WORLD ACADEMIC FRONTIERS
世界学术研究前沿丛书

Environmental Pollution and Control

环境污染及其防治

"世界学术研究前沿丛书"编委会
THE EDITORIAL BOARD OF
WORLD ACADEMIC FRONTIERS

中国出版集团公司
世界图书出版公司
广州 · 上海 · 西安 · 北京

图书在版编目（CIP）数据

环境污染及其防治：英文 / “世界学术研究前沿丛书”编委会编. —广州：世界图书出版广东有限公司，2017.8
ISBN 978-7-5192-2459-2

Ⅰ. ①环… Ⅱ. ①世… Ⅲ. ①环境污染—污染防治—英文 Ⅳ. ①X5

中国版本图书馆CIP数据核字(2017)第040977号

Published by arrangement with Scientific Research Publishing
Through Wuhan Irvine Culture Company

书　　名：环境污染及其防治
Huanjing Wuran Ji Qi Fangzhi
编　　者：“世界学术研究前沿丛书”编委会
责任编辑：康琬娟
出版发行：世界图书出版广东有限公司
地　　址：广州市海珠区新港西路大江冲25号
邮　　编：510300
电　　话：（020）84460408
网　　址：http://www.gdst.com.cn/
邮　　箱：wpc_gdst@163.com
经　　销：新华书店
印　　刷：广州市德佳彩色印刷有限公司
开　　本：787 mm×1092 mm　1/16
印　　张：56.25
插　　页：4
字　　数：1030千
版　　次：2017年8月第1版　　2017年8月第1次印刷
国际书号：ISBN 978-7-5192-2459-2
定　　价：598.00元

Preface

Environmental pollution is the introduction of contaminants into the natural environment that causes adverse change. Pollution can take the form of chemical substances or energy, such as noise, heat or light. Pollutants, the components of pollution, can be either foreign substances/energies or naturally occurring contaminants. Pollution is often classed as point source or nonpoint source pollution. Pollution control is a term used in environmental management. It means the control of emissions and effluents into air, water or soil. Without pollution control, the waste products from overconsumption, heating, agriculture, mining, manufacturing, transportation and other human activities, whether they accumulate or disperse, will degrade the environment. In the hierarchy of controls, pollution prevention and waste minimization are more desirable than pollution control.[1]

In the present book, thirty-five literatures about environmental pollution and its prevention published on international authoritative journals were selected to introduce the worldwide newest progress, which contains reviews or original researches on environmental pollution and human health, air pollution, water pollution, soil pollution and so on. We hope this book can demonstrate advances in environmental pollution and its prevention as well as give references to the researchers, students and other related people.

March 9, 2017

[1]From Wikipedia: https://en.wikipedia.org/wiki/Pollution.

Selected Authors

Saffa Riffat, Department of Architecture and Built Environment, University of Nottingham, University Park, Nottingham, UK.

Nick Hanley, Department of Geography and Sustainable Development, School of Geography and Geosciences, University of St. Andrews, St Andrews, Scotland, UK.

Carmen E. Pavel, King's College London, Strand Building, Strand Campus, London, UK.

Kirstin I. Conti, Governance and Inclusive Development, Amsterdam Institute for Social Science Research(AISSR), University of Amsterdam, Nieuwe Actergracht 166, Amsterdam, The Netherlands.

Cornelis J. van Leeuwen, Copernicus Institute for Sustainable Development and Innovation, Utrecht University, Heidelberglaan, Utrecht, The Netherlands.

Esther Rind, Centre for Research on Environment, Society and Health (CRESH), Institute of Geography, School of GeoSciences, University of Edinburgh, Edinburgh, UK.

Michael E. Scheurer, Department of Pediatrics, Section of Hematology-Oncology and Dan L. Duncan Cancer Center, Baylor College of Medicine, Houston, Texas, USA.

Martyn N. Futter, Department of Aquatic Sciences and Assessment, Swedish University of Agricultural Sciences, Uppsala, Sweden.

Henriette Naims, Institute for Advanced Sustainability Studies e.V., Berliner Strasse 130, Potsdam, Germany.

Yunbi Xu, Institute of Crop Science, Chinese Academy of Agricultural Sciences, Beijing, China.

Contents

Chapter 1

Future Cities and Environmental Sustainability

Saffa Riffat, Richard Powell, Devrim Aydin

Department of Architecture and Built Environment, University of Nottingham, University Park, Nottingham NG7 2RD, UK

Abstract: Massive growth is threatening the sustainability of cities and the quality of city life. Mass urbanisation can lead to social instability, undermining the capacity of cities to be environmentally sustainable and economically successful. A new model of sustainability is needed, including greater incentives to save energy, reduce consumption and protect the environment while also increasing levels of citizen wellbeing. Cities of the future should be a socially diverse environment where economic and social activities overlap and where communities are focused around neighbourhoods. They must be developed or adapted to enable their citizens to be socioeconomically creative and productive. Recent developments provide hope that such challenges can be tackled. This review describes the exciting innovations already being introduced in cities as well as those which could become reality in the near future.

Keywords: Future Cities, Sustainability, Urbanisation, Environment, Innovations

1. Introduction

Throughout history, cities have been at the heart of human development and technological advancements[1]. Although an element of planning can be discerned even in the earliest cities they have often evolved in response to the changing needs and aspirations of their inhabitants. Some cities have survived for millennia, including Rome, Athens, Cairo, Alexandra, Baghdad and Beijing, and are still flourishing. Other, once mighty, cities have disappeared, their ruins being unearthed by present-day archaeologists. A fascinating example, built by the Khmer civilisation, is Ankor Wat[2] in present day Cambodia, which boasted features very relevant to the design of future cities. Notable achievements by the ancient Khmer engineers were the control and distribution of water through a sophisticated canal system irrigating agriculture within the city bounds supplying citizens with ample food. The fundamental problems that the Khmer solved were the prevention of flooding by Monsoon rains, and storing water for the subsequent periods of drought. Despite its success over 8 centuries, Ankor Wat collapsed in 1431[3]. Climate change resulting in extended droughts is considered to be a contributory factor, which even the excellence of Khmer engineering was unable to counter.

Tenochtitian, the Aztec capital in what is now Mexico, was built in a lake bordered by swamps. The flow of water was controlled to provide land for building and irrigate fields, the so-called floating gardens. The city districts were connected by both causeways and canals. The Aztec engineers also had to separate the brackish water of the lake from spring water from nearby hills for drinking. In 1519, at the time of the Spanish conquest, Tenochtitian, with an estimated population of 200,000 to 300,000, and was one of the largest cities in world. Although conquered it did not collapse like Ankor Wat, but was developed by the Spanish into what is now Mexico City, with a population of 21m.

Although in designing cities of the future we have a much greater range of technologies than our ancestors, we must not make the hubristic mistake of assuming that these will ensure our success. The words of George Santayana are apposite, "Those who do not remember the past are condemned to repeat it." We need an enlightened approach to design cities of the future, learning from the experience of the past and applying the advanced technologies of the present.

2. Tomorrow's Cities: Chaotic or Strategic? Lessons from the Immediate Past

First and foremost, future cities must serve their citizens, combining increased prosperity for all with desirable life styles. These aims must be achieved without detriment to people who live in other regions; for example they must not export carbon emissions by importing goods manufactured by fossil fuel and feedstock dependent processes or create pollution elsewhere. This is not to say that a city cannot import feedstock or energy intensive goods from outside its borders; rather the energy and material contents of imports must be balanced by those of exports. To this end, future cities must adopt wide scale utilisation of renewable energy, waste management/minimisation, water harvesting/recycling, landscape/ biodiversity to enhance the natural environment, use of green transport systems, applications of innovative material/construction methods (low/zero carbon buildings) and local food production.

While such aspirations would have been familiar to city designers in antiquity, their modern counterparts can draw upon newer technologies such as integrated smart management control systems based on wireless sensor networks, which by detailed monitoring can turn wasteful cities into sustainable cities. Technologies will need to be tailored to particular geographic, climatic and cultural conditions, but all will have a similar philosophy of turning buildings from passive entities to active, adaptive and adaptable spaces that takes advantage of the surrounding environment for heat, cooling, light and electricity. A key to achieving low carbon cities is understanding how best to select and integrate various technologies from the many available, to optimise performance for different building types, climates, cultures and socio-economic conditions. A strategic approach will be required to achieve a sustainable city to ensure that it functions efficiently as a whole. But the planning parameters should not be so centralised and rigid that they do not allow different community designs and architecture styles to find expression.

Several impacts of overpopulation in urban areas are already appreciable as summarized in **Table 1** with possible mitigation strategies. First of all concerns of food and water security is arising in many cities. As these cities expand, agricultural land is converted into residential and industrial areas. For instance in Conception, a Chilean City with a population of 500,000, 1734 hectares of wetlands and 1417 hectares of agricultural land and forests were transformed into residential areas between 1975-2000[4]. In Accra, Ghana, it is estimated that 2600 hectares of agricultural land is converted every year where Chinese and Indonesian cities have the similar pattern[4]. In the future agriculture will be challenged to meet the

Table 1. Impacts of global urbanisation and mitigation strategies.

Impacts ⟹	*Mitigation Strategies*
High traffic density	✓ Efficient public transport ✓ Compact city design
High amount of waste	✓ Recycling
Urban warming	✓ Increasing green space, ✓ Using reflective materials
Increasing Air pollution	✓ CO_2 capture, ✓ Filtering exhaust gases, ✓ Increasing efficiency of industrial processes/vehicles
Increasing energy consumption/sinking resources	✓ Using renewable sources, ✓ Achieving low energy buildings, ✓ Increasing efficiency of devices/ processes
Lack of biodiversity/natural habitat	✓ Increasing green space, ✓ Developing animal/plant protection areas
Sinking water resources	✓ Water purification ✓ Desalination ✓ Rainwater harvesting
Rising food demand/ poverty	✓ Vertical farming ✓ Artificial food production ✓ Greening the deserts
Land shortage for housing	✓ Constructing multifunctional buildings, ✓ Creative architectural designs
Weak Social cohesion	✓ Improving sociocultural environment ✓ Increasing the number of organisations-events that bring people together

demand of a population that is projected to grow and to urbanize. This indicates that more food will be demanded by a population of net food buyers; and food demand will have to be met by rural and peri-urban areas or by food imports.

On the other hand, in many emerging cities, people are obligated to live in more marginal regions. They have less adaptive capacity, low incomes and no assets. Additionally less legal and financial protection, no insurance, no land tittle... Indeed urban areas are more attractive than rural areas for many people in terms of job opportunities, improved living conditions, multicultural environment and dynamic life. In fact developed countries are already highly urbanized due to the opportunities they have, and the United Nations estimates that the urban populations of Africa, Asia, and Latin America will double over the next 30 years, from 1.9 billion in 2000 to 3.9 billion in 2030. At that point, over 60% of the world's population will live in cities[5].

Cities with the growing urban mass will turn to a more resource, land, food and energy demanding consumers and they should be productive to be genuinely sustainable. Unproductive urban areas will probably face with poverty, inequality of individuals, pollution, illnesses and external economical dependency. Productivity is clearly desirable in emerging cities as it increases competitiveness thereby prosperity and sustainability of any city. More productive cities are able to increase output with the same amounts of resources, generating additional real income that can raise living standards through more affordable goods and services[6]. More specifically, the generated extra income and municipal revenue will enable any city to provide more, better services, including housing, education and health services, social programmes and expanded infrastructure networks to support both productive and leisure activities. Indeed, the productivity of the city is directly influences the citizen well-being and socio-economic status.

Urban productivity is the measure of how efficient a city transforms inputs into outputs. Gross domestic product (GDP) per capita is commonly used as a proxy for urban productivity, with a city's GDP measuring local production of goods and services and the population serving as a proxy for inputs related to human capital. GDP is an important measure of sustainability of developing cities. The report by UN demonstrates that GDP of emerging cities is significantly influenced by the national development level. The importance of national comparative advantage is illustrated by the fact that, while 22 of the top 30 largest urban areas (by population) were located in emerging or developing economies in 2008, only seven emerging economy cities ranked among the top 30 in terms of urban GDP. The group included Mexico City, São Paulo, Buenos Aires, Moscow, Shanghai, Mumbai and Rio de Janeiro, but no Middle Eastern or African cities (See: **Figure 1**). The average GDP per capita of these emerging/developing country cities tends to be substantially smaller than that of developed cities[6].

The recent report by the United-Nations (2014) estimate 3.9 billion urban population in 2014 itself. The urban population will be increase about 2.5 billion, which means by 2050 UN estimates about 6.4 billion population residing in urban

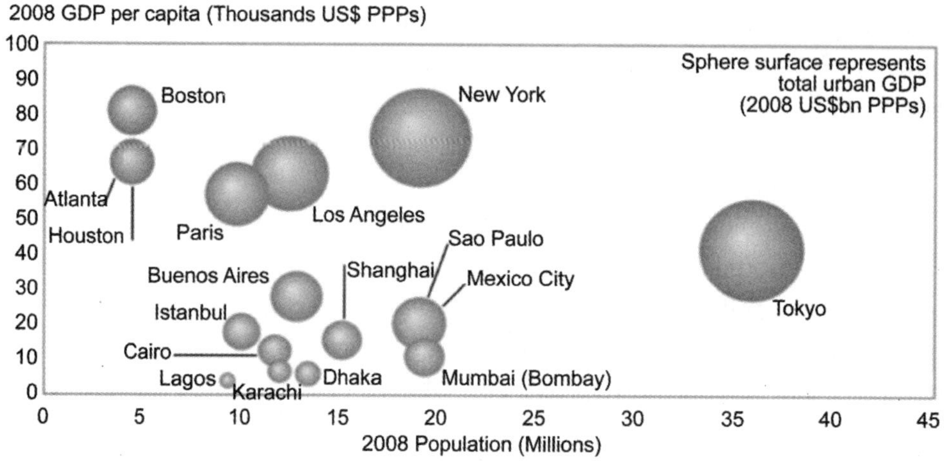

Figure 1. Population, GDP per capita and total GDP for selected metropolitan areas (2008)[6].

areas. Increasing urban population will lead to a significant increase of urban energy consumption and urban emissions. Higher population may also cause an increase in urban density (number of people per unit area). Key strategies are needed for minimizing energy consumption, efficient use of land, sustainable food production and transport. In a recent study Singh and Kennedy developed a tool for predicting future energy consumption and CO_2 emissions based on electricity, heating and transportation in urban areas[7]. The tool was applied to 3646 urban areas and three projections were used in the analysis for the years 2020 and 2050. In medium and high projection it is assumed that the number of people per unit area (km^2), (urban density), will increase 1% and 2%, whereas in low projection no urban density increase was assumed. The predicted emissions and energy consumption in the years 2020 and 2050 with the baseline of 2000 are given in **Figure 2(a)** and **2(b)**. The results revealed that, for the high projection, CO_2 emissions based on electricity usage will be doubled in 2020 and increase more than four times in 2050. According to the developed tool, heating sector based CO_2 emissions and energy usage will not affected from the urban density and will slightly increase in all cases. On the other hand the increase in urban density may dramatically increase the transportation sourced emissions and energy consumption.

While some cities in the past have been planned, at least in part, they have often grown chaotically with minimal strategic design and service provision, especially when under pressure from population growth often driven by migrations

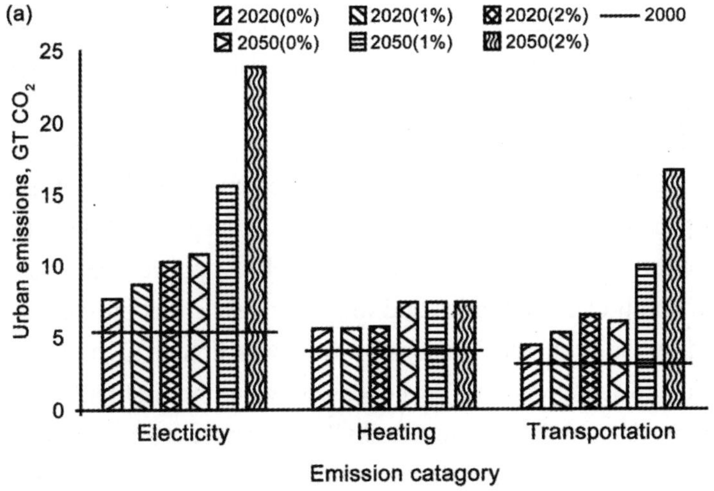

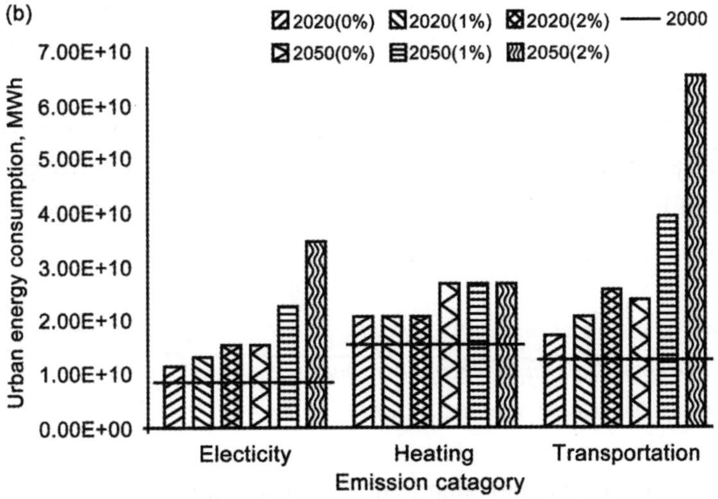

Figure 2. Predicted sectoral urban (a) emissions, (b) energy consumption by 2020 and 2050 based on percent urban density decline[7].

from the countryside during periods of gricultural mechanisation and industrialisation. The large industrial cities of 19th century Britain and the favelas of 20th century Rio de Janeiro are obvious examples. Lack of sanitation, overcrowding, disease, child labour and little provision for education resulted in the horrendous living conditions described in the novels of Charles Dickens and the factual reports of investigative journalist Henry Mayhew[8]. In Britain a century and a half of social reform, legislation and regulation has been required to heal the wounds of past piece-meal urban growth. Even now Britain suffers from a considerable 19th and early 20th century house stock, which whilst impractical to replace, is also technically challenging to refurbish to the modern standards of energy efficiency necessary to reduce national carbon emissions.

To avoid the chaos of the past future cities must be planned. But city wide planning, especially on grand scale, has a legacy of suspicion. Like other European nations after the Second World, Britain, embarked upon a combination of slum clearance in bomb-damaged cities and initiating new cities on greenfield sites, in part inspired by the "concrete" visions of the Swiss-French architect Le Corbusier (Charles-Édouard Jeanneret-Gris), who was active in the first half of the 20th century. Tower blocks and "streets in the sky" constructed using reinforced concrete and industrial engineering-type processes proved to be a disaster, both technically, because of lax quality standards, and socially, because of social isolation and creation of conditions conducive to petty crime. In Britain, 60 years later, these structures are regularly demolished to be replaced by more traditional low rise dwellings built to modern standards. Even in 1945 John Betjeman, the Poet Laureate, who was a staunch defender of historic buildings, warned about overpowering development in his poem "The Planster's Vision"[9] where the second verse begins with the ironic lines,

"*I have a vision of the future, chum,*
The workers' flats in fields of soya beans
Tower up like silver pencils, score on score..."

The key word is "chum" that implies an insensitivity for humans as individuals. On the other hand, influenced by the British planning failures of the 1950s and 60s, heir to the British Throne Prince Charles, is highly critical of "modern architecture" and strongly advocates a return to local vernacular architecture for new communities, a philosophy he has applied to the development of Poundbury in southern England[10][11]. Charles' ideas have been dismissed as an anti-progressive pastiche of the traditional. But, in some respects, Charles is returning to the earlier vision of British urban planner, Ebenezer Howard the creator of the garden city movement[12]. Letchworth was the first garden city followed by Welwyn Garden City, both built in the early 20th century to the north of London, recognised the importance of an integrated transportation system, but, not surprisingly for the time they were built, did not foresee the rapid growth of automobile ownership.

The garden city philosophy was adopted in other countries.

In summary, inspired city planning working to a well-defined strategy with a regard for human values can avoid the chaos created by the uncontrolled growth of cities during rapid industrialisation. Engineering has much to offer future city design but it must be tempered by a respect for citizen's aspirations. The designs for future cities must be flexible, responding to evolving technologies and cultural changes. With rapid global urbanisation the challenges are immense; but to learn what works new cities must be built to new designs. We shall need to accept that success will be accompanied by failure, from which we shall, and must, learn.

3. Review of the Visions for Future Cities

Innovative visions are needed in emerging cities to reduce the impact on the environment while creating places that increase social cohesion, or accelerating human interaction in education, health and employment to improve the quality of life for an ever greater percentage of our world population. The technological advancements should be fully utilized to realize these visions and goals. For instance, temperature, pollution, water systems, waste management systems, radiation, traffic, air pollution and other components can be monitored through wireless sensor networks for achieving the greatest efficiency[13]. These systems can help detect leaks and problem areas quickly, potentially saving electricity and other precious resources. In order to save additional resources, cities can consider grassroots initiatives, like farmer's markets and community-supported agriculture. Urban farming is a simple change, since dirt beds can be put nearly anywhere and grow food locally[13]. Organizing community carpools and encouraging people to recycle waste and use reusable bags for shopping can make huge impacts as well. A staggering 75% of solid waste is recyclable, but steps need to be made to encourage more recycling to happen, as 70% is still thrown into the trash[13]-[15]. Cities can become also more sustainable and attractive by adding open space. Hiking trails, activity centres, and parks can draw people into the city and reduce waste.

Cities are vital to the future global economy. For instance 41% of the UK's population lives in the country's ten largest urban areas[16]. However, cities are struggling with climate change, changes in population and demographics, congestion and healthcare, and pressure on key resources[17][18]. In future there will be a large market for innovative technologies/approaches to create efficient, attractive and resilient cities[18][19].

Recent research has been focused on the development of a data platform for power, heat and cooling usage in cities and individual usage patterns in domestic, commercial and industrial buildings[20][21]. There is a lack of information in the rapidly changing energy market. Solutions are required to better handling of cost, supply and demand of energy in cities and towns. With macro-level energy data, cities can invest in new innovations, provide more focused geographic support to areas where energy supply is lacking, and gain better decision-making evidence on issues such as targeted building retrofitting and fuel poverty[21].

Responding to the rapid urban development and challenges, future cities have become a pressing issue due to the impacts of global warming problems. This inevitably requires identifying prioritizing and structuring new design and managerial tools to improve their environmental, urban and fiscal sustainability.

Emerging cities should also develop local and national policies to retain highly qualified individuals. Currently in developing world, the proportion of cities making effort to retain talented and visionary individuals is alarmingly low. Asia could count as an exception where half of the cities are putting effort to retain talent. In China, Chongqing has developed an ambitious training programme to support the transition of rural migrants from manual-based to skill-based types of work; by 2009, nearly one-third of migrants had benefited from the scheme[22]. Dubai is also promoting education especially in the fields of engineering and information technologies[23].

Some cities in developing countries have embraced the model of world-class innovation clusters, such as California's Silicon Valley or Boston's Route, to become "high-tech hubs"[6]. Those that have met with success in this endeavour, such as India's Bangalore, owe it to the same basic factors: the presence of top-quality academic and research institutions as well as substantial public and corporate investment. However, low infrastructure development rate and unbalanced distribution of benefits of growth across all the population are signalling threat for these regions. Quality of life is rapidly emerging as a major asset in any efforts to attract and retain creative minds and businesses. It is not surprising that Toronto, San

Francisco or Stockholm are regularly ranked among the top performing cities in the world, since they are found as performing particularly well in a wide range of both economic and quality of life indicators including crime, green areas, air quality and life satisfaction. Except more developed nations, Singapore, with a similar balance of quality of life attributes, also ranks among the top world cities and the highest among developing countries[6].

Inspiring from the above given successful examples, each city should develop its own strategic future vision for realizing the basic concepts, with the aim of maximizing an integrated total of environmental, social and economic values. When setting out the future vision, both a back casting approach of looking back from a desirable future to the present and a forecasting approach of looking forward from the present to the future are essential to enhance feasibility. Moreover, it is important to set the vision in a way that fully embodies each city's diverse and unique features that arise from its natural and social characteristics. Each city is required to tackle the challenges of the environment and aging society, and is further encouraged to take on additional challenges in areas that can enhance their originality and comparative advantages in cooperation with other cities in the same nation and abroad. It will be important to gather world-wide wisdom by absorbing information on other cites' successes from all over the world, as this will help integrate a variety of efforts in different fields and realize synergistic effects. By accumulating successes, cities are expected to break away from subsidies and acquire self financing independence, establishing financially and socially autonomous models[6].

The European "Smart Cities & Communities Initiative" of the Strategic Energy Technology Plan (SET-Plan) promotes 40% reduction of greenhouse gases in the urban environment by 2020, which could be achieve with sustainable and efficient production, conversion and use of energy. Yet the domestic sector will increasingly become the leading energy sector as more people around the world aspire to higher living standards, which will drive the demand for air conditioning and electric power. Zero energy buildings (ZEB)/Zero carbon buildings (ZCB), therefore, expected to have a vital role to achieve sustainable and smart cities. Kylili and Fokaides define ZEBs as buildings that have zero carbon emissions on an annual basis[24]. The required ZEB aspects as part of future's smart cities are demonstrated by the Kylili, and Fokaides as given in **Figure 3**.

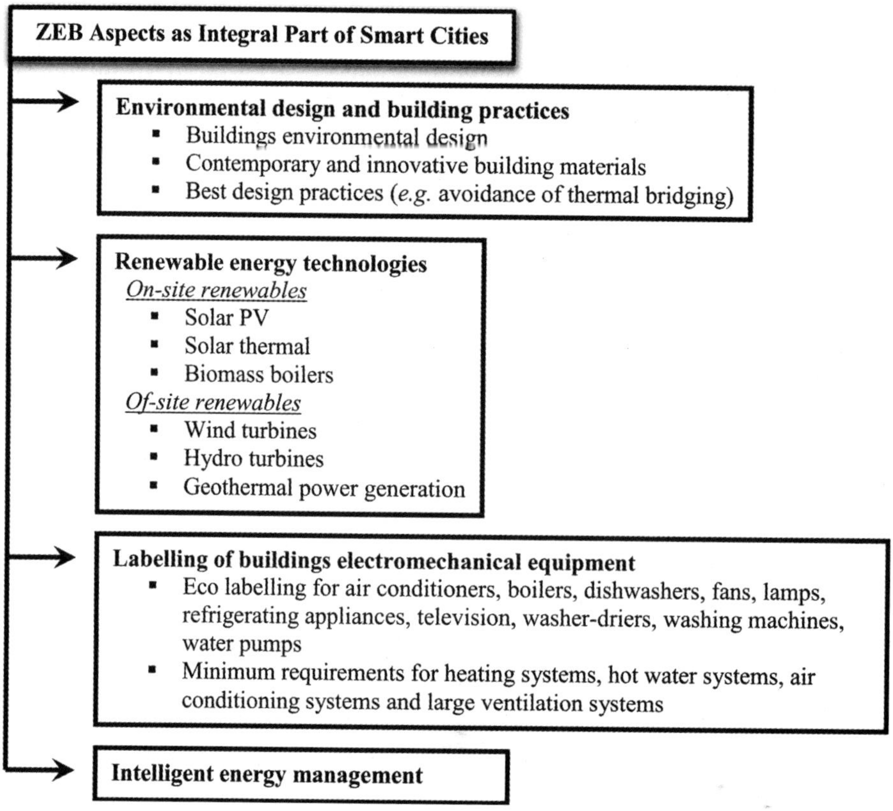

Figure 3. ZEB aspects as integral part of smart cities. Adapted from[24].

Various designs for future cities have been mooted, some more adventurous than others. Some are actually being built. All aspire to being carbon neutral and sustainable, exploiting the latest technologies for construction, renewable energy, recycling and transportation. Recently the British Government has announced plans for new garden cities in the UK which emphasised the development of new communities adapted to local needs[25]. The aspirational "wish list" harks back to Howard and, although arguably obvious, it does express what is expected of a future British garden city:

- Strong vision, leadership and community engagement

- Land value capture for the benefit of the community

- Community ownership of land and long-term stewardship of assets

- Mixed-tenure homes and housing types that are affordable for ordinary people
- A strong local jobs market in the Garden City itself, with a variety of employment opportunities within easy commuting distance of homes
- Beautifully and imaginatively designed homes with gardens, combining the very best of town and country living to create healthy homes in vibrant communities
- Generous green space linked to the wider natural environment, including a surrounding belt of countryside to prevent sprawl, well connected, biodiverse public parks, and a mix of public and private networks of well-managed, high quality gardens, tree-lined streets and open spaces
- Opportunities for residents to grow their own food, including generous allotments
- Strong local cultural, recreational and shopping facilities in walkable neighbourhoods
- Integrated and accessible low-carbon transport systems—with a series of settlements linked by rapid transport providing a full range of employment opportunities

Garden cities built along these lines will largely exploit existing technologies, an approach already adopted elsewhere. The Zero Carbon Building (ZCB) in Hong Kong, is located at the heart of Kowloon Bay, the upcoming vibrant premier business district. Covering a total area of 14,700 m^2 comprising a 3-storey Zero Carbon Building and a landscape area[26] it both showcases state-of-the art eco-building design and technologies to the construction industry locally and internationally and raises community awareness of low carbon living in Hong Kong. To achieve zero carbon emissions, ZCB adopts an integrated design where the ZCB building and its surrounding woodland must be seen as a single entity. Nevertheless, in addition to the sustainability of future's construction we should also consider the "visuality" and "functionality" of buildings. An intriguing example, Hong Kong Polytechnic University's 15-storey Jockey Club Innovation Tower competed last year[27]. It prospers the diversity, expresses the dynamism and creativity of university life with creating a fascinating turban area. While the tower provides multi-functional usage and is visually attractive, its unique geometry covers less land space than its contenders. The building is a showcase of future high rise construction. Future cities could evolve by progressively adding more buildings following the same principles, each designed for its intended function, residential, offices etc.

The principles incorporated into Hong Kong's ZCB and Jockey Club Innovation Tower can also be seen in Swedish developments. Malmö, Sweden's largest city has undergone economic changes replacing its tradition heavy industry with small and medium size companies. Kjellgren Kaminsky in combination with builders Höllviksnäs Förvaltnings AB, won an open competition for passive houses in April 2009 which have now been built. The buildings have a number of measures for ecological sustainability using a combination of wind, geothermal and solar energy. The original biodiversity of the local area has been maintained and especial attention has been applied to rainwater collection and sewage treatment.

Hammarby Sjöstad (Hammarby Lake City) is a new district in Stockholm built on a previously industrial and harbor area. Hammarby is meant to provide 10,000 apartments for 25,000 inhabitants and occupies 200 hectares of land, close to the city centre. The required environmental impact of the project was limited to no more than half that of the best projects built at the end of the 1990s; in the long term, the energy demand should not exceed 60kWh/m^2 per year of which not more than 20kWh/m^2 per year should be electric energy[28]. As with the Malmo development, energy, waste and water systems have been designed for sustainability. A similar development, Beddington Zero Energy Development (BedZED) in London, was completed in 2002 comprises 82 affordable dwellings and commercial site (offices, workspaces) spread on approximately 2500 m^2. The project is a conspicuous example of urban development as it addresses many challenges such as combining workspace with housing, matching with dense urban population, achieving zero carbon standards and increasing comfort level[29].

Japan is also actively developing sustainable "eco" cities, of which a particularly interesting example is the Kitakyushu Eco-Town project[30]. Like the Swedish examples, its development is a response to the decline in highly polluting heavy industry, which contaminated the local, land, sea and air in the 1960s. The target is to reverse this environmental damage by creating a sustainable community through a partnership of the government, commercial organisations and citizens. A key aspect is local recycling of discarded items from bottles to bicycles. Furthermore, all Eco-Town companies must allow their facilities to be inspected by citizens in order to eliminate public distrust and anxiety concerning potential pollution.

The developments described above are based essentially on established technologies following principles that can be applied readily elsewhere to achieve urban sustainability in the near future. They are targeted at relatively modest sized communities typically adjacent or within existing conurbations.

In parallel with these projects, far more ambitious, schemes have been initiated that are creating completely new sustainable cities on virgin ground, especially in states with strong central, governments and with considerable national wealth earned from the sale of fossil fuels. A good example, of a future community is Masdar City in Abu Dhabi (UAE), a project to create the world's first low carbon/zero waste sustainable city[31][32]. Completely powered by renewable energy, and covering an area of more than seven square kilometres, Masdar City will have the capacity to house 40,000 residents, and host a range of businesses and institutions employing 50,000+ people. But, it is intended to be more than just a demonstration of the practicality of using renewable energy technologies. Masdar City will host a vibrant, innovative, community of academics, researchers, start-up companies and financiers—all focused on developing renewable energy and sustainability technologies.

Another interesting project, Silk City in Kuwait, will be completed in 2023 and will include 30 communities grouped into four main districts; Finance city, Leisure city, Ecological City and the Educational-Cultural city. Silk City will become a new urban centre accommodating 750,000 residents in over 170 thousand residential units. This $132 billion project will create a modern and sustainable oasis, providing hundreds of thousands of jobs and investment opportunities within the world's tallest tower "Burj Mubarak al-Kabir" located in Finance City[33].

King Abdullah Economic City is another representative of the future city concept aiming to have a positive impact on the socio-economic development of Kingdom of Saudi Arabia[34]. The first stage of the city was finished in 2010 and it will be fully completed in 2020. It will consist of several zones enabling industrial, educational, business and residential activities over an area of 173 km^2. Energy/carbon, water, waste, ecology/biodiversity and pollution prevention have been adopted as key parameters in the design of the city[34][35]. It is also aimed to create up to one million jobs for the youthful population of the country, with where 40% are under 15[36].

In response to its considerable environmental problems, a result of its recent industrial growth and need to meet the aspirations of its increasingly wealthy population, China has initiated the construction of many cities based on sustainable designs. In contrast to Europe and Japan, China is able to build on green field sites, an example is Tianjin Eco City in China[37]. Although its development has not been without problems[38] it does appear to be growing at a viable pace[39]. The stated intention is to move one hundred million people into new cities in the next decade, especially in the western part of the country.

Azerbaijan is developing Khazar Islands, a sustainable \$100bn city in central Asia on the Caspian Sea, which, when complete in 2020 to 2025 will have 1 m inhabitants. Amenities provided in the city will include; cultural centres and university campuses[40]. The prestigious \$2bn Azerbaijan Tower, intended to be the world's tallest, presumably trying to outdo Burj Mubarak al-Kabir. A Formula 1 circuit will also be included. All buildings will be capable of withstanding magnitude 9.0 earthquakes.

While the new cities described above are ambitious they are based on existing or emerging technology and, in principle at least, can be completed within the next decade, designs for far more futuristic cities have also been mooted, siting them underground[41]-[43], underwater[44], floating on the sea[45][46] or even in the sky[47][48].

Arguably the development of the underground city has already started. In London, where real estate is very expensive, wealthy property owners are digging downwards to expand their living space thus avoiding planning regulations. The London Crossrail scheme shows that large underground spaces can be created, but as Harris notes, quoting London's Road Task Force, why not put major roads into new tunnels "...leaving the surface, with its sunlight and trees, for public spaces"[41]? Maybe in localities such as London, where the underlying clay is conducive to excavation, a present day city can evolve into a future city by digging downwards rather growing upwards?

An alternative option for London is Sure Architecture's "Endless (Vertical) City" envisages a 55 storey tower designed for London site which will be a self-contained community complete with areas dedicated to parks[49]. Two ramps wind around the exterior essentially providing "vertical" streets since London does not have the space to accommodate further horizontal streets.

With Japan's lack of building land and susceptibility to earthquakes it is perhaps not surprising that a Japanese company, Shimizu Corporation, has proposed building self-sufficient cities under the sea called "Ocean Spirals"[50]–[52]. A city with typically 5000 inhabitants will be contained within a 500 m diameter water-tight sphere, at or near the ocean surface, and connected by a huge spiral to the ocean floor as much as 4000 m below. Aquaculture would be practised in the surrounding sea to produce food sustainably and fresh water would be obtained by desalination. Shimizu claims the first city, costing £16bn, could be ready by 2030, having taken just 5 years to build...and the price of further cities would be reduced as numbers increased.

In contrast to Shimizu, Architect Vincent Callebaut has designed the "Lily-pad" city, capable of accommodating 50,000 people floating on the ocean surface[53][54]. The city integrates a range of renewable energies (solar, thermal, photovoltaic and wind). Intriguingly, since these floating cities float near a coast or travel around the world following the ocean currents, they would avoid the problems of sea level rise resulting from climate change[53][54].

The Venus Project, proposed by US inventor, Jacque Fresco, is another circular city comprising a central dome containing the cybernetic systems that maintain core automated city functions[55][56]. Fresco goes way beyond developing a sustainable city. He wishes to create an utopian, technological civilisation without money that avoids the ills of all previous forms of economic and political systems... capitalism, government, fascism, communism, socialism and democracy. Fresco considers that by creating the ideal environment for humans it will naturally eliminating violence, greed, and the inequalities that presently afflict us. His philosophy seems to be in a tradition that can be traced back to Plato and Thomas

More. The ideas espoused are beguiling, but are they achievable? Could they survive in a world where the pursuit of power and wealth is the prime objective of some individuals, whether ostensibly justified by nationalism, religious belief, or political creed? Indeed, to fully buy into the Venus Project requires a strong belief in its philosophy.

Even more fanciful than London's "endless City" and inspired by the form of the lotus flower[48], is Tsvetan Toshkov's, "'City in the sky' which he claims is a concept embodying an imaginary tranquil oasis above the mega-developed and polluted city, where one can escape from the everyday noise and worries." Although a delightful exercise in creating a utopia away from the strains of modern city life, the engineering stresses within the proposed structure raise questions about its practicality.

Despite the ambitious, indeed grandiose, designs of future cities requiring considerable planning, rapid urban renewal may become vital in response to natural disasters notably earthquakes and hurricanes. While nobody would wish such misfortunate on any city with the human tragedies engendered, the opportunity presented to rebuild a devastated city to both improve its sustainability and to reduce the risk of future disaster cannot be overlooked, not least as an honour to those who have suffered. Two examples are the Wenchuan and Qingchuan districts of Sichuan Province, severely damaged by the 2008 earthquake, which are now in the reconstruction process.

These areas suffered because buildings were not earthquake resistant. Reconstruction has been difficult and a large number of temporary shelters that are neither durable nor thermally comfortable have been built in an attempt to meet the urgent needs of those affected. A research team led by Prof. Zhu Jingxiang of the School of Architecture at The Chinese University of Hong Kong (CUHK) has developed an integrated light-structure system for the reconstruction of New Bud Primary School at Xiasi village in Sichuan's Jiange County[57]. With the support of the Hong Kong Dragon Culture Charity Fund and the CUHK New Asia Sichuan Redevelopment Fund, the new school was completed and in operation in just two weeks. The building is safe and durable, and the cost of construction is low. It also looks attractive and features good thermal performance and a high energy-saving capacity. Maybe inspired, elegant, but eminently practical designs to rebuild shattered communities rapidly and sustainably will be more important and helpful to humanity than some of the grandiose schemes presently on drawing boards?

4. Building Future Cities

Of course, new cities, based on existing modern technology, are already being designed and built. China is responsible for about half of global construction work and will build 400 new cities and towns within the next 20 years[58]. China is also moving rapidly towards implementing a low carbon economy and have recently selected 5 provinces and 8 cities for low carbon demonstration[58][59]. However, city design and technology must continue to develop, not least because more than 80% of the world's Global Warming Potential (GWP) is created in cities[60][61] and by 2050, 66% of the world's population will be urban[62]. Accordingly, advanced construction methods and materials will be needed for sustainability in future's cities. Robotic/digital design based technologies coupled with 3D printing of prefabricated modules will reduce construction times, minimise energy consumption and eliminate wasted material, all contributing to lower costs[63][64]. This technology will be an important step for redesign/reconstruction of cities towards sustainability. Basically with the 3D printing, robotic arms with three axis freedom of movement can construct the building, based on the architectural design, which is coded into the controller of the 3D printer. 3D printed buildings provide aesthetics while minimizing the constructional defects which is generally an issue in conventional buildings[65][66].

A return to timber as a major building material is especially attractive since each cubic meter of wood can store half tonne of carbon[67]. Can we make buildings that work like trees and cities like forests? New cities will exploit new materials that will deliver greater functionality. For example nano-materials already offer opportunities for advances in sensors[68] and smart polymers[69]. However, it is just as important that future cities are constructed from materials that are completely recyclable and sustainable[70]. Where virgin feedstock is required it must be taken from renewable sources, which in many instances will be biomass-based[71]. For health and safety reasons manufacturing processes are currently located at distances from major conurbations. In the future processes are required that are low hazard and can be integrated into cities, close to workers homes. The newly emergent disciplines of Green Chemistry and Green Engineering are addressing the development of future manufacturing industry[72].

With buildings being responsible for almost half of all energy consumption and carbon emissions in Europe, new build properties are becoming much more energy efficient and their environmental footprint is being reduced[73][74].

The Energy Performance of Buildings Directive, EPBD (EU, 2010) requires that all new buildings shall be "nearly zero energy buildings" (nZEB) by the end of 2020[75]. EPBD is not limited to new buildings, but also covers retrofit of existing buildings because these constitute the majority. Accordingly, building materials are in the spotlight as they have a large influence on building energy consumption, carbon emissions, urban warming and comfort level. Solid wood has been used as a building material for thousands of years, appreciated for being a lightweight, easy reusable and naturally regrown resource (See: **Figure 4**). Today, "wooden construction" is innovative and on the rise: timber is again regarded as an ideal green building material[76]–[78]. Recently a 30 storey tower, has been designed by Michael Green for Vancouver, Canada. Once built, it will be the tallest wooden construction, overtaking its competitors Forte Building, Melbourne and Stadthaus, London[78].

Since timber is one of the few materials that has the capacity to store carbon in large quantities over a long period of time, some of the historically negative environmental impact of urban development and construction could be avoided. As seen in **Figure 5**, while 1 kg wood can store 9 kg CO_2, the rest of building construction materials positively contribute to the CO_2 emissions, particularly the aluminium has a significant footprint on the environment by releasing 27 kg CO_2 per kg[79][80].

Nanotechnology is also expected to have a wide range of usage in future's buildings. In the last decade this technology has been used in the building industry to improve the structural, mechanical, hygienic, aesthetic and energy-related properties of building materials. Nanomaterials can be either added to the building materials or used as coatings. For instance applying nano scale coatings of titanium dioxide breaks down the dirt as and provides a self-cleaning effect when it is applied to windows, frame, glazing or roof tiles[81][82].

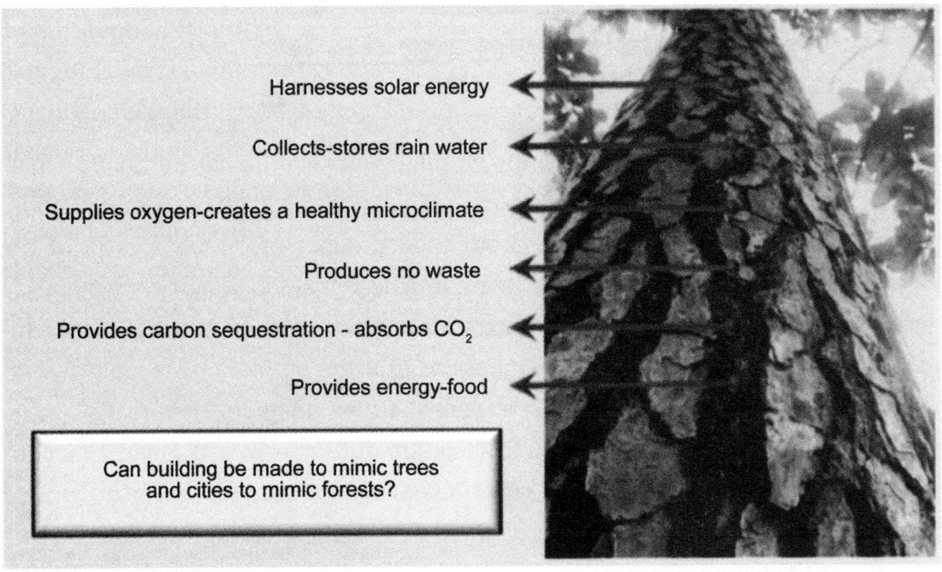

Figure 4. Benefits of using wood as building construction material. Legend—Wood is the only material with a negative CO_2 balance; each cubic metre of wood sequestrates on average 0.8 to 0.9 tonnes CO_2[79][80]. Building with wood could play a vital role in reducing air pollution and global warming. Being a natural material it will not produce any waste and can be recycled. Wood can also be an energy source for future cities.

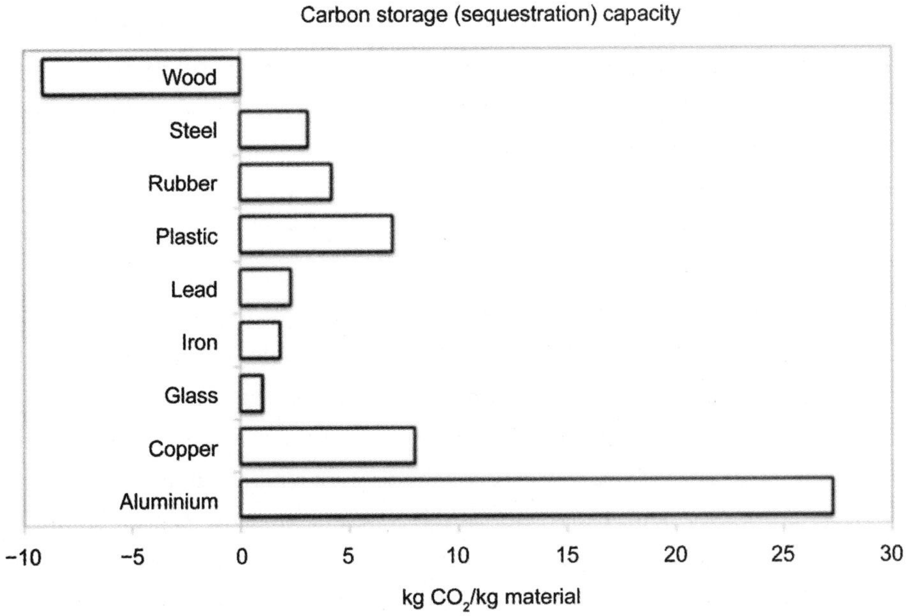

Figure 5. Carbon storage capacity of building construction materials[79][80].

5. Feeding Future Cities

According to WHO 50 percent of the 7 bn global population is currently living in cities requiring a land area for farming equivalent to half of South America to produce their food[83]. In the next 40 years there will be 3 billion more people[84][85] to feed implying 50% more food production[83]. Since 80% of the world's population is predicted to be living in cities by 2050, seemingly generating a conflict between using land for agriculture and for cities if the extra food production is obtained via traditional agriculture. But the problem of feeding the inhabitants in future cities may be less severe than we imagine. Historically, some cities at least integrated agriculture into their structure... Ankor Wat and Tenochtitian were mentioned above. During WW1 and WW2 the gardens and spare ground within British and German cities were turned over to the growing vegetables. Even Einstein cultivated an allotment in WW1, although he was reprimanded for it being untidy. Even the USA increased its food production in WW2 by promoting victory gardens. With the collapse of the Soviet Union in 1990, Cuba lost its supplies of fertilizers and agrichemicals precipitating a crisis in food production. To survive, Cubans turned to intensive urban agriculture to augment their food supplies, an activity which continues to this day. Ironically, when people are restricted to a diet of smaller amounts of freshly grown local food less in quantity than previously, their general level of health improves, an effect clearly evident in both 1940s Britain and 1990s Cuba. In a recent paper Thebo *et al.* suggest, based on satellite data, that urban agriculture already contributes significantly to the global food supply since an area within 20 km of cities equivalent to the 28 EU states combined is already being used for agricultural activities[86]. The detailed analysis is summarised in **Figure 6**;

Martellozzo *et al.* suggest the potential for vegetable growing within urban areas would require roughly one third of the total global urban area to meet the global vegetable consumption of urban dwellers. But the urban area available and suitable for urban agriculture varies considerably depending upon the nature of the agriculture performed. They reluctantly conclude that the space required is regrettably the highest where need is greatest, *i.e.*, in more food insecure countries. They note that smaller urban areas offer the most potential as regards physical space[87].

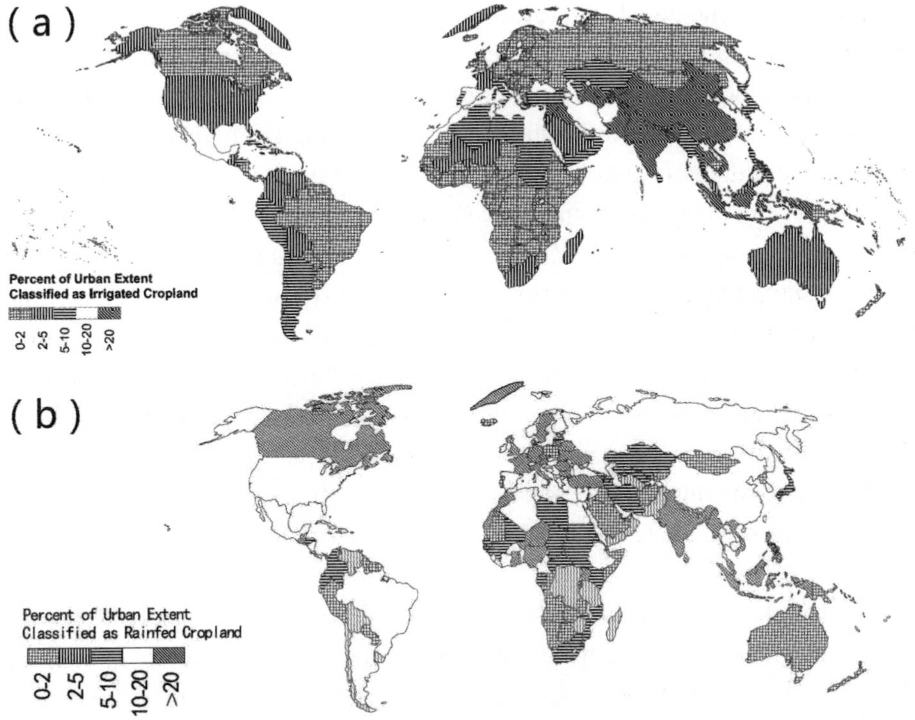

Figure 6. Schematic illustration of allocated urban area for irrigated and rainfed croplands[86]. Percent of urban area classified as (a) irrigated cropland, (b) rainfed cropland by country. Legend Proportion of irrigated cropland tend to be higher in regions having larger urban extend area used for irrigated cropland. However proportion of rainfed cropland is more dependent on regional climate patterns

In the developed world urban food growing is becoming popular perhaps for three reasons: firstly by the middle classes the appreciation that urban food cultivation can re-establish the link between food production and consumption, especially for children, encouraging them to adopt a more healthy diet; to supply free, fresh food for those in poverty and perhaps already relying upon food banks; and ironically for high end restaurants. One example of such community organisations world-wide is York Edible[88] in the city of York, UK.

To reduce their environmental impact future urban dwellers will increasingly grow food within, or at least in the immediate hinterlands, of their cities to avoid the CO_2 emissions associated with food transportation especially over transcontinental distances[89]. It is estimated that each 1 Calorie of consumed food uses currently 10 Calories of oil[90]–[92]. But where ground is at a premium, food production might be integrated into future cities by "Vertical Farming," *i.e.* multi-tier city farms in the form of glass protected skyscrapers or high rise towers that grow the maximum amount of on a minimum land area[93]–[95]. Although one dedicated vertical farm could feed up to 50,000 people[96], it is still likely that it will be beneficial for all buildings in future to have space reserved for food production. With the recent developments in photovoltaic (PV) technology it will be also possible to design vertical farms self-sufficient and completely sustainable. The primary energy consumption of vertical farms is for lighting (creating mimic sunlight) and water pumping for irrigation. Al-Chalabi conducted a study evaluating the sustainability of skyscrapers for vertical farming with different building dimensions as given in **Table 2**. According to the study results, for the vertical farms with a floor area less than 500 m^2, the available space on roof/façade is enough to install required the required number of PV panels. But once the floor area exceeds 500 m^2 the space on roof/façade is insufficient[97].

In March 2014, the world largest vertical farm was opened in Michigan (USA) with 17 million plants in plant racks using LED light to mimic sunlight[98][99]. The American National League of Cities is promoting urban agriculture[100] as a part of its remit to make cities more sustainable.

The most ambitious schemes for vertical farms will take a long time to realise, if ever. But some more modest examples already exist, for example in Singapore[101][102], Sky Greens has constructed a four storey building using traditional growing systems comprising soil based potted plants on a series of conveyor belts which migrate the plants near the windows maybe once or twice an hour so that every plant gets same amount of sunlight during the day. The technology increases food production by a factor of ten compared to that of traditional farming

Table 2. Optimisation model for the vertical farm. Adapted from[97].

Dimensions of building		Energy demand (one month timeline)				Energy supply	Feasible
Length/width (m)	Area/floor(m^2)	Water pumping required (kWh)	Light required (kWh)	Total required (kWh)	PV required (number of panels)	PV available on roof/façade (number of panels)	PV available-PV required
10	100	148	0	148	4	593	Yes
20	400	591	0	591	15	1289	Yes
22.5	506	748	57946	58694	1398	1479	Yes
25	625	923	137388	138311	3294	1675	No
28	784	1158	257393	258551	6165	1920	No
30	900	1329	352350	353679	8421	2088	No

on an equal land area[102]. Other vertical farms have been built in Korea, Japan, the USA and Sweden[98]–[100][102]. Singapore, one of the most densely populated countries, is considering a futuristic "floating vertical farm" designed by Forward Thinking Lab of Barcelona[103]. The system basically consists of looping towers that could float in local harbours, providing new space for year-round crops. The concept is inspired in part by floating fish farms that have been in use locally since the 1930s[103].

The flip side of producing and consuming food is that it creates human waste that must be treated to avoid pollution. Although human faeces and urine have been used historically as fertiliser thus creating an "eco-cycle" (**Figure 7**), it

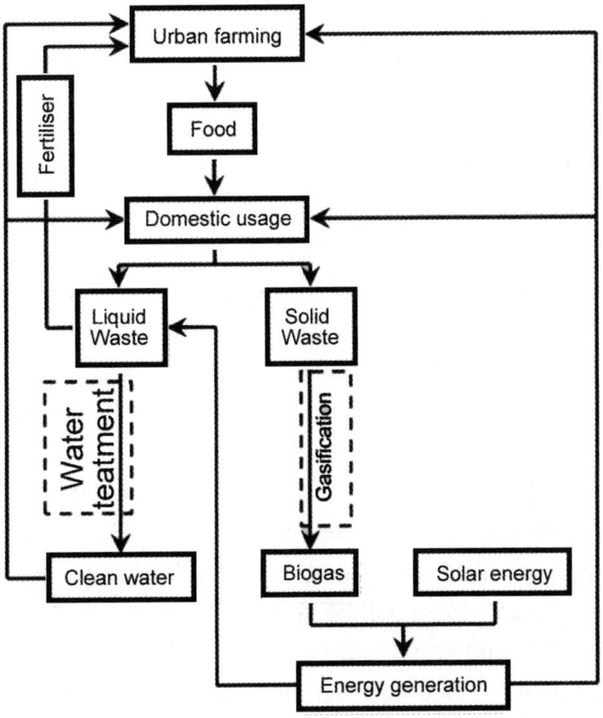

Figure 7. Eco-Cycle describing the recycling of solid waste and sewage for food, water and energy production for sustainability of future cities. Adapted from[105]. Legend Advanced water treatment systems for clean water production and advanced systems for gasification of solid waste for energy generation could allow considerable amount of water and energy savings. These could be reused for domestic needs and urban farming for food production. Additionally required fertilizer for urban farming can also be produced from sewage with processes such as "Pearl Process".

has long been discouraged in developed countries because of food contamination by pathogens. This especially applies to uncooked foods such as salads. Progressive build-up of toxic, heavy metals in the soil and thus plants is also a long term problem. But merely treating sewage and discharging the resulting effluent to rivers or the sea loses valuable nutrients, notably phosphorus, and also nitrogen and potassium, which have to be replaced from unsustainable sources. Vancouver based, Ostara has developed the "Pearl Process®" that recovers phosphorus and nitrogen plus magnesium from waste water to produce a slow release fertiliser called "Crystal Green®", which has a low heavy metal content[104]. The first European plant has recently been installed in Slough UK to treat water from a local industrial estate. Although Crystal Green is presently sold for conventional agriculture, technology of this type will be essential for sustainable urban agriculture. Energy input, required to operate the process, can potentially be obtained from renewable sources, especially solar[105].

6. Transport for Future Cities

To overcome rising traffic problems, cities should be compactly structured with improved accessibility, and have a well-designed transport network.

In future cities effective transportation will play a key role. For the health and well-being of citizens walking and cycling between their homes, workplaces, shops and other locations is already being encouraged. Undoubtedly these self-propelled systems will be integrated into future cities, avoiding the modern day perils of mixing pedestrians, cyclists and powered vehicles. For distances and occasions where self-propelled travel is impractical then future cities will need to strike a balance between mass public transport (buses and trams), individually-hired vehicles (taxis and rental cars) and individually owned vehicles. Various technologies already in development will impinge upon the choices made, such as self-driving vehicles[106][107], electric vehicles[108][109] and Aero-Mobil[110]. Aero-Mobil is a flying car that integrates existing infrastructure used for automobiles and planes. As a car it can fit into any standard parking space, uses regular gasoline, and can be used in road traffic just like any other car. As a plane it can use any airport in the world, but can also take off and land using any grass strip or paved surface just a few hundred meters long. It is now finalised and has been in regular flight-testing program in real flight conditions since October 2014. The Aero-Mobil is built from advanced composite material includes its body shell, wings, and wheels. According to company authorities the final product will include all the standard avionics and, an autopilot plus an advanced parachute deployment system[110].

Undoubtedly safety will be improved over current transport systems, but a serious debate is required concerning citizens' freedom to own their vehicles against potentially more energy efficient mass transport. The driverless, for-hire vehicle summoned by an app may find increasing acceptance. At present road vehicles are dual purpose, being used for both modes but in the future a distinction may be drawn between intra-city transport and inter-city transport[111][112]. For long distances the speeds of trains may increase so journey times rival those of intercity aircraft, but operate with superior energy efficiency, for example maglev trains[113][114] exploiting superconductivity and the "Aero-train" that is part train and part aircraft[115].

Traffic and transportation are growing problems for all cities. In Europe, people are wasting 10 to 60[116] hours in traffic jams each year, while their vehicles are contributing significantly to global warming by emitting carbon dioxide emissions and to air pollution by emitting nitrogen oxides and carbon particles. Therefore traffic management and monitoring systems are currently already being applied in many large cities and people are strongly encouraged to use public transport instead of personal vehicles. Even though some steps have been taken more radical, innovative will be needed. Perhaps "futuristic" solutions can no longer be viewed as engineers' fantasies but essential to avoid increasing urban transport problems.

Chinese engineers have allowed a Hongqi Q3 car to navigated itself through traffic to a destination 286 km away guided by cameras and sensors[117]. It is clear that computers can be safer drivers than human beings. They can react quicker, they can look in all directions at once and they don't get distracted. They won't speed or cut people up. Principally, self-driving cars will be connected with a wireless network similar to the internet or telephone network and all cars will be travelling on major roads under control of satellite and roadside control systems[118]. A traffic jam will be predicted before it even happens by using roadside sensors, GPS and other advanced software.

An alternate innovative design for future transport is the Aero-train which is partly train and partly aircraft. The vehicle is designed with wings and flies on an air cushion along a concrete track using wing and ground effects. This minimises the drag effect allowing the aero-train to consume less energy whilst reaching higher speeds than the conventional trains[115].

Another imaginative idea, first proposed by Robert M. Salter in the 1970s, is the evacuated tube transport (ETT) where a vehicle occurs in a vacuum to eliminate air resistance and friction,[119]. The ETT system envisions super-conducting maglev trains operating at speeds of up to 6,500 km/h (4,039 mph) on international trips—that's New York to Beijing in two hours! Although the proponents say that ETT could be 50 times more efficient than electric cars or trains it is only a concept that is the subject of ongoing research[120]. But the achievement of ETT would revolutionise future long distance transportation.

7. Administering Future Cities

According to the UN report 70% of the world's population will live in urban areas by 2050[61][62] with the number of cities expected to exceed 2000 by 2030, compared to 1551 in 2010[121]. Whilst there are 43 "large cities" with populations between 5 and 10 million in 2014, there are expected to be 63 by 2030[62]. The UN estimates that there will be more than 40 mega-cities worldwide by 2030, each with a population of at least 10 million, compared to 28 today. It is projected that Delhi, Shanghai and Tokyo will each have more than 30 million people by 2030, and will be the world's largest urban agglomerations[62]. This massive global growth of urban areas will requires developments in administrative systems to ensure that technological advances described in previous sections truly deliver improved living conditions for all urban dwellers. Key challenges for future cities will be productivity, sustainability, liveability and good governance[122]–[125] as summarized in **Table 3**.

Although the well-established scientific basis for global warming is well established, its full impact appears to be several decades in the future, action is

Table 3. Challenges for future cities and desired objectives-principles to overcome these challenges[122]-[125].

Challenges ⇨	Objectives ⇨	Principles
Productivity	o Improve labour and capital productivity	
	o Integrate land use and infrastructure	Efficiency
	o Improve the efficiency of urban infrastructure	
Sustainability	o Protect and sustain our nature land built environments	Value for money
	o Reduce greenhouse gas emissions and improve air quality	Innovation
	o Manage our resources sustainably	Adaptability
	o Increase resilience to climate change, emergency events and natural hazards	Resilience
Liveability	o Facilitate the supply of appropriate mixed income housing	Equity
	o Support affordable living choices	
	o Improve accessibility and reduce dependency on private vehicles	Affordability
	o Support community wellbeing	Subsidiarity
Good Governance	o Improve the planning and management of our cities	Integration
	o Streamline administrative processes	
	o Evaluate progress	Engagement

required now to ameliorate its effects by identifying, prioritizing, and structuring new design and managerial tools to improve urban environmental and fiscal sustainability[126]. Despite the vociferous assertions of those denying man-made global warming, the ill effects of "local warming", known as the "urban heat island effect" (UHIE), are already manifest in large cities, especially in the tropics and serves as early warning of what is likely to happen worldwide as global warming becomes more pronounced. UHIE refers to the urban temperature being higher than that of the surrounding countryside, 3 K being not a typical, a combination of solar heat absorption by buildings, roads etc., heat emitted by vehicles, and the conversion of electrical to thermal energy, for example by air conditioners.

The UHIE does not just cause discomfort for urban inhabitants, it is also a killer. Various studies of temperature related excess mortality using historical data have shown that during heat waves above a threshold temperature deaths increase significantly with each further degree rise. Not surprisingly, the young, old and those with serious medical conditions, a most vulnerable[127].

If a city experiences a heat wave of lasting longer than 5 days the UHIE results in excess mortality that raises steeply with every extra degree of excess temperature during night time; the correlation is weaker with the daytime temperature. Not surprisingly, the young, the old and those with existing medical conditions are most at risk.

Mitigation technologies such as increasing green urban space and biodiversity, use of reflective materials, decrease of anthropogenic heat levels and use of low temperature natural sinks (such as ground or water bodies) aiming to counter the impact of the phenomenon are rapidly being developed and applied in real projects[128]. Rehan provided a detailed framework, including several measures that will diminish the accumulation of heat in urban areas and mitigate their UHIE by a set of planning actions as a strategy to cool the cities. The framework is given in **Figure 8**[129]. Richer urbanites can in principal, offset the effects of the UHIE merely by turning up their air conditioning or installing more powerful units. But such would be socially reprehensible since it would increase urban temperatures

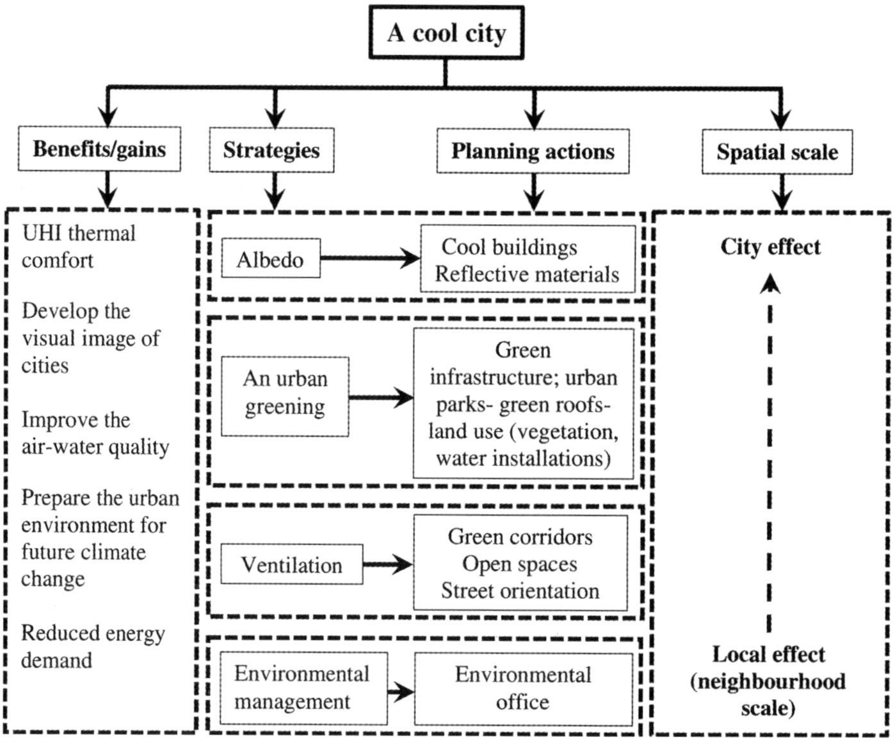

Figure 8. Cool city-Framework[129].

further to the disadvantage of poor who cannot afford a/c or the power to run it. Nevertheless, to protect the vulnerable it may be necessary to build air conditioned refuges where they can be sent when local temperatures are high.

Administrators are already aware of the need to incorporating UHIE mitigation as cities are further developed and is required in temperate regions as well as the topics. For example Public Health England has recently published an excellent guide the adverse effects of high temperatures and methods to combat them, both short and long term[130].

Although excess mortality caused by the UHIE is occurring now, it will be exacerbated by the rise in global temperatures projected by the Intergovernmental Panel on Climate Change (IPCC)[128][131]. Those who argue strongly that man-made global is a myth and therefore nothing needs to be done to mitigate it, must face the consequence if the world follows their lead and they are wrong people, especially the poor and vulnerable will die. The adverse effect of increasing temperatures is based on sound research and historical data. It is not a theory derived from a computer model. Planning and building "cool" cities must be a primary objective of administrations now to minimise deaths.

In large cities excess mortality from attributable to high temperatures is exacerbated by air pollution, notably NO_x from internal combustion engines that ozone produced by the sunlight-induced reaction of oxygen with unburnt hydrocarbons. Indeed, separating excess urban mortality arising from pollution and high temperatures is problematical. Fortunately mitigations for both tend to be the same... for example by reducing fossil fuel combustion to supply electric power by increasing renewables, reducing the number of internal combustion engine vehicles coupled with the provision of better public transport and "greening" cities by planting more vegetation capable of both cooling the air and absorbing airborne pollutants.

Originally, "green" infrastructure was identified with parkland, forests, wetlands, greenbelts, or floodways in and around cities that provided improved quality of life or "ecosystem services" such as water filtration and flood control[132][133]. Now, green infrastructure is more often related to environmental or sustainability goals that cities are trying to achieve through a mix of natural approaches. Examples of "green" infrastructure and technological practices include green[134], blue, and white roofs[132] (See: **Figure 9**); hard and soft permeable surfaces; green alleys and streets; urban forestry; green open spaces such as parks and wetlands; and adapting buildings to better cope with floods and coastal storm surges[132].

The climate adaptation benefits of green infrastructure are generally related to its ability to moderate the expected increases in extreme precipitation or temperature. Benefits include better management of storm-water runoff, lowering incidents of combined storm and sewer overflows (CSOs), water capture and conservation, flood prevention, accommodation of natural hazards (e.g., relocating out of floodplains), reduced ambient temperatures and urban heat island (UHI) effects, and defense against sea level rise (with potential of storm-surge protection measures). The U.S. Environmental Protection Agency (EPA) has also identified green infrastructure as a contributor to improving human health and air quality, lowering energy demand, reducing capital cost savings, increasing carbon storage, expanding wildlife habitat and recreational space, and even increasing land-values by up to 30%[132][135].

Green infrastructure (GI) in urban areas can ameliorate the warming effects of climate change and the UHIE. In a study performed by Gill *et al.* (2007)[136][137], it is found that that increasing the current area of GI in Greater Manchester,

Figure 9. A view of an urban green roof in Chicago, USA[134].

UK by 10% (in areas with little or no green cover) would result in a cooling by up to 2.5°C under the high emissions scenarios based on UKCP02 predictions. The potential benefits of increased green infrastructure/green space on reducing UHIE are presented below:

- Trees and shrubs provide protection from both heat and UV radiation by direct shading (both of buildings and outdoor spaces).
- Evapotranspiration reduces the temperature in the area around vegetation by converting solar radiation to latent heat.
- Lower temperatures caused by both evapotranspiration and direct shading lead to a reduction in the amount of heat absorbed (and therefore emitted) by low albedo man-made urban surfaces[136][138].

These changes can only be achieved by a city administration that has both the appropriate legal power and the will to serve all its city's inhabitants, the poor as well as the rich.

Much of what has been discussed above has ignored the differing sizes of conurbations.

Cities face different impacts, depending upon their sizes and levels of development. Small cities of upper income nations are facing with population decline as a result of the migration to larger cities for better job opportunities and higher life standards. Diminishing manpower makes it difficult for small cities to compete globally in terms of economy and productivity. On the other hand large cities in developed world are facing with the impacts of aging infrastructure and population. Increasing population creates inequality and social cohesion inside the cities while job opportunities become more competitive[125].

In contrast to developed nations, small cities in developing countries are faced with the impacts of weak economies and weak urban governance. Due to their inadequate infrastructure and buildings, such cities lack the resilience to survive natural disasters such as earthquake or flood is very low. This Survival is threatened and in many cases many people are forced to vacate their homes [See

Figure 10(b)]. Even large cities in the developing world with inadequate/insufficient transport as well as poor housing stock, are threatened by demographic challenges resulting in social and economical inequality. Environmental pollution is probably the most significant problem facing these cities, a result of the rapid industrialisation. But the latter potentially creates the wealth that can enable developing world cities to overcome their growing pains provided it is harnessed for benefit of all and is not siphoned off by corruption.

In many cities of China [See **Figure 10(a)**] and India, environmental pollution, particularly P.M 2.5 (e.g. oil smoke, fly ash, cement dust) level is sharply rising and threatening the human health as P.M 2.5 particles have the ability to penetrate deep into the lungs and can cause severe health problems[139].

	Upper Income Nations	Developing World
Small Cities	o Population decline o Competing globally o Economic restructuring	o Weak economies o Inadequate infrastructure for provision of basic public services o Weak urban governance
Large Cities	o Ageing infrastructure & population o Inequality and social cohesion o Increasing competition for global leadership	o Inadequate transport infrastructure and severe congestion o Population housing and slums o Demographic challenges o Environmental problems o Social and economic inequality

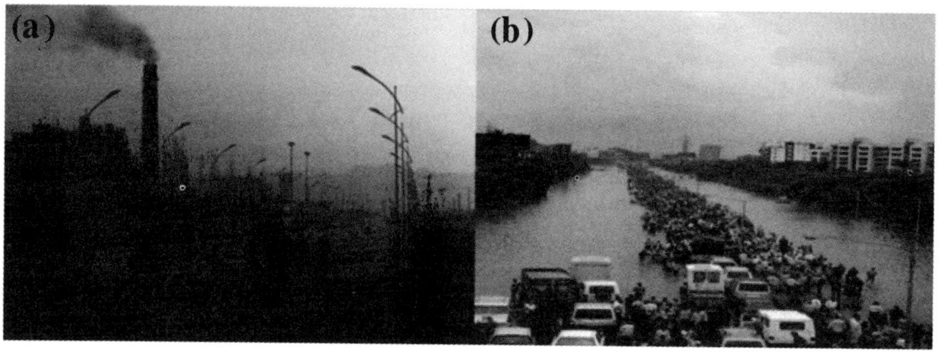

Figure 10. Impacts of global urbanization; (a) Flue gas emitting from a factory polluting the air in an urban area. (b) People vacating their houses after a flood disaster. Table represents the specific impacts of global urbanization depending on the size and development level of a city.

The severe impacts of rapid urbanisation can be ameliorated by applying creative design to infrastructure. Ecosystems like multifunctional units will provide several uses rather than a single functionality thereby saving energy, time and cost. For instance garden plots can serve as water management system while providing food for citizens. Similarly multifunctional buildings could save time for people while allowing efficient use of land[140]. Significant advances in computer simulation provided tools that enable us to evaluate current conditions and requirements thus modelling future scenarios. This phenomenon will have increasing importance in future cities to monitor existing conditions for efficient use of capital and natural resources or controlling traffic flow through wireless sensor networks[141]-[143]. In addition it will allow modifying energy usage or household waste of urban dwellings with real time feedback[144]-[147]. Republic of Korea has already put this technology into practice in city of Songdo, where traffic, waste and energy usage are monitored[140]. Similarly in Rio de Janeiro there is a high-tech centre where public safety responses to natural disasters or building collapses are quickly identified[146][147]. The recent earthquake in Nepal demonstrated that, this kind of technological centre could save many lives with timely intervention during disasters.

Technically, highly automated management systems are very attractive, but they have potential downsides. The amassing of large amounts of data about individuals' daily lives is already creating grave concern, both via the internet and CCTV. Increasing data collection could offer the potential for city authorities to exercise greater control over citizen's lives. Technology must be tempered by democratic safeguards if individual liberties are not to be infringed. The vulnerability of a highly networked city to a physical or a cyberattack on data centres must be minimised. A fascinating concept currently being pursued is the adaption of the "block-chain" algorithm underlying "bitcoin" to the administration of organisations to increase transparency and to minimise corruption[148].

For example recently "Ethereum" launched is a block-chain platform that allows secure systems to be developed with transactions permanently and transparently recorded[149].

Large corporations already experimenting Ethereum, include UBS and Barclays. "BoardRoom" is a block-chain dapp (distributed app) founded by Dobson who claims that it can be used to run large organisations by "collaborative decision making"[150].

These developments could potentially be just important to the operation of modern cities as the new engineering technologies.

8. Socio-Economic Development and Prosperity of Future Cities

Emerging cities should be where human beings find satisfaction of basic needs and essential public goods. Where various products can be found in sufficiency and their utility enjoyed. Future cities should also be the habitats where ambitions, aspirations and other immaterial aspects of life are realized, providing contentment and happiness and increasing the prospects of individual and collective well-being. However in many developing cities, prosperity is absent or restricted to some groups or only enjoyed in some parts of the city[6].

Low purchase power contrarily increasing expenses could socioeconomically pressurize individuals and minimize their social subsistence. This situation will turn citizens from productive and creative individuals to the ones just trying to survive. "Liveable" cities should support affordable living choices, provide citizens options to have a social status and life conditions independent than their income. Cities also should be compact structured with improved accessibility, they should include natural habitats allowing biodiversity and socialisation of individuals and should have a well-designed transport network which will eliminate the need for private vehicles to overcome the rising traffic problem in growing cities. Besides they should offer a profusion of public goods, develop actions/policies for a sustainable use and more importantly should enable equitable access to "commons"[3] in order to ensure well-being of citizens.

The future urban configurations should concentrate on efficient use of resources and opportunities that could help to achieve prosperity and citizen well-being in five dimensions as defined below and illustrated in **Figure 11**.

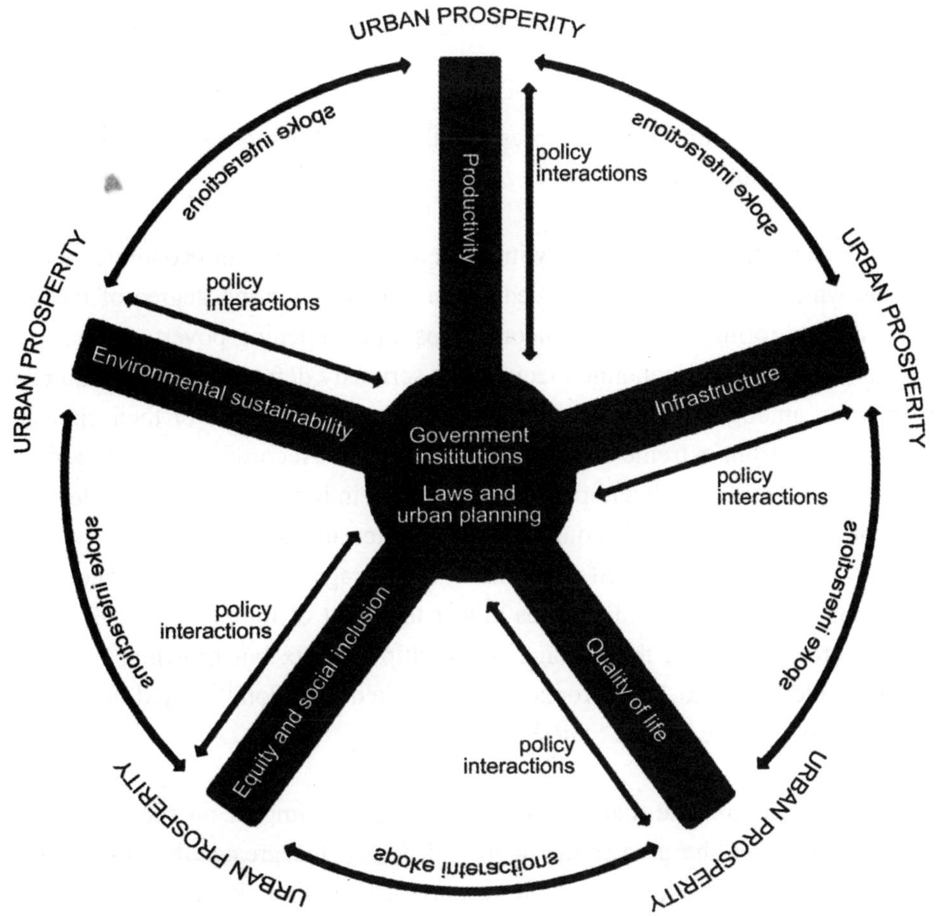

Figure 11. Wheel of urban prosperity[6].

- Contribute to economic growth through productivity, generating the income and employment that afford adequate living standards for the whole population.

- Deploy the infrastructure, physical assets and amenities—adequate water, sanitation, power supply, road network, information and communications technology etc.—required to sustain both the population and the economy.

- Provide the social services—education, health, recreation, safety and security etc.—required for improved living standards, enabling the population to maximize individual potential and lead fulfilling lives.

- Minimize poverty, inequalities and segments of the population live in abject poverty and deprivation.

- Protect the environment and preserve the natural assets for the sake of sustainable urbanization.

The past few decades have witnessed a notable surge in economic growth, but one which has been accompanied by an equally daunting degree of inequity under various forms, with wider income gaps and deepening poverty in many cities across the world. Economic inequality is seriously detrimental to the equitable distribution among individuals of opportunities to pursue a life of their choosing and be spared from extreme deprivation in outcomes. According to recent reports, income gaps between rich and poor are expanding in both developed and developing countries[6][151]. In OECD countries, inequalities are as steep as they have been for over 30 years. In advanced economies, the average income of the richest 10% of the population is about nine times higher than that of the poorest 10%. In Europe's Nordic countries, the average is a multiple of six but growing, compared with multiples of 10 in Italy, Korea and the United Kingdom[151], and up to 14 in Israel, Turkey and the United States[6].

Cities must realize that equity has a significant impact on socio-economic performance, since the greater the degree of equity, the greater the chances of a wider, more efficient use of available resources, including skills and creative talent[152] Urban prosperity thrives on equity, which involves reduction in barriers on individual/collective potential, expansion of opportunities, and strengthening of human agency[6][153] and civic engagement. Cities generate wealth, but the problem is the unequal distribution of it. Despite considerable increases in productivity (e.g. GDP per capita) along with reductions in extreme poverty, inequality as a whole is growing in most parts of the world—a process that undermines urban life quality[154]. In many cities, the population and local experts concur that inequalities are becoming steeper which could be a threat for emerging cities in terms of their sustainability and well-being of citizens.

9. Conclusions

The "Future Cities" topic employs a multidisciplinary approach to address the urban development challenges facing emerging cities. This can integrate environmental technologies, comprehensive urban development, fiscal sustainability and good governance, to provide emerging cities with a set of tools in order to improve the quality of life globally.

New-born babies in developed countries are projected to have a life expectancy of 80+ years[155], with the majority living in cities, increasing yet further the demand for energy, water, food, housing and other services. However, cities are struggling with climate change, changes in population and demographics, congestion, healthcare, and pressure on key resources. In the future innovative technologies/approaches will create considerable market opportunities to transform existing conurbations into the efficient, attractive and resilient cities of the future.

Nevertheless, simply applying innovative technologies alone will not guarantee the combination of sustainability and acceptable living standards for future cities... good governance and management will also play a pivotal role. This can only be provided by utilizing technological advancements optimally whilst also developing short and long term management, organization and development strategies to realize the desired objectives.

Abbreviations

BedZED: Beddington zero energy development; EPBD: energy performance of buildings directive; ETT: evacuated tube transport; EU: European Union; GWP: global warming potential; IPCC: intergovernmental panel on climate change; PV: Photovoltaic; UA: urban area; UHI: urban heat island; UHIE: urban heat island effect; UK: United Kingdom; UN: united nations; ZCB: zero carbon building; ZEB: zero energy building.

Competing Interests

The authors declare that they have no competing interests.

Authors' Contributions

DA and RP drafted the manuscript. SR has supervised the presented research, done the final revision of the manuscript and given final approval of the version to be published. All authors read and approved the final manuscript.

Acknowledgements

The authors wish to gratefully acknowledge the various reference resources benefitted for preparing this paper. The valuable suggestions of the anonymous reviewers in improving the quality of the manuscript are also greatly appreciated.

Source: Riffat S, Powell R, Aydin D. Future cities and environmental sustainability[J]. Future Cities & Environment, 2016, 2(1):1−23.

References

[1] Mumford L (1961) The City in History: Its origins, its transformations and its prospects. Harcourt, Brace and World Inc., New York.

[2] Hirst KK (2015) Khmer Empire Water Management System. http://archaeology.about.com/od/transportation/qt/Khmer-Empire-Water-Management-System.htm. Accessed 15 August 2015.

[3] UNESCO World heritage list: Angkor. http://whc.unesco.org/en/list/668/. Accessed 18 August 2015.

[4] Matuschke I (2009) Rapid urbanization and food security: Using food density maps to identify future food security hotspots. International Association of Agricultural Economists Conference, Beijing, China, August 16-22, 2009. http://www.fao.org/fileadmin/user_upload/esag/docs/RapidUrbanizationFoodSecurity.pdf. Accessed 12 August 2015.

[5] Allenby B, Fink J (2005) Toward Inherently Secure and Resilient Societies. Science 309:1034–1036. doi:10.1126/science.1111534. http://science.sciencemag.org/content/sci/309/5737/1034.full.pdf. Accessed 10 August 2015.

[6] United Nations Human Settlements Programme (UN-HABITAT) (2013) State of the World's Cities 2012/2013. Prosperity of cities. Routledge, Taylor & Francis, New York, USA, file:///C:/Users/ezxda4/Downloads/3387_alt.pdf. Accessed 01 February

2016. ISBN 978-92-1-132494-5.

[7] Singh S, Kennedy C (2015) Estimating future energy use and CO2 emissions of the world's cities. Environmental Pollution 203:271–278.

[8] Mayhew H (2010) London Labour and the London Poor. OUP Oxford, ISBN 0191501476, 9780191501470. Provides a modern selection from the original volumes.

[9] Betjeman J (1958) John Betjeman's Collected Poems. John Murray, London.

[10] HRH Prince Charles (1989) A Vision of Britain: A Personal View of Architecture. Doubleday, ISBN-13: 978-0385269032.

[11] http://poundbury.org.uk/. Accessed 18 August 2015.

[12] Howard E (1902) Garden Cities of Tomorrow. S. Sonnenschein & Co. Ltd., London.

[13] http://www.sustainablecities.org.uk/sustainable-cities/ Accessed18 August 2015.

[14] http://www.mrra.net/wp-content/uploads/Why-Recycle.pdf. Accessed 02 Fubruary 2015.

[15] United Nations Human Settlements Programme (UN Habitat) (2010) Solid Waste Management In the World Cities: Water and Sanitation in the World's Cities. Earth-scan publishing, London, UK. mirror.unhabitat.org/pmss/getElectronicVersion.aspx?nr=2918&alt=1. Accessed 02 February 2016. ISBN 978-1-84971-169-2.

[16] Pointer G (2005) Focus on People and Migration. Chapter 3: The UK's major urban areas. http://www.ons.gov.uk/ons/rel/fertility-analysis/focus-on-people-and-migration/december-2005/focus-on-people-and-migration—focus-on-people-and-migration—chapter-3.pdf. Accessed 02 February 2016.

[17] https://connect.innovateuk.org/web/future-cities-special-interest-group/definition. Accessed 03 February 2016.

[18] Technology Strategy Board (2013) Solutions for Cities: An analysis of the feasibility studies from the Future Cities Demonstrator Programme; ARUP. http://publications.arup.com/~/media/Publications/Files/Publications/S/Solutions_for_Cities_An_analysis_of_the_Feasibility_Studies_from_the_Future_Cities_Demonstrator_Pro-gramme.ashx. Accessed 03 February 2016.

[19] Clark L (2012) Technology Strategy Board opens "Future Cities" design contest. http://www.wired.co.uk/news/archive/2012-06/12/future-cities-competition. https://connect.innovateuk.org/web/future-cities-specialinterest-group/definition. Accessed 04 February 2016.

[20] Special Report of The Intergovernmental Panel on Climate Change (IPCC) (2011). Renewable Energy Sources and Climate Change Mitigation. Summary for Policy makers and Technical Summary. ISBN 978-92-9169-131-9. https://www.ipcc.ch/pdf/special-reports/srren/SRREN_FD_SPM_final.pdf.
Accessed 04 February 2016.

[21] European Technology Platform on Renewable Heating and Cooling (2011) 2020 –

2030 – 2050 Common Vision for the Renewable Heating & Cooling sector in Europe. Publications Office of the European Union, Luxembourg. doi:10.2788/20474. ftp://ftp.cordis.europa.eu/pub/etp/docs/rhc-vision_en.pdf. Accessed 04 February 2016. ISBN 978-92-79-19056-8.

[22] Phills JA, Deiglmeier K and Miller DT (2008) Rediscovering Social Innovation. Stanford Social Innovation Review. http://www.ssireview.org/articles/entry/rediscovering_social_innovation/. Accessed 04 February 2016.

[23] UN-Habitat: Scaling New Heights (2010) New Ideas in Urban Planning. Urban World, Vol. 1, Issue, 4, Nairobi. http://docplayer.net/7307781-Urban-world-scaling-new-heights-new-ideas-in-urban-planning-volume-1-issue-4.html. Accessed 01 February 2016.

[24] Kylili A, Fokaides PA (2015) European smart cities: The role of zero energy buildings. Sustainable Cities and Society 15:86–95.

[25] UK Department for Communities and Local Government (2014) Locally led Garden Cities, ISBN: 978-1-4098-4204-0. https://www.gov.uk/government/uploads/system/uploads/attachment_data/file/303324/20140414_Locally-led_Garden_Cities_final_signed.pdf. Accessed 28 July 2015.

[26] http://zcb.hkcic.org/Eng/index.aspx. Accessed 21 July 2015.

[27] http://www.zaha-hadid.com/architecture/jockey-club-innovation-tower/. Accessed 23 July 2015.

[28] Future Communities (2015) Hammarby, Sjostad, Stockholm, Sweden, 1995 to 2015; building a green city extension.http://www.futurecommunities.net/case-studies/hammarby-sjostad-stockholm-sweden-1995-2015. Accessed 14 August 2015.

[29] http://www.zedfactory.com/zed/. Accessed 12 August 2015.

[30] http://www.hkip.org.hk/plcc/download/Japan.pdf. Accessed 29 July 2015.

[31] Nader S (2009) Paths to low-carbon economy—The Masdar example. Energy Procedia 1:3951–3958. doi:10.1016/j.egypro.2009.02.199.

[32] http://www.masdar.ae/en/masdar-city/the-built-environment. Accessed 21 July 2015.

[33] Goldschein E (2011) Kuwait is building a $132 billion city around a skyscraper with an "Arabian Nights" Theme. Business Insider. http://www.businessinsider.com/kuwait-madinat-al-hareer-skyscraper-2011-12?IR=T. Accessed 15 July 2015.

[34] http://www2.kaec.net/. Accessed 06 July 2015.

[35] Moser S, Swain M, Alkhabbaz MH (2015) King Abdullah Economic City: Engineering Saudi Arabia's post-oil future. Cities 45: 71–80. http://dx.doi.org/10.1016/j.cities.2015.03.001.

[36] Thorold C (2008) New cities rise from Saudi Desert. BBC News.http://news.bbc.co.uk/1/hi/world/middle_east/7446923.stm. Accessed 02 July 2015.

[37] http://www.tianjinecocity.gov.sg/. Accessed 28 June 2015.

[38] Kaiman J (2014) China's eco-cities: Empty of hospitals, shopping centres and hospitals. The Guardian. http://www.theguardian.com/cities/2014/apr/14/china-tian-jin-eco-city-empty-hospitals-people. Accessed 21 June 2015.

[39] Feldman J (2014) China's lofty dream of eco-friendly super cities at major crossroads. The Huffington Post. http://www.huffingtonpost.com/2014/07/17/wuhan-china-environmental_n_5579019.html. Accessed 05 June 2015.

[40] http://www.avestaconcern.com/en/project/7. Accessed 19 June 2015.

[41] Harris S (2015) Going underground: Cities of the future. The Engineer. http://www.theengineer.co.uk/blog/going-underground-cities-of-the-future/1019844.article. Accessed 13 June 2015.

[42] Kaliampakos D, Benardos A, Mavrikos A, Panagiotopoulos G (2015) The Underground Atlas Project. Tunnel. Underg. Space Technol. Article in press. http://dx.doi.org/10.1016/j.tust.2015.03.009.

[43] Good A (2014) The city of the future could lie below your feet. USC News. https://news.usc.edu/71414/the-city-of-the-future-could-lie-below-your-feet/. Accessed 01 July 2015.

[44] Nuwer R (2013) Will we ever…live in underwater cities. BBC Future. http://www.bbc.com/future/story/20130930-can-we-build-underwater-cities. Accessed 08 July 2015.

[45] http://www.seasteading.org/floating-city-project/. Accessed 12 July 2015.

[46] Gamble J (2014) Has the time come for floating cities. The Guardian. http://www.theguardian.com/cities/2014/mar/18/floating-cities-proposals-utopiansci-fi. Accessed 26 June 2015.

[47] Robinson J (2014) The city in the sky: Ambitious blueprint for London tower block that could house thousands of people—as well as schools, offices, shops and even parks. Daily Mail.http://www.dailymail.co.uk/news/article-2735522/City-sky-Ambitious-tower-block-house-thousands-people-schools-offices-shops-parks.html. Accessed 23 June 2015.

[48] http://www.hrama.com/skycity/. Accessed 26 May 2015.

[49] http://www.sure-architecture.com/. Accessed 02 June 2015.

[50] Withnall A (2014) Japanese construction firm says this "Ocean Spiral" is the underwater city of the future. The Independent. http://www.independent.co.uk/news/world/asia/japanese-construction-firm-says-this-ocean-spiral-is-the-underwater-city-of-the-future-9882532.html. Accessed 23 June 2015.

[51] O'Callaghan J (2014) A 21st century ATLANTIS: Floating spheres that house entire cities and sink to the seabed in extreme weather could be built by 2030. Daily Mail. http://www.dailymail.co.uk/sciencetech/article-2847244/A-21st-century-ATLANTIS

-Floating-spheres-house-entire-cities-sink-seabed-extreme-weather-built-2030.html. Accessed 22 June 2015.

[52] http://www.shimz.co.jp/english/theme/dream/oceanspiral.html. Accessed 23 June 2015.

[53] http://vincent.callebaut.org/page1-img-lilypad.html. Accessed 28 June 2015.

[54] Chapa J (2008) Lilypad: Floating City for Climate Change Refugees. Inhabitat. http://inhabitat.com/lilypad-floating-cities-in-the-age-of-global-warming/. Accessed 29 June 2015.

[55] https://www.thevenusproject.com/en/. Accessed 01 July 2015.

[56] Tomorrow's cities (2013) How the Venus Project redesigning the future. BBC News. http://www.bbc.co.uk/news/technology-23799590. Accessed 01 July 2015.

[57] Jordana S (2010) NewBud eco-school/Zhu Jingxiang architects. Arch Daily. http://www.archdaily.com/82039/newbud-eco-school-zhu-jingxiang-architects/. Accessed 22 June 2015.

[58] Bullivant L (2012) Master planning futures. Routledge Publishing, Taylor & Francis Group, New York.

[59] Kamal-Chaoui L, Leman E, Rufei Z, Urban Trends and Policy in China (2009) OECD Regional Development Working Papers. OECD publishing. http://www.oecd.org/china/42607972.pdf. doi:10.1787/225205036417. Accessed 23 June 2015.

[60] An Assessment of the Intergovernmental Panel on Climate Change (2007) Climate Change 2007: Synthesis Report. https://www.ipcc.ch/pdf/assessment-report/ar4/syr/ar4_syr.pdf. Accessed 28 June 2015.

[61] The International Bank for Reconstruction and Development/The World Bank (2010) Cities and Climate Change: An Urgent Agenda. Urban Development Series Knowledge Papers. http://siteresources.worldbank.org/INTUWM/Resources/340232-1205330656272/CitiesandClimateChange.pdf. Accessed 18 June 2015.

[62] The Department of Economic and Social Affairs of the United Nations (2014) World Urbanization Prospects: The 2014 Revision. United Nations, New York. http://esa.un.org/unpd/wup/Highlights/WUP2014-Highlights.pdf. Accessed 21 June 2015. ISBN 978-92-1-151517-6.

[63] Gillman O (2015) The villas created using 3D printers: £100,000 five storey homes made using construction waste in China, Daily Mail. http://www.dailymail.co.uk/news/article-2917025/The-villas-created-using-3D-printers-100-000-five-storey-homes-using-construction-waste-China.html. Accessed 13 June 2015.

[64] Berman B (2012) 3-D printing: The new industrial revolution. Business Horizons 55:155–162. doi:10.1016/j.bushor.2011.11.003.

[65] http://www.contourcrafting.org/. Accessed 21 June 2015.

[66] http://gadgets.ndtv.com/laptops/news/new-giant-3d-printer-can-build-a-house-in-24-

hours-470564. Accessed 21 June 2015.

[67] Gold S, Rubik F (2009) Consumer attitudes towards timber as a construction material and towards timber frame houses – selected findings of a representative survey among the German population. Journal of Cleaner Production 17:303–309. doi:10.1016/j.jclepro.2008.07.001.

[68] Su S, Wu W, Gao J, Lu J, Fan C (2012) Nanomaterials-based sensors for applications in environmental monitoring. J Mater Chem 22:18101–18110. doi:10.1039/C2JM33284A.

[69] Xia L, Xie R, Ju X, Wang W, Chen Q, Chu L (2013) Nano-structured smart hydrogels with rapid response and high elasticity. Nature Communications 4:2226. doi:10.1038/ncomms3226.

[70] United Nations Environment Programme (2009) Critical metals for future sustainable technologies and their recycling potential. Sustainable innovation and technology transfer industrial sector studies. http://www.unep.fr/shared/publications/pdf/DTIx 1202xPA-Critical%20Metals%20and%20their%20Recycling%20Potential.pdf. Accessed 12 August 2015.

[71] UK Department of Energy & Climate Change (2013) Government response to the consultation on proposals to enhance the sustainability criteria for the use of biomass feedstocks under the renewables obligation (RO). https://www.gov.uk/government /uploads/system/uploads/attachment_data/file/231102/RO_Biomass_Sustainability_c onsultation_-_Government_Response_22_August_2013.pdf. Accessed 06 June 2015.

[72] EPA Region 2 (2012) Unleashing Green Chemistry and Engineering in Service of a Sustainable Future, Final Report. New York. http://www.njbin.org/attachments/368_ Unleashing%20Green%20Chemistry%20Report%20-%20Final.pdf. Accessed 28 July 2015.

[73] Buildings Performance Institute Europe (BPIE) (2011) Europe's buildings under the microscope: A country-by-country review of the energy performance of buildings. ISBN: 9789491143014, Belgium. http://www.europeanclimate.org/documents/LR_ %20CbC_study.pdf. Accessed 23 July 2015.

[74] Power A (2010) Housing and sustainability: demolition or refurbishment? Urban Design and Planning 163:205–216. doi:10.1680/udap.2010.163.4.205.

[75] Directive 2010/31/eu of the european parliament and of the council (2010) Official Journal of the European Union. http://eur-lex.europa.eu/legal-content/en/TXT/?uri= celex%3A32010L0031. Accessed 23 July 2015.

[76] Mallo MFL, Espinoza O (2015) Awareness, perceptions and willingness to adopt cross-laminated timber by the architecture community in the United States. Journal of Cleaner Production 94:198–210. doi:10.1016/j.jclepro.2015.01.090.

[77] Cuadrado J, Zubizarreta M, Pelaz B, Marcos I (2015) Methodology to assess the environmental sustainability of timber structures. Construction and Building Materials 86:149–158. http://dx.doi.org/10.1016/j.conbuildmat.2015. 03.109. Accessed 22

July 2015.

[78] Cathcart-Keays A (2014) Wooden skyscrapers could be the future of flat-pack cities around the world. The Guardian. http://www.theguardian.com/cities/2014/oct/03/-sp-wooden-skyscrapers-future-world plyscrapers. Accessed 25 July 2015.

[79] Lehmann S (2013) Low-carbon construction: the rise, fall and rise again of engineered timber systems. 12th International Conference on Sustainable Energy technologies (SET-2013), Hong Kong.

[80] Dodoo A, Gustavsson L, Sathre R (2014) Lifecycle carbon implications of conventional and low-energy multi-storey timber building systems. Energy and Buildings 82:194–210. http://dx.doi.org/10.1016/j.enbuild.2014.06.034.

[81] European Comission Research Innovation (2015) Nano in Energy/ Environment. http://ec.europa.eu/research/industrial_technologies/nano-in-energy-environment_en.html. Accessed 25 July 2015.

[82] Quagliarini E, Bondioli F, Goffredo GB, Cordoni C, Munafò P (2012) Self-cleaning and de-polluting stone surfaces: TiO2 nanoparticles for limestone. Construction and Building Materials 37:51–57. doi:10.1016/j.conbuildmat.2012.07.006.

[83] http://www.verticalfarm.com/. Accessed 16 June 2015.

[84] United Nations Department of Economic and Social Affairs, Population Division (2004) World Population to 2300. United Nations, New York. http://www.un.org/esa/population/publications/longrange2/WorldPop2300final.pdf. Accessed 15 June 2015.

[85] United Nations, Department of Economic and Social Affairs, Population Division (2015) World Population Prospects: The 2015 Revision, Key Findings and Advance Tables. United Nations, New York, Working Paper No. ESA/P/ WP.241. http://esa.un.org/unpd/wpp/Publications/Files/Key_Findings_WPP_2015.pdf. Accessed 12 June 2015.

[86] Thebo AL, Drechsel P, Lambin EF (2014) Global assessment of urban and peri-urban agriculture: irrigated and rainfed croplands. Environmental Research Letters 9:114002. doi:10.1088/1748-9326/9/11/114002.

[87] Martellozzo F *et al* (2014) Urban agriculture: a global analysis of the space constraint to meet urban vegetable demand. Environmental Research Letters 9:064025. doi:10.1088/1748-9326/9/6/064025.

[88] http://www.edibleyork.org.uk/edibleinitiatives/communitygrowing/. Accessed 12 June 2015.

[89] Angotti T (2015) Urban agriculture: long-term strategy or impossible dream? Lessons from Prospect Farm in Brooklyn, New York. Public Health 129:336–341. http://dx.doi.org/10.1016/j.puhe.2014.12.008.

[90] Lott MC (2011) 10 Calories in, 1 Calorie Out – The Energy We Spend on Food. Scientific American. http://blogs.scientificamerican.com/plugged-in/2011/08/11/10-calories-in-1-calorie-out-the-energy-we-spend-on-food/. Accessed 21 June 2015.

[91] Church N (2005). Why our food is so dependent on oil. PowerSwitch. http://www.po werswitch.org.uk/portal/index.php?option=content&task=view&id=563. Accessed 15 February 2016.

[92] Heller MC, Keoleian GA (2000) Life Cycle-Based Sustainability Indicators for Assessment of the U.S. Food System. Center for Sustainable Systems, University of Michigan, Ann Arbor, MI, Report No. CSS00-04. http://css.snre.umich.edu/css_doc /CSS00-04.pdf. Accessed 23 June 2015.

[93] Technology quarterly, Q4, 2010 (2010) Does it really stack up. The Economist. http://www.economist.com/node/17647627. Accessed 15 June 2015.

[94] Ramankutty N, Evan AT, Monfreda C, Foley JA (2000) Farming the planet: geographic distribution of global agricultural lands in the year 2000. Global Biogeochem. Cycles 22:GB1003. doi:10.1029/2007GB002952.

[95] Despommier D (2013) Farming up the city: the rise of urban vertical farms. Trends In Biotechnology 31:388–389. doi:10.1016/j.tibtech.2013.03.008.

[96] Heath T, Shao Y (2014) Vertical farms offer a bright future for hungry cities. The Conversation. http://theconversation.com/vertical-farms-offer-a-bright-future-for-hu ngry-c ities-26934. Accessed 11 June 2015.

[97] Al-Chalabi M (2015) Vertical farming: Skyscraper sustainability? Sustainable Cities and Society 18:74–77. doi:10.1016/j.scs.2015.06.003.

[98] http://agritecture.com/post/52866684629/check-out-this-vertical-farming-outfit-in-mi chigan. Accessed 05 June 2015.

[99] Marks P (2014) Vertical farms sprouting all over the world. New Scientist. http:// www.newscientist.com/article/mg22129524.100-vertical-farms-sprouting-all-over-th e-world.html#.VTJLq9JViko. Accessed 02 June 2015.

[100] Kisner C (2011) Developing a Sustainable Food System. City Practise Brief. National League of Cities, Centre for Research and Innovation, Washington. http://www. nlc.org/documents/Find%20City%20Solutions/Research%20Innovation/Sustainabilit y/developing-a-sustainable-food-system-cpb-mar11.pdf. Accessed 14 June 2015.

[101] http://www.skygreens.com/about-skygreens/. Accessed 15 June 2015.

[102] http://bigthink.com/think-tank/vertical-farming-will-help-us-meet-the-challenges-of-t omorrow. Accessed 15 June 2015.

[103] http://www.forwardthinkingarchitecture.com/. Accessed 13 June 2015.

[104] http://www.ostara.com/about. Accessed 01 June 2015.

[105] http://acoulstock.com/resources-and-sustainability/creating-sustainable-districts/. Accessed 07 June 2015.

[106] Thielman S (2015) Nevada clears self-driving 18-wheeler for testing on public roads. The Guardian. http://www.theguardian.com/technology/2015/may/06/nevada-self-

driving-trucks-public-roads-daimler-inspiration. Accessed 17 June 2015.

[107] Griffiths S (2014) Self-driving cars to hit British roads next month: Four cities will host trial projects featuring driverless shuttles to smart roads. Daily Mail. http://www.dailymail.co.uk/sciencetech/article-2860451/Self-driving-cars-hit-British-roads-year-Four-cities-host-trial-projects-featuring-driverless-pods-smart-roads.html. Accessed 12 June 2015.

[108] Morais H, Sousa T, Soares J, Faria O, Vale Z (2015) Distributed energy resources management using plug-in hybrid electric vehicles as a fuel-shifting demand response resource. Energy Conversion and Management 97:78–93. doi:10.1016/j.enconman.2015.03.018.

[109] Merrill J (2015) Are e-cars the future of motoring? Find out on a long, but not long enough, drive up the electric highway. The Independent. http://www.independent.co.uk/life-style/motoring/motoring-news/are-ecars-the-future-of-motoring-find-out-on-a-long-but-not-long-enough-drive-up-the-electric-highway-9955940.html. Accessed 21 June 2015.

[110] http://www.aeromobil.com. Accessed 06 July 2015.

[111] Gota S, Fabian B (2009) Emissions from India's Intercity and Intracity Road Transport. Clean Air Initiative for Asian Cities Center. http://www.Indiaen-vironment-portal.org.in/files/274555.pdf. Accessed 14 July 2015.

[112] School of Public and Environmental Affairs, Indiana University (2011) The Future of Intercity Passenger Transportation. V600 Capstone Course, Indiana. http://www.In-diana.edu/~cree/pdf/Future%20of%20Intercity%20Passenger%20Transport%20Report.pdf. Accessed 12 July 2015.

[113] Culpan, D (2015) Japan's maglev train breaks world speed record. Wired Technology; 2015. http://www.wired.co.uk/news/archive/2015-04/17/japan-maglev-train-world-speed-record. Accessed 13 July 2015.

[114] Technology quarterly; Q2 (2013) Reinventing the Train; ideas coming down the track. The Economist; 2013. http://www.economist.com/news/technology-quarterly/21578516-transport-new-train-technologies-are-less-visible-and-spread-less-quickly. Accessed 23 July 2015.

[115] Japan unveils levitating high-speed electric aero train (2015) Inhabitat. http://inhabitat.com/japan-unveils-levitating-high-speed-electric-aero-train/aero-train-4/. Accessed 19 July 2015.

[116] Collins N (2014) Drivers spend 30 hrs each year in gridlock. The Telegraph. http://www.telegraph.co.uk/news/uknews/road-and-rail-transport/10673424/Drivers-spend-30-hrs-each-year-in-gridlock.html. Accessed 23 July 2015.

[117] Nasowitz D (2011) Driverless car drives 175 miles on busy chinese expressway, no gps necessary. Popular Science. http://www.popsci.com/cars/article/2011-08/chinese-driverless-car-travels-over-175-miles-no-gps-required. Accessed 24 July 2015.

[118] Manzalini A (2015) Enabling the self-driving car. Network Computing. http://www.networkcomputing.com/cloud-infrastructure/enabling-the-self-driving-car/a/d-id/1319538. Accessed 01 August 2015.

[119] Salter RM (1972) The Very High Speed Transit System. RAND Corporation, Santa Monica, CA. http://www.rand.org/pubs/papers/P4874. Accessed 02 August 2015.

[120] http://www.et3.com/. Accessed 29 July 2015.

[121] UN Data available at: http://esa.un.org/unpd/wup/CD-ROM/Default.aspx. Accessed 26 July 2015.

[122] Australian Government Department of Infrastructure and Transport (2011) Our Cities, Our Future: A national Urban Policy for a Productive, Liveable and Sustainable future. Department of Infrastructure and Transport, Canberra. https://www.Infrastruc-ture.gov.au/infrastructure/pab/files/Our_Cities_National_Urban_Policy_Paper_2011.pdf. Accessed 22 June 2015.

[123] https://www.consultaustralia.com.au/docs/default-source/infrastructure/Tomorrow_s_Cities_Today-web.pdf?sfvrsn=0. Accessed 28 June 2015.

[124] Australian Sustainable Built Environmental Council (2013) Snapshot of Australian Cities and Urban Policy Landscape. ASBEC Cities and Regions Policy Task Group; 2013. http://www.asbec.asn.au/wordpress/wp-content/uploads/2007/09/130328-Cit-ies -briefing-paper-ASBEC-Mar-2013-ST-V8.a.pdf. Accessed 28 June 2015.

[125] Moir E, Moonen T, Clark G (2014) The future of cities: What is the global agenda? UK Government Office of Science. https://www.gov.uk/government/uploads/system/uploads/attachment_data/file/377470/future-cities-global-agenda.pdf. Accessed 28 June 2015.

[126] El Sioufi M (2010) Climate Change and Sustainable Cities: Major Challenges Facing Cities and Urban Settlements in the Coming Decades. International Federation of Surveyors. https://www.fig.net/resources/monthly_articles/2010/june_2010/june_2010_el-sioufi.pdf. Accessed 05 July 2015.

[127] Ekamper P, van Poppel F, van Duin C, Garssen J (2009) 150 Years of temperature-related excess mortality in the Netherlands. Demographic Research 21:385–426. doi:10.4054/DemRes.2009.21.14. http://www.demographic-research.org/Volumes/Vol21/14/. Accessed 28 July 2015.

[128] McCarthy MP, Best MJ, Betts RA (2010) Climate change in cities due to global warming and urban effects. Geophysics Research Letters 37:L09705. doi:10.1029/2010GL042845.

[129] Rehan RM (2014) Cool city as a sustainable example of heat island management case study of the coolest city in the world. HBRC J. (in press). doi:10.1016/j.hbrcj.2014.10.002.

[130] England PH (2015) Heatwave plan for England Making the case: the impact of heat on health – now and in the future. PHE publications, London. https://www.gov.uk/

government/uploads/system/uploads/attachment_data/file/429572/Heatwave_plan_Making_the_case_-_2015.pdf. Accessed 28 July 2015.

[131] Edenhofer O, Pichs-Madruga R, Sokona Y, Farahani E, Kadner S, Seyboth K, Adler A, Baum I, Brunner S, Eickemeier P, Kriemann B, Savolainen J, Schlömer S, von Stechow C, Zwickel T, Minx JC, PCC (2014) Climate Change 2014: Mitigation of Climate Change. Contribution of Working Group III to the Fifth Assessment Report of the Intergovernmental Panel on Climate Change. Cambridge University Press, Cambridge, United Kingdom and New York, NY, USA. https://www.ipcc.ch/pdf/assessment-report/ar5/wg3/ipcc_wg3_ar5_summary-for-policymakers.pdf. Accessed 26 July 2015.

[132] The Center for Clean Air Policy (2011) The Value of Green Infrastructure for Urban Climate Adaptation. http://www.grabs-eu.org/downloads/Value_GI_Urban_Adaptation_CCAP_Feb2011.pdf. Accessed 03 February 2016.

[133] McMahon E (2000) Looking Around: Green Infrastructure. Planning Commission Journal, Burlington, Vermont, No. 37. http://landcarecentral.org/References/EMcMahon%20PCJ%20Green%20Infrastructure%20Article.pdf. Accessed 03 February 2016.

[134] Novy M (2013) Green Roofs. Schaumburg's Sustainable Future. https://futureofschaumburg.wordpress.com/green-design/green-roofs/. Accessed 03 February 2016.

[135] EPA Wet Weather (2015) Managing Wet Weather with Green Infrastructure: Action Strategy. http://nepis.epa.gov/Exe/ZyPDF.cgi/P1008SI8.PDF?Dockey=P1008SI8.PDF. Accessed 04 February 2016.

[136] Forest research (2015) Green infrastructure and the urban heat island. Benefits of green infrastructure evidence note. http://www.forestry.gov.uk/pdf/urgp_evidence_note_004_Heat_amelioration.pdf/$FILE/urgp_evidence_note_004_Heat_amelioration.pdf. Accessed 05 February 2016.

[137] Gill SE, Handley JF, Ennos AR, Pauleit S (2007) Adapting cities for climate change: the role of green infrastructure. Built Environment 33(1):115–133.

[138] Dimoudi A, Nikolopoulou M (2003) Vegetation in the urban environment: microclimatic analysis and benefits. Energy and Buildings 35:69–76.

[139] Marshall J (2013) PM 2.5. PNAS 110(22):8756. www.pnas.org/cgi/doi/10.1073/pnas.1307735110. Accessed 22 July 2015.

[140] Carter T (2013) Smart cities: The future of urban infrastructure. BBC News. http://www.bbc.com/future/story/20131122-smarter-cities-smarter-future. Accessed 22 July 2015.

[141] Lau SP, Merrett GV, Weddell AS, White NM (2015) A traffic-aware street lighting scheme for Smart Cities using autonomous networked sensors. Computers & Electrical Engineering 45:192–207.

[142] Avelar E, Marques L, dos Passos D, Ricardo M, Dias K, Nogueira M (2015) Intero-

perability issues on heterogeneous wireless communication for smart cities. Computer Communications 58:4–15.

[143] Zahurul S, Mariun N, Grozescu IV, Tsuyoshi H, Mitani Y, Othman ML, Hizam H, Abidin IZ (2016) Future strategic plan analysis for integrating distributed renewable generation to smart grid through wireless sensor network: Malaysia prospect. Renewable and Sustainable Energy Reviews 53:978–992.

[144] Wakefield J (2013) What if...you could design a city. BBC News. http://www.bbc.co.uk/news/technology-21032725. Accessed 05 August 2015.

[145] Carter P, Rojas B, Sahni M (2011) Delivering next-generation citizen services: Assessing the environmental, social and economic impact of intelligent X on future cities and communities. IDC White Paper. http://www.cisco.com/web/strategy/docs/scc/whitepaper_cisco_scc_idc.pdf. Accessed 07 August 2015.

[146] Wakefield J (2013) Tomorrow's cities: Rio de Janeiro's bid to become a smart city. BBC News. http://www.bbc.co.uk/news/technology-22546490. Accessed 05 August 2015.

[147] Wakefield J (2013 Tomorrow's cities: Do you want to live in a smart city? BBC News. http://www.bbc.co.uk/news/technology-22538561. Accessed August 06 2015.

[148] Aron J (2015) Automatic world. New Scientist p18-19. http://njscience.com/resources/journals/New%20Scientist/NS3.pdf.Accessed 07 August 2015.

[149] https://www.ethereum.org/. Accessed 30 July 2015.

[150] http://boardroom.to/. Accessed 30 July 2015.

[151] Reuben A (2015) Gap between rich and poor "keeps growing". BBC News; 2015. http://www.bbc.co.uk/news/business-32824770. Accessed 01 August 2015.

[152] Oyelaran-Oyeyinka B, Sampath PG (2010) Latecomer Development: Innovation and Knowledge for Economic Growth. Routledge, London and New York.

[153] Glaeser EL, Resseger MG and Tobio K (2008) Urban Inequality. National Bureau of Economic Research, Working Paper 14419. http://www.nber.org/ papers/w14419.pdf. Accessed 02 August 2015.

[154] Kratke S (2011) The Creative Capital of Cities: Interactive Knowledge Creation and The Urbanization Economies of Innovation. Wiley-Blackwell, Chichester. ISBN 978-1-4443-3621-4.

[155] United Nations, Department of Economic and Social Affairs, Population Division (2013) World Population Ageing 2013. New York: United Nations, Working Paper No. ST/ESA/SER.A/348. http://www.un.org/en/development/desa/population/publications/pdf/ageing/WorldPopulationAgeing2013.pdf. Accessed 08 August 2015.

Chapter 2

A Conceptual Approach to a Citizens' Observatory - Supporting Community-Based Environmental Governance

Hai-Ying Liu[1], Mike Kobernus[1], David Broday[2], Alena Bartonova[1]

[1]Norwegian Institute for Air Research (NILU), Instituttveien 18, 2027 Kjeller, Noway
[2]Division of Environmental, Water and Agricultural Engineering, Faculty of Civil & Environmental Engineering, Technion, Israel Institute of Technology, Haifa 3200003, Israel

Abstract: In recent years there has been a trend to view the Citizens' Observatory as an increasingly essential tool that provides an approach for better observing, understanding, protecting and enhancing our environment. However, there is no consensus on how to develop such a system, nor is there any agreement on what a Citizens' Observatory is and what results it could produce. The increase in the prevalence of Citizens' Observatories globally has been mirrored by an increase in the number of variables that are monitored, the number of monitoring locations and the types of participating citizens. This calls for a more integrated approach to handle the emerging complexities involved in this field, but before this can be achieved, it is essential to establish a common foundation for Citizens' Observatories and their usage. There are many aspects to a Citizens' Observatory. One view is that its essence is a process that involves environmental monitoring, information gathering, data management and analysis, assessment and reporting systems. Hence, it requires the development of novel monitoring technologies and of advanced data management strategies to capture, analyse and survey the data, thus facilitating their exploitation for policy and society. Practically, there are many challenges in implementing the Citizens' Observatory approach, such as ensuring effective citizens' participation, dealing with data privacy, accounting for ethical and security requirements, and taking into account data standards, quality and reliability. These concerns all need to be addressed in a concerted way to provide a stable, reliable and scalable Citizens' Observatory programme. On the other hand, the Citizens' Observatory approach carries the promise of increasing the public's awareness to risks in their environment, which has a corollary economic value, and enhancing data acquisition at low or no cost. In this paper, we first propose a conceptual framework for a Citizens' Observatory programme as a system that supports and promotes community-based environmental governance. Next, we discuss some of the challenges involved in developing this approach. This work seeks to initiate a debate and help defining what is the Citizens' Observatory, its potential role in environmental governance, and its validity as a tool for environmental research.

Keywords: Citizens' Observatory, Citizen Science, Environmental Governance, Environmental Monitoring, Top-down and Bottom-up Approach,
Public Participation

1. Background

The word "environment" is derived from the old French, 'environ' which means encircle (en viron = in circle) or surround[1]. For us, the environment is literally the immediate surroundings within a circumference, with citizens at its centre. As such, this is a citizen and sensor centric perspective, and requires that we enable citizens to observe with their own senses and with sensors what is happening within their immediate circumference. This observational circle follows each individual as he navigates in his surroundings.

Recently, it has been increasingly suggested that the key to protecting our environment is to engage the average citizens, not only highly active environmentalists. Although our political, economic and administrative structures are designed to tackle our environmental concerns via large-scale policies and strategic decisions, these often leave citizens as unengaged and silent observers[2]. A major drawback of the traditional approaches to environmental monitoring (e.g., Earth Observation through satellites and in-situ observations through monitoring networks) is the sparsely collected data and their inherent remoteness from the citizens' experience of the environment[3]. It is no longer sufficient to develop and provide passive lists of environmental indices or reports and inform citizens about changes in their environment. There is a need to engage citizens to find out how they can inform the community, and to empower citizens to improve their own health and wellbeing through actively making informed choices via the Citizens' Observatory (CO) process[4]–[7]. Involving citizens at the local level by developing knowledge pools can help to create an atmosphere of active participation and generate a sustainable movement that can build over time[2]. Citizens have expectations to interact and participate in the decision making processes, and to be engaged in a dialogue about their communities, preferences and future. According to sociological research, the recent increase in pro-active participants of social IT media, with a particular focus on environmental issues, results from a shift from materialism to post-materialism[8] where more and more people are showing increasing interest in renewable and sustainable life style, which are the key objectives of the 'Environmental Governance'[9]. Developing a CO is a crucial step in bridging the gap between Environmental Governance and the public.

There is no a globally agreed and understood definition of Environmental Governance[10]. It can be interpreted in many different ways. In principle, Environmental Governance comprises the rules, practices, policies and institutions that shape how humans interact with the environment[11]. In this paper, 'Environmental Governance' refers to the processes of decision-making involved in the control and management of the environment for the purpose of attaining environmentally-sustainable development. Good Environmental Governance takes into account the role of all actors that impact the environment. From governments to Non-Governmental Organizations (NGOs), the private sector and civil society, the individual and the citizen groups, cooperation is critical to achieving effective governance that can help us move towards a more sustainable future[11]. A CO for supporting community-based environmental governance may be defined as the participation of citizens in monitoring the quality of the environment they live in, with the help of one or more of the following: (1) mobile devices of everyday utility; (2) specialized static and/or portable environmental and/or wearable health sensors, and (3) personal, subjective and/or objective observations, information, annotation and exchange routes, coming from social media technologies or other similar platforms[12]. In this context, the key aspect of Citizens' Observatories (COs) is the direct involvement of ordinary citizens, and not just that of scientists/professionals in data collection as well as harnessing the citizens' collective intelligence, *i.e.*, the distributed information, experience and knowledge embodied within individuals and communities, to meet gaps that many areas of environmental management are still suffering from. Namely, CO should enable citizens' participation in environmental monitoring, and contribute to environmental governance by providing relevant data and information that can help decision-makers make sound decisions. This can be advanced by providing citizens with a voice and supporting them with knowledge of their environment and as a consequence of raising their awareness.

The environmental concept of CO was first introduced in the project 'Eye on Earth'[13] with the European Environmental Agency (EEA) creating the first 'official' environmental portal that includes a CO on air, noise, nature, coral reefs and water quality. In addition, the EU has funded five CO-related projects under the FP7 topic ENV.2012.6.5-1 "Developing community-based environmental monitoring and information systems using innovative and novel earth observation applications" at the end of 2012, including (1) "Citclops-Citizens' Observatory for coast and ocean optical monitoring", 2012-2015[14]; (2) "Omniscientis-Odour monitoring and information system based on citizens and technology innovative sensors", 2012–2014[15]; (3) "CITI-SENSE—Development of sensor-based Citizens' Observatory Community for improving quality of life in cities", 2012–2016[16]; (iv) "WeSenseIt—Citizen Observatory of Water", 2012–2016[17]; and (v) "COBWEB—Citizen Observatory Web", 2012–2016[18]. On this basis it is expected that CO will have an increasing importance in supporting environmental governance and other applications over the next years.

As participants in a major EU FP7 project (CITI-SENSE)[16] as well as from work undertaken across several EU-funded projects, e.g., ENVIROFI, 2011– 2013[19] and HENVINET, 2007-2011[20], which contained core components that were heavily based on the CO concept[21][22], we have gained some insight on best practices for a CO programme. These have been implemented in the ongoing Citizens' Observatory work in CITI-SENSE. The role of the Citizens' Observatory in the project is to harmonise various independent/local Citizens' Observatories and to develop coherent understanding of COs with regards to the overall project objectives.

While we agree that there might be various perspectives to each aspect of a CO, we have developed an initial concept that we believe can be applied to many CO initiatives in general. To this end, we propose a framework to support and influence community and policy priorities and associated decision making in environmental stewardship. Informed by existing approaches in CO-related environmental governance support, we propose a structural work system that enables effective citizens' participation, data collection and interpretation, and information dissemination. Further, we review and discuss the main challenges faced by a CO programme in support of environmental governance. The aim of this paper is, therefore, to provide a platform for debate that should lead to a more comprehensive understanding of what a Citizens' Observatory is, and how it can support environmental research and governance.

2. Current Citizens' Observatories

To put CO into perspective as an instrument to support community-based environmental decision making, it is useful to get a sense of the variety of COs within environmental media and relevant aspects[23]. A wealth of CO-related initiatives (e.g., Citizen Science, Community-Based Monitoring (CBM), Volunteered Geographic Information (VGI), Volunteered Environmental Monitoring (VEM), etc.) can be found around the globe. The Waterkeeper Alliance, for example, which includes the Riverkeeper, Lakekeeper, Baykeeper, and Coastkeeper programmes, which works towards the goals of ecosystem and water quality protection and enhancement has over 200 programmes in 15 nations[23]–[25]. The majority of these are located in the USA, Australia, India, Canada, and Russia[26]–[29]. A review of the academic- and non-academic-based literature indicates that these nations are among those leading many CBM initiatives and that by all indications the movement of 'Citizen Science' is increasing[21]. Kerr *et al.* (1994)[30] indicated a near tripling of new monitoring programmes with citizens' engagement between 1988 and 1992, all related to water monitoring. Pretty (2003)[31] reported that since the 1990s, up to 500,000 new local population groups were established in varying environmental and social contexts. A review in 2006 showed that the increase of CBM has been particularly dramatic in the USA and Canada[32]. The cause for this rise has been attributed to an increase in public knowledge and concern about anthropogenic impacts on natural ecosystems[33]–[35] and recent public and NGOs concern about governmental monitoring of the environment and ecosystems[36]. In addition to COs involving the general public, private individuals and NGOs play many roles[33]–[36], the critical role of various institutions for observing earth and conserving the environment[16][17][19][37], for achieving citizen participation in environmental monitoring[16][17][19][38], for facilitating science-policy dialogue[20][39], and for setting up Citizen Science as a discipline in its own right[40][41], etc., is increasing as well[42].

In Europe, several ongoing national and international community-based environmental monitoring programmes currently exist, e.g., the EEA project 'Eye on Earth'[13][43][44], the European Mobile and Mobility Industries Alliance (EMMIA) project Citi-Sense-MOB[45][46], and the EU FP7 funded five CO-related projects (WeSenseIt[17], Omniscientis[15], COBWEB[18], Citclops[14], and CITI-SENSE[16]. According to Wiggins and Crowston (2011)[47], the recent decades have seen a growing emphasis on 'scientifically sound practices and measurable goals of public education'. Some of the well-known projects were and are focused on nature and biodiversity, for example, The Open Air Laboratories[OPAL[48], The Big Butterfly Count[49], and Citizens' Network for the Observation of Marine Biodiversity (COMBER)[50]]. However, there are many more CO-related programmes, encompassing different models of Citizen Science and within the environmental sciences these span a diverse range of subject (e.g., biodiversity, water, air, climate change, agriculture, disaster, etc.). To promote debate on the CO definition, concept and practices, we provide a brief review of nine programmes (Additional file 1): Citclops, CITI-SENSE, Citi-Sense-MOB, COBWEB, Eye on Earth, Omniscientis, Waterkeeper Alliance, WeSenseIt, The Big Butterfly Count. We focus on (i) the aim/purpose of each programme; (ii) its geographic scope; (iii) project duration; (iv) target groups; (v) monitoring parameters; (vi) data collection and interpretation, visualization and information dissemination technologies. These six properties determine the potential of the programmes for supporting informed decision-making. These programmes can be classified into:

- International programmes whose objectives are to develop Citizens' Observatories using innovative earth observation technologies (air, water, odour, biodiversity, etc.), e.g., CITI-SENSE, WeSenseIt, COBWEB, Citclops, Omniscientis.

- International programmes whose objectives focus on enabling greater access to and sharing of environmental and societal data, e.g., Eye on Earth.

- National and/or international programmes whose objectives are on creating community-based environmental monitoring in varying environmental and social contexts towards the goal of ecosystem, biodiversity and environmental quality protection, e.g., the Waterkeeper Alliance programmes, The Big Butterfly Count, Citi-Sense-MOB.

From the information in Additional file 1, we have identified the following characteristics that seem to be vital for the Citizens' Observatories: (i) A CO should involve citizens as active partners in environmental monitoring and decision-making, since this is central for protecting and enhancing our environment; (ii) CO-related environmental monitoring should target an array of natural resources and/or a range of environmental components; (iii) Generally, the involvement of citizens in CO has multiple purposes, with education and raising public awareness being the most common objectives associated with a CO[45][51]–[53]; (iv) There is value in CO as a way to bring community groups together. CO, like other forms of civic engagement, can build social capital within the community[53][54]; (v) Evaluation of the effectiveness of a CO as well as of the public involvement in environmental decision-making is generally lacking. There are a number of questions about its potential as a democratizing force in environmental policy and management[52][53][55]. However, given the many contending conceptions within democratic theory (e.g., direct, representative, participatory, minimal, deliberative, aggregative, etc.)[56], it should be noted that this aspect is a complex subject with no "one size fits all". Nevertheless, a CO in the environmental domain has shown its potential role to address issues of environmental equity and to improve social justice[57].

3. Conceptual Framework of Citizens' Observatory in Support of Environmental Governance

3.1. Definition of Citizens' Observatory

There is no clear definition of CO available yet. In the broadest sense, a CO for supporting community-based environmental governance may be defined as 'the citizens' own observations and understanding of environmentally-related problems, and in particularly as reporting and commenting on them'. As such, the CO promotes communicates and supports sharing of technological solutions (e.g., sensors, mobile apps, web portals) and community participatory governance methods (e.g., aided by various social media streams) among citizens. A CO is also open and democratic, enabling the possibility for anyone who is interested or willing to contribute and participate in earth observation and environmental conservation[58][59]. It also promote a more active role for the community with regards to understanding the environment, since citizens are traditionally considered to be merely consumers of information services at the very end of the information chain[4] and not as data providers. This definition reveals three core components that underpin some of its objectives, *i.e.*, raising the citizens' environmental awareness; enabling dialogue among citizens, scientists and policy/decision makers and supporting data exchange among citizens, scientists and other stakeholders.

3.2. Citizens' Observatory in Support of Environmental Governance

We believe that the above three components of CO can explain the major links between Citizens' Observatory and environmental governance, and in fact as the three pillars that sustain a Citizens' Observatory for supporting environmental governance. In the context of Citizens' Observatories contribution to environmental governance, it is important to recognise that citizens are not monolithic, with CO stakeholders/user groups[59] including individual or groups of volunteers, scientists, government authorities, emergency services, etc. Hence, various stakeholder actors in a CO have different behaviours, intentions, interrelations, agendas, interests, as well as influence, resources and power on decision-making and political processes[60][61].

3.3. Raising Awareness

Information is available to us in a myriad of ways and from many sources: newsprint, radio, television, online portals and mobile device. In fact, there is so much information that it is sometimes hard to keep track of what we need, or even to really understand what we need to know. Recently awareness grew that "it is no longer sufficient to develop passive lists or report to 'inform' citizens of changes in our environment. We need to engage with citizens and ask how they can 'in-form' us" (Prof. Jacqueline McGlade, Executive Director, European Environment Agency[2][62]). At the recent 2013 Green Week conference, the European Environment commissioner, Janez Potočnik, reinforced this when he stated that "We have learned that public awareness is of key importance for the implementation of existing air policy, as well as for the success of any future air pollution strategy"[63]. Clearly, getting the useful 'message' across to the public, in the right way, and thereby effectively raising public awareness, is critical. The first criterion therefore is to determine who we would like to get the message to, and to target those users in a way that ensures a certain level of interest.

In previous projects (e.g., ACCENT, 2009–2011[64], HENVINET[20], and ENVIROFI[19]) we have attempted to engage users through various campaigns, including mass emailing, printed media such as brochures, online video presentations and workshops in the Café Scientifique format. These methods generated a sufficiently moderate number of public users interested in knowing more about the project but ultimately did not create a self-sustaining community of users that are willing to engage/participate for a long period in a community forum that is based on social network platform(s). Hence, while it could be argued that we were moderately successful, it was clear that we did not really create a viable, sustainable community. What was missing was the emphasis on knowledge transfer. Raising awareness is not just about alerting the public or recruiting users; it is just as much about helping those users understand the problems and concerns so that they can make informed decisions of their own. While these platforms did include expert users who could answer questions about relevant environmental issues, this does not automatically translate into true knowledge transfer. An additional factor is to ensure that the communities' opinions, thoughts, questions, etc., are not only heard, but are valued. For this, we need to provide a platform that support a dialogue among the users in a CO programme.

Furthermore, to facilitate citizens play an active role in the data collection process (e.g., via portable sensors and smart phones, information and communication technology), as well as harness the citizens' collective intelligence (e.g., using apps and social medias), a self-sustaining community of users should be formed, which exchange data/information and knowledge, reach the expert who could answer questions about relevant environmental issues, and disseminate information to understand environmental issues. The Chinese Proverb "tell me and I'll forget; show me and I may remember; involve me and I'll understand", does apply in this context.

3.4. Enabling Dialogues

Successful multi-stakeholder dialogues are critical to ensuring a deeper level of interest of the stakeholders, especially the general public. At the most basic level, these can take the form of peer-to-peer as well as public-to-expert. Yet, any discussion forum needs to have a comprehensive and consistently active membership drawn from a multidisciplinary volunteer 'workforce'. Nothing kills a communication portal quicker than low levels of active participation. Only if regular activity of a varied group of users is achieved one can likely see a sustained growth over time, as more people begin to participate than fall away. It cannot be overemphasized that this is not a place for passive participation and that static information portals guarantee a quick demise. Social media applications can be employed as a platform for initiating dialogue but in itself, this is ultimately insufficient. It is critical to move to a more advanced level, since multi-stakeholder dialogues are more than just question and answer or discussion forum style communication. They must include technology based information gathering and exchange systems, including sensors, smart-phones, personal subjective observations, etc. These will create a much broader canvas for information gathering and for data exchange.

3.5. Data Exchange

Data exchange is much more than just pushing data to users, and it goes beyond the sharing of ideas or questions. In a Citizens' Observatory context, this must include a variety of Volunteered Geographical Information (VGI) observation types, in addition to personnel observations on an array of topics, such as physical wellbeing, perceived environmental effects and even just personal opinions. The key for this is user that is encouraged to provide data inputs regularly, and finds value in the way that this information is used. The user's peers should also find value in these data and be further encouraged to make their own observations available. An important aspect is that all data, not just electronic sensor data, has a geo-temporal marker.

Public users are now in a position to use micro-sensors in increasing numbers due to advances in technology and lowering costs. An individual might purchase sensors for different reasons and tie them to a network that collect, store and disseminate data. Such platforms become increasingly possible. However, while this results in more data being generated it does not necessarily engage the users, who might be entirely passive data providers. For example, pollen data is generally very limited, so major generalizations are often made about the prevalence of pollen in any given area. If individuals reported the presence of particular types of pollen in a specific area, this could be of great interest to others who also have an allergic reaction to that particular pollen. Therefore engaging users in providing personal observations on their perception of the environment can have beneficial consequences for others, which will further encourage others to participate and share their own observations. Finally, presenting information that combines heterogeneous data sources which includes VGI data allows the stakeholders, in particular public users, to see how their individual contributions add to the value chain, ultimately creating a reinforcing mechanism that will help to create a self-sustaining community.

3.6. Citizens' Observatory Framework

Based upon the definition of the CO we have given and our understanding of how a Citizens' Observatory may supports environmental governance, we propose that a Citizen's Observatory comprises four aspects, which we refer to as the CO framework (**Figure 1**), as follows: (i) Collaborative participation process; (ii) Two data layers: hard layer comprising data generated from sensors and the soft layer comprising data generated from citizens; (iii) Two-directional approach: top-down and bottom-up and (iv) Two-way interactive communication model.

In the following sections we elaborate each of these concepts in turn.

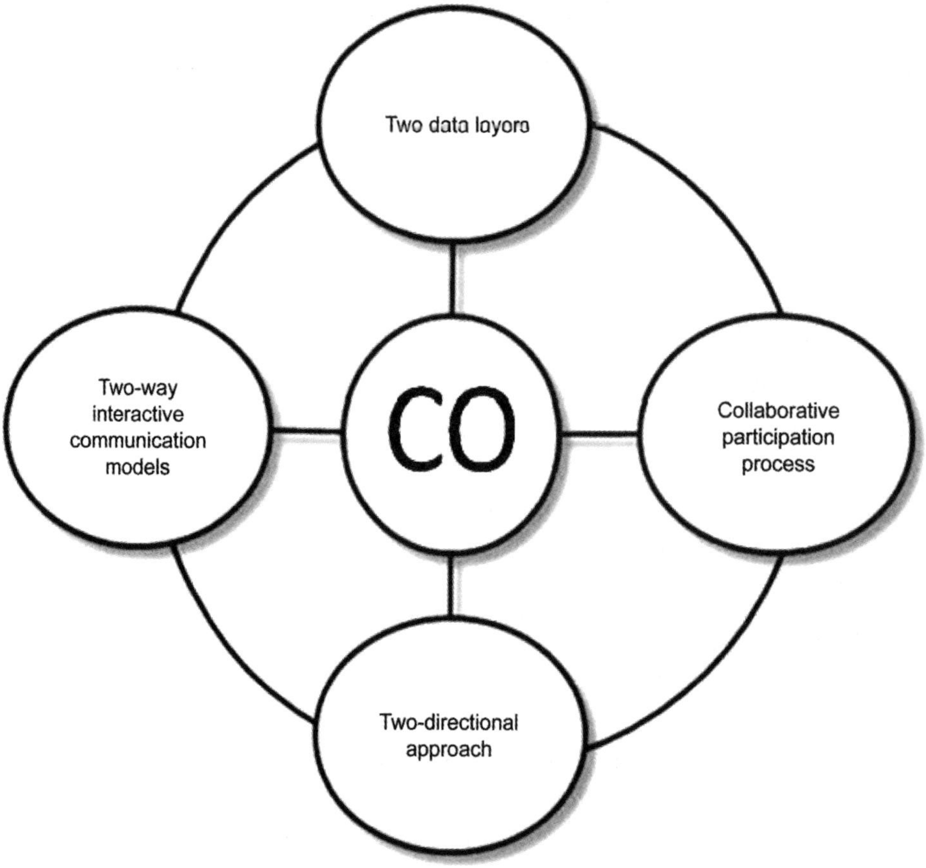

Figure 1. Conceptual frameworks to a Citizens' Observatory. Grouped as follows: (i) Collaborative participation process; (ii) Two data layers: hard layer and soft layer; (iii) Two-directional approach: top-down and bottom-up, and (iv) Two-way interactive communication models.

3.7. Collaborative Participation Process

Citizen participation should be considered throughout the entire chain of monitoring-data-assessment-reporting within an environmental monitoring programme. Key to this is ensuring that citizens are both motivated and equipped to influence the decision making process. This approach will also enable and motivate the citizens to change their personal behaviour and priorities in order to improve their environments. For example, indoor air quality in schools can be improved considerably if all children and staff take off their shoes before entering the classroom. This change in behaviour directly affects the environment and is possible due to the active participation of the citizens.

Moving away from the traditional one-way transfer of knowledge between scientists and citizens is important. Collaborative participation demands that the citizens not only consume information, but also provide it, leading to the joint production of knowledge (where multiple forms of expertise, for example from researchers, practitioners and the public) are valued equally in the production of knowledge[65]. Fernandez-Gimenez *et al.*[66] have posited that a collaborative participation process in environmental monitoring can lead to shared environmental understanding among diverse participants, build trust internally and credibility externally, foster social learning and community-building, and advance adaptive management[66].

3.8. Two Data Layers

Many citizens and volunteers are equipped today with Global Positioning System (GPS) enabled devices, digital cameras and numerous other resources, turning citizens into potential resources for creating and sharing publicly relevant information[66]. In other words, citizens become part of the information chain as data suppliers, not just consumers. In addition to collecting existing data and information from relevant programmes and projects, the two data layer approach can be applied in a CO programme, utilizing citizens as mobile sensors as well as implementing physical sensors. The hard layer (physical sensor layer) includes static and portable devices for sensing and transferring environmental information via mobile devices. The objective here is to have a large number of sensors providing spatial patterns and temporal evolution of the changing environment and real-time information for decision-making. For example, in several current projects (e.g., CITI-SENSE, Citi-Sense-MOB), static sensors will be installed in parallel with existing stationary monitoring networks, while portable sensors will be carried by citizens[16][45]. The soft layer (Human layer) is harnessing citizens' own observations of their surroundings/environment. This includes social trends and social activity, online participation in public forums (such as Facebook page, Twitter account, LinkedIn group, etc.), participative Geotagging and sharing of their own subjective/objective observations on their perceived environment.

3.9. Top-down and Bottom-up Approaches

Top-down and bottom-up are strategies of information processing and knowledge ordering, mostly involving software, but also other humanistic and scientific theories[67]. In many cases top-down is used as a synonym of analysis or decomposition, and bottom-up of synthesis[68]. Both top-down and bottom-up approaches exist in CO-related programmes (e.g., CITI-SENSE, Citi-Sense-MOB). To meet the challenges in the data coverage, a combined top-down and bottom-up approach is often used[68]. Top-down approaches are typically research-led (expert) and often start with the formulation of visions of future direction. At the same time, a broad variety of bottom-up initiatives are taken by different public groups (citizens) who develop and try out new approaches to meet the challenges as they see them. Most of these initiatives are not guided by broad future visions and focus on specific aspects[69][70]. Accordingly, Citizens' Observatories can be seen as a combination of top-down and a multi-layer bottom-up approach. This can be defined and interpreted differently. The method proposed here, involves a combined top-down and bottom-up approach in a CO programme and can interpret the use of citizen science as a two-way data connection between the researchers and citizen scientists who work from opposite approaches (Top-down and bottom-up)[71][72]. The current scientific knowledge and policy analysis (top-down gathered knowledge) are combined with local knowledge, experiences and perceptions (bottom-up collected information through ordinary citizens' observatories). This approach creates a platform for exchange of information in two directions, and an involvement and engagement with all stakeholders which is crucial if a sustainable CO programme is to take place.

For multiple location-based case studies in a CO programme (e.g., CITI-SENSE[16][73]), the combined top-done and bottom-up approach can be defined by the following: (i) The key components of the top-down CO approach includes the definition of the COs goals, selecting and applying the necessary standards, protocols, sampling designs and methodologies, wherever these have one identical purpose in various case studies within one environmental domain. (ii) Crucial to the multiple layer bottom-up COs approach are heterogeneous data sources that can accommodate multiple data standards, conflicting requirements from diverse user groups and a definite need to develop methodologies that are able to integrate diverse systems and ultimately synthesize data of many types and formats. Together, these two approaches aim to minimize the differences, and maximize what is similar, among multiple systems, enabling both individual case study data analysis and integrated data analysis to be performed[21]. **Figure 2** presents the top-down and bottom-up process in this context. The top-down approach is to ensure individual case studies within a CO programme have coordinated data types and acquisition, so that they can be analysed in an integrated manner at a later stage. The bottom-up approach enables the integration and synthesizing of results from individual case studies, these results arising from multiple and potentially conflicting needs.

In addition, the top-down approach may be considered to be management driven, and the bottom-up approach may be considered as driven by the needs of the client/s. The ultimate goal is that both approaches should be merged. It is important to understand that these are not separate approaches and that they should be run in parallel, not individually.

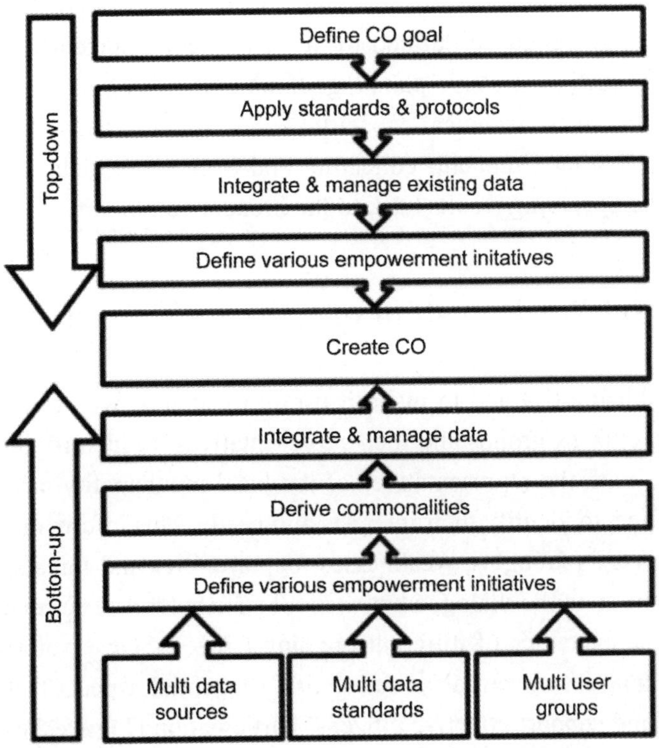

Figure 2. A top-down and bottom-up approach have the same goals via different paths, but both allow for individual case study data analysis and integrated data analysis.

3.10. Two-Way Interactive Communication Model

In order to harness environmental data and knowledge to effectively and efficiently management of environmental issues, a CO which will enable citizens and communities to take on a new role in the environmental management chain needs to be developed: a shift from the traditional one-way communication paradigm in which citizens are passive information receivers[74], into a two-way communication model, in which citizens become active stakeholders in information capturing, evaluation and communication[4]. As a result, citizens will become important in two ways: (i) as data providers through the direct involvement of user communities in the data provision and collection process; and (ii) by solving consensus tasks, e.g., by collecting multiple assessments from citizens[75] and gathering information on the environment.

To realize this, there is the need for understanding the citizens' demographics[76] to develop a CO platform that meets their needs. Beyond demographics, community needs should be defined by the community stakeholders themselves through efforts such as strategic planning, community visioning, design charrettes, etc.[77][78]. Furthermore, technical capacity need to be built as well for facilitating citizens observing environment, collecting and exchanging data, communicating and visualizing observing results back to the broader community.

4. Structural Work System of Citizens' Observatory

To establish a CO and to make it useful to society, we need to collaborate with citizens, citizens groups and their representatives, to identify their needs and concerns, and with the representatives of the local municipality or environmental protection office, to identify their interests and needs. This information can then be cast into a SWOT (strengths, weaknesses, opportunities and threats) analysis approach[79] to promote the dialogue between all stakeholders. The review of the varieties and characteristics of different ongoing Citizens' Observatories that focus on environmental issues reveals a set of five sequential aspects that underlie the CO skeleton and support effective citizens' participation (**Figure 3**): (A) Citizens' participation in identifying what a CO can offer to provide information and knowledge in response to public concerns. This is achieved mainly by a dialogue among

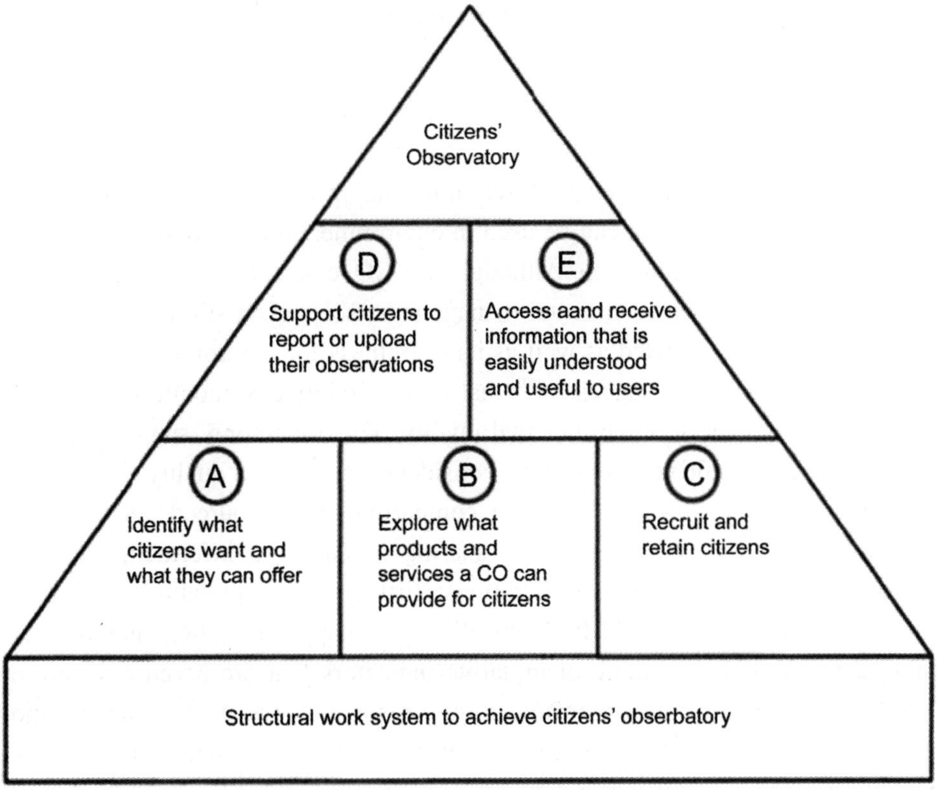

Figure 3. Sequential aspects of a Citizens' Observatory programme.

the stakeholders; (B) Citizens' participation in exploration of what products and services a CO can provide for the citizens. This involves systematizing and structuring citizens-created content to make it appealing for use by citizens during their normal daily life; (C) Recruitment and retaining of citizens to participate in and contribute to environmental governance: further clarify the purpose, scope and expected impact of the CO, identify motivations that will promote citizens to contribute to and take part in the CO, and encourage public participation in data collection and interpretation; (D) Obtaining public participation in the relevant decision making and/or in changing their related personal priorities and behaviour by gaining access to environmental data, knowledge and experience, and by using tools that can support citizens to report or upload their objective/subjective observations; (E) Providing tools to access and receive timely information on relevant environmental issues in a manner that is both easily understood and useful to the users.

5. Challenges and Development Needs

5.1. Challenges

The essence of a Citizens' Observatory lies in public wide-spread engagement in collecting data that can be used in environmental decision-making, which is relevant to public concerns. Although the CO concept is becoming a more common practice for environmental management, scepticism still exists about the quality of the data collected as well as its usefulness for environmental policy[52]. Furthermore, it has been suggested[80] that improved citizens' monitoring can even have adverse effects on environmental quality. This dichotomy suggests that the following areas should receive careful consideration: (i) Data quality (*i.e.*, accuracy and uncertainty)—especially when comparing crowd-sourced and reference data; (ii) Data privacy and security—sharing of data and information requires strong ethical and security considerations; (iii) Data interpretation—qualitative indicators such as "quality of life", "wellbeing", "happiness", etc., should be developed in parallel with more quantitative indicators that are based not only on individual perception, but on an integrated sensor network; (iv) Systematisation and structuring of citizens-created content and feedback—establishing a viable model(s) to support decisions and empower the public[81]; (v) Involving and maintaining a broad spectrum of society—implementing various location-specific and target group-tailored tools in recruiting and sustaining citizens' participation in environmental monitoring[82].

Whereas we recognize the critical importance of the above issues, which must be solved to ensure the viability of the CO model, we believe that the value it carries and its pluralistic and democratic foundations override many of the reservations currently still associated with it.

5.2. Development Needs

In terms of ensuring a usable CO, some key challenges that we have faced during the course of several projects where we developed a CO component include the following: (i) We need to adequately promote the CO platform and tools, to raising awareness, recruiting and sustaining citizens' participation. The old adage, "if you build it, they will come", does not apply. (ii) We need a good understanding of citizens' demographics in order to develop the CO platform to meet their needs, especially as they change; (iii) We need to build a long lasting infrastructure that uses open standards, is easily exploitable through an open Application Programming Interface (API), can be widely accessed, extended and maintained, and is seen as a generic environmental enabler rather than a project specific outcome; (iv) We need to address and evaluate Citizens' Voice and Accountability (CV & A) in the social and political context in which Citizens' Observatories are embedded[82]-[84], to actively promote the CV & A concepts as important dimensions of good environmental governance[84], to address CO's potential role to influence environmental equity and to improve social justice[75]; and (v) we need to develop particular channels and mechanisms that can underpin the sound environmental-social-political actions in which Citizens' Observatories are addressed, in a manner which facilitates citizens to influence environmental governing priorities and processes.

Surveying technology evolves quickly but issues relating to data collection and analysis always prevail. The latter include: (i) Building needed technical capacity and overcoming the 'digital divide' for environmental monitoring, data exchange, visualizing and communicating results back to the broader users[83][85]; (ii) Managing and analysing increasing data volumes, variety and velocity[86]; (iii) Reducing measurement uncertainties; (iv) Developing reliable and fast quality assurance/quality control (QA/QC) tools that can work in real-time; and (v) Increasing need for interdisciplinary use of data, integration of different types of data. Whereas solutions toward many of these potential limiting factors already exist, progress in computational tools that can process large volumes of data and enable analysis of large volume of data is foreseen[87]. Decision support methods and tools (e.g., Aguila[88]; the Numerical Unit Spread Assessment and Pedigree (NUSAP) system[89][90]) can be used to deal, to some extent, with the inherent data uncertainty. Expert elicitation can be used to deal with some aspect of uncertainty by consulting experts as a means to derive preliminary estimates for information[91]-[95]. And more traditional methods can be used to overcome the 'digital divide', by making data available in other methods, such as web, or even TV and Radio (e.g., high pollen warnings, dust, etc.).

Furthermore, data privacy issues need to be addressed in a spatial-temporal data mining context. In the Geographic Privacy-Aware Knowledge Discovery and

Delivery (GeoPKDD) project, Giannotti and Pedreschi (2010)[96] investigated various scientific and technological issues of mobility data, open problems, and roadmap. They found that privacy issues related to Information Communication Technologies (ICT) can only be addressed through an alliance of technology, legal regulations and social norms. In the meanwhile, increasingly sophisticated privacy-preserving data mining techniques are being studied and need to be further developed. The final aim is to achieve appropriate levels of anonymity by means of controlled transformation of data and/or patterns but with limited distortion, to avoid the undesired side effects on privacy, to preserve the possibility of discovering useful patterns and trends.

In addition, an issue with data quality and its use in shaping environmental policy is the gap between science and policy[95][96] caused by poor timing, ambiguous results and lack of relevant data[97]. Addressing these concerns requires approaches that are both interdisciplinary and engages scientists with societal needs, developmental needs and the implementation of a variety of novel methods and tools to bridge the communication gap[98]. In this regard, we believe Citizens' Observatories provide the possibility by addressing several of the concerns mentioned, such as increased spatial resolution, up-to-the-minute data coverage and improved environmental awareness leading to a stronger public voice.

6. Conclusions

In this paper, we lay the groundwork for a debate on the conceptual framework for developing a Citizens' Observatory. Based upon the review of different ongoing COs and of CO-related programmes in the environmental domain, we have identified key elements and qualities which are essential for a CO programme: (i) Be a unique virtual place to gather and share data from a variety of sources: novel sensor-technologies, open environmental data from public and national sources, and personal perceptions and textual/graphical contribution; and (ii) Extract and make use of relevant citizens-related data and provide multimodal services for citizens, communities and authorities.

Based upon our experience in Citizens' Observatory from the CITI-SENSE and Citi-Sense-MOB projects, we posit that citizens observing and understanding environment related problems, as well as reporting and commenting on them within a dedicated platform, is the key to a successful CO implementation.

To better understand the links between Citizens' Observatory and environmental governance, we first propose three pillars: (i) Raising awareness; (ii) Enabling dialogue; and (iii) Data exchange. In addition, we suggest a CO framework which provides: (i) A collaborative participation process; (ii) Two data layers: a hard layer and a soft layer; (iii) Two-directional approach: top-down and bottom-up; and (iv) A two-way interactive communication model. With these processes in place, citizens will be in a position to participate actively in environmental surveillance in a way that will benefit them in a timely manner.

Current CO programmes attempt demonstrate the main aspects needed to effectively address citizens' participation. These include participation in data collection, data interpretation and information delivery. Alternatively, this can be expressed as: A) Identifying what citizens want and what citizens can offer; B) Exploring what products and services a CO can provide for the citizens; C) Recruiting and retaining citizens to participate in and contribute to environmental governance; D) Providing tools that support citizens to report their observations, inference and concerns; and E) Supplying tools to access/receive timely information on the environment in a manner that is both easily understood and useful.

We believe that achieving these milestones will facilitate the CO objective of engaging citizens and stakeholders in participating in environmental surveillance. This, in turn, will contribute to better informed decisions, contribute to improved quality of life, and ensure that the interest of people in the environment and its impact on human health and wellbeing continues to grow. By raising a debate on this topic we hope to further the understanding and potential of Citizens' Observatory and their wider acceptance in environmental monitoring.

6. Additional File

Additional file 1: Overview of nine Citizens' Observatories programmes with their aim, location, period, target groups, monitoring parameter(s), data collection, and communication and visualization methodologies. For acronyms please see the text. http://www.biomedcentral.com/content/supplementary/1476-069X-13-

107-S1.docx.

7. Abbreviations

CBM: Community-Based Monitoring; Citclops: Citizens' observatory for coast and ocean optical monitoring; CITI-SENSE: Development of sensor-based Citizens' Observatory Community for improving quality of life in cities; Citi-Sense-MOB: Mobile services for environment and health citizen's observatory; CO: Citizens' observatory; CO_2: Carbon dioxide; COs: Citizens' observatories; COBWEB: Citizen observatory web; Copernicus: Global monitoring for environment and security; CVA: Citizens' Voice and Accountability; DALYs: The disability-adjusted life years; EEA: European environmental agency; EMMIA: The European mobile and mobility industries alliance; ENVIROFI: The environmental observation web and its service applications within the future internet; GeoPKDD: Geographic Privacy-aware knowledge discovery and delivery; GEOSSS: Global earth observation system of systems; GO: The global observatory; GPS: Global Positioning System; HENVINET: Health and environmental network systems; ICT: Information communication technology; NUSAP: Numerical unit spread assessment and pedigree; Omniscientis: Odour monitoring and information system based on citizen and technology innovative sensors; QA/QC: Quality assurance/quality control; SDI: Spatial data infrastructure; SEIS: Shared Environmental Information System; VEM: Volunteered Environmental Monitoring; VGI: Volunteered geographical information; SWOT: Strengths, weaknesses, opportunities and threats; WeSenseIt: Citizen water observatories; WNBR: The world network of biosphere reserves.

Competing Interests

The authors declare that they have no competing interests.

Authors' Contributions

HYL planned this work. HYL and MJK wrote the manuscript. DB contributed content and provided editorial input. AB is the project leader. All authors approved the final version.

Acknowledgements

The ideas presented here evolved from work undertaken in the context of studies funded under the projects HEVINET, ENVIROFI, COST Action TD 1202-Mapping and citizen sensor, CITI-SENSE and Citi-Sense-MOB. HENVINET is a coordination action funded by the EU FP6 Programme with the contract no. GOCE-CT-2006-037019. ENVIROFI is funded by the EU FP7 Programme under the FI.ICT-2011.1.8 Work Programme. COST Action TD 1202 is an intergovernmental framework for European Cooperation in Science and Technology funded by the EU within information and Communication Technologies programme. CITI-SENSE is a Collaborative Project partly funded by the EU FP7-ENV-2012 under grant agreement no. 308524. Citi-Sense-MOB is a collaborative project partly funded by The European Mobile and Mobility Industries Alliance (EMMIA). The initial proof of this manuscript benefited from the suggestions and comments from Sonja Grossberndt (NILU) and Erik Skjetne (Statoil). We would like to thank the referees Oral Saulters and Marcos Engelken-Jorge for their very valuable comments. We want to make a special acknowledgement to Fintan Hurley (IOM) for his suggestions, as well as Mark Nieuwenhuijsen (CREAL), Britt Ann Kåstad Høiskar (NILU) and William Lahoz (NILU) for their inputs. And finally, thank you to the members of ENVIROFI, CITI-SENSE, Citi-Sense-MOB and Cost Action TD 1202, who contributed to our understanding and ideas.

Source: Liu H Y, Kobernus M, Broday D, *et al.* A conceptual approach to a citizens' observatory-supporting community-based environmental governance[J]. Environmental Health A Global Access Science Source, 2014, 14(1):1–13.

References

[1] Sadik S: What is environment? http://shadmansadik.blogspot.no/search/label/Environmental%20 Geography.

[2] McGlade J: Global citizen observatory—The role of individuals in observing and understanding our changing world. http://www.eea.europa.eu/media/speeches/global-citizen-observatory-the-role-of-individuals-in-observing-and-understanding-our-

changing-world.

[3] Lanfranchi V, Ireson N, When U, Wrigley SN, Fabio C: Citizens' Observatories for Situation Awareness in Flooding. In Proceedings of the 11th International Conference on Information Systems for Crisis Response and Management (ISCRAM 2014): 18-21 May 2014. Edited by Hiltz SR, Pfaff MS, Plotnick L, Shih PC. Pennsylvania, USA: University Park; 2014:145−154.

[4] Copernicus: Global monitoring for environment and security. http://www.copernicus.eu.

[5] GEOSS: Global earth observation system of systems. http://www.epa.gov/geoss.

[6] GO: The global observatory. http://www.globalobservatory.eu.

[7] SEIS: Shared environmental information system. http://ec.europa.eu/environment/seis.

[8] Inglehart R: Modernization and Postmodernization: Cultural, Economic, and Political Change in 43 Societies. New Jersey: Princeton University Press; 1997.

[9] UNEP: Environmental governance. http://www.unep.org/environmentalgovernance/.

[10] Stakeholder Forum for Our Common Future: International environmental governance—A briefing paper. http://www.stakeholderforum.org/publications/reports/IEG-SFpaper.pdf.

[11] UNEP: Environmental governance. http://www.unep.org/pdf/brochures/EnvironmentalGovernance.pdf.

[12] Karatzas KD: Participatory environmental sensing for quality of life information services. In Information Technologies in Environmental Engineering—New Trends and Challenges. Edited by Golinska P, Fertsch M, Marx-Gomez J. Heidelberg: Springer; 2011:123−133.

[13] EEA: Eye on Earth. http://www.eyeonearth.org.

[14] CitCLOPS: Citizens' observatory for coast and ocean optical monitoring. http://www.citclops.eu.

[15] Omniscientis: Odour monitoring and information system based on citizen and technology innovative sensors. http://www.omniscientis.eu.

[16] CITI-SENSE: Development of sensor-based citizens' observatory community for improving quality of life in cities. http://www.citi-sense.eu.

[17] WeSenseIt: Citizen water observatories. http://www.wesenseit.com.

[18] COBWEB: Citizen observatory web. http://cobwebproject.eu.

[19] ENVIROFI: The environmental observation web and its service applications within the future internet. http://www.envirofi.eu.

[20] HENVINET: Health and environmental network systems. http://www. henvinet.eu.

[21] Liu H-Y, Bartonova A, Pascal M, Smolders R, Skjetne E, Dusinska M: Approaches to integrated monitoring for environmental health impact assessment. Environ Health 2012, 11:88.

[22] Liu H-Y, Bartonova A, Neofytou P, Yang Y, Kobernus M, Negrenti E, Housiadas C: Facilitating knowledge transfer: decision support tools in environment and health. Environ Health 2012, 11:S17.

[23] Pfeffer MJ, Wagenet LP: Volunteer environmental monitoring, knowledge creation and citizen-scientist interaction. http://knowledge.sagepub.com/view/hdbk_enviroso ciety/n16.xml.

[24] Conrad CC, Hilchey KG: A review of citizen science and community-based environmental monitoring: issues and opportunities. Environ Monit Assess 2011, 176: 273–291.

[25] Waterkeeper alliance. http://waterkeeper.org/.

[26] Keough HL, Blahna DJ: Achieving integrative, collaborative ecosystem management. Conserv Biol 2006, 20:1373–1382.

[27] Pattengill-Semmens CV, Semmens BX: Conservation and management applications of the reef volunteer fish monitoring program. Environ Monit Assess 2003, 81:43–50.

[28] Savan B, Morgan A, Gore C: Volunteer environmental monitoring and the role of the universities: The case of citizen's watch. Environ Manag 2003, 31:561–568.

[29] Sultana P, Abeyasekera S: Effectiveness of participatory planning for community management of fisheries in Bangladesh. J Environ Manag 2008, 86:201–213.

[30] Kerr M, Ely E, Lee V, Mayio A: A profile of volunteer environmental monitoring: National survey results. Lake Reserv Manag 1994, 9:1–4.

[31] Pretty J: Social capital and the collective management of resources. Science 2003, 302:142–148.

[32] Lawrence A: No personal motive? Volunteers, biodiversity, and the false dichotomies of participation. Ethics Place Environ 2006, 9:279–298.

[33] Conrad C: Towards meaningful community-based ecological monitoring in Nova Scotia: Where we are versus where we would like to be. Environment 2006, 34: 25–36.

[34] Conrad C, Daoust T: Community-based monitoring frameworks: Increasing the effectiveness of environmental stewardship. Environ Manag 2008, 41:356–358.

[35] Whitelaw G, Vaughan H, Craig B, Atkinson D: Establishing the canadian community monitoring network. Environ Monit Assess 2003, 88:409–418.

[36] Pollock RM, Whitelaw GS: Community based monitoring in support of local sustainability. Local Environ 2005, 10:211–228.

[37] Earth Institute centre for environmental sustainability. http://eices. columbia.edu/.

[38] Community science institute. http://communityscience.org/.

[39] The Institute for environmental science and policy. http://iesp.uic.edu/.

[40] Vespucci Institute: Vespucci institute on citizen science and VGI. http://povesham. wordpress.com/2014/07/12/vespucci-institute-on-citizen-science-and-vgi/.

[41] Wikipedia: Citizen cyberscience centre. http://en.wikipedia.org/wiki/Citizen_Cyber science_Centre.

[42] Ostrom E: Governing the commons: The Evolution of Institutions for Collective Action. Cambridge: Cambridge University Press; 1990.

[43] New eye on earth global mapping and information service now live. http://www. eea.europa.eu/media/newsreleases/new-eye-on-earth-global

[44] EyeOnEarth—Network. http://network.eyeonearth.org/.

[45] Citi-Sense-MOB: Mobile services for environment and health citizen's observatory. http://www.citi-sense-mob.eu.

[46] Castell N, Kobernus M, Liu H-Y, Schneider P, Lahoz W, Berre AJ, Noll J: Mobile technologies and services for environmental monitoring: The Citi-Sense-MOB approach. Urban Clim. In Press.

[47] Wiggins A, Crowston K: From conservation to crowdsourcing: A typology of citizen science. In Proceedings of the 11th International Conference on Autonomous Agents and Multiagent Systems (AAMAS 2012): 4-8 June 2012; Valencia, Spain. Edited by IEEE Computer Society. 2011:1−10.

[48] The open air laboratories (OPAL). http://www.sei-international.org/projects?prid=237.

[49] The big butterfly count. http://www.bigbutterflycount.org.

[50] COMBER: Citizens' network for the observation of marine biodiversity. http://www. comber.hcmr.gr/.

[51] Gouveia C, Fonseca A, Camara A, Ferreira F: Promoting the use of environmental data collected by concerned citizens through information and communication technologies. J Environ Manag 2004, 71:135−154.

[52] Hanahan RA, Cottrill C: A comparative analysis of water quality monitoring programs in the southeast: lessons for tennessee. http://www.academia.edu/1612982 /A_Comparative_Analysis_of_Water_Quality_Monitoring_Programs_in_the_Southe ast_Lessons_for_Tennessee.

[53] Nerbonne JF, Nelson KC: Volunteer macro-invertebrate monitoring in the United States: Resource mobilization and comparative state structures. Soc Nat Resour 2004, 17:817−839.

[54] Buytaert W, Zulkafli Z, Grainger S, Acosta L, Alemie TC, Bastiaensen J, Bièvre BD,

Bhusal J, Clark J, Dewulf A, Foggin M, Hannah DM, Hergarten C, Isaeva A, Karpouzoglou T, Pandeya B, Paudel D, Sharma K, Steenhuis T, Tilahun S, Hecken GV, Zhumanova M: Citizen science in hydrology and water resources: opportunities for knowledge generation, ecosystem service management, and sustainable development. Front Earth Sci 2014, 2:1–21.

[55] Kaufmann J: Three view of associationalism in 19th century America: An empirical investigation. Am J Sociol 1999, 104:1296–1345.

[56] Fore LS, Paulsen K, O'Laughlin K: Assessing the performance of volunteers in monitoring streams. Freshw Biol 2001, 46:109–123.

[57] Fung A: Democratic theory and political science: a pragmatic method of constructive engagement. Am Polit Sci Rev 2007, 101(3):443–458.

[58] ELLA: Citizen Participation in Latin America: innovations to strengthen governance. http://ella.practicalaction.org/sites/default/files/ 130516_CitPar_GOV_GUIDE.pdf.

[59] Holohan A: Community, competition and citizen science: voluntary distributed computing in a globalized world. Farnham, Surrey: Ashgate Publishing Limited; 2013.

[60] Fraser EDG, Dougill AJ, Mabee WE, Reed M, McAlpine P: Bottom up and top down: Analysis of participatory processes for sustainability indicator identification as a pathway to community empowerment and sustainable environmental management. J Environ Manag 2006, 78:114–127.

[61] Brugha R, Varvasovsky Z: Stakeholder analysis: a review. Health Policy Plan 2000, 15:239–246.

[62] Potočnik J: Only one air. europa.eu/rapid/press-release_SPEECH-13-516_en.doc.

[63] Potočnik J: Raising air pollution awareness 'of key importance'. http://www.airqualitynews.com/2013/06/10/raising-air-pollution-awareness-of-key-importance/.

[64] ACCENT: Atmospheric composition change—the european network of excellence. http://www.accent-network.org/.

[65] Phillipson J, Liddon A: Common knowledge? An exploration of knowledge transfer. http://www.relu.ac.uk/news/briefings/RELUBrief6%20Common%20Knowledge.pdf.

[66] Fernandez-Gimenez ME, Ballard HL, Sturtevant VE: Adaptive management and social learning in collaborative and community-based monitoring: a study of five community-based forestry organizations in the western USA. Ecol Soc 2008,13:4.

[67] Goodchilda MF, Glennona JA: Crowdsourcing geographic information for disaster response: a research frontier. Int J Digit Earth 2010, 3:231–241.

[68] Linhart J, Papp L: Bridging the gap between bottom-up and top-down e-participation approaches—e-participation as active citizenship. http://www.vitalizing-democracy.org/site/downloads/591_265_EDEM2010_paper_20100402_revision.pdf.

[69] Wikipedia: Top-down and bottom-up design. http://en.wikipedia.org/wiki/Top-down

_and_bottom-up_design.

[70] Extreme Citizen Science blog: Citizen cyberscience summit 2014: Day 1—policy and citizen science. http://uclexcites.wordpress.com/2014/03/05/citizen-cyberscience-summit-2014-day-1-policy-and-citizen-science/.

[71] Airqualitynews: Raising air pollution awareness 'of key importance'. http://www.airqualitynews.com/2013/06/10/raising-air-pollution-awareness-of-key-importance.

[72] Ciravegna F, Huwald H, Lanfranchi V, Wehn de Montalvo U: Citizen observatories: the WeSenseIt vision. http://inspire.ec.europa.eu/events/conferences/inspire_2013/schedule/submissions/261.pdf.

[73] Berre AJ: Citizens observatories: The CITI-SENSE project. http://www.gepw8.noa.gr/files/presentations/splinter1/1%20-%20Arne%20Berre.pdf.

[74] Wynne B: Misunderstood misunderstanding: social identities and public uptake of science. Public Underst Sci 1992, 1:281−304.

[75] Kamar E, Hacker S, Horvitz E: Combining human and machine intelligence in large-scale crowdsourcing. In Proceedings of the 11th International Conference on Autonomous Agents and Multiagent Systems (AAMAS 2012): 4-8 June 2012; Valencia, Spain. Edited by International Foundation for Autonomous Agents and Multiagent Systems. 2012:467−474.

[76] ORACLE: Meeting citizen expectations in new ways. http://www.oracle.com/us/products/applications/meeting-citizen-exp-wp-1560499.pdf.

[77] collective action problem and the potential of social capital. Local Environ 2010, 5(2):153−169.

[78] Stout M: Delivering an MPA emphasis in local governance and community development through service learning and action research. JPAE 2013, 19(2):217−238.

[79] Wikipedia: SWOT analysis. http://en.wikipedia.org/wiki/SWOT_analysis.

[80] Goeschl T, Jürgens O: Environmental quality and welfare effects of improving the reporting capability of citizen monitoring schemes. J Regul Econ 2012, 42:264−286.

[81] Engelken-Jorge M, Moreno J, Keune H, Verheyden W, Bartonova A, CITI-SENSE consortium: Developing citizens' observatories for environmental monitoring and citizen empowerment: challenges and future scenarios. In Proceedings of the Conference for E-Democracy and Open Governement (CeDEM14): 21-23 May 2014; Danube University Krems, Austria. Edited by Parycek P, Edelmann N. 2014:49-60.

[82] Fernandez-Gimenez ME, Ballard HL, Sturtevant VE: Adaptive management and social learning in collaborative and community-based monitoring: a study of five community-based forestry organizations in the western USA. Ecol Soc 2008,13:4.

[83] Department for International Development: Citizens' voice and accountability—evaluation mozambique country case study (final report). http://diplomatie.belgium.be/en/binaries/evaluation_cva_mozambique_en_tcm312-64788.pdf.

[84] Foresti M, Sharma B, O' Neil T, Evans A: Evaluation of citizens' voice and accountability evaluation framework. http://diplomatie.belgium.be/en/binaries/evaluation_cva_framework_en_tcm312-64813.pdf.

[85] Brabham DC: Crowdsourcing the public participation process for planning projects. Plan Theory 2009, 8(3):242–262.

[86] Zikopoulos IBMP, Eaton C, Zikopoulos P: Understanding Big Data: Analytics for Enterprise Class Hadoop and Streaming Data. McGraw-Hill Professional; 2011.

[87] Manovich L: Trending: The promises and the challenges of big social data. http://manovich.net/content/04-projects/065-trending-the-promises-and-the-challenges-of-big-social-data/64-article-2011.pdf.

[88] van der Sluijs JP, Craye M, Funtowicz S, Kloprogge P, Ravetz J, Risbey J: Experiences with the NUSAP system for multidimensional uncertainty assessment in model based foresight studies. Water Sci Technol 2005, 52:133–144.

[89] Knol A: Health and the environment: assessing the impacts, addressing the uncertainties. PhD thesis. Utrecht University, Institute for Risk Assessment Sciences; 2010.

[90] van der Sluijs JP, Craye M, Funtowicz S, Kloprogge P, Ravetz J, Risbey J: Combining quantitative and qualitative measures of uncertainty in model-based environmental assessment: the NUSAP system. Risk Anal 2005, 25:481–492.

[91] Adams SM: Establishing causality between environmental stressors and effects on aquatic ecosystems. Hum Ecol Risk Assess 2003, 9:17–35.

[92] Knol AB, Slottje P, van der Sluijs JP, Lebret E: The use of expert elicitation in environmental health impact assessment: a seven step procedure. Environ Health 2010, 9:19.

[93] Risbey JS, Kandlikar M: Expressions of likelihood and confidence in the IPCC uncertainty assessment process. Clim Chang 2007, 85:19–31.

[94] Cohen Y, Cohen A, Hetzroni A, Alchanatis V, Broday DM, Gazit Y, Timar D: Spatial decision support system for Medfly control in citrus. Comput Electron Agric 2008, 62:107–117.

[95] Cohen A, Cohen Y, Broday D, Timar D: Performance and acceptanceevaluation of a knowledge-based SDSS for Medfly area-wide control.J Appl Entomol 2008, 132:734–745.

[96] Giannotti F, Pedreschi D: Mobility, Data Mining and Privacy: Geographic Knowledge Discovery. New York: Springer; 2010.

[97] Brownson RC, Roer C, Ewing R, McBride TD: Researchers and policymakers: Travellers in parallel universes. Am J Prev Med 2006, 30(2):164–172.

[98] Bartonova A: How can scientists bring research to use: the HENVINET experience. Environ Health 2012, 11(Suppl 1):S2.

Chapter 3

Remoteness from Sources of Persistent Organic Pollutants in the Multi-Media Global Environment

Recep Kaya Göktaş[1,2*], Matthew MacLeod[2]

[1]Department of Environmental Engineering, Kocaeli University, Umuttepe Yerles, kesi, 41380, İzmit, Kocaeli, Turkey
[2]Department of Environmental Science and Analytical Chemistry (ACES), Stockholm University, Svante Arrhenius väg 8, SE 11418, Stockholm, Sweden

Abstract: Quantifying the remoteness from sources of persistent organic pollutants (POPs) can inform the design of monitoring studies and the interpretation of measurement data. Previous work on quantifying remoteness has not explicitly considered partitioning between the gas phase and aerosols, and between the atmosphere and the Earth's surface. The objective of this study is to present a metric of remoteness for POPs transported through the atmosphere calculated with a global multimedia fate model, BETR-Research. We calculated the remoteness of regions covering the entire globe from emission sources distributed according to light emissions, and taking into account the multimedia partitioning properties of chemicals and using averaged global climate data. Remoteness for hypothetical chemicals with distinct partitioning properties (volatile, semi-volatile, hydrophilic, low-volatility) and having two different half-lives in air (60-day and 2-day) are presented. Differences in remoteness distribution among the hypothetical chemicals are most pronounced in scenarios assuming 60-day half-life in air. In scenarios with a 2-day half-life in air, degradation dominates over wet and dry deposition processes as a pathway for atmospheric removal of all chemicals except the low-volatility chemical. The remoteness distribution of the low-volatility chemical is strongly dependent on assumptions about degradability on atmospheric aerosols. Calculations that considered seasonal variability in temperature, hydroxyl radical concentrations in the atmosphere and global atmospheric and oceanic circulation patterns indicate that variability in hydroxyl radical concentrations largely determines the seasonal variability of remoteness. Concentrations of polybrominated diphenyl ethers (PBDEs) measured in tree bark from around the world are more highly correlated with remoteness calculated using our methods than with proximity to human population, and we see considerable potential to apply remoteness calculations for interpretation of monitoring data collected under programs such as the Stockholm Convention Global Monitoring Plan.

Keywords: BETR, Multimedia Modeling Long-Range Transport Remoteness, POPs

1. Introduction

The potential of a chemical to pose an environmental and/or human health risk as a result of long-range transport is a defining property of a persistent organic pollutant (POP). As such, long-range transport potential is identified in the Stockholm Convention as one of the screening criteria for determining whether a substance should be classified as a POP, and subject to global regulation. According to Annex D of the convention text (S.C, 2009), measured levels of a chemical "in locations distant from its release" should be supplied as evidence of a chemical's long-range transport potential during the screening process. Therefore, in the planning and design of measurement campaigns, and also, when analyzing measurement data from monitoring networks, it is critical to determine how distant or remote sampling locations are from pollution sources.

In a pioneering study, von Waldow *et al.* (2010) introduced a quantitative metric of the remoteness of regions at a global or regional scale from emissions of pollutants to the atmosphere, called remoteness index. Von Waldow *et al.*'s (2010) remoteness index is calculated from atmospheric transport modeling of a suite of volatile tracers with different atmospheric half-lives emitted at a constant rate on a specified geographical distribution. The results of the tracer modeling are fit to a non-linear function that provides a location-specific parameter that indicates the remoteness of locations in the model domain from the emissions. Westgate *et al.* (2010) subsequently proposed three methods to quantify the remoteness of air sampling sites. They define pertingency index values calculated for sampling sites as a measure of the proximity (1/remoteness) of the sites to spatially distributed emissions. Two of the methods proposed by Westgate *et al.* (2010) are based on the geographical distance of sources from sampling sites, and one method uses Lagrangian trajectory modeling. They applied these methods to determine the proximity of the sampling sites in the Global Atmospheric Passive Sampling (GAPS) study to estimated global emissions of polycyclic aromatic hydrocarbons (PAHs). Similar methods using Lagrangian trajectory modeling to quantify the remoteness of sampling sites in studies were later used by Westgate and Wania (2011) and Westgate *et al.* (2013).

In the approach proposed by Westgate *et al.* (2010), the location of a sampling site and also the specific time period for sampling are required to obtain the air-mass trajectories that end at the site. While, in principle, the whole globe can be divided as a collection of cells and the air-mass back trajectories for all cells can be calculated, it is more practical to use this method to quantify remoteness of a limited number of predetermined sampling sites to a given emissions distribution to avoid extensive computations. Von Waldow *et al.*'s (2010) method, on the other hand, used a simplified global atmospheric circulation model to calculate long-term average circulation patterns at steady-state as the basis of the remoteness index. Von Waldow *et al.*'s (2010) method, therefore, produces time-independent remoteness values unlike the methods based on air-mass back-trajectory calculations.

Neither the von Waldow *et al.* (2010) nor the Westgate *et al.* (2010) approaches to quantify remoteness take into account the properties of specific pollutants. In this study, we propose a method to quantify remoteness from global sources of POPs that considers the multimedia fate and transport properties of chemical pollutants. We use a global multimedia fate model to derive remoteness values of regions covering the entire globe for a set of hypothetical pollutants assuming a representative generic global emissions distribution. Running the fate model in steady-state mode yields long-term average remoteness values similar to the von Waldow *et al.* (2010) approach. However the global multimedia fate model is capable of dynamic simulations, and therefore, seasonal variations in the remoteness of locations can be analyzed with our method when time-dependent remoteness values are of interest.

Our method of quantifying remoteness is based on the transfer efficiency concept introduced by MacLeod and Mackay (2004). Transfer efficiency is a model-based metric of the contribution of emissions distributed in a defined geographical pattern to the flux of chemical that reaches a specific receptor area. In MacLeod and Mackay's (2004) analysis, the source area was the continent of North America and the receptor area was the Great Lakes. In this study, the whole globe is considered as both the source and the receptor.

This study presents an investigation of the relationship between the multimedia partitioning properties of pollutants and the remoteness of regions from pollution sources. Hypothetical POPs that represent different categories of partitioning properties are used in the analysis. A case study with polybrominated diphenyl ethers (PBDEs) is included as an evaluation of the methodology. The proposed approach for quantifying remoteness allowed us to conduct a critical analysis of the 2-day-half-life-in-air criteria that is used to screen for long-range transport potential in the Stockholm Convention and under various national regulations as it is applied to pollutants with different multimedia partitioning properties.

2. Methods

2.1. Global Multimedia Fate Model-BETR-Global

We used the global multimedia fate model BETR-Global (MacLeod *et al.*, 2011) to calculate the global transfer efficiency values for the hypothetical and real organic chemicals analyzed in this study. BETR-Global is a member of the Berkeley-Trent (BETR) family of spatially resolved multimedia fate models. In BETR models, the environment is conceptualized as a collection of interconnected regional mass-balance models. Each region has seven compartments: upper atmosphere, lower atmosphere, vegetation, freshwater, ocean, soil, and freshwater sediments. The regions are connected by bi-directional flows of air and water between the atmosphere, ocean and fresh water compartments. Chemical mass balance equations for each of the compartments in each of the model regions are defined, and numerical solution of the resulting set of equations provides the spatial and temporal distribution of the simulated chemical.

Since its first introduction (MacLeod *et al.*, 2005), BETR-Global has been applied to analyze many different global-scale chemical pollution problems (e.g. Armitage *et al.*, 2009a; Armitage *et al.* 2009b; Gusev *et al.*, 2010; Lamon *et al.*, 2009; Li *et al.*, 2015; Wöhrnschimmel *et al.*, 2012). These applications have built confidence in the model structure and assumptions by conducting model evaluations where modeled and measured concentrations of chemicals are compared. Lamon *et al.*'s (2009) study showed satisfactory agreement between the dynamic model results and long-term monitoring data for globally distributed concentrations of PCB 28 and PCB 153 in air. Woöhrnschimmel *et al.* (2012) modeled the global fate of α- and β-HCH and evaluated the model performance by comparing the model results with measured concentrations in air and ocean water. Recently, two new software implementations of BETR-Global has been published (MacLeod *et al.*, 2011). BETR-Global 2.0 is coded in Visual Basic for Applications and runs as a Microsoft Excel macro (https://sites.google.com/site/betrglobal). BETR-Research is a re-implementation of BETR-Global in the Python programming language (http://betrs.sourceforge.net). BETR-Research is an open-source project aimed at researchers who would like to modify the model code and add new capabilities to the model according to their specific research interests. Some of the modifications to BETR-Research since its first introduction include increasing the resolution of the global data sets to model the global environment on 3.75° × 3.75° spatial resolution (in addition to the base resolution of 15° × 15°), implementing a fast numerical solver, and adding new code to track mass fluxes between compartments. In this study, the newest version of BETR-Research that allows fast simulations with high spatial resolution was used. Code introduced to the model to track mass fluxes between compartments (Woöhrnschimmel *et al.*, 2013) was critical in calculating the transfer efficiency values.

2.2. Quantifying Remoteness

Our proposed multimedia remoteness measure is based on the global distribution of transfer efficiency values calculated by the BETR-Research model. The definition of transfer efficiency used in this study is a slight modification of the definition given by MacLeod and Mackay (2004). In this study, transfer efficiency is calculated by dividing the rate of contaminant flux to a selected target environmental compartment (including emissions) by the rate of global emissions:

$$\text{Transfer Efficiency}\left(\text{TE}\right) = \frac{\text{Rate of contaminant flux to the target environmental compartment}\left(\text{mol/h}\right)}{\text{Rate of global emissions}\left(\text{mol/h}\right)} \quad (1)$$

Our remoteness measure is then calculated by transforming the TE values so that lower values indicate less remoteness while higher values indicate more remoteness. This is achieved by defining remoteness as equal to −log(TE). The resultant dimensionless remoteness metric is called pTE:

$$\text{pTE} = -\log\left(\text{TE}\right) \quad (2)$$

In this study, the focus is on POPs that are emitted into, transported through and measured in the atmosphere. Therefore, the target environmental compartment in Equation (1) is the lower atmosphere compartment, and the transfer efficiency to a model region is calculated by summing all the atmospheric fluxes into the lower atmosphere compartment of that region, *i.e.*, the sum of the input fluxes from the lower atmosphere compartments of the neighboring regions, input flux from the upper atmosphere of the model region, and the direct emissions to the lower air compartment. This approach is compatible with the remoteness concept that has been proposed in previous studies (von Waldow *et al.*, 2010; Westgate *et al.*, 2010), which were also primarily interested in atmospheric pollutants. However, in principle, the remoteness measure proposed in this study (pTE) can be used to quantify remoteness of any model compartment and could consider transport in water as well as air.

2.3. Simulation Experiments

2.3.1. Remoteness Analysis with Hypothetical Chemicals

Wania (2006) and Gouin and Wania (2007) proposed a categorization of chemicals according to their airewater and octanoleair partition coefficients, KAW and KOA, respectively. This categorization defines four types of POPs according to their global transport behavior (**Figure S1**). The "flyers" are volatile chemicals that do not deposit onto surface compartments. At the other extreme, "single-hoppers" have low volatility and are predominantly associated with aerosols in the atmosphere and irreversibly deposit onto surface compartments. Between these two extremes are "multiple hoppers" that are semi-volatile chemicals that are readily exchanged between the atmosphere and the surface compartments, and "swimmers" that tend to be present in the water phase and transport via oceans. In this study, four hypothetical chemicals (volatile, semi-volatile, hydrophilic and low-volatility) that belong distinctly to each of these categories (flyers, multiple-hoppers, swimmers and single hoppers, respectively) are defined by assigning appropriate partitioning properties between air, water and octanol. **Table 1** lists the partition coefficient values assigned to each of the hypothetical chemicals.

Annex D of the Stockholm Convention (S.C, 2009) states that for an organic pollutant to be classified as persistent, its half-life in water should be longer than 2 months, or its half-life in soil and/or sediment should be longer than 6 months. In this study, all the hypothetical chemicals are assigned these limit values as their half-lives in water, soil and sediment compartments. Half-life in vegetation is assumed to be the same as half-life in water. We considered two different scenarios for degradation half-lives in air, where chemical degradation in air is assumed to occur only by reaction with hydroxyl radicals. In the first scenario, the hypothetical chemicals are assumed to have a 60-day half-life in air when exposed to hydroxyl radicals in the gas phase at global-average concentrations. This scenario simulates the properties of POPs that are very persistent in air (e.g. similar to hexocholorobenzene). In a second scenario, the hypothetical chemicals have a 2-day halflife in air when exposed to hydroxyl radicals in the gas phase at global-average concentrations, which is the limit value given in the Stockholm Convention (S.C, 2009, Annex D) when screening for chemicals with long-range transport potential. The half-lives of hypothetical chemicals used in model simulations to analyze our proposed remoteness metric are given in **Table 1**.

2.3.2. Remoteness Distribution of PBDEs

Salamova and Hites (2013) reported concentrations of PBDEs in tree bark from 12 locations around the globe. They showed a correlation between concen

Table 1. Properties of hypothetical chemicals and PBDEs used in model simulations.

	(Flyer) volatile	(Multi-hopper) semi-volatile	(Swimmer) hydrophilic	(Single-hopper) low-volatility	BDE-47*	BDE-99*	BDE-209*
			Partition coefficients				
log KOW	6	6	2	8	6.53	7	9.97
log KAW	3	−2	−4	−4	−3.12	−3.37	−4.81
log KOA	3	8	6	12	10.4	11.3	16.8
			Degradation half-lives (days)				
in Air			60 (1st scenario); 2 (2nd scenario)		11	19	318
in Water			60		192	354	1583
in Vegetation			60		192	354	1583
in Soil			180		385	708	3167
in Sediment			180		1155	2125	9500

*: BDE-47, -99, -209 properties are compiled from Schenker *et al.* (2008).

trations of PBDEs in tree bark and human population in nearby areas. As shown in **Table 1**, PBDEs have properties that place them between the semi-volatile and the low-volatility hypothetical chemicals considered in our calculations. Therefore the dataset from Salamova and Hites (2013) offers an opportunity to evaluate our pTE calculations by investigating the correlation between concentrations of PBDEs in tree bark and remoteness values calculated for the sampling sites. In Salamova and Hites (2013), on average, 80% of total PBDEs measured in tree-bark samples was comprised of BDE-47, -99, and -209. We calculated the global remoteness distributions of these three PBDE congeners using the property values in **Table 1** and assuming they are emitted to air according to the night-light emission scenario defined below.

2.3.3. Emission Scenario

For all calculations presented here we used an emission scenario that assumes the global distribution of pollutant emissions follows the same pattern as nighttime light emissions into space. This approach was originally proposed by von Waldow *et al.* (2010) and was also used in subsequent studies that applied BETR-Global to model pollutants whose usage is related with economic activity and population density (e.g. MacLeod *et al.*, 2011; Wöhrnschimmel *et al.*, 2013). **Figure S2** in the Supporting Information shows the global emissions distribution used in the simulations.

3. Results

3.1. Global Remoteness Distributions of POPs

Figure 1 shows the global distribution of pTE for the four hypothetical chemicals when the half-life in air is 60 days. The minimum, the maximum and the mean pTE values for each type of chemical are specified in the figure, and histograms illustrate the frequency of occurrence of different pTE values for each hypothetical chemical. The distribution of remoteness follows the distribution of emissions and also the global atmospheric circulation patterns. Visual comparison of pTE distributions and the histograms indicate that a higher fraction of the global atmosphere at surface level is more remote for the hydrophilic and low-volatility chemicals compared to the volatile and semi-volatile chemicals. The minimum, the maximum and the mean pTE values increase in the order of Volatile < Semi-Volatile

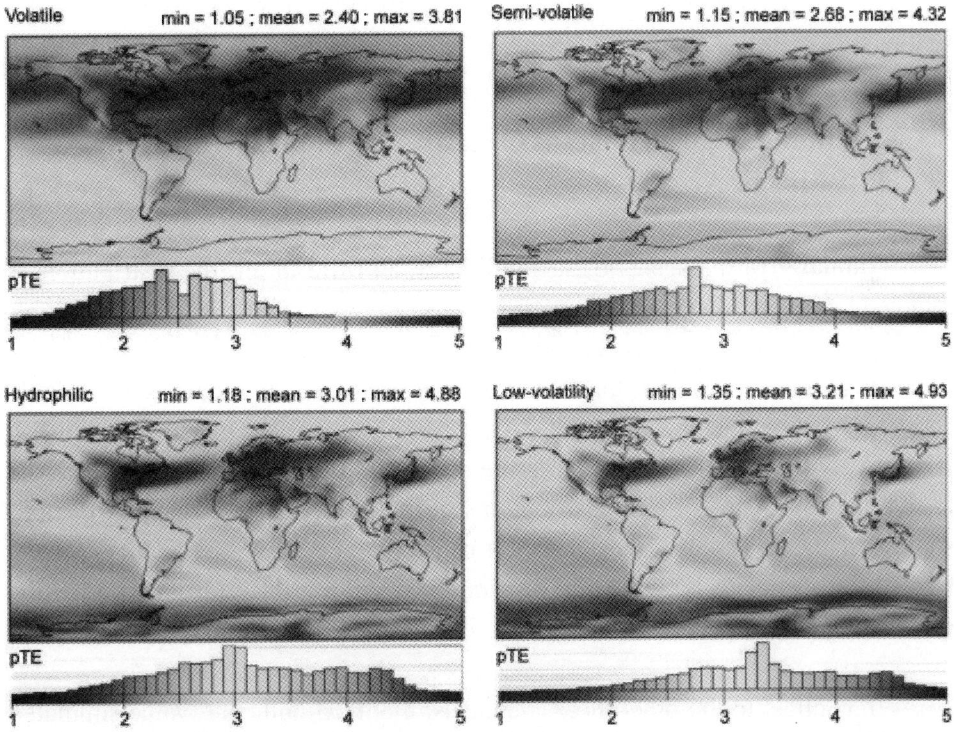

Figure 1. Global distribution of remoteness (pTE) obtained from steady-state simulations of four hypothetical chemicals when half life in air is 60 days. Histograms illustrate the frequency of occurrence of specified pTE values for each of the four chemicals.

< Hydrophilic < Low-Volatility. The same trend in the distribution of pTE values can be seen in the histograms.

3.2. Correlation of Remoteness Values and Tree-Bark Concentrations for PBDEs

The global remoteness distribution maps for PBDE-47, PBDE-99 and PBDE-209, and the average remoteness distribution map for PBDEs are plotted in **Figures S3** and **S4**, respectively, and the pTE values at the sampling locations of Salamova and Hites (2013) are given in **Table S1**. Atmospheric transport of the PBDEs and the low volatility chemical shown in **Figure 1** is mostly determined by the model's description of aerosol transport, so the remoteness distributions of these chemicals are very similar. The correlations between concentrations of PBDEs in tree bark with population and with the pTE values of the sampling locations are shown in **Figure 2**. The correlation of the logarithm of concentration values with pTE ($r^2 = 0.48$, $p = 0.012$) is stronger than it is with population ($r^2 = 0.35$, $p = 0.045$).

3.3. Remoteness for Persistent Organic Pollutants with 2-Day Half-Life in Air

Similarly to the scenario for the 60-day half-life in air, the distribution of remoteness in the 2-day half-life in air scenario for the hypothetical chemicals follows the distribution of emissions and also the global atmospheric circulation patterns (**Figure 3**). However, the maximum pTE values for volatile, semi-volatile and hydrophilic pollutants are much higher for this scenario, *i.e.*, 8.6, 8.6, and 8.8, respectively (note the different pTE scale in **Figure 3** compared to **Figure 1**). The histograms and maps of pTE distributions in the 2-day half-life scenario demonstrate that the pTE value distribution is nearly identical for the volatile and the semi-volatile chemicals, while the distribution of pTE values for the hydrophilic chemical includes higher values.

In contrast to the other three chemicals, the maximum pTE value simulated for the low-volatility chemical in the 2-day half-life in air scenario is 5 (**Figure 3**), which is nearly identical to the maximum pTE value of 4.9 in the 60-day half-life in air scenario illustrated in **Figure 1**. Thus, in the 2-day half-life scenario locations

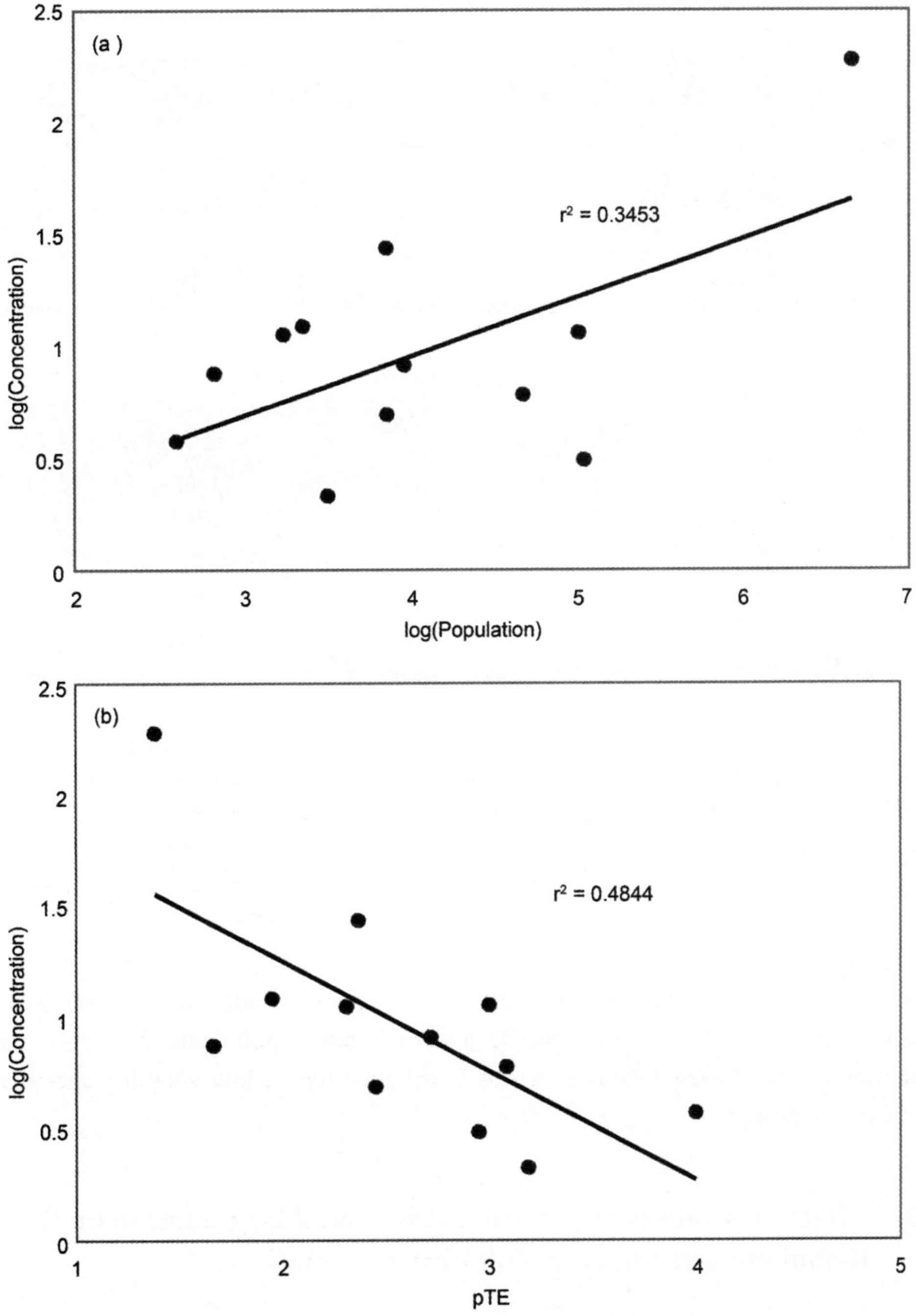

Figure 2. Regressions of the concentrations of total PBDE in tree-bark samples measured by Salamova and Hites (2013) with (a) population ($r^2 = 0.35$, $p = 0.045$); and (b) pTE ($r^2 = 0.48$, $p = 0.012$).

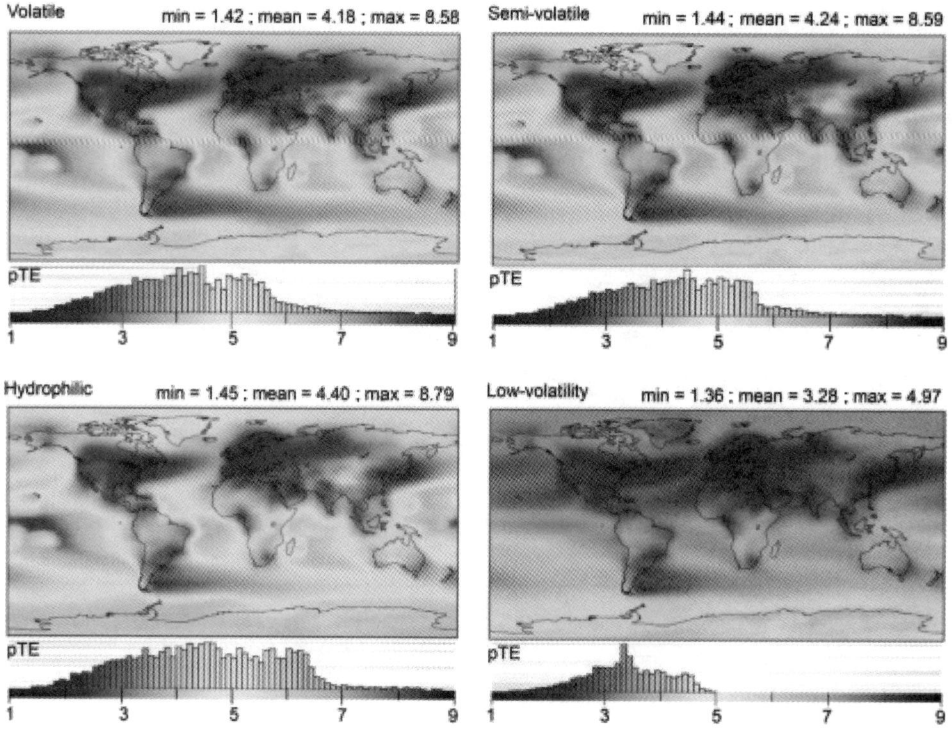

Figure 3. Global distribution of remoteness (pTE) obtained from steady-state simulations of the four hypothetical chemicals when half life in air is 2 days. Histograms illustrate frequency of occurrence of specified pTE values for each of the four chemicals.

far from source areas in the global environment are less remote for the low-volatility chemical than they are for the other 3 substances. Comparing **Figure 1** and **Figure 3** and taking note of the different scales indicates that when the 60-day half-life in air was shortened to 2-day, most locations on the earth became more remote for the volatile, semi-volatile and hydrophilic pollutants. However, the shorter half-life affected the remoteness distribution for the low-volatility chemical much less strongly.

3.4. Effect of Atmospheric Particle Associated Degradation on the Remoteness of the Low-Volatility Chemical

In BETR-Global, one of the base assumptions is to model atmospheric degradation only in the gas phase. Under this assumption, substances that are sorbed to atmospheric particles are not degraded. The model simulations presented above are based on this assumption, which would most strongly affect the low-volatility chemical that is highly associated with aerosols in the atmosphere. There are high uncertainties associated with calculating the amount of chemical sorbed to aerosols and the degradation of sorbed chemicals (Scheringer, 2009). In order to analyze the sensitivity of the remoteness calculations to the quantification of the degradation rate within the aerosols, we designed a simulation scenario that represents the assumption on the other extreme: degradation rate in aerosols being equal to gas-phase degradation rate. We therefore repeated the model simulations for the 2-day half-life in air scenario, but applied that half-life to chemical present in the bulk air compartment consisting of both the gas phase and aerosols. The resultant pTE distribution for the low-volatility chemical is compared with the base case in **Figure 4**. Remoteness values calculated for the low-volatility chemical when degradation in air is simulated both in the gas phase and the particle phase are more highly variable geographically, and conform more closely to results for the other three hypothetical chemicals. Repeating the simulations assuming degradation in both the gas phase and aerosols does not significantly affect the remoteness distribution for the other hypothetical chemicals that is illustrated in **Figure 3** (**Figure S5** and **Figure S6**).

3.5. Comparison of the Global pTE Distribution of the Volatile Chemical with the Remoteness Index Distribution of von Waldow *et al.* (2010)

In order to further evaluate the proposed method for quantifying remoteness from pollution sources, the remoteness index distribution obtained by von Waldow *et al.* (2010) was compared with the pTE distribution for the hypothetical volatile chemical. **Figure 5(a)** is a reproduction of von Waldow *et al.*'s (2010) remoteness index distribution for the ECON emission scenario, which also used night-light emissions to represent the geographical distribution of emissions of chemicals. von Waldow *et al.* (2010) presented the remoteness index values after classifying them as deciles of earth's total surface area. In order to facilitate comparison of pTE distributions with remoteness index distributions, we calculated the percentile ranks of model regions according to their pTE values. **Figure 5(b)**, **Figure 5(c)** shows the percentile rank distributions of pTE values of the volatile chemical for the 60-day air half-life scenario and for the 2-day air half-life scenario, respectively. The pTE distribution for the 60-day air half-life case follows a very similar

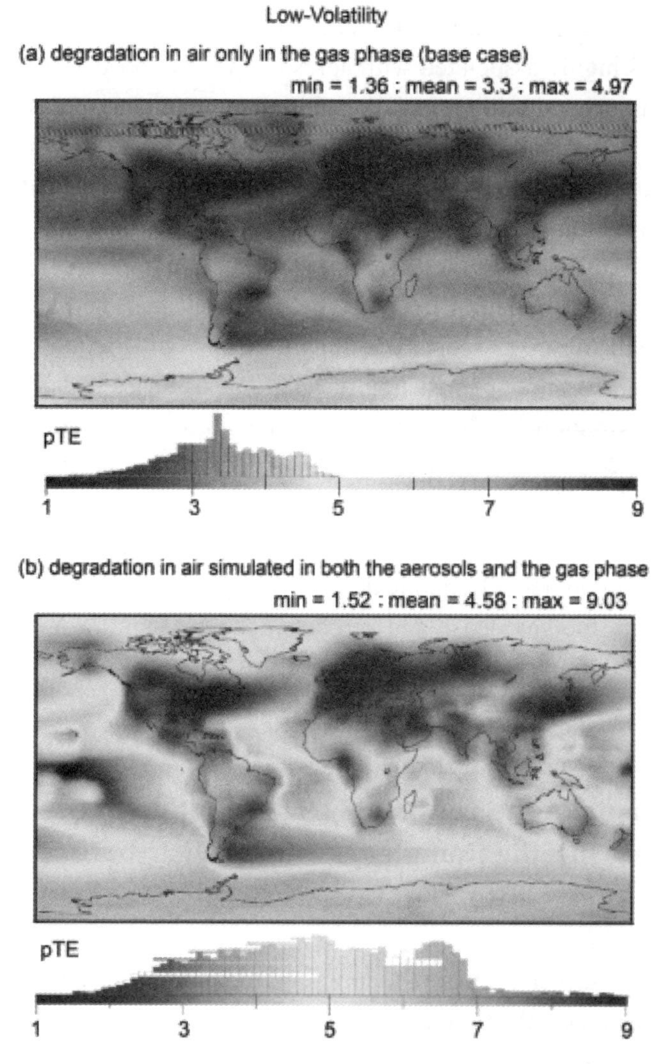

Figure 4. Global distribution of steady-state pTE values for the low-volatility chemical when half-life in air is 2 days. In case (a), degradation in air is simulated only in the gas phase. In case (b), degradation in air is simulated in both aerosols and the gas phase. Histograms illustrate frequency of occurrence of specified pTE values for each case.

pattern to the remoteness index distribution of von Waldow *et al.* (2010). In the case of 2-day air half-life scenario, although a similar pattern is still distinguishable, pTE values are less homogenous.

We calculated the correlation between RI and pTE distributions based on their raw values presented in **Figure S7**. The Pearson correlation coefficient (r)

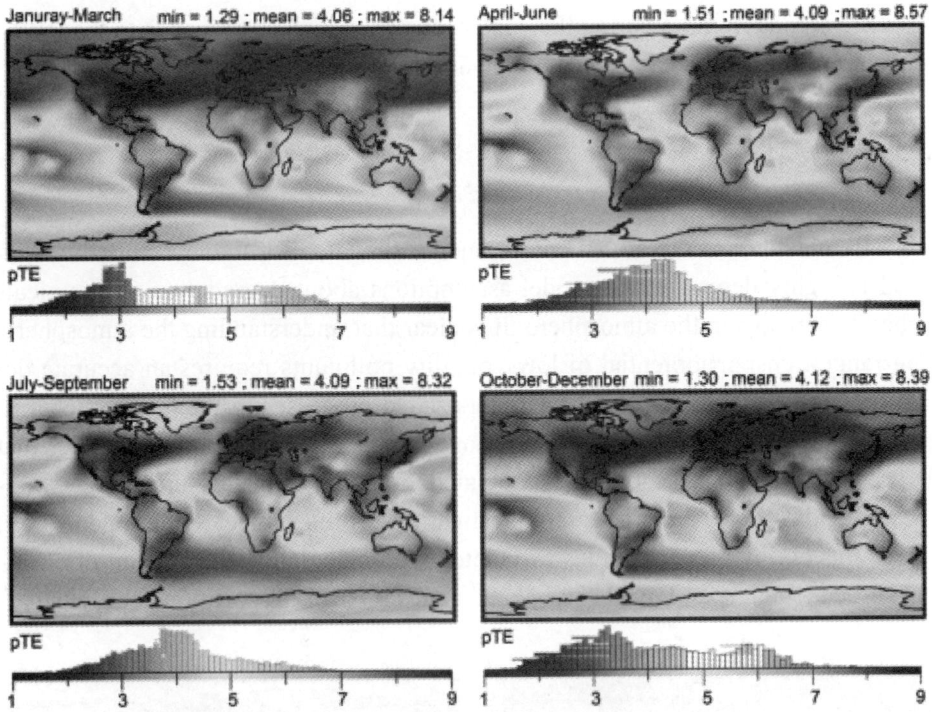

Figure 5. (a) Percentile ranks of von Waldow *et al.*'s (2010) remoteness index distribution; and the percentile ranks of pTE distributions for the volatile chemical for both scenarios with (b) 60-day half-life in air, and (c) 2-day half-life in air.

between RI and pTE values are 0.88 and 0.75 for the 60-day air half-life and the 2-day air half-life cases, respectively, providing a quantitative measure of the similarity between the RI and pTE distributions (**Figure S8**).

3.6. Dynamic Global Remoteness Distribution for the Volatile Chemical

BETR-Global is capable of dynamic simulations, and therefore, seasonally variable remoteness distributions can be obtained by the proposed method. Using the steady-state model simulation results as initial condition, we performed dynamic global simulations for the hypothetical volatile chemical assuming a 2-day half-life in air. The simulation time period was 5 years and the fifth year's results are presented here interpreted in terms of pTE. **Figure 6** shows the seasonal (three-month) averages of the pTE values for the volatile chemical. The summer and the winter months are visually distinguishable as the Arctic and the whole northern hemisphere are less remote during the winter months, which corresponds to the lowest concentrations of hydroxyl radical in the atmosphere of the northern hemisphere.

4. Discussion

Remoteness of the global environment calculated for the low-volatility chemical is highly dependent on model assumptions about degradation of chemicals sorbed to aerosols in the atmosphere. It is clear that understanding the atmospheric long-range transport potential of low-volatility pollutants requires an accurate description and parameterization of aerosol-related processes in the model. Scheringer (2009), in a review paper on long-range atmospheric transport, concludes that the degradation rates of particle-associated organic chemicals are "largely unknown" and calls for experimental investigations of the subject. Scheringer (2009) also points out high uncertainties associated with calculating the amount of chemical sorbed to aerosols. BETR-Global calculates aerosol-air partition coefficients

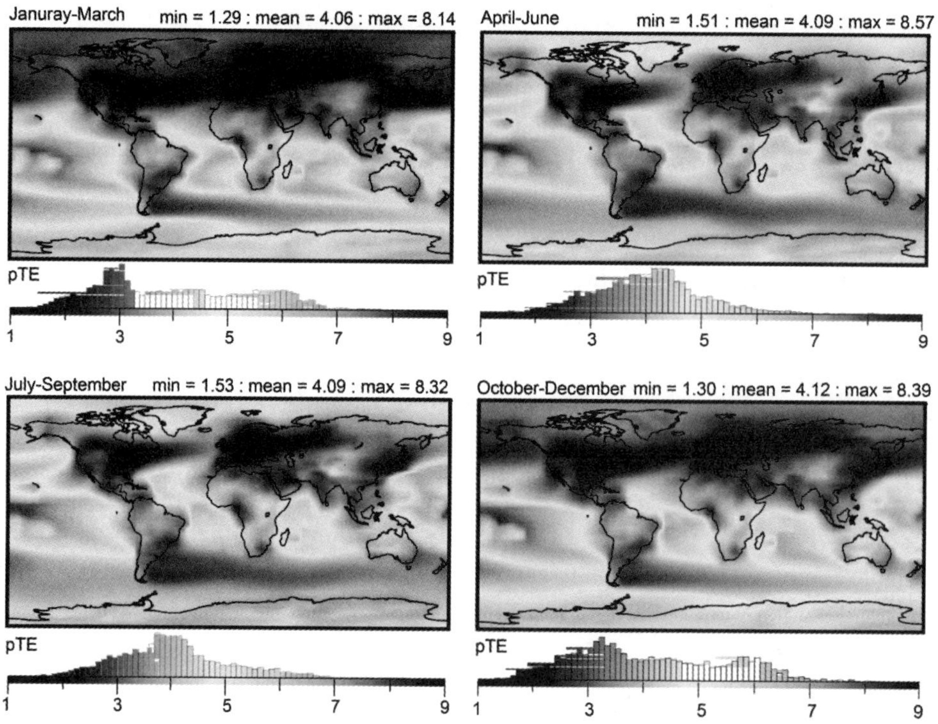

Figure 6. The dynamic global distribution of pTE values for the volatile chemical assuming an average half life in air of 2-day. Histograms illustrate frequency of occurrence of specified pTE values for each season.

from KOA values using the relationship recommended by MacLeod *et al.* (2010). Another critical parameter that influences the modeled pTE distributions of the low-volatility chemical is the life-time of aerosols in the atmosphere before being deposited on the surface. We conducted model experiments to determine the average global time scales for removal from atmosphere via wet and dry deposition using hydrophilic and hydrophobic tracers. In BETR-Global, the global average time-scale for the sum of wet and dry deposition is about 3 days for highly hydrophilic pollutants, whereas it is about 19 days for low-volatility chemicals sorbed to aerosols. In our scenarios that assume degradation half-lives in air of 60 days, these deposition processes can compete with degradation in air, and the results reflect the different partitioning behaviors of the hypothetical chemicals. However, when the degradation half-life in air is set to be 2 days, degradation dominates over wet and dry deposition processes as a pathway for atmospheric removal of all chemicals except the low-volatility chemical in the scenario where degradation in air occurs only in the gas phase. In that scenario, deposition is the dominant removal process for the low-volatility chemical. The three PBDE congeners simulated in this study (BDE-47, BDE-99, BDE-209) have different physico-chemical properties (**Table 1**). However, their partitioning properties are similar to the low-volatility chemical's and they all belong to the single-hopper region in Wania's (2006) chemical space (**Figure S1**). The pTE distributions of these PBDE congeners are mostly determined by the lifetime of aerosols in the atmosphere resulting in similar pTE distributions (**Figure S3**). Continued research and development is required to improve the description of atmospheric deposition, atmospheric degradation and aerosol-air partitioning processes in BETR-Global, and to provide a better interpretation of the pTE distributions for low-volatility chemicals.

All the analysis presented in this study is focused on pollutants emitted to the atmosphere and the target environmental compartment was chosen as the lower atmosphere. This study design places attention on transport of chemicals from source regions to the atmosphere in remote locations that is in contact with the surface, and it allowed us to evaluate the proposed method within the same context of the previous studies, such as von Waldow *et al.*'s (2010) remoteness index. However, the pTE method can be used to analyze remoteness of other environmental media (such as water, soil and vegetation) from pollutants emitted to air, water or soil. This type of an analysis is a potential future application of the pTE method introduced in this study.

In the scenario assuming a 60-day half-life in air for the hypothetical chemicals, the volatile chemical is distributed most efficiently in the global atmosphere (**Figure 1**). In this scenario, even the most remote region for the volatile chemical (pTE = 3.81) is less remote than 23% of the model regions for the low-volatility chemical. The behaviors of the semi-volatile and the hydrophilic chemicals are in-between these two extremes, however, the difference in the remoteness distributions of the volatile and the semi-volatile chemicals is small. The global mean pTE values of the volatile and the semi-volatile chemicals are 2.4 and 2.7 respectively, and only 1.4% of model regions are more remote for the semi-volatile chemical than the most remote region for the volatile chemical. This result suggests that the semi-volatile chemical can be transported to the lower air of regions far from sources nearly as efficiently as the volatile substance. Being more likely to be deposited onto surface compartments of regions after reaching the atmosphere of a remote region, the semi-volatile chemical can cause additional exposure and exposure-related effects. On the other hand, in this model scenario, the remoteness distribution of the hydrophilic chemical is similar to the low-volatility chemical. The global mean pTE values of the hydrophilic and the low-volatility chemicals are 3.0 and 3.2 respectively. And, few model regions are more remote for the low-volatility chemical than the most remote regions for the hydrophilic chemical. When emitted to the atmosphere, the hydrophilic chemical tends to be scavenged by precipitation. Then, its global transport will be mainly through freshwaters and oceans. This behavior is not distinguishable for the hydrophilic chemical in the analysis presented here since this study is focused on transport in the atmosphere and the atmosphere as a target compartment.

The remoteness index (RI) introduced by von Waldow *et al.* (2010) is an established quantitative measure of remoteness from sources of atmospheric pollutants. RI takes into account the emissions distribution and the global circulation patterns; but it is not dependent on physicochemical properties of pollutants. We compared the pTE distribution of the volatile chemical with the RI distribution for the same emission scenario used in this study (**Figure 5**). Although the pTE distribution of the volatile chemical changes when the modeled half-life in air is changed, the resultant pTE distribution maps are comparable indicating that pTE can be used as a quantitative measure of remoteness. Comparison of **Figure 5(b)**, **Figure 5(c)** indicates that when the half-life in air is longer, the pollutant is more homogeneously distributed within the northern hemisphere since the inter-hemispheric transfers become more significant as a limitation on the global atmospheric transfer efficiencies. Most of the emission sources are located in the northern hemisphere, and the volatile chemical that is not degraded or transferred to the other hemisphere is relatively well-mixed on the hemispheric scales in **Figure 5(b)** (60-day half-life in air). In contrast, in **Figure 5(c)** (2-day half-life in air), there are stronger concentration gradients on hemispheric scales, and the global remoteness distribution closely reflects atmospheric circulation patterns emanating from source regions.

The stronger correlations between PBDE concentrations in tree-bark and pTE compared to population indicate that metrics of remoteness have considerable untapped potential to help interpret measurements of persistent organic contaminants in the global environment. The correlation coefficient ($r^2 = 0.48$) indicates that 48% of the variability in the logarithm of PBDE concentrations in tree bark can be explained by variability in pTE under the assumption that PBDEs are emitted with the same geographical distribution as nighttime light emissions to space, and illustrates that the model's description of transport pathways for these chemicals is reasonable. Using measured concentrations of chemicals in different media to determine correlations with the pTE values calculated for various target environmental compartments other than the atmosphere would be a very interesting analysis. Comparing seasonally varied concentrations with dynamic pTE distributions is also possible and might provide insight into seasonally varying concentrations at remote locations. The remoteness metric has considerable potential to be used in interpretation of monitoring data collected under programs such as the Stockholm Convention Monitoring Plan. In cases such as the one for the PBDEs, where remoteness and measured concentrations are highly correlated, more detailed modeling using alternative emission reduction scenarios will make it possible to formulate strategies to target sources of chemicals for emission reductions and to estimate the effects of these actions on pollutant concentrations at remote locations.

Acknowledgements

We thank Harald von Waldow for providing the raw data of the remoteness index maps published in von Waldow *et al.* (2010). Funding for this research was provided by the Swedish Research Council FORMAS project number 216-2011-

427 and the Swedish Research Council Vetenskapsrådet through contract grant number 2011−3921.

Appendix A. Supplementary Data

Supplementary data related to this article can be found athttp://dx.doi.org/10.1016/j.envpol.2015.12.058. http://www.sciencedirect.com/science/article/pii/S0269749115302724.

Source: Recep Kaya Göktaş, Matthew MacLeod. Remoteness from sources of persistent organic pollutants in the multi-media global environment [J]. Environmental Pollution:barking, Essex, 2016.

References

[1] Armitage, J.M., MacLeod, M., Cousins, I.T., 2009a. Comparative assessment of the global fate and transport pathways of long-chain perfluorocarboxylic acids (PFCAs) and perfluorocarboxylates (PFCs) emitted from direct sources. Environ. Sci. Technol. 43 (15), 5830e5836.

[2] Armitage, J.M., MacLeod, M., Cousins, I.T., 2009b. Modeling the global fate and transport of perfluorooctanoic acid (PFOA) and perfluorooctanoate (PFO) emitted from direct sources using a multispecies mass balance model. Environ. Sci. Technol. 43 (4), 1134e1140.

[3] Gouin, T., Wania, F., 2007. Time trends of arctic contamination in relation to emission history and chemical persistence and partitioning properties. Environ. Sci. Technol. 41, 5986e5992.

[4] Gusev, A., MacLeod, M., Shatalov, V., Bartlett, P., Hollander, A., Gong, S., Lammel, G., 2010. Global and Regional Modelling of POPs. United Nations, Economic Commission for Europe, Task Force on Hemispheric Transport of Air Pollution.

[5] Lamon, L., von Waldow, H., MacLeod, M., Scheringer, M., Marcomini, A., Hungerbuehler, K., 2009. Modeling the global levels and distribution of polychlorinated biphenyls in air under a climate change scenario. Environ. Sci. Technol. 43 (15), 5818e5824.

[6] Li, L., Liu, J.G., Hu, J.X., 2015. Global inventory, long-range transport and environmental distribution of dicofol. Environ. Sci. Technol. 49 (1), 212e222.

[7] MacLeod, M., Mackay, D., 2004. Modeling transport and deposition of contaminants

to ecosystems of concern: a case study for the Laurentian Great Lakes. Environ. Pollut. 128 (1e2), 241e250.

[8] MacLeod, M., Riley, W.J., McKone, T.E., 2005. Assessing the influence of climate variability on atmospheric concentrations of polychlorinated biphenyls using a global-scale mass balance model (BETR-global). Environ. Sci. Technol. 39 (17), 6749e6756.

[9] MacLeod, M., Scheringer, M., Götz, C., Hungerbühler, K., Davidson, C.I., Holsen, T.M., 2010. "Deposition from the Atmosphere to Water and Soils with Aerosol Particles and Precipitation." Handbook of Chemical Mass Transport in the Environment. CRC Press, pp. 103e135.

[10] MacLeod, M., von Waldow, H., Tay, P., Armitage, J.M., Wöhrnschimmel, H., Riley, W.J., McKone, T.E., Hungerbuhler, K., 2011. BETR global e a geographically-explicit global-scale multimedia contaminant fate model. Environ. Pollut. 159 (5), 1442e1445.

[11] S.C, 2009. Stockholm Convention on Persistent Organic Pollutants (POPs) as Amended in 2009: Text and Annexes. UNEP.

[12] Salamova, A., Hites, R.A., 2013. Brominated and chlorinated flame retardants in tree bark from around the Globe. Environ. Sci. Technol. 47 (1), 349e354.

[13] Schenker, U., Soltermann, F., Scheringer, M., Hungerbuhler, K., 2008. Modeling the environmental fate of polybrominated diphenyl ethers (PBDEs): the importance of photolysis for the formation of lighter PBDEs. Environ. Sci. Technol. 42 (24), 9244e9249.

[14] Scheringer, M., 2009. Long-range transport of organic chemicals in the environment. Environ. Toxicol. Chem. 28 (4), 677e690.

[15] von Waldow, H., MacLeod, M., Scheringer, M., Hungerbuhler, K., 2010. Quantifying remoteness from emission sources of persistent organic pollutants on a global scale. Environ. Sci. Technol. 44 (8), 2791e2796.

[16] Wania, F., 2006. Potential of degradable organic chemicals for absolute and relative enrichment in the arctic. Environ. Sci. Technol. 40 (2), 569e577.

[17] Westgate, J.N., Shunthirasingham, C., Oyiliagu, C.E., von Waldow, H., Wania, F., 2010. Three methods for quantifying proximity of air sampling sites to spatially resolved emissions of semi-volatile organic contaminants. Atmos. Environ. 44 (35), 4380e4387.

[18] Westgate, J.N., Sofowote, U.M., Roach, P., Fellin, P., D'Sa, I., Sverko, E., Su, Y., Hung, H., Wania, F., 2013. "In search of potential source regions of semi-volatile organic contaminants in air in the Yukon Territory, Canada from 2007 to 2009 using hybrid receptor models. Environ. Chem. 10 (1), 22e33.

[19] Westgate, J.N., Wania, F., 2011. On the construction, comparison, and variability of airsheds for interpreting semivolatile organic compounds in passively sampled air. Environ. Sci. Technol. 45 (20), 8850e8857.

[20] Wöhrnschimmel, H., MacLeod, M., Hungerbuhler, K., 2013. Emissions, fate and transport of persistent organic pollutants to the arctic in a changing global climate. Environ. Sci. Technol. 47 (5), 2323e2330.

[21] Wöhrnschimmel, H., Tay, P., von Waldow, H., Hung, H., Li, Y.F., MacLeod, M., Hungerbuhler, K., 2012. Comparative assessment of the global fate of alpha- and beta-hexachlorocyclohexane before and after phase-out. Environ. Sci. Technol. 46 (4), 2047e2054.

Chapter 4

A Legal Conventionalist Approach to Pollution

Carmen E. Pavel

King's College London, Strand Building, Strand Campus, London WC2R 2LS, UK

Abstract: There are no moral entitlements with respect to pollution prior to legal conventions that establish them, or so I will argue. While some moral entitlements precede legal conventions, pollution is part of a category of harms against interests that stands apart in this regard. More specifically, pollution is a problematic type of harm that creates liability only under certain conditions. Human interactions lead to harm and to the invasion of others' space regularly, and therefore we need an account of undue harm as a basis of assigning legal protections (rights) and obligations (duties) to different agents, which creates standards for holding those agents responsible for harm. Absent such positive standards with respect to pollution at the domestic or international level, it does not make sense to hold agents responsible. This fact has two fundamental implications. First, contrary to what some defenders of environmental justice argue, we cannot hold people responsible for polluting without a system of legal rights in place that assigns entitlements, protections, and obligations, and second, contrary to what opponents of environmental regulation claim, the lack of moral entitlements to pollute creates room for quite extensive legal restrictions on people's ability to pollute for the sake of the environment and human health. Indeed the scope of those restrictions is wide and open-ended.

1. Introduction

There is no right to pollute. The advocates of environmental regulation proclaim this loudly, and they are right to do so. They are right in the sense that there is no pre-political, natural right to use our liberty to harm other people. But they are wrong to conclude that therefore those who pollute bear full responsibility for the effects of their activities. While they bear some responsibility, how much depends on regulations established by communities that define and protect both rights to engage in activities that may cause pollution, and to be protected against the harms of pollution.[1]

Thus, there are no moral entitlements with respect to pollution (for or against) prior to legal conventions that establish them, or so I will argue. While some moral entitlements precede legal conventions, pollution is part of a category of harms against interests that stands apart in this regard. This fact has two fundamental implications. First, contrary to what some defenders of environmental justice argue, we cannot hold people responsible for polluting without a system of legal rights in place that assigns entitlements, protections, and obligations, and second, contrary to what opponents of environmental regulation claim, the lack of moral entitlements to pollute creates room for quite extensive legal restrictions on people's ability to pollute for the sake of the environment and human health. Indeed the scope of those restrictions is wide and open-ended.

First I describe the thought experiment of a common pasture to provide a model for how to think about pollution and boundary crossing in the absence of a developed legal system. The purpose of the thought experiment is to show that absent clear legal rules, we lack rules of thumb for knowing how much interference is justified, if any, with pollution causing activities, and how much protection from harm, if any, people are entitled to. The following section defends the model by drawing a distinction between different kinds of harms and interests against being harmed. I then present a contrast with existing approaches to pollution that presuppose strong pre-existing moral entitlements against being harmed or against being interfered with one's actions. I consider some classic libertarian views on pollution as well as more standard accounts provided by Robert Goodin, Peter Singer and Simon Caney. The final sections defend legal conventionalism as a politically negotiated solution to the problem of pollution and draw further implications.

2. Entitlements in the Common Pasture

Pollution is a specific type of harm. In the absence of positive laws, we have no basis to judge whether someone has imposed undue harm that constitutes a rights violation on others. To capture this point, imagine a little isolated village of a couple dozen families. They all have houses adjoining a large pasture that they use in common to raise livestock. The main water source of the village is a river that meanders through the pasture. In the beginning, the villagers raise a small number of cattle, horses and sheep. The pasture is large enough so that the effects of their use are hardly noticeable.

But as the village grows, and as the use of the pasture diversifies, so does the pressure on common resources. One villager builds a blacksmith shop next to the river so he can have ready access to the water. Another villager opens a business tanning cowhides to sell leather locally and to far-off villages. The waste from these shops is sometimes dumped in the river, harming the health of some of the people in the village who use the water downstream. More animals means more manure that also seeps into the river, creating additional health risks for the village inhabitants. The ones affected start to complain and ask that something be done.

There are different ways to think about what the best solutions are to the village's emerging problems. Some villagers will demand a stop to the use of commons resources as a dumping ground. Others will argue that no interference with people's industrious use of the land is warranted. Both such reactions seem rash and unwarranted. At this stage of the social life, it is premature to know how much interference with some people's actions is justified for the sake of preventing harm to others. The villagers cannot claim rights to pollute or against being polluted, or so I will argue. Talk of rights, obligations, and responsibilities will not make sense until the villagers gather evidence, evaluate competing interests, and make new rules to solve the problems confronting them.

Perhaps they will choose to keep the pasture in common, but create rules about its use, such as how many cattle each person is allowed to graze, where to dump the waste, and so on. Or perhaps they will choose to divide up plots of land among the villagers and adopt a strict 'no spillage' rule from one plot to the next that also applies to common areas, so that villagers who produce waste will have to find a remote dumping ground. In the process of coming up with solutions to their problems, the villagers will have to define what counts as harm, the rules for permissions and restrictions, and to elaborate on the consequences of not obeying those rules and on the mechanism for addressing conflicts. The rules and their arbitration system will produce judgments about whether one is entitled to engage in certain activities, what kind of precautions one must take so that those activities do not have harmful effects on others, and what redress measures are available in cases of non-compliance.

Whatever their options, the villagers will have created positive rules that (further) regulate their common life together. If they can enforce those laws with the backing of a legal system, they will have created in effect rights and obligations relating to polluting activities. Perhaps in some instances they will allow certain kinds of pollution (cows using the pasture) if the outcome (the seepage into the river) is below a certain harmful threshold for human health. Or perhaps they will prohibit certain kinds of activities completely, such as dumping chemicals into the water stream, because they pose the risk of severe damage to human health. The process of creating common rules to define and manage harm will generate permissions/liberty rights to pollute and protections/claim rights against being polluted. The lines drawn will be different for different activities, and the villagers will use them as the basis for holding people responsible for their actions.

Notice a few things about this thought experiment. First, there are no obvious, uniquely correct ways of assigning permissions and protections. While some solutions will make more sense than others, there are various right ways of addressing the problems that arise in the village. Commons create opportunities for multiple equilibria.[2] The equilibrium the villagers will settle on will depend on prior patterns of compliance, history, and local norms about appropriate behavior.

Second, this means that the balance of permissions and restrictions will be different in different villages even under similar circumstances. A justifiable balance will be reached when all the interests at stake are given their due and the procedures for reaching agreements are acceptable to most inhabitants in the community.[3]

The variation may very well be wider than what would be reasonable to expect under the circumstances. For example, some villages will choose not to regulate environmental pollution at all, because say, those affected downstream are of lower socio-economic standing, and their problems do not count for much in the eyes of the village elites. Or they will choose not to regulate because they believe that while there are bad things happening in the village, they are not caused by pollution, but by the evil eye, or the sinfulness of the locals.

At the other extreme, some villages might choose to impose drastic restrictions which in effect end up prohibiting productive occupations and activities, and therefore closing economic opportunities for some, with repercussions to all. Such restrictions would deprive individual villagers of occupations that the community as a whole has an interest in being pursued. Furthermore, the restrictions would interfere with a general liberty interest everyone has in having a choice whether to pursue the prohibited occupations and activities.[4] This interest is affected every time options are closed or made prohibitively expensive by regulation. In between these two extremes, lies a range of sensible, moderate approaches that take everybody's interest into account and attempt a delicate balancing act between reducing harm from environmental damage and allowing people to engage in productive activities.

Third, in the common pasture there is no pre-political standard for assigning blame, responsibility for compensation, or entitlements to being compensated for harm caused by pollution. This is true even if villagers enjoy various pre-political, moral rights, such as rights to property or rights to bodily integrity, that offer protections against standard threats such as theft or physical assault. A moral right against physical assault protects a fundamental interest that people have in physical security, and there is no balancing interest served by people being allowed to assault each other. The case of harm caused by pollution is different. The difference consists in the fact that pollution can affect a person's fundamental interest in bodily integrity or in her property, but the case for restricting it must be balanced against legitimate interests people have in engaging in some of the pollution-causing activities. There are examples of pollution that come close to physical assault, such as when someone poisons a water source with the sole intent of causing harm, and there is no other, legitimate interest at stake, but the pollution problems the villagers confronted were (and typically are) not of this kind. What explains this difference?

3. Legal Conventionalism and Pollution

Regulating pollution entails regulating a class of harms with the same structure. This class comprises tragedy of the commons problems and other scenarios in which people have an interest in engaging in activities that cumulatively have a tendency to cause harm, but prohibiting these activities causes harm to the people who have an interest in engaging in them. To clarify people's rights and responsibilities in these situations, communities must resort to the balancing of conflicting interests. In most cases, there is no obvious way to strike the balance between allowing activities that people have an interest in pursuing, and limiting the effects those activities have on others, as I will show below. Therefore communities must rely on legal conventions about what is permissible, what counts as harm, what kinds of protections are needed, and for whom or what (people? animals? the rainforest?).[5]

Not all rights are like this. Some moral rights protect interests that are so important no balancing is required to determine if we enjoy those protections as legal rights. I will assume for the sake of the argument a standard account of moral rights. According to this account, moral rights are rights that protect vital welfare interests against standard threats.[6] We have pre-political rights to bodily integrity, to property, and to the liberty to shape our lives as we see fit.[7] We require legal conventions to adequately determine the proper boundaries of those rights, that is, to specify them correctly. We do not require legal conventions to tell us whether we have those rights in the first place. Entitlements to pollute and protections against pollution are different. In the process of assessing the proper balance between conflicting interests, communities both create and specify rights and obligations with respect to pollution, usually via specifying other rights whose exercise is impeded by pollution.

Conventionalism is the view that certain phenomena arise out of conventions, that is, explicit or implicit agreements, promises, contracts or decisions made by a community.[8] By extension, legal conventionalism is the view that certain duties and rights, such as in our case permission to pollute or protections against pollution, arise out of legal conventions, that is legal decisions made by the community or the relevantly situated people in that community (judges, legislators). One can be a thoroughgoing legal conventionalist and assert that all rights and responsibilities arise this way. I endorse no such sweeping view here. On the contrary, I take on the view that we have rights and obligations to treat each other in certain ways by virtue of our common humanity that both precede and transcend political communities. Among them are the right to bodily integrity and certain (but not all) forms of property in material objects. Such a position is compatible with a restricted legal conventionalism, one that sees certain rights and obligations arising solely out of processes of social agreement, widely construed. My view is also compatible with, but does not require, social agreement arising out of a state's legal institutions. Legal conventions can arise in communities with dispersed and decentralized systems for the creation of legal rules. Communities whose legal systems are based on common/customary and treaty law, such as the international community, qualify.

Pollution as a type of harm is a violation of people's interests. While all harms are violations of interests, not all harms are important enough violations of someone's interest to warrant either moral opprobrium or legal censure. This is indeed the distinction that Joel Feinberg draws in Harm to Others. The interests of different people will inevitably come into conflict. Legally we can try to minimize the harm by making judgments about the relative importance of different kinds of interests and the trade-offs we are ready to bear.[9] And because there is no manual that assigns exact weights to various interests, we must rely on the fallible judgment of the community legislators. The implication of this legal conventionalism is that (1) not all harms are rights violations, and (2) the law cannot always aim to (and it would be inadvisable to) reduce all harms to zero. In fact, 'virtually every kind of human conduct can affect the interest of others for better or worse to some degree,' and without fine tuning our assessments of the types and degree of harm that merits proscription, we would create a society with a large, illegitimate and inappropriate amount of interference into people's lives.[10]

One implication of the idea that not all harms are rights violations is that for the class of harms that are considered rights violations according to legal conventions, moral responsibility and legal responsibility are entangled. 'Wrong' or 'harmful' is that which goes above a legal threshold. Thresholds set standards for allocating permissions, protections and liabilities. This is the case with activities that have a tendency to cause harm, but whose prohibition would also cause harm to the people who have an interest in engaging in them.[11] The standard for responsibility for harm is set by the legislator who judges the relative importance of conflicting interests. The point is not simply that there are pre-political rights against pollution that lack specificity. Rather, harms caused by pollution are of a different kind, and rights to be protected arise only out of positive conventions that create, define, and specify them.

The common pasture example seeks to clarify our intuitions about pollution as precisely the kind of harm for which the standard for moral responsibility cannot be disentangled from legal responsibility as judged by a set of positive rules. Pollution is produced by people pursuing their interests through a variety of activities that individually or in the aggregate have a tendency to cause harm to the people who are affected by them. On the one hand, there is nothing fundamentally wrong or rights violating (in the moral sense) in driving cars, running factories or producing energy.[12] On the other hand, many of these activities have consequences that damage common resources, such as the air we breathe and the water we drink. An interest in a healthy life is a fundamental human interest that can be invaded by activities with negative environmental consequences. Protections, restrictions, and permissions create legal entitlements and obligations related to pollution causing activities, and they can be determined after a careful balancing of these different interests. Standards for imputing harms and responsibility must be based on such legal entitlements and prohibitions.

The balancing of interests will vary with activity and type of effects. Some will be judged so harmful as to justify outright prohibition, regardless of the benefits. The use of asbestos in construction materials, or lead in paint are among the activities that have been prohibited due to their serious consequences to human health. For other kinds of harmful pollution, the effects will be so small as to lack a justification for restrictions, even when the interest of engaging in those activities is relatively minor. Smoke crossing fences from neighbors barbecuing in their backyard is such a case. And we can also distinguish between different contexts or spaces in which the activity in question takes place. Barbecuing indoors has different effects than smoking outdoors.

Other existing moral and political rights will serve as inputs to help structure reasoning about what constitutes harmful pollution and how to apportion responsibility for its effects. For example, preexisting moral rights may serve as a justification that the community takes action to rectify perceived harm. Or they may serve to characterize the harm being caused. Additionally, there will be a variety of moral ideals beside rights that will serve as guideposts in reasoning about the boundaries of pollution, such as fairness, proportionality, equality, inclusiveness, and human flourishing. Moral reasoning is an unescapable part of law and public policy, and this is as true of pollution regulation as is of other policy areas.[13]

Nonetheless, the fact that we have some moral rights, and a moral language to describe and reason about harms more generally, does not entail that we have moral entitlements with respect to pollution (rights to pollute or rights against pollution). Yet this is not what prominent views on environmental justice argue. On the contrary, some assert strong, pre-political entitlements to protections against pollution, and pre-political benchmarks for the responsibility to compensate.

4. Against Conventionalism?

Libertarians draw strong invisible lines around private property and believe that few, if any, encroachments on individuals' right to use their land as they see fit are ever justified. Property rights as natural rights are considered relatively inviolable because they protect an extensive sphere of personal freedom from social intrusion and because such protection is essential for individual welfare and for economic and social progress.[14] Murray Rothbard, a prominent advocate of libertarianism, says that 'the central core of the libertarian creed, then, is to establish the absolute right to private property of every man: first, in his own body, and second, in the previously unused natural resources which he first transforms by his labor.'[15] Property rights should not be limited or redistributed for short-term gains. Not only would such imitation violate individual rights, but they would also reduce the ability of individuals to flourish and to engage in a mutually beneficial system of social cooperation.

Rothbard makes clear that private property regulation is unwelcomed: 'since the libertarian also opposes invasion of the rights of private property, this also means that he just as emphatically opposes government interference with property rights or with the free market economy through controls, regulations, subsidies, or prohibitions'.[16] Ostensibly, regulations aimed at curbing pollution are part of unjustified 'invasions' of private property.

The language of 'absolute' property rights would seem to license individual and industrial polluters alike to soil the water, dirty the air, and harm other people's bodies.

It would be easy, but mistaken, to infer from this strong defence of private property rights and opposition to social regulation that libertarians are in favor of allowing individuals and corporations to use their land and their natural resources as they see fit, and therefore engage in unlimited, unregulated pollution.

This is not the kind of license that libertarians believe property rights grant people. Here is Rothbard again: 'The vital fact about air pollution is that the polluter sends unwanted and unbidden pollutants—from smoke to nuclear radiation to sulfur oxides—through the air and into the lungs of innocent victims, as well as onto their material property. All such emanations which injure person or property constitute aggression against the private property of the victims.'[17] Far from giving blanket permission to property owners to pollute to their heart's desire, libertarian rights create quite stringent constraints on the permissible ways in which people can enjoy their property. Such restrictions are so severe, that even very small, non-consensual boundary crossing of smoke, air particles, or chemicals can be considered as an unjustified interference with one's property and are therefore liable to be made illegal. It would seem that libertarian property rights are at once too permissive and too restrictive. This is a point made forcefully both by David Sobel and Matt Zwolisnki recently.[18]

How then to reconcile the idea of property as unfettered license with the idea of property as a shield, including a shield against pollution and bodily harm? Libertarians believe the solution lies again with property rights. Fully privatizing natural resources, by extending the existing system of property rights in land to all existing common resources, such as mineral reserves, highways, air, and water, including the oceans, will lead to a system that encourages the internalization of costs, and the reduction or elimination of harmful effects on others. The authors of the sophisticated Free Market Environmentalism, Terry L. Anderson and Donald R. Leal, describe the free market approach to pollution as one that 'establishes rights to clean air or rights to dump into the air, and allows the holders of those rights to bargain over the optimal mix of competing uses, in this case clean air and garbage disposal'.[19] Thus they would object to my interpretation of the primitive common pasture example on the grounds that because property rights are not fully privatized, individuals lack incentives to fully internalize the costs they impose on others.

Many pollution and environmental problems are the result of tragedy of the commons scenarios in which land or resources are typically held in common and end up overexploited and under-protected. Anderson and Leal are right to argue (with convincing examples) that more adequately specified property rights regimes for those commons would lead a long way toward reducing the environmental impact of overuse. It is an underappreciated feature of private property rights and markets that they encourage people to internalize externalities, that is, to bear the costs of the negative effects their actions have on third parties.[20] Yet, a large problem looms in the libertarian treatment of pollution.

There are serious limits to the ability of any system of private property rights to encourage property owners to internalize the effects of pollution and to prevent it from crossing property boundaries. Think of the air or ocean water. Although various libertarians have proposed imaginative ways of creating property rights in the air and homesteading the ocean, and are encouraged by technologically forward environmental entrepreneurs with visionary proposals, it is hard to imagine such solutions to be feasible.[21] As long as the air circulates freely, so will pollution. The implication of this limitation is momentous: when it comes to pollution, a system of property rights must be complemented with restriction on the appropriate use of both private and common property to minimize harm caused to innocent third parties.

Think for example of various forms of intrusion on our bodies and property that Rothbard labels as 'aggression.' To describe any non-consensual property boundary crossing as aggression and to therefore call for it to be prohibited is to take the idea of property as a shield too far. The air a person next to you exhales, which as David Friedman pointed out, contains carbon dioxide, can be considered an unwanted interference if it touches your body.[22] The absolutist character of this understanding of libertarian property rights seems to preclude reasonable compromises. Yet assessing what counts as aggression, and what is the threshold above which pollution is damaging and must be curtailed, is why we need conventions in the first place. There is nothing else that could replace conventions for this purpose. Depending on the type of harm and the relative interests that must be balanced against restrictions, the scope of regulation can be quite wide and is indeed open ended. Therefore libertarians must make peace with much more extensive intrusion in property rights for the sake of the environment and human health than they have been ready to accept so far.

Rothbard has granted that not all pollution can be zero, and has argued that we should distinguish between 'visible and tangible' air pollution, which interferes with possession and the use of property, and 'invisible or insensible', which cannot be counted as interference.[23] But what if invisible turns out to be harmful? More generally, who is to make the decision that interference with property is harmful? People will have very different understanding of harm, and in the absence of legally sanctioned norms of right conduct communities will not be able to coordinate their expectations about what counts as undue interference with other people's body or property.

5. Rights and Compensation

Robert Nozick was aware of the dilemma facing strong property rights. He crystalized a central insight of libertarian views: that property rights act as a moral side constraint on permissible action. 'The rights of others', he said, 'determine the constraints upon your actions'.[24] He thus embraced a view of natural rights which reflects the Kantian principle of respecting individuals as ends and not merely as means. Expressing this respect requires us to not trespass certain boundaries around them (material property, physical body).[25] Physical aggression is one way to disrespect individuals and fail to take their rights as side constrains on one's actions seriously. So is stealing from them or interfering with the use of their property.

Nozick was also among the first to realize how inflexible this rule is, and how implausible its implications. The natural rights theorist that insists on strong property rights has no way of selecting a threshold measure below which certain harms or right violations are permitted.[26] And such a system would lead to a world unfriendly to liberty, because it would severely limit the ability of anyone to engage in boundary crossing with significant social benefits. In his view, a society that prohibited actions of unintended harm or intended, but small, border crossings 'would ill fit a picture of a free society as one embodying a presumption in favor of liberty,' presumably because it would prohibit many occupations, hobbies, and activities in which people would like to be at liberty to engage.[27]

Nozick's answer to this problem is to permit boundary crossing and harming with compensation. Since consent is often impossible to negotiate beforehand or too costly when it involves minor crossings and large numbers of actors, Nozick seeks to establish a principle of compensation according to which people can engage in inadvertent, unplanned boundary crossing or harming, when the harm is minor and the benefits large.[28] Thus he proposes what we would call a modified natural rights approach to work around the more absolutist implications of taking rights as side constraints.

What Nozick misses, however, is the fact that there is a prior question that he assumes settled, that in fact is not. His question is how to deal with the fact that boundary crossing and harming, while rights violating, is necessary both to secure liberty and to provide socially beneficial outcomes. However, the prior question that a legal conventionalist is in a better position to broach, is what kinds of actions constitute boundary crossings and harms at all? Thus the 'threshold problem' is not about determining the threshold above which rights violations are permitted, but the threshold above which an activity counts as rights violating.

Pollution is directly addressed by Nozick and presents an interesting case of the difference between Nozick's modified natural rights approach and legal conventionalism. Since a society cannot prohibit all polluting activities, it should prohibit only those activities 'whose benefits are greater than their social costs'. Furthermore, the test for whether an activity produces more benefits than costs is 'whether those who benefit from it are willing to pay enough to cover the cost of compensating those ill affected by it'.[29] For example, if air transport imposes noise pollution on homes surrounding the airport, then airlines should compensate those who suffer the pollution's effects.[30]

Suffering 'ill effects' is ambiguous between having rights violated and suffering merely unwanted, unpleasant effects. The prior questions that the legal conventionalist can solve but the natural rights theorist cannot are: does any noise pollution trigger compensation? What level of noise pollution is considered harmful and thus rights violating? Unless we answer these questions, anyone living on a busy street where car traffic creates noise or indeed, anyone with neighbors you can hear coming in and out of their homes would be entitled for compensation. Therefore regulation is needed to establish the level of noise that is harmful and the rights people have against noise pollution above that level.

Compensation is thus owed, if at all, not for noise pollution simpliciter, but for noise pollution above the level established by regulation, which would take into account the noise produced from normal human activity and the interests of those engaging in noise-producing actions. Legal conventions would thus establish liberty rights that people have in engaging in noise producing activities, and claim rights people have against excessive levels of noise.

An instructive example is Hinman v. Pacific Air Transport (1936). A property owner living next to a newly opened military airport sought injunctive relief against planes flying over his property.[31] Hinman argued that the airplanes trespassed on his property and as such were in violation of his rights. The majority opinion in the Supreme Court decision argued that although the air above one's property has been, since Roman law, considered one's property, this was a poorly specified property boundary that must be revisited in light of the needs of modern societies. Hinman can at most claim the rights in the air that he makes use of, say, by erecting structures on the land, but not above it. Thus planes flying above his head did not commit any rights violations because the air space into which the planes few were not Hinman's property to begin with, or if was, such property title was the result of incorrectly specified property boundaries. The case helped to define (or in this case redefine) rights and therefore what counts as harm or right violations.

Libertarians thus either wrongly assume, in more extreme versions, that boundary crossing such as pollution is forbidden, or that it is a rights violation that must be compensated. Both the more extreme Rothbardian version and the more flexible Nozickian version share the idea that the benchmark for rights violations is pre-political. Nozick allows conventionalism in the specification of the level of compensation or the allocation of such compensation to victims of rights violations, but not necessarily in deciding whether rights violations have occurred to begin with. Whether individuals are entitled to compensation and at what level the compensation should be set is related but conceptually distinct from whether rights violations have occurred.

6. Responsibility, Harm and the Environment

There are strong similarities between libertarian views of property rights and a very different defense of claim rights against pollution as human rights. Both either end up with implausible restrictions on people's actions, or with pre-determined benchmarks for compensation. Robert Goodin is in the former category. He argues that to permit pollution is to 'permit the impermissible'.[32] Rights, quotas, and permissions of any kind resemble medieval indulgences because they give 'sinners' a pass to do wrong. The assumption behind the analogy with indulgences is that any amount of pollution is wrong. The implication of this view is that the proper way to deal with this kind of sin is not to create pollution permits or rights, but to prohibit it altogether. Although he does not exactly endorse a 'zero emissions' standard, he believes most arguments against such a standard fail.[33]

Goodin presupposes that the standard for harm is straightforward and devoid of entanglement with legitimate, competing interests. This kind of approach to pollution is too one-sided, since it considers the harmful effects of pollution to the exclusion of the interests people have in engaging in pollution-causing activities. While Goodin emphasizes that pollution causes great harms, he fails to account for the fact that outright prohibition of polluting activities also causes great harms.[34]

For Goodin, moral and legal responsibility can come apart because we can pass judgments about the value of nature and the costs of environmental damage. Nature 'provide(s) a context... in which to set our lives', and environmental despoliation 'deprives us of that context'.[35] But it is not enough to point to the fact that certain activities can cause environmental despoliation in order to pass judgment on whether they are wrong tout court. To have a context for our lives, we must also engage in life-sustaining pursuits. We must combine our skills with earth's natural resources productively and with a view to improving our wellbeing. There are limits to how much of it we should be able to enjoy given the costs. But the costs cannot be zero.

The many discussions on behalf of the polluter pays principle (PPP) reveal the same problematic assumption. PPP provides a standard of responsibility for harm. Its defenders claim that developed nations have caused most of the atmospheric pollution, and are therefore responsible for cleaning it up. Peter Singer says:

So, to put it in terms a child could understand, as far as the atmosphere is concerned, the developed nations broke it. If we believe that people should contribute to fixing something in proportion to their responsibility for breaking it, then the developed nations owe it to the rest of the world to fix the problem with the atmosphere.[36]

Others echo this attitude. Henry Shue says that those in developing nations who face the unequal burdens of environmental damage without their consent are entitled to compensation from those that have imposed the costs.[37] The partial destruction of the ozone layer and global warming impose unequal burdens on the developing world. 'Those societies whose activities have damaged the atmosphere ought, according to the first principle of equity, to bear sufficiently unequal burdens henceforth to correct the inequality that they have imposed', Shue says.[38] Although he seeks to distinguish this 'principle of equity' from PPP, both endorse the idea that compensation is proportional to the damage caused.

PPP has been adopted by international agreements as the basis on which to judge the allocation of responsibility to correct the effects of global warming. The Organization for Economic Co-operation and Development (OECD) and the European Union and Council of Ministers have affirmed the principle in their resolutions. It is now part of the common language in which people discuss how to allocate the obligations to reduce environmental pollution.

There are numerous problems with PPP, and Simon Caney has exposed the most serious ones. In particular, he has drawn attention to the fact that the damage to the environment was caused by earlier generations and yet new generations are required to bear the cost. PPP places blame on the wrong agent.[39] Another potential problem is that, for a long time, developed nations were unaware of the negative effects of emissions on a global scale. One cannot hold people responsible if they are not aware of the harmful effects of their actions. Ignorance can mitigate responsibility.[40]

The analysis offered here identifies a different, more fundamental problem with PPP. As I have argued earlier, polluters, strictly speaking, do not have to pay if they are permitted to engage in pollution-causing activities within limits. They are only responsible if they produce effects above a threshold defined in the law. In order to know how to divide responsibility, we need, as Simon Caney puts it, 'an account of persons' entitlements'.[41] Moreover, if the common pasture case is right, that account must come from legal conventions. It is people's legal rights and responsibilities that determine protections against pollution and the level of compensation that polluters must pay.

It is inadequate, therefore, to think of protections against being polluted and permissions to pollute only in terms of the harms or costs that pollution imposes, or of the environment only as a resource that is depleted or abused rather than one that is essential to human survival. So, for instance, it won't do to think of the atmosphere as 'a giant global sink into which we can pour our waste gases', as Singer does.[42] It is more like a common pasture from which we take things out and we put things back in. Zero damage to the pasture is not a plausible benchmark, so the question becomes how much damage are we willing to live with and how to weight the cost of the damage against the benefits of using the pasture.

7. Caney's Solution

Caney has helped shift the debate surrounding the issue of allocating responsibilities when he observed that 'to make the claim that someone should pay also requires an account of what their entitlement is'.[43] He then proceeded to give an account of such entitlements. In 'Climate Change, Human Rights and Moral Thresholds', he describes the effects of climate change as violations of human rights.[44] Climate change affects three particular human rights: the right to life, health and subsistence. For example, climate change increases the frequency of events such as tornadoes, flooding, and storm surges that lead to direct loss of life. In Caney's view, 'all persons have a human right that other people do not act so as to create serious threats to human health'.[45] The implication of this view is that 'if, as argued above, climate change violates human rights, then it follows that compensation is due to those whose rights have been violated'.[46]

Elsewhere, Caney reaches the same conclusion: '[P] ersons have the human right not to suffer from the disadvantages generated by global climate change'.[47] In contrast to the PPP, his view does not rest on the assumption that climate change is human caused. Even if climate change would not be anthropogenic, the human rights not to suffer from the effects of climate change would hold. Caney then adds an 'ability to pay' principle: the most advantaged have a duty to reduce their greenhouse gas emissions and to address the ill effects of clime change.[48]

This account of person's entitlements is incomplete. There cannot be such a sweeping, general human right 'that other people do not act so as to create serious threats to human health', as Caney claims. This claim is too broad. People have rights against standard threats to wellbeing, but they do not have rights against all threats, even if such threats are serious. They cannot have 'the human right not to suffer from the disadvantages generated by global climate change', because focusing on harm alone does not offer guidance for entitlements related to activities with complex effects where multiple interests are at stake. There cannot be a 'human right to not be polluted'.

Caney himself notes that his account of the division of responsibilities is incomplete, because we still need to ascertain what counts as a fair pollution quota.[49] But, of course, the fact that other people enjoy permissions or rights to pollute by virtue of their quota affects whether those who suffer the effects of their actions have legitimate complaints against them, and thus 'a human right to not be polluted'. Or to put it differently, if people have human rights against being polluted, how can anyone else have a right to engage in activities that violate human rights? The kind of balancing Caney would like to engage in, according to which we first ignore people's human rights not to be harmed, and then we balance that against people's entitlements to pollute, contains one step too many. People's rights against being polluted are determined in the process of determining people's rights to pollute. Legal protections from the harmful effects of pollution are inextricably tied to permissions to engage in activities to pollute.

There is a larger methodological point at stake in this debate. Caney believes that in order to have an account of persons' entitlements, we require a background theory of justice.[50] This is the spirit in which he offers his account of human rights against the damaging effects of climate change and of the obligations different agents incur to reduce the damage. But the kind of judgments for fine-tuning different trade-offs required by considering the relative importance of different interests and harms in the case of pollution cannot be the result of abstract moral theorizing. Only communities armed with a proper understanding of the relationships between those interests in context can pass those kinds of judgments. This means that entitlements against pollutions and permissions to pollute must be the result of positive law, not of abstract theorizing about peoples' interests.

Perhaps a human right such as the one Caney deploys would be consistent with varying levels of acceptable pollution. In Caney's defense, one might say that perhaps we can consider the existence of such a right to offer a simplistic threshold, which needs to be further refined by a political or common law process. But then it is unclear what is gained by saying we have such a right in the first place. There cannot be a right apart from the marking and making of the threshold, which defines liberties and protections depending on where the line is drawn. The fact that we have a general right to liberty, or a general interest in being protected against harm, does not tell us where to draw the line.

To sum up, Caney is right that in order to assign responsibility for the damaging effects of climate change we require an account of entitlements. But the idea of a human right against pollution is not a helpful way to specify what those entitlements are, and it precludes the conventionalist legal approach proposed here, which defends the idea that entitlements must result from balancing conflicting interests.

8. Conventionalism and Complex Balancing

The two general frameworks that rely on natural rights and human rights to set standards for people's entitlements are not in themselves objectionable. Yet as frameworks that rely on pre-political rights they cannot determine the scope of rights and restrictions with respect to various environmental harms. They can at most serve as inputs into political processes that determine the complex relationships between conflicting interests and harms to those interests, but by themselves, natural or human rights cannot offer an adequate characterization of liberties and claim rights related to pollution.

The clearest case for legal conventionalism is the case of air particle pollution. From the simple fact that some agent is releasing particles in the air and that those particles cross some physical property boundary or are inhaled by persons we cannot deduce that any rights violations has occurred, contrary to certain libertarian and human rights views about the scope of protection pre-political rights confer against pollution.

The process for determining if any rights violations have indeed occurred consists of assessing and balancing the nature of the interests at stake (for producing air particles as well as against letting them cross certain physical boundaries), how weighty such interests are, what kind of harm is produced if any, how to balance all of these considerations taking into account the existing background of rights and obligations, and so on. In many cases, such judgments will depend heavily on expert analysis about patterns of diffusion of air particles, projections and modeling about future air quality in light of existing trends of emission, likely effects on human health, and on personal property of various sorts. Without such a complex, explicit, publicly endorsed process, our judgments are likely to be premature and highly arbitrary. We would lack a good sense about what the threshold for harm is and how to balance freedom and welfare-enhancing rights to engage in polluting producing activities with protections for human health and property.

Chloroform offers another instructive case. Is an air concentration of 1000 parts per million (ppm) harmful to human health?[51] And if so, do people have rights against chloroform being used at all, only in certain concentrations (under 1000ppm), or no right at all against companies and individuals that use chloroform? Correspondingly, do companies and private individuals enjoy complete freedom to use it, are allowed to use it only in certain quantities/concentrations, or not at all? Notice that we cannot answer any of these questions based on pre-political property rights or some general human rights against being harmed. We would need to know much more about what its uses are, who has stakes in those uses, if it is toxic and what levels of toxicity are tolerable if at all.

It turns out chloroform is ubiquitous and widely used. Chloroform is used to chlorinate drinking water, swimming pools, wastewater, and is also used in industrial processes in paper mills. It is released into the ambient air as a result of these uses. Based on existing analyses, chloroform is categorized as a toxic air pollutant with significant effects on human health. According to the EPA data, at higher than 40,000ppm, exposure may result in death, between 1,500 and 30,000ppm it produces anesthesia, and at lower concentrations of fewer than 1,500ppm it results in dizziness, tiredness, headaches and other health effects.[52] Estimates for the threshold of chronic exposure are in the neighborhood of 0.05ppm to 0.1ppm.[53] The background level of chloroform in the ambient air in the early 1990's was at 0.00004ppm.[54] So while 1000ppm concentration in the ambient air is unlikely based on projections derived from past emission patterns, localized concentrations can exceed dangerous levels for certain population subgroups, such as workers in paper mills. Restrictions on water chlorination, industrial paper mills and other industries that rely heavily on chlorination are justified based on estimated thresholds of harm, and unjustified if they go beyond such estimates.

This example goes at the heart of legal conventionalism, the view that I defend here. Legal conventionalism opposes any view that argues that pollution rights and restrictions are in some sense natural, or that natural property rights (or the boundaries of such rights as specified by either original acquisition or positive laws) provide functional guidelines for determining permissions or restrictions against pollution. Pre-political property rights exist and can inform how pollution rights and restrictions are distributed, but they do not determine such rights and restrictions. Only a political or legal process can.

I contrast conventionalism understood as a social process of rule generation with the view that such rules can be in principle under-stood as natural or arising out of pre-political norms and principles. Conventionalism in this sense is broad and covers most types of legislative and judicial action directed at regulating environmental harms. My use is thus a departure from the narrow use of conventionalism in the literature on social norms, which restricts conventions to describe simple coordination problems.[55]

Legal conventionalism as an approach for pollution regulation is the right approach even if it creates the problems that public choice theorists make vivid, such as political capture, rent seeking, and collective action dilemmas facing large numbers of people that are subject to diffuse effects. What legal conventionalism implies in the face of these challenges is not abandoning a publicly negotiated solution due to the high costs of politicization, but legal conventionalism plus efforts to minimize political capture and rent seeking.

9. Implications

Legal conventionalism is no endorsement of the status quo. Peter Singer may well be right that there is no 'ethical basis' for the current levels of pollution.[56] Although departing in important respects from Singer's view, the conventional approach defended here is compatible with quite extensive restrictions on different agents' abilities to pollute. It also brings a necessary corrective to a view such as Singer's. In order to know what the ethical basis for acceptable levels of pollution is, we must rely on legal conventions of a certain kind, which balance complex combinations of interests rather than on any pre-political intuitions about harm. But not all conventions are made equal.

Thus an important and related question follows given that we have to rely on imperfect, fallible publics, legislators, an acceptable range of legal rules? Some communities will err on the side of being too cautions while others will rest content with massive unregulated harm. Therefore we need an account of legitimacy for legal rules that create entitlements related to pollution, and for the institutions and procedures according to which such rules develop, for both local communities and the globe as a whole. I do not have the space to develop such an account here, but a few preliminary remarks are in order.

First, while there can be legitimate variation across communities and across issue areas in the acceptable ways of balancing conflicting interests when generating positive rules regarding pollution, some possible trade-offs are beyond the pale. Those based in superstition, mistaken beliefs or ignorance of widely available scientific data raise question about the legitimacy of the rules as a whole, regardless of their content. This means that there will be epistemic constraints on the legitimacy of a given legal regime.

Second, in addition to being epistemically defective, legal regimes can fall short by failing to take into account the relevant interests at stake, or the relevant contributions to harm. Communities in which the political or economic elites have an interest in engaging in pollution causing activities may ignore the legitimate interests of significant portions of the population. In doing so, these communities will fail to include in the political calculus of the best rule relevant moral data. Such regimes are morally defective.

Third, and related, regimes whose procedures do not encourage or allow the emergence of information about the relevant interests affected by polluting activities are also problematic. These regimes fall short by not being sufficiently inclusive. Democratic representative procedures are typically judged to do well according to this procedural requirement, but so may other institutional alternatives.

Consequently, the approach defended here departs in important respects from pure conventionalism, the idea that whatever decision a community makes provides the legitimate legal, and therefore moral, standard to assess claims related to pollution. Rather, defensible legal conventionalism must respond to epistemic, moral and procedural constraints. Here too there are trade-offs in fulfilling these requirements. Imagine a global forum that is as inclusive as possible. The likely result is that different constituencies will have vastly different and conflicting interests, not all of which can be adequately aggregated in a policy position. This does not mean that inclusiveness must be abandoned, but that perhaps there is an optimal amount of inclusiveness for the purposes of decision-making. Optimality will bring its own costs.

Whatever constraints these three requirements place on policy and law-making, individual and collective agents cannot be held responsible in the absence of a set of clear rules that define restrictions, permissions, obligations and rights. Whether such regimes are already in place and are legitimate is, of course, a different matter. Some could point to the Kyoto treaty as a positive agreement that spells out the obligations of different nations of cutting carbon emissions and the permissions they enjoy based on a trading scheme.

But others claim that Kyoto is seriously defective as an international legal regime for failing to assign responsibility to developing countries such as China and India who are on target to become the largest polluters in the world. In the eyes of its critics this fact affects the legitimacy of the regime, and as a result, the acceptability of the entitlements it generates.[57] I do not know whether this criticism is correct or not, but it is difficult to evaluate partly because our ideas about the content of the epistemic, moral and procedural requirements at the global level are still uncertain compared to those at the domestic level. Perhaps we need an account of legitimate global environmental institutions before we can generate legitimate global rules, if we have to insist, as I think we must, that legitimate rules come from legitimate institutions.

This is however beyond the scope of this paper. The main contribution of the argument offered here is to bring clarity with respect to the responsibility for global climate change by describing moral responsibility as a function of legal responsibility. Society evolves, and so does our understanding of what activities people are entitled to engage in or not. When it comes to harm generated by new technologies or by new activities that people have a legitimate interest in engaging, but that also generate harmful effects, we do not have good guidance from pre-political norms or rules of thumb about how to balance conflicting interests. Legal rights are necessary to pass judgment on the right way to balance different interests. Without legal rights in that special group of cases, we cannot make sense of the extent and limits of our liberties.

Acknowledgments

For useful feedback I thank Mark Budolfson, Peter Stone, David Schmidtz, Jacob Levy, David Boonin, Hillel Steiner, Guido Pincione, Ralf Bader, Steve Wall, Josef Sima, the audiences at the European Political Science Association Meeting in Edinburgh, Scotland (June 2014) and Western Political Association Meeting in Hollywood, California, (March 2013), and two anonymous referees. This article was written with the support of a grant from the John Templeton Foundation. The opinions expressed in it are those of the author and do not necessarily reflect the views of the John Templeton Foundation.

Source: Pavel C E. A Legal Conventionalist Approach to Pollution[J]. Law & Philosophy, 2016:1–27.

References

[1] By pollution I mean the release of fumes, air particles, chemicals, and contaminants into the air, land, and water with the potential to damage property, the environment, and human health.

[2] Elinor Ostrom discusses the many practical solutions people in different parts of the world have adopted to deal with commons problems, Governing the Commons: The Evolution of Institutions for Collective Action (Cambridge University Press, 1990).

[3] We need not agree what the interests are, how to give each interest its due, and what procedural hurdles policies must pass in order to agree that policies that fail to protect significant interests of the people affected, or which are adopted arbitrarily, will not be justifiable on one or more dimension.

[4] Joel Feinberg, Harm to Others (Oxford University Press, USA, 1987), 206. In Feinberg's words: 'Whenever a person's interest in X is thwarted, say by legal prohibitions against anyone doing, pursuing or possessing X's, an interest in liberty is also impeded, namely, the interest in having a choice whether to do, possess or pursue X or not'.

[5] Intellectual property rights have a similar structure. Although people's views about the desirability of intellectual property rights cover the full spectrum, a conventional approach seems to offer the best solution to balancing the authors' interest in benefiting from their ideas and creations with the public's interest in putting those ideas to different and productive uses.

[6] James Nickel, Making Sense of Human Rights, 2nd ed. (Wiley-Blackwell, 2007); Charles R. Beitz, The Idea of Human Rights (Oxford University Press, USA, 2009); Henry Shue, Basic Rights (Princeton University Press, 1996).

[7] Property rights are pre-political if one adopts the original appropriation justification of property claims.

[8] For a solid introduction to conventions and conventionalism see Michael Rescorla, 'Convention', The Stanford Encyclopedia of Philosophy (Spring 2011 Edition), Edward N. Zalta (ed.), URL =http://plato.stanford.edu/archives/spr2011/entries/convention/ (accessed July 8, 2014). As distinct from conventionalism that arises out of agreement, a form of conventionalism is generated by conventional use (the value of money, or the meaning of words), meaning shared practices that over time imprint meaning or value to objects or activities.

[9] Feinberg, Harm to Others, 35.

[10] Ibid., 12.

[11] Ibid., 203.

[12] Ibid., 230.

[13] I thank an anonymous referee for emphasizing this point.

[14] Robert Nozick, Anarchy, State, and Utopia (Basic Books, 1977), 26–53; Milton Friedman, Capitalism and Freedom: Fortieth Anniversary Edition, 40 (Chicago: University of Chicago Press, 2002), 7–21.

[15] Murray Newton Rothbard, For a New Liberty: The Libertarian Manifesto (Ludwig von Mises Institute, 1978), 47.

[16] Ibid., 27–28.

[17] Ibid., 319.

[18] David Sobel, 'Backing Away from Libertarian Self-Ownership,' Ethics 123, no. 1 (2012): 32; Matt Zwolinski, 'Libertarianism and Pollution,' in The Routledge Companion to Environmental Ethics, ed. Benjamin Hale and Andrew Light (Routledge Press, 2015), 3, 9–11.

[19] Terry L. Anderson and Donald R. Leal, Free Market Environmentalism, Revised edition (New York, NY: Palgrave Macmillan, 2001), 126.

[20] David Schmidtz, 'The Institution of Property,' Social Philosophy and Policy 11, no. 02 (1994): 42–62; Carol M. Rose, 'Liberty, Property, Environmentalism,' Social Philosophy and Policy 26, no. 02 (2009): 1–25.

[21] Anderson and Leal, Free Market Environmentalism, 107–142. For a measured assessment of the limits and possibilities of such proposals, see Daniel H. Cole, 'Clearing the Air: Four Propositions about Property Rights and Environmental Protection,' Duke Environmental Law & Policy Forum 10, no. 1, 103–130.

[22] David Friedman, The Machinery of Freedom: Guide to a Radical Capitalism, 2nd edition (La Salle, IL: Open Court, 1989), 168; Matt Zwolinski, 'Libertarianism and Pollution,' 11.

[23] Murray Rothbard, 'Law, Property Rights and Air Pollution,' Cato Journal 2, no. 1: 82.

[24] Nozick, Anarchy, State, and Utopia, 29.

[25] Ibid., 31.

[26] Ibid., 75.

[27] Ibid., 78; David Sobel, 'Backing Away from Libertarian Self-Ownership,' 36–37.

[28] Nozick, Anarchy, State, and Utopia, 57–58, 65–85.

[29] Ibid., 79.

[30] Ibid., 80.

[31] Injunctive relief is a court ordered prohibition of an act.

[32] Robert E. Goodin, 'Selling Environmental Indulgences,' Kyklos 47, no. 4 (1994): 578.

[33] Ibid., 575–576.

[34] See also Feinberg, Harm to Others, 227 on this point.

[35] Goodin, 'Selling Environmental Indulgences,' 578.

[36] Peter Singer, 'One Atmosphere,' in Climate Ethics: Essential Readings, ed. Stephen Gardiner et al. (Oxford University Press, USA, 2010), 190.

[37] Henry Shue, 'Global Environment and International Inequality,' International Affairs 75, no. 3 (1999): 533–534.

[38] Ibid., 534.

[39] Simon Caney, 'Cosmopolitan Justice, Responsibility, and Global Climate Change,' Leiden Journal of International Law 18, no. 04 (2005): 756; See also Eric A. Posner and David Weisbach, Climate Change Justice (Princeton, NJ: Princeton University Press, 2010), especially 99–118.

[40] Caney, 'Cosmopolitan Justice, Responsibility, and Global Climate Change,' 751.

[41] Ibid., 765.

[42] Singer, 'One Atmosphere,' 188.

[43] Caney, 'Cosmopolitan Justice, Responsibility, and Global Climate Change,' 765.

[44] Simon Caney, 'Climate Change, Human Rights and Moral Thresholds,' in Human Rights and Climate Change, ed. Humphreys Stephen (Cambridge University Press, 2009), 163-166. Simon Caney, 'Climate Change, Human Rights and Moral Thresholds,' in Human Rights and Climate Change, ed. Humphreys Stephen (Cambridge University Press, 2009), 163–166.

[45] Ibid., 166.

[46] Ibid., 171.

[47] Caney, 'Cosmopolitan Justice, Responsibility, and Global Climate Change,' 768.

[48] Ibid., 769.

[49] Ibid., 770.

[50] Ibid., 765.

[51] Parts per million is a standard measure of air particle concentration.

[52] EPA data http://www.epa.gov/ttn/atw/hlthef/chlorofo.html accessed Nov. 13, 2014.

[53] These estimates are concentrations 'at or below which adverse health effects are not likely to occur.' Idem.

[54] The EPA has not established a reference concentration for chloroform. A reference concentration is 'An estimate (with uncertainty spanning perhaps an order of magni-

tude) of a continuous inhalation exposure of a chemical to the human population through inhalation (including sensitive subpopulations), that is likely to be without risk of deleterious noncancer effects during a lifetime.' http://www.epa.gov/ttn/atw/hlthef/hapglossaryrev.html#rfc accessed Nov 13, 2014.

[55] The dominant use of the term conventionalism in the philosophical literature has been heavily influenced by David Lewis's account, according to which conventions are solutions to simple coordination problems that arise in social life, such as which side of the street to drive on. See for example David Lewis, Convention: A Philosophical Study (Blackwell, 2002), 5–51; Cristina Bicchieri, The Grammar of Society: The Nature and Dynamics of Social Norms (New York: Cambridge University Press, 2005), 34-41; Geoffrey Brennan et al., Explaining Norms (Oxford: Oxford University Press, 2013), 14–19.

[56] Singer, 'One Atmosphere,' 197.

[57] Richard B. Steward and Jonathan Baert Wiener, Reconstructing Climate Policy: Beyond Kyoto (Washington, DC: AEI Press, 2007), 83–95; Posner and David Weisbach, Climate Change Justice; Cass R. Sunstein, 'The World vs. the United States and China? The Complex Climate Change Incentives of the Leading Greenhouse Gas Emitters,' UCLA Law Review 55, no. 6 (2008): 1682.

Chapter 5

Environmental Impacts of Farm Land Abandonment in High Altitude/Mountain Regions: A Systematic Map of the Evidence

Neal R. Haddaway[1], David Styles[2], Andrew S. Pullin[1*]

[1]Centre for Evidence-Based Conservation, School of the Environment and Natural Resources and Geography, Bangor University, Bangor LL57 2UW, UK
[2]School of the Environment and Natural Resources and Geography, Bangor University, Bangor LL57 2UW, UK

Abstract: Background: Environmental impacts of farm land abandonment can be viewed as either an opportunity for ecological restoration to a state prior to agricultural establishment, or as the loss of an on-going process of land management and an associated threat to biodiversity. Whether land abandonment poses an ecological opportunity or threat depends upon the agricultural history and the presence of ecological systems that depend upon regular management for their existence. In Europe, many ecosystems have developed in the presence of agriculture and the loss of continued management resulting from land abandonment can have significant negative ecological impacts. Around 56 percent of the utilised agricultural area (UAA) of the EU is classified as 'less-favourable areas' and much of this is mountainous. The small-scale and extensively managed farmlands that are common in mountain areas are particularly vulnerable to marginalisation and abandonment. The work herein will form the first systematic synthesis of the evidence of impacts of farm land abandonment in mountain areas across the globe. Methods: This review will take the form of two interrelated systematic maps, cataloguing the existing evidence across a wide range of variables such as setting, methodology, scale, measured outcomes etc. Mapping will be undertaken both at abstract-level at a coarse scale and at full text-level at a finer scale. Literature databases, organisational web sites, and search engines will be used to collate all of the available literature regarding the impacts of agricultural land abandonment. All studies investigating farmland abandonment in mountainous regions with an appropriate comparator and measuring an appropriate outcome will be included. Outcomes will be coded in a partly iterative process but will include; natural hazards (fire-/flood risk, land/mud slides), soil (fertility, erosion), water (chemistry, eutrophication, sediment load, hydrology), ecosystem functioning (biodiversity, abundance, invasive species presence), socio-economics (e.g. health, wellbeing, employment). The systematic map outputs will be in the form of searchable databases of relevant and obtainable (full text only) literature, coded by subject, methodology and study design, and internal validity.

Keywords: Agriculture, Abandonment, Mountains, Alpine, Remote, Farming, Socio-Economic Impacts, Environmental Impacts

1. Background

Farm land abandonment can be simply defined as the cessation of agricultural activities on a given surface of land, yet there is no common precise definition of agricultural farmland abandonment in the literature[1]. Farm land abandonment occurs when income or resource generation cease to be economically viable or sustainable and the possibilities of adapting via changes in farming practices have been expended[2]. According to a study by Ramankutty and Foley[3], global abandonment of croplands has oc-curred over an estimated 1.47 million·km^2 between 1700 and 1992. Meanwhile, Pointereau *et al.*[1] estimate that 9.09Mha of agricultural land have been abandoned across 20 European countries between 1990 and 2000. Data cited for France for the period 1992 and 2003 show that grassland represented 57% of abandoned agricultural land; cropland 30% and vinyards and hedges/groves each 6%. However, the lack of a standardised definition of abandoned agricultural land, and the difficulty of matching this to available datasets, means that accurate estimates of abandoned area are lacking.

Land abandonment has a number of well-studied drivers, including environmental (e.g. reductions in soil fertility), economic (e.g. market globalisation) and socio-political (e.g. rural depopulation) causes[4]. The environmental impacts of farm land abandonment can be viewed as either an opportunity for ecological restoration to a state prior to agricultural establishment, or as the loss of an ongoing process of land management and an associated threat to biodiversity. Whether land abandonment poses an ecological opportunity or threat depends upon the agricultural history and the presence of systems that depend upon regular management for their existence. In Europe, many ecosystems have developed in the presence of agriculture and the loss of continued management resulting from land abandonment can have significant negative ecological impacts[4]. Pointereau *et al.*[1] suggest that abandonment of intensive agriculture often results in ecological benefits for the affected parcel of land, whilst abandonment of low intensity agricultural is more likely to result in a negative ecological impact owing to the role of such agriculture in maintaining systems classified as "high nature value" (HNV). From a soci-economic perspective, the abandonment of agricultural land is usually regarded as detrimental owing to implied loss of employment and income in rural areas.

Around 56 percent of the utilised agricultural area (UAA) of the EU is classified as 'less-favourable areas' by the Common Agricultural Policy (CAP). According to MacDonald *et al.*[2], much of this is mountainous, and a report in 2004 identified mountainous regions as constituting 39.9 percent of the area of the 15 Member States at the time[5]. Mountain areas, however, are difficult to define. For the purposes of examining farm land abandonment, mountainous areas are defined by their unfavourable topography, remoteness and extreme climate. Mountainous areas are typically described by elevation and/or slope, but this can vary significantly between countries. For example, Austria defines mountain areas as being above 700m or above 500m if slope is greater than 20 percent, whilst Spain more strictly defines them as being above 1000m, over 20 percent slope and a 400m elevation gain relative to surrounding land. Some definitions include low altitude areas where temperature contrasts reflect those in the high altitude Alps, such as Sweden and Finland. Other definitions use ruggedness assessed from satellite imagery e.g.[6].

The small-scale and extensively managed farmlands that are common in mountain areas are particularly vulnerable to marginalisation and abandonment[7]. A report from the Cross-Compliance Network identified mountainous areas as key areas at threat from farmland abandonment[8]. The causes of farmland abandonment in mountainous areas are expanded upon in more detail in Pointereau *et al.*[1] to include; steep slope, distance from the farm to the field, low accessibility, poor soils, land used as alpine pastures, small farms, high cultivation costs and small field size.

Resilience and adaptability in farming systems in mountain regions is limited for a number of reasons, including remoteness, climate and physical constraints, and the aversion to risk-taking, traditional cultural values and limited skill sets often held by the local population[2]. Limitations to the adaptability of mountain regions have been compounded by the historical paucity of agricultural research in these areas and a bias towards lowland regions e.g.[9].

A limited review of CAB Abstracts focusing on land abandonment was published in 2007[10]. A systematic review is currently underway on the subject of land abandonment in the Mediterranean[11]. A conceptual review of several case studies of land abandonment and EU policies responding to the problem for mountain areas was published in 2000[2]. The work herein will form the first systematic synthesis of the evidence of impacts of farm land abandonment in mountain areas across the globe. This systematic map of the literature will identify and catalogue all available evidence from a wide variety of sources, including the grey literature. Here we set out our methodology.

Objective of the Review

Primary Question

The primary question of this systematic review will be; What are the environmental impacts of farm land abandonment in high altitude/mountain regions?

This review will take the form of a systematic map, cataloguing the existing evidence across a wide range of variables such as setting, methodology, scale, measured outcomes etc. Mapping will be undertaken at two levels. Coarse-scale mapping will be undertaken on all identified abstracts, whilst fine-scale mapping will be undertaken on all available full texts.

The map databases that we will produce will catalogue the focus and location of relevant articles on the subject of agricultural land abandonment in high altitude/mountain regions. We anticipate that these maps will form a vital resource for researchers to identify subsets of this literature for further systematic review, and to identify knowledge gaps in the primary research.

The question has the following components:

Population: All mountainous* agricultural lands (global scope).

Exposure: Abandonment of agricultural land management. This definition is in accordance with that of Coppola[12] and Pointereau *et al.*[1] and specifies the cessation of all agricultural activity.

Comparator: Before-after land abandonment (temporal comparator), or un-abandoned nearby surrogate (spatial comparator).

Outcome: All outcomes relating to environmental and socio-economic impacts, including but not restricted to; natural hazards (fire-/flood risk, land/mud slides), soil (fertility, erosion), water (chemistry, eutrophication, sediment load, hydrology), ecosystem functioning (biodiversity, abundance, invasive species presence), socio-economics (household income, gender equity, health, wellbeing, employment).

Due to the difficulties in defining a mountainous region and the differences in definition between institutions, we will include any studies that make reference to a study site that is mountainous (including synonyms, e.g. uplands) or that has limited accessibility (e.g. of farming machinery) due to topography (*i.e.* altitude or slope).

By mapping the literature at abstract AND full text levels we hope to identify significant details about the availability of certain groups of studies. Scoping has suggested that a potentially large body of research has been carried out in the Loess Hilly Plateau in China on soil erosion. Preliminary scoping suggests that many of these articles may be published in Chinese language journals with restricted access; a potentially systematic limitation to the synthesis of the entire body of available evidence. Comparisons between the two levels of maps will highlight these potential deficiencies in the full text map and will allow users of the full text map to avoid outcomes that may be particularly susceptible to such restrictions.

No language restrictions will be employed in this systematic map. However, searches will be undertaken only using English search terms, since inclusion of all languages in such a global study would be impractical. In order to identify whether evidence exists that may have been missed by this limitation, during coding we will extract author email addresses. Following full-text coding we will then invite authors to participate in a simple online survey asking several questions: (1) if they are aware of non-English language research and research that may have gone un-catalogued by databases searched herein that has been published in this topic within their country/language/research area; (2) requesting that they give examples of such studies against which the search can be tested. The results of this survey will be compared with the two maps to identify where potentially missed studies may lie. To our knowledge, such involvement of primary researchers has not previously been undertaken in CEE systematic reviews and will strengthen any conclusions made concerning knowledge gaps in the evidence base.

2. Methods

2.1. Search Strategy

2.1.1. Search Terms

Scoping was undertaken in order to identify suitable relevant key terms to be included in the finalised search string. These terms include aspects of the exposure (farm land abandonment) and the population (high altitude/mountain regions) and are displayed in **Table 1**. Outcome terms were not included in the search string because of the size of returns based only on exposure and population terms, which was deemed to be manageable. Furthermore, the aim of the map is to document

Table 1. Summary of outputs from scoping study for search terms using web of knowledge

	Search string	WoK hits
Exposure terms	[(grassland OR farm* OR cropland OR agricultur* OR land OR *field OR pasture) AND (destock* OR abandon*)]	26,325
	AND	
Population terms	("high altitude" OR "higher altitude" OR "high ground" OR "higher ground" OR *alpine OR montane OR mount* OR elevat* OR highland OR hill* OR upland OR plateau OR mesa OR tableland OR slope OR aspect OR remote* OR massif OR sierra OR steep OR rugged)	2,579

*indicates a wildcard in search term [*i.e.* any character(s) permitted].

the available literature, including the forms of outcomes measured in the evidence base. Outcome documentation will therefore be an iterative process, and all relevant outcomes will be coded. No language restrictions will be put in place: automated language translation software will be used to complement the review teams' abilities.

2.1.2. Databases

The search aims to include the following online databases which cover the breadth and depth of available literature on the topic:

1) ISI Web of Knowledge (inc. ISI Web of Science and ISI Proceedings)

2) Science Direct

3) Directory of Open Access Journals

4) Copac

5) Agricola

6) CAB Abstracts

7) CSA Illumina/Proquest

8) GreenFile

Where databases cannot accept the full search strings detailed in **Table 1**, search strings will be modified according to the database help files. All database searches and outcomes will be recorded in a Search Record Appendix.

2.2. Search Engines

The following internet search engines will be used to identify relevant grey literature. The first 150 hits from each engine will be screened (based on sorting by relevance of results where possible).

Google Scholar http://scholar.google.co.uk/.

Scirus http://www.scirus.com/.

Dogpile http://www.dogpile.co.uk/.

Where search engines cannot accept the full search strings detailed in **Table 1**, search strings will be modified according to the search engine help files. All search engine searches and outcomes will be recorded in a Search Record Appendix.

2.3. Specialist Sources

The following specialist organisations will be searched for relevant grey literature using manual searches of their websites and automatic search facilities using key terms (such as abandon*).

Alterra http://www.wageningenur.nl/en/Expertise-Services/Research-Insttutes/alterra.htm.

Centre for Ecology and Hydrology http://www.ceh.ac.uk/.

National Farmers Union http://www.nfuonline.com/home/.

Global Environment Centre http://www.gec.org.my/.

Greenpeace http://www.greenpeace.org.uk/.

Joint Nature Conservation Committee http://jncc.defra.gov.uk/.

Macaulay Land Use Research Institute http://www.macaulay.ac.uk/.

National Soil Resources Institute http://www.cranfield.ac.uk/sas/nsri/.

Natural England http://www.naturalengland.org.uk/.

Royal Society for the Protection of Birds http://www.rspb.org.uk/.

Society for Ecological Restoration http://www.ser.org/.

DEFRA http://www.defra.gov.uk/.

Environment Agency http://www.environment-agency.gov.uk/.

PBL Netherlands http://www.pbl.nl/en/.

German Federal Ministry of Ag http://www.bmelv.de/EN/Homepage/homepage_node.html.

Thunen Institute http://www.ti.bund.de/en/.

ETH Zurich http://www.ethz.ch/index_EN.

European Environment Agency http://www.eea.europa.eu/.

EC Ag and Rural Dev site http://ec.europa.eu/agriculture/.

IEEP http://www.ieep.eu/.

JRC Institute for Env Sustainability http://ies.jrc.ec.europa.eu/.

JRC Institute for Prospective Tech Studies http://ipts.jrc.ec.europa.eu/.

United Nations Environment Programme http://www.unep.org/.

Food and Agriculture Organisation http://www.fao.org/index_en.htm.

Convention on Biological Diversity http://www.cbd.int/convention/.

World Wildlife Fund http://www.wwf.org.uk.

Associations des Populations des Montagnes du Monde http://www.mountainpeople.org.

Mountain Partnership http://www.mountainpartnership.org.

The International Centre for Integrated Mountain Development http://www.icimod.org.

Where organisational website search facilities cannot accept the full search strings detailed in **Table 1**, search strings will be modified according to the search help files (where provided), or a small subset of key terms will be searched individually. All organisational website searches and outcomes will be recorded in a Search Record Appendix.

2.4. Search Comprehensiveness Assessment

The comprehensiveness of the above search strategies will be assessed in a number of ways. Firstly, key bibliographies from relevant reviews e.g.[2] will be compared to the search results to check that all relevant articles have been identified through searches. Secondly, search results will be compared with a list of includable studies, identified by subject experts prior to the review (see **Table 2**). We will post questions on social media (www.academia.edu, www.researchgate.net and www.linkedin.com) to alert the research community to this systematic map

Table 2. List of key includable articles identified by subject experts for checking the comprehensiveness of the search strategy.

1. Cammeraat, E.L.H., A. Cerda, and A.C. Imeson, Ecohydrological adaptation of soils following land abandonment in a semi-arid environment. Ecohydrology, 2010. 3(4): p. 421–430.
2. Catorci, A., G. Ottaviani, and S. Cesaretti, Functional and coenological changes under different long-term management conditions in Apennine meadows (central Italy). Phytocoenologia, 2011. 41(1): p. 45–58.
3. Cocca, G., *et al.*, Is the abandonment of traditional livestock farming systems the main driver of mountain landscape change in Alpine areas? Land Use Policy, 2012. 29(4): p. 878–886.
4. Deleglise, C., G. Loucougaray, and D. Alard, Effects of grazing exclusion on the spatial variability of subalpine plant communities: A multiscale approach. Basic and Applied Ecology, 2011. 12(7): p. 609–619.
5. Durak, T., Long-term trends in vegetation changes of managed versus unmanaged Eastern Carpathian beech forests. Forest Ecology and Management, 2010. 260(8): p. 1333–1344.
6. Ferlan, M., *et al.*, Comparing carbon fluxes between different stages of secondary succession of a karst grassland. Agriculture Ecosystems & Environment, 2011. 140(1–2): p. 199–207.
7. Fonderflick, J., *et al.*, Avifauna trends following changes in a Mediterranean upland pastoral system. Agriculture Ecosystems & Environment, 2010. 137(3–4): p. 337–347.
8. Garcia-Ruiz, J.M. and N. Lana-Renault, Hydrological and erosive consequences of farmland abandonment in Europe, with special reference to the Mediterranean region -A review. Agriculture Ecosystems & Environment, 2011. 140(3–4): p. 317–338.
9. Gellrich, M., *et al.*, Agricultural land abandonment and natural forest re-growth in the Swiss mountains: a spatially explicit economic analysis. Agriculture, Ecosystems & Environment, 2001. 18(1): p. 93–108.
10. Kampmann, D., *et al.*, Agri-environment scheme protects diversity of mountain grassland species. Land Use Policy, 2012. 29(3): p. 569–576.
11. Knapp, B.A., A. Rief, and J. Seeber, Microbial communities on litter of managed and abandoned alpine pastureland. Biology and Fertility of Soils, 2011. 47(7): p. 845–851.
12. Lesschen, J.P., L.H. Cammeraat, and T. Nieman, Erosion and terrace failure due to agricultural land abandonment in a semi-arid environment. Earth Surface Processes and Landforms, 2008. 33(10): p. 1574–1584.
13. Marriott, C.A., *et al.*, Impacts of extensive grazing and abandonment on grassland soils and productivity. Agriculture Ecosystems & Environment, 2010. 139(4): p. 476–482.
14. Nikolov, S.C., Effects of land abandonment and changing habitat structure on avian assemblages in upland pastures of Bulgaria. Bird Conservation International, 2010. 20(2): p. 200-213.
15. Nunes, A.N., *et al.*, SOIL EROSION AND HYDROLOGICAL RESPONSE TO LAND ABANDONMENT IN A CENTRAL INLAND AREA OF PORTUGAL. Land Degradation & Development, 2010. 21(3): p. 260–273.
16. Obrist, M.K., *et al.*, Response of bat species to sylvo-pastoral abandonment. Forest Ecology and Management, 2011. 261(3): p. 789–798.
17. Peco, B., *et al.*, Effects of grazing abandonment on functional and taxonomic diversity of Mediterranean grasslands. Agriculture Ecosystems & Environment, 2012. 152:p.27–32.
18. Tocco, C., *et al.*, Does natural reforestation represent a potential threat to dung beetle diversity in the Alps? Journal of Insect Conservation, 2013. 17(1): p. 207–217.
19. Uematsu, Y., *et al.*, Abandonment and intensified use of agricultural land decrease habitats of rare herbs in semi-natural grasslands. Agriculture Ecosystems & Environment, 2010. 135(4): p. 304–309.
20. Waesch, G. and T. Becker, Plant diversity differs between young and old mesic meadows in a central European low mountain region. Agriculture Ecosystems & Environment, 2009. 129(4): p. 457–464.
21. Zimmermann, P., *et al.*, Effects of land-use and land-cover pattern on landscape-scale biodiversity in the European Alps. Agriculture Ecosystems& Environment, 2010. 139(1–2): p. 13–22.

and to request that subject experts submit studies that they feel may not be readily accessible or catalogued by the most common academic databases. In addition, authors of unobtainable articles will be contacted by email to request the submission of other pertinent articles in addition to the unobtainable literature. These studies will be used in addition to the articles highlighted in **Table 2** to test the comprehensiveness of our search strategy.

2.5. Study Inclusion Criteria

Study selection according to the predefined inclusion criteria detailed below will proceed according to a three stage, hierarchical process: titles, abstracts and finally full texts will be assessed against the inclusion criteria. If there is any doubt over the presence of a relevant inclusion criterion (or if information is absent) the articles will be retained for assessment at a later stage. Title and abstract-level assessment will not assess the presence of a comparator, which is typically not explicit. Since titles and abstracts in grey literature do not conform to scientific standards, assessment will proceed immediately to full text assessment. Consistency checks will be undertaken using a subset of 100 abstracts by two reviewers independently of one another. Screening decisions will then be compared using a Kappa test of agreement[13]. A score of greater than 0.6 indicates substantial agreement. Any disagreements will be discussed and any terms that need redefining or expansion will be adapted accordingly.

The following aspects of the systematic review question will form inclusion criteria when assessing potentially relevant literature:

Relevant population(s): Any high altitude or mountainous region, any region with restricted access due to ruggedness, any region with agricultural difficulties or limits on agricultural advancement or adaptability due to slope, altitude or ruggedness (global scope).

Types of exposure/intervention: Abandonment of agricultural land or reinstating of agricultural activity in agricultural land following abandonment.

Types of comparator: Before land abandonment and/or an un-abandoned control site.

Types of outcome: All outcomes, including but not restricted to; soil chemistry (including carbon and GHG flux), soil erosion, water chemistry, hydrology, natural hazards, biological diversity and abundance, presence of invasive species, socioeconomics (including health, wellbeing, employment).

Types of study: Both observational and experimental field studies. Experimental field studies (*i.e.* simulated abandonment) must investigate continued abandonment over a period in excess of one year.

2.6. Map Coding

Mapping will be undertaken in two stages to produce two interrelated databases. One database will map studies using abstracts at a coarse scale (*i.e.* study location, population descriptor, measured outcome, study design, and comparator type).

A second database will expand on this information by extracting summary details for each study where a full text is available, producing a fine-scale. Coding of full texts will be undertaken using key words describing various aspects of study design and setting. Key variables of interest were identified through scoping activities and discussion with subject experts. Coding options within these key variables were then compiled in a partly iterative process, expanding the range of options as they were encountered during scoping. The finalised coding tool for the full text map is displayed in **Table 3**.

Studies may be coded with multiple keywords within each coding variable where appropriate, for example one study in multiple countries. The coding will be undertaken by one reviewer, with a subset of 10% of the coding carried out independently by a second reviewer and cross checked. Discrepancies will be discussed and coding moderated accordingly to reflect any clarification.

2.7. Critical Appraisal of Study Internal Validity

Coding will be used to describe the internal validity (IV) of each included study. This will be assessed using the following coding variables; study length, study timescale, comparator appropriateness, comparator type, replication, sources

Table 3. Coding tool for the systematic map.

Coding variables	Details/examples
Author	
Full reference	
Publication type	e.g. book chapter, journal paper, conference paper, thesis, organisation report
Holding institution	Organisation/body holding access to article
Article access issues	*i.e.* open access, subscription only
Study year	Time period of experimentation/observation
Study length	Time over which study undertaken
Study timescale	Period between intervention and study
Study description	Brief description of study
Intervention description	Full description of intervention and final state
Intervention time period	Years intervention in place
Comparator description	Full description of comparator
Comparator appropriateness	Brief description of how well matched the comparator is to the intervention population
Comparator type	*i.e.* spatial, temporal, both
Replication	Unit of replication (e.g. patch, farm, landscape)
Spatial scale	*i.e.* landscape scale, single farm, multiple farm, whole farm, within field
Sources of potential bias	Brief description of potential sources of bias in study results
Methodological detail	Level of methodological detail; low (very little detail, significant information missing), medium (some detail missing but generally sufficient), high (very high level of detail, no obvious information lacking)
Study country/ies	
Study region	
Mountain descriptor	Quoted description of mountain type, e.g. alpine
Altitude	
Farming system	e.g. organic farming, conventional farming, integrated farming, intensive grassland, extensive grassland, tillage, ploughing, non inversion tillage, minimal tillage
Broad outcome group	*i.e.* soil, water, natural hazard, ecosystem functioning
Outcome focus	e.g. water chemistry, butterfly
Measured outcome	e.g. total suspended solids, Simpson's diversity index
Experimental design	*i.e.* observation, experimentation
Additional details	*i.e.* multiple outcomes studied, multiple articles of one study, multiple experiments in one article

of potential bias and methodological detail. A judgment based on this critical appraisal will be made by placing each study into one of three categories; low, high, or unclear IV. For each study a short descriptive explanation for this judgement will be given for transparency.

2.8. Systematic Map Database

The systematic map outputs will be in the form of two databases of studies (at abstract and full text levels) that will describe the nature and location of evidence on the review topic. These databases will be easily searchable and freely accessible. The maps may form the basis for further primary research by identifying key knowledge gaps, and may also form the basis for further secondary research as a starting point for the synthesis of information in focused systematic reviews.

Competing Interest

The authors declare that they have no competing interest.

Authors' contributions

NH, DS and AP conceived the review question. NH undertook pilot research. NH drafted the protocol text with support from DS and AP. All authors read and approved the final manuscript.

Acknowledgements

The authors thank the European Commission for funding this research. We also thank Nicola Randall for advice on systematic mapping practicalities.

Source of Support

This research is undertaken as part of a project funded by the European

Commission's Joint Research Centre through Service Contract Number 153172-012 A08 GB. All statements/comments within this document belong to the authors and do not necessarily represent the views of the European Commission.

Source: Haddaway N R, Styles D, Pullin A S. Evidence on the environmental impacts of farm land abandonment in high altitude/mountain regions: a systematic map[J]. Environmental Evidence, 2013, 2(18):1−19.

References

[1] Pointereau P, Coulon F, Girard P, Lambotte M, Stuczynski T, Sanchez Ortega V, Del Rio A, Anguiano E, Bamps C, Terres J: Analysis of farmland abandonment and the extent and location of agricultural areas that are actually abandoned or are in risk to be abandoned. Ispra: European Commission-JRC-Institute for Environment and Sustainability; 2008.

[2] MacDonald D, Crabtree J, Wiesinger G, Dax T, Stamou N, Fleury P, GutierrezLazpita J, Gibon A: Agricultural abandonment in mountain areas of Europe: environmental consequences and policy response. J Environ Manage 2000, 59:47−69.

[3] Ramankutty N, Foley JA: Estimating historical changes in global landcover: croplands from 1700 to 1992. Global Biogeochem Cy 1999,13:997−1027.

[4] Hobbs RJ, Cramer VA: Why old fields? Socioeconomic and ecological causes and consequences of land abandonment. In Old fields: dynamics and restoration of abandoned farmland. Edited by Cramer VA, Hobbs RJ. Washington DC: Island Press; 2007:1−14.

[5] Schuler M, Stucki E, Roque O, Perlik M: Mountain Areas in Europe: Analysis of mountain areas in EU member states, acceding and other European countries. European Commission contract No 2002.CE.16.0.AT.136. Final report, January. Stockholm: Nordregio, Nordic Centre for Spatial Development; 2004.

[6] Körner C, Paulsen J, Spehn EM: A definition of mountains and their bioclimatic belts for global comparisons of biodiversity data. Alp Bot 2011, 121:73−78.

[7] Baldock D, Beaufoy G, Brouwer F, Godeschalk F: Farming at the margins: abandonment or redeployment of agricultural land in Europe. London: Institute for European Environmental Policy The Hague; 1996.

[8] Moravec J, Zemeckis R: Cross compliance and land abandonment. In A research paper of the Cross-Compliance Network (Contract of the European Community's Sixth Framework Programme, SSPE-CT-2005-022727), Deliverable D17 of the Cross-Compliance Network. 2007.

[9] Jodha NS: Mountain agriculture: the search for sustainability. J Farm Syst Res Exten-

sion 1990, 1:55–75.

[10] Rey Benayas J, Martins A, Nicolau JM, Schulz JJ: Abandonment of agricultural land: an overview of drivers and consequences. CAB Reviews: Perspect Agric Vet Sci Nutr Natural Resour 2007, 2:1–14.

[11] Plieninger T, Gaertner M, Hui C, Huntsinger L: Does land abandonment decrease species richness and abundance of plants and animals in Mediterranean pastures, arable lands and permanent croplands? Env Evid 2013, 2:1–7.

[12] Coppola A: An economic perspective on land abandonment processes, working paper n. 1/2004. presented at the AVEC Workshop on "Effects of Land Abandonment and Global Change on Plant and Animal Communities. Anacapri; 2004.

[13] Cohen J: A coefficient of agreement for nominal scales. Educ Psychol Meas 1960, 20:37–46.

Chapter 6

The Challenges of Water, Waste and Climate Change in Cities

S. H. A. Koop[1,2], C. J. van Leeuwen[1,2]

[1]KWR Water cycle Research Institute, Groningen haven 7, 3433 PE Nieuwegein, The Netherlands

[2]Copernicus Institute for Sustainable Development and Innovation, Utrecht University, Heidelberglaan 2, 3584 CS Utrecht, The Netherlands

Abstract: Cities play a prominent role in our economic development as more than 80% of the gross world product (GWP) comes from cities. Only 600 urban areas with just 20% of the world population generate 60% of the GWP. Rapid urbanization, climate change, inadequate maintenance of water and wastewater infrastructures and poor solid waste management may lead to flooding, water scarcity, water pollution, adverse health effects and rehabilitation costs that may overwhelm the resilience of cities. These megatrends pose urgent challenges in cities as the cost of inaction is high. We present an overview about population growth, urbanization, water, waste, climate change, water governance and transitions. Against this background, we discuss the categorization of cities based on our baseline assessments, *i.e.* our City Blueprint research on 45 municipalities and regions predominantly in Europe. With this bias towards Europe in mind, the challenges can be discussed globally by clustering cities into distinct categories of sustainability and by providing additional data and information from global regions. We distinguish five categories of sustainability: (1) cities lacking basic water services, (2) wasteful cities, (3) water-efficient cities, (4) resource-efficient and adaptive cities and (5) water-wise cities.

Many cities in Western Europe belong to categories 3 and 4. Some cities in Eastern Europe and the few cities we have assessed in Latin America, Asia and Africa can be categorized as cities lacking basic water services. Lack of water infrastructures or obsolete infrastructures, solid waste management and climate adaptation are priorities. It is concluded that cities require a long-term framing of their sectoral challenges into a proactive and coherent Urban Agenda to maximize the co-benefits of adaptation and to minimize the cost. Furthermore, regional platforms of cities are needed to enhance city-to-city learning and to improve governance capacities necessary to accelerate effective and efficient transitions towards water-wise cities. These learning alliances are needed as the time window to solve the global water governance crisis is narrow and rapidly closing. The water sector can play an important role but needs to reframe and refocus radically.

Keywords: Water Governance Capacities, Sustainability Transitions, Urban Agenda, Smart Cities, City Blueprints, Circular Economy, HABITAT III

1. Introduction

1.1. Population Growth

The world population is projected to increase by more than one billion people within the next 15 years, reaching 8.5 billion in 2030, and to increase further to 9.7 billion in 2050 and 11.2 billion by 2100 (UN 2015). Approximately 60% of the global population lives in Asia (4.4 billion), 16% in Africa (1.2 billion), 10% in Europe (738 million), 9% in Latin America and the Caribbean (634 million), and the remaining 5% in Northern America (358 million) and Oceania (39 million). Population growth patterns are different across the globe. **Figure 1** shows population growth in some world regions from the year 1600 to 2100 (Klein Goldewijk *et al.* 2010; UN 2015a). Many countries in Africa are still growing exponentially. This implies that their claims on resources also increase rapidly. In Western Europe and India? (India, Bangladesh, Nepal, Sri Lanka, Bhutan, Pakistan, Afghanistan and Maldives), population growth is gradually levelling off (logistic growth), while in China growth will soon decline due to the one-child family policy. This policy was introduced in 1979 to halt the rapid growth in the Chinese population, and it included late marriage and childbearing (delaying the start of reproduction)

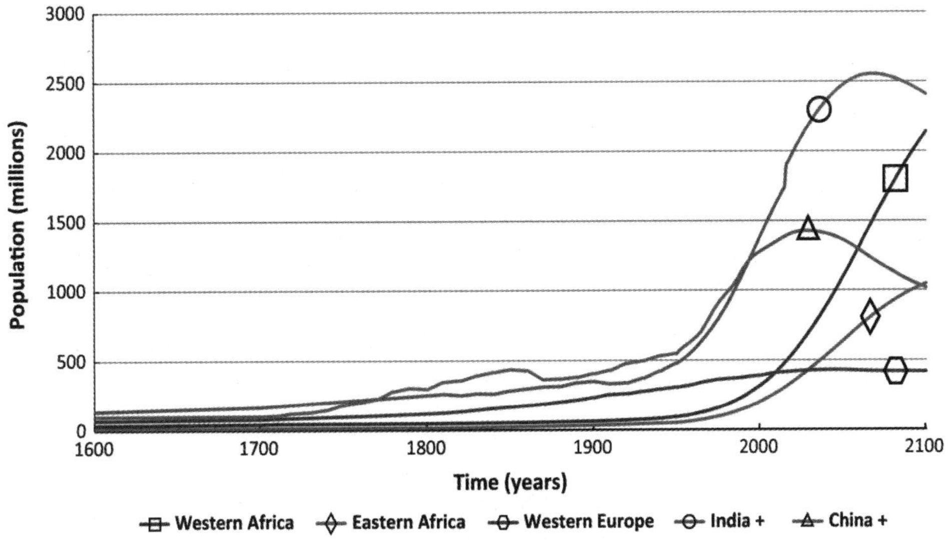

Figure 1. Total population estimations of India? (India, Bangladesh, Nepal, Sri Lanka, Bhutan, Pakistan, Afghanistan and Maldives), China? (China, Hong Kong, Macao and Mongolia), Eastern Africa, Western Africa and Western Europa based on the HYDE 3.1 database (Klein Goldewijk *et al.* 2010), and the UN medium variant of the world population predictions (UN 2015a).

as well as the restriction on family size to just one child per family with high penalties for infringement (Hesketh *et al.* 2005). In fact, the maximum population densities in China+ (China, Hong Kong, Macao and Mongolia), Western Europe and India+ are expected in the year 2026 (1428 million), 2045 (424 million) and 2069 (2554 million), respectively (UN 2015a).

Urbanization will continue in both the more developed and the less developed regions so that, by 2050, urban dwellers will likely account for 86% of the population in the more developed regions and for 64% of that in the less developed regions. Overall, the world population is expected to be 67% urban in 2050 (UN 2012). Thus, urban areas of the world are expected to absorb all the population growth over the next decades.

1.2. City Blueprint Methodology

The development of the City Blueprint methodology to assess the sustainability of integrated water resources management (IWRM) in municipalities and regions started in 2011 (Van Leeuwen *et al.* 2012). A baseline assessment was developed as part of the strategic planning process in cities as described in the training modules developed in the SWITCH project (managing water for the city of the future). The assessment was kept as short, clear and simple as possible. The strategic planning process and the role of the City Blueprint are provided in **Figure 2**. The indicators in the City Blueprint are based on the 3 Ps (People, Planet and Profit) in the water cycle (Van Leeuwen *et al.* 2012, Koop and Van Leeuwen 2015a). Use has been made of several other assessment frameworks (Van Leeuwen *et al.* 2012), including the Green City Index (2015). Similar assessment schemes have been published by SDEWES (2015).

The sustainability of IWRM is assessed in an interactive process involving the most important IWRM actors (Philip *et al.* 2011). This interactive approach has been used for the assessment of all cities, except for Rotterdam, Ankara and London. For these cities, an extensive literature search was completed. For all other cities, a comprehensive questionnaire was completed by municipalities and regions, which takes them a few days (European Commission 2015a). The City Blueprint offers cities a threefold benefit: (1) an interactive quick scan of their

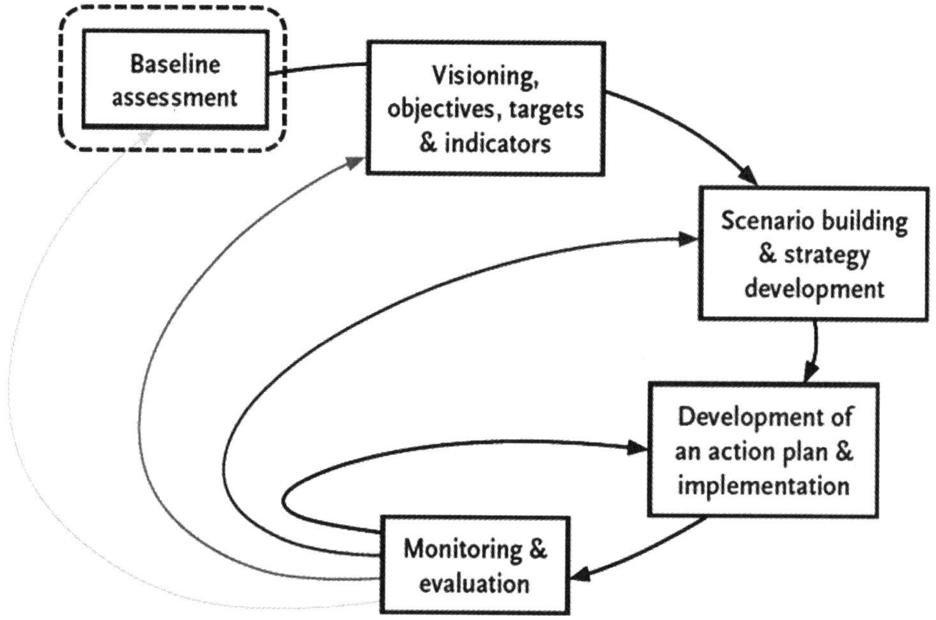

Figure 2. Function of the City Blueprint (red box) in the strategic planning process for IWRM according to SWITCH (Philip *et al.* 2012).

own water cycle, (2) access to best practices in other cities (Koop *et al.* 2015) and (3) participation in an international platform (European Commission 2015a). After the completion of the questionnaire, a radar chart of all 25 performance indicators (the City Blueprint) and the Blue City Index (BCI) are provided both varying from 0 (concern) to 10 (no concern). This initiative has been scaled up to an action under the flag of the European Innovation Partnership on Water of the European Commission (European Commission 2015a) in the framework of the European Blueprint for water (European Commission 2012). The City Blueprint provides municipalities and regions with a practical and broad framework to define steps towards realizing a more sustainable and resilient water cycle in collaboration with key stakeholders.

1.3. Outline of This Study

The aim of this study is to present an overview of the challenges of water, waste and climate adaptation in cities and to link the City Blueprint activities to major developments such as: the challenges of urbanization (Section 2), water governance (Section 3) and transitions in cities (Section 4). In Section 5, we summarize our work on City Blueprints and discuss their role in learning alliances of cities (Section 6). Concluding remarks are provided in Section 7.

2. Urbanization and the Dynamics of the City

2.1. Homes

Most people live in cities. There are more than 400 big cities (urban areas with more than one million inhabitants) and 23 megacities (metropolitan areas with a population of more than 10 million). Most of these megacities are in Asia (UN 2012). The United Nations (UN) estimates that 54% of all people live in cities, and by 2050, this will increase to 66% (UN 2015a). In developed countries, this percentage is even higher (more than 80%). Global urbanization is taking place at a high speed. In 1970, for example, there were only two megacities (Tokyo and New York); in 1990, there were 10; in 2011, there were 23, and by 2025, there will be 37 megacities. Tokyo, the largest megacity, will grow from 37 million to about 40 million people in 2025 (UN 2012).

The United Nations (UN 2015a) estimates that between 2015 and 2050 the world population will grow from 7.32 to 9.55 billion. At the same time, the population in cities will increase from 3.96 to 6.34 billion, while the number of people living in rural areas will decline. Due to population growth and migration from rural areas to cities, approximately 190,000 people per day will need to find a new place to live. In other words, over the next 40 years, we will build approximately 3000 big cities with a population size of Amsterdam. It should be noted that there are major differences in the rate of population growth and urbanization in different parts of the world (UN 2012, 2015a). Developing countries account for 93% of the urbanization globally, 40% of which is the expansion of slums. By 2030, the urban population in Africa and Asia will double (UNESCO 2015a).

2.2. Work

Cities play a prominent role in economic development. More than 80% of the gross world product (GWP) comes from cities. Only 600 urban areas with just 20% of the world population generate 60% of the GWP (Dobbs *et al.* 2011). Cities are therefore also job generators and centres of communication, innovation and creativity. They also play a large part in social and cultural matters (European Commission 2011; BAUM 2013). Cities can also take the lead in sustainable development as they offer many economies of scale per head of the population in terms of raw material use, energy consumption, waste recycling and transport (BAUM 2013).

The continued acceleration of change or 'rapidification' of our planet, our life and the global economy is higher than ever (Francis 2015). The transformation of China due to urbanization and industrialization is taking place on a scale 100 times greater and ten times faster as compared to UK a century ago (Dobbs *et al.* 2012). This comes with unavoidable consequences (Van Leeuwen 2008). To illustrate this, two examples are provided: one looking back and the other forward: first of all, a backward glance based on the turnover of the chemical industry over the last 10 years. In 2003, the production of the chemical industry was roughly equally divided between Europe, North America and the rest of the world. In 10 years' time sales, figures have almost doubled to €3156 billion, but the hub has shifted to Asia with a share of 57% in 2013 (CEFIC 2014).

The second example has been taken from a report by Dobbs *et al.* (2012). Emerging cities create opportunities. That is why entrepreneurs increasingly focus on cities with great economic growth potential as in 2025 one billion new consumers are expected. This will create many opportunities and is undoubtedly the main reason why people are moving on such a massive scale to the city. There will be an acceleration in the shift of the economic hub from the old developed countries to the new developing or transitioning countries, particularly in Asia. The population growth of Chinese cities until 2025 and the accompanying growth in gross domestic product (GDP) and drinking water supply needs has been estimated at 30.9%, 39.7% and 25.6%, respectively, whereas growth in European cities has been estimated at 1.8%, 5.7% and 1.7%, respectively. This means that after about five centuries (since the discovery of America), Asia will again become the global economic epicentre (Dobbs *et al.* 2012).

2.3. Challenges of Urbanization

The concentration of homes and employment in cities also has its downside. Cities currently take up about 2% of the land surface on Earth, but account for 60%–80% of the energy consumption and 75% of global CO_2 emissions (UN 2013a). Roughly the same percentage will also apply to the use of raw materials (e.g. metals, wood, plastics) for infrastructure, houses, cars and numerous other consumer items. Cities are concentrated centres of production, consumption and waste (Grimm *et al.* 2008; Bai 2007). Ecological studies of cities have shown that they sometimes exceed their environmental footprint by a factor 10–150 (Doughty and Hammond 2004). This creates enormous pressure not only on water supply, solid waste recycling and wastewater treatment (Grant *et al.* 2012), but also on nature and the built environment too, including soil, air and water pollution (UN 2013a; Hoekstra and Wiedman 2014). Water pollution reduces the availability of healthy water (Schwarzenbach *et al.* 2006; WHO 2008; Van Leeuwen and Vermeire 2007). Cities are therefore becoming increasingly dependent on rural areas for the supply of energy, water, building materials and food, as well as for the removal of waste and waste substances (OECD 2015a; UN 2014). A summary of some of the challenges in cities is provided in **Figure 3**.

The consequences of urbanization extend to areas far beyond the city, areas which are vital to supply cities with important 'ecosystem services' (OECD 2015a).

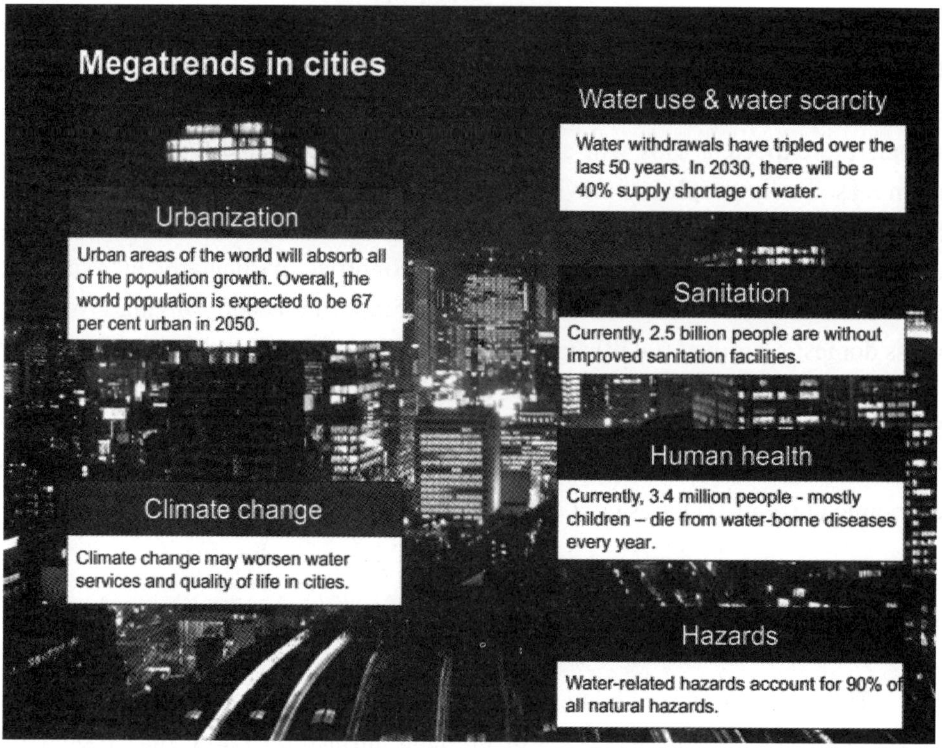

Figure 3. Megatrends pose urgent challenges in cities (Van Leeuwen 2013).

An example is provided by the megacity Istanbul, where water is now supplied from a basin by a 180-km-long pipeline (Van Leeuwen and Sjerps 2015a). Habitat preservation, *i.e.* the conservation of the forests surrounding Istanbul—vital habitats for the water supply—is extremely important for the future of Istanbul (Atelier Istanbul 2012).

2.4. Water Challenges in the City

Drinking water consumption in cities makes up a small fraction of the total water footprint. For example, people in the Netherlands use about 2300m^3 of water per person per year of which 67% is for agriculture, 31% is used in industry, while only 2% makes up household water (Van Oel *et al.* 2009). This means that water challenges in cities need to be solved predominantly by actors outside the traditional water sector. In fact, half of all cities with populations greater than 100,000 are located in water-scarce basins. In these basins, agricultural water consumption accounts for more than 90% of all freshwater depletions (Hunger and Döll 2008; Richter *et al.* 2013). In a critical analysis, Richter *et al.* (2013) point out that nearly all water used for domestic and industrial purposes is eventually returned to a water body. For instance, toilets are flushed and purified wastewater as well as cooling water in power plants is often returned to rivers. Because much of this water is not consumed, efforts to reduce urban water use or to recycle water with the aim to alleviate water scarcity *per se*, hardly makes any difference. In total, the domestic, industrial and energy sectors account for less than 10% of global water consumption (Richter *et al.* 2013; Hoekstra *et al.* 2012). Of course, proper urban use and reuse of water, as well as adequate sanitation, contribute significantly to pollution reduction, local water availability, as well as to energy efficiency, energy and nutrient recovery.

Hoekstra *et al.* (2012) estimate that agriculture accounts for 92% of the global blue water footprint. Land, energy and climate studies have shown that the livestock sector plays a substantial role in deforestation, biodiversity loss and climate change. Livestock also significantly contributes to humanity's water footprint, water pollution and water scarcity (Jalava *et al.* 2014; Hoekstra 2014). Furthermore, the Food and Agriculture Organization of the United Nations (FAO) estimates that 32% of all food produced in the world was lost or wasted in 2009 (Lipinski *et al.* 2013; FAO 2011a). Therefore, consumers, *i.e.* citizens, can play a major role in the reduction in the global water footprint by both reducing the fraction of animal products in their diets and by curbing their food waste.

With a changing climate comes a greater demand for proactive adaptation processes, as well as knowledge of how adaptation policies and measures could be implemented successfully. Accidents often lead to major policy changes. In 1953, almost 2000 people drowned in the Netherlands. As a result of this catastrophe, a long-term plan was devised, the Delta Plan, with a Delta Fund, and a Delta Commissioner appointed, reporting directly to the Dutch Minister-President (Delta programme 2013). Another example of a reactive adaptation policy can be observed in the city of Melbourne. Melbourne is a city of extremes: floods due to excessive rainfall, but drought too. A 10-year period of drought has recently come to an end. This has forced the city to take rigorous measures: (1) the construction of a costly desalination plant as backup for drinking water supply, (2) rainwater harvesting and (3) the reuse of wastewater (Van Leeuwen 2015). Melbourne has become 'water sensitive' or water-wise (Brown *et al.* 2009), and the citizens 'do their bit', e.g. by limiting water use and installing rainwater tanks on a wide scale to make good use of the rain when it does fall.

Disasters quickly raise awareness, whether that be about defending against flooding or dealing with drought (Koop and Van Leeuwen 2015a, b). Hence, adaptation measures are mainly reactive (Amundsen *et al.* 2010; Reckien *et al.* 2015), ad hoc, and often ineffective and expensive (UNEP 2013). Globally, the main challenge is to move from reactive measures to proactive transitions, by taking bold decisions based on a cohesive long-term process as shown in **Figure 2**.

2.5. Solid Waste and Water

Cities generate massive amounts of solid waste. Poor waste management, ranging from non-existing collection systems to ineffective disposal, causes air, water and soil contamination. Open and unsanitary landfills contribute to contamination of drinking water and increase infection and transmit diseases. Managing solid waste is another challenge of urban areas of all sizes, from megacities to the small towns and large villages (UN-Habitat 2010).

Plastics easily enter rivers and ultimately oceans. Jambeck *et al.* (2015) calculated that 275 million metric tons of plastic waste was generated in 192 coastal countries in 2010. Approximately 1.7%–4.6% of this plastic enters oceans (Jambeck *et al.* 2015). Plastic waste does not readily biodegrade but degrades into smaller pieces that affect marine ecosystems (Derraik 2002). The plastics form 'soups' in five major ocean gyres: two in the Pacific, one in the Indian and two in the Atlantic and affect many marine animals by ways of ingestion (Zarfl *et al.* 2011; McFedries 2012). Also consumer products contribute to the emission of microplastics to surface water such as cosmetics and personal care products, cleaning agents, paint and coatings (Van Wezel *et al.* 2015). Recently, a detailed study was made for the river Rhine, one of the largest European rivers. Microplastics were found in all samples, with 892,777 particles per km^2 on average. These microplastics concentrations were diverse across the river, reflecting various sources and sinks such as wastewater treatment plants, tributaries and weirs (Mani *et al.* 2015).

Recycling leads to substantial resource savings (EMF 2014, 2015a) and to significant reductions in greenhouse gas (GHG) emissions. GHG emissions from open dump landfilling are about 1000kg CO_2-eq. tonne^{-1} of solid waste, whereas this can be largely reduced to 300kg CO_2-eq. tonne^{-1} for conventional landfilling. Actually, it can even be a net sink of carbon when most material is recycled or the energy is recovered (Manfredi *et al.* 2009). The global GHG emissions of solid waste disposal sites are estimated to be approximately 5%–20% of the global anthropogenic methane emission, which is equal to about 1%–4% of the total anthropogenic GHG emissions (IPCC 2006).

The order of preference of managing waste also known as the Lansink's ladder has been laid down in the Dutch Environmental Management Act (VROM 2001) and subsequently across Europe, as the waste hierarchy in the Waste Framework Directive (2008). The waste hierarchy is a preference order from: prevention, preparing for reuse, recycling, other recovery (e.g. energy recovery) and disposal.

Recently, the European Commission announced a plan for the circular economy. One of the reasons is that Europe currently loses around 600 million tonnes of materials contained in waste each year, which could potentially be re-cycled or reused. On average, only 40% of the waste produced by EU households is recycled ranging from 5% in some areas to 80% in others. Turning waste into a resource is an essential part of increasing resource efficiency and part of this circular economy package (European Commission 2015b; EMF 2015b).

Solid waste data in many cities are largely unreliable. Available data show that cities can improve on their solid waste management as waste collection rates for cities in low- and middle-income countries range from 10% in peri-urban areas to 90% in commercial city centres (UN-Habitat 2010). Even in Europe, recycling rates are rather low (EEA 2013). As the sustainability of IWRM in municipalities and regions is intrinsically linked to proper solid waste management, it was decided to include the following three indicators in the improved City Blueprint framework (Koop and Van Leeuwen 2015a), *i.e.* solid waste collected (the per capita non-industrial solid waste that is collected; kg/cap./year), solid waste recycled (% of collected non-industrial solid waste that is recycled or composted) and solid waste energy recovery (% of collected non-industrial waste that is incinerated with energy recovery). This information has been gathered for 45 municipalities and regions (Koop and Van Leeuwen 2015b, c).

2.6. The Cost of Urban Water Infrastructure

Cities need to protect their citizens against water-related disasters (e.g. droughts and floods), to guarantee water availability and high-quality groundwater, surface water and drinking water. Cities also need to have adequate infrastructure in response to climate, demographic and economic trends (OECD 2015a). The cost of urban infrastructure is high. The UNEP (2013) estimates that for the period 2005–2030 about US$ 41 trillion is needed to refurbish the old (in mainly developed countries) and build new (mainly in the developing countries) urban infrastructures. The cost of the water infrastructure (US$ 22.6 trillion) is estimated at more than that for energy, roads, rail, air and seaports put together. The wastewater infrastructure is responsible for the largest share of this 22.6 trillion. The report also warns that 'Sooner or later, the money needed to modernise and expand the world's urban infrastructure will have to be spent. The demand and need are too great to ignore. The solutions may be applied in a reactive, ad hoc, and ineffective fashion, as they have been in the past, and in that case the price tag will probably be higher than US$ 40 trillion'.

To support projected economic growth between now and 2030, McKinsey (2013) has estimated that the investments on global infrastructure need to increase by nearly 60% from the US$ 36 trillion spent on infrastructure over the past 18 years. Therefore, an investment of US$ 57 trillion over the next 18 years is necessary. This is approximately 3.5% of anticipated global GDP. These figures do not account for the cost of addressing the large maintenance and renewal backlogs and infrastructure deficiencies in many economies (McKinsey 2013). Cashman and Ashley (2008) have estimated the required annual expenditure on water and sanitation infrastructure for high-, middle- and low-income countries at, respectively, 0.35%–1.2%, 0.54%–2.60% and 0.71%–6.30% of the annual GDP.

Water goals have big costs but also big returns. Conservative estimates of global investments in a post-2015 water for sustainable development and growth agenda have been estimated (UN University 2013). Between 1.8% and 2.5% of the annual global GDP is needed for implementation of water-related sustainable development goals. This would also generate a minimum US$ 3108 billion in additional economic, environmental and social benefits, *i.e.* a net annual benefit of US$ 734 billion.

2.7. Time Is Running out

In many countries, awareness of the urban challenges is low. Nevertheless, there are developments which cannot be ignored:

- The UN (2012) estimates that in 2025 about 2 billion people will have an absolute water shortage and that two-thirds of the world population will be affected by water scarcity. Estimates for 2030 assume 40% more demand for water than is actually available (2030 Water Resources Group 2009).

- The world population growth and immigration will take place mainly in cities (UN 2012).

- Many cities lie in high-risk areas (UN 2012, 2013b). It is estimated that two-thirds of the world's largest cities will be vulnerable to rising sea levels. At the same time, many delta cities suffer from severe land subsidence. Consequently, the vulnerability of cities to both marine and fluvial flooding is expected to increase (Molenaar *et al.* 2015). It is predicted that the frequency, intensity and duration of extreme precipitation events will increase, as well as the frequency and duration of droughts (EEA 2012; Jongman *et al.* 2014).

- Large areas of productive agricultural land is fed by groundwater which is becoming increasingly depleted (UNEP 2007).

- Sea water intrusion, salinization of irrigated land, erosion and desertification are growing problems affecting global water and food security (FAO 2011b; UNESCO 2015a).

- Wastewater treatment in Asia and Africa is sparse, and nutrient emissions are projected to double or triple within 40 years as a result of rapid urbanization (**Figure 1**). This will strongly enhance eutrophication, biodiversity loss, and threaten fisheries, aquaculture, tourism, and drinking water (Ligtvoet *et al.* 2014).

- Adequate sanitation remains a challenge for 2.5 billion people and lack of improvement will continue to lead to mortality, particularly among children (WHO 2008).

Sustainable water management is a major challenge. This is probably also the reason why the World Economic Forum (2014) ranked the water crisis and water-related risks as major global risks in terms of both probability and impact. Water is also high on the agenda of many other international organizations, such as the Organisation for Economic Cooperation and Development (OECD 2011a), the UN, the World Health Organization (WHO) and the Food and Agriculture Organization (FAO 2011b).

2.8. Benefits of Smart Adaptation

The cost of preventable accidents in urban areas is high, and smart coherent transitions in cities are likely to prevent both human and capital losses. For instance, the overall economic impacts of water scarcity and drought events in the past 30 years were estimated at € 100 billion in the European Union (EU). From 1976–1990 to the following 1991–2006 period, the average annual impact doubled, rising to € 6.2 billion per year in the most recent years. The price tag of the exceptional European heat wave in 2003 was estimated at € 8.7 billion and caused up to 70,000 excess deaths over a four-month period in Central and Western Europe (EEA 2012).

Assets can be directly damaged by droughts, floods and severe storms. Floods are the most prevalent natural hazard in Europe. In a recent analysis, it was estimated that EU floods cost €4.9 billion a year on average from 2000 to 2012, a figure that could increase to €23.5 billion by 2050 (Jongman *et al.* 2014). In addition, large events such as the European floods in 2013 are likely to increase in frequency from an average of once every 16 years to a probability of every ten years by 2050. A well-known example is the City of Copenhagen. During a two-hour thunderstorm, 150 mm of rain fell in the city centre on 2 July 2011. Sewers were unable to handle this amount of water, and many streets were flooded and sewers overflowed into houses, basements and onto streets, thereby flooding the city. The first estimate of the damage was €700 million (EEA 2012), but a more indepth review showed that the damage was actually nearly €1 billion (Leonardsen 2012). Hurricane Katrina was one of the deadliest hurricanes ever to hit the USA. An estimated 1836 people died. Total property damage from Katrina was estimated at US$ 81 billion, which was nearly triple the damage inflicted by Hurricane Andrew in 1992 (Zimmerman 2012). Casualties, pollution and social stress are more difficult to quantify financially, but in general it can be concluded that the real costs of flooding in cities are seriously underestimated.

There is an increasing amount of information and evidence on the impacts of climate change and also on adaptation. However, information on the costs of inaction (future losses as a result of non-adaptation) remains limited, and there is an even larger gap for the costs of adaptation (EEA 2007, 2012). Preliminary estimates suggest that benefits often exceed costs. Taking advantage of opportunities related to urban renewal as well as designing multi-purpose solutions will often result in adaptation benefits exceeding the costs. More recent information shows that cost of inaction is significant. The global expected losses of the asset management industry as a result of climate change are valued at US$ 4.2 trillion (Economist Intelligence Unit 2015). An example of smart adaptation is provided by the City of Copenhagen. The cost of inaction for climate adaptation in Copenhagen has been valued at €4–€4.7 billion, and the climate adaptation cost at €1.3–€1.6 billion, resulting in future savings of €2.6–€3.2 billion (Leonardsen 2012).

The economic gain from materials savings alone is estimated at over a trillion US$ a year. A shift to innovative reusing, remanufacturing and recycling products could lead to significant job creation. For instance, 1000,000 jobs have been created by the recycling industry in the EU alone (EMF 2014). These figures may even rise when also wastewater utilities will be considered as 'profit centres', *i.e.* as sources of energy (Grant *et al.* 2012; Van Leeuwen and Bertram 2013) and nutrients (Van Leeuwen and Sjerps 2015b), as phosphate is on the EU list of critical raw materials (European Commission 2014).

The consequence of all these developments, the short-term framing of many politicians and the long-term existence ('generation time') of cities (**Table 1**), may be perceived as a recipe for disaster. Cities require an integrated long-term framing of their plans and actions (proactive transitions) as there will not be a second

Table 1. Generation times for some species (modified after Van Leeuwen and Vermeire 2007).

Species	Generation time
Bacteria	≈0.1 days
Green algae (Chlorella sp.)	≈1 day
Water fleas (Daphnia sp.)	≈10 days
Snails (Lymnaea sp.)	≈100 days
Rats	≈1 year
Politicians	≈5 years
Man	≈25 years
Cities	>100 years

chance to plan and build cities in a smart, sustainable, flexible and adaptive manner. Time is pressing, and the reality is increasingly becoming more a matter of now or never, of make or break.

3. Water Governance

3.1. What Is Water Governance?

To tackle the challenges of water in the city, it is necessary to take numerous aspects, interests and actors into account (Philip *et al.* 2011). These can be brought together under the heading of water governance. Hofstra (2013) considered a number of definitions. The Water Governance Centre (2012) and the OECD (2011a) have adopted the definition of the Global Water Partnership (GWP) on governance: 'the range of political, social, economic and administrative systems that are in place to develop and manage water resources, and the delivery of water services, at different levels of society and for different purposes'. According to the GWP, water governance covers the mechanisms, processes and institutions by which all stakeholders—government, the private sector, civil society, pressure groups—on the basis of their own competences, can contribute their ideals, express their priorities, exercise their rights, meet their obligations and negotiate their differences. Recently, the OECD (2015b) adopted the following definition of water governance: the range of political, institutional and administrative rules, practices and processes (formal and informal) through which decisions are taken and implemented, stakeholders can articulate their interests and have their concerns considered, and decision-makers are held accountable for water management.

Driessen *et al.* (2012) carried out an analysis of various governance models. The authors differentiated between central, decentral, public-private and interactive governance, as well as self-governance. Lange *et al.* (2013) elaborated this further in a multidimensional approach in which a distinction was made between political processes (politics), institutional structures (polity) and policy content (policy). The recently published UN guidelines on water governance (UNDP 2013) set out four dimensions: the economic, social, political and ecological dimensions, in which the UNDP makes no distinction between the political and administrative dimension but combines both aspects in the political dimension.

Governance is the work of people and is all about 'who does what?' According to Kuijpers *et al.* (2013), the term actually covers three essentially different aspects, *i.e.*:

1) Governing: holding responsibility for and directing the management of a water or other system;

2) Managing: ensuring adequate capacity and overseeing the operation, etc. of a managed water or other system;

3) Supervising: exercising influence and intervening in the water or other system for the purpose of its management.

3.2. Governance Gaps and Capacities

An OECD study on water governance in 17 OECD countries (OECD 2011a) revealed that obstacles can be found at several levels. The OECD listed seven of them (**Figure 4**). The biggest challenges, according to the OECD, are institutional fragmentation, ambiguous legislation, poor implementation of multi-layered governance, as well as matters such as limited capacity at local level, unclear allocation of roles and responsibilities, fragmented financial management and uncertain

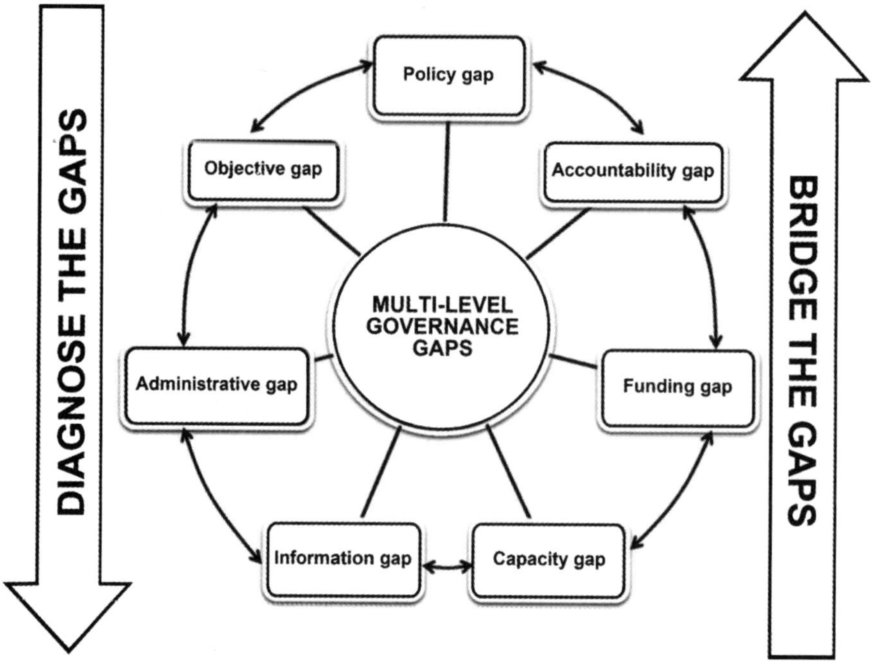

Figure 4. OECD multi-level governance framework (OECD 2015b).

allocation of resources. Often there are also no long-term strategic plans and insufficient resources to be able to measure performance. This leads to weak accountability and little transparency. All these challenges are often rooted in inadequately coordinated goals and insufficient steering of the interactions between stakeholders, the actors in the water cycle. In short, many plans sprouting in various directions, but, all in all, they do not add up to a clearly signposted route heading in a common sustainable direction. Recently, the OECD published their principles on water governance as well as a review of water governance in 48 cities (OECD 2015b, c). One of the conclusions is that building adequate governance capacities is a premise for sustainable futures of cities (OECD 2015c).

4. Transitions

By the year 2100, the total world population is estimated at 11 billion and about 80% will live in cities (UN 2015a). This raises questions about options we may have to make our cities more sustainable and resilient, particularly with regard to water. How can we successfully transform our cities with future generations (our children and grandchildren) in mind? Transitions are understood as multilevel, multiphase processes of structural change in societal systems; they realise themselves when the dominant structures in society (regimes) are put under pressure by external changes in society, as well as endogenous innovation (Loorbach 2010). Gleick (2003) talked about soft-path solutions. Three important considerations are raised by Loorbach (2010):

1) All societal actors exert influence and thus direct social change, being aware of the opportunities as well as the restrictions and limitations of directing;

2) Top-down planning and market dynamics only account for part of societal change; network dynamics and reflexive behaviour account for other parts;

3) Steering of societal change is a reflexive process of searching, learning and experimenting.

Examples of transition practices are provided by Loorbach and Rotmans (2010). Strategies, actors and resources are discussed by Farla *et al.* (2012) and Gupta *et al.* (2010), whereas Markard *et al.* (2012) provide a review on the conceptual framework of sustainability transitions. Many international organizations address these issues too (BAUM 2013; OECD 2011a, 2015a, b; European Commission 2011, 2012). Practical guidance on the governance of transitions is provided by UNDP (2013), OECD (2015b), and in training modules (Philip *et al.* 2011).

Frijns *et al.* (2013) discuss the future challenges in the Dutch water sector such as (1) unstable economy, (2) citizen centric, (3) changing demographics, (4) sustainability, (5) raw material shortages/prices, (6) NBIC convergence (the convergence and growing importance of nano, bio, information and cognitive technologies), (7) transsectoral innovation, (8) shifts in governance, (9) the city and (10) social networks. When this is scaled up to a global approach, there will be numerous added factors, including social and cultural differences in policy formulation and, especially, in the areas of implementation and enforcement. In practical terms, how to deal with corruption and how to communicate with people who are hungry, living under appalling conditions and are still illiterate too? With no pretence to be complete, seven points for successful transitions can be brought forward:

1) Develop a shared long-term vision.

2) Stakeholder participation: involve civil society, the commercial sector along with other stakeholders.

3) SMART transitions with a focus on co-benefits.

4) Not only technology development.

5) Make data accessible and applicable.

6) Carry out a thorough cost-benefit analysis and remove financial barriers.

7) Monitor implementation.

4.1. Develop a Shared Long-Term Vision

Developing a long-term vision together is an important prerequisite to bring about change. This can be summarized as participative scenario planning and back-casting. This approach aims to envision a coherent future picture for the long term together with the actors/stakeholders involved and from that, by working backwards (backcasting) to arrive at a plan of action for that period (*i.e.* for the short term). This process begins by involving the most relevant actors (open and inclusive development), and doing so as early as possible in the process (Van Leeuwen and Vermeire 2007). There are many actors in IWRM, as described in the excellent training modules of SWITCH (Philip *et al.* 2011), the guide for water governance (UNDP 2013) and the OECD (2011a; 2015b).

4.2. Stakeholder Participation: Involve Civil Society, the Commercial Sector along with Other Stakeholders

Governance is a concept that has emerged in political, environmental and sustainability studies in response to a growing awareness that the authorities are no longer the only relevant actors when it comes to managing society's public affairs (Lange *et al.* 2013). This is reflected in the European Green City Index (2009) in Europe that was commissioned by Siemens. This index shows how sustainable European cities are. This study of 30 European cities showed a surprisingly strong correlation between the green city index and the voluntary participation index. In the notes to this report, it is also concluded that achieving the CO_2 reduction targets in London had more to do with the involvement of the people and businesses than the authorities. It provides a good example of the opportunities available for achieving ambitious goals in IWRM. The process is supported by a common interest and a 'broadly accepted' purpose among the parties involved (Kuijpers *et al.* 2013).

4.3. SMART Transitions with a Focus on Co-Benefits

Today, the consequences of short-term governance are particularly clear in the fragmentation of urban development and transitions. Far more coherence is needed between urban, regional and national policies (UN-Habitat 2013; OECD 2015a). According to the OECD (2011a; 2015a, b), water governance often shows many gaps (**Figure 4**). In some countries, even at central level, sometimes ten or more ministries are actively concerned with water policy. This is worrying when you realize that 21 of the 33 cities which in 2015 will have more than 8 million inhabitants are along the coast (UN 2013b).

Ideally, cities should develop a cohesive set of long-term objectives that should be SMART: Specific (target a specific area for improvement), Measurable (quantify or at least suggest an indicator of progress), Assignable (specify who will do it), Realistic (state what results can realistically be achieved, given available resources), Time-related (specify when the result(s) can be achieved). Very often clear objectives are not set and—as a result—many cities are neither smart nor future proof. The cost of inaction (or ad hoc sectoral action) is generally very high (Economist Intelligence Unit 2015; UNEP 2013; Leonardsen 2012).

Governance of cities is never simple (**Figure 4**). It is a matter of cooperation in complexity. Transparency, accountability and participation are the criteria for good governance. In the development of a long-term vision for a city with different stakeholders, there will be differences of outlook, interests, short-term and long-term perspectives, 'generation times', planning horizons, investments and returns. The transitions in infrastructure, in particular, need to be flexible and adaptive, because, as indicated above, the investments are huge and, in principle, must create value (Kuijpers *et al.* 2013). Colliding short- and long-term interests will threaten the success of the process. Long-term goals are often not served by short-term political thinking as cities have long generation times (**Table 1**).

Over the past 20 years, a different view of the role of government has evolved, both in government itself and in society. To an increasing degree, government sees for itself only a legislating and facilitating role. Under this new political and social philosophy, government is operating more at arm's length and new initiatives are increasingly being developed by society. It is said, however, that steering is necessary, but it no longer needs to be government which arranges and decides on everything (Lange *et al.* 2013; Kuijpers *et al.* 2013). All these actors, government included, need sufficient expertise at their disposal. Local stakeholder needs 'knowledge receptors' too in order to properly manage or co-manage these complex governance processes. Both the City of Amsterdam (Van Leeuwen and Sjerps 2015a, b) and Melbourne (Van Leeuwen 2015) are examples of adequate water governance at local level. The secret of Melbourne's success was the transparent governance structure that has been set up in a reaction to the 'Millennium Drought' and success has come from many organizations working together to a common goal (Van Leeuwen 2015). Amsterdam has a long tradition in water management, and its current focus is on the integration of water, energy and material flows (Van der Hoek *et al.* 2015).

In the development or reconstruction of cities, optimal use should be made by exploring options for win-win's or co-benefits for the different issues that need to be addressed in cities. For instance, road reconstruction can be combined with the renewal or installing of water distribution networks, sewer systems, and the creation of blue and green space. This would save a lot of time, money and nuisance for citizens (**Figure 5**; **Table 2**). **Figure 5** represents a simplified city in which nine urban sectoral agendas are shown: ICT (information and communications technology), energy and transport (European Commission 2013), solid waste, green and blue space, water supply, wastewater, climate adaptation, houses and factories. Governance is considered to be a horizontal issue linked with all other agendas in a city. At a recent public consultation, the European Commission has decided for an upgraded and more holistic Smart Cities and Communities policy to

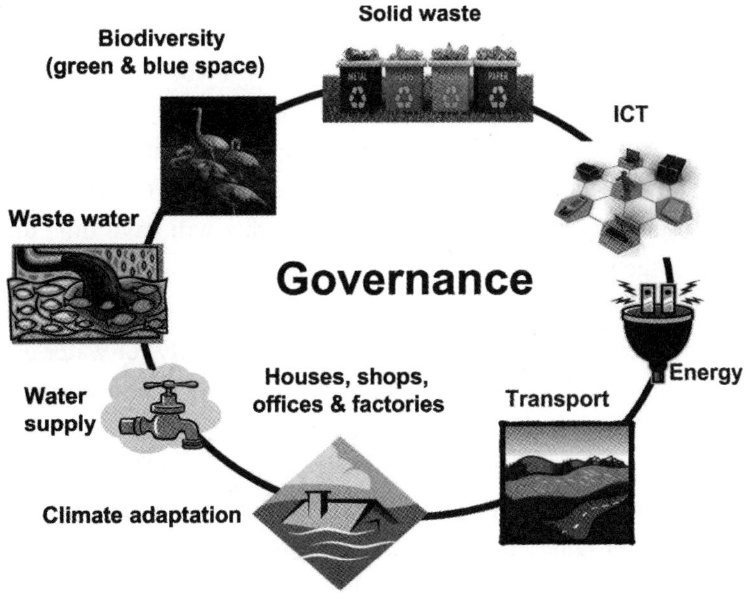

Figure 5. Simplification of a city. The red items ICT, transport and energy are part of the EU Smart City Policy (European Commission 2013). Governance is considered to be a horizontal activity. Recently, water and waste have been included in the EU policy on smart cities (European Commission 2015c).

Table 2. Illustration of the relevance of co-benefits of integration in city planning as part of a cohesive long-term strategy for cities.

Policy	Number of issues (n)	Number of P. I.[a]	Issues addressed	Interactions addressed	Missed P. I.	Missed P. I. (%)
Smart cities[b]	9	36	3	3	33	92
Smart cities[c]	9	36	6	15	21	58
SMARTER cities[d]	9	36	9	36	0	0

The total number (n) of issues in cities is nine (**Figure 5**). Governance is considered to be a horizontal aspect interacting with all other issues in cities. [a]P. I. is the total number of potential interactions. The number of potential interactions is calculated as follows: P. I. = 1/2n 9 (n − 1), [b]Issues addressed are ICT, transport and energy (European Commission 2013), [c]Issues addressed are ICT, transport, energy, waste (taken as solid waste and wastewater) and water (European Commission 2015c), [d]Example of a cohesive integral Urban Agenda addressing all nine topics in a city.

better integrate and connect energy, transport, water, waste and ICT (European Commission 2015c). From **Table 2**, it can be demonstrated that a smart city policy addressing only ICT, transport and energy can be considered as a maximization of missed opportunities in cities as more than 90% of the potential interactions or win-wins between these sectoral agendas are not explored. The recent decision to include also waste and water is a step forward, but still many opportunities (58%; **Table 2**) are not explored, including climate adaptation in cities, which is another omission. The obvious conclusion is that smarter cities need to develop a cohesive long-term plan and integrate/combine agendas as this will save time and money and better serves the needs of their taxpayers.

Often, there are governance gaps and barriers, not only for water governance (OECD 2015b, c), but also for all other urban adaptation and mitigation plans (Reckien *et al.* 2015), making smart long-term transitions, easier said than done. Nevertheless, inspiring examples are provided by the city of Melbourne on water and climate adaptation (Van Leeuwen 2015), by the city of Hamburg on energy efficiency and the introduction of the water cycle concept in city planning (Van Leeuwen and Bertram 2013), and by the city of Amsterdam on the integration of water, energy and material flows (Van der Hoek *et al.* 2015; Van Leeuwen and Sjerps 2015).

4.4. Not Only Technology Development

The recent attention devoted to the complex issue of water governance follows a general shift in the focus on 'technical' infrastructure-driven solutions to demand-driven solutions which underline the role of institutions, along with economic and social processes (OECD 2011a, 2015b; Van Someren and Van Someren-Wang 2013). According to European Commissioner Hahn (BAUM 2013) 'technology is important to implement an intelligent city concept, to create new business opportunities, to attract investments and to generate employment. But technology alone would not bring about any wonders. Good governance and the active involvement of citizens in the development of new organisation models for a new generation of services and a greener and healthier lifestyle are also important'. At the global level, there seems to be a greater need for smart implementation of state-of-the-art technologies, *i.e.* communities of practice, rather than in the development of new technologies for two reasons: (1) developing countries account for 93% of urbanization globally, 40% of which is the expansion of slums (UN 2015b), and (2) major improvements in urban water cycle services can be obtained by cleverly combining best practices in cities as clearly demonstrated in a study of 11 municipalities and regions (Van Leeuwen 2013). Therefore, it is important to speed up implementation by investing in smart demonstration projects on water, waste and climate mitigation and adaptation with affordable and adaptive state-of-the-art technologies (CCS 2008). Good water governance is critical to manage water-related risks at an acceptable cost and in a reasonable time frame so that the next generation does not inherit liabilities and costs from either inaction or poor decisions taken today (OECD 2015c). This is the real challenge for the upcoming HABITAT III conference (UN-HABITAT 2015).

4.5. Make Data Accessible and Applicable

Utilities in general obtain a lot of information on their water and wastewater services. One of the recommendations of the OECD (2011a) is to create, update and harmonize information systems and databases in order to share water policy at river basin, national and international levels. Most of the data for the baseline assessments (City Blueprints) of cities have been collected and provided by the cities or their utilities (Koop and Van Leeuwen 2015a, b). The collection of data is time-consuming, both for the utility and for the scientists who gather these data in order to provide baseline assessments of IWRM. Some of this knowledge is collated and held by water management actors including the utility operators and the different levels of environmental authorities; all of which may have their own distinct reference points and definitions (EEA 2014).

Benchmarking improves performance by identifying and applying best demonstrated practices to operations and sales. The objectives of benchmarking are (1) to determine what and where improvements are called for, (2) to analyse how other organizations achieve their high-performance levels and (3) to use this information to improve performance. Benchmarking networks collect data from their members. The European Environment Agency (EEA 2014) observed that the data policies for benchmarking networks are defined by their members and that results are often presented in an aggregated or anonymous form, preventing individual plants/utilities to be identified directly. Often, the underlying data are considered confidential (EEA 2014). In order to meet the enormous water challenges as described above, this policy needs to change. Transparency and accountability are crucial for utilities, and certainly utilities paid by the taxpayers. These asymmetries of information (quantity, quality, type, scale and confidentiality) between different stakeholders are one of the key coordination gaps in (water) policy (**Figure 4**). Secondly, there is the problem of scale. Given that cities are becoming increasingly important, then it is necessary to have harmonized and up-to-date data at city level (urban hydro-informatics). Applicable knowledge that is understandable for all stakeholders is necessary to enhance public engagement and well-informed decisions.

4.6. Carry out a Thorough Cost-Benefit Analysis and Remove Financial Barriers

To start at the end: scarce financial resources need not necessarily be an obstacle. On the contrary, limited resources often inspire creativity and foster cooperation between public and private investors, as well as the involvement of civil society. Civil society underpins urban development and will strive for cost-effective operations in cities with a maximum of cost-saving options (**Table 2**). It is primarily all about three things: communication, involvement and ownership. The decisive factor is that through transparency, inspiring confidence and specifying the tangible benefits, private individuals will get behind a common ideal. This will enable civil society to strongly identify with the city and urban society. Ordinary people will then feel involved as individuals and support developments with their time and money (BAUM 2013). Groups of people, private institutions, societies, clubs, religious communities, charitable organizations, pressure groups, *i.e.* non-governmental organizations (NGOs), should not be overlooked (Philip *et al.* 2011). Financial limitations are therefore not always an obstacle but often provide the impetus for creative solutions because it is then necessary to look for ways to link up with other interests and solutions (**Table 2**). Further to which, a thorough cost-benefit analysis of various promising solutions is required. Often it turns out that these solutions are also more affordable when considered over the longer term. Institutional investors—pension funds, insurance companies and mutual funds—are able to invest in high yield, smart and sustainable infrastructures (OECD 2011b). It is therefore mainly a matter of making transparent long-term plans which will create value.

4.7. Monitor Implementation

It was once said by the American delegation during the negotiations on the

European REACH regulation that legislation is only as good as its implementation and enforcement. That also applies to city planning. Furthermore, continuous monitoring is necessary for learning, maintaining flexibility and securing continuous improvement.

5. City Blueprints

5.1. Results

The City Blueprint provides municipalities and regions with a practical and broad framework to define steps towards realizing a more sustainable and resilient water cycle in collaboration with key stakeholders. This assessment methodology has been applied to 45 municipalities and regions, mainly in Western Europe (Van Leeuwen *et al.* 2015b). Detailed reviews are available for Malmö (Mottaghi *et al.* 2015), Rotterdam (Van Leeuwen *et al.* 2012), Hamburg (Van Leeuwen and Bertram 2013), Amsterdam (Van Leeuwen and Sjerps 2015b) and Istanbul (Van Leeuwen and Sjerps 2015a). Detailed reviews of cities outside Europe are available for Dar Es Salaam (Van Leeuwen and Chandy 2013), Ho Chi Minh City (Van Leeuwen *et al.* 2015a) and Melbourne (Van Leeuwen 2015).

Recently, the City Blueprint approach was critically reviewed to better separate cities' IWRM performance from general trends and pressures that can hardly be influenced directly (Koop and Van Leeuwen 2015a). The Trends and Pressures Framework (TPF) comprises indicators for social, environmental and financial classes and these indicators have been scaled from 0 to 4 points, where a higher score represents a higher urban pressure or concern. The following ordinal classes, expressed as 'degree of concern', have been used: 0–0.5 points (no concern), 0.5–1.5 (little concern), 1.5–2.5 (medium concern), 2.5–3.5 (concern) and 3.5–4 (great concern). In this way, a TPF is provided that depicts the most relevant topics that either hamper sustainable IWRM or, on the contrary, pose opportunity windows (Koop and Van Leeuwen 2015a). The results for 45 municipalities and regions are provided in **Figure 6**.

The performance-oriented set of indicators of the City Blueprint Framework (CBF) provides a snapshot of the current IWRM performance. The Blue City Index® or BCI is the geometric mean of 25 indicators which varies from 0 to 10

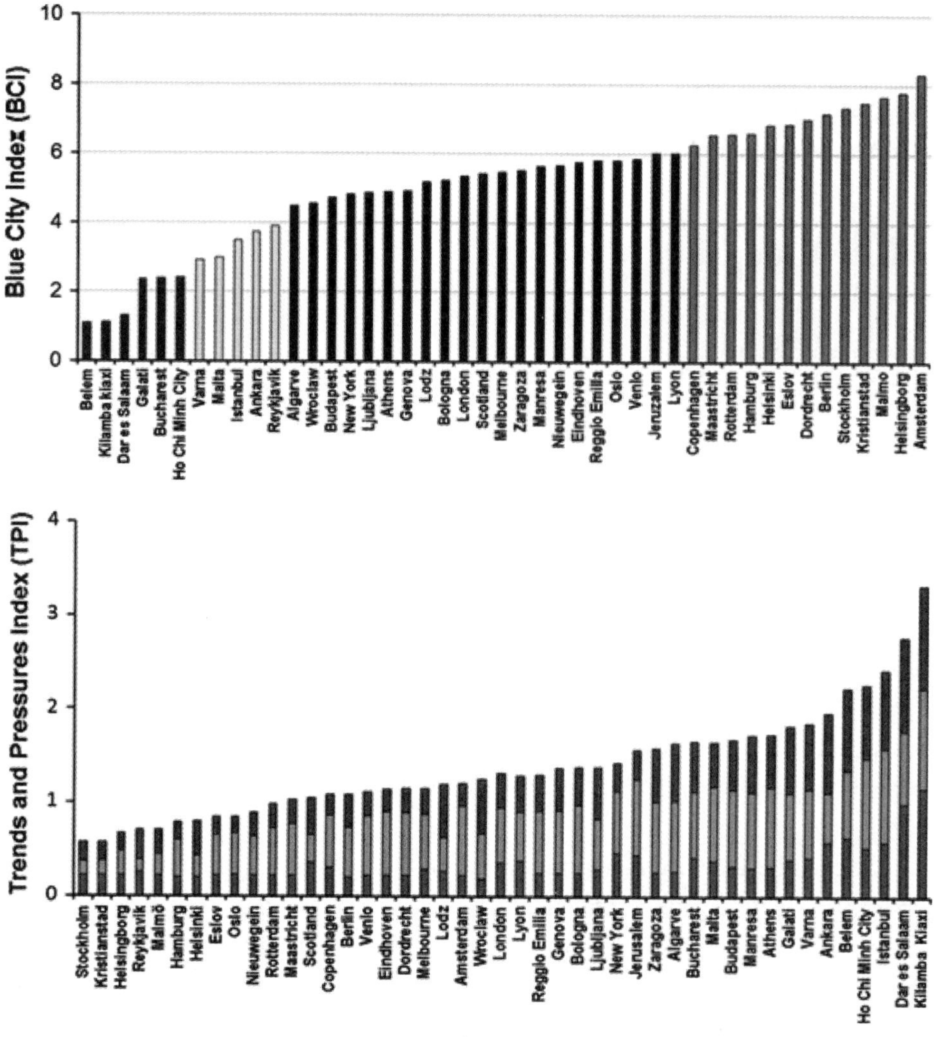

Figure 6. Results of the City Blueprint analysis of 45 municipalities and regions in 27 different countries. Bottom TPI (arithmetic average of 12 indicators), where green, red and blue represent the share of the environmental, financial and social indicators, respectively, to the overall TPI. Top BCI (geometric mean of 25 indicators) of the City Blueprint according to Koop and Van Leeuwen (2015a, b, c).

(Koop and Van Leeuwen 2015a). The BCIs for 45 municipalities and regions are also provided in **Figure 6**. The CBF consists of 25 indicators divided over the following seven categories: water quality, solid waste treatment, basic water services, wastewater treatment, infrastructure, climate robustness and governance. The indicator scores of each city are shown in a spider diagram (Koop and Van Leeuwen 2015a). The methodology is summarized in a simple brochure (Van Leeuwen and

Elelman 2015), two publications (Koop and van Leeuwen 2015a, b) and in a detailed report (Koop and Van Leeuwen 2015c).

The indicator scores may facilitate sharing of knowledge, experiences and best practices between cities (Van Leeuwen 2013). The potential performance improvement (PPI) for each indicator is the maximum indicator score minus the actual score. The PPI may guide cities in their transitions towards more sustainable IWRM and innovative urban planning, leapfrogging arrangements that have locked-in many cities (Brown *et al.* 2009; OECD 2015a).

5.2. Categorization of Different Levels of Sustainability in Cities

Although our City Blueprint research is focussed on the performance of IWRM in European cities, we have tried to include also other geographical regions. The selection of cities is therefore not random at all, but regionally biased towards Western Europe. With these limitations in mind, the challenges on water, waste and climate change can be discussed globally by clustering cities into distinct categories of sustainability and by providing additional data and information for various global regions. The categorization of cities is based on hierarchical clustering with the squared Euclidean distances for all 25 indicators (Koop and Van Leeuwen 2015b) and provided in **Table 3**.

5.3. Regional Challenges

The geographical distribution of municipalities and regions and their categorization is shown in **Figure 7**. Basic information on regions and cities is provided in **Table 4**. As stated before, the selection of cities is not random at all, but regionally biased towards Western Europe. Therefore, further research of cities in other global regions is needed. With this limitation in mind, the following general observations can be made.

- The challenges of water, waste and climate change development vary from one region to another.

Table 3. Categorization of different levels of sustainable IWRM in cities (Koop and Van Leeuwen 2015b).

IWRM category	Description
Cities lacking basic water services (BCI 0–2)	Access to potable drinking water of sufficient quality and access to sanitation facilities are insufficient. Typically, water pollution is high due to a lack of wastewater treatment (WWT). Solid waste production is relatively low but is only partially collected and, if collected, almost exclusively put in landfills. Water consumption is low, but water system leakages are high due to serious infrastructure investment deficits. Basic water services cannot be expanded or improved due to rapid urbanization. Improvements are hindered due to governance capacity and funding gaps
Wasteful cities (BCI 2–4)	Basic water services are largely met but flood risk can be high and WWT is poorly covered. Often, only primary and a small portion of secondary WWT is applied, leading to large-scale pollution. Water consumption and infrastructure leakages are high due to the lack of environmental awareness and infrastructure maintenance. Solid waste production is high, and waste is almost completely dumped in landfills. Governance is reactive, and community involvement is low
Water-efficient cities (BCI 4–6)	Cities implementing centralized, well-known, technological solutions to increase water efficiency and to control pollution. Secondary WWT coverage is high, and the share of tertiary WWT is rising. Water-efficient technologies are partially applied; infrastructure leakages are substantially reduced, but water consumption is still high. Energy recovery from WWT is relatively high, while nutrient recovery is limited. Both solid waste recycling and energy recovery are partially applied. These cities are often vulnerable to climate change, e.g. urban heat islands and drainage flooding, due to poor adaptation strategies, limited storm water separation and low green surface ratios. Governance and community involvement has improved
Resource-efficient and adaptive cities (BCI 6–8)	WWT techniques to recover energy and nutrients are often applied. Solid waste recycling and energy recovery are largely covered, whereas solid waste production has not yet been reduced. Water-efficient techniques are widely applied, and water consumption has been reduced. Climate adaptation in urban planning is applied, e.g. incorporation of green infrastructures and storm water separation. Integrative, centralized and decentralized as well as long-term planning, community involvement and sustainability initiatives are established to cope with limited resources and climate change
Water-wise cities (BCI 8–10)	There is no BCI score that is within this category so far. These cities apply full resource and energy recovery in their WWT and solid waste treatment, fully integrate water into urban planning, have multi-functional and adaptive infrastructures, and local communities promote sustainable integrated decision-making and behaviour. Cities are largely water self-sufficient, attractive, innovative and circular by applying multiple (de)centralized solutions

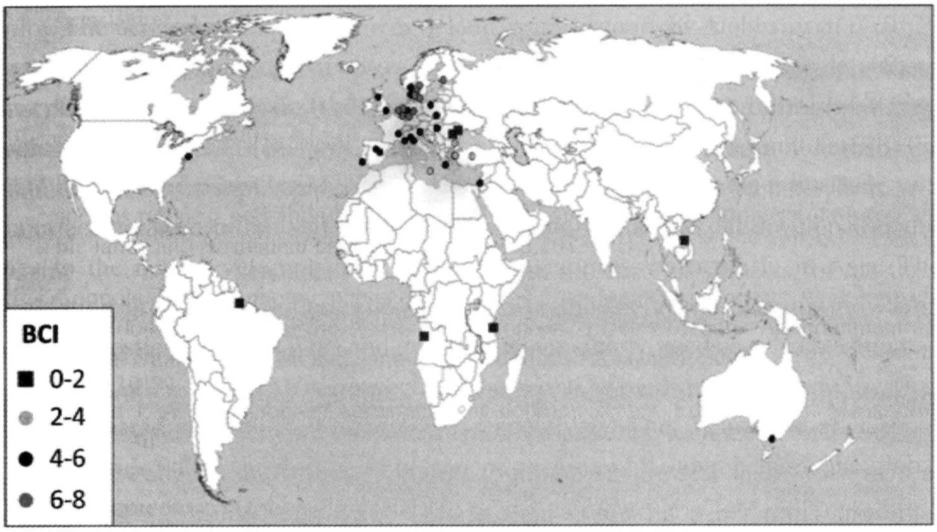

Figure 7. Municipalities and regions assessed with the City Blueprint. Red, orange, black and blue represent municipalities and regions with a geometric BCI between 0–2 (cities lacking basic water services), 2–4 (wasteful cities), 4–6 (water-efficient cities), and 6–8 (resource-efficient and adaptive cities), respectively (Koop and Van Leeuwen 2015b). Most cities are from north-western Europe. Cities outside Europe are: Ankara and Istanbul (Turkey), Jerusalem (Israel), Kilamba Kiaxi (Angola), Dar es Salaam (Tanzania), Ho Chi Minh City (Vietnam), Bele´m (Brazil), Melbourne (Australia) and New York City (USA).

- South-east Asia. Rapid population growth and rapid socio-economic changes place increasing pressure on natural resources (Dobbs *et al.* 2012; Green City Index 2015). Excessive water abstraction, land subsidence, decline in groundwater level, saline water intrusion and pollution can be observed in Ho Chi Minh City and many other cities in south-east Asia (Van Leeuwen *et al.* 2015a). This is in line with observations by UNESCO (2015a, b). Solid waste collection and recycling (Jambeck *et al.* 2015) as well as water infrastructure upgrading are major challenges as well (Van Leeuwen *et al.* 2015a).

- Africa. By 2030, the urban population in Africa and Asia will double (UNESCO 2015a). Dar es Salaam in Tanzania is among the ten fastest growing cities in the world (Green City Index 2015). Little more than half of the population in Dar es Salaam has access to some form of sanitation, but the wastewater generated by 15% of the city residents who are connected to the sewer system is discharged into the sea untreated (Van Leeuwen and Chandy 2013). There is also no regular waste collection and many residents simply

Table 4. Basic socio-economic context and IWRM performance in different world regions.

	North-Western Europe [Amsterdam]	Eastern Europe and Turkey [Istanbul]	Africa [Dar es Salaam]	Australia [Melbourne]	South-east Asia [Ho Chi Minh City]	Latin America [Belem]	North America [New York]
Social economic context							
Urbanization rate (%)[a]	0.9 [1.0]	2.0 [2.0]	3.7 [5.4]	1.31 [1.6]	2.9 [2.9]	1.4 [1.2]	1.3 [1.0]
GDP (US$ 2013/capita)[b]	58,334 [47,651]	10,777 [10,745]	2148 [690]	52,311 [64,157]	9977 [1896]	9058 [10,958]	38,645 [52,839]
Poverty rate (% pop. < 2US$)[c]	0.4 [0.4]	1.0 [2.6]	34.4 [33.6]	1.4 [1.4]	11.9 [12.5]	3.4 [6.8]	2.2 [1.7]
Water							
Access to drinking water (%)[d]	99.9 [100.0]	99.3 [100.0]	89.1 [60.0]	100.0 [100.0]	95.5 [84.0]	97.3 [70.0]	98.7 [100.0]
Access to sanitation (% of urban pop.)[e]	98.4 [100.0]	96.0 [95.0]	47.7 [56.0]	100.0 [100.0]	88.6 [12.0]	88.9 [7.0]	96.0 [100.0]
Secondary WWT (%)[f]	85.4 [99.0]	47.8[j] [35.0]	– [5.0]	88.0 [100.0]	– [6.0]	– [7.7]	68.3 [72.0]
Nutrient recovery (%)	45.2[k] [100.0]	5.4[l] [11.8]	– [0.0]	– [0.0]	– [0.0]	– [0.0]	– [72.0]
Energy recovery (%)	56.2[k] [100.0]	39.7[l] [2.1]	– [0.0]	– [90.0]	– [0.0]	– [0.0]	– [72.0]
Leakages (%)	12.9[k] [5.4]	30.7[l] [25.0]	– [30.0]	– [11.0]	– [23.0]	– [48.0]	– 8.0]
Storm water separation (%)	55.7[k] [82.8]	47.1[l] [70.0]	– [20.0]	– [100.0]	– [5.0]	– [30.0]	– [40.0]
Solid waste							
Solid waste collected (kg/cap/year)[g]	546.6 [600.0]	360.4 [419.0]	195.3 [365.0]	640.0 [640.0]	281.3 [296.0]	331.7 [383.2]	617.9 [730.0]
Solid waste recycled (%)[g]	29.6 [99]	11.4 [1.0]	3.0 [5.0]	30.3 [42.0]	10.2 [5.0]	5.35 [0.0]	18.0 [46.0]
Climate robustness							
CO_2 emissions (metric tons per capita)[h]	8.51 [–]	5.9 [–]	0.8 [–]	11.8 [–]	2.8 [–]	2.9 [–]	11.7 [–]
Climate adaptation (–)	7.5[k] [10]	4.5[l] [4]	– [2]	– [8]	– [7]	– [0]	– [10]
Governance							
Management and action plans (–)	7.5[k] [7]	5.6[l] [5]	– [2]	– [10]	– [7]	– [0]	– [10]
Public participation (%)[i]	41.2 [44.0]	23.3 [8.0]	– [5.0]	– [44.0]	– [5.0]	– [5.0]	– [40.0]

Provided are averages for countries in these regions as well as one city, assessed with the City Blueprint approach (Koop and Van Leeuwen 2015c)

[a] CIA (2014); [b] IMF (2013); [c] World Bank (2015a); [d] World Bank (2015b); [e] World Bank (2015c); [f] OECD (2013); [g] Waste Atlas (2015); [h] World Bank (2015d); [i] EFILWC (2012); [j] Secondary WWT of Ukraine, Moldavia, Bosnia, Montenegro and Macedonia are missing and are not included in the average of Eastern Europe; [k] Average of all North-Western European cities ($n = 21$) assessed by the City Blueprint approach; [l] Average of all Eastern European cities ($n = 9$) assessed by the City Blueprint approach

burn their rubbish (Koop and Van Leeuwen 2015c). Based on other cities assessed in Africa (Green City Index 2015), the challenges of Dar es Salaam are no exception. The security of water, food and energy are major challenges, and sustainable development is perhaps more important for Africa than other regions of the world (UNESCO 2015a).

- Australia. Melbourne is the only city in this world region that has been assessed with the City Blueprint approach (Van Leeuwen 2015). The challenges of Melbourne under a changing and uncertain climate became apparent during the 'Millennium drought', a decade long period of extreme dry conditions across southern Australia throughout the 2000s. Melbourne scores highly in areas such as water efficiency, wastewater efficiency, energy recovery, and climate change commitments related to heat and water scarcity. Nearly 30% of the houses in Melbourne have installed rainwater tanks and plans to increase the use of storm water have recently been published. Energy efficiency of buildings, nutrient recovery (especially phosphate) from wastewater and sewage sludge recycling are topics for improvement. The same holds for the production and proper handling of solid waste. Moreover, the emissions of GHGs in Australia are relatively high (**Table 4**).

- Latin America. Bele'm is the only city in this world region that has been assessed with the City Blueprint. Flooding is a very serious concern in Bele'm. Urban environmental concerns such as traffic congestion, land use policies, waste disposal and air quality are immediate concerns to the majority of Latin America's residents, simply because 81% of the population already lives in cities (Green City Index 2015). Access to sanitation and drinking water are challenges in several cities in Latin America (UNESCO 2015a, b). According to UNESCO (2015a), a major priority for Latin America is to build the formal institutional capacity to manage water resources and bring sustainable integration of water resources management and use into socio-economic development and poverty reduction. Another priority is to ensure the full realization of the human right to water and sanitation in the context of the post-2015 development agenda. Provided that Bele'm is a representative sample of a city in Latin America, these observations are fully supported by the City Blue-print analysis, as the BCI of

Bele'm is 1.1. In other words, the challenges of Bele'm expressed as PPI are nearly nine points.

- North America. New York is the only city in this world region that has been assessed with the City Blueprint approach. Parts of the North-American continent suffer from droughts, whereas in 2012, New York suffered from hurricane Sandy. Sandy's impacts included the flooding of the New York City subway system, many suburban communities and many road tunnels entering Manhattan. Sandy damaged 200,000 homes and was blamed for 117 US deaths. The total damage in New York was estimated at more than $19 billion (Toro 2013). The USA emits double the average amount of GHGs, while their BCI is about average (World Bank 2015d; Koop and Van Leeuwen 2015c). New York is vulnerable to extreme weather because the urban soil is largely sealed with impermeable concrete, asphalt and stone (NYC 2010). Rainwater can hardly infiltrate and forms large amounts of runoff which may result in urban drainage flooding and amplifies the impact of extreme weather which happened in 2012. Furthermore, New York produces a lot of solid waste and can improve on solid waste recycling, sewage sludge recycling, sewer maintenance and green space (Koop and Van Leeuwen 2015c). UNESCO (2015a) concludes that increasing resource use efficiency, reducing waste and pollution, influencing consumption patterns and choosing appropriate technologies are the main challenges facing both Europe and North America.

- Europe. The only continent for which an adequate number of municipalities and regions have been assessed using the City Blueprint shows a high variation in IWRM performance (**Figure 6**, **Figure 7**; **Table 4**). The differences between Western and Eastern Europe is striking, part of which can be explained by non-existing, badly maintained or outdated water infrastructure and technology in Eastern Europe. The overall conclusion of UNESCO (2015a) as quoted for North America also holds for Europe. Upgrading and renewing existing infrastructures remain a challenge and are illustrated by the high leakage rates (>40%) in some European cities and fully support the conclusions of the OECD (2015a).

- Until now, none of the cities can be categorized as water-wise cities (Koop and Van Leeuwen 2015a, b).

Our research shows that cities with a high BCI are those cities with high ambitions to improve IWRM, with an active civil society (involvement in voluntary work), in countries with greater prosperity (high GDP) and high governmental effectiveness (Koop and Van Leeuwen 2015b). Similar conclusions have been provided by Reckien *et al.* (2015) in an empirical analysis of urban adaptation and mitigation plans in European cities. Our work is mainly based on an analysis of European cities. There is a great need to assess more cities, especially in other world regions, as a starting point for sustainability transitions and to monitor their progress on the implementation of the Sustainable Development Goals for better urban futures (UN-Habitat 2015).

6. City-to-City Learning

Our work on City Blueprints shows that results can be used for a variety of purposes to:

- Aid in the evaluation and compare outcomes with other cities;
- Translate knowledge and educate;
- Raise/improve awareness (particularly in communicating with the public);
- Enable informed decision-making, *i.e.* stimulate proactive transitions;
- Refine parts of the assessment, with tailor-made in-depth studies and advanced models, if necessary;
- Monitor progress;
- Stimulate the exchange of best practices (Koop *et al.* 2015).

An important result from our work is that the wide variation in the way cities deal with their water, wastewater, solid waste and climate adaptation offers key insights for improving their resilience and sustainability, provided that cities share their best practices (Van Leeuwen 2013; Frijns *et al.* 2013). Theoretically, if cities would share their best practices, the BCI can reach a maximum value of 10 (Van Leeuwen 2013). It also shows that cities that currently perform well can still improve. Of course, this is ultimately the responsibility of the cities themselves. These challenges are too often not taken up, because people are waiting for new technological breakthroughs and fail to make use of existing knowledge and technologies. Therefore, we have three recommendations:

1) Cities require a long-term framing of their sectoral challenges into a proactive and coherent Urban Agenda to maximize the co-benefits and to minimize their cost.

2) Cities are encouraged to participate in learning alliances to actively share knowledge and experiences on implementation of state-of-the-art technologies (city-to-city learning). This is the most efficient way to improve IWRM (Van Leeuwen 2013; Koop and Van Leeuwen 2015b). Recently, a compendium of best practices has been completed that can help cities to choose among options to improve their performance on water, waste and climate adaptation (Koop *et al.* 2015).

3) Given the megatrends and water challenges in cities, existing technologies and innovations should be better embedded in urban planning. This is mainly a governance challenge (OECD 2015a). As developing countries account for 93% of urbanization globally, 40% of which is the expansion of slums (UNESCO 2015a), new affordable technologies need to be developed. These new and efficient technologies can gradually be introduced in the transition process allowing these cities to leapfrog towards waterwise cities.

7. Concluding Remarks

It has been attempted to shed light on growth and the limits to growth, with particular emphasis on water. Freshwater scarcity is a major challenge (FAO 2011b; UNEP 2012, 2013; World Economic Forum 2014; UNESCO 2015a). The UN (2012) estimates that in 2025 about 2 billion people will have an absolute water shortage and that two-thirds of the world population will be affected by water scarcity. Estimates for 2030 assume 40% more demand for water than is actually available (2030 Water Resources Group 2009). It means that the window we have for solutions is narrow and rapidly closing.

In the Netherlands, excellent drinking water is readily available by turning on a tap and safety is provided by the Delta Programme, while the history of that too lies in the flood disaster of 1953. Water safety and water security are not a matter of course. Actually, there is not a water crisis but a water governance crisis which now and in the very near future will become manifest in cities (OECD 2011a; Engel *et al.* 2011; European Commission 2011, 2015b). The solutions must also come from cities. Cities, as global change makers, must make the difference. And they can too, because there are already many good initiatives (C40Cities 2015; Philip *et al.* 2011; World Future Council 2014).

According to the European Commission (2013), smart cities are cities that focus on ICT, energy and transport. This definition was recently broadened to include water and waste (European Commission 2015a). Unfortunately, the proposed policy is still not cohesive, but fragmented and will lead to many missed opportunities for cities that are lost in sectoral agenda's and mists of techno-optimism. With the urgency of the water governance crisis, it is time that we cannot afford to lose. The European Commission can take the lead in the development of a practical coherent long-term European Urban Agenda, e.g. an EUA-2050, with cities and based on the needs of cities (European Commission 2015a, c). Such an initiative may also lead to improved visibility and a better image of Europe for the European citizens, which is a political priority for Europe. An Urban Agenda is even more needed in the rest of the world, where the challenges of water, waste and climate change are much greater than in Europe (**Figure 7**; **Table 4**). There is a need to move towards smarter cities:

- Smarter cities are cities with a coherent long-term social, economic and ecological agenda.

- Smarter cities are water-wise cities that integrate their sectoral agendas on water, wastewater, energy, solid waste, transport, ICT, climate adaptation and nature into a forward-looking, coherent Urban Agenda to maximize co-benefits and to minimize the cost.

- Smarter cities implement a circular economy (EMF 2014, 2015a; European Commission 2015b), focus on social innovation (Science Communication Unit 2014) and, last but not least, greatly improve on governance (OECD 2011a, 2015a, b).

Inaction can be overcome by setting up learning alliances of cities. Globally, we need regional platforms to exchange challenges, policies and best practices between cities. International organizations (e.g. OECD, UN, WHO, FAO, and the European Commission), the scientific community, the private sector, utilities (e.g. transport, water, waste, energy and telecom utilities), the civil society, city planners, architects, coordination providers, and last but not least, all the mayors in the world, are in a remarkably privileged position to contribute to the solutions of these urgent challenges in our cities.

Water utilities have much expertise and an extensive water consumers network. There are many opportunities for the water sector as a whole and the drinking water sector in particular, but under a number of conditions which can be summarized as the three Rs: 'Reframe, Refocus, Radically'.

1) Reframe. The Netherlands' drinking water sector has achieved a great deal but is faced by challenges such as salinization and groundwater depletion. There are also promising opportunities for nutrient recovery and energy conservation and production (More'e *et al.* 2013; Frijns *et al.* 2012). Nevertheless, water challenges require a broader framing as water is more than just drinking water (Van Oel *et al.* 2009; Van Someren and Van Someren-Wang 2013).

2) Refocus. In view of the declining level of government involvement, there will be major opportunities for initiatives launched by civil society and the private sector. Participative scenario development and the implementation of sustainability processes in the city—a highly complex environment—make it necessary that the focus be placed primarily on governance. The extensive expertise of the technology and drinking water sectors will be vital for this. But success will not be achieved by looking to technology alone (European Commission 2011; OECD 2011a, 2015a, b; BAUM 2013).

3) Radically. It has been attempted here to give an impression of the speed at which global change is taking place, both economically and ecologically. The challenges are high: urbanization at a rate of 190,000 people per day, the shift in the labour market (e.g. the exodus of businesses and employment from Europe), and the safety of cities in relation to climate change and water security (World Economic Forum 2014; UNESCO 2015a). The same holds for the challenges of irrigation, *i.e.* food security (UNEP 2007, 2012; FAO 2011b). This together with the high costs for water infrastructure and its maintenance make water a high priority, where procrastination, *i.e.* the avoidance of doing tasks which need to be accomplished, will not do (UNEP 2013; Cashman and Ashley 2008; UN University 2013). Mahatma Gandhi has raised this too: 'The difference between what we do and what we are capable of doing would suffice to solve most of the world's problems'.

Acknowledgements

This paper provides a broad overview of water, waste and climate change in cities. It is an update of a previous paper on water in the city at the University of Utrecht (Van Leeuwen 2014). Our work has been financed by KWR Watercycle Research Institute in the context of Watershare®: sharing knowledge in the water sector (http://www.watershare.eu/). It is a contribution to the European Innovation Partnership on Water of the European Commission and more specifically to the City Blueprint Action Group (European Commission 2015b), coordinated by both Dr. Richard Elelman of Fundacio' CTM Centre Tecnolo`gic and NETWERC H_2O and Prof. Dr. C.J. van Leeuwen (KWR Watercycle Research Institute). The authors would like to thank Prof. dr. Wim van Viersen, Prof. dr. Annemarie van Wezel, Ir. MBA Idsart Dijkstra and Ir. Jos Frijns (KWR Watercycle Research Institute, the Netherlands), as well as Prof. Dr. Peter Driessen (Utrecht University, the Netherlands) for their practical and stimulating contributions. Last but not least, we would like to thank Richard Elelman, the members of the City Blueprint Action Group, and all partners of the EU BlueSCities project and all cities involved, for their dedication, voluntary contributions and discussions related to the work described in this manuscript. The European Commission is acknowledged for funding TRUST in the 7th Framework Programme under Grant Agreement No. 265122 and for BlueSCities in H2020-Water under Grant Agreement No. 642354.

Source: Koop S H A, Leeuwen C J V. The challenges of water, waste and climate change in cities[J]. Environment Development & Sustainability, 2016:1–34.

References

[1] Amundsen, H., Berglund, F., & Westskog, F. (2010). Overcoming barriers to climate change adaptation—A question of multilevel governance? Environment and Planning C: Government and Policy, 28, 276–289.

[2] Atelier Istanbul. (2012). http://vimeo.com/41973779. Accessed September 15, 2014.

[3] Bai, X. (2007). Industrial ecology and the global impacts of cities. Journal of Industrial Ecology, 11, 1-6. BAUM. (2013). Intelligent cities: Routes to a sustainable, efficient and livable city. Hamburg: Bundes-deutscher Arbeitskreis fu¨r Umweltbewusstes Management.

[4] Brown, R. R., Keath, N., & Wong, T. H. F. (2009). Urban water management in cities, historical, current and future regimes. Water Science and Technology, 59, 847–855.

[5] C40Cities. (2015). Powering climate action: Cities as global changemakers. London: C40 Cities.

[6] Cashman, A., & Ashley, R. (2008). Costing the long-term demand for water sector infrastructure. Foresight, 10(3), 9–26.

[7] CCS. (2008). The European carbon dioxide capture and storage (ccs) project network. https://ec.europa.eu/energy/sites/ener/files/documents/ccs_project_network_booklet.pdf. Accessed December 15, 2015.

[8] CEFIC. (2014). Facts and figures 2014. Brussels: The European Chemical Industry Council.

[9] CIA Central Intelligence Agency. (2014). The world factbook. Urbanization. https://www.cia.gov/library/publications/the-world-factbook/fields/2212.html. Accessed December 21, 2015.

[10] Delta programme. (2013). http://www.rijksoverheid.nl/onderwerpen/deltaprogramma. Accessed September 15, 2013, in Dutch.

[11] Derraik, J. G. B. (2002). The pollution of the marine environment by plastic debris: A review. Marine Pollution Bulletin, 44, 842–852.

[12] Dobbs, R., Remes, J., Manyika, J., Roxburgh, C., Smit, S., & Schaer, F. (2012). Urban world: Cities and the rise of the consuming class. Washington, DC: McKinsey Global Institute.

[13] Dobbs, R., Smit, S., Remes, J., Manyika, J., Roxburgh, C., & Restrepo, A. (2011). Urban world: Mapping the economic power of cities. Washington, DC: McKinsey Global Institute.

[14] Doughty, M., & Hammond, G. (2004). Sustainability and the built environment at and beyond the city scale. Building and Environment, 39(10), 1223–1233.

[15] Driessen, P. P. J., Dieperink, C., Van Laerhoven, F., Runhaar, H. A. C., & Vermeulen, W. J. V. (2012). Towards a conceptual framework for the study of shifts in modes of environmental governance—Experiences from the Netherlands. Environmental Policy and Governance, 22, 143–160.

[16] Economist Intelligence Unit. (2015). The cost of inaction: recognizing the value at risk from climate change. London: The Business Intelligence Unit. The Economist. http://www.economistinsights.com/financialservices/analysis/cost-inaction. Accessed October 18, 2015.

[17] EEA. (2007). Climate change: the costs of inaction and the costs of adaptation. Technical report No 13/2007. Copenhagen: European Environment Agency.

[18] EEA. (2012). Urban adaptation to climate change in Europe: Challenges and opportunities for cities together with supportive national and European Policies. (EEA Report 2/2012). Copenhagen: European Environment Agency.

[19] EEA. (2013). Recycling rates in Europe. Copenhagen: European Environment Agency. http://www.eea.europa.eu/about-us/what/public-events/competitions/waste-smart-competition/recycling-rates-in-europe/image_view_fullscreen. Accessed December 15, 2015.

[20] EEA. (2014). Performance of water utilities beyond compliance. Sharing knowledge bases to support environmental and resource-efficiency policies and technical improvements. EEA Technical report No 5/2014. Copenhagen: European Environment Agency.

[21] EFILWC. (2012). Quality of life in Europe: Impact of the crisis. Luxembourg: European foundation for the improvement of living and working conditions.

[22] EMF. (2014). Towards the Circular Economy vol.3: accelerating the scale-up across global supply chains. Ellen MacArthur Foundation. http://www.ellenmacarthurfoundation.org/business/reports/ce2012. Accessed June 2, 2014.

[23] EMF. (2015a). Towards a circular economy. Business rationale for an accelerated transition. Ellen MacArthur Foundation. http://www.ellenmacarthurfoundation.org/assets/downloads/TCE_EllenMacArthur-Foundation_9-Dec-2015.pdf. Accessed December 15, 2015.

[24] EMF. (2015b). Growth within: a circular economy vision for a competitive Europe. Ellen MacArthur Foundation. http://www.ellenmacarthurfoundation.org/publications/growth-within-a-circular-economy-vision-for-a-competitive-europe. Accessed December 15, 2015.

[25] Engel, K., Jokiel, D., Kraljevic, A., Geiger, M., & Smith, K. (2011). Big cities: Big water: Big challenges: Water in an urbanizing world. Berlin: World Wildlife Fund.

[26] European Commission. (2011). Cities of Tomorrow. Challenges, vision, ways for-

ward. European Union. Regional Policy. Brussels: European Commission.

[27] European Commission. (2012). Communication from the commission to the European parliament, the council, the European economic and social committee and the committee of the regions. A blueprint to safeguard Europe's water resources. COM (2012)673 final.

[28] European Commission. (2013). European innovation partnership on smart cities and communities: Strategic implementation plan. Brussels: European Commission.

[29] European Commission. (2014). The European critical raw materials review. MEMO/14/377 26/05/2014. Brussels. http://europa.eu/rapid/press-release_MEMO-14-377_en.htm. Accessed June 2, 2014.

[30] European Commission. (2015a). European innovation partnership on water. City blueprints action group, Brussels, Belgium. http://www.eip-water.eu/City_Blueprints. Accessed June 30, 2015.

[31] European Commission. (2015b). Closing the loop—An EU action plan for the Circular Economy. Brussels, Belgium. http://ec.europa.eu/priorities/jobs-growth-investment/circular-economy/docs/communication-action-plan-for-circular-economy_en.pdf. Accessed December 17, 2015.

[32] European Commission. (2015c). Results of the public consultation on the key features of an EU Urban Agenda. SWD(2015) 109 final/2. Brussels. http://ec.europa.eu/regional_policy/en/conferences/cities-2015/. Accessed September 30, 2015.

[33] European green city index. (2009). Assessing the environmental impact of Europe's major cities. Siemens: A research project conducted by the Economist Intelligence Unit.

[34] FAO. (2011a). Global food losses and food waste—Extent, causes and prevention. Rome: The Food and Agriculture Organization of the Unities Nations.

[35] FAO. (2011b). The state of the world's land and water resources for food and agriculture: Managing systems at risk. Rome: The Food and Agriculture Organization of the Unities Nations.

[36] Farla, J., Markard, J., Raven, R., & Coenen, L. (2012). Sustainability transitions in the making: A closer look at actors, strategies and resources. Technological Forecasting and Social Change, 79(6), 991–998.

[37] Francis, H. F. (2015). Encyclical letter LAUDATO SI' of the Holy Father Francis on care for our common home. Rome: Vatican Press. http://w2.vatican.va/content/francesco/en/encyclicals/documents/papa-francesco_20150524_enciclica-laudato-si.html. Accessed December 17, 2015.

[38] Frijns, J., Bu¨scher, C., Segrave, A., & Van der Zouwen, M. (2013). Dealing with future challenges: A social learning alliance in the Dutch water sector. Water Policy, 15, 212–222.

[39] Frijns, J., Middleton, R., Uijterlinde, C., & Wheale, G. (2012). Energy efficiency in

the European water industry: Learning from best practices. Journal of Water and Climate Change, 3(1), 11–17.

[40] Gleick, P. H. (2003). Global freshwater resources: soft-path solutions for the 21st century. Science, 302, 1524–1528.

[41] Green City Index. (2015). http://www.siemens.com/entry/cc/en/greencityindex.htm. Accessed December 10, 2015.

[42] Grant, S. B., Saphores, J. D., Feldman, D. L., Hamilton, A. J., Fletcher, T. D., Cook, P. L. M., *et al.* (2012). Taking the "waste" out of "wastewater" for human water security and ecosystem sustainability. Science, 337, 681–686.

[43] Grimm, N. B., Faeth, S. H., Golubiewski, N. E., Redman, C. L., Wu, J., Bai, X., & Briggs, J. M. (2008). Global change and the ecology of cities. Science, 319, 756–760.

[44] Gupta, J., Termeer, C., Klostermann, J., Meijerink, S., van den Brink, M., Jong, P., & Bergsma, E. (2010). The adaptive capacity wheel: a method to assess the inherent characteristics of institutions to enable the adaptive capacity of society. Environmental Science & Policy, 13(6), 459–471.

[45] Hesketh, T., Lu, L., & Xing, Z. W. (2005). The effect of China's one-child family policy after 25 years. New England Journal of Medicine, 353(11), 1171–1176.

[46] Hoekstra, A. Y. (2014). Water for animal products: A blind spot in water policy. Environmental Research Letters, 9, 091003.

[47] Hoekstra, A. Y., Mekonnen, M. M., Chapagain, A. K., Mathews, R. E., & Richter, B. D. (2012). Global monthly water scarcity: Blue water footprints versus blue water availability. PLoS ONE, 7(2), e32688. doi: 10.1371/journal.pone.0032688.

[48] Hoekstra, A. Y., & Wiedman, T. O. (2014). Humanity's unsustainable environmental footprint. Science, 344(6188), 1114–1117.

[49] Hofstra, M. (2013). Water governance, a framework for better communication. Water Governance, 1, 9-13. Hunger, M., & Do¨ll, P. (2008). Value of river discharge data for global-scale hydrological modeling. Hydrology and Earth Systems Science, 12(3), 841–861.

[50] IMF International Monetary Fund (2013). http://www.imf.org/external/pubs/ft/weo/2013/01/weodata/weoselco.aspx?g=2001&sg=All?countries. Accessed December 21, 2015.

[51] IPCC. (2006). CH4 emissions from solid waste disposal. Background paper expert group CH4 emissions from solid waste disposal. Geneva: Intergovernmental Panel on Climate Change. http://www.ipccnggip.iges.or.jp/public/gp/bgp/5_1_CH4_Solid_Waste.pdf. Accessed December 21, 2015.

[52] Jalava, M., Kummu, M., Pokka, M., Siebert, S., & Varis, O. (2014). Diet change—A solution to reduce water use? Environmental Research Letters, 9, 074016.

[53] Jambeck, J. R., Geyer, R., Wilcox, C., Siegler, T. R., Perryman, M., Andrady, A., *et al.* (2015). Plastic waste inputs from land into the ocean. Science, 347, 768–771.

[54] Jongman, B., Hochrainer-Stigler, S., Feyen, L., Aerts, J. C. J. H., Mechler, R., Botzen, W. J. W., *et al.* (2014). Increasing stress on disaster-risk finance due to large floods. Nature Climate Change, 4, 264–268.

[55] Klein Goldewijk, K., Beusen, A., & Janssen, P. (2010). Long-term dynamic modeling of global population and built-up area in a spatially explicit way: HYDE 3.1. The Holocene, 20(4), 565–573.

[56] Koop, S. H. A., & Van Leeuwen, C. J. (2015a). Assessment of the sustainability of water resources management: A critical review of the City Blueprint approach. Water Resources Management, 29(15), 5649–5670.

[57] Koop, S. H. A., & Van Leeuwen, C. J. (2015b). Application of the improved City Blueprint framework in 45 municipalities and regions. Water Resources Management, 29(13), 4629–4647.

[58] Koop, S. H. A. & Van Leeuwen, C. J. (2015c). Towards sustainable water resources management: Improving the city blueprint framework. Report KWR 2015.025. KWR Watercycle Research Institute, Nieuwegein, the Netherlands. http://www.eip-water.eu/City_Blueprints. Accessed August 7, 2015.

[59] Koop, S., Van Leeuwen, K., Bredimas, A., Arnold, M., Makropoulos, M. & Clarens, F. (2015). Compendium of best practices for water, waste water, solid waste and climate adaptation. http://www.bluescities.eu/wp-content/uploads/2015/12/D2_3_Formatted_v7.pdf. Accessed December 16, 2015.

[60] Kuijpers, C., Nap, R., & de Bruijn, P. (2013). Good governance in local groundwater management. By coincidence or not? Water Governance, 1, 19–26. (in Dutch).

[61] Lange, P., Driessen, P. P. J., Sauer, A., Bornemann, B., & Burger, B. (2013). Governing towards sustainability—Conceptualizing modes of Governance. Journal of Environmental Policy & Planning, 15(3), 403–425.

[62] Leonardsen, L. (2012). Financing adaptation in Copenhagen. http://resilient-cities.iclei.org/fileadmin/sites/resilient-cities/files/Webinar_Series/Webinar_Presentations/Leonardsenfinancing_adaptation_in_Copenhagen_ICLEI_sept_2012.pdf. Accessed May 12, 2015.

[63] Ligtvoet, W., Hilderink, H., Bouwman, A., Puijenbroek, P., Lucas, P., & Witmer, M. (2014). Towards a world of cities in 2050. An outlook on water-related challenges. Background report to the UN-Habitat Global Report. Bilthoven: Netherlands Environmental Assessment Agency.

[64] Lipinski, B., Hanson, C., Lomax, J., Kitinoja, L., Waite, R., & Searchinger, T. (2013). Reducing Food Loss and Waste. Working Paper, Installment 2 of Creating a Sustainable Food Future. Washington, DC: World Resources Institute. http://www.worldresourcesreport.org.

[65] Loorbach, D. (2010). Transition management for sustainable development: A prescriptive, complexity-based governance framework. Governance, 23, 161–183.

[66] Loorbach, D., & Rotmans, J. (2010). The practice of transition management: Examples and lessons from four distinct cases. Futures, 42, 237–246.

[67] Manfredi, S., Tonini, D., Christensen, T. H., & Scharff, H. (2009). Landfilling of waste: Accounting of greenhouse gases and global warming contributions. Waste Management and Research, 27(8), 825–836.

[68] Mani, T., Hauk, A., Walter, U., & Burkhardt-Holm, P. (2015). Microplastics profile along the Rhine river. Scientific Reports. doi: 10.1038/srep17988.

[69] Markard, J., Raven, R., & Truffer, B. (2012). Sustainability transitions: An emerging field of research and its prospects. Research Policy, 41, 955–967.

[70] McFedries, R. (2012). Littered with 'plastic soup'. http://worldmaritimenews.com/archives/141427/%EF%BF%BClittered-with-plastic-soup/. Accessed December 15, 2015.

[71] McKinsey. (2013). Infrastructure productivity: How to save $1 trillion a year. London: McKinsey & Company.

[72] Molenaar, A., Aerts, J., Dircke, P., & Ikert, M. (2015). Connecting Delta Cities: Resilient cities and climate adaptation strategies. Rotterdam: Connecting Delta Cities.

[73] More'e, A. L., Beusen, A. H. W., Bouwman, A. F., & Willems, W. J. (2013). Exploring global nitrogen and phosphorus flows in urban wastes during the twentieth century. Global Biogeochemistry Cycles, 27(3), 836–846.

[74] Mottaghi, M., Aspegren, H., & Jo¨nsson, K. (2015). The necessity for re-thinking the way we plan our cities with the focus on Malmo¨/Towards urban-planning based urban runoff management. Vatten, 2015(1), 37–43.

[75] NYC. (2010). City of New York: NYC green infrastructure plan. A sustainable strategy for clean waterways. http://www.nyc.gov/html/dep/html/stormwater/nyc_green_infrastructure_plan.shtml. Accessed March 25, 2015.

[76] OECD. (2011a). Water governance in OECD countries: A multi-level approach. Paris: Organization for Economic Cooperation and Development.

[77] OECD. (2011b). International Futures Programme Pension funds investment in infrastructure. A survey. Paris: Organization for Economic Cooperation and Development.

[78] OECD. (2013). Environment at a glance 2013. OECD indicators. OECD Publishing. Paris: Organisation for Economic Co-operation and Development. doi:10.1787/9789264185715-en. http://www.oecd-ilibrary.org/environment/environment-at-a-glance-2013_9789264185715-en. Accessed April 7, 2015.

[79] OECD. (2015a). Water and cities: Ensuring sustainable futures. Paris: Organisation for Economic Cooperation and Development.

[80] OECD. (2015b). OECD Principles on water governance. Paris: Organisation for Economic Cooperation and Development.

[81] OECD. (2015c). Water Governance in Cities: GOV/RDPC (2015)17. Paris: Organisation for Economic Cooperation and Development.

[82] Philip, R., Anton, B., & van der Steen, P. (2011). SWITCH training kit. Integrated urban water management in the city of the future. Module 1. Strategic planning, ICLEI, Freiburg. http://www.switchtraining.eu/. Accessed May 15, 2013.

[83] Reckien, D., Flacke, J., Olazabal, M., & Heidrich, O. (2015). The Influence of drivers and barriers on urban adaptation and mitigation plans—An empirical analysis of european cities. PLoS ONE, 10(8), e0135597. doi:10.1371/journal.pone.0135597.

[84] Richter, B. D., Abell, D., Bacha, E., Brauman, K., Calos, S., Cohn, A., *et al.* (2013). Tapped out: How can cities secure their water future? Water Policy, 15, 335–363.

[85] Schwarzenbach, R. P., Escher, B. I., Fenner, K., Hofstetter, T. B., Johnson, C. A., Von Gunten, U., & Wehrli, B. (2006). The challenge of micropollutants in aquatic systems. Science, 313, 1072–1077.

[86] Science Communication Unit (2014). Science for Environment Policy. In-depth Report: Social Innovation and the Environment. Report produced for the European Commission DG Environment. Bristol: University of the West of England.

[87] SDEWES. (2015). The Sustainable Development of Energy, Water, and Environment Systems (SDEWES) Index. Zagreb: International Centre for Sustainable Development of Energy, Water and Environment Systems. http://www.sdewes.org/sdewes_index.php. Accessed December 15, 2015.

[88] Toro, R. (2013). Hurricane Sandy's impact. http://www.livescience.com/40774-hurricane-sandy-s-impact-infographic.html. Accessed March 24, 2015.

[89] UN. (2012). World urbanization prospects: The 2011 revision. New York: United Nations.

[90] UN. (2013a). Sustainable cities. Facts and figures. http://www.un.org/en/sustainablefuture/cities.asp. Accessed October 15, 2015.

[91] UN. (2013b). Disaster-resilient societies. http://www.un.org/en/sustainablefuture/disasters.asp. Accessed October 15, 2015.

[92] UN. (2014). The United Nations world water development report. Water and energy Vol. 1. New York: United Nations Development Programme.

[93] UN. (2015a). World urbanization prospects: The 2015 revision. New York: United Nations.

[94] UN. (2015b). The United Nations world water development report 2015: Water for a sustainable world. Paris: UNESCO.

[95] UN University. (2013). Catalyzing water for sustainable development and growth. Framing water within the post development agenda: Options and considerations.

Hamilton: United Nations University Institute for Water, Environment and Health.

[96] UNDP. (2013). User's guide on assessing water governance. Oslo: United Nations Development Programme.

[97] UNEP. (2007). Fourth Global Environment outlook: Environment for development. Geneva: United Nations Environment Programme.

[98] UNEP. (2012). Fifth global environment outlook: Environment for development. Geneva: United Nations Environment Programme.

[99] UNEP (2013). City-level decoupling: Urban resource flows and the governance of infrastructure transitions. A report of the working group on cities of the International Resource Panel. Swilling M., Robinson B., Marvin S., & Hodson, M. Nairobi: United Nations Environment Programme.

[100] UNESCO. (2015a). The United Nations world water development report Water for a sustainable world. Paris: United Nations Educational, Scientific and Cultural Organization.

[101] UNESCO. (2015b). Facing the Challenges. Paris: United Nations World Water Assessment Programme, United Nations Educational, Scientific and Cultural Organization.

[102] UN-Habitat. (2010). Solid waste management in the world's cities. London: United Nations Human Settlements Programme.

[103] UN-Habitat. (2013). The state of European cities in transition. Taking stock after 20 years of reform. Nairobi: United Nations Settlement Programme.

[104] UN-Habitat. (2015). http://unhabitat.org/habitat-iii-conference/. Accessed December 16, 2015.

[105] Van der Hoek, J. P., Struker, A., & De Danschutter, J. E. M. (2015). Amsterdam as a sustainable European metropolis: Integration of water, energy and material flows. Urban Water Journal. doi:10.1080/1573062X.2015.1076858

[106] Van Leeuwen, C. J. (2008). The China environment yearbook 2005. Book review. Environmental Science and Pollution Research, 15, 354–356.

[107] Van Leeuwen, C. J. (2013). City Blueprints: Baseline assessment of sustainable water management in 11 cities of the future. Water Resources Management, 27, 5191–5206.

[108] Van Leeuwen, K. (2014). Water in the city: Inauguration speech. Utrecht: University of Utrecht Faculty geosciences. (in Dutch).

[109] Van Leeuwen, C. J. (2015). Water governance and the quality of water services in the city of Melbourne. Urban Water Journal. doi: 10.1080/1573062X.2015.

[110] Van Leeuwen, C. J., & Bertram, N. P. (2013). Baseline assessment and best practices in urban water cycle services in the city of Hamburg. Bluefacts. International Journal of Water Management 2013,10–16.

http://www.bluefacts-magazin.de/heftarchiv/2013/.

[111] Van Leeuwen, C. J., & Chandy, P. C. (2013). The City Blueprint: Experiences with the implementation of 24 indicators to assess the sustainability of the urban water cycle. Water Science and Technology. Water Supply, 13(3), 769–781.

[112] Van Leeuwen, C. J., Dan, N. P., & Dieperink, C. (2015a). The challenges of water governance in Ho Chi Minh City. Integrated Environmental Assessment and Management. doi:10.1002/ieam.1664.

[113] Van Leeuwen, C.J., & Elelman, R. (2015). E-Brochure of the City Blueprint. Available at: http://www.eipwater.eu/City_Blueprints.

[114] Van Leeuwen, C. J., Frijns, J., Van Wezel, A., & Van De Ven, F. H. M. (2012). City blueprints: 24 indicators to assess the sustainability of the urban water cycle. Water Resources Management, 26, 2177–2197.

[115] Van Leeuwen, C. J., Koop, S. H. A., & Sjerps, R. M. A. (2015b). City Blueprints: Baseline assessments of water management and climate change in 45 cities. Environment, Development and Sustainability. doi:10.1007/s10668-015-9691-5.

[116] Van Leeuwen, C. J., & Sjerps, R. M. A. (2015a). The City Blueprint of Amsterdam. An assessment of integrated water resources management in the capital of the Netherlands. Water Science and Technology Water Supply, 15(2), 404–410.

[117] Van Leeuwen, K., & Sjerps, R. (2015b). Istanbul: The challenges of integrated water resources management in Europa's Megacity. Environment, Development and Sustainability. doi:10.1007/s10668-015-9636-z.

[118] Van Leeuwen, C. J., & Vermeire, T. G. (Eds.). (2007). Risk assessment of chemicals. An introduction (2nd ed.). Dordrecht: Springer Publishers.

[119] Van Oel, P. R., Mekonnen, M. M., & Hoekstra, A. Y. (2009). The external water footprint of the Netherlands: geographically-explicit quantification and impact assessment. Ecological Economics, 69(1), 82–92.

[120] Van Someren, T. C. R. & Van Someren-Wang, S. (2013). Strategic innovation: the creation of the water companies of the future. 65e Vakantiecursus Drinkwater en Afvalwater. Delft: Technical University Delft.

[121] Van Wezel, A., Caris, I., & Kools, S. A. E. (2015). Release of primary microplastics from consumer products to wastewater in The Netherlands. Environmental Toxicology and Chemistry. doi:10.1002/etc.3316.

[122] VROM. (2001). General policy on waste. Fact sheet. The Hague: Ministry of Housing, Spatial Planning and the Environment. http://www.wbcsdcement.org/pdf/tf2/01 GenPolonWaste.pdf. Accessed December 14, 2015.

[123] Waste Atlas. (2015). http://www.atlas.d-waste.com/. Accessed December 20, 2015.

[124] Waste Framework Directive. (2008). Directive 2008/98/EC of the European Parliament and of the Council of 19 November 2008 on waste and repealing certain Direc-

tives. Brussels: EUR-Lex—32008L0098—EN.

[125] Water Resources Group. (2009). Charting our water future. West Perth, USA: Economic framework to inform decision making.

[126] WHO. (2008). Safer water, better health: Costs, benefits and sustainability of interventions to protect and promote health. Geneva: World Health Organization.

[127] World Bank. (2015a). Poverty gap at $2 a day (PPP) (%). http://data.worldbank.org/indicator/SI.POV. GAP2/countries/1W?display=default. Accessed December 21, 2015.

[128] World Bank. (2015b). Improved water source, urban (% of urban population with access). http://data.worldbank.org/indicator/SH.H2O.SAFE.UR.ZS. Accessed December 21, 2015.

[129] World Bank. (2015c). Improved sanitation facilities, urban (% of urban population with access). http://data.worldbank.org/indicator/SH.STA.ACSN.UR. Accessed December 21. 2015.

[130] World Bank. (2015d). CO2emissions (metric tons per capita). http://data.worldbank.org/indicator/EN.ATM.CO2E.PC. Accessed January 4, 2016.

[131] World Economic Forum. (2014). Global risks 2013 (9th ed.). Geneva: World Economic Forum.

[132] World Future Council. (2014). Regenerative urban development: A roadmap to the city we need: 3rd Future of Cities Forum. Hamburg: World Future Council.

[133] Zarfl, C., Fleet, D., Fries, E., Galgani, F., Gerdts, G., Hanke, G., & Matthies, M. (2011). Microplastics in oceans. Marine Pollution Bulletin, 62, 1589–1591.

[134] Zimmerman, K. A. (2012). Hurricane Katrina: facts, damage & aftermath. Livescience, August 27, 2015. http://www.livescience.com/22522-hurricane-katrina-facts.html. Accessed October 15, 2015.

Chapter 7

Economic Evaluation of the Air Pollution Effect on Public Health in China's 74 Cities

Li Li[1,2], Yalin Lei[1,2], Dongyan Pan[3], Chen Yu[1,2], Chunyan Si[1,2]

[1]School of Humanities and Economic Management, China University of Geosciences, Beijing, China
[2]Key Laboratory of Carrying Capacity Assessment for Resource and Environment, Ministry of Land and Resources, Beijing, China
[3]Central University of Finance and Economics, Beijing, China

Abstract: Air deterioration caused by pollution has harmed public health. The existing studies on the economic loss caused by a variety of air pollutants in multiple cities are lacking. To understand the effect of different pollutants on public health and to provide the basis of the environmental governance for governments, based on the dose-response relation and the willingness to pay, this paper used the latest available data of the inhalable particulate matter (PM_{10}) and sulphur dioxide (SO_2) from January 2015 to June 2015 in 74 cities by establishing the lowest and the highest limit scenarios. The results show that (1) in the lowest and highest limit scenario, the health-related economic loss caused by PM_{10} and SO_2 represented 1.63% and 2.32% of the GDP, respectively; (2) For a single city, in the lowest and the highest limit scenarios, the highest economic loss of the public health effect caused by PM_{10} and SO_2 was observed in Chongqing; the highest economic loss of the public health effect per capita occurred in Hebei Baoding. The highest proportion of the health-related economic loss accounting for GDP was found in Hebei Xingtai. The main reason is that the terrain conditions are not conducive to the spread of air pollutants in Chongqing, Baoding and Xingtai, and the three cities are typical heavy industrial cities that are based on coal resources. Therefore, this paper proposes to improve the energy structure, use the advanced production process, reasonably control the urban population growth, and adopt the emissions trading system in order to reduce the economic loss caused by the effects of air pollution on public health.

Keywords: Air Pollution, The Public Health Effect, The Economic Loss, 74 Cities, China

1. Introduction

Air pollution mainly refers to human activities or the natural processes that cause a certain substance to continuously enter the atmosphere at a sufficient concentration to endanger the health and cause environmental pollution. There are many types of air pollutants, the primary of which are total suspended particulates (TSP), inhalable particulate matter (PM_{10}), fine particulate matter ($PM_{2.5}$), SO_2 and NO_x, among others (Yu *et al.* 2008). After inhalation of harmful pollutions, humans may develop respiratory disease and can suffer from serious diseases, such as tracheitis, bronchitis, asthma, lung disease and lung cancer, for many years. The energy consumption structure in China is mainly based on coal resources, and the rapid growth of motor vehicles in cities results in increasingly more serious air pollution in China's large cities; in addition, the questions regarding public health and air pollution have garnered widespread attention (Wilkinson and Smith 2007).

The main air pollutions in China are PM_{10}, SO_2 and NO_x (Chen *et al.* 2001). In the 12th Five-Year period, China clearly proffered the target to reduce the total SO_2 emissions by 8% and increase the ratio of the urban air quality to achieve the second level of 8% (Ministry of Environmental Protection of the People's Republic of China 2011). According to an environmental analysis reported by the Asian Development Bank in 2013, the Chinese government was taking measures to control air pollution; however, of the world's 10 most seriously polluted cities, 7 cities were in China. Of China's 500 major cities, less than 1% met the standards of the

World Health Organization (Zhang 2012). In 2015, the World Health Organization released a report stating that at least one in eight people died of air pollution globally. Air pollution has become the world's largest environmental health risk (Huanqiunet 2015).

Since January 1, 2013, the Ministry of Environmental Protection has monitored the air quality index of Beijing, Tianjin, Hebei, the Yangtze River Delta, the Pearl River Delta region, the municipality directly under the central government, the provincial capital cities and the cities specifically designated in the state plan, which are collectively called the 74 cities, in brief (China's National Environmental Monitoring Centre 2013a). The concentrations of pollutants such as PM_{10} and SO_2 have been monitored since November 2014 (China's National Environmental Monitoring Centre 2013b). As World Bank (1997) didn't identify the health effects of NO_x and NO_x wasn't included in the dose-response relation of Ho and Jorgenson (2007), thus to quantitatively evaluate the economic loss due to the effects of air pollution on public health in China, this paper analyzes PM_{10} and SO_2 based on the latest available data from January 2015 to June 2015, uses the method of foreign study on China's economic loss due to air pollution effects on public health for reference (Wang and Smith 1999; Ho and Nielsen 2007) and estimates the economic loss caused by the effects of air pollution on public health in the 74 cities. An evaluation of the health-related economic loss can provide a basis for the government to develop and initiate preventative measure for controlling air pollution. At the same time, these findings can also improve the environmental protection awareness of the local government and the public.

2. Literature Review

In recent decades, industrialization and urbanization have experienced rapid development, which has resulted in increasing air pollution. According to the World Bank, there is a close relationship between air pollution and public health. There is a positive relation between the concentration of air pollutants and respiratory diseases, lung function loss, chronic bronchitis and premature death (World Bank SEPA 2007). The evaluation of health-related economic loss caused by air pollution has become a hot topic for scholars and institutions.

2.1. Research Progress on the Economic Loss Regarding to Public Health Impacts Caused by Air Pollution

Ridker (1967) calculated the economic loss associated with different diseases which caused by air pollution in the USA in 1958 by using the human capital method. The results showed that the economic loss related to the effects on public health was 80.2 billion dollars in the USA. This study hailed the beginning of the calculation of health related economic loss caused by air pollution.

Employing a survival analysis and the data from a 14- to 16-year mortality follow-up of 8111 adults in the six cities in the U.S., Dockery *et al.* (1993) estimated the associations between particulate air pollution and daily mortality rates. Their results confirmed that the mortality rate was associated with the level of air pollution. Using data from 1994 to 1995 in Hong Kong, Wong *et al.* (1999) determined that adverse health effects were evident at the current ambient concentrations of air pollutants. Samet *et al.* (2000) recognized an association between daily changes in the concentration of ambient particulate matter and the daily number of deaths (mortality) in the United States. Wong *et al.* (2002) used Poisson regression to estimate the associations between daily admissions and the levels of PM_{10} and SO_2 in Hong Kong and London. The results confirmed that air pollution caused detrimental short-term health effects.

Using the collective data regarding to PM_{10} and SO_2 from January 1999 to September 2000, Kaushik *et al.* (2006) assessed the ambient air quality status in the fast growing urban centres of Haryana state, India. Adopting the daily data during 2008 and 2009 in Beijing, Xu *et al.* (2014b) confirmed that short-term exposure to particulate air pollution was associated with increased ischemic heart disease (IHD) mortality.

In 1981, the concept, theory and method of environmental pollution economic loss assessment were put forward and discussed in the congress of the National Symposium on Environmental Economics (Xia 1998). Thereafter, the economic loss associated with environmental pollution was of interest to scholars. Gao *et al.* (1993) adopted the GEE (Generalized Estimation Equation) to study the relationship between TSP in Haidian District, Beijing and low air pollution. Using the two methods (ecology and time series) and the data of 1992 in Shenyang, Xu *et al.* (1996) determined that total mortality, chronic obstructive pulmonary disease (COPD), cardiovascular disease and pollution levels were significantly correlated. Jing and Ren (2000) conducted an epidemiological survey on adults who were older than 25 years using the multiple logistic regression analysis. The results showed that 6 types of respiratory system diseases or symptoms appeared with an increasing frequency as air pollution levels increased. Chen and Hong (2002) quantitatively evaluated the air pollution in Shanghai based on the risk evaluation method and found that the health effects caused by SO_2 exhibited a gradually declining trend. Chen *et al.* (2010) evaluated the health impacts of particulate air pollution on urban populations in 113 Chinese cities, and it was estimated that the total economic cost of the health impact was approximately 341.4 billion Yuan, 87.79% of which was attributable to premature deaths. Chen *et al.* (2015) employed a Poisson regression model to estimate residents' health benefits in two scenarios: environmentally controlled scenario 1 and environmentally controlled scenario 2. Scenario 2 showed a potentially higher reduction of emissions and greater health benefits than scenario 1. Xu *et al.* (2014a) used the established model between PM_{10} and thermal environmental indicators to evaluate the PM_{10}—related health risk in Beijing.

Certain scientific institutions also focused more on the economic loss associated with public health effects caused by air pollution. The World Health Organization estimated that the total loss globally caused by air pollution-related disease was 0.5% (Murray and Lopea 1997) in 1997. In the same year, the World Bank systematically studied the health effects caused by air pollution in China (World Bank 1997). The U.S. Environmental Protection Agency estimated that the economic benefits of health and ecological improvement in the United States from 1990 to 2010 were as high as \$6–50 trillion, most of which could be attributed to the decrease in the number of deaths caused by air pollution (U.S. EPA 1999). The World Health Organization reported that 80% of the world's cases of heart disease and stroke deaths were due to air pollution, and a total of 7 million people in the world died of air pollution in 2014 (Huanqiunet 2014). In 2015, the World Health Organization released data that at least 1 in every 8 people died of air pollution throughout the world. Air pollution has become the world's largest environmental health risk (Huanqiunet 2015).

2.2. Research Progress on the Method Used to Evaluate the Economic Loss Associated with Public Health Effects Caused by Air Pollution

The previous studies regarding to the economic loss caused by the effects of air pollution on public health generally included the determination of economic loss using the contents of the environmental pollution assessment, the public health impact assessment and a choice of methods. Generally, the methods used were as follows.

1) Modified human capital method

Ridker (1967), Dockery *et al.* (1993), Wang *et al.* (2005), Jia *et al.* (2004), Wan *et al.* (2005), Han *et al.* (2006), Zhang *et al.* (2008), Shang *et al.* (2010), Han (2011), and Shen *et al.* (2014) quantitatively estimated the economic loss in different regions and obtained different results.

2) Illness cost method

Air pollution led to changes in the disposable income of people, particularly, an increase in medical expenses. Medical expenses became a recognized fact, and they also became a very heavy burden on civilians. Based on the above views, certain scholars obtained conclusions by analysing the illness costs caused by air pollution. These scholars include Chen *et al.* (2010), Zmirou *et al.* (1999), Hedley *et al.* (2008), Patankar and Trivedi (2011), Brandt *et al.* (2014), and Yan (2012).

3) Willingness to pay

Willingness to pay is an indirect evaluation method which constructs a simulated market to reveal people's willingness to pay for certain environmental goods, in order to evaluate the value of environmental quality. Researchers included Carlsson and Martinsson (2001), Wang and John (2006), Koop and Tole (2004), Pascal *et al.* (2013), Yaduma *et al.* (2013), Ami *et al.* (2014), Istamto *et al.* (2014), Cai and Yang (2003), Peng and Tian (2003), Cai *et al.* (2007), Zhou *et al.* (2010), and Zeng *et al.* (2015).

2.3. Literature Summary

1) In the studies of the economic loss caused by air pollution, domestic and foreign researchers studied the qualitative relationship between and quantitative analysis of air pollution and its health effect. However, generally, previous studies were solely based on a country, a city or a type of air pollutant; the health effects of many types of air pollutants in a city of a typical city are moderate.

2) Regarding to the method for evaluating the economic loss caused by the effects of air pollution on public health, the deficiencies of the modified human capital method were that the life prediction of the society may not be reasonable, and the different choices of the discount rate would have a large impact on the evaluation results. The disadvantage of the illness cost method is that it may underestimate the illness value. Additionally, the method's other disadvantage was that the individual may have a willingness to pay.

From the economic perspective, the willingness to pay method is the most reasonable method because it can reveal the value of all goods and utilities, and it can completely evaluate the economic values of environmental resources, which is currently being widely recognized and accepted. Thus, this paper utilizes the willing to pay method to evaluate the economic loss associated with public health effects caused by air pollution in 74 cities.

3. Methods and Data

3.1. The Dose-Response Relationship and the Willingness to Pay

To study the economic loss related to public health effects caused by air pollution, it is necessary to consider the types of public health effects and to establish the relation between the concentration of air pollutants and the effect on public health, which is called a dose-response. In different studies, the dose-response relationship is different. The indexes of public health effects caused by air pollution, which the World Bank put forward, included premature deaths, hospitalization and emergency caused by respiratory diseases, the number of restriction days caused by health problems related to the inhalation of particulate matter, lower respiratory tract infections, childhood asthma, asthma, chronic bronchitis, respiratory symptoms, and chest discomfort. This paper used the dose-response relationship of Ho and Jorgenson (2007) for reference and assumed that all the people in the 74 cities were exposed to the same concentrations of PM_{10} and SO_2. The dose-response relationship is shown in the Equation (1).

$$HE_{xrh} = DR_{xh} \times C_{rx} \times POP_r \tag{1}$$

where HE_{xrh} is the h-th type of public health effect caused by the air pollutant x (including PM_{10} and SO_2) in the region r. DR_{xh} is the dose-response coefficient of the air pollutant x (unit: the number of the people suffer with the concentration of the air pollutant increasing by $1\mu g/m^3$) and the h-th type of public health. C_{rx} is the concentration of the air pollutant x in the region r. POP_r is the number of the people in the region r. Ho and Jorgenson (2007) used the survey data of Beijing and Anqing in 1997 to estimate the economic loss caused by the health effects of Chinese residents using the willingness to pay method, the population of was were 6.53 million and 0.35 million respectively. Based on the loss value estimation of Ho and Jorgenson (2007), this paper modified it, as shown in **Table 1**. The total economic loss of the 74 cities is obtained by adding up all of the economic loss types relating to the health effects, as shown in the Equations (2) and (3):

$$HEV_{xrh} = V_{xh} \times HE_{xrh} \tag{2}$$

$$THEV = \sum_r \sum_x \sum_h HEV_{xrh}$$

where HEV_{xrh} is the economic loss of the h-th type of public health effect caused by the air pollutant x (including PM_{10} and SO_2) in the region r. V_{xh} is the economic loss value of the h-th type of public health effect caused by the air pollutant x (including PM_{10} and SO_2). THEV is the total economic loss related to public health effects caused by the air pollutant. In calculating total health-related economic loss, the paper adds up eight effects of PM_{10} on public health and three effects of SO_2 on public health together, which may appear double counting. As there is little literature of this issue, the paper hasn't analyzed it.

Table 1. The dose-response relationship and the loss value estimation of the public health effects.

	Coefficient of the dose-response relationship		Loss value estimation (yuan, the price in 2002)	Loss value estimation (yuan, the mean price from January 2015 to June 2015)
	The lower limit scenario	The highest limit scenario	Ho and Jorgen-son's estimation	Modified estimation
		Effects of PM_{10} on public health		
Premature death	1.3	2.6	370,000	528,370.2
Hospitalization caused by respiratory disease	12	12	1751	2500.5
Emergency	235	235	142	202.8
Number of restriction days	18,400	57,500	14	20
Lower respiratory tract infections and childhood asthma	23	23	80	114.2
Asthma	1770	2608	2.5	3.57
Chronic bronchitis	61	61	48,000	68,545.33
Respiratory symptoms	49,820	183,000	3.7	5.28
		Effects of SO_2 on public health		
Premature death	1	2.6	370,000	528,370.2
Chest discomfort	10,000	10,000	6.2	8.85
Lower respiratory tract infections and childhood asthma	5	5	6.2	8.85

Source: Ho and Jorgenson (2007).

3.2. Data

The environmental data of 74 cities during the period from January 2015 to June 2015 were reported by China's National Environmental Monitoring Centre (2015a, b, c, d, e, f), which mainly contained the monthly mean concentrations of

PM_{10} and SO_2 (**Figure 1, Figure 2**). The number of the population and GDP in 74 cities were obtained from Askcinet (2015).

Figure 1. Monthly mean concentration of SO_2 in China's 74 cities.

Figure 2. Monthly mean concentration of PM_{10} in China's 74 cities.

From **Figure 1** and **Figure 2**, Taiyuan, Shenyang and Yinchuan are the top three cities with the highest monthly mean concentration of SO_2. Baoding, Zhengzhou and Xingtai are the top three cities with the highest monthly mean concentration of PM_{10}. According to the ambient air quality standard GB3095-2012, the paper compared Chongqing, Baoding and Xingtai with Beijing and found that: (1) For SO_2, Beijing and Chongqing achieved the first level of national standards from January 2015 to June 2015. Baoding and Xingtai achieved the first level of national standards from April to June 2015, and they achieved the second level of national standards from January to March 2015. (2) For PM_{10}, Beijing achieved the second level of national standards from January 2015 to June 2015. Chongqing achieved the second level of national standards from February 2015 to June 2015, which didn't achieve the national standards in January 2015. Baoding and Xingtai achieved the second level of national standards from April to June 2015, but they didn't achieve the national standards from January to March 2015. Overall, the monthly mean concentration of SO_2 and PM_{10} in the 4 cities appeared a downward trend.

4. Results and Discussions

4.1. The Total Economic Loss Associated with Public Health Effects Caused by Air Pollution in 74 Cities

The dose-response relationship and the loss value estimation of the public health effects in different cities vary; therefore, this paper establishes different scenario parameters for the lowest limit scenario and the highest limit scenario in order to evaluate the total economic loss related to the effects of air pollution on public health in 74 cities.

1) The lowest limit scenario

As shown in **Table 2**, there were 84,917 premature deaths caused by PM_{10} and SO_2. There were 646,282 hospitalizations caused by respiratory disease, 12.66 million emergencies, more than 990 million restriction days and 1.23 million lowest respiratory tract infections and occurrence of childhood asthma due to PM_{10}. However, the economic loss caused by the effects of SO_2 on public health was less than that of PM_{10}.

Table 2. The total lowest economic loss caused by the effects of air pollutants on public health in 74 cities.

	Coefficient of the dose-response relation	Cases	Loss value estimation (yuan, the mean price from January 2015 to June 2015)	Economic loss (million yuan)
Effects of PM_{10} on public health				
Premature death	1.3	70,014	528,370.2	36,993.31
Hospitalization caused by respiratory disease	12	646,282	2500.5	1616.03
Emergency	235	12,656,359	202.8	2566.71
Number of restriction days	18,400	990,965,979	20	19,819.32
The lowest respiratory tract infections and childhood asthma	23	1,238,707	114.2	141.46
Asthma	1770	95,326,619	3.57	340.32
Chronic bronchitis	61	3,285,268	68,545.33	225,189.78
Respiratory symptoms	49,820	2,683,148,101	5.28	14,167.02
Total economic loss of the public health effect caused by PM_{10}	300,833.95			
Effects of SO_2 on public health				
Premature death	1	14,903	528,370.2	7874.30
Chest discomfort	10,000	149,030,000	8.85	1318.92
The lowest respiratory tract infections and childhood asthma	5	74,515	8.85	0.66
Total economic loss caused by the effects of SO_2 on public health effect	9193.88			
Total economic loss related to public health effects	310,027.82			
Economic loss related to public 598 yuan health effects per capita				
GDP in the 74 cities	18,987,854			
Total economic loss of the public health effects accounting for the GDP	1.63%			

This paper calculated that the total health-related economic loss caused by the air pollutant in 74 cities was approximately 310 billion yuan, explaining approximately 1.63% of the 74 cities' GDP, which was higher than the result of Wei *et al.* (2012). The total economic loss of the public health effect caused by PM_{10} was 300.8 billion yuan, explaining approximately 97.03% of the total economic loss, which was the major economic loss and was consistent with the result of Zhang (2012). The economic loss caused by chronic bronchitis, which was approximately 225.2 billion yuan. It was the largest in the total economic loss, explaining approximately 72.64% of the total economic loss. The result was different from that of Chen *et al.* (2010), who determined that the economic loss caused by premature death was the largest.

2) The highest limit scenario

Table 3 showed there were 178,776 premature deaths caused by PM_{10} and SO_2. There were 646,282 hospitalizations caused by respiratory disease, 12.66 million emergencies, more than 3090 million restriction days and 1.23 million lower respiratory tract infections and childhood asthma caused by PM_{10}. Similarly, the economic loss of the public health effect caused by SO_2 was also less than that of PM_{10}.

The total economic loss of the public health effect caused by the air pollutant in 74 cities was approximately 439.8 billion yuan in the highest limit scenario, representing approximately 2.32% of the GDP in 74 cities, a slight difference from the result of Wei *et al.* (2012). The total highest economic loss of the public health effect caused by PM_{10} was 418 billion yuan, explaining approximately 95.04% of the total economic loss, which was the major economic loss and was also consistent with the result of Zhang (2012). The economic loss caused by chronic bronchitis was also the largest in the total economic loss, approximately 225.2 billion yuan, explaining approximately 51.24%. The result was also different from that of Chen *et al.* (2010).

4.2. The Economic Loss Caused by Effect on the Public Health Effect in the Major Cities

To further understand the health-related economic loss caused by the air

Table 3. The total highest economic loss of the public health effect caused by the air pollutant in 74 cities.

	The coefficient of the dose-response relation	Cases	the loss value estimation (yuan, the mean price from January 2015 to June 2015)	economic loss (million yuan)
PM_{10}'s public health effect				
Premature death	2.6	140,028	528,370.2	73,986.62
Hospitalization caused by respiratory disease	12	646,282	2500.5	1616.03
Emergency	235	12,656,359	202.8	2566.71
The number of restriction days	57,500	3,096,768,683	20	61,935.37
Lower respiratory tract infections and childhood asthma	23	1,238,707	114.2	141.46
Asthma	2608	140,458,656	3.57	501.44
Chronic bronchitis	61	3,285,268	68,545.33	225,189.78
Respiratory symptoms	183,000	9,855,802,940	5.28	52,038.64
The total economic loss of the public health effect caused by PM_{10} SO_2's public health effect	417,976.05			
Premature death	2.6	38,748	528,370.2	20,473.29
Chest discomfort	10,000	149,030,000	8.85	1318.92
Lower respiratory tract infections and childhood asthma	5	74,515	8.85	0.66
The total economic loss of the public health effect caused by SO_2	21,792.86			

Continued

The total economic loss of the public health effect	439,768.91
Economic loss related to public health effects per capita	848 yuan
The GDP in 74 cities	18,987,854
The total economic loss of the public health effect accounts for the GDP	2.32%

pollutant, this paper estimated the health effects of air pollutants and the economic loss in the major cities from January 2015 to June 2015.

1) The lowest limit scenario

From **Figure 3**, Chongqing's public health economic loss was 17 billion yuan, ranking the first among the 74 cities, followed by Beijing, Baoding, and Tianjin. From the regional perspective, there were 4 municipalities and 4 cities in Hebei Province in the top 10 cities with the highest economic loss. There were 7 cities in North China excluding Chongqing, Shanghai and Chengdu in the top 10 cities with the highest economic loss.

As shown in **Figure 4**, there were 7 cities in Hebei Province in the top 10 with the highest economic loss. Zhengzhou and Ji'nan also ranked in the top 10 cities due to their poor air quality. Urumqi's health-related economic loss was not high; however, because of its low population, the resulting health economic loss per capita was higher.

Figure 5 shows that the proportion of the health-related economic loss accounting for GDP in Xingtai was the highest, which was 10.15%. In the 10 cities with the highest proportion of health-related economic loss accounting for GDP, there were 9 cities in the Hebei province. Suqian in Jiangsu province had a higher health-related economic loss and a lower GDP, ranking the tenth; therefore, the proportion of its health-related economic loss accounting for GDP was higher,

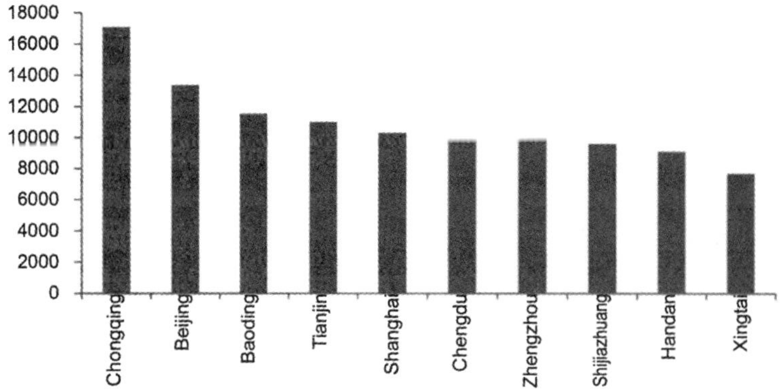

Figure 3. The 10 cities with the highest economic loss (million yuan).

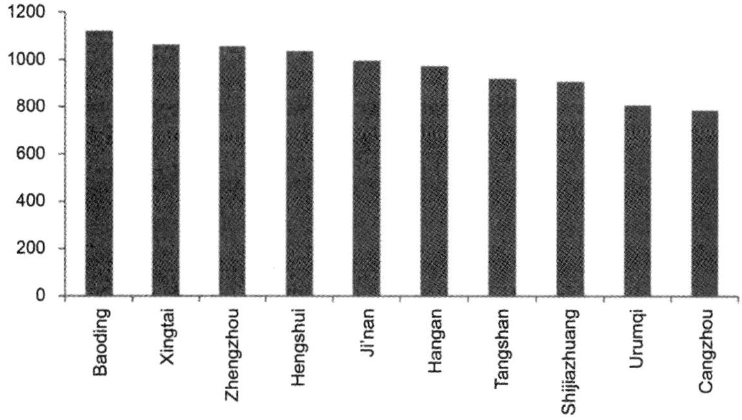

Figure 4. The 10 cities with the highest health-related economic loss per capita (yuan/per person).

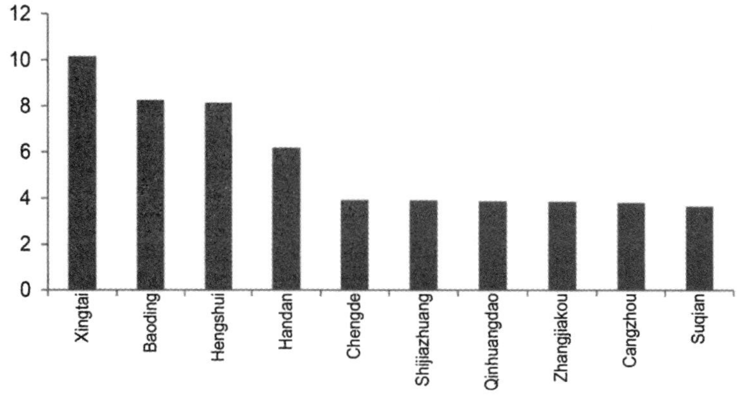

Figure 5. The 10 cities with the highest proportion of health-relate economic loss accounting for GDP (%).

ranking in the top 10. The economic losses of the public health effect of 4 municipalities ranked in the top 10. Although their GDPs were higher, the proportions of their health-related economic loss accounting for GDP were relatively lower, which did not rank in the top 10.

2) The highest limit scenario

In the highest limit scenario, the cities ranking in the top 10 with the highest economic loss in **Figure 6** were the same as those in the lowest limit scenario in **Figure 3**. However, the rankings of Zhengzhou and Chengdu were different in **Figure 3** and **Figure 4**.

The cities ranking in the top 10 with the highest economic loss per capita in **Figure 5** in the highest limit scenario were the same as in the lowest limit scenario in **Figure 4**. However, as shown in the **Figure 7** in the following, the economic loss per capita in the top 10 cities in the highest limit scenario was higher than that in the lowest limit scenario. For example, the economic loss per capita in Baoding was 1599.9yuan/per person, higher by yuan than that in the lowest limit scenario.

The 10 cities with the highest proportion of health-related economic loss accounting for GDP in **Figure 8** in the highest limit scenario were the same as in the lowest limit scenario in **Figure 3**. However, the rankings of Shijiazhuang, Chengde, Zhengjiukou and Qinhuangdao and the proportions of the health-related economic loss accounting for GDP were different in **Figure 5** and **Figure 8**.

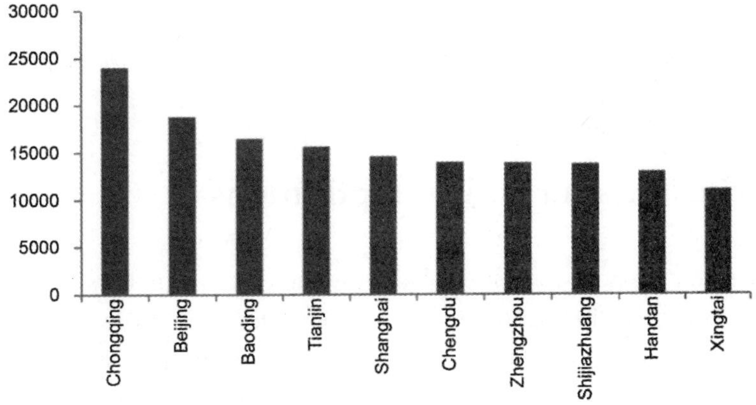

Figure 6. The 10 cities with the highest economic loss (million yuan).

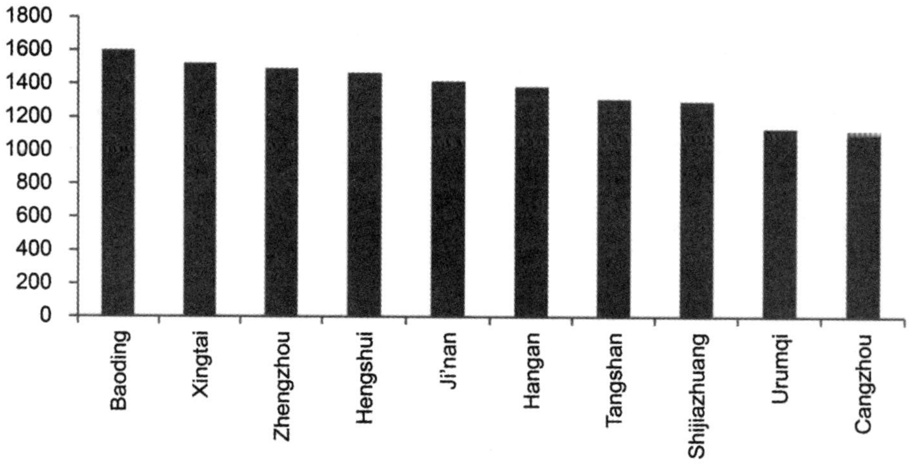

Figure 7. The 10 cities with the highest health-related economic loss per capita (yuan/per person).

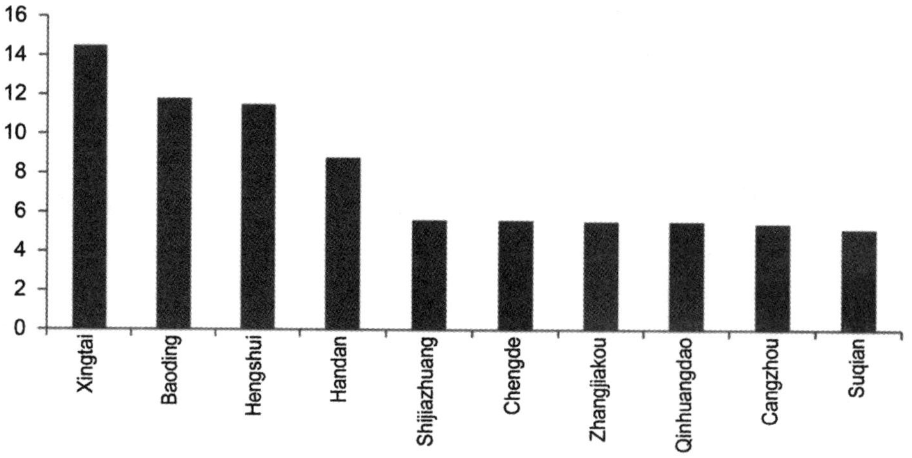

Figure 8. The 10 cities with the highest proportion of the health-related economic loss accounting for GDP (%).

5. Conclusions and Policy Implications

Based on the dose-response relationship and the willingness to pay method, this paper evaluated the health-related economic loss caused by air pollution in China's 74 cities using the latest available data regarding to PM_{10} and SO_2 from January 2015 to June 2015, by establishing lowest and highest limit scenarios. The conclusions and policy implications are as follows.

6. Conclusions

1) For the whole 74 cities

The health-related economic loss caused by PM_{10} was larger than that caused by SO_2 in the lowest and highest limit scenarios, and the economic loss associated with chronic bronchitis caused by PM_{10} was the largest in all the losses. Thus, PM_{10} has become the main air pollutant in 74 cities, and it is necessary to focus on the issue of chronic bronchitis caused by PM_{10}.

2) For the major cities

In the lowest and highest limit scenarios, the health-related economic loss in Chongqing, Beijing, Baoding, Tianjin and other major cities was larger than in other cities. The health-related economic loss per capita in Baoding, Xingtai and Zhengzhou was higher than in other cities. Regarding to the proportion of the health-related economic loss accounting for GDP, there were 9 cities in the Hebei Province included in the top 10 cities with the highest loss. It was evident that the air pollution was serious in North China, particularly in the Hebei Province, except in Shanghai in East China and in Chongqing and Chengdu in the southwest of China.

7. Policy Implications

1) According to the results of this paper, Hebei Province is a typical polluted area.

Energy consumption structure in Hebei Province is mainly composed of coal resource. Main pollutants from coal combustion are PM_{10} and SO_2, etc. Thus, to reduce air pollution, the coal-based energy structure needs to be improved, and new energies and advanced production processes need to be utilized. Furthermore, strengthening the development of technology and equipment; improving combustion technology, combustion devices and the fuel utilization rate; and reducing the additional pollutants generated by fuel burning will help reduce air pollution.

2) Pollutant concentration is an important factor affecting the loss of health effects.

The government should adopt the emissions trading system, limit the pollutant discharges of enterprises, grasp the influence of the enterprises on environmental pollution, strengthen the management of enterprises, and develop a number of administrative regulations conducive to environmental protection to ensure the implementation of environmental protection measures.

3) From the calculation results, the urban population is also an important factor affecting the loss of health effects. The government should reasonably control the urban population and improve people's awareness of environmental protection. Accelerating the transfer of industries and upgrading may reduce the non-household population in order to reduce the side-effects caused by population growth and improve the urban environment and public health.

Here the paper uses the data from January to June 2015 to calculate the health-related economic loss in 74 cities. If the paper uses a long term data, for example, which covers from January to December 2015, there may be a larger health-related economic loss (assuming the population in 74 cities is constant). However, the health-related economic loss from January to December 2015 may not be twice as much as that from January to June 2015. As it can been seen from **Figure 1** and **Figure 2**, the monthly mean concentration data of SO_2 and PM_{10} from January to June 2015 appeared a downward trend. In the future, data (if complete and available) can be combined using a geographic information system and other new tools to determine the economic loss of caused by the effects of air pollution on public health in the typical resource-based regions or cities and to provide references for environmental management and sustainable development.

Authors' Contributions

LL and YL designed the research and methodology; DP, CY collected the data and compiled all the data and literature; LL and CS finished the experiment and calculation; YL and LL analyzed the results and put forward the policies; YL and LL revised the manuscripts and approved the manuscripts; YL will be responsible for the future questions from readers as the corresponding authors. All authors read and approved the final manuscript.

Acknowledgements

The authors express their sincere thanks for the support from the National Natural Science Foundation of China under Grant No. 71173200 and the support from the Development and Research Center of China Geological Survey under Grant Nos. 1212011220302 and 12120114056601, Key Laboratory of Carrying Capacity Assessment for Resource and Environment, Ministry of Land and Resources (Chinese Academy of Land and Resource Economics, China University of Geosciences Beijing) under Grant No. CCA2015.08.

Competing Interests

The authors declare that they have no competing interests.

Source: Li L, Lei Y, Pan D, *et al.* Economic evaluation of the air pollution effect on public health in China's 74 cities[J]. Springerplus, 2016, 5(1):1−16.

References

[1] Ami D, Aprahamian F, Chanel O (2014) Willingness to pay of committed citizens: a field experiment. Ecol Econ 105:31-39 Askcinet (2015) The people's number and GDP ranking in 74 cities in the first half of 2015. http://www.askci.com/data/2015/07/23/15652fovg.shtml. Accessed on 23 Jul 2015.

[2] Brandt S, Perez L, Künzli N (2014) Cost of near-roadway and regional air pollution-attributable childhood asthma in Los Angeles County. J Allergy Clin Immunol 5:1028−1035.

[3] Cai YP, Yang ZF (2003) Health loss estimation of air pollution in the township industrial enterprises in Tianjin. J Saf Environ 1:33−36.

[4] Cai CG, Chen G, Qiao XC (2007) Comparison of the value evaluation method on the two fractional conditions of the single boundary and two boundary conditions—taking the investigation on health hazard caused by air pollution in Beijing city as an example. China Environ Sci 1:39−43.

[5] Carlsson F, Martinsson P (2001) Willingness to pay for reduction in air pollution: a multilevel analysis. Environ Econ Policy Stud 1:17–27.

[6] Chen BH, Hong CJ (2002) Quantitative assessment on the health effect of SO2 pollution in Shanghai City. J Environ Health 1:11–13.

[7] Chen BH, Hong CJ, Kan HD (2001) Methodological research on the health-based risk assessment on air pollution. J Environ Health 2:67–69.

[8] Chen RJ, Chen BH, Kan HD (2010) A health-based economic assessment of particulate air pollution in 113 Chinese cities. China Environ Sci 3:410–415.

[9] Chen J, Li W, Cheng HG (2015) Evaluation of emission reduction potentials of key air pollutants and health benefits for residents of Beijing. Res Environ Sci 7:1114–1121.

[10] China's National Environmental Monitoring Centre (2013a) Monthly air quality report in January 2013 in 74 cities. http://www.cnemc.cn/publish/106/news/news_33883.html. Accessed on 07 Feb 2013.

[11] China's National Environmental Monitoring Centre (2013b) Monthly air quality report in November 2014 in 74 cities. http://www.cnemc.cn/publish/106/news/news_43865.html. Accessed on 21 Dec 2013.

[12] China's National Environmental Monitoring Centre (2015a) Monthly air quality report in January 2015 in 74 cities. http://www.cnemc.cn/publish/106/news/news _44259.html. Accessed on 05 Feb 2015.

[13] China's National Environmental Monitoring Centre (2015b) Monthly air quality report in Febuary 2015 in 74 cities. http://www.cnemc.cn/publish/106/news/news_44345.html. Accessed on 17 Mar 2015.

[14] China's National Environmental Monitoring Centre (2015c) Monthly air quality report in March 2015 in 74 cities. http://www.cnemc.cn/publish/106/news/news_44415.html. Accessed on 14 Apr 2015.

[15] China's National Environmental Monitoring Centre (2015d) Monthly air quality report in April 2015 in 74 cities. http://www.cnemc.cn/publish/106/news/news_44867.html. Accessed on 25 May 2015.

[16] China's National Environmental Monitoring Centre (2015e) Monthly air quality report in May 2015 in 74 cities. http://www.cnemc.cn/publish/106/news/news_45103.html. Accessed on 16 May 2015.

[17] China's National Environmental Monitoring Centre (2015f) Monthly air quality report in June 2015 in 74 cities. http://www.cnemc.cn/publish/106/news/news_45341.html. Accessed on 14 Jul 2015.

[18] Dockery DW, Pope CA, Xu X (1993) An association between air pollution and mortality in six US cities. N Engl J Med 24:1753–1759.

[19] Gao J, Xu XP, Li BL (1993) Investigation on the relationship between air pollution

and death in Haidian District, Beijing. Chin J Prev Control Chronic Non-Commun Dis 5:207–210.

[20] Han Q (2011) The health damage caused by the particulate matters in air pollutants in Beijing—human capital method. North Environ 11:150–152.

[21] Han MX, Guo XM, Zhang YS (2006) Human capital loss of urban air pollution. China Environ Sci 4:509–512.

[22] Hedley AJ, McGhee SM, Barron B (2008) Air pollution: costs and paths to a solution in Hong Kong—understanding the connections among visibility, air pollution, and health costs in pursuit of accountability, environmental justice, and health protection. J Toxicol Environ Health Part A 9–10:544–554.

[23] Ho MS, Jorgenson DW (2007) Sector allocation of emissions and damages in clearing the air: the health and economic damages of air pollution in China. The MIT Press, Cambridge.

[24] Ho MS, Nielsen CP (2007) Clearing the air: the health and economic damages of air pollution in China. The MIT Press, Cambridge.

[25] Huanqiunet (2014) The World Health Organization reported that a total of 7 million people died from air pollution in 2012. http://world.huanqiu.com/exclusive/2014-03/4929416.html. Accessed on 25 Mar 2014.

[26] Huanqiunet (2015) WHO: air pollution has become the world's largest environmental health risk. http://health.huanqiu.com/health_news/2015-06/6581336.html?referer=huanqiu. Accessed on 02 Jun 2015.

[27] Istamto T, Houthuijs D, Lebret E (2014) Willingness to pay to avoid health risks from road-traffic-related air pollution and noise across five countries. Sci Total Environ 497:420–429.

[28] Jia L, Guttikunda SK, Carmichael GR (2004) Quantifying the human health benefits of curbing air pollution in Shanghai. J Environ Manage 1:49–62.

[29] Jing LB, Ren CX (2000) Relationship between air pollution and acute and chronic respiratory diseases in Benxi City. J Environ Health 5:268–270.

[30] Kaushik CP, Ravindra K, Yadav K (2006) Assessment of ambient air quality in urban centres of Haryana (India) in relation to different anthropogenic activities and health risks. Environ Monit Assess 1–3:27–40.

[31] Koop G, Tole L (2004) Measuring the health effects of air pollution: to what extent can we really say that people are dying from bad air? J Environ Econ Manage 1:30–54.

[32] Ministry of Environmental Protection of the People's Republic of China (2011) National environmental protection "12th Five-Year Plan". http://gcs.mep.gov.cn/hjgh/shierwu/201112/t20111221_221595.htm. Accessed on 21 Dec 2011.

[33] Murray CJ, Lopea AD (1997) Global mortality, disability, and the contribution of risk

factors: global burden of disease study. Lancet 349:1436–1442.

[34] Pascal M, Corso M, Chanel O (2013) Assessing the public health impacts of urban air pollution in 25 European cities: results of the Aphekom project. Sci Total Environ 449: 390–400.

[35] Patankar AM, Trivedi PL (2011) Monetary burden of health impacts of air pollution in Mumbai, India: implications for public health policy. Public Health 3:157–164.

[36] Peng XZ, Tian WH (2003) Study on the willingness to pay for the economic loss of air pollution in Shanghai City. World Econ Pap 2:32–44.

[37] Ridker RG (1967) Economic costs of air pollution: studies in measurement. Praeger, New York.

[38] Samet JM, Zeger SL, Dominici F (2000) The national morbidity, mortality, and air pollution study. Part II: morbidity and mortality from air pollution in the United States. Res Rep Health Eff Inst 2:5–79.

[39] Shang YH, Zhou DJ, Yang J (2010) Study on the economic loss of the human health caused by air pollution. Ecol Econ 1:178–179.

[40] Shen XW, Wang YH, Zhang WX (2014) The economic loss of the human health caused by air pollution in Kunming. China Market 46:124–126.

[41] U.S. EPA (1999) Benefits and costs of the clean air act. U.S. EPA Office of Air and Radiation, Washington, DC.

[42] Wan Y, Yang HW, Masui T (2005) Health and economic impacts of air pollution in China: a comparison of the general equilibrium approach and human capital approach. Biomed Environ Sci 6:427.

[43] Wang H, John M (2006) Willingness to pay for reducing fatal risk by improving air quality: a contingent valuation study in Chongqing, China. Sci Total Environ 1:50–57.

[44] Wang X, Smith K (1999) Near-term health benefits of greenhouse gas reductions: a proposed assessment method and application to two energy sectors of China. World Health Organization, Geneva.

[45] Wang Y, Zhao XL, Xu Y, Li Y (2005) Estimation on the economic loss of air pollution in Shandong Province. Urban Environ Urban Ecol 2:30–33.

[46] Wei YM, Wu G, Liang QM (2012) China Energy Security Report (2012): Study on energy security. Science Press, Beijing.

[47] Wilkinson P, Smith KR (2007) A global perspective on energy: health effects and injustices. The Lancet 370(9591):965–978.

[48] Wong TW, Lau TS, Yu TS (1999) Air pollution and hospital admissions for respiratory and cardiovascular diseases in Hong Kong. Occup Environ Med 10:679–683.

[49] Wong CM, Atkinson RW, Anderson HR (2002) A tale of two cities: effects of air pol-

lution on hospital admissions in Hong Kong and London compared. Environ Health Perspect 1:67.

[50] World Bank (1997) Clear water, blue skies: China's environment in the new century. World Bank, Washington, DC World Bank SEPA (2007) Cost of pollution in China. World Bank, World Bank.

[51] Xia G (1998) Economic measurement and study on the loss of environmental pollution in China. China Environmental Sciences Press, Beijing.

[52] Xu ZY, Liu YQ, Yu DQ (1996) Impact of air pollution on mortality in Shenyang. Chin J Public Health 1:61–64.

[53] Xu LY, Yin H, Xie XD (2014a) Health risk assessment of inhalable particulate matter in Beijing based on the thermal environment. Int J Environ Res Public Health 12: 12368–12388.

[54] Xu MM, Guo YM, Zhang YJ (2014b) Spatiotemporal analysis of particulate air pollution and ischemic heart disease mortality in Beijing, China. Environ Health 1:1–12.

[55] Yaduma N, Kortelainen M, Wossink A (2013) Estimating mortality and economic costs of particulate air pollution in developing countries: the case of Nigeria. Environ Resour Econ 3:361–387.

[56] Yan J (2012) Residents' breathing, the hospital cost of the disease of the circulatory system caused by air pollution in Lanzhou. In: Food safety and healthy life conference in Gansu in 2012, pp 61–64.

[57] Yu F, Guo XM, Zhang YS (2008) Evaluation on health economic loss caused by air pollution in 2004. J Environ Health 12:999–1003.

[58] Zeng XG, Xie F, Zong Q (2015) The behavior choice and willingness to pay to reduce the health risk of PM2. 5—taking the residents in Beijing as an example. China Popul Resour Environ 1:127–133.

[59] Zhang QF (2012) Toward an environmentally sustainable future country environmental analysis of the People's Republic of China. China Financial and Economic Publishing House, Beijing.

[60] Zhang GZ, Chun R, Nan ZR (2008) Study on the effects of air pollution on human health and economic loss in Lanzhou. J Arid Land Resour Environ 8:120–123.

[61] Zhou J, Wang Y, Ren L (2010) Willingness to pay for the improvement of air quality in Shandong typical city. J Environ Health 6:507–510.

[62] Zmirou D, Deloraine AMD, Balducci F, Boudet C (1999) Health effects costs of particulate air pollution. J Occup Environ Med 10:847–856.

Chapter 8

Are Income-Related Differences in Active Travel Associated with Physical Environmental Characteristics? A Multi-Level Ecological Approach

Esther Rind[1], Niamh Shortt[1], Richard Mitchell[2], Elizabeth A Richardson[1], Jamie Pearce[1]

[1]Centre for Research on Environment, Society and Health (CRESH), Institute of Geography, School of GeoSciences, University of Edinburgh, Edinburgh, UK
[2]Centre for Research on Environment, Society and Health (CRESH), Institute of Health and Wellbeing, University of Glasgow, Glasgow, UK

Abstract: Background: Rates of active travel vary by socio-economic position, with higher rates generally observed among less affluent populations. Aspects of both social and built environments have been shown to affect active travel, but little research has explored the influence of physical environmental characteristics, and less has examined whether physical environment affects socio-economic inequality in active travel. This study explored income-related differences in active travel in relation to multiple physical environmental characteristics including air pollution, climate and levels of green space, in urban areas across England. We hypothesised that any gradient in the relationship between income and active travel would be least pronounced in the least physically environmentally-deprived areas where higher income populations may be more likely to choose active transport as a means of travel. Methods: Adults aged 16+ living in urban areas (n = 20,146) were selected from the 2002 and 2003 waves of the UK National Travel Survey. The mode of all short non recreational trips undertaken by the sample was identified (n = 205,673). Three-level binary logistic regression models were used to explore how associations between the trip being active (by bike/walking) and three income groups, varied by level of multiple physical environmental deprivation. Results: Likelihood of making an active trip among the lowest income group appeared unaffected by physical environmental deprivation; 15.4% of their non-recreational trips were active in both the least and most environmentally-deprived areas. The income-related gradient in making active trips remained steep in the least environmentally-deprived areas because those in the highest income groups were markedly less likely to choose active travel when physical environment was 'good', compared to those on the lowest incomes (OR = 0.44, 95% CI = 0.22 to 0.89). Conclusions: The socio-economic gradient in active travel seems independent of physical environmental characteristics. Whilst more affluent populations enjoy advantages on some health outcomes, they will still benefit from increasing their levels of physical activity through active travel. Benefits of active travel to the whole community would include reduced vehicle emissions, reduced carbon consumption, the preservation or enhancement of infrastructure and the presentation of a 'normalised' behaviour.

Keywords: Active Travel, Urban Areas, Physical Environment, Health Inequality, Ecological Analysis

1. Background

There is growing evidence that active travel (walking or cycling for non-recreational purposes, including trips undertaken for commuting, business, shopping etc.) can contribute significantly to levels of overall physical activity[1][2], with associated benefits for health[3]. Even short bouts of activity have been shown to contribute to physical and mental well-being[3][4]. Yet, the average annual distance actively travelled per person in the UK actually decreased by 28% (from 306 to 221 miles) since the 1970s[5] (though this decline may be slowing or even reversing[6]). For many local journeys, walking or cycling could be a reasonable alternative to motorised transport and, in addition to conveying health benefits to the participant, would contribute to a reduction in traffic-related air pollution, accidents, and carbon use, as well as helping to normalise physical activity[2][7]. These environmental and health-related co-benefits emphasise the potential importance of creating an environment conducive to active transport[7].

The environmental and social determinants of health behaviours, including active travel, have been conceptualised using ecological models[8][9]. These comprise 'layers' of influence including individual biological (e.g. age), psychological (e.g. attitudes towards physical activity), intrapersonal (e.g. social support) influences, and broader social and physical environmental factors[10]. Prior research has explored aspects of the social environment, such as levels of social capital and perceived safety of an area[10][11] but research on the physical environment has been largely focused on examining built environment features, including functional patterns (e.g. street connectivity), safety issues (e.g. heavy traffic), aesthetic components (e.g. maintenance of green spaces), and destination accessibility (e.g. proximity to shops)[12]. Less is known about how active travel patterns may be related to 'natural' physical environmental factors such as air pollution, climate and green space.

A small study of residents from the San Francisco Bay area revealed that exogenous factors such as topography, darkness and rainfall had stronger associations with walking and cycling than did established characteristics of the built environment including street connectivity, land use mix, and proximity to retail facilities[13]. Other results from the US indicate that active transport is higher in areas with access to national parks, forests and blue space, and that greater participation in cycling is associated with aspects of moderate climate, topography, and low levels of air pollution[14]. The latter has been a concern to on-road cyclists, a group more likely to be exposed to harmful levels of air pollutants[15] potentially increasing the risk of respiratory and cardiovascular morbidity and mortality[16]. Physical features including pleasant views of gardens, roadside greenery and other green spaces, as well as low air pollution, may also encourage older people to walk to destinations such as shops or community services[17].

Some research has also reported a socio-economic gradient in active travel, with higher levels often found in the most socio-economically deprived groups[18]. However, there is a less than consistent relationship between active travel and socio-economic position; it appears to vary by place and time[19]–[21]. Where higher levels of active travel are found among more deprived populations, it has been attributed to a lack of material resources including car access, resulting in greater dependency on modes of active travel[18][22]. Although there have been calls for a more integrated analysis of all determinants of active transport[23], including the social and the physical environment[24][25], little is known about whether, and how, the income-related social gradient in active travel patterns may be related to physical environment. A better understanding might aid the development of tailored interventions aiming to increase levels of physical activity in particular population subgroups, by considering their socio-economic circumstances as well as characteristics of their local environment.

In previous research we explored the association between physical activity and multiple aspects of the physical environment using an index of multiple environmental deprivation (MEDIx)[26] derived for the year 2001. The index consisted of aspects of the environment that are both health damaging (air pollution, proximity to industry and cold climate) and health enabling (green space and UV levels)[27]. Results demonstrated respondents were most likely to engage in active travel, and specifically walking for transport, in the most physically-deprived environments. The 'choice' to engage in active travel is not made in isolation, but rather reflects broader socio-ecological environments, alongside individual characteristics. For the most economically deprived populations, the affordability of mechanised transport may constrain an individual's choice and as such, render levels of active travel within this population group unrelated to broader physical environment. Yet, a more conducive physical environment might encourage those less dependent on active transport to choose walking or cycling over mechanised transport. If this hypothesis is correct, we would expect the social gradient in active transport to be reduced in areas with better physical environments. It is likely that the most deprived populations will still have to actively travel, but a better quality environment may encourage more affluent groups to actively travel, and hence reduce the inequality. The relationship between multiple aspects of physical environment and socio-economic inequalities in active travel remains unexplored.

To address this knowledge gap, we explored income-related differences in active travel in relation to physical environmental disadvantage in urban areas across England. We expected the income-related gradient in active travel to be less pronounced in the least environmentally-deprived areas where, in comparison with more environmentally-deprived areas, more affluent individuals may be more likely to choose active travel.

2. Methods

2.1. Survey Data

Active travel data were taken from the National Travel Survey (NTS), a nationally-representative cross-sectional survey first commissioned by the Ministry of Transport in 1965 to monitor long-term changes in individual travel patterns in Great Britain[28]. The full sampling and interview methodology are described elsewhere[29]. In short, face-to-face interviewing was used to collect key socio-economic, demographic and travel-related characteristics of participants. A sub-group of individuals completed a travel diary recording trips undertaken over the course of a week. To boost the sample size for statistical analysis, we pooled data from the survey waves 2002 and 2003 (n = 42,817, including 33,717 adults aged 16+ and 9,100 children <16 years) which matched the timeframe of the available environmental data. Whilst acknowledging the age of these data, these survey years were selected to closely match the time period for which the measure of physical environment was available. Our final sample included all participants of the diary subsample (age 16+) with full information on active travel, living in urban areas (n = 20,146).We opted to include those aged 16 and 17 in our definition of adults as they are potentially working people, and are normally making independent journeys and travel decisions at these ages. Those aged over 17 are also eligible to drive in the UK. Note that ethical approval was not required for the analysis of this publicly available, anonymised secondary dataset.

2.2. Active Travel

We defined active travel as walking or cycling for non-recreational purposes, including for commuting, business, education, shopping and any other personal business. We measured such active travel at 'trip' level. In this case a trip is defined as a one-way course of travel from one place to another with a single main purpose. First we identified all trips (active and mechanised) with a non-recreational purpose. We then selected trips for which there might reasonably be a 'choice' over mode: we excluded trips that were so short as to be almost certainly walked or biked, defined as less than 1/10th of a mile (160m), and trips that were so long that active travel was likely only by cycling enthusiasts, defined as more than 5 miles (8km). These thresholds were defined based on the distribution of active travel mode within the data; almost all trips less than 1/10th mile (160m) were active, but almost none more than 5 miles (8km) were active. We considered separate analysis for cycling only, but numbers were too low. This approach identified 205,673 trips of interest. The resulting binary outcome variable then assessed whether a trip of interest was made actively (*i.e.* by walking or cycling) or not (*i.e.* by mechanised means). The mode of travel referred to the main mode, *i.e.* that which was used to complete most of the journey.

2.3. Trip-, Individual-, Household- and Area-Level Covariates

We selected a range of covariates known to be associated with active travel. All models were adjusted for trip distance (in miles), age group (16 to 29, 30 to 49, 50 to 69, and 70+), sex (male/female), ethnicity (White/non-White), self-reported walking difficulties (yes/no), car access (yes/no), bicycle access (yes/no) and household income (<£25,000, £25,000 to £49,999, and £50,000+).

Previous research has shown that residents living in socio-economically deprived neighbourhoods were more likely to actively travel than their less deprived counterparts[18]. Hence, we included the Carstairs Deprivation Index 2001, a well-established and robust area-level measure of socioeconomic deprivation including low social class, lack of car ownership, overcrowding and male unemployment[30]. We were concerned about the inclusion of car ownership in the area-level measure of deprivation since it is known to be skewed by urbanity and may also reflect alternative transport options (active or public transport) in the neighbourhood. However, there was no other measure available for the whole UK, on a consistent basis for this time period. Although our sample was urban residents only, we included a measure of urban settlement size to adjust our analysis for increased public transport demand and provision in more densely populated urban areas[31][32]. Based on their area of residence, each household of the survey was assigned to an urban category including very large (population >250,000), large (population >10,000 to 250,000) and smaller (population over 3,000 to 10,000) urban areas[29]. Other correlates, plausibly associated with active travel, were also explored. These included interview season, employment status, and having children <16 years in the household. These were not significantly associated with active travel ($p > 0.05$) and were thus excluded from further analysis.

The appropriate weights for trip-level analyses were provided by the NTS and were applied to all models. These accounted both for the drop-off in the number of trips recorded by participants over the course of the week and non-response of households to the survey[29].

2.4. Measuring Environmental Deprivation

Physical environment was measured using the multiple environmental deprivation index (MEDIx) for UK Census Area Statistic (CAS) wards (n = 10,654). The development of this indicator is described elsewhere[27][32]. In short, MEDIx assesses environmental dimensions both detrimental (air pollution, proximity to industry and cold climate measures) and beneficial (green space availability and UVB measures) to health. MEDIx scores range from −2 (least deprived) to +3 (most deprived). Due to small numbers at the extreme ends of the MEDIx scale, we combined categories MEDIx +2 and +3 into one group, and MEDIx categories −1 and −2 into another, leaving four categories of environmental deprivation. For each of the NTS respondents a CAS ward of residence identifier was obtained which made it possible to link information on individual active travel patterns to physical environmental deprivation. For reasons of confidentiality, survey respondents living in wards with an individually-identifiable combination of environmental characteristics were excluded from this study [n = 4538 (18.4%)]. In comparison to the original dataset, the final dataset (n = 20,146) comprised more adults (+4.1%) living in very large urban areas with a population of >250,000, but fewer observations (−2.4%) in urban areas with a population of >10,000 to 250,000. All other socio-demographic differences between the included and excluded participants were ≤±2.0%.

Finally, we also explored adjusting for region of residence, and in particular for residence in London given its unique urban structure and public transport infrastructure. Analyses showed, however, that i) relationships between active trips and income did not seem to vary significantly by residence in Greater London but also ii) that there were collinearities between this variable and MEDIx score, Carstairs score and urban category. We therefore opted not to adjust for region.

3. Statistical Analysis

3.1. Descriptive Statistics

First, we tested univariate associations between active trips and all covariates using χ^2-tests. We then explored the relationship between income and active trips across different categories of environmental deprivation using a graph.

3.2. Multivariate, Multi-Level Models

We ran a three-level binomial logistic random-intercept model, predicting the choice of active mode for a trip, where trips (level 1) were nested within individuals (level 2, nested in CAS wards (level 3). The trip weight included a household weight which allowed us to run a three-level model. Models were run first without covariates and were then adjusted. We initially stratified the analysis by MEDIx category to explore the nature of any interactions, following an approach previously applied in health-related research e.g.[33][34]. A full non-stratified model was then run, including interaction terms for income group and MEDIx to establish formally whether income-related differences in active trip mode varied significantly (at a significance level of $p < 0.05$) by physical environmental deprivation. The lowest income group (less than £25,000) and environments with an intermediate level of environmental deprivation (MEDIx 0) were used as reference categories. This choice was based on associations between household income, active trips and environment observed at the descriptive statistics stage. We checked the impact of our reference category choices by repeating analyses using alternatives, but there were no substantive differences. All statistical analyses were run in Stata/IC 12.1[35].

4. Results

Characteristics of the Sample in Relation to Active Trips

About 13% of the 205,673 trips were active (**Table 1**). Further characteristics of the study sample and univariate associations with trip mode are summarised in **Table 1**.

Table 1. Characteristics of the study sample in relation to making a trip of interest (0.1 to 5 miles) by active means, including adults aged 16+ from urban areas, National Travel Survey 2002 and 2003.

	$^{+}n_{weighted}$	%	$^{+}n_{weighted}$	% active	Pearson chi^2	p-value
Age group					2600.0	<0.001
16–29	55,047	26.8	6,166	11.2		
30–49	88,960	43.3	10,120	11.4		
50–69	41,651	20.3	8,985	21.6		
70+	20,015	9.7	2,382	11.9		
Total	205,673	100.0	27,652	13.4		
Sex					119.7	<0.001
Males	93,521	45.5	13,618	14.6		
Females	112,152	54.5	14,034	12.5		
Total	205,673	100.0	27,652	13.4		
Ethnicity					38.4	<0.001
White	189,176	92.0	25,239	13.3		
Non-White	16,486	8.0	2,408	14.6		
Total	205,662	100.0	27,647	13.4		
Missing	11	<0.1%	5			
Walking difficulties					371.7	<0.001
Yes	22,148	10.8	2,015	7.9		
No	183,505	89.2	25,633	14.0		
Total	205,654	100.0	27,648	13.4		
Missing	19	<0.1%	4			
Car access					8800.0	<0.001
Yes	175,691	85.4	18,736	10.7		
No	29,982	14.6	8,916	29.7		
Total	205,673	100.0	27,652	13.4		
Bike access					1100.0	<0.001
Yes	80,576	39.2	14,224	17.7		
No	125,088	60.8	13,424	10.7		
Total	205,665	100.0	27,648	13.4		
Missing	8	<0.1%	4			

Continued

Journey distance					21000.0	<0.001
<1 mile	20,291	9.9	1,047	5.2		
1 to <2 miles	72,991	35.5	20,392	27.9		
2 to <3 miles	50,474	24.5	4,208	8.3		
3 to <5 miles	61,918	30.1	2,006	3.2		
Total	205,673	100.0	27,652	13.4		
Household income					849.9	<0.001
less than £25,000	99,289	48.3	15,603	15.7		
£25,000 to £49,999	74,631	36.3	8,802	11.8		
£50,000 and over	31,753	15.4	3,248	10.2		
Total	205,673	100.0	27,652	13.4		
Socio-economic deprivation					783.5	<0.001
1 (Carstairs tertiles)	67,033	32.6	6,978	10.4		
2 (least deprived)	70,578	34.3	9,887	14.0		
3 (most deprived)	68,062	33.1	10,788	15.8		
Total	205,673	100.0	27,652	13.4		
Urban classification					221.3	<0.001
population over 3k to 10k	10,740	5.2	1,323	12.3		
population over 10k to 250k	88,373	43.0	12,876	14.6		
population over 250k	106,561	51.8	13,453	12.6		
Total	205,673	100.0	27,652	13.4		
Environmental deprivation					93.6	<0.001
(MEDIx* category)						
−2/−1 (least deprived)	19,441	9.5	2,590	13.3		
0	53,786	26.2	7,755	14.4		
+1	101,770	49.5	13,554	13.3		
+2/+3 (most deprived)	30,676	14.9	3,753	12.2		
Total	205,673	100.0	27,652	13.4		

†Results weighted for sample and trip bias (see Methods). *Multiple Environmental Deprivation Index, capturing small-area exposure to multiple health-related environmental characteristics including air pollution, proximity to industry, cold climate, green space and UVB.

Whilst all of the bivariate associations were significant, this is likely to have been a consequence of the large sample size. Unsurprisingly, not having walking problems, having a bike, and not having a car appeared strongly related to making an active trip. Trip distance was also strongly negatively associated with active mode. Of particular interest for this analysis was the negative association between household income group and active trip choice. However, associations between MEDIx and active trips were more modest, with no clear gradient.

(**Figure 1**) presents the relationship between active trip mode and income across the categories of environmental deprivation. These are unadjusted values, but in fact give a very clear picture of the results obtained from the adjusted models. An income-related gradient in active trips was clear across all of the MEDIx categories, with the lowest income group always reporting the highest levels of active trips. **Figure 1** suggests that choosing active trip mode was barely affected by environmental deprivation for those in the lowest income group. Indeed, the proportion of non-recreational trips being made actively in the least and most environmentally deprived areas was the same, at 15.4%. However, **Figure 1** also suggests that the middle and higher income groups did show some sensitivity to

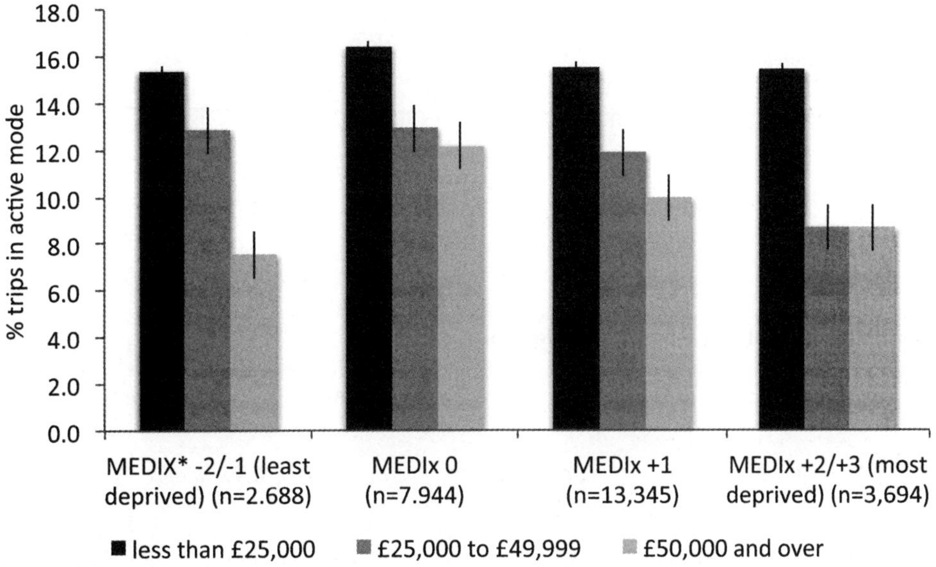

Figure 1. Percentage of trips of interest (0.1 to 5 miles) made by active means, stratified by income group and environmental deprivation. *Multiple Environmental Deprivation Index, capturing small-area exposure to multiple health-related environmental characteristics including air pollution, proximity to industry, cold climate, green space and UVB.

environmental deprivation. The middle income group was increasingly less likely to choose an active mode for their trips as environmental deprivation worsened. The high income group followed this pattern too, but only as MEDIx score worsened from 0 to +2/+3. In the least environmentally deprived areas (MEDIx −2/−1) the high income group showed a marked and significantly lower likelihood to choose an active mode. **Table 2** presents results from stratified models which confirm that adjustment for covariates did not alter the observed associations. Compared to the lowest income group, the odds of choosing to make an active trip reduced with increasing environmental deprivation among those in the middle income group. Yet, they were also significantly lower among the highest income group living in the least environmentally deprived areas (OR = 0.44,

Table 2. Odds ratios for the likelihood of trips of interest (0.1 to 5 miles) being made by active means in the middle and highest income groups, relative to the lowest income group, stratified by environmental deprivation.

	Groups of environmental deprivation											
	MEDIx^ −2/−1			MEDIx 0			MEDIx +1			MEDIx +2/+3		
	(least deprived)									(most deprived)		
	OR	95% CI		OR	95% CI		OR	95% CI		OR	95% CI	
Unadjusted models												
Less than £25,000	1.00			1.00			1.00			1.00		
£25,000 to £49,999	0.60	0.45	0.81	0.79	0.63	0.98	0.68	0.58	0.81	0.53	0.39	0.71
£50,000 and over	0.32	0.18	0.56	0.66	0.48	0.90	0.51	0.40	0.64	0.43	0.29	0.62
Adjusted for individual variables & contextual covariates												
Less than £25,000	1.00			1.00			1.00			1.00		
£25,000 to £49,999	0.97	0.60	1.56	0.92	0.73	1.17	0.81	0.67	0.99	0.68	0.48	0.95
£50,000 and over	0.44	0.22	0.89	0.87	0.62	1.21	0.62	0.48	0.81	0.57	0.36	0.89

Bold = significant at $p < 0.01$; italics = significant at $p < 0.05$. ^Multiple Environmental Deprivation Index, capturing small-area exposure to multiple health-related environmental characteristics including air pollution, proximity to industry, cold climate, green space and UVB.

95% CI = 0.22 to 0.89). Not surprisingly, the full model showed no evidence of a significant interaction in the association between household income and active trips, by environmental deprivation (results not shown).

5. Discussion

Regardless of levels of physical environmental deprivation, those in the lowest income groups had greater odds of making active non-recreational trips. Contrary to our expectations however, the income-related gradient in making active trips remained as steep in the least environmentally-deprived areas as in the most environmentally-deprived areas. This reflects the fact that those in the highest income groups were markedly less likely to choose active travel when the physical environment was good. Within the limits of a cross-sectional study design, the results suggest that improving physical environment is unlikely to mitigate socio-economic differences in active travel.

To our knowledge this was the first study assessing variation in income-related inequalities in active travel by multiple characteristics of the physical environment. The particular advantage of the MEDIx score is that it enables us to look not only at negative aspects of physical environment, such as air pollution and cold climate, but positive aspects such as green space. Our results suggest that where environment is adverse, those whose incomes may enable them to choose not to actively travel, are indeed more likely to do so. Yet, the most benign or favourable physical environment does not seem to encourage active travel among the same higher income groups.

Previous research had suggested that neither the social[11] nor the built-environment[20] are major factors in determining active transport behaviours in socio-economically deprived populations. For example, in a study on active travel amongst a deprived urban population group in the UK, Ogilvie *et al.*[21] concluded that measures of the local environment did not explain much of the variance in active travel. Their research points to the fact that such population groups 'may simply have adapted to adverse conditions in their local environment'[21]. It may be that the higher levels of active travel among those on low incomes in our study were also evidence of 'adapting' to such adverse conditions. Such 'adaptation' raises questions of risk and harm. Literature on active travel has discussed whether the benefits of active travel always outweigh the associated costs and risks, in particular focussing on active travel in adverse environments. We know that risks of active transport include increased mortality and morbidity from traffic accidents[16], the mental drain of having to rely on such forms of transport in stressful environments[23] as well as exposure to air pollution[7] which is related to cardiovascular mortality and respiratory conditions[36]. Gatrell[37] recently highlighted that all modes of transport, including walking and cycling, have 'dis-benefits'. Previous research has also shown that health is worse in areas with higher levels of multiple environmental deprivation[27]. It is pertinent to question whether higher levels of active travel in areas of extreme environmental deprivation are wholly health-enabling. However, whilst there may be risks associated with active travel, the estimated health benefits substantially outweigh the risks relative to car driving[16]. Nevertheless, policies aiming to increase active transport also need to consider the risks involved, particularly for the most income-deprived populations in the most environmentally-deprived areas.

The low levels of active travel among the affluent in good physical environments is intriguing and of concern. It may reflect the ubiquity of motorised transport among socio-economically advantaged groups[6]. It is also possible that the combination of higher income and MEDIx score −2/−1 is capturing residence in environments which are different somehow to those with the same MEDIx score but occupied by those on middle or lower incomes. Whilst everyone in the sample lived in an urban area, and we adjusted for size of settlement, these crude measures cannot capture the other environmental characteristics which may influence active travel such as walkability, or the nature and number of destinations.

Does a low level of active travel for the more advantaged really matter? Whilst the affluent are more likely to reach recommended levels of physical activity through recreational activity and to retain a whole host of other health advantages over more deprived populations, we believe this is an important issue for at least two reasons. First, we are not just interested in the health of those who do or do not participate in active travel. Population health reflects the social and physical environment that everyone shares and contributes to. We know that the benefits of active travel to the non-participating community include reduced vehicle emissions, reduced carbon consumption, the preservation or enhancement of infrastructure and the presentation of a 'normalised' behaviour. If one group does not participate in active travel, this affects the health of others. Second, physical activity may reduce the risk of health outcomes which are not strongly socially patterned; mental health and wellbeing for example. Thus, just because a more affluent group seems systematically less likely to participate in active travel and they may on some health measures, be relatively advantaged, this does not mean that their health could not or should not be improved. Public health has a duty to maximise population health, not just minimise inequalities. We suggest therefore, that policy options to reduce private transport must target the more affluent. The related co-benefits will be felt by the whole of the population through the reduction of broader environmental concerns.

Key strengths of this paper include its use of the nationally representative National Travel Survey (NTS), designed to measure travel behaviour, the use of a robust multiple environmental deprivation measure which captures both 'good' and 'bad' physical environments, a large sample size, adjustment for a good range of potential confounders as identified by the literature and the use of multilevel models to allow for sample design. Alongside these however, were several limitations. The data used in the paper were relatively old. This was necessary because whilst the NTS is a regularly repeated survey, the measure of physical environmental deprivation was only available in 2001. It was essential to measure the environment and behaviour at about the same time. It is plausible that both physical environments, and active travel behaviour have changed since this time period and indeed, recent work in the UK has suggested some minor increases in this behaviour, but that socio-economic differences remain stark[6]. Whilst more up-to-date data would be desirable, it does not appear that there has been a substantial shift either in the levels, or the socio-economic distribution, of active travel sufficient to warrant our results irrelevant.

Effect sizes did not change substantially with adjustment for individual- and area-level confounders. However, other factors not included in this study may be more important in explaining the relationship between household income and active travel. These could include individual psychological measures such as personal attitudes, perceptions, motivations and preferences related to active travel[24]. Furthermore, it is possible that these factors are patterned by individual-level socio-demographic covariates included in our analysis. The extent to which residual confounding remains in our results is thus not clear.

The study modelled trips in which the main form of travel was active, rather than the mode of travel on each leg of each trip. This approach was taken because of the computational complexity of trying to model all legs on 205,673 trips. However, we acknowledge that mixed-mode journeys, for example walking to public transport, then taking it, then walking from the terminal to the destination, will not have been well handled. This was a very large study and what we have lost in detail by modelling trips, is arguably offset by the gain in modelling at a population scale, and with a range of physical environments.

The area-based measure of socio-economic deprivation deployed was not ideal as it included car ownership levels. However, the Carstairs Index is one of the most widely used and tested measures in the UK and it is known to be effective at identifying socio-economic situations likely to affect health and related behaviours. Use of the Carstairs Index is highly unlikely to have affected our substantive findings.

Although our outcome measure was based on self-reported travel mode, which may make it vulnerable to over-reporting of active travel levels[38], the prevalence of active transport in our study sample was relatively low (~13%) , and was similar to comparable findings from the US[39]. There is a substantial literature comparing self-reported levels of physical activity with those measured by objective devices such as accelerometers, and this literature often finds discrepancies between such measurements[38]. However, the literature exploring the validity or reliability of travel diaries is much smaller. Panter *et al.* have explored this issue in a recent study and their work suggested diaries led to overestimation of active commuting[40]. We have no reason to think the travel diary data from the NTS are abnormally unreliable or lack external validity, nor is there reason to believe that their quality would vary by type of physical environment the respondent lived in. However, we acknowledge that the data are based on self-report which may be subject to bias and inaccuracy.

Our study used cross-sectional data, which does not allow the inference of causality. Natural experiments have been shown to be better suited to establish determinants, rather than correlates, of physical activity[41]. However, natural experiments are designed to study particular behaviours (e.g. participation in active travel) before and after plausibly-related exogenous changes (e.g. new traffic infrastructure), and the results are therefore frequently specific to a particular locality and not necessarily transferrable to a broader population. In this paper we were interested in inequalities in active travel at the population-level. In terms of assessing health-relevant longitudinal changes in the physical environment it would be particularly challenging to collect comparable periodical data on a variety of measures such as captured by our indicator of environmental deprivation[27].

There have been attempts in the UK to apply a health framework to transport policies, with the aim of integrating perspectives related to economy, regeneration, social justice and health. Yet, UK transport policy has largely focused on improving conditions for motorised transport, neglecting interests of those participating in active travel, particularly in areas characterised by economic decline and social exclusion[42]. Changing broader social and environmental conditions for the whole of society might result in creating time, space and capacities for individuals to reconsider the feasibility of active transport. It might also mitigate the socially patterned risks involved for those who do actively travel[42].

6. Conclusions

This research found that the likelihood of choosing an active travel mode for short trips was relatively high for low income people in the most environmentally-disadvantaged areas, and relatively low for high income people in the least environmentally deprived areas. This suggests that physical environment, as measured in this work, is not a strong influence on socio-economic inequalities in active travel. Nevertheless improvements in the physical environment may mitigate the risks for those who actively travel and continue to encourage such forms of transport in the face of increasing car ownership.

Competing Interests

The authors declare that they have no competing interests.

Authors' Contributions

ER, NS, JP and RM contributed to the design of the study. EAR was responsible for the data acquisition. ER and EAR carried out the statistical analysis and ER wrote the first draft of the paper. RM drafted the revised version of the paper. All authors contributed to the interpretation of the results, critically reviewed and revised versions of the manuscript. All authors read and approved the final manuscript.

Acknowledgements

This research was supported by the Economic and Social Research Council [RES-000-22-3974] and the European Research Council [ERC-2010-StG Grant 263501]. The development of MEDIx was supported by the Joint Environment and Human Health Programme [NE/E008720/2] which was funded by the UK's Natural Environment Research Council, the Department for the Environment Food and Rural Affairs, the Environment Agency, the Ministry of Defence and the Medical Research Council.

Source: Rind E, Shortt N, Mitchell R, *et al.* Are income-related differences in active travel associated with physical environmental characteristics? A multi-level ecological approach[J]. International Journal of Behavioral Nutrition & Physical Activity, 2015, 12(1):1−10.

References

[1] Sahlqvist S, Song Y, Ogilvie D. Is active travel associated with greater physical activity? The contribution of commuting and non-commuting active travel to total physical activity in adults. Prev Med. 2012;55(3):206–11.

[2] Mackett RL, Brown B. Transport, Physical Activity and Health: Present knowledge and the way ahead. London: Department for Transport; 2011.

[3] Warburton DE, Nicol CW, Bredin SS. Prescribing exercise as preventive therapy. Can Med Assoc J. 2006;174(7):961–74.

[4] Garrard J. Active Transport: Adults. An Overview of Recent Evidence. Australia:

VicHealth; 2009.

[5] Townsend N, Bhatnagar P, Wickramasinghe K, Scarborough P, Foster C, Rayner M. Physical Activity Statistics 2012. London: British Heart Foundation; 2012.

[6] Goodman A. Walking, cycling and driving to work in the English and welsh 2011 census: trends.Socio-economic patterning and relevance to travel behaviour in general. PLoS One. 2012;8(8), e71790.

[7] de Nazelle A, Nieuwenhuijsen MJ, Antó JM, Brauer M, Briggs D, Braun-Fahrlander C, *et al.* Improving health through policies that promote active travel: A review of evidence to support integrated health impact assessment. Environ Int. 2011;37(4): 766–77.

[8] Sallis JF, Cervero RB, Ascher W, Henderson KA, Kraft MK, Kerr J. An ecological approach to creating active living communities. Annu Rev Public Health. 2006;27: 297–322.

[9] Spence JC, Lee RE. Toward a comprehensive model of physical activity. Psychol Sport Exerc. 2003;4(1):7–24.

[10] Bauman AE, Reis RS, Sallis JF, Wells JC, Loos RJF, Martin BW. Correlates of physical activity: why are some people physically active and others not? Lancet. 2012;380(9838):258–71.

[11] Caspi CE, Kawachi I, Subramanian SV, Tucker-Seeley R, Sorensen G. The social environment and walking behavior among low-income housing residents. Soc Sci Med. 2013;80:76–84.

[12] Pikora T, Giles-Corti B, Bull F, Jamrozik K, Donovan R. Developing a framework for assessment of the environmental determinants of walking and cycling. Soc Sci Med. 2003;56(8):1693–703.

[13] Cervero R, Duncan M. Walking, bicycling, and urban landscapes: evidence from the San Francisco Bay area. Am J Public Health. 2003;93(9):1478–83.

[14] Zahran S, Brody SD, Maghelal P, Prelog A, Lacy M. Cycling and walking: Explaining the spatial distribution of healthy modes of transportation in the United States. Transp Res Part D: Transp Environ. 2008;13(7):462–70.

[15] Kingham S, Longley I, Salmond J, Pattinson W, Shrestha K. Variations in exposure to traffic pollution while travelling by different modes in a low density, less congested city. Environ Pollut. 2013;181:211–8.

[16] Hartog JJ, Boogaard H, Nijland H, Hoek G. Do the health benefits of cycling outweigh the risks? Cien Saude Colet. 2011;16(12):4731–44.

[17] Franke T, Tong C, Ashe MC, McKay H, Sims-Gould J. The secrets of highly active older adults. J Aging Stud. 2013;27(4):398–409.

[18] Turrell G, Haynes M, Wilson L-A, Giles-Corti B. Can the built environment reduce health inequalities? A study of neighbourhood socioeconomic disadvantage and

walking for transport. Health Place. 2013;19:89–98.

[19] Stokes G, Lucas K. National Travel Survey Analysis. Working Paper No 1053. Oxford: Transport Studies Unit. School of Geography and the Environment; 2011.

[20] Adams J. Prevalence and socio-demographic correlates of "active transport" in the UK: analysis of the UK time use survey 2005. Prev Med. 2010;50(4):199–203.

[21] Ogilvie D, Mitchell R, Mutrie N, Petticrew M, Platt S. Personal and environmental correlates of active travel and physical activity in a deprived urban population. Int J Behav Nutr Phys Act. 2008;5(1):43.

[22] Chapman R, Howden-Chapman P, Keall M, Witten K, Abrahamse W, Woodward A, *et al.* Increasing active travel: aims, methods and baseline measures of a quasi-experimental study. BMC Public Health. 2014;14(1):935.

[23] Bostock L. Pathways of disadvantage? Walking as a mode of transport among low-income mothers. Health Soc Care Community. 2001;9(1):11–8.

[24] Panter JR, Jones A. Attitudes and the environment as determinants of active travel in adults: what Do and Don't We know? J Phys Act Health. 2010;7(4):551–61.

[25] National Institue for Health and Clinical Excellence (NICE). Promoting and Creating Built or Natural Environments that Envourage and Support Physical Activity. London: NHS; 2008.

[26] Shortt NK, Rind E, Pearce J, Mitchell R. Integrating environmental justice and socio-ecological models of health to understand population-level physical activity. Environ Planning A. 2014;46:1479–95.

[27] Richardson EA, Mitchell R, Shortt NK, Pearce J, Dawson TP. Developing summary measures of health-related multiple physical environmental deprivation for epidemiological research. Environ Planning A. 2010;42(7):1650–68.

[28] National Travel Survey data. User guidance 2002–2012. London: Department for Transport; 2012.

[29] Hayllar O, McDonnell P, Mottau C, Salathiel D. National Travel Survey Technical Report. London: NatCen Social Research; 2004.

[30] Carstairs V, Morris R. Deprivation and Health in Scotland. Aberdeen: Aberdeen University Press; 1991.

[31] Halcrow Group Ltd. Land use and transport: settlement patterns and the demand for travel. Stage 2 Background Technical Report PPRO/04/07/13. London: Commission for Integrated Transport (CfIT); 2009.

[32] Richardson EA, Mitchell RJ, Shortt NK, Pearce J, Dawson TP. Evidence-based selection of environmental factors and datasets for measuring multiple environmental deprivation in epidemiological research. Environ Heal. 2009;8(Supp 1):S18.

[33] Mitchell R, Popham F. Effect of exposure to natural environment on health inequali-

ties: an observational population study. Lancet. 2008;372(9650):1655–60.

[34] Bauman AE, Sallis JF, Dzewaltowski DA, Owen N. Toward a better understanding of the influences on physical activity: The role of determinants, correlates, causal variables, mediators, moderators, and confounders. Am J Prev Med. 2002;23(2):5–14.

[35] StataCorp. Stata Statistical Software: Release 12. College Station: TX: StataCorp LP; 2011.

[36] Allender S, Foster C, Scarborough P, Rayner M. The burden of physical activity-related ill health in the UK. J Epidemiol Community Health. 2007;61(4):344–8.

[37] Gatrell AC. Therapeutic mobilities: walking and 'steps' to wellbeing and health. Health Place. 2013;22:98–106.

[38] Prince SA, Adamo KB, Hamel ME, Hardt J, Gorber SC, Tremblay M.A comparison of direct versus self-report measures for assessing physical activity in adults: a systematic review. Int J Behav Nutr Phys Act. 2008;5(1):56.

[39] Kim D, Kawachi I. Contextual Determinants of Obesity: An Overview. In: Pearce J, Witten K, editors. Geographies of Obesity. Environmental Understanding of the Obesity Epidemic. Farnham, Burlington: Ashgate; 2010. p. 39–54.

[40] Panter J, Costa S, Dalton A, Jones A, Ogilvie D. Development of methods to objectively identify time spent using active and motorised modes of travel to work: how do self-reported measures compare? Int J Behav Nutr Phys Act. 2014;11:116.

[41] Ogilvie D, Bull F, Cooper A, Rutter H, Adams E, Brand C, *et al.* Evaluating the travel, physical activity and carbon impacts of a 'natural experiment' in the provision of new walking and cycling infrastructure: methods for the core module of the iConnect study. BMJ Open. 2012;2(1).

[42] Coyle E, Huws D, Monaghan S, Roddy G, Seery B, Staats P, *et al.* Transport and health—a five-country perspective. Public Health. 2008;123(1):e21–3.

Chapter 9

Association of Traffic-Related Hazardous Air Pollutants and Cervical Dysplasia in an Urban Multiethnic Population: A Cross-Sectional Study

Michael E. Scheurer[1*], Heather E. Danysh[1], Michele Follen[2], Philip J. Lupo[1]

[1]Department of Pediatrics, Section of Hematology-Oncology and Dan L. Duncan Cancer Center, Baylor College of Medicine, Houston, Texa
[2]Department of Obstetrics and Gynecology, Paul L. Foster School of Medicine, Texas Tech University Health Sciences Center, El Paso, Texas

Abstract: Background: Human papillomavirus (HPV) infection is a necessary cause in the development of cervical cancer; however, not all women infected with HPV develop cervical cancer indicating that other risk factors are involved. Our objective was to determine the association between exposure to ambient levels of common traffic-related air toxics and cervical dysplasia, a precursor lesion for cervical cancer. Methods: The study sample consisted of women enrolled in a Phase II clinical trial to evaluate diagnostic techniques for cervical disease in Houston, Texas. The current assessment is a secondary data analysis in which cases were defined as women diagnosed with cervical dysplasia, while those without cervical dysplasia served as controls. Residential census tract-level estimates of ambient benzene, diesel particulate matter (DPM), and polycyclic aromatic hydrocarbons (PAHs) were used to assess exposure. Census tract-level pollutant estimates were obtained from the United States Environmental Protection Agency. Multivariable logistic regression was used to estimate prevalence odds ratios (aOR) and 95% confidence intervals (CI) adjusted for age, race/ethnicity, education, smoking status, and HPV status. Results: Women in the highest residential exposure categories for benzene and DPM had an increased prevalence of cervical dysplasia compared to the lowest exposure category (Benzene: aOR [95% CI] for high exposure = 1.97 [1.07–3.62], very high exposure = 2.30 [1.19–4.46]. DPM: aOR [95% CI] for high exposure = 2.83 [1.55–5.16], very high exposure = 2.10 [1.07–4.11]). Similarly, women with high residential exposure to PAHs had an increased prevalence of cervical dysplasia (aOR [95% CI] = 2.46 [1.35–4.48]). The highest PAH exposure category was also positively associated with cervical dysplasia prevalence but was not statistically significant. Assessment of the combined effect of HAP exposure indicates that exposure to high levels of more than one HAP is positively associated with cervical dysplasia prevalence (p for trend = 0.004). Conclusions: Traffic-related HAPs, such as benzene, DPM, and PAHs, are not as well-regulated and monitored as criteria air pollutants (e.g., ozone), underscoring the need for studies evaluating the role of these toxicants on disease risk. Our results suggest that exposure to traffic-related air toxics may increase cervical dysplasia prevalence.

Keywords: Benzene, Cervical Dysplasia, Diesel Particulate Matter, Hazardous Air Pollutants, Polycyclic Aromatic Hydrocarbons

1. Background

Cervical cancer is the third most common cancer and the fourth leading cause of cancer-related mortality among women worldwide[1]. The majority (>85%) of cervical cancer cases and deaths occur in the developing world[1], largely due to limited screening programs which allow for the detection of cervical dysplasia (precancerous lesions) and early stage cervical cancer[2]–[4]. Infection with human papillomavirus (HPV) has been established as a necessary cause in the development of cervical cancer; however, most women infected with HPV do not develop cervical cancer[5][6]. Lifestyle factors, such as sexual behavior and smoking cigarettes[7][8], use of oral contraceptives[9][10], high parity[10][11], as well as co-infections, such as human immunodeficiency virus infection[12], are associated with the development of cervical cancer. Taken together these findings suggest that genetic and/or environmental factors may play a role in the development of this malignancy[13].

Air pollution is largely composed of automobile emissions, a complex mixture of compounds many of which are known to adversely affect human health. Benzene, polycyclic aromatic hydrocarbons (PAHs), and diesel particulate matter (DPM) are components of automobile emissions and have been designated by the United States Environmental Protection Agency (U.S. EPA) as hazardous air pollutants (HAPs)[14]. HAPs are toxic substances known or suspected to be carcinogenic. In fact, the International Agency for Research on Cancer has classified benzene and DPM as carcinogenic compounds, and many PAHs as possible, probable, or known carcinogens[15]-[17]. Some studies have suggested that exposure to traffic-related air pollutants, such as benzene, PAHs, and DPM, is associated with an increased risk for several types of cancer, including lung cancer[18][19], brain cancer[20][21], and leukemia[22]-[24]. The carcinogenic impact of exposure to traffic-related air pollution is an ever-increasing public health concern as urbanization continues to rise, and a greater proportion of the population is exposed to higher levels of HAPs.

Exposure to traffic-related HAPs may play a role in the development of cervical cancer. Occupational exposure to diesel engine exhaust has previously been shown to be associated with an increased risk of cervical cancer among women[25]. In addition, a recent study found an increased risk of cervical cancer in Danish women with higher residential concentration levels of nitrogen oxides (NO_x), a component of automobile engine emissions[21]. The objective of the current study is to evaluate the association between levels of traffic-related HAPs and cervical dysplasia, a precursor lesion for cervical cancer, in a multi-ethnic sample of women receiving cervical cancer screening and diagnostic services in Houston, Texas.

2. Methods

2.1. Study Population

The study population consisted of women attending the colposcopy clinics at The University of Texas MD Anderson Cancer Center and Lyndon B. Johnson General Hospital in Houston, Texas, between 2000 and 2004, and enrolled in a multi-center Phase II clinical trial (registered at www.Clinical-

Trials.gov as NCT00511615) to evaluate fluorescence and reflectance spectroscopy for diagnosing cervical disease. The detailed data collection methods for this trial have been previously described[26]. Briefly, women had to be 18 years of age or older, have no history of hysterectomy, and not be pregnant to be eligible for enrollment in the parent trial. Of those who were eligible and enrolled in the parent trial, only women who had available information on HPV infection status and residence (*i.e.*, address) at the time of diagnosis were eligible for our study, a secondary data analysis using data collected during the parent trial. After applying all eligibility criteria, 736 women were included in the current assessment. Cases (n = 173) included women who were diagnosed with cervical intraepithelial neoplasia (CIN) I, CIN II, or CIN III. Women who had a normal Papanicolaou test result (*i.e.*, not diagnosed with cervical dysplasia or cervical cancer) served as control subjects (n = 563). Details on the laboratory methods used to test for HPV status as well as those used to confirm cervical dysplasia disease status have been previously described[26]. Epidemiologic data were obtained from a risk factor interview conducted with each patient at the time of enrollment on the parent trial. Women received the standard treatment according to their colposcopy results. The Institutional Review Boards at the University of Texas MD Anderson Cancer Center and Harris Health System approved the protocol for the parent trial. Informed consent was obtained from all subjects.

2.2. Ambient Air Pollutant Concentration Levels

Annual concentration estimates of benzene, DPM, and PAHs were obtained for each census tract from the U.S. EPA's 1999 Assessment System for Population Exposure Nationwide (ASPEN). ASPEN is a computer simulation model used in the National-Scale Air Toxics Assessment conducted by the U.S. EPA[27]. The ASPEN model was derived from the U.S. EPA's Industrial Source Complex Long Term model designed to model air pollutant dispersion. To generate annual pollutant concentration estimates, ASPEN uses information on meteorological conditions, the location and height of pollutant release, rate of release, and pollutant deposition, reactive decay, and transformation properties. Other epidemiological studies have used pollutant concentration estimates from ASPEN to evaluate the effect of HAPs on disease risk[23][28][29].

Residential pollutant levels were determined based on the ASPEN estimates for the census tract of the subject's residence at the time of the clinic visit. Addresses were geocoded and mapped to determine residential census tracts using ArcGIS software (Esri, Redlands, California). Benzene, DPM, and PAH levels from ASPEN were categorized using the distribution of the residential HAP levels among the controls: low or reference exposure (<25th percentile), medium exposure (25th−74th percentiles), high exposure (75th−89th percentiles), and very high exposure (≥90th percentile). In addition, a HAP composite score was created to assess the combined effect of exposures to multiple HAPs on cervical dysplasia. To create the HAP composite score, each exposure category of each HAP was assigned a score: 0 = low, 1 = medium, 2 = high, and 3 = very high. The scores from the three HAPs were then summed for each subject based on their residential census tract in order to generate the overall HAP composite score, which could range from 0−9.

2.3. Statistical Analysis

Descriptive statistics were generated to characterize the demographic variables among the case and control groups, which included computing means and standard deviations (SD) for continuous variables and frequency distributions for categorical variables. Potential differences between the case and control groups in the distribution of each of the demographic variables were tested for statistical significance using t tests for continuous variables and χ^2 tests for categorical variables. Correlations between levels of benzene, DPM, and PAHs were determined using Spearman's rank correlation. We used multivariable logistic regression to assess the associations between HAP levels and cervical dysplasia. Adjusted prevalence odds ratios (aOR) and 95% confidence intervals (CI) were estimated for each association. Covariates were selected a priori and included age, race/ethnicity, years of completed education, HPV infection status[30][31], and smoking status. All multivariable regression models were adjusted for these covariates. Univariable logistic regression analyses were conducted to evaluate the associate between cervical dysplasia and demographic variables, including age, race/ethnicity, smoking status, and completed years of education, variables which also serve as covariates in the adjusted analyses. These results were included in Additional file 1: **Table S1**.

3. Results

There were no statistically significant differences between the case and control groups in regard to racial makeup, completed years of education, smoking status, HPV infection status, and the screening institution (**Table 1**). Approximately half

Table 1. Characteristics of cervical dysplasia cases and controls from colposcopy clinics in Houston, Texas, 2000–2004.

Characteristic	Cases (n = 173)	Controls (n = 563)	P-value
Age in years, mean ± SD	37.0 ± 11.0	43.5 ± 11.8	<0.001
Race/ethnicity, n (%)			
Non-Hispanic White	90 (52.0)	296 (52.6)	0.180
Non-Hispanic Black	36 (20.8)	85 (15.1)	
Hispanic	39 (22.5)	137 (24.3)	
Other	8 (4.6)	45 (8.0)	
Education in years, mean ± SD	13.1 ± 1.98	13.4 ± 3.57	0.376
Marital status, n (%)			
Never	45 (26.0)	102 (18.2)	0.038
Married/partnered	85 (49.1)	331 (58.9)	
Divorced/widowed/separated	43 (24.9)	129 (23.0)	
Smoking status, n (%)			
Never	108 (62.4)	389 (69.1)	0.101
Ever	65 (37.6)	174 (30.9)	
HPV status, n (%)			
Positive	129 (74.6)	408 (72.5)	0.587
Negative	44 (25.4)	155 (27.5)	
Histology, n (%)			
CIN I (mild dysplasia)	141 (81.5)		
CIN II (moderate dysplasia)	19 (11.0)		
CIN III (severe dysplasia)	13 (7.5)		
Clinic, n (%)			
MD Anderson Cancer Center	141 (81.0)	481 (85.4)	0.153
Lyndon B. Johnson General Hospital	33 (19.0)	82 (18.6)	

Abbreviations: SD, standard deviation; HPV, human papillomavirus; CIN, cervical intraepithelial neoplasia.

of the sample was non-Hispanic white (52.5%), with a quarter being Hispanic (24.2%) and fewer being non-Hispanic black (16.0%). This racial/ethnic distribution approximates the distribution of the population in the Houston area. The mean (SD) completed years of education was 13.3 (3.3) years. Approximately one-third of the sample (32.5%) had a history of smoking. Most of the subjects (both cases and controls) were positive for HPV infection (73.0%); 27.0% were HPV negative. The majority of the sample (84.4%) was screened at the colposcopy clinic at the University of Texas MD Anderson Cancer Center, while the remainder (15.6%) was screened at the Lyndon B. Johnson General Hospital. Cases were younger ($p < 0.001$) and more likely to have never been married ($p = 0.04$) compared to controls. Cases primarily had CIN I (81.5%) while the remainder of the cases had moderate to severe dysplasia (18.5%). All controls were confirmed to not have cervical dysplasia.

The distributions of selected census tract-level HAP concentration levels among the control group are presented in **Table 2**. The mean benzene exposure level among the controls was 1.917μg/m^3, with the lowest (<25th percentile) and highest (≥90th percentile) exposure categories representing those living in census tracts with an estimated average benzene level <1.415μg/m^3 and ≥2.977μg/m^3, respectively. Similarly, the mean level of DPM and PAHs among the controls was 2.015μg/m^3 and 0.011μg/m^3, respectively. The lowest and highest exposure categories included those living in census tracts with an estimated average DPM level <1.413μg/m^3 and ≥2.794μg/m^3, respectively. For PAHs, the lowest exposure category was <0.007μg/m^3 and the highest exposure category was ≥0.018μg/m^3.

Scatterplots of benzene and each of the other HAPs (*i.e.*, DPM and PAHs) are presented in **Figure 1** to assess correlations between the pollutant levels. Overall, benzene levels are highly correlated with levels of DPM ($\rho = 0.91$, $p < 0.0001$) and

Table 2. Distributions of hazardous air pollutants.

Pollutant (μg/m$_3$)*	Mean	25th percentile	50th percentile	75th percentile	90th percentile
Benzene	1.917	1.415	1.760	2.261	2.977
DPM	2.015	1.413	1.724	2.175	2.794
PAHs	0.011	0.007	0.009	0.013	0.018

Abbreviations: DPM, diesel particulate matter; PAHs, polycyclic aromatic hydrocarbons. *All concentrations are presented as μg/m^3.

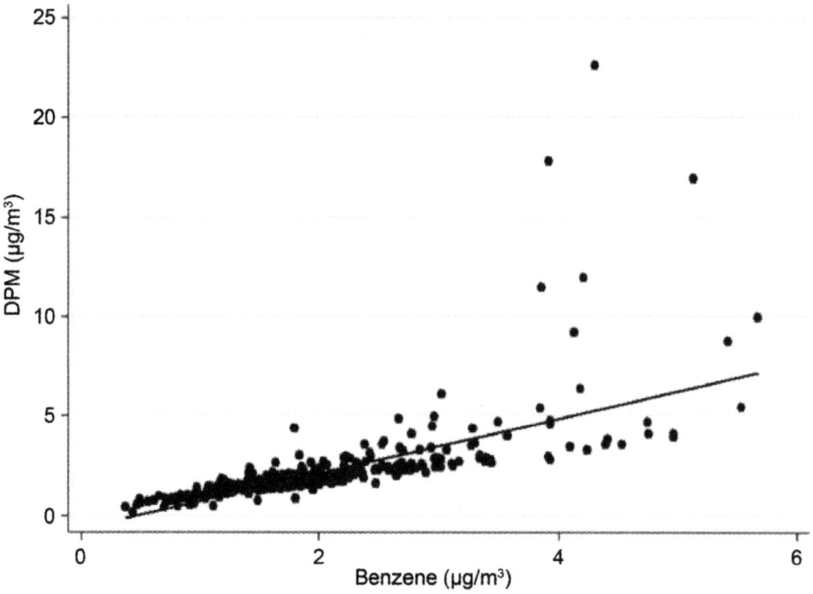

(a)

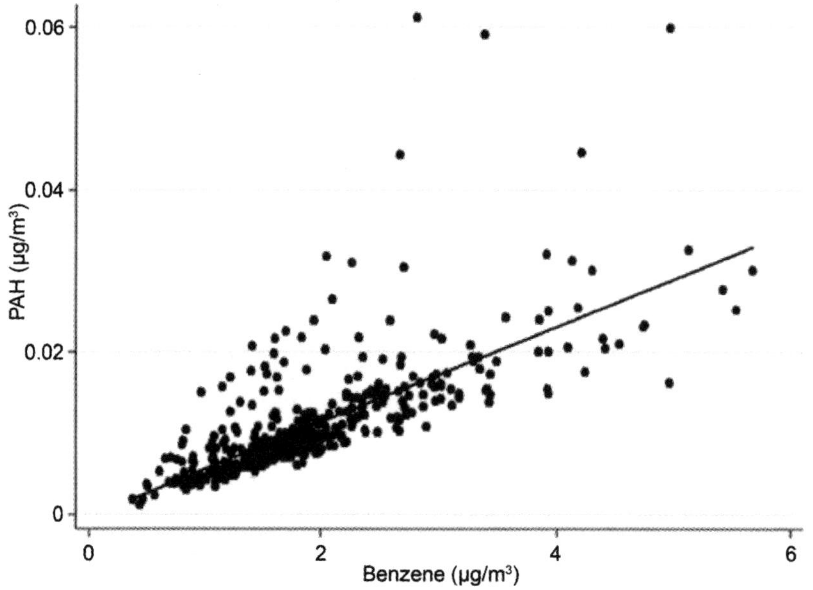

(b)

Figure 1. Correlations between traffic-related hazardous air pollutant levels. Scatterplots of (a) diesel particulate matter (DPM) and benzene, (b) polycyclic aromatic hydrocarbons (PAHs) and benzene from the 1999 United States' Environmental Protection Agency's (U.S. EPA) Assessment System for Population Exposure Nationwide's (ASPEN) model for Texas census tracts included in the current assessment.

PAHs (ρ = 0.80, p < 0.0001). However, these correlations were not as strong when restricted to those in the "very high" exposure category. Specifically, when restricting the correlation analysis to those in the very high benzene exposure category (n = 83) the correlations with DPM (ρ = 0.66, p < 0.0001) and PAHs (ρ = 0.65, p < 0.0001) are attenuated.

The aORs and 95% CIs of the associations between residential exposure to benzene, DPM and PAH's and the prevalence of cervical dysplasia are presented in **Table 3**. Residential exposure to benzene was associated with cervical dysplasia (aOR medium exposure = 1.39, 95% CI: 0.83–2.32; aOR high exposure = 1.97, 95% CI: 1.07–3.62; and aOR very high exposure = 2.30, 95% CI: 1.19–4.46) after adjusting for age, race/ethnicity, years of education, HPV infection status, and smoking. We also observed a trend with increasing exposure (p for trend = 0.005). Residential exposure to DPM is positively associated with cervical dysplasia when comparing high and very high exposure to low exposure (aOR for high exposure = 2.83, 95% CI: 1.55–5.16; aOR for very high exposure = 2.10, 95% CI: 1.07–4.11). Compared to low exposure, women living in census tracts with high levels of PAHs had an increased prevalence of cervical dysplasia (aOR = 2.46, 95% CI: 1.35–4.48). The effect estimates for medium or very high exposure to PAHs suggest an increased prevalence of cervical dysplasia; however, these associations are not statistically significant. When assessing the combined effect of HAP exposure as defined using the HAP composite score, there is a positive association with cervical dysplasia prevalence when comparing women with a higher (score of 3+) HAP composite score to those with a low composite score (score <3) (aOR for medium score = 1.78, 95% CI: 1.11–2.84; aOR for high score = 2.30, 95% CI: 1.28–4.13). All regression analyses were repeated on a restricted sample that excluded cases with mild dysplasia, which yielded similar results (data not shown), and therefore we retained the grouping of mild and progressive cervical disease in our final models[30][32]. Moreover, regression analyses were repeated stratifying the sample on HPV infection status, and the overall results were similar regardless of HPV status (data not shown).

4. Discussion

In one of the first studies of its kind, our results indicate traffic-related HAPs are associated with cervical dysplasia. Specifically, women living in census

Table 3. Associations between selected traffic-related hazardous air pollutants and cervical dysplasia.

Pollutant	Prevalence aOR* [95% CI]
Benzene	
Low (<25th percentile)	1.00
Medium (25th–74th percentile)	1.39 [0.83, 2.32]
High (75th–89th percentile)	1.97 [1.07, 3.62]
Very high (≥90th percentile)	2.30 [1.19, 4.46]
	P for trend = 0.005
DPM	
Low (<25th percentile)	1.00
Medium (25th–74th percentile)	1.41 [0.84, 2.37]
High (75th–89th percentile)	2.83 [1.55, 5.16]
Very high (≥90th percentile)	2.10 [1.07, 4.11]
	P for trend = 0.002
PAHs	
Low (<25th percentile)	1.00
Medium (25th–74th percentile)	1.30 [0.77, 2.20]
High (75th–89th percentile)	2.46 [1.35, 4.48]
Very high (≥90th percentile)	1.88 [0.94, 3.73]
	P for trend = 0.006
HAP composite	
Low (score 0–2)	1.00
Medium (score 3–6)	1.78 [1.11, 2.84]
High (score 7–9)	2.30 [1.28, 4.13]
	P for trend = 0.004

Abbreviations: aOR, adjusted odds ratio; CI, confidence interval; DPM, diesel particulate matter; PAHs, polycyclic aromatic hydrocarbons; HAP, hazardous air pollutant. *Odds ratios adjusted for age, race/ethnicity, education, smoking status, and status of human papilloma virus (HPV) infection.

tracts with high levels of benzene, DPM, or PAHs were approximately two to three times more likely to be diagnosed with cervical dysplasia compared to women living in census tracts with relatively low pollutant levels. We also observed a statistically significant dose-response relationship between each pollutant and cervical dysplasia prevalence (p for trend ≤ 0.006 for benzene, PAHs, and DPM). These findings are notable as cervical dysplasia in conjunction with chronic HPV infection is a precursor of cervical cancer. In fact, 18.5% of our cases had medium to severe dysplasia (CIN II or higher), lesions associated with an increased risk of progression to cervical cancer, and the majority of the women in our sample were positive for HPV infection.

While previous studies have not evaluated the role of air pollution on the risk of cervical dysplasia, our findings are in keeping with previous reports of an increased risk of cervical cancer after exposure to traffic-related pollutants and HAPs[21][25]. In a recent assessment by Raaschou-Nielsen *et al.* using data from the Danish Diet Cancer and Health cohort, residential exposure to NO_x was associated with an increased risk of cervical cancer (incidence rate ratio = 2.45, 95% CI: 1.01−5.93)[21]. Another study conducted in Sweden indicated that occupational exposure to diesel engine emissions was associated with an increased incidence of cervical cancer (standardized incidence ratio = 1.48, 95% CI: 1.17−1.84)[25]. Our study builds upon these assessments by including a multi-ethnic population and evaluating cervical dysplasia rather than cervical cancer. In addition, there is considerable variation in HAP concentration levels between the HAP exposure categories defined in our analysis, which also includes HAP levels positively associated with adverse health risks[33].

Our findings are consistent with the role of cigarette smoking on cervical cancer risk. A large body of epidemiological evidence supports the association between cigarette smoking and cervical cancer risk[34]−[36]. In fact, when assessing smoking as a predictor of cervical cancer among women who have a confirmed HPVpositive status, those who were current or ever smokers had a two times greater risk of developing cervical cancer compared to never smokers (OR = 2.17, 95% CI: 1.46−3.22)[8]. Even passive exposure to cigarette smoke (*i.e.*, second-hand smoke) may increase the risk of cervical cancer[37]. These previously reported associations are of note, as cigarette smoke contains carcinogenic compounds similar to those found in automobile emissions (e.g., benzene and PAHs), and, consequently, the effect of exposure to traffic-related air pollutants may be analogous to the effect of smoking on cervical cancer risk.

While the mechanism underlying the association between these pollutants and cervical dysplasia is unclear, there are certain biological pathways that may lead to dysplasia after exposure to HAPs. For instance, these HAPs may have genotoxic effects on the cervix. This is supported by evidence of high levels of PAHs in cervical mucus and cervical tissue among smokers[38]. PAHDNA adducts have also been found in the cervical tissue of smokers[39]. These adducts have genotoxic properties, which may lead to cancer. It is unknown if benzene or DPM has been identified in cervical tissue, however, these compounds are also known to have genotoxic effects[15][16]. Another potential mechanism may be related to oxidative stress, as these pollutants have been shown to form reactive oxygen species, leading to DNA damage[40]. These pollutants may also play a role in an altered immune environment leading to a state of chronic inflammation[41][42]. Ultimately, it is suspected that these pollutants may lead to a state of DNA damage and/or inflammation, which may facilitate viral integration, resulting in enhanced cervical carcinogenesis. For instance, some evidence suggests that PAHs may actually play a role in the proliferation and persistence of HPV[43].

There are several limitations to consider when interpreting the results presented in this report. First, the cross-sectional study design and the inclusion of prevalent cases of cervical dysplasia make it difficult to assess temporality in terms of the timing of exposure and the development of disease. However, it is not uncommon to evaluate prevalent cases when assessing an asymptomatic condition that requires an invasive screening technique for detection[26]. Second, ASPEN yields area-based pollutant concentration estimates and therefore exposure misclassification is a possibility. Although we used an area-based exposure measure there are few sources of population-based exposure assessments of HAPs. Additionally, ASPEN pollutant concentration estimates have been used extensively in other assessments of HAPs and adverse health outcomes[23][28][29]. Lastly, it is not clear if the women in this assessment later developed more progressive cervical disease or cervical cancer; however, cervical dysplasia is an important precursor to cervical cancer and therefore it is important to understand risk factors for cervical dysplasia.

To our knowledge, this study is the first to evaluate the association of air pollution exposure and predictors of cervical dysplasia in a racially diverse sample of women. A growing body of evidence shows that there are disparities in air pollution exposure among racial minorities and those with a low socioeconomic status[44]-[46]. Racial minorities are more likely to live in urban areas with higher concentrations of traffic-related air pollution, therefore, increasing the risk of these groups to related health effects, such as cervical dysplasia. In addition, Hispanic and black women are less likely to receive Pap screening services, have a higher incidence of cervical cancer, and have poorer survival after a cervical cancer diagnosis compared to white women in the U.S.[47]-[51]. The vulnerability of racial and ethnic minority groups to air pollution exposure in combination with lower adherence to screening is a pressing public health concern.

The results from this study have pertinent implications for health in the global context. Cervical cancer incidence and mortality is highest in the developing world, specifically in Africa and Central and South America[1][52]. This is largely due to the high prevalence of HPV infection in conjunction with limited access to screening services to detect precancerous lesions and early stage cervical cancer, as well as the lack of effective dissemination of the HPV vaccine and targeted vaccine delivery programs[4][52][53]. Moreover, due to increasing urbanization, specifically in Latin American cities, a larger portion of the population living in developing regions is being exposed to high levels of air pollution largely resulting from rapidly growing roadway and traffic density in urban areas[54][55]. The increased prevalence of exposure to traffic-related HAPs in these vulnerable populations may lead to an even higher incidence of cervical cancer.

5. Conclusions

This is the first study to evaluate the association between exposure to traffic-related HAPs and cervical dysplasia prevalence in a multi-ethnic sample of women. Our results suggest that women with high residential exposure to benzene, DPM, or PAHs have an increased prevalence of cervical dysplasia compared to women with relatively low exposure to these pollutants. Our findings highlight the need to continue to identify novel risk factors that contribute to cervical disease in conjunction with HPV infection. In addition, our findings may also be important in future cervical cancer prevention efforts as public health strategies are directed toward the detection, management, and prevention of cervical dysplasia.

6. Additional File

Additional file 1: **Table S1**. Associations between demographic variables and cervical dysplasia: Univariable analyses. http://www.biomedcentral.com/content/supplementary/1476-069X-13-52-S1.docx.

Abbreviations

aOR: Adjusted prevalence odds ratio; ASPEN: Assessment System for Population Exposure Nationwide; CI: Confidence interval; CIN: Cervical intraepithelial neoplasia; DNA: Deoxyribonucleic acid; DPM: Diesel particulate matter; HAPs: Hazardous air pollutants; HPV: Human papillomavirus; NOx: Nitrogen oxides; PAHs: Polycyclic aromatic hydrocarbons; SD: Standard deviation; U.S. EPA: United States Environmental Protection Agency.

Competing Interests

The authors declare that they have no competing interests.

Authors' Contributions

MES and PJL conceived and designed the study. PJL carried out the primary data analysis with the assistance of HED. MES and HED drafted the initial manuscript. MF was the PI of the parent study and provided input regarding study design and the assessment of cervical dysplasia. All authors contributed to revisions of the manuscript and read and approved the final manuscript.

Acknowledgements

This work was supported by the National Cancer Institute (R03 CA143965 awarded to MES and P01 CA082710 awarded to MF).

Source: Scheurer M E, Danysh H E, Follen M, *et al.* Association of traffic-related hazardous air pollutants and cervical dysplasia in an urban multiethnic population: a cross-sectional study[J]. Environmental Health A Global Access Science Source, 2014, 13(1):1–8.

References

[1] Jemal A, Bray F, Center MM, Ferlay J, Ward E, Forman D: Global cancer statistics. CA Cancer J Clin 2011, 61(2):69–90.

[2] Gustafsson L, Ponten J, Zack M, Adami HO: International incidence rates of invasive cervical cancer after introduction of cytological screening. Cancer Causes Control 1997, 8(5):755–763.

[3] Mathew A, George PS: Trends in incidence and mortality rates of squamous cell carcinoma and adenocarcinoma of cervix-worldwide. Asian Pac J Cancer Prev 2009, 10(4):645–650.

[4] Parkin DM, Almonte M, Bruni L, Clifford G, Curado MP, Pineros M: Burden and trends of type-specific human papillomavirus infections and related diseases in the latin america and Caribbean region. Vaccine 2008, 26(Suppl 11):L1–L15.

[5] Bosch FX, Lorincz A, Munoz N, Meijer CJ, Shah KV: The causal relation between human papillomavirus and cervical cancer. J Clin Pathol 2002, 55(4):244–265.

[6] Schiffman MH, Castle P: Epidemiologic studies of a necessary causal risk factor: human papillomavirus infection and cervical neoplasia. J Natl Cancer Inst 2003, 95(6):E2.

[7] Au WW: Life style, environmental and genetic susceptibility to cervical cancer. Toxicology 2004, 198(1–3):117–120.

[8] Plummer M, Herrero R, Franceschi S, Meijer CJ, Snijders P, Bosch FX, de Sanjose S, Munoz N: Smoking and cervical cancer: pooled analysis of the IARC multi-centric case-control study. Cancer Causes Control 2003, 14(9):805–814.

[9] Moreno V, Bosch FX, Munoz N, Meijer CJ, Shah KV, Walboomers JM, Herrero R, Franceschi S: Effect of oral contraceptives on risk of cervical cancer in women with human papillomavirus infection: the IARC multicentric case-control study. Lancet 2002, 359(9312):1085–1092.

[10] Hildesheim A, Herrero R, Castle PE, Wacholder S, Bratti MC, Sherman ME, Lorincz AT, Burk RD, Morales J, Rodriguez AC, Helgesen K, Alfaro M, Hutchinson M, Balmaceda I, Greenberg M, Schiffman M: HPV co-factors related to the development of cervical cancer: results from a population-based study in Costa Rica. Br J Cancer 2001, 84(9):1219–1226.

[11] Thomas DB, Ray RM, Koetsawang A, Kiviat N, Kuypers J, Qin Q, Ashley RL,

Koetsawang S: Human papillomaviruses and cervical cancer in Bangkok. I. Risk factors for invasive cervical carcinomas with human papillomavirus types 16 and 18 DNA. Am J Epidemiol 2001, 153(8):723–731.

[12] Denny LA, Franceschi S, de Sanjose S, Heard I, Moscicki AB, Palefsky J: Human papillomavirus, human immunodeficiency virus and immunosuppression. Vaccine 2012, 30(Suppl 5):F168–F174.

[13] de Freitas AC, Gurgel AP, Chagas BS, Coimbra EC, do Amaral CM: Susceptibility to cervical cancer: an overview. Gynecol Oncol 2012, 126(2):304–311.

[14] Technology transfer network: air toxics web site. http://www.epa.gov/ttnatw01/allabout.html.

[15] IARC: IARC monographs on the evaluation of carcinogenic risks to humans. Volume 46. Diesel and gasoline engine exhausts and some nitroarenes. IARC Monogr Eval Carcinog Risks Hum 1989, 46:1–458.

[16] IARC: IARC monographs on the evaluation of carcinogenic risks to humans. Volume 100F. Chemical agents and related occupations. IARC Monogr Eval Carcinog Risks Hum 2012, 100(Pt F):9–562.

[17] IARC: IARC monographs on the evaluation of carcinogenic risks to humans. Volume 82. Some traditional herbal medicines, some mycotoxins, naphthalene and styrene. IARC Monogr Eval Carcinog Risks Hum 2002, 82:1–556.

[18] Pope CA 3rd, Burnett RT, Thun MJ, Calle EE, Krewski D, Ito K, Thurston GD: Lung cancer, cardiopulmonary mortality, and long-term exposure to fine particulate air pollution. JAMA 2002, 287(9):1132–1141.

[19] Raaschou-Nielsen O, Bak H, Sorensen M, Jensen SS, Ketzel M, Hvidberg M, Schnohr P, Tjonneland A, Overvad K, Loft S: Air pollution from traffic and risk for lung cancer in three Danish cohorts. Cancer Epidemiol Biomarkers Prev 2010, 19(5):1284–1291.

[20] Boeglin ML, Wessels D, Henshel D: An investigation of the relationship between air emissions of volatile organic compounds and the incidence of cancer in Indiana counties. Environ Res 2006, 100(2):242–254.

[21] Raaschou-Nielsen O, Andersen ZJ, Hvidberg M, Jensen SS, Ketzel M, Sorensen M, Hansen J, Loft S, Overvad K, Tjonneland A: Air pollution from traffic and cancer incidence: a Danish cohort study. Environ Health 2011, 10:67.

[22] Vinceti M, Rothman KJ, Crespi CM, Sterni A, Cherubini A, Guerra L, Maffeis G, Ferretti E, Fabbi S, Teggi S, Consonni D, De Girolamo G, Meggiato A, Palazzi G, Paolucci P, Malagoli C: Leukemia risk in children exposed to benzene and PM10 from vehicular traffic: a case-control study in an Italian population. Eur J Epidemiol 2012, 27(10):781–790.

[23] Whitworth KW, Symanski E, Coker AL: Childhood lymphohematopoietic cancer incidence and hazardous air pollutants in southeast Texas, 1995–2004. Environ

Health Perspect 2008, 116(11):1576–1580.

[24] Visser O, van Wijnen JH, van Leeuwen FE: Residential traffic density and cancer incidence in Amsterdam, 1989–1997. Cancer Causes Control 2004, 15(4):331–339.

[25] Boffetta P, Dosemeci M, Gridley G, Bath H, Moradi T, Silverman D: Occupational exposure to diesel engine emissions and risk of cancer in Swedish men and women. Cancer Causes Control 2001, 12(4):365–374.

[26] Pham B, Rhodes H, Milbourne A, Adler-Storthz K, Follen M, Scheurer ME: Epidemiologic differentiation of diagnostic and screening populations for the assessment of cervical dysplasia using optical technologies. Gend Med 2012, 9(1 Suppl):S36–S47.

[27] The ASPEN Model. http://www.epa.gov/ttn/atw/nata/aspen.html.

[28] Lupo PJ, Symanski E, Waller DK, Chan W, Langlois PH, Canfield MA, Mitchell LE: Maternal exposure to ambient levels of benzene and neural tube defects among offspring: Texas, 1999-2004. Environ Health Perspect 2011, 119(3):397–402.

[29] Reynolds P, Von Behren J, Gunier RB, Goldberg DE, Hertz A, Smith DF: Childhood cancer incidence rates and hazardous air pollutants in California: an exploratory analysis. Environ Health Perspect 2003, 111(4):663–668.

[30] Nye MD, Hoyo C, Huang Z, Vidal AC, Wang F, Overcash F, Smith JS, Vasquez B, Hernandez B, Swai B, Oneko O, Mlay P, Obure J, Gammon MD, Bartlett JA, Murphy SK: Associations between methylation of paternally expressed gene 3 (PEG3), cervical intraepithelial neoplasia and invasive cervical cancer. PLoS One 2013, 8(2):e56325.

[31] Tomita LY, D'Almeida V, Villa LL, Franco EL, Cardoso MA: Polymorphisms in genes involved in folate metabolism modify the association of dietary and circulating folate and vitamin B-6 with cervical neoplasia. J Nutr 2013, 143(12):2007–2014.

[32] Grimm C, Watrowski R, Polterauer S, Baumühlner K, Natter C, Rahhal J, Heinze G, Schuster E, Hefler L, Reinthaller A: Vascular endothelial growth factor gene polymorphisms and risk of cervical intraepithelial neoplasia. Int J Gynecol Cancer 2011, 21(4):597–601.

[33] Sexton K, Linder SH, Marko D, Bethel H, Lupo PJ: Comparative assessment of air pollution-related health risks in Houston. Environ Health Perspect 2007, 115(10): 1388–1393.

[34] Appleby P, Beral V, de Gonzalez Berrington A, Colin D, Franceschi S, Goodill A, Green J, Peto J, Plummer M, Sweetland S: Carcinoma of the cervix and tobacco smoking: collaborative reanalysis of individual data on 13,541 women with carcinoma of the cervix and 23,017 women without carcinoma of the cervix from 23 epidemiological studies. Int J Cancer 2006, 118(6):1481–1495.

[35] Gadducci A, Barsotti C, Cosio S, Domenici L, Riccardo Genazzani A: Smoking habit, immune suppression, oral contraceptive use, and hormone replacement therapy use and cervical carcinogenesis: a review of the literature. Gynecol Endocrinol 2011,

27(8):597–604.

[36] Roura E, Castellsague X, Pawlita M, Travier N, Waterboer T, Margall N, Bosch FX, de Sanjose S, Dillner J, Gram IT, Tjonneland A, Munk C, Pala V, Palli D, Khaw KT, Barnabas RV, Overvad K, Clavel-Chapelon F, Boutron-Ruault MC, Fagherazzi G, Kaaks R, Lukanova A, Steffen A, Trichopoulou A, Trichopoulos D, Klinaki E, Tumino R, Sacerdote C, Panico S, Bueno-de-Mesquita HB: Smoking as a major risk factor for cervical cancer and pre-cancer: results from the EPIC cohort. Int J Cancer 2013.

[37] Trimble CL, Genkinger JM, Burke AE, Hoffman SC, Helzlsouer KJ, Diener-West M, Comstock GW, Alberg AJ: Active and passive cigarette smoking and the risk of cervical neoplasia. Obstet Gynecol 2005, 105(1):174–181.

[38] Pratt MM, Sirajuddin P, Poirier MC, Schiffman M, Glass AG, Scott DR, Rush BB, Olivero OA, Castle PE: Polycyclic aromatic hydrocarbon-DNA adducts in cervix of women infected with carcinogenic human papillomavirus types: an immunohistochemistry study. Mutat Res 2007, 624(1-2):114–123.

[39] Melikian AA, Sun P, Prokopczyk B, El-Bayoumy K, Hoffmann D, Wang X, Waggoner S: Identification of benzo[a]pyrene metabolites in cervical mucus and DNA adducts in cervical tissues in humans by gas chromatography-mass spectrometry. Cancer Lett 1999, 146(2):127–134.

[40] Pilger A, Rudiger HW: 8-Hydroxy-2'-deoxyguanosine as a marker of oxidative DNA damage related to occupational and environmental exposures. Int Arch Occup Environ Health 2006, 80(1):1–15.

[41] Sorensen M, Autrup H, Moller P, Hertel O, Jensen SS, Vinzents P, Knudsen LE, Loft S: Linking exposure to environmental pollutants with biological effects. Mutat Res 2003, 544(2–3):255–271.

[42] Wang F, Li C, Liu W, Jin Y: Effect of exposure to volatile organic compounds (VOCs) on airway inflammatory response in mice. J Toxicol Sci 2012, 37(4):739–748.

[43] Alam S, Conway MJ, Chen HS, Meyers C: The cigarette smoke carcinogen benzo[a]pyrene enhances human papillomavirus synthesis. J Virol 2008, 82(2): 1053–1058.

[44] Gray SC, Edwards SE, Miranda ML: Race, socioeconomic status, and air pollution exposure in North Carolina. Environ Res 2013, 126:152–158.

[45] Gwynn RC, Thurston GD: The burden of air pollution: impacts among racial minorities. Environ Health Perspect 2001, 109(Suppl 4):501–506.

[46] Brochu PJ, Yanosky JD, Paciorek CJ, Schwartz J, Chen JT, Herrick RF, Suh HH: Particulate air pollution and socioeconomic position in rural and urban areas of the Northeastern United States. Am J Public Health 2011, 101(Suppl 1):S224–S230.

[47] Bazargan M, Bazargan SH, Farooq M, Baker RS: Correlates of cervical cancer screening among underserved Hispanic and African-American women. Prev Med2004,39(3):465–473.

[48] Howlader N, Noone AM, Krapcho M, Garshell J, Neyman N, Altekruse SF, Kosary CL, Yu M, Ruhl J, Tatalovich Z, Cho H, Mariotto A, Lewis DR, Chen HS, Feuer EJ, Cronin KA (Eds): SEER Cancer Statistics Review, 1975-2010. National Cancer Institute. Bethesda, MD. http://seer.cancer.gov/csr/1975_2010/, based on November 2012 SEER data submission, posted to the SEER web site, April 2013.

[49] del Carmen MG, Avila-Wallace M: Effect of health care disparities on screening. Clin Obstet Gynecol 2013, 56(1):65–75.

[50] Howe HL, Wu X, Ries LA, Cokkinides V, Ahmed F, Jemal A, Miller B, Williams M, Ward E, Wingo PA, Ramirez A, Edwards BK: Annual report to the nation on the status of cancer, 1975-2003, featuring cancer among U.S. Hispanic/Latino populations. Cancer 2006, 107(8):1711–1742.

[51] Niccolai LM, Julian PJ, Bilinski A, Mehta NR, Meek JI, Zelterman D, Hadler JL, Sosa L: Geographic poverty and racial/ethnic disparities in cervical cancer precursor rates in Connecticut, 2008-2009. Am J Public Health 2013, 103(1):156–163.

[52] Forman D, de Martel C, Lacey CJ, Soerjomataram I, Lortet-Tieulent J, Bruni L, Vignat J, Ferlay J, Bray F, Plummer M, Franceschi S: Global burden of human papillomavirus and related diseases. Vaccine 2012, 30(Suppl 5):F12–F23.

[53] Sankaranarayanan R: HPV vaccination: the promise & problems. Indian J Med Res 2009, 130(3):322–326.

[54] Bell ML, Davis DL, Gouveia N, Borja-Aburto VH, Cifuentes LA: The avoidable health effects of air pollution in three Latin American cities: Santiago, Sao Paulo, and Mexico City. Environ Res 2006, 100(3):431–440.

[55] Molina MJ, Molina LT: Megacities and atmospheric pollution. J Air Waste Manag Assoc 2004, 54(6):644–680.

Chapter 10

Chemical Content and Estimated Sources of Fine Fraction of Particulate Matter Collected in Krakow

Lucyna Samek, Zdzislaw Stegowski, Leszek Furman, Joanna Fiedor

Faculty of Physics and Applied Computer Science, AGH University of Science and Technology, 30 Mickiewicza Ave., 30-059 Krakow, Poland

Abstract: The monitored level of pollution remains high in Krakow, Poland. Alerts regarding increased levels of pollution, which advise asthmatics, the elderly, and children to limit their exposure to open air, continue to be issued on numerous days. In this work, seasonal variations in $PM_{2.5}$ (particulate matter containing particles with aerodynamic diameter no higher than 2.5μm) concentrations are shown. An increasing trend is reported, which is enhanced during the colder seasons. The mean $PM_{2.5}$ concentrations in Krakow exceeded the target value of 25μg/m^3 specified for 2015 in the spring, autumn, and winter seasons. For this reason, particulate matter pollution is of special concern. Elemental concentrations as well as the presence of black carbon (BC) and black smoke (BS) in $PM_{2.5}$ samples were determined. Seasonal variations of Cl, K, Ca, Ti, Mn, Fe, Cu, Zn, Br, Rb, Sr, and Pb concentrations were observed whereas V, Cr, Ni, BC, and BS concentrations did not significantly change with the time of year. Seven factors were identified by the positive matrix factorization (PMF) technique, and one was non-identified. They were attributed to the following sources of pollution: steel industry, traffic (diesel exhaust), traffic (gasoline exhaust, brake wear), road dust, construction dust, combustion (biomass, coal), and non-ferrous metallurgical industry. The last, non-identified source, could be attributed to secondary aerosols. It is worth to mention that combustion shows significant seasonal variations with a high impact in winter. The reported results of the completed studies may significantly aid in solving air quality issues in the city by highlighting major sources of air pollution.

Keywords: Particulate Matter, Energy Dispersive X-Ray Spectrometry, Positive Matrix Factorization

1. Introduction

According to European Directives, the concentrations of NO_x, CO_2, and particulate matter significantly exceed the specified limit values (Ostro *et al.* 2015; Kim *et al.* 2004; Lim *et al.* 2011). Increased levels of air pollution can negatively affect human health, especially prolonged exposure to polluted air may cause respiratory and cardio-vascular diseases. The overall mortality and morbidity can also be influenced (Brunekreef and Holgate 2002; Anenberg *et al.* 2010; Samek 2016). A component of air pollution which is given special attention is $PM_{2.5}$—a number of research groups has reported major and trace elements concentrations in $PM_{2.5}$ (Cuccia *et al.* 2013; Yu *et al.* 2013; Moreno *et al.* 2006; Samek *et al.* 2015, 2016; Zhang *et al.* 2015, Terrouche *et al.* 2016). Due to the fact that elemental composition is determined by unique factors, its determination may lead to the identification of sources of air pollution. Properly identified sources can then improve the efforts to minimize pollution levels in different cities. Since chemical content of air particulate matter determines its toxicity, receptor models are used for source identification and apportionment based on the concentrations of chemical species in $PM_{2.5}$ (Mazzei *et al.* 2008; Laupsa *et al.* 2009; Querol *et al.* 2007; Masiol *et al.* 2014).

Krakow is located in a valley in Southern Poland. The city is characterized by a high level of particulate matter pollution. Steel and non-ferrous metallurgical industries are located within the city. Other sources of PM pollution are also present, they include traffic (from diesel or gasoline exhausts), brake, and tire wear. Additionally, the air in Krakow is influenced by long-range transport of aerosols (e.g., secondary aerosols) (Samek 2012; Samek *et al.* In press). These studies identified potential sources. However, the time series of source contributions was not identified in these studies due to limited number of samples, short duration of sampling, or discontinuous sampling. A long period of daily $PM_{2.5}$ data is needed in order to obtain a precise source identification. The present study was designed to fill this gap.

The aim of this work is to present results of chemical analyses of $PM_{2.5}$ samples collected during a 1-year period (2014/2015) in an urban area of Krakow. Major and trace element, black carbon, and black smoke concentrations as well as seasonal variations were determined. Additionally, source identification was performed and seasonal variations of likely $PM_{2.5}$ sources were unveiled. The application of a full year continuous data could not only improve the efficiency of positive matrix factorization (PMF) analysis, but also help to perform the time series analysis of various sources.

In this work, PMF is applied as described by Paatero (Paatero 1997). The sources are classified as natural or anthropogenic. Natural sources include suspended soil and road dust, sea salt, forest fires as well as long-range transported dust (Marconi *et al.* 2014). The following are a part of anthropogenic sources: industry, traffic, combustion of biomass, or coal for residential heating.

2. Experimental

2.1. Sampling

Samples were collected by the Voivodship Inspectorate of Environmental Protection in Krakow. The site selected for the study was an urban area of Krakow, specifically, the southeastern part of the city (district Kurdwanow). The major local sources of pollution are municipal emissions, industry, and traffic. Traffic in the city is dense with frequent traffic jams. Factories are located at a distance of about 10km from the sampling site. Additionally, a power plant is located in the southern area of the city. The Upper Silesian industry area can be found approximately 80km to the west from Krakow. Moreover, the zinc industry is situated about 50km to the north of the city. Twenty-four-hour $PM_{2.5}$ fraction samples were collected between February 1, 2014 and January 31, 2015 with the use of a low volume sampler with a flow rate of 2.3m^3/h. Quartz (46.2mm) filters were used as a support. Overall, 194 samples were collected during the entire year.

2.2. Chemical Analysis

$PM_{2.5}$ concentrations were determined by the Voivodship Inspectorate of Environmental Protection in Krakow. Concentrations of the following elements were quantified: Cl, K, Ca, Ti, V, Cr, Mn, Fe, Ni, Cu, Zn, Br, Sr, Rb, and Pb. Samples of $PM_{2.5}$ were analyzed with the use of a multifunctional energy dispersive X-ray fluorescence spectrometer as thin samples. The instrument is a micro-beam X-ray fluorescence spectrometer with capillary X-ray optics, a broad X-ray beam from a molybdenum secondary target for XRF analysis of bulk samples, and a total reflection X-ray technique. The molybdenum tube is the source of X-rays. The tube has the power of 2kW. The excited X-rays were detected by a Si(Li) detector with resolution of 170eV at an energy of 5.9keV. Data collection was completed using the Canberra system. The measurements were carried out under the following conditions: voltage of 55kV, current of 30mA, measurement time of 10,000s, and under atmospheric air. In order to calculate the concentrations of different elements in the filters, the spectrometer was calibrated using thin-film standards (Micromatter, USA). The calibration was verified by the analysis of the NIST Standard Reference Material (2783-Air particulate matter on filter media). **Table 1** presents certified and measured elemental concentrations of the NIST Standard Reference Material. The XRF spectra were quantitatively analyzed with the use of the QXAS package (Vekemans *et al.* 1994).

Table 1. Measured and certified concentrations of elements in NIST SRM2783.

Element	Measured values (ng)	Certified values (ng)
K	6011 ± 4500	5280 ± 520
Ca	18,780 ± 3500	13,200 ± 1700
Ti	1384 ± 130	1490 ± 240
Cr	115 ± 67	135 ± 25
Mn	319 ± 66	320 ± 12
Fe	27,111 ± 678	26,500 ± 1600
Cu	398 ± 25	404 ± 42
Zn	2077 ± 63	1790 ± 130
Pb	244 ± 25	317 ± 54

The presence of black carbon (BC) and black smoke (BS) was determined by UV-VIS spectroscopy. Spectroscopic measurements were performed using a Varian Cary 50-Bio UV-VIS spectrophotometer (Agilent). Transmittance was recorded at the 880nm wavelength, triplicate for each sample. Additionally, transmission spectra were collected in the range of 200–1000nm. All measurements were carried out in reference to air. Concentrations of BC and BS were calculated according to formulas listed by Quincey P. (Quincey 2007).

2.3. Statistical Analysis

Source apportionment analysis was carried out using the positive matrix factorization receptor model (version PMF5.0) developed by the United States Environmental Protection Agency (US EPA).

PMF requires two inputs to run, namely concentration and its uncertainty. In this work, if the concentration was less than or equal to the detection limit (DL) provided, the uncertainty was calculated as five sixths DL and the concentration as one half DL (Pollisar 1998). Missing data was substituted with median values, and the corresponding uncertainties were replaced by four times the median values. $PM_{2.5}$ concentration was included as a total variable, and all the species were characterized as “strong,” “weak,” or “bad” depending on the signal to noise ratio.

The input data for the PMF model included 194 samples with 18 species (concentration of $PM_{2.5}$, 15 elements, BC and BS). Concentrations of three elements, V, Sr, and Rb, were excluded because many of the corresponding values were below the detection limit and they were characterized by low signal to noise values.

3. Results and Discussion

Table 2 shows concentrations of $PM_{2.5}$, BC, BS, and major and trace elements in $PM_{2.5}$ during different seasons of the 2014/2015 period. The lowest concentration of $PM_{2.5}$ was observed in the summer. It was significantly below the target value specified by the EU Directive (2008). Concentrations reported during spring and autumn were slightly higher than the target value whereas those observed in the winter were as high as twice the target value. In Poland, in the

Table 2. Concentrations of chemical species in Krakow during different seasons of the year (concentrations of $PM_{2.5}$, BC, BS are in μg/m^3, and rest of species in ng/m^3).

Element	Spring	Summer	Autumn	Winter
$PM_{2.5}$	31 ± 23	12.7 ± 4.6	30 ± 21	57 ± 39
Cl	674 ± 1408	<DL	361 ± 704	3609 ± 3548
K	255 ± 411	17 ± 36	49 ± 115	507 ± 537
Ca	264 ± 380	59 ± 144	36 ± 72	169 ± 221
Ti	13 ± 13	8.4 ± 7.5	6.8 ± 5.3	7.7 ± 7.3
V	1.7 ± 1.3	1.7 ± 1.6	1.7 ± 1.3	1.6 ± 1.3
Cr	3.8 ± 2.7	3 ± 2	6.5 ± 5.1	6.0 ± 4.4
Mn	5.9 ± 4.5	3.5 ± 2.4	5.1 ± 4.7	7.7 ± 8.7
Fe	261 ± 217	102 ± 90	195 ± 139	267 ± 219
Ni	1.9 ± 1.0	1.7 ± 1.0	0.9 ± 0.8	0.6 ± 0.2
Cu	8.5 ± 9.1	2.9 ± 2.2	5.5 ± 4.6	12 ± 11
Zn	67 ± 52	18 ± 10	59 ± 55	118 ± 89
Br	11 ± 9	3.3 ± 1.5	9.8 ± 8.5	25 ± 17
Rb	1.5 ± 1.0	0.9 ± 0.5	1.4 ± 1.1	2.3 ± 1.3
Sr	1.9 ± 1.1	1.7 ± 1.1	1.1 ± 0.8	0.9 ± 0.9
Pb	20 ± 17	5.4 ± 3.8	18 ± 16	43 ± 32
BC	2.4 ± 1.3	2.7 ± 1.2	2.8 ± 1.2	2.6 ± 1.3
BS	10.4 ± 6.5	12.0 ± 5.6	12.6 ± 6.2	11.4 ± 6.1

St.dev.—variability of concentrations during the measured period.

Regulation of the Minister of the Environment (2012), the limit value of $PM_{2.5}$ concentration was 25μg/m^3—the same as target value to be met in 2015 (phase I) as well as target value equal to 20μg/m^3 to be met in 2020 (phase II). Strong seasonal variations were noted for concentrations of Cl, K, Br, Pb, Cu, and Zn. Cl and K can originate from combustion of coal and/or biomass. Ratios of Cu to Zn were in the range of 0.09–0.16 depending on the season. A Cu to Zn ratio equal to 0.3 indicative of traffic was reported by (Mazzei *et al.* 2008). Ratios of Cu to Pb were 0.43 and 0.46 in spring and summer, respectively. A higher value of 0.69 was reported in the summer, and a low number of 0.28 was found in the winter. These elements present in ratios in the range of 2.3–3.0 were identified as indicators of traffic by (Mazzei *et al.* 2008). These results suggest that a source of Pb other than traffic also exists. The Zn to Pb ratio in winter was equal to 2.74 and in other seasons of the year it was in the range 3.28–3.35. The ratio of Pb to Br was equal to 1.64–1.85, and it remained constant during all seasons of the year. Concentrations of Ca and Ti were the highest in spring and the lowest in autumn. These elements can be related to construction dust created at prevalent construction sites in Krakow. Other elements such as V were present in constant concentrations during the year. Ni concentrations in spring and summer were twice as large as those reported in winter and autumn. V and Ni are indicators of traffic. Black carbon and black smoke had the highest values in autumn and were higher than those found in Ghent and Amsterdam but were similar to those noted in Barcelona. During the remaining seasons, the values were comparable between the various cities. Based on the assumption that black carbon is a tracer for primary emissions, mostly derived from traffic, these results suggest that the influence of traffic on air pollution is higher in Krakow than in Amsterdam and Ghent and on a similar level as in Barcelona (Viana *et al.* 2007).

Figure 1 presents the factor profiles. **Figure 2** and **Figure 3** show source contributions in microgams per cubic meter and percent, respectively. A positive matrix factorization model was used for source identification and apportionment.

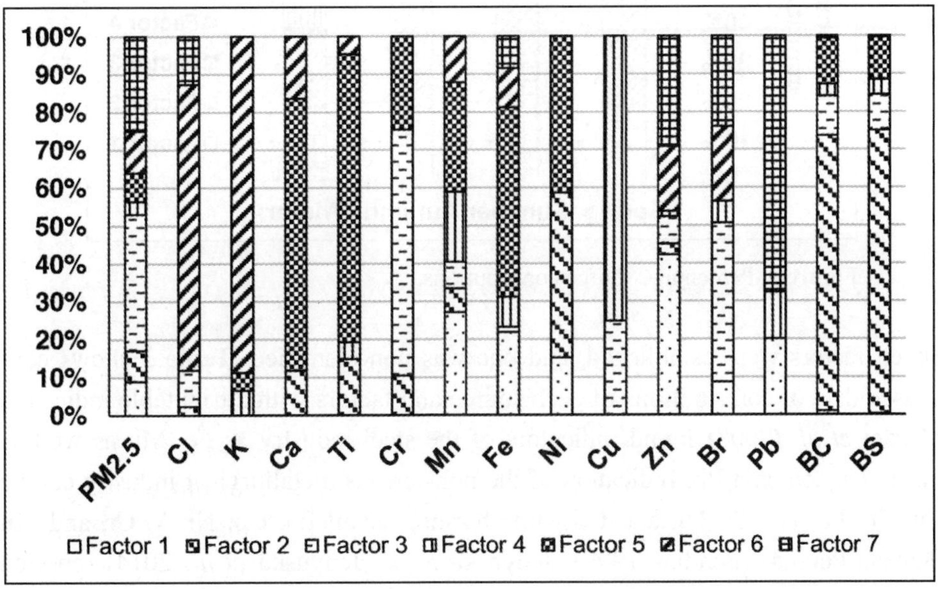

Figure 1. Factor profiles of fine particulate matter. For attributions to factors, see **Table 3**.

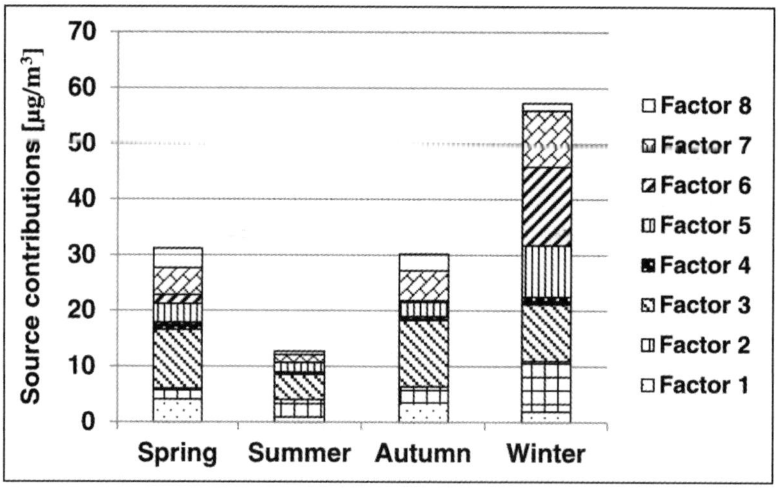

Figure 2. Source contributions in micrograms per cubic meter. For attributions to factors, see **Table 3**.

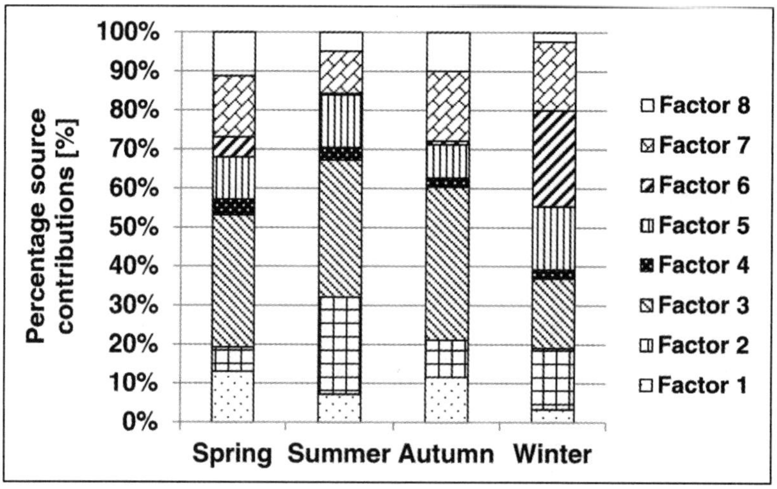

Figure 3. Percentage source contributions.

Seven factors were established, and one was non-identified. **Table 3** shows condensed data of contributions of each source and factors with attributable indicators. Mazzei *et al.* (2008) found indicators of the steel industry as Fe, Mn as well as traffic Cu, Zn, and Pb. Indicators of the non-ferrous metallurgical industry can be Cu, Zn, Pb, As, Cr, Ni, and Co while burning of oil Ba, Co, Ni, V, Cr, and Mn (Kabata-Pendias, Pendias 1999). Jedynska *et al* (Jedynska *et al.* 2014) reported that EC can be an indicator of diesel emissions. Yu *et al.* (2013) published that

Table 3. Factors with attributed sources and their contribution to PM mass concentrations.

Factor number	Attributed source	Indicators	Source contributions (%)			
			Spring	Summer	Autumn	Winter
Factor 1	Steel industry	Mn, Fe, Zn, Br, Pb	13	11.7	7	3.3
Factor 2	Traffic (diesel exhaust)	Ni, BC, BS	6.2	24.9	9.5	15.7
Factor 3	Traffic (gasoline exhaust)	Cr, Cu, Br, BC, BS	33.7	35	39	17.8
Factor 4	Road dust	Cu	4.0	3.2	2.2	2.3
Factor 5	Construction dust and/or soil dust	Ca, Ti, Cr, Mn, Fe, Ni	10.8	13	8.5	10
Factor 6	Combustion coal and/or biomass	Cl, K,	5	0.3	0.9	24.6
Factor 7	Non-ferrous metallurgical industry	Zn, Br, Pb	15.7	11	18	18
Factor 8	Non-identified source		11.2	5	9.9	2.4

biomass burning indicator was K as well as fossil fuel combustion indicators were Cl, V, Ni, As, and Pb. Confirmation of this finding also is the paper of Viana (Viana *et al.* 2008).

In a previous study (Samek *et al.* In press), fine fraction principal component analysis (PCA) and multilinear regression analysis (MLRA) were performed. The contribution of municipal emissions combined with industry was estimated to be 49.2% during winter, that of traffic was at 37.8%. In the present study, the contribution of traffic in winter was similar at 36.5% (15.7% diesel exhaust, 17.8% gasoline exhaust, 3% road dust) while that of combustion was equal to 25%; the steel industry contributed 3.4% and the non-ferrous metallurgical industry added 18%. The previously mentioned study (Samek *et al.* In press) reported a mean contribution of traffic equal to 53% (40%–60%) in summer. In this study, it was found to be equal to 62.9% (24.9% diesel exhaust, 35% gasoline exhaust, 3% road dust). The cited work stated the contribution of industry in summer to be equal to 18% (in the range of 5%–40%). In this study, the corresponding contribution was at 20.3% (3.3% steel industry, 17% non-ferrous metallurgical industry). PMF results were consistent with the previously completed research. This research provided an increased number of factors indicated as pollutants as well as a more complete analysis than previous ones. This study complements previous investigations performed during a more limited time period (1 month in summer and one in winter) by delivering new valuable information.

4. Conclusions

Chemical characterization of $PM_{2.5}$ fraction collected in Krakow was reported. Certain elements such as Cl, K, Br, Pb, Cu, and Zn show strong seasonal variations of concentrations whereas others—V, Ni, BC, BS—do not exhibit such deviations. Eight contributing factors were identified, and a single pollution source was attributed to each one. Strong seasonal variations were associated with combustion, with the peak value reported in winter. Contributions of traffic were constant throughout the year. This study is consistent with a previously published work regarding $PM_{2.5}$. It involved an extended period of time, a larger number of analyzed samples, and, as a result, more detailed information. This data is particularly relevant to the identification of pollution sources in Krakow. However, further analyses need to be performed in order to determine the impact of the described pollution on human health.

Source: Samek L, Stegowski Z, Furman L, *et al.* Chemical content and estimated sources of fine fraction of particulate matter collected in Krakow[J]. Air Quality Atmosphere & Health, 2016:1−6.

References

[1] Anenberg SC, Horowitz LW, Tong DQ, West JJ (2010) An estimate of the global burden of anthropogenic ozone and fine particulate matter on premature human mortality using atmospheric modeling. Environ Health Persp 118:1189−1195. doi:10.1289/ehp.0901220.

[2] Brunekreef B, Holgate ST (2002) Air pollution and health. Lancet 360: 1233−1242.

[3] Cuccia E, Massabo D, Ariola V, Bove MC, Fermo P, Piazzalunga A, Prati P (2013) Size-resolved comprehensive characterization of airborne particulate matter. Atmos Environ 67:14−26. doi:10.1016/j. atmosenv.2012.10.045.

[4] EU Directive 2008/50/EC of the European Parliament and the Council of 21 May

2008.

[5] Jedynska A, Hoek G, Eeftens M, Cyrys J, Keuken M, Ampe C, Beelen, Cesaroni G, Forastiere F, Cirach M, de Hoogh K, de Nazelle A, Madsen C, Declercq C, Eriksen K,T, Katsouyanni K, Akhlaghi H, M, Lanki T, Meliefste K, Nieuwenhuijsen M, Oldenwening M, Pennanen A, Raaschou-Nielsen O, Brunekreef B, Kooter I, M (2014) Spatial variations of PAH, hopanes/steranes and EC/OC concentrations within and between European study areas. Atmos Environ 87:239–248. doi:10.1016/jatmosenv.2014.01.026.

[6] Kabata-Pendias A, Pendias H (1999) Biogeochemistry of trace elements. PWN, Warsaw (In Polish).

[7] Kim JJ, Smorodinsky S, Lipsett M, Singer BC, Hodgson AT, Ostro B (2004) Traffic-related air pollution near busy roads: the East Bay Children's Respiratory Health Study. Am J Resp Crit Care 170:520–526.

[8] Laupsa H, Denby B, Larssen S, Schaug J (2009) Source apportionment of particulate matter (PM2.5) in an urban area using dispersion, receptor and inverse modelling. Atmos Environ 43:4733-4744. doi:10. 1016/j.atmosenv.2008.07.010.

[9] Lim J, Jeong J, Lee J, Moon J, Chung Y, Kim K (2011) The analysis of PM2.5 and associated elements and their indoor/outdoor pollution status in an urban area. Indoor Air 21:145–155. doi:10.1111/j.1600-0668.2010.00691.x.

[10] Marconi M, Sferlazzo DM, Becagli S, Bommarito C, Calzolai G, Chiari M, Sarra A, Ghedini C, Gómez-Amo JL, Lucarelli F, Meloni D, Monteleone F, Nava S, Pace G, Piacentino S, Rugi F, Severi M, Traversi R, Udisti R (2014) Saharan dust aerosol over the Central Mediterranean Sea: PM10 chemical composition and concentration versus optical columnar measurements. Atmos Chem Phys 14: 2039–2054. doi:10.5194/acp-14-2039-2014.

[11] Masiol M, Squizzato S, Rampazzo G, Pavoni B (2014) Source apportionment of PM2.5 at multiple sites in Venice (Italy): spatial variability and the role of weather. Atmos Environ 98:78-88. doi:10.1016/j. atmosenv.2014.08.059.

[12] Mazzei F, Alessandro AD, Lucarelli F, Nava S, Prati P, Valli G, Vecchi R (2008) Characterization of particulate matter sources in an urban environment. Sci Total Environ 401:81–89. doi:10.1016/j.scitotenv.2008.03.008.

[13] Moreno T, Querol X, Alastuey A, Viana M, Salvador P, de la Campa AS, Artinano B, de la Rosa J, Gibbons W (2006) Variations in atmospheric PM trace metal content in Spanish towns: illustrating the chemical complexity of the inorganic urban aerosol cocktail. Atmos Environ 40: 6791–6803. doi:10.1016/j.atmosenv.2006.05.074.

[14] Ostro B, Tobias A, Karanasiou A, Samoli E, Querol X, Rodopoulou S, Basagana X, Eleftheriadis K, Diapouli E, Vratolis S, Jacquemin B, Katssouyanni K, Suner J, Forastiere F, Stafoggia M (2015) The risk of acute exposure to black carbon in southern Europe: results from the MED-PARTICLES project. Occup Environ Med 72:123–129. doi:10.1136/oemed-2014-102184.

[15] Paatero P (1997) Least squares formulation of robust nonnegative factor analysis. Atmos Environ 37:23–35. doi:10.1016/S0169-7439(96) 00044-5.

[16] Pollisar AV (1998) Atmospheric aerosol over Alaska: 2. Elemental Composition and Sources J Geophys Res 10:19045–19057. doi: 10.1029/98JD01212.

[17] Querol X, Viana M, Alastuey A, Amato F, Moreno T, Castillo S, Pey J, de la Rosa J, Sanchez de la Campa S, Artinano B, Salvador P, Garcia Dos Santos S, Fernandez-Patier R, Moreno-Grau S, Negral L, Minguillon MC, Monfort E, Gil JI, Inza A, Ortega LA, Santameria JM, Zabalza J (2007) Source origin of trace elements in PM from regional background, urban and industrial sites of Spain. Atmos Environ 41:7219–7231.
doi:10.1016/j.atmosenv.2007.05.022.

[18] Quincey P (2007) A relationship between Black Smoke Index and black carbon concentration. Atmos Environ 41:7964–7968. doi:10.1016/j. atmosenv.2007.09.033.

[19] Regulation of the Minister of the Environment (2012) on the levels of certain substances in the air 24 August 2012, Dz.U. 1031.

[20] Samek L (2012) Source apportionment of the PM10 fraction of particulate matter collected in Krakow, Poland. Nukleonika 57(4):601-606 Samek L, Furman L, Kawik T, Welnogorska K (2015) Application of X-ray fluorescence method for elemental analysis of PM2.5 fraction. Nukleonika 60(3):621–626.

[21] Samek L (2016) Overall human mortality and morbidity due to exposure to air pollution. Int J Occup Med Env 29: 417–426. doi:10.13075/ ijomeh.1896.00560.

[22] Samek L, Gdowik A, Ogarek J, Furman L (2016) Elemental composition and rough source apportionment of fine particulate matter in Krakow, Poland. Environ Prot Eng (in press).

[23] Terrouche A, Ali-Khodja H, Kemmouche A, Bouziane M, Derradji A, Charron A (2016) Identification of sources of atmospheric particular matter and trace metals in Constantine, Algeria. Air Qual Atmos Health 9:69-82.
doi:101007/s11869-014-0308-1.

[24] Vekemans B, Janssens K, Vincze L, Adams F, Van Espen P (1994) Analysis of X-ray spectra by iterative least squares (AXIL): new developments. X-Ray Spectrom 23:278–285. doi:10.1002/xrs. 1300230609.

[25] Viana M, Maenhaut W, ten Brink HM, Chi X, Weijrs E, Querol X, Alastuey A, Mikuska P, Vecela Z (2007) Comparative analysis of organic and elemental carbon concentrations in carbonaceous aerosols in three European cities. Atmos Environ 41:5972–5983. doi:10.1016/j.atmosenv.2007.03.035.

[26] Viana M, Kuhlbusch TAJ, Querol X, Alastuey A, Harrison RM, Hopke PK, Winiwarter W, Vallius m, szidat S, Prevot ASH, Hueglin C, Bloemen H, Wahlin P, Vecchi R, Miranda AI, Kasper-Giebl A, Maenhaut W, Hitzenberger R (2008) Source apportionment of particulate matter in Europe: a review of methods and results. Aerosol Sci 39:827–849. doi:10.1016/j.jaerosci.2008.05.007.

[27] Yu L, Wang G, Zhang R, Zhang L, Song Y, Wu B, Li X, An K, Chu J (2013) Characterization and source apportionment of PM2.5 in an urban environment in Beijing. Aerosol Air Qual Res 13:574–583. doi:10.4209/aaqr.2012.07.0192.

[28] Zhang N, Han B, He F, Xu J, Niu C, Zhou J, Kong S, Bai Z, Xu H (2015) Characterization, health risk of heavy metals, and source apportionment of atmospheric PM2.5 to children in summer and winter: an exposure panel study in Tianjin, China. Air Qual Atmos Health 8:347–357. doi:10.1007/ s11869-014-0289-0.

Chapter 11

A Systematic Review on Status of Lead Pollution and Toxicity in Iran; Guidance for Preventive Measures

Parissa Karrari[1,2], Omid Mehrpour[1,2], Mohammad Abdollahi[3*]

[1]Medical Toxicology and Drug Abuse Research Center (MTDRC), Pasdaran Avenue, Birjand University of Medical Sciences, Birjand, Iran
[2]Department of Clinical Toxicology and Forensic Medicine, Faculty of Medicine, Birjand University of Medical Sciences, Ghaffari Avenue, Birjand, Iran
[3]Department of Toxicology and Pharmacology, Faculty of Pharmacy, and Pharmaceutical Sciences Research Center, Tehran University of Medical Sciences, Tehran, Iran

Abstract: Lead is an old environmental metal which is presented everywhere and lead poisoning is an important health issue in many countries in the world including Iran. It is known as a silent environmental disease which can have life-long adverse health effects. In children, the most vulnerable population, mental development of children health effects is of the greatest influence. Low level lead exposure can significantly induce motor dysfunctions and cognitive impairment in children. The sources of lead exposure vary among countries. Occupational lead exposure is an important health issue in Iran and mine workers, employees of paint factories, workers of copying centers, drivers, and tile making factories are in higher risk of lead toxicity. Moreover lead processing industry has always been a major of concern which affects surface water, drinking waters, and ground waters, even water of Caspian Sea, Persian Gulf and rivers due to increasing the number of industries in vicinity of rivers that release their waste discharges into river or sea. In addition, lead contamination of soil and air especially in vicinity of polluted and industrialized cities is another health problem in Iran. Even foods such as rice and fishes, raw milk, and vegetables which are the most common food of Iranian population are polluted to lead in some area of Iran. Adding lead to the opium is a recently health hazard in Iran that has been observed among opium addicts. There are few studies evaluated current status of lead exposure and toxicity in the Iranian children and pregnant women which should be taken into account of authorities. We recommend to identify sources, eliminate or control sources, and monitor environmental exposures and hazards to prevent lead poisoning.

Keywords: Iran, Lead, Poisoning, Pollution, Toxicity

1. Background

Lead is an old environmental xenobiotic metal which is presented everywhere[1] and its chemical properties make a wide spectrum of applications possible for lead. Lead is used in more than 900 industries, including mining, smelting, refining, battery manufacturing and so on[2]. It is one of the most abundant natural substances[3] and is the fifth highest metal used throughout the world. In Iran, application of lead dates back to 5,000 years ago[4] and previous Iranian scientific such as Haly abba (10th century), Rhazea (865–952 CE) knew about concept of lead poisoning. Iranian people used lead for different purposes such as facial powder, painting, and traditional tile brick glazing. Industries such as mining had not been modernly managed until 1930, thus there were many lead exposure in the workers. Lead toxicity is of the major concerns to public health due to widespread persistence of lead in the environment[5]. In the history of medicine; lead poisoning has been a well-known disease. The first article about lead poisoning was published in 1848[4]. Although lead toxicity has been relatively controlled in industries but it is still the most common environmental toxicity in the United States of America (U.S.)[6] and it is an important health issue in countries such as Iran[7]. Over the past decades; efforts have been made to reduce its exposure[5]. The activity related to the workers safety and occupational health has been started from 1946 in Iran[4].

Signs and symptoms of lead poisoning included hearing loss, anemia, renal failure, and weakened immune system, and Low birth weights, still births and miscarriages, premature births, and increased urine and blood lead levels (BLL) are the most common reports[8].

BLL provides the best parameter of recent exposure to this metal[9]. Normal BLL is less than 30μg/dL, whereas acceptable BLL ranges between 30 and 49.9μg/dl while high BLL refers to higher than 49. 9μg/dL[9]. The World Health Organization (WHO) expresses the limit for BLL as 1.9μmol/L (40μg/dL) for men, and 1.4μmol/L (30μg/dL) for women of child-bearing age[3].

There are so many papers about hazards and adverse effects of lead on human worldwide and meanwhile the incidence, prevalence and sources of its contamination is clear. But this is the first comprehensive review about lead toxicity in Iran. The fact is that information on lead contamination in Iran are incomplete and dispersed. In this review, all papers published about sources and hazards of lead exposure in Iran during the past two decades have been gathered and criticized to reach a conclusion about exact risk of this metal in a large country with highest rate of import and export in the Middle East.

2. Methods

We looked up the terms lead, Pb, toxicity, poisoning, exposure, source and Iran in all bibliographical databases such as TUMS digital library, PubMed, Scopus and Google Scholar. This review includes relevant articles published between 1990 and 2011.

3. Findings and Discussions

3.1. Occupational Lead

Lead is a toxic heavy metal for human that is recognized as an environmental and occupational hazard. However, in industry, it is a useful metal and is still being used in various industries in Iran, for example, in producing of lead bullets, in battery manufacturing, lead refinery industry, and is used as a smelter metal for purifying gold and silver. The workers who work in these factories can be easily exposed to the dusts or fumes of lead. Occupational lead poisoning has been a human health hazard for more than two centuries[10]. While, acute lead poisoning is rare, subacute and chronic intoxication (occupational) are not uncommon in cities where industries or mines are located. As reviewed by Mañay *et al.* (2008) in Uruguay, it was revealed that exposed workers with lead from different manufacturing industries such as battery factories, foundries, wire factories, etc., showed that almost 60% of BLL of tested cases were above 40μg/dL[11]. It has been reported that lead exposure in Brazil occurs mainly in battery plants (recycling plants and lead-acid battery producing). Also lead in pigments, plastic and ceramics, and rubber industries are other concerns in Brazil. Moreover, small recovery battery workshops and medium size secondary smelting plants have been found responsible for the most occupational lead poisoning cases in the Brazil[12]. Similarly, it has been reported that almost 95% of lead poisoning among US adults comes from occupational exposure[7].

In Iran, people or workers of industrial cities such as Tehran, Mashhad, Isfahan, Tabriz, Zanjan, and Arak are at greater risks. Some years ago when leaded gasoline was highly used in Iran, professional drivers such as taxi drivers or bus drivers were found highly-exposed through polluted air[13] but this is not the case in the most recent years. Lead toxicity in drivers may be influenced by availability of lead dust, and spreading in air breathing[3]. Thus, considering the similar countries in the world, people of cities with major motor vehicle traffic and air pollution may be still at risk of lead pollution[13].

Another reported occupational lead exposure is workers of copying centers[9] and employees of paint production factories[3][14]. Inorganic lead compounds are widely used in paint and pigment industries[6] and inhalation seems the most probable route of exposure[9] to lead. Besides, lead naphthenate oxalate is also used as a drying agent in the paint[3] and thus painters seems to have higher BLL but it has not been studied in Iran yet.

Kalantary *et al.* compared the BLL in workers of Zinc melting factory of Dandi Zanjan with healthy men who were living around the factory and found that BLL in factory workers were higher than that of controls[15]. In another study carried out in Zanjan city on the workers of a lead refinery industry and two control groups, the mean concentrations of hair lead in the lead refinery workers (case group), the staff (control group A) and the citizens (control group B) were 131.7 ± 93.4μg/g, 21.1 ± 13.2μg/g, and 27.9 ± 14.1μg/g, respectively. The mean hair lead concentration in the case group was more than normal range (0–30μg/g). The mean hair lead level in the citizens who used gas vehicles was statistically higher than who had not used it (36.9 ± 12.2μg/g vs. 16.6 ± 4.9μg/g)[16].

Tabrizizadeh *et al.* evaluated the relationship between the prevalence of oral complications and BLL in workers employed in Koushk lead mines from Yazd province and compared BLL with a control group. They found that factory workers had higher BLL than controls and in the meantime neurologic disorder, chronic fatigue, existence of lead line, mucous pigmentation, gingivitis, tongue burning, taste sense reduction and dimethylformamide (DMF) were higher among workers, although the BLL in most of mine workers was in normal limits[17]. Moreover in a study conducted for determination of BLL on workers of lead and zinc mine in Kooshk City, it was revealed that BLL in 45.7% of workers were more than permissible limit[18]. In another study in mine workers exposed to lead and zinc in Arak city, it was shown that the mean scores of physical complaints, anxiety, and aggression scales were significantly higher in the case group than the control. They concluded that oxidative stress induced by lead results in mental disorders and thus mine workers suffered from more psychological disorders should be in greater care[19]. Moreover, evaluating BLL in Welders of a car company in suburb of Tehran revealed that BLL in those who smoke more than seven cigarettes per day was significantly higher than those who smoke less than seven cigarettes per day or no smoking group, also the hemoglobin concentrations in frequently cigarette users was significantly lower than that of the non-smokers or less cigarette users. They concluded, cigarette smoking in occupationally lead-exposed workers makes them in higher risk of lead as well as inhibition of hemoglobin synthesis[20]. In another study, lead concentration in urine of urban service workers of Tehran were compared with control group, and the results showed that lead levels in 77. 1% of the urine samples were higher than Health and Safety Executive (HSE)-recommended limits (643.86 ± 353.73). Also, mean urine lead levels in smokers were significantly higher than that of non-smokers in case group[21]. Yartirah *et al.* in a study on workers of refinery in Kermanshah found that those workers had higher blood and urine lead levels in comparison to control group.

Also lead concentration among those who worked with tin was higher than others. In the meantime, there was a correlation between increase of lead level and increase of age or cigarette smoking[22]. Actually, workers who are involved in glazing the traditional tiles are easily exposed to the lead. In a study conducted by Balali-Mood *et al.* in Mashhad to determine the prevalence of lead intoxication and its complication in traditional tile workers, they concluded that lead toxicity in these workers is not uncommon and the toxic effects of lead were more often found on the teeth, central nervous system and peripheral nervous system[7][23]. Lead is also a significant occupational hazard in ceramic industries and it is still used in ceramic industries in many of Asian countries. Lead glaze is commonly used for hand-crafted pottery in Iran to produce certain colors and help to prevent cracking. Inhalation of airborne lead and ingestion of lead through contaminated hands are generally the common sources of lead absorption in lead-glazed ceramics workers. In Iran, the Lalejin city in Hamadan province is the main center of hand-crafted pottery. This city has about 15,000 residents and about a hundred glazing workshops. Lead, copper, zinc and magnesium are used in these glazing workshops without preventive measures against heavy metal toxicity and the workers of these workshops are at great risk of lead toxicity. Some reports documented lead poisoning coming from these potteries glazing[24].

It has been pointed out that tetraethyl lead is added to petrol for reducing flammability so gasoline station workers are another group at risk of lead toxicity[25]. Repetitive stopping of numerous vehicles, that are coming and going along the days, contaminated floor in gasoline station and workers clothes, makes this group exposed to the lead[26].

It should be considered that battery plant workers, solder ammunitions, workers involved welding and tile making factories, painters, car radiators, manufacturing or use of cable and wires, ceramic ware with lead glaze and tin cans, lead smelting plants and steel plants are other main jobs with possible occupational lead toxicity[2][7][14].

Beside this, workers of battery manufactories were found at great risk of lead toxicity in developing countries such as Uruguay and Brazil[11][12]. In a study performed on 105 workers who were exposed to lead in a car battery manufacturer in Mashhad, Iran, in 2006, it was revealed that all of these workers had lead intoxication with mean BLL of 32.2 ± 13.7μg/dl[27].

In addition to this, inhalation of fumes from burned car batteries, and ingestion of flaking paint are other occupational sources of lead poising[28].

Direct contact of oral mucosa with the lead in breathing air[17] or difficulty of environment and work conditions as risk factors, and smoking[20] may be reasons of lead toxicity in these workers and more important of that is work location. Of course, age, duration of employment and smoking habit[21][22] have direct effect in toxicity.

3.2. Air as a Source of Lead Exposure

Composition of settled dust is similar to air suspended particulates, so it can be a marker of pollutants such as heavy metal contamination in the air. In China, heavy metals were determined in dust of roads, tunnels, urban parks, playgrounds, children's nurseries and households[29]. Humans can be exposed to heavy metals such as lead in dust through several routes including inhalation, ingestion and skin. In dusty environments, it is estimated that adults could ingest up to 100mg dust every day. Children are even exposed to greater amounts of dust than adults due to play behavior and hand-mouth pathway[30]. Exposure to lead for general population comes mainly from airborne dusts containing lead particles and from food or water contaminated by lead, of which 15%–30% is inhaled and 70%–86% is ingested[13]. Each year, 200 million tons of man-made waste products are released into the air, and 50% of this data belong to burning internal engine[31] and the main reason is due to leaded gasoline that is still used as an automobile fuel[13].

Tehran is one of the crowded cities of the world with vehicular terrific, which leaded gasoline is still used although prohibited. Beside from geographical aspect, Tehran is surrounded on three sides by mountains, with no continuous flow of air, and makes this city the highest polluted area[13]. Six types of materials have been known as the major air pollutants which make more than 90% of air pollutants. These include carbon monoxide, nitrogen oxide, hydrocarbons, sulfur oxide, suspended solids, and lead. Those who live in south and central part of Tehran had the highest BLL[32] and those who live in downtown and busy streets are in higher risk of lead toxicity[13] in comparison to those who live in suburb. In a study conducted by Farzin *et al.* (2008) for obtaining the usual value of Pb, Cd, and Hg in normal human blood of 101 volunteers resident in Tehran, it was found that BLL in normal volunteers living in Tehran were 123.75 ± 56.42 and their results showed significantly higher content of Pb in blood of males compared to females (138.11 ± 65.43 and 101.84 ± 51.38μg/dL, respectively)[33].

Another study in Kerman one of industrial cities in center of Iran on lead concentration of gasoline station air indicated that the mean value of all the stations was higher than the control one, but because of lack of inversion phenomenon in this city, lead concentration in all the gasoline stations was lower than the leads TLV (total lead value)[26].

High population, vehicles, urban activities, and around industries have made Yazd a city in center of Iran to face serious air lead pollution[34]. It has been estimated that 60%–90% of lead in airborne dust of the ambient and 10%–50% of lead in the blood of the non-occupationally-lead-exposed population can be attributed to lead in gasoline[13]. Evaluation of suspended air particles and their composition in central area of this city showed that it is higher than national standard[34].

Also it has been reported that air of Zanjan a city in which major lead and zinc factories are located its around is full of heavy metals[35]. Beside, in a study conducted in Tehran, about 40% of randomly selected children had higher BLL which clearly showed importance of screening test for lead poisoning in the population[36]. The measurement of heavy metals in atmospheric precipitation shows the effects of anthropogenic sources in air quality. The heavy metals concentration can be used as air pollution index[24]. Some European countries such as Bulgaria were lucky in full banning import and use of leaded gasoline. The lead levels in Varna, the third largest city, decreased up to 63-fold in year of 1996–2007[37]. It seems it is the time to be so serious in policies to decrease heavy metal air pollution in Iran such as full prohibition of using leaded gasoline and advises to use filters to reduce the amount of pollutants produced in factories, relocating factories to outside of cities, getting rid of old automobiles and regulating automobile engines that all are undergoing.

3.3. Water as a Source of Lead Exposure

Heavy metals processing industry has always been a major of concern which affect surface water, drinking waters, ground waters and rivers contamination. There are several sources of water around us which make concern about our future life. All of these sources should be examined for the presence of lead when determining a person's total lead exposure and risk.

Based on WHO standard, concentration of lead in drinking water was limited to 0.01mg/L, and based on drinking water standard in Iran, upper limit of the concentration of lead in drinking water announced to 0.05mg/L[38][39]. Lead exposure from drinking water has been a topic of public prevention programs in European countries[40]. To assess the present state of drinking water contamination with lead, a free examination of lead in drinking water was offered in cooperation with local public health departments for private households in Germany. In the screening part of that project, 2,901 tap water collected during 2005–2007 which of those, 7.5% had lead concentration of more than 10μg/L (recommended limit of the WHO) and 3.3% had concentrations above the present limit of the German drinking water ordinance (25μg/L)[40].

Recently the problem received attention in the US when report of drinking water at schools was published[41]. Beside, several European countries are known to have significant numbers of building with elevated concentration of lead in drinking water, such as UK, Austria, and Germany[42]–[44]. In a study conducted in UK (1996), it was found that 17% of households had water lead concentration of 10μg/L (48.3nmol/L) or more in 1993 in comparison with 49% of households in 1981. Meanwhile, tap water lead remained the main cause of raised maternal BLL accounted for 62% and 76% of cases whom maternal BLL were above 5 and 10μg/dL (0.24 and 0.48μmol/L), respectively[42].

In another study conducted in Austria (2002), the collecting data of the upper floors showed significantly higher lead concentration compared to the lower floors, which indicates that in Viennese drinking water, house installations were the major causes of lead contamination, but in comparison to other European countries the percentage of samples exorbitant the guideline levels (50μg/L as current value and 10μg/L as target value) was lower.

Typically, lead gets into tap water after it leaves the water treatment plant, so its monitoring is difficult and somehow impossible to estimate such exposures to lead and other metals, because contamination occurs when the distribution system is not monitored[43].

Lead contamination of drinking water is also a major concern in Iran. In a study carried out to determine heavy metals in water sources of Hamadan city (West of Iran) in 1994, 90 water samples were analyzed and the results showed that the mean concentration of lead were 0. 514mg/L, which are higher than the standard levels and the authors concluded that these pollutants are mainly sourced from industrial waste and/or fuel consumption. Authors suggested authorities to force factories to restructure their wastewater treatment plant[45].

Another study conducted in Ahwaz city (South-West of Iran) to evaluate the corrosion and leakage potentials of some important heavy metals (Pb, Cd, Zn, Cu, Fe, Mn) using the USEPA (United States Environmental Protection Agency) standard procedure. They selected 76 sampling points including raw water intakes, treatment effluents, and tap waters in Ahwaz distribution network. The results from six rounds of tests showed a lead concentration of 8.47μg/L in drinking water. Furthermore, the data indicated high corrosion potential in Ahwaz drinking water distribution network and the leakage of lead and other heavy metals into the network closely associated with the corrosion phenomenon[46].

Shah-Mansouri *et al.* (2003) evaluated trace metals in the drinking water distribution system in Zarin Shahr and Mobareke of Isfahan province. They found that the average concentration of lead in water distribution system of Zarin Shahr were 5.7μg/L and in Mobareke were 7.83μg/L. They discovered that lead concentration was zero at the beginning of the water samples from the municipal drinking water distribution system for both cities. Also they showed corrosion by lead that was the result of dissolution of the galvanized pipes and brass facets. Lead concentration in over 10% of the water samples of Zarin Shahr exceeded the drinking water standard level[47]. In another survey conducted in Isfahan city to evaluate the leakage of heavy metals from the polypropylene pipes and PVC which are used in the water distribution system, the mean lead concentration in old and new PVC pipes was higher than other pipes. Mean leakage of lead was higher in polypropylene (PP) pipes produced in manufacturing plants. Lead leakage was lower than

Iranian standards, but exceeds than EPA standards or WHO guidelines in PP pipes produced in manufacturing plants[48]. Thus, it seems use of these types of pipes in the water distribution systems may increase lead concentration in drinking water. In the meantime, metallic structure and inappropriate plastic production are potential factors in contamination of network drinking water with heavy metals. A high protein of lead concentration in municipal drinking water may be related to dissolution of the brass facets and galvanized pipes[47] and leakage of heavy metals from the PVC and PPpipes used in distribution system[48].

Ground water resources in arid and semiarid regions are very important[49]. Groundwater is contaminated by agricultural, industrial and municipal activities[50]. In US, more than half of the consuming water for population and one-third of water for agriculture is supported by groundwater[51]. Thus lead contamination of this groundwater can cause health problems for a country.

Sanitary landfill is one of the potential factors in contamination of ground water. Utilizing the groundwater closed to the landfill and lack of insulating the landfills floor, makes easy movement of current contamination into groundwater. Ebrahimi *et al.*, evaluated pH and metal concentration of area's groundwater near the municipal solid waste landfill of Yazd city in center of Iran and compared with the ground waters far from them. It was found that the ground waters pH in downstream was significantly lower than upstream and both case and control groundwater were contaminated with lead at the same amount[49]. Moreover in a survey on chemical quality of groundwaters in Zarin Shahr city, it was found that the mean concentration of lead and cadmium exceeded than standard levels. The authors concluded that the water wells are polluted due to high discharge rate of agricultural and industrial wastewater[50].

It is essential to control and treat the wastewater appropriately and also to monitor the groundwater to prevent the aquifer pollution. In another study looking for heavy metals in soils, water, and vegetables of Shahnama region in Shahroud city, it was revealed that mean concentration of lead in water samples were 7.55mg/L, which was much higher than standard value (0.71mg/L). The authors concluded that use of synthetic fertilizers, unsanitary disposal of sewage and fossil fuel combustion has made water, soil and plants of the region polluted with heavy metals[52].

In the study by Mohammadian *et al.* (2008), water wells close to Zanjan zinc and lead smelting plant was examined and a lead concentration of 53% in water wells was found that is higher than standard values of WHO[53].

Collection and storage of roof rainwater (Cisterns) in rural areas are traditionally done from long time ago in Iran and many other countries. Many residents in rural areas of Turkaman Sahra located in Golestan province in North of Iran are providing part of drinking and municipal water by this way[54]. In a study detecting probable contamination resources in cisterns in this province, it was found that lead concentration in 51% of samples were higher than reference level. Any of water cisterns in this province was unfavorable for drinking due to lead contamination. Lead pollution in the roof rainwater maybe due to infiltration surface and agricultural waters and precipitation of air pollution and high lead content in these water indicate the need for some form of treatment[54][55].

3.4. Soil as a Source of Lead Exposure

Soil contaminated with lead is not only a major concern in developing countries, but also it is a health problem in western countries. Rabito *et al.* recently reported the high incidence (61%) of lead above recommended levels in soil and dust samples of New Orleans in the US. Most notably children and around residences were concerned about potential health risks to the lead-contaminated soil and dust in that area[56].

Plant and soil surface are the major sink for airborne lead in the environment and may take a contribution to dietary lead intake[57]. Hereby, in the following paragraphs, we are going to illustrate lead contamination potential of soil in industrial areas, vehicular traffics, and near shore areas.

The rapid industrialization in developing countries frequently causes a high anthropogenic emission of heavy metals into the soil[58]. Hamadan province is located in West of Iran with 1.75 million inhabitants, and semiarid climate. Although agriculture is a major habit of these people but this city has becoming industrial in the recent years[58]. Jalali *et al.* researched about contamination of lead in industrial areas in this province. They revealed that industrial soils were contaminated with lead to some extent. Application of sewage sludge, fertilizers, and pesticide in agriculture[58], and industrial activities such as opencast mining and smelting[59], and failure to complete recycling of city refuses or discharge of municipal waste urban in soil had a serious environmental impact on this area and contributed to a continuous accumulation of heavy metals in soil[58].

Zanjan province located in North-West of Iran has been considered as a traditional mining region since ancient time[59] and there are several studies reporting heavy metal contamination in vicinity of lead and zinc mines[53][59][60]. Parizanganeh *et al.* studied heavy metal pollution in superficial soils surrounding industrial area in Zanjan and found wide spread heavy metal contamination of soil[59]. In another study, soil, water and vegetables of Shahnama region in Shahroud city was examined and the mean concentration of lead in soil was found 81.12 (μg/g), which was much higher than standard value (0.2−1 μg/g). The authors concluded that use of synthetic fertilizers, unsanitary disposal of sewage and fossil fuel combustion has resulted in pollution of water, soil and plants of the region with heavy metals[52].

Pollution caused by traffic activities and exhaust product of leaded gasoline, are one of the major source of contamination by lead in urban environment[57][61]. Expositing to contaminated surface soil with lead, through indoor and outdoor inhalation of lead in dust and ingestion of lead deposited within houses[61]. Farsam *et al.* studied lead deposition on plant leaves in Tehran and reported that older leaves have higher lead level[57]. In overall, their results tended them to the conclusion that major cause of lead contamination in downtown plants is vehicle exhaust and low rain fall[57]. In another study conducted to investigate the concentration of lead, cadmium, copper and zinc in different sites of the Sari-Ghaemshahr road in Iran[61] which is one of the crowded roads in North of Iran, the soil samples were collected along the sampling section with different distances from the road edge of both sides of the road. Their results showed high amount of lead in nearest distance to the road. Of course, amount of heavy metals is basically dependent on wind[62], traffic intensity, and tire wear[63]. Thus, the highest value of lead in nearest distances could be because of emissions from vehicle exhausts.

In another study, concentration of heavy metals (Pb, AL, Cu, Ni, Zn) in near shore sediments in alongshore direction of the Iranian coast of Caspian Sea was examined[64]. The results showed that concentrations reflected metal loading from anthropogenic sources located at and in the vicinity of the sampling sites[64]. Metal discharged into coastal areas of marine environments is likely to be scavenged by particles and removed to the sediments. So sediments, become large repositories of toxic heavy metals. In overall, although those experiments give information concerning possible enrichment of the soil with heavy metal, but the severity of pollution depends also on the proportion of their mobile and bioavailable form which determines their mobilization capacity and behavior in the environment[58].

3.5. Fish

Heavy metals have a high resistance against degeneration (stable pollution)[65]. Thus, their amount in fish may be increased even several times either in water or air, due to bioaccumulation though fish is often at the top aquatic food chain[8].

Fish as human food is considered as a good source of protein, polyunsaturated fatty acids (omega-3), calcium, iron, zinc and generous supply of minerals and vitamins[66]. The demand for fish as a source of protein is on arise. During the last few decades, great attention has been paid to the possible hazards of heavy metal poisoning in human due to the consumption of contaminated fish. Based on our statistics its consumption during last 20 years, its increased up to 5kg per capita in Iran[67]. Here we are going to illustrate 4 different parts of Iran that measurement of lead in fish was conducted in the last decade.

3.6. Caspian Sea

The Caspian Sea with 386,400Km^2, with 5 major inlets and no outlet acting as a watershed reservoir is the largest lake surrounded by 5 countries, Azerbaijan, Iran, The Russian Federal, Kazakhstan and Turkmenistan[8]. Determination of lead in the most consumed fishes in Caspian Sea in different studies[8][66][68] revealed existence of exposure to lead.

Shokrzadeh *et al.* (2004) measured the amount of lead in five species of most consumed fishes in five fishery areas of the Caspian Sea region and their results showed that Rutilus frsii kutum fish had the highest concentration of lead compared to other fishes. Rutilus frsii kutum is living in depth of water, and in this depth, concentration of heavy metals in animals tissue are much higher in comparison to other parts of water[68]. Thus this kind of fish has the most concentration of lead in comparison to the rest. Although lead concentration of Rutilus frsii kutum fish was at standard levels (less than 0.5ppm) but it was a significant increase in its lead concentration comparing to year of 1997 with value of 0. 07ppm[68]. Clupeonella delicatulu is another kind of consumed fish from Caspian Sea in North of Iran, and its calcium was replaced with heavy metals from contaminated water because it is consumed with bone, it can cause lead toxicity in humans[68]. Gorgan coast is located in southeastern of Caspian Sea. This coast is one of the most important ecosystems in the North of Iran[66]. Large amount of pesticide and chemical fertilizers containing heavy metals that are used by agricultural industry of this region, are brought via surface run off from farm to river and increases lead concentration in Caspian Sea and its fishes[66]. Tabari *et al.* (2010) conducted a study on Rutilus frsii kutum, Cyprinus carpio, Mugila auratus in Gorgan coast and reported increasing hazard of lead concentration in fishes, water and the sediment[66]. Also Eslami *et al.* (2011) measured lead level in Rutilus frisii kutum from Tajan River, one of the significant rivers of Caspian Sea water basin[8] and reported existence of lead toxicity. In above investigations that carried out in one region, it is clear that although the observed heavy metals concentrations were below the recommended limit, but existence of pollution cannot be ignored[66][68]. In Eslami *et al.* study, they found nearly all non-essential metals (Cd, Ni, Pb) in Rutilus frisii kutum fish higher than limits for fish proposed by FAO/WHO and EU[8].

Besides, Rutilus frisii kutum was the common fish among these studies, this fish is a very valuable commercial fish in that region, with a high demand, due to good taste and kitchen customs, and is consumed in all year around, the average annual catch of Kutum in Iran was about 96,000 Rials in 1991–2001[69]. In relation to this, spreading lead toxicity through fish and fishery product consumption would be catastrophic.

Increasing the number of factories and industries in vicinity of rivers and receiving effluent discharges which end to sea, presence of thousands of vehicular traffics spreading heavy metals like lead into atmosphere, raining back to earth and sea, leaking petrol from petrol port during exiting or transmission of oil, and land-locked body of water boarded by five countries are all common reasons which makes Caspian sea in priority to any measures to reduce environmental pollution[8][66][68].

3.7. Kor River

The Kor River is the longest freshwater river in Fars Province in South of Iran, it is approximately 50 km long, 15–20 km wide and nearly 20 m depth and it originates from Zagros Mountain. The Kor River is used for irrigation of rice paddies and homesteads, as the supply of drinking water and industrial water needs, and for hydroelectric energy production[67]. Every year the entrance of factory waste such as Shiraz Petrochemical Complex, Marvdasht sugar cube factory, and Charmineh factory and other industrial units into the Kor increase[70]. In a study in Fars province, Ebrahimi *et al.*[67] studied on lead concentration in (Cyprinus Carpri and Capoeta sp.) from Kor River indicating presence of maximum amount of lead higher than the permissible levels for human consumption[67]. Also it was revealed that lead toxicity in the fishes can induce pathological changes in blood cells, liver and kidney of fishes and these changes were significantly higher in highly polluted area[67].

3.8. Persian Gulf

The Persian Gulf is an extension of the Arabian Sea, positioned in the heart of the Middle East. It connects with the Gulf of Oman and the Arabian Sea through the Strait of Hormuz, and it's approximately 990 km long. The Persian Gulf is certainly one of the most vital strategic bodies of water on the planet, as gas and oil from Middle Eastern countries flow through it, supplying most of the world's energy needs. The Persian Gulf has been subject to inputs of heavy metals from different sources, and it has been estimated that oil contamination in the Persian Gulf represents 4. 7% of the total oil pollution in the world (National Research Council, 1985)[71]. This oil pollution has increased even more after the wars occurred around Persian Gulf, about 11 million oil barrels were discharged into the Persian Gulf[72].

Shahriari *et al.*[65] reported that in edible tissue of Lutjans Coccineus and Tigeratooh Croaker in Persian Gulf, concentration of lead in 27% of collected samples was more than acceptable limit of WHO[65].

Ashraf in a study evaluated lead level in the kidney and heart tissues of Epinephelus Microdon collected from the Persian Gulf, Eastern province of Saudi Arabia, and it was found that the average lead (3.19 ± 2.03ppm) concentrations of heart tissues is exactly high[73].

Raissy *et al.* (2011) in a study determined the concentrations of lead and 3 other toxic metals in lobster (Panulirus homarus) muscles from the Persian Gulf. Lead concentrations in muscle samples were 379–1,120μg/kg, with means of 629. 4μg/kg. Lead in the edible muscle tissue, was above the acceptable level and showed a health risk for consumers[74].

3.9. Farmed Fishes

Pourmoghaddas *et al.* in a study measured lead concentration in Cyprinus carpio (farmed fishes) and Lutjans Coccineus and Tigeratooh Croaker (from Persian Gulf) species which are the most consuming fishes in Isfahan city[75] and found a mean concentration of 0.48ppm for lead in the Cyprinus carpio fishes. It was also revealed that lead concentration in 27% of collected cases were more than upper limit in WHO[75]. High lead concentration in farmed fishes maybe due to limited sources of pound water and even there are situations that replace sewage agriculture instead of fresh water in some regions of this city. Also in this study, it was found that lead concentration in the farmed fishes is more than those collected from Persian Gulf. Although lead concentration in Caspian Sea fish in comparison to this region was two times more[75].

Determination of heavy metals in fish in different parts of Iran depends on concentration of the metals in water and exposure period[67], geographical area, quality of water source, distance of industrial units to coast, legal rules in disposal of sewage effluent, type of fish, type of organ tissue, condition of laboratory experimental[65] and other environmental factors, such as salinity, pH, water hardness, and temperature[67].

3.10. Rice

Lead is an unnecessary metal for human body, and any amount of it would be harmful[76] but it is accumulated in rice that is the most popular food among Asian people probably causing silent toxicity demonstrating itself as insufficiency in different tissues and organs[53][77]. Every year, factories and industrial units, and city sewers cause the pollution of agricultural land by adding large resources of contaminated water containing heavy metals[56]. Irrigation of farmland and crop-land with this water can cause potential harm for human[70].

The lead content in rice samples from various countries ranged from 1.6 to 58.3 ng/g and the average content was 15.7 ng/g[78].

Jahedkhaniki *et al.* (2005) determined the lead contents in rice in the North of Iran. They collected samples from four areas of Qaemshahr region in North of Iran (Mazandaran province) at harvesting time of rice. Their results showed that average concentrations of lead in rice was 2.23 ± 18mg/kg dry weight, which was upper than the FAO/WHO limits. Also the weekly intake of lead from rice was upper than the maximum weekly intake recommended by WHO/FAO[77].

Bakhtiarian *et al.* (2001) evaluated the effect of the Kor River's pollution on the lead and cadmium content of the Korbal rice samples in Fars Province in South of Iran[70]. Comparison of the pollution level of the Korbal and Gilan rice samples (which were cultivated with unpolluted water) indicated a significant difference and confirmed the significant effect of the pollution of the river on the lead and cadmium content of the Korbal rice samples. The reason is the entrance of drainage water from different factories like petrochemical factory, charmineh factory, and other industrial units and also entrance of Marvdasht and Zarghan sewer system wastes into the Kor River that is used for cultivation of the rice[70].

Shakerian *et al.* (2012) investigated the lead content of several commercially available brands of rice grains in central Iran. The results showed that lead concentration in rice grains ranged from 0.0405 to 0.1281ppm dry weight and its average concentration was 0.068 ± 0.0185ppm. They found that lead concentration in the sampled rice grains was lower in comparison with their upper limits (0.2ppm)[79].

In another survey by Malakootian *et al.* (2011) about determination of lead concentration in imported Indian rice, the result indicated that weekly intake of heavy metal by rice was below the provisional tolerable weekly intake recommended by WHO/FAO[76], but their results were against that of Bakhtiarian *et al.* who studied in South of Iran and Jahedkhaniki who studied in North of Iran which reported higher lead concentrations in rice samples[70][77]. The most important anthropogenic sources of soil pollution to metal are industrial sludge, effluent discharging, using super phosphate fertilizers, burying the non-ferrous waste in land and closing the agricultural fields to zinc mine and lead or refining factories[80]. These metals accumulate in agricultural products and enter to food chain. In overall health risk of lead intake through rice is high in Iran and might be even increased with consumption of vegetable, fish, etc.[77].

3.11. Vegetables

Food safety is a major public concern worldwide. Vegetables constitute essential components of diet such as vitamins, fiber, mineral, and other nutrients. In a study, determining of heavy metals in soils, water, and vegetables of Shahnama region in Shahroud city indicated that mean concentration of lead in soil samples was 23.99 (μg/g) that was much higher than standard value (0.1−10μg/g). The authors correlated it to use of synthetic fertilizers, unsanitary disposal of sewage and fossil fuel combustion, water, soil and plants of the region that are polluted with heavy metals[52]. Heavy metals such as lead are easily absorbed by soil but have no toxicity for plants[81]. There is evidence suggesting that vegetables cultivation vary in uptake of pollutants[82]. Generally, plants translocate larger quantities of metals into their leaves rather than to their fruits and seed[82]. The amount of lead in the soil is important due to the direct transmission of lead, and also due to forming water-soluble forms of lead by streams of water or rain[82], furthermore the effect of lead in the water or air is directly transmitted[82]. These amounts maybe hazardous if the vegetables are taken in large quantities. Nonetheless, all these metals have toxic potential, but the detrimental impact become apparent only after decades of exposure.

Contamination of vegetables with heavy metals may be due to irrigation with contaminated water[45][52][83]−[85], addition of fertilizers and metal-based pesticides, industrial emissions, transportation, harvesting process, and storage at the point of sale[83].

There are several studies indicating that irrigation with polluted water is the main source of lead in vegetables[45][83][84]. In another study about determination of lead in cultural vegetables in suburb in Shahroud city, it was shown that surplus water of urban and industrial facilities are main reason of rising lead content to above the standard zone in cultural vegetables[52].

The accumulation of heavy metals varies greatly both between species and cultivars[82]. Lettuce is one of those vegetables that has high capacity in absorption and storage of lead[81] through contaminated soil by sewage or dust deposited on plants exposed to polluted air[81][83]. In Malakootian *et al.* (2009) study, the mean concentration of lead in lettuce imported to Kerman city from Dezfool, Jahrom, Yazd and Varamin cities were lower than WHO guideline while the lead level in lettuce in Turkey and Kenya was higher than that of used in Kerman[81].

Tea is the most popular beverage in Iran and the presence of lead in tea has been a concern in the recent years. A study by Ebadi *et al.* (2005) done in Gilan province (North of Iran) on green leaf of tea cultivated in Lahijan and Fuman cities indicated that green leaf of tea in this region had very high amount of lead. In explanation, this region is a transit automobile way for import and export inside the country[82]. In another survey which was carried out on consumed black tea in Tehran, eleven types of the most widely consumed brands of dry black tea were purchased from local market of Tehran, and the results indicated that lead concentration in black tea was higher than the permissible limit for human food[86] keeping in mind that tea is not always used in the fresh green leaf like other vegetables. Matsuura *et al* reported that after making tea, 80% of lead content is reduced in comparison to dry tea[87].

3.12. Raw Milk

Milk is one of the important selective foods to nourish infant and other age groups. Many reports indicate the presence of heavy metals in milk[88][89]. In Tajkarimi *et al.* study that lead residue in raw milk from different parts of Iran was assessed, the cities of Isfahan, Tehran, and West Azerbaijan showed higher levels of contamination[88]. The reason is that these regions are more industrialized than others[88]. This result is so critical especially in Isfahan state, because a new infant milk formula plant has been established there[88]. In another survey carried out in Yazd province on raw milk, the lead content in samples were less than limit of FAO/WHO standard[90].

In China, the lead contamination in meat, eggs and milk-based products increased during last decade[91]. One of the probable reasons for this rising in other countries maybe the wide use of leaded gasoline during the recent years[88] where one of the most important sources of food lead contamination is water[92]. However, there are several factors which effect lead content in raw milk such as range of contamination of what cows graze and drink, geographical condition of pasture, distance of stable to industrial area, vehicular area, climate weather, different seasons, and type of soil[88][89].

3.13. Other Foods

Bread is the most important food of Iranian people and due to immense side effects of long-term exposure of people to contamination, being lead toxicity in daily life food seems a serious problem[93]. WHO in 1998 announced that maximum limit of lead content of bread is 0.1ppm[93]. Khabnadideh *et al.* (2004), evaluated lead concentration of collected breads from various parts of Shiraz to indicate whether bread ingredients are lead-contaminated. They found that in lead polluted area, lead level in salt and water applied by all bakeries were below the standard level (0.05ppm), but this level for flour samples were higher than limits[93]. This results indicate that to decrease lead contamination of bread, it is necessary to usually monitor the internal (bread ingredient, machine related to bakery) and external (distance between bakery and petrol station or to cross road) condition of bakeries[93]. In another study conducted in Finland, it was found that mean and median lead contents of all breads were 14 and 8μg/kg. The collected samples showed a very high variation of lead contents. Also in that study, the lead content of Finnish breads was much lower than that in the late 1970s[94].

One of the sources of heavy metals and trace elements entrance to body is their release from manufacturing apparatus solving in food materials due to low pH of food[95]. pH is an important factor that effects the concentration of the cadmium and lead of the solution, because an increase of pH causes a decrease in the solubility of the lead and cadmium compounds[70].

Lemon juice and tomato paste with an average pH value of 2.3 and 4.6 are at risk of lead pollution. Poormo-ghaddas *et al.* (1998, 2001) evaluated lead concentration in lemon juices and tomato paste in Isfahan city, Iran[95][96]. They found that method of preparation (non-standard apparatuses like metallic machine-made) has a significant effect in increasing the concentration of toxic elements. In handmade juice and tomato paste, concentration of lead was in normal range where in metallic machine-made lemon juices tomato paste samples, the lead was 58% and 93% higher than normal levels[95][96].

Peanut is a kind of nut that grow in shell underground and widely eaten by people and its residue is used to richen farm animal foods[97]. Contamination of peanut, make a trouble for human and animals because it enters into human food chain through different ways. Rahbar *et al.* reminded in his survey that there was a quite high lead and cadmium levels in the peanut, although it depends on food habitants, geographical situation, and level of contamination environment[97].

3.14. Medications

From a long time ago people prefer to consume herbal medicines and even doctors are in believe that herbal medicines have no side effects[98]. One of these side effects is lead toxicity which is not far from mind by rapid industrialization. Several sources of lead contamination are estimated for medical products and drugs, such some oral herbal drops which is available in markets. Asghari *et al.* examined 10 different oral herbal medicines in the market of Iran and found existence of heavy metals but they were below acceptable recommended intake[99]. It should be noticed that patients can be in higher risk when they use medicines for long time and their organs might not be functioning adequately to detoxify heavy metals. Of course this is not restricted to Iran and can be a concern of many countries of the world especially when trend to use herbal medicines is increasing in the world. The existence of heavy metals in herbals medicines have been confirmed in different countries[100][101]. In a study, Obi *et al.* (2006) evaluated existence of heavy metals in herbal drugs of Nigeria. They found that 100% of the collected samples contained elevated amounts of heavy metals. These data alert us to the possibility of heavy metals toxicity from herbal products in the public that should be studied in-depth[101].

Make up products are another source of lead exposure. Traditional eye make ups such as powder of Surma and powders of Kohl, which are used in Middle East countries, contain lead and due to the long time contact with skin and eye mucosa they can cause lead toxicity in the users[102]. In a study, Malakootian *et al.* (2010) evaluated the amount of lead in kohl in Kerman city. He found that mean concentration of lead in measured samples was 254.5μg/g with range of 3.2−1219.4μg/g. Also it was found that plant-based kohl samples had lower amount of lead in comparison to mineral-based ones[102]. Existence of lead in tooth amalgam is another catastrophic event in Iran. Amalgam has been used for tooth restoration for decades and it is still used heavily in dentistry. In an study, Mortazavi *et al.* (2000) evaluated substantial amounts of heavy metals and found that lead and cadmium exist in the commercial amalgam which was available in Iran at year of 2000[103].

3.15. Opium

Exposure to lead is usually considered only when a patient's history points to well-known traditional sources of lead, although the incidence of lead poisoning has declined, but the presence of new forms of non-occupational poisoning poses new problems[104]. The earliest report of this strange source of lead backed to 1973 related to father and his middle aged daughter who were diagnosed as lead poisoning due to ingestion of home-made opium[105]. Additionally, acute lead poisoning as a result of self-injection of lead and opium pills, crushed and suspended in water, has been reported[105]. Inorganic lead poisoning due to intravenous or inhalation abuse of lead-contaminated heroin has been reported since 1989[106]. Other examples include adulterated marijuana, methamphetamine and Indian herbal medicines[107]–[109]. In some parts of the US, illegally distilled alcohol (moonshine) is an important source of lead exposure[110].

Recently, there have been few reports about lead poisoning as a consequence of opium addiction in Iran[14][111]–[115]. Also, researchers reported the presence of lead in opium in the South-East Iran[116]. In the study of Salehi *et al.* in 2009, BLL in opium addicted patients was measured in comparison to healthy controls. The results showed that BLL in opium addicts had a range of 7.2–69.9μg/dl with a mean of 8.6 ± 3.5μg/dl that was significantly higher than that of controls[114]. Aghaee-Afshar *et al.* reported that lead existed in 10 opium samples collected from various sources with a mean concentration of 1.88 ± 0.35ppm[116] that might be harmful in chronic consumption by addicts. Informal and often illegal laboratories refine opium into a sticky, brown paste, which is pressed into bricks and sun dried[112]. This process results in introduction of impurities such as lead into the products, but it is still unknown whether it is added to opium during the process of preparation or it is added by dealers or smugglers to increase the opium weight for more profit[104][116]. In adults, absorption of lead via the respiratory tract is the most prevalent route of opium abuse in Iran[104] since it has higher bioavailability[117] with an average absorption of approximately 40%[118]. The heat of smoking opium can affect the amount of lead absorbed in blood while other methods of consumption such as oral may have not that much effect on the opium lead and thus blood absorption of lead can be higher in these methods[119]. Also, it must be noted that several symptoms of lead poisoning are similar to that of opium abuse such as constipation, nausea, irritability, anorexia and various other neuropsychiatric symptoms. The diagnosis of lead poisoning is based on an elevated BLL, which is defined as equal to or greater than 25μg/dL[104]. Many of toxic effects of lead is reversible if lead poisoning is identified early but high BLL and delay in treatment may lead to irreversible symptoms like motor neuron defects[111]. Unrecognized lead poisoning in drug abusers presenting with symptoms of abdominal pain can lead to misdiagnosis and unnecessary gastrointestinal evaluations or abdominal surgeries such as appendectomy[2], decreased consciousness[120], and even paralysis of four limbs[111]. Adding lead to the opium is a recently health hazard in Iran[121]. Thus, lead poisoning should be considered in patients with a history of opium abuse who present with non-specific clinical manifestations[104]. Finally, it would be noticed that substance addicts may have an elevated BLL in comparison to healthy subjects and thus screening of BLL would be helpful.

3.16. Children Exposure

Pediatric lead poisoning is still an important public health problem for millions of children in the world. In South and Central America, 33%–34% of children have BLL above 10μg/dL (0. 48μmol/L) as compared with 7% in North America[122]. Another survey showed that one in every 20 children in the US has toxic blood levels[112]. In China, because up to 23% of populations (377 million) are children, lead exposure is still a serious public concern. Published data from this country showed that children's BLL are higher than other developed countries due to its heavy metal pollution[123]. In another study, the mean BLL of Chinese children was 62.31μg/L and (9.2%) of 3,624 children's BLL were above 100μg/L. Taking Chinese medicinal herbs, substitutes of breast milk and puffed foods, seem the main risk factors[124].

Lead has many toxic effects on human health. In children, the most vulnerable population, mental development of children health effects is of greatest effect. Low level lead exposure can significantly induce motor dysfunctions and cognitive impairment in pediatric, especially if the exposures occurs before the age of six. Recent surveys showed that cognitive impairment is associated with BLLs < 10mg/dL among pediatrics[125]–[127]. Lanphear *et al.* observed cognitive effects in children aged 6–16 years with BLLs < 5mg/dL in USA[125]. Two other cross-sectional studies, in Detroit and Mexico City evaluated children at age 7 years and found an inverse relationship with BLLs < 10mg/dL and cognitive development[126][127]. Moreover, several cohort studies have provided more data that prenatal exposure to lead is associated with child cognitive development[128]. Because children and fetus are in a rapid growth course, they absorbed heavy metal in food content more than adults, so they are in exposure of greater danger than others.

Published studies indicate that children exposed to contaminated water, soils, dust, and air particulates may ingest a significant amount of lead and other toxic metals through the hand-mouth activity or the inhalation of lead dust. After banning of leaded gasoline, the US focused on paints, and the lead-based paint in old houses was considered as the main exposure resource for children lead poisoning. In the US, the main source of lead exposure in children is believed lead-based house paint and the contaminated dust[129].

There are very few studies conducted on lead exposure and its related factors in Iranian children. Faranoush *et al.* (2003) studied 320 students who were randomly selected in two areas in Semnan city and it was revealed that 78.8% of the children had BLLs > 10μg/dl, also, in 5% of them toxic levels of lead was observed (Pb > 20μg/dl)[130]. In another study, it was reported that 32% of randomly selected students in a polluted district in Tehran, capital of Iran had BLLs of more than 10μg/dl which clearly showed importance of screening test for lead poisoning in the population. Also in this study it was found that BBL in boys was 1.6 times more than girls[36]. In another study, two groups of 7−11 years old children, from a lead mining area (Angooran, Zanjan Province, Iran) and 36 from control area, were selected to assess BLL and grown parameters such as weight and height. The mean BLLs in case group were 36.97μg/dL which was significantly higher than controls (13.35μg/dL). Also there was no significant difference in growth parameters, including weight and height, in the children of two groups[131], suggesting that the BLL was not correlated with growth parameters of children in lead mining area. In another survey, the average intelligence quotient (IQ) in 64 children living in high lead area of Zanjan province (Center of Iran) was 86. 64 ± 9.68 with 40 (62.5%) having less than normal level. In comparison in the other group, the average IQ was 91.98 ± 10.26 with 24 of the children (38.7%) having less than normal level. The IQ of pediatric living in high-lead area was significantly lower than children living in low-lead area[132]. Moreover, Dehghan *et al.* evaluated BLL of children with age of 2−12 years old in Yazd city (Center of Iran) and showed that 93.1% of children have higher BLL than standard values[133]. In another study conducted in Mashhad, East of Iran, 32 children aged 3−7 years whose parents were lead-exposed workers were randomly selected and studied. All of the children had BLL of above the standard (more than 100μg/L). Duration of fathers' exposure to lead at work was 9.14 ± 5.63 years. BLL was 163.81 ± 57.19μg/L and urine lead concentration was 97 ± 48.12mg/L. The children whose parents worked at battery plant manufacturing had higher BLL (217mg/L) comparing to children of tile workers (151mg/L)[134].

Although, in some countries, addition of compounds containing lead to toys has been banned in the current decade, but there are some reports which indicating lead toxicity from contaminated toys worldwide. In 2006, a child aged 4 years died of acute lead poisoning which was the first child lead-poisoning death since 2000 in the US. The autopsy revealed a heart-shaped metallic charm in the abdomen that was found to have a lead content of 99.1%[135].

In 2007, a series of recalls of pediatrics' toys that were suspected lead contamination was issued in the US[128]. In Iran, results of some surveys showed that plastic toys and other PVC products manufactured for children in some area are contaminated with lead[136]. In Iran comparing to other countries, studies on lead exposure and lead toxicity in the children are too few. In fact sources of lead exposure in Iranian children are unknown. In recognizing that there is no safe BLL for pediatric and chelating agents have limited value in decreasing the harmful effect of lead poisoning or even cost-benefit. The government should control or eliminate lead hazards in children's environment before they are exposed.

3.17. Pre- and Post-Natal Exposure

There is a strong relation between umbilical cord and maternal BLL which proves the transfer of lead from mother to fetus[128]. Golmohammadi *et al.* (2007) evaluated lead concentration of specimens of maternal blood, new born, cord blood, and colostrums in polluted area of Tehran and compared with non-polluted areas. Their data revealed an association between mean concentrations in blood lead of mothers and newborns and between mean concentrations of colostrums lead and newborn BLL in both area. The lead concentration of mother blood, newborn cord blood and colostrums in polluted area were significantly higher than non-polluted area. The mean BLL of mothers, cord blood of newborns and colostrums were 7.6 ± 4.1, 5.9 ± 3, and 4.2 ± 2.5μg/dl, respectively in the non-polluted area and they were 9.1 ± 8.4, 6.5 ± 5.2, and 5.8 ± 5.5μg/dl, respectively, in the polluted area[137]. Moreover, Pourjafari *et al.* (2007) evaluated the fetal deaths rate among progenies of workers at two high risk occupations (lead mine and dyehouses) of Hamadan city (West of Iran) and the results were compared with general population. The rates of abortions plus stillbirths among their wives' pregnancies were 13.15% and 13.30%, respectively. Fetal death rates were significantly higher than general population that suggests the idea that long-term genetic consequences occur following working in lead mine and dye-houses[138].

In addition, Vigeh *et al.* (2010) conducted a study to clarify the effects of lead on fetal premature rupture of the membranes (PROM). They measured BLL in 332 women with age range of 16–35 years, during their early pregnancy period. They found that BLL of PROM deliveries were significantly higher than non-PROM deliveries (4.61 ± 2.37g/dl versus 3.69 ± 1.85g/dl) and suggested that lead can increase the risk of PROM in pregnant women even with mean BLL less than 5g/dl[139].

Moreover, Vige *et al.* (2011) studied the effect of lead on occurrence of preterm labor. The BLL of mothers who delivered preterm babies was significantly higher than that of those who delivered full-term babies (4.46 ± 1.86 versus 3.43 ± 1.22μg/dl). This suggests that adverse pregnancy outcomes may occur at BLL even below the current acceptable level[140].

Norouzi *et al.* (2010) conducted a study to determine concentration of lead in the milk of women living in the vicinity of a metal smelter area. Their results showed that mean level of lead in milk nulliparous women was significantly higher than multiparous women (70.64 versus 23.73g/l). Also, they found that milk lead level of women with age of 24 or less was significantly higher than age greater than 24 years old[141].

In addition, cohort surveys indicate that prenatal exposure is associated with pediatric cognitive development[128]. Hu *et al.* (2006) evaluated fetal lead toxicity during trimesters of pregnancy as predictors of infant neurodevelopment. They revealed that exposure to lead during the first trimester may have a greater effect on adverse neurodevelopment later in life than the second or third trimesters[142]. The Mexico City Prospective Lead survey[143] showed that higher maternal BLL at third trimester of pregnancy, especially around week 28, was correlated to reduction of intellectual child development. Wasserman *et al.*[144] studied the impact of pre and postnatal lead exposure to early intelligence and found that pre and postnatal exposures to lead that occur during the first 7 years of life are independently accompanied with small reduction in later IQ scores[144].

4. Conclusion

Several metal chelators can be used to prevent lead poisoning after occurrence of exposure or can save life in persons with very high BLL but none of them are suitable in reducing lead burden in chronic lead exposure[128]. Moreover, chelators are not always available in all countries or if available they are too expensive[145] and are not included by health insurance companies and most importantly they have limited value in decreasing the sequel of lead poisoning. Also some clinical trials demonstrated no developmental benefit in the group that received succimer after 3 and 7 years treatment[146][147]. These findings highlight the importance of undertaking further precautions and designing programs to prevent lead contamination. Identification of sources, elimination or control of sources, and monitoring environmental exposures and hazards can be used to prevent lead poisoning. Hopefully Department of Public Health and Environment from World Health Organization has made special attention to update guidelines on the prevention and management of lead poisoning started since 2011. The first meeting was held 11−13 July 2011 at WHO Geneva, Switzerland where the corresponding author of this paper attended as an adviser. The meeting tried to recognize world problem of lead poisoning country by country to reach a global conclusion about protocols and measures to prevent lead poisoning. One of the conclusions of the meeting was to ask all countries to extend their studies to recognize source of lead exposure and extent of toxicity in their people by general screening and finally conducting systematic reviews country by country. Taken collectively, authors recommend the following actions to be considered in Iran.

1) Full prohibition of use of leaded gasoline and alternating with naturally obtained ethanol as a clean fuel.

2) Authorities should introduce measures for assessing and controlling occupational exposure to lead and exposure of workers' families. People who work with lead should be taught about the hazards of lead and how to minimize their own exposure as well as exposure to their families.

3) Authorities should conduct systematic country-wide studies to identify, document and map the important environmental sources of lead exposure in their populations.

4) Authorities should work towards elimination of the non-essential use of lead in household and consumer products (including paint, toys, solder in food cans, lead batteries, cosmetics, traditional medicines or remedies, ceramics used in connection with food, cosmetics, etc.). Where water is supplied through lead pipes, these pipes should be replaced with safer materials or water filters should be used to reduce the lead content of the water. In homes where painted with leaded paint, the paint should either be removed or stabilized using appropriate safe measures.

5) Public health authorities should establish and implement a BLL screening

program in populations known or suspected to be at risk of lead exposure. Reference laboratories can be set up to regularly measurement of lead in biological and environmental samples. In this regard, setting national limits for concentration of lead in air, water, food and soil that are based on the protection of human health in vulnerable populations should be taken into priority. Special attention to children would need establishing national children's environmental health centers to provide effective and rapid diagnosis, treatment and increased awareness of lead poisoning and other environmental threats.

6) The removal, remediation and/or replacement of lead-contaminated soil in communities where lead contamination is high should be considered by authorities.

7) Where there is a significant risk of exposure to lead from environmental and other sources, parents and pregnant women should be taught about the hazards of lead and how to reduce or prevent exposure. Generally, education of all public at lead-polluted area should be taken into priority through media.

As a final point, this review confirms that chronic lead toxicity should be concerned for Iranian population because sources of lead pollution in air, waters, soil, food, and etc. exist. Although many studies have been conducted on environmental lead contamination in Iran but there are very few studies evaluated current status of lead exposure and toxicity in the Iranian children and pregnant women. Screening of lead exposures especially in children and prevention strategies should be a priority for Iranian authorities. We recommend identify sources, eliminate or control sources, and monitor environmental exposures and hazards to prevent lead poisoning.

Competing Interests

The authors declare that they have no competing interests.

Acknowledgements

This paper is the outcome of an in-house financially non-supported study.

Authors' Contributions

PK and OM collected data and drafted the manuscript. MA gave the idea, designed and supervised the study, and edited the manuscript. All authors contributed equally. All authors read and approved the final manuscript.

Author's Information

Mohammad Abdollahi is an Adviser to the World Health Organization (WHO) for providing WHO guidelines on the prevention and management of lead poisoning. He is doing his best to integrate identification of lead toxicity in human and its hazards to the environment.

Source: Karrari P, Mehrpour O, Abdollahi M. A systematic review on status of lead pollution and toxicity in Iran; Guidance for preventive measures.[J]. Daru Journal of Pharmaceutical Sciences, 2011, 20(2):703–712.

References

[1] Malekirad AA, Oryan S, Fani A, Babapor V, Hashemi M, Baeeri M, Bayrami Z, Abdollahi M: Study on clinical and biochemical toxicity biomarkers in a zinc-lead mine workers. Toxicol Ind Health 2010, 26(6):331–337.

[2] Mohammadi S, Mehrparvar A, Aghilinejad M: Appendectomy due to lead poisoning: a case-report. J Occup Med Toxicol 2008, 17(3):23.

[3] Abdollahi M, Sadeghi Mojarad A, Jalali N: Lead toxicity in employees of a paint factory. MJIRI 1996, 10:203–206.

[4] Azizi MH, Azizi F: Lead poisoning in the world and Iran. Int J Occup Environ Med 2010, 1:81–87.

[5] Karimooy HN, Mood MB, Hosseini M, Shadmanfar S: Effects of occupational lead exposure on renal and nervous system of workers of traditional tile factories in Mashhad (northeast of Iran). Toxicol Ind Health 2010, 26(9):633–638.

[6] Landrigan PJ, Todd AC: Lead poisoning. West J Med 1994, 161(2):153–159.

[7] Balali-Mood M, Shademanfar S, Rastegar Moghadam J, Afshari R, Namaei Ghassemi M, Allah Nemati H, Keramati MR, Neghabian J, Balali-Mood B, Zare G: Occu-

pational lead poisoning in workers of traditional tile factories in Mashhad, Northeast of Iran. Int J Occup Environ Med 2010, 1:1.

[8] Eslami S, Hajizadeh Moghaddam A, Jafari N, Nabavi SF, Nabavi SM, Ebrahimzadeh MA: Trace element level in different tissues of Rutilus frisii kutum collected from Tajan River, Iran. Biol Trace Elem Res 2011, 143(2):965–973.

[9] Abdollahi M, Ebrahimi-Mehr M, Nikfar S, Jalali N: Monitoring of lead poisoning in simple workers of a copying center by flame atomic absorption spectroscopy. MJIRI 1996, 10:69–72.

[10] Staudinger KC, Roth VS: Occupational lead poisoning. Am Fam Physician 1998, 57(4):719–726. 731–2.

[11] Mañay N, Cousillas AZ, Alvarez C, Heller T: Lead contamination in Uruguay: the "La Teja" neighborhood case. Rev Environ Contam Toxicol 2008, 195:93–115.

[12] Paoliello MM, De Capitani EM: Occupational and environmental human lead exposure in Brazil. Environ Res 2007, 103(2):288–297.

[13] Abdollahi M, Shohrati M, Nikfar S, Jalali N: Monitoring of lead poisoning in bus drivers of Tehran. Irn J Med Sci 1995, 20:29–33.

[14] Masoodi M, Zali MR, Ehsani-Ardakani MJ, Mohammad-Alizadeh AH, Aiassofi K, Aghazade R, Shavakhi A, Somi MH, Antikchi MH, Yazdani S: Abdominal pain due to lead-contaminated opium: a new source of inorganic lead poisoning in Iran. Arch Iran Med 2006, 9:72–75.

[15] Kalantari S, Khoshi AH, Mohebbi MR, Fooladsaz K: Investigation of blood lead levels and its toxicity in workers of zinc melting factory of Dandi, Zanjan, Iran. J Zanjan Univ Med Sci Health Serv 2009, 17(66):79–86.

[16] Pirsaraei SR: Lead exposure and hair lead level of workers in a lead refinery industry in Iran. Indian J Occup Environ Med 2007, 11(1):6–8.

[17] Tabrizizadeh M, Boozarjomehri F, Akhavan Karbasi MH, Maziar F: Evaluation of the relationship between blood lead level and prevalence of oral complication in Koushk lead mine workers, Yazd province. J Dent Tehran Univ Med Sci 2006, 19(1):91–99.

[18] Aminpour MR, Barkhordari A, Ehrampoush MH, Hakimian AM: Blood lead levels in workers at Kooshk lead and zinc mine. J Shahid Sadoughi Univ Med Sci Health Serv 2008, 16(2):24–30.

[19] Malekirad AA, Fani A, Abdollahi M, Oryan S, Babapour V, Shariatzade SMA, Davodi M: Blood-urine and cognitive -mental parameters in mine workers exposed to lead and zinc. AMUJ 2011, 13(4):106–113.

[20] Shahrabi J, Dorosti AR: Study of blood lead levels, hemoglobin & plasma ascorbic acid in a car company welders. Iran J Occup Health 2006, 3(1–2):50–55.

[21] Meshkinian A, Asilian H, Nazmara Sh, Shahtaheri DJ: Determination of Lead in the

environment and in the urban services workers in Tehran municipality district. J Sch Public Health Inst Public Health Res 2003, 1(3):31–40.

[22] Yartireh HA: Determination of blood and urine lead level among workers of Kermanshah refinery in 1994. Sci Med J 2001, 31:60–65.

[23] Herman DS, Geraldine M, Venkatesh T: Evaluation, diagnosis, and treatment of lead poisoning in a patient with occupational lead exposure: a case presentation. J Occup Med Toxicol 2007, 2:7.

[24] Shiri R, Ansari M, Ranta M, Falah-Hassani K: Lead poisoning and recurrent abdominal pain. Ind Health 2007, 45(3):494–496.

[25] Mirsattari SG: Urine lead levels in service station attendants exposed to tetraethyl lead. J Res Med Sci 2001, 6(3):151–154.

[26] Yaghmaie B, Faghihi Zarandi A, Bazrafshani M, Arjomand Tajaddini A: Study of lead concentration in the air of gasoline station of Kerman city. J Kerman Univ Med Sci 1995, 2(2):66–70.

[27] Keramati MR, Nemai Ghasemi M, Balali-Mood M: Correlation between iron deficiency and lead intoxication in the workers of a car battery manufacturer. J Birjand Univ Med Sci 2009, 16(1):51–58.

[28] Ibrahim AS, Latif AH: Adult lead poisoning from a herbal medicine. Saudi Med J 2002, 23(5):591–593.

[29] Leung AO, Duzgoren-Aydin NS, Wong H: Heavy metals concentrations of surface dust from e-waste recycling and its human health implications in southeast China. Environ Sci Technol 2008, 42(7):2674–2680. 1.

[30] Centers for Disease Control and Prevention: Preventing lead poisoning in young children. Atlanta, GA: Centers for Disease Control; 2005.

[31] Abdollahi M, Zadparvar L, Ayatollahi B, Baradaran M, Nikfar S, Hastaie P, Khorasani R: Hazard from carbon monoxide poisoning for bus drivers in Tehran, Iran. Bull Environ Contam Toxicol 1998, 61(2):210–5.

[32] Kebriaeezadeh A, Abdollahi M, Sharifzadeh M, Mostaghasi R: Lead levels in the inhabitants of Tehran city districts. Pajouhandeh 1997, 2(5):72–67.

[33] Farzin L, Amiri M, Shams H, Ahmadi Faghih MA, Moassesi ME: Blood levels of lead, cadmium, and mercury in residents of Tehran. Biol Trace Elem Res 2008, 123 (1–3):14–26.

[34] Naddafi K, Ehrampoush MH, Jafari R, Nabizadeh M, Younesian M: Complete evaluation of suspended air particles and their composition in the central area of Yazd city. J Shahid Sadoughi Univ Med Sci Health Serv 2008, 16(1):21–25.

[35] Farahmandkia Z, Mehrasbi MR, Sekhawatju MS, Hasanalizadeh AS, Ramezanzadeh Z: Study of heavy metals in the atmospheric deposition in Zanjan, Iran. Iran J Health Environ 2010, 2(4):240–249.

[36] Zaman T, Hosseinzadeh H: Lead poisoning in a highly polluted district of Tehran in high school children. Iran J Pediatr 1999, 4:207–212.

[37] Chuturkova R, Iossifova Y, Clark S: Decrease in ambient air lead concentrations in Varna, Bulgaria, associated with the introduction of unleaded gasoline. Ann Agric Environ Med 2010, 17(2):259–261.

[38] WHO: WHO Guidelines for drinkingwater quality. Geneva: World Health Organization; 2006:35–38.

[39] Karbasi M, Karbasi E, Saremi A, Ghorbani Zade Kharazi H: Determination of heavy metals concentration in drinking water resources of Aleshtar in 2009. YAFT-E 2010, 12(1):65–70.

[40] Fertmann R, Hentschel S, Dengler D, Janssen U, Lommel A: Lead exposure by drinking water: an epidemiologial study in Hamburg, Germany. Int J Hyg Environ Health 2004, 207(3):235–244.

[41] Renner R: Out of plumb: when water treatment causes lead contamination. Environ Health Perspect 2009, 117(12):A542–A547.

[42] Watt GC, Britton A, Gilmour WH, Moore MR, Murray GD, Robertson SJ, Womersley J: Is lead in tap water still a public health problem? An observational study in Glasgow. BMJ 1996, 313(7063):979–981. 19.

[43] Zietz BP, Lass J, Suchenwirth R, Dunkelberg H: Lead in drinking water as a public health challenge. Environ Health Perspect 2010, 118(4):A154–A155.

[44] Haider T, Haider M, Wruss W, Sommer R, Kundi M: Lead in drinking water of Vienna in comparison to other European countries and accordance with recent guidelines. Int J Hyg Environ Health 2002, 205(5):399–403.

[45] Karimpour M, Shariat MA: A study of heavy metals in drinking water network in Hamadan city in 1994. Sci J Hamadan Univ Med Sci Health Serv 2007, 7(17): 47–44.

[46] Savari J, Jaafarzadeh N, Hassani AH, Shams Khoramabadi GH: Heavy metals leakage and corrosion potential in Ahvaz drinking water distribution network. Water Wastewater J 2008, 18(64):16–24.

[47] Shahmansouri MR, Poormoghadas H, Shams Khorramabadi GH: A study of Leakage of trace metals from corrosion of the municipal drinking water distribution system. J Res Med Sci 2003, 8(3):34–30.

[48] Tashauoei HR, Hajian Nejad M, Amin MM, Karakani F: A study on leakage of heavy metals from the PVC and polypropylene pipes used in the water distribution system in Isfahan. Health Syst Res 2010, 6(3):373–382.

[49] Ebrahimi A, Ehrampoush MH, Ghaneian MT, Davoudi M, Hashemi H, Behzadi S: The survey chemical quality of ground water in the vicinity of sanitary landfill of Yazd in 2008. Health Syst Res 2010, 6:1048–1056.

[50] Ebrahimi A, Amin MM, Hashemi H, Foladifard R, Vahiddastjerdi M: A survey of groundwater chemical quality in Sajad Zarinshahr. Health Syst Res 2011, 6:918–926.

[51] Legg TM, Zheng Y, Simone B, Radloff KA, Mladenov N, González A, Knights D, Siu HC, Rahman MM, Ahmed KM, McKnight DM, Nemergut DR: Carbon, metals, and grain size correlate with bacterial community structure in sediments of a high arsenic aquifer. Front Microbiol 2012, 3:82.

[52] Nazemi S, Khosravi A: A study of heavy metals in soil, water and vegetables. Knowledge Health 2011, 5(4):27–31.

[53] Mohammadian M, Nouri J, Afshari N, Nassiri J, Nourani M: Investigation of heavy metal concentrations in the water wells close to Zanjan zinc and lead smelting plant. Iran J Health Environ 2008, 1:51–56.

[54] Zafarzadeh A: The determination of water chemical quality of cisterns in rural areas of Golestan province. J Gorgan Univ Med Sci 2006, 8:51–54.

[55] Yaziz MI, Gunting H, Sapari N, Ghazali AW: Variations in rainwater quality from roof catchments. Water Res 1989, 23:761–765.

[56] Rabito FA, Iqbal S, Perry S, Arroyave W, Rice JC: Environmental lead after Hurricane Katrina: implications for future populations. Environ Health Perspect 2012, 120(2):180–184.

[57] Farsam H, Zand N: A preliminary study of lead deposition on plant leaves in Tehran. Iran J Publ Health 1991, 20(1–4):27–34.

[58] Jalali M, Khanlari ZV: Enviromental contamination of Zn, Cd, Ni, Cu and Pb from industrial areas in Hamadan Province, western Iran. Environ Geol 2008, 55: 1537–1543.

[59] Parizanganeh A, Hajisoltani P, Zamani A: Assessment of heavy metal pollution in surficial soils surrounding Zinc Industrial Complex in Zanjan-Iran. Procedia Environ Sci 2010, 2:162–166.

[60] Chehregani A, Noori M, Yazdi HL: Phytoremediation of heavy-metal-polluted soils: screening for new accumulator plants in Angouran mine (Iran) and evaluation of removal ability. Ecotoxicol Environ Saf 2009, 72(5):1349–1353.

[61] Masoudi SN, Ghajar Sepanlou M, Bahmanyar MA: Distribution of lead, cadmium, copper and zinc in roadside soil of Sari-Ghaemshahr road, Iran. Afr J Agric Res 2012, 7(2):198–204.

[62] Piron-Frenet M, Bureau F, Pineau A: Lead accumulation in surface roadside soil: its relationships to traffic density and meteorological parameters. Sci Total Environ 1994, 144:297–304.

[63] Ozkan M, Gurkan R, Ozkan A, Akcay M: Determination of manganese and lead in roadside soil samples by FAAS with ultrasound assisted leaching. J Anal Chem 2005, 60:469–474.

[64] Parizanganeh A, Lakhan VC, Jalalian H: A geochemical and statistical approach for assessing heavy metal pollution in sediments from the southern Caspian coast. Int J Environ Sci Tech 2007, 4(3):351–358.

[65] Shahriari A: Determination of heavy metals (Cd, Cr, Pb, Ni)in edible tissues of Lutjans Coccineus and Tigeratooh Croaker in Persian Gulf-2003. J Gorgan Univ Med Sci 2005, 7(16):67–65.

[66] Tabari S, Saravi SS, Bandany GA, Dehghan A, Shokrzadeh M: Heavy metals (Zn, Pb, Cd and Cr) in fish, water and sediments sampled form Southern Caspian Sea, Iran. Toxicol Ind Health 2010, 26(10):649–656.

[67] Ebrahimi M, Taherianfard M: Concentration of four heavy metals (cadmium, lead, mercury, and arsenic) in organs of two cyprinid fish (Cyprinus carpio and Capoeta sp.) from the Kor River (Iran). Environ Monit Assess 2010, 168:575–585.

[68] Shokrzadeh M, Ebadi AG, Heidari R, Zare S: Measurement of lead, cadmium and chromium in five species of mostconsumed fish in Caspian sea. Int J Biol Biotech 2004, 1:673–675.

[69] Dorafshan S, Heyrati FP: Spawning induction in Kutum Rutilus frisii kutum (Kamenskii, 1901) using carp pituitary extract or GnRH analogue combined with metoclopramide. Aquacult Res 2006, 37:751–755.

[70] Bakhtiarian A, Gholipour M, Ghazi-Khansar M: Lead and cadmium of Korbal rice in Northern Iran. Iran J Publ Health 2001, 30(3–4):129–132.

[71] National Research Council: Oil in the sea. Inputs fates and effects. Washington DC: National Academy Press; 1985.

[72] Price ARG, Shepparrd CRC: The Gulf: past, present, and possible future states. Marine Pollut Bull 1991, 22:222–227.

[73] Ashraf W: Accumulation of heavy metals in kidney and heart tissues of Epinephelus microdon fish from The Arabian Gulf. Environ Monit Assess 2005, 101(1–3): 311–316.

[74] Raissy M, Ansari M, Rahimi E: Mercury, arsenic, cadmium and lead in lobster (Panulirus homarus) from the Persian Gulf. Toxicol Ind Health 2011, 27(7):655–659.

[75] Pourmoghaddas H, Shahryari A: The concentration of lead, chromium, cadmium, nickel and mercury in three speices of consuming fishes of Isfahan city. Health Syst Res 2010, 6(1):30–35.

[76] Malakootian M, Yaghmaeian K, Meserghani M, Mahvi AH, Daneshpajouh M: Determination of Pb, Cd, Cr and Ni concentration in Imported Indian Rice to Iran. Iran J Health Environ 2011, 4(1):77–84.

[77] Jahedkhaniki GR, Zazoli MA: Cadmium and lead contents in rice (Oryza sativa) in the North of Iran. Int J Agr Biol 2005, 7(6):1026–1029.

[78] Lin HT, Wong SS, Li GC: Heavy metal content of rice and Shellfish in Taiwan. J

Food Drug Anal 2004, 12:167–174.

[79] Shakerian A, Rahimi E, Ahmadi M: Cadmium and lead content in several brands of rice grains (Oryza sativa) in central Iran. Toxicol Ind Health 2012. Epub ahead of print.

[80] Zazouli MA, Shokrzadeh M, Izanloo H, Fathi S: Cadmium content in rice and its daily intake in Ghaemshahr region of Iran. Afr J Biotechnol 2008, 7:3686–3689.

[81] Malakootian M, Aboli M, Ehrampoosh M: Determination of lead level in Lettuce in Kerman. Tolooe Behdasht 2009, 8(1–2):62–67.

[82] Ebadi AG, Zare S, Mahdavi M, Babaee M: Study and measurement of Pb, Cd, Cr and Zn in green leaf of tea cultivated in Gillan province of Iran. Pak J Nutr 2005, 4:270–272.

[83] Maleki A, Zarasvand MA: Heavy metals in selected edible vegetables and estimation of their daily intake in Sanandaj, Iran. Southeast Asian J Trop Med Public Health 2008, 39(2):335–340.

[84] Jahangir-Touyserkani A, Tavakolian F, Valaie N: Determination of the lead, cadmium, copper in the vegetables, grown in south part of Tehran. Pajouhandeh 2004, 9(38): 105–108.

[85] Asadi M, Bazargan N: Survey of heavy metal in the waste water of a canal after irrigation of vegetable farms in south of Tehran. Iran J Public Health 1994, 23(4–1): 44–35.

[86] Malakootian M, Mesreghani M, Danesh Pazhoo M: A survey on Pb, Cr, Ni and Cu concentration in Tehran consumed black tea. Sci J Rafsanjan Univ Med Sci Health Serv 2011, 10:138-143.

[87] Matsuura H, Hokura A, Katsuki F, Itoh A, Haraguchi H: Multielement determination and speciation of major-to-trace elements in black tea leaves by ICP-AES and ICP-MS with the aid of size exclusion chromatography. Anal Sci 2001, 17(3): 391–398.

[88] Tajkarimi M, Ahmadi-Faghih M, Poursoltani H, Salahnejad A, Motallebi AA, Mahdavi H: Lead residue levels in raw milk from different regions of Iran. Food Control 2008, 19(5):495–498.

[89] Caggiano R, Sabia S, D'Emilio M, Macchiato M, Anastasio A, Ragosta M, Paino S: Metal levels in fodder, milk, dairy products, and tissues sampled in ovine farms of Southern Italy. Environ Res 2005, 99(1):48–57.

[90] Yasaei-Mehrgrdy GHR, Ezzatpanah H, Yasini-Ardakani SA, Dadfarnia SH: Assessment of lead and cadmium levels in raw milk from various regions of Yazd province. Food Tech Nutr 2010, 7:35–42.

[91] Wang M, Wang Z, Ran L, Han H, Wang Y, Yang D: Study on food contaminants monitoring in China during 2000–2001. Wei Sheng Yan Jiu 2003, 32(4):322–326.

[92] Renner R: Out of plumb: when water treatment causes lead contamination. Environ Health Perspect 2009, 117(12):A542–7.

[93] Khabnadideh S, Mokhtarifard A, Namavar-Jahromi B, Malekpour MB: Lead detection in bread ingredients of fifth district of Shiraz city in 2000. Hakim Res J 2004, 7:17–21.

[94] Tahvonen R, Kumpulainen J: Lead and cadmium contents in Finnish breads. Food Addit Contam 1994, 11(5):621–631.

[95] Poormoghaddas H, Javadi I, Eslamieh R: Cadmium, chromium, mercury and lead concentration in lemon joices. J Res Med Sci 1998,3:114–118.

[96] Poormoghaddas H, Javadi I, Eslamieh R: Study of the toxic trace elements cadmium, lead, and mercury in tomato paste in Isfahan city. Sci J Hamedan Univ Med Sci Health Serv 2002, 9:40–46.

[97] Rahbar N, Nazari Z: Level of lead and cadmium in peanut. KAUMSJ (FEYZ) 2004, 7:71–77.

[98] Eisenberg DM: Advising patients who seek alternative medical therapies. Ann Intern Med 1997, 127(1):61–69.

[99] Asghari G, Palizban AA, Tolue-Ghamar Z, Adeli F: Contamination of cadmium, lead and mercury on Iranian herbal medicines. Tabriz J Pharm Sci 2008, 1:1–8.

[100] Abou-Arab AAK, Soliman Kawther M, El Tantawy ME, Ismail Badeaa R, Naguib K: Quantity estimation of some contaminants in commonly used medicinal plants in the Egyptian market. Food Chem 1999, 67: 357–363.

[101] Obi E, Akunyili DN, Ekpo B, Orisakwe OE: Heavy metal hazards of Nigerian herbal remedies. Sci Total Environ 2006, 369(1–3):35–41.

[102] Malakootian M, Pourshaaban Mazandarany M, Hossaini H: Lead levels in powders of surma (Kohl) used in Kerman. J Kerman Univ Med Sci 2010, 17:167–174.

[103] Mortazavi VS, Fathi MH: Tooth restoration with Amalgam; treatment or tragedy. J Dent School 2000, 18:32–40.

[104] Soltaninejad K, Fluckiger A, Shadnia SH: Opium addiction and lead poisoing. J Subst Use 2011, 16(3):208–212.

[105] Chino M, Moriyama K, Saito H, Morn T: The amount of heavy metal derived from domestic sources in Japan. Water Air Soil Pollut 1991, 57–58(1). 57:829–837.

[106] Antonini G, Palmieri G, Millefiorini E, Spagnoli LG, Millefiorini M: Lead poisoning during heroin addiction. Ital J Neurol Sci 1989, 10(1):105–108.

[107] Dunbabin DW, Tallis GA, Popplewell PY, Lee RA: Lead poisoning from Indian herbal medicine (Ayurveda). Med J Aust 1992, 157(11–12):835–836.

[108] Norton RL, Burton BT, McGirr J: Blood lead of intravenous drug users. J Toxicol Clin Toxicol 1996, 34(4):425–430.

[109] Busse F, Omidi L, Timper K, Leichtle A, Windgassen M, Kluge E, Stumvoll M: Lead poisoning due to adulterated marijuana. N Engl J Med 2008, 358(15):1641–1642.

[110] Morgan BW, Barnes L, Parramore CS, Kaufmann RB: Elevated blood lead levels associated with the consumption of moonshine among emergency department patients in Atlanta, Georgia. Ann Emerg Med 2003, 42(3):351–358.

[111] Beigmohammadi MT, Aghdashi M, Najafi A, Mojtahedzadeh M, Karvandian K: Quadriplegia due to lead-contaminated opium-case report. Middle East J Anesthesiol 2008, 19(6):1411–1416.

[112] Jalili M, Azizkhani R: Lead toxicity resulting from chronic ingestion of opium. West J Emerg Med 2009, 10(4):244–246.

[113] Afshari R, Emadzadeh A: Case report on adulterated opium-induced severe lead toxicity. Drug Chem Toxicol 2010, 33(1):48–49.

[114] Salehi H, Sayadi AR, Tashakori M, Yazdandoost R, Soltanpoor N, Sadeghi H, Aghaee-Afshar M: Comparison of serum lead level in oral opium addicts with healthy control group. Arch Iran Med 2009, 12:555–558.

[115] Fatemi R, Jafarzadeh F, Moosavi S, Afshar Amin F: Acute lead poisoning in an opium user: a case report. GHFBB 2008, 1(3):139–142.

[116] Aghaee-Afshar M, Khazaeli P, Behnam B, Rezazadehkermani M, Ashraf-Ganjooei N: Presence of lead in opium. Arch Iran Med 2008, 11(5):553–554.

[117] Mahaffey KR: Quantities of lead producing health effects in humans: sources and bioavailability. Environ Health Perspect 1977, 19:285–295.

[118] Froutan H, Kashefi Zadeh A, Kalani M, Andrabi Y: Lead toxicity: a probable cause of abdominal pain in drug abusers. MJIRI 2011, 25:16–20.

[119] Hayatbakhsh Abbasi MM, Ansari M, Shahesmaeili A, Qaraie A: Lead serum levels in opium-dependent individuals. Addic Health 2009, 1(2):106–110.

[120] Moharari RS, Khajavi MR, Panahkhahi M, Mojtahedzadeh M, Najafi A: Loss of consciousness secondary to lead poisoning-case reports. Middle East J Anesthesiol 2009, 20(3):453–455.

[121] Karrari P, Mehrpour O, Balali-Mood M: Iranian crystal: a misunderstanding of the crystal-meth. J Res Med Sci 2012, 17(2):96–97.

[122] Jones RL, Homa DM, Meyer PA, Brody DJ, Caldwell KL, Pirkle JL, Brown MJ: Trends in blood lead levels and blood lead testing among US children aged 1 to 5 years, 1988–2004. Pediatrics 2009, 123(3):e376–e385.

[123] Wang S, Zhang J: Blood lead levels in children, China. Environ Res 2006, 101(3):412–418.

[124] Shi H, Jiang YM, Li JY, Liu F, Wang H, Yu F, Yang H: Environmental lead exposure among children in Chengdu, China, 2007–2009. Biol Trace Elem Res 2011, 143(1):

97–102.

[125] Lanphear BP, Dietrich K, Auinger P, Cox C: Cognitive deficits associated with blood lead concentrations <10 microg/dL in US children andadolescents. Public Health Rep 2000, 115(6):521–529.

[126] Chiodo LM, Jacobson SW, Jacobson JL: Neurodevelopmental effects of postnatal lead exposure at very low levels. Neurotoxicol Teratol 2004, 26(3):359–371.

[127] Kordas K, Canfield RL, López P, Rosado JL, Vargas GG, Cebrián ME, Rico JA, Ronquillo D, Stoltzfus RJ: Deficits in cognitive function and achievement in Mexican first-graders with low blood leadconcentrations. Environ Res 2006, 100(3):371–386.

[128] Meyer PA, Brown MJ, Falk H: Global approach to reducing lead exposure and poisoning. Mutat Res 2008, 659(1–2):166–175.

[129] Lin GZ, Peng RF, Chen Q, Wu ZG, Du L: Lead in housing paints: an exposure source still not taken seriously for children lead poisoning in China. Environ Res 2009, 109(1):1–5.

[130] Faranoush M, Malek M, Ghorbani R, Rahbar M, Safaei Z: Study of the blood lead levels and related factors in the 6-11 years old children in Semnan. Koomesh 2003, 4(3):79–86.

[131] Mahram M, Mousavinasab N, Dinmohammadi H, Soroush S, Sarkhosh F: Effect of living in lead mining area on growth. Indian J Pediatr 2007, 74(6):555–559.

[132] Mahram M: Intelligence quotient level of the children living high lead areas in Zanjan province. Behbood 2004, 7(4):36–42.

[133] Dehghan L, Ghane A: Study of the blood lead levels in the 2-12 years old children in Yazd.[MD dissertation]. Iran,Yazd: Faculty of Medicine, Yazd Univ Med Sci; 2000: 70–85.

[134] Deldar K, Nazemi E, Balali-Mood M, Emami SA, Mohammadpour AH, Tafaghodi M, Afshari R: Effect of Corandrum sativum L. extract on lead excretion in 3-7 year old children. J Birjand Univ Med Sci Health Serv 2008, 15(3):11–19.

[135] CDC: Death of a child after ingestion of a metallic charm-Minnesota. MMWR Morb Mortal Wkly Rep 2006, 55:340–341.

[136] Pourmoghaddas H, Pishkar AR, Kavehzade FM: Perecentage of toxic trace elements; Pb, Cr, and Cd in certain plastic toys, Isfahan city. J Shahid Sadoughi Univ Med Sci Health Serv 2006, 14:59–64.

[137] Golmohammadi T, Ansari M, Nikzamir A, Safary R, Elahi S: The effect of maternal and fetal lead concentration on birth weight: polluted versus non-polluted areas of Iran. Tehran Univ Med J 2007, 65(8):74–78.

[138] Pourjafari H, Rabiee S, Pourjafari B: Fetal deaths and sex ratio among progenies of workers at lead mine and dye-houses. Genet Third Millennium 2007, 5(2):1057–1060.

[139] Vigeh M, Yokoyama K, Shinohara A, Afshinrokh M, Yunesian M: Early pregnancy blood lead levels and the risk of premature rupture of the membranes. Reprod Toxicol 2010, 30(3):477–80.

[140] Vigeh M, Yokoyama K, Seyedaghamiri Z, Shinohara A, Matsukawa T, Chiba M, Yunesian M: Blood lead at currently acceptable levels may cause preterm labour. Occup Environ Med 2011, 68(3):231–4.

[141] Norouzi E, Bahramifar N, Ghasempouri SM: Determination concentration of lead in breast in lactating women in region industrial zarinshahr and effect on infant. J Isfahan Med School 2010, 28(112):640–646.

[142] Hu H, Téllez-Rojo MM, Bellinger D, Smith D, Ettinger AS, Lamadrid-Figueroa H, Schwartz J, Schnaas L, Mercado-García A, Hernández-Avila M: Fetal lead exposure at each stage of pregnancy as a predictor of infant mental development. Environ Health Perspect 2006, 114(11):1730–5.

[143] Schnaas L, Rothenberg SJ, Flores MF, Martinez S, Hernandez C, Osorio E, Velasco SR, Perroni E: Reduced intellectual development in children with prenatal lead exposure. Environ Health Perspect 2006, 114(5):791–7.

[144] Wasserman GA, Liu X, Popovac D, Factor-Litvak P, Kline J, Waternaux C, LoIacono N, Graziano JH: The Yugoslavia Prospective Lead Study: contributions of prenatal and postnatal lead exposure toearly intelligence. Neurotoxicol Teratol 2000, 22(6): 811–8.

[145] Nikfar S, Khatibi M, Abdollahiasl A, Abdollahi M: Cost and utilization study of antidotes: an Iranian experience. Int J Pharmacol 2011, 7:46–49.

[146] Rogan WJ, Dietrich KN, Ware JH, Dockery DW, Salganik M, Radcliffe J, Jones RL, Ragan NB, Chisolm JJ Jr, Rhoads GG: The effect of chelation therapy with succimer on neuropsychological development in children exposed to lead. N Engl J Med 2001, 344(19):1421–6.

[147] Dietrich KN, Ware JH, Salganik M, Radcliffe J, Rogan WJ, Rhoads GG, Fay ME, *et al*: Effect of chelation therapy on the neuropsychological and behavioral development of lead-exposed children after school entry. Pediatrics 2004, 114(1):19–26.

Chapter 12

Coal Mining in Northeast India: An Overview of Environmental Issues and Treatment Approaches

Mayuri Chabukdhara[1], O. P. Singh[2]

[1]Department of Environmental Biology and Wildlife Sciences, Cotton College State University, Guwahati, Assam, India
[2]Department of Environmental Studies, North Eastern Hill University, Shillong, Meghalaya, India

Abstract: Northeast India has a good deposit of sub-bituminous tertiary coal. The northeast Indian coals have unusual physico-chemical characteristics such as high sulfur, volatile matter and vitrinite content, and low ash content. In addition, many environmental sensitive organic and mineral bound elements such as Fe, Mg, Bi, Al, V, Cu, Cd, Ni, Pb, and Mn etc. remain enriched in these coals. Such characteristics are associated with more severe environmental impacts due to mining and its utilization in coal based industries. Environmental challenges include large scale landscape damage, soil erosion, loss of forest ecosystem and wildlife habitat, air, water and soil pollution. Several physical and chemical methods are reported in literature for the removal of mineral matter, total sulfur and different forms of sulfur from high sulfur coal in northeast India. This paper may help different researchers and stakeholders to understand current state of research in the field. Initiatives may be taken towards sustainable use of coal resources by adopting innovative clean technologies and by implementing effective control measures and regulatory policies.

Keywords: Northeast India, Sub-Bituminous Coal, Environmental Issues, Innovative Technologies, Management and Regulatory Policies

1. Introduction

Coal is the most important and abundant fossil fuel in India. With increased population, growing economy and a quest for improved quality of life, energy demand in India is rising. Mining is not only fulfilling the increasing energy demand of industry, but also plays an important role in the economic development of the country (Chaulya and Chakraborty 1995). Power sector is the largest consumers of coal followed by iron, steel and cement segments in the last four decades (**Figure 1**). Other smaller consumers include fertilizer, textile (including jute and jute products), paper and the brick industry. Coal mining and its utilization is associated with substantial environmental challenges as it creates significant and often irreversible impacts upon the terrestrial and aquatic environment.

Open cast or surface mining is dominant in India and it not only alters the nature of groundwater-surface water interactions but also contributes to major air pollutants to the atmosphere and results in dramatic changes in the landscape. Most coal mining districts in India have been declared as critically polluted areas (CPAs) by MoEF in 2009 (CSE 2012).

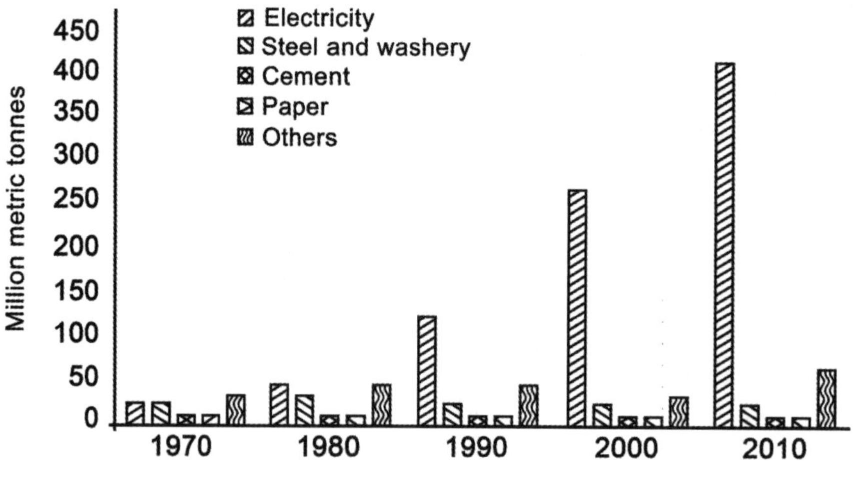

Figure 1. Consumption of raw coal by different industries in India. Source: India Energy Book 2012.

In northeast India, coal mining was initiated by Medlicott in 1869 and 1874 (Sarma 2005a, b). The Cenozoic coals in the northeast states of India with its unusual physico-chemical characteristics have been playing an important role in the Indian economy for the last few decades (Baruah 2009; Saikia *et al.* 2014a). Due to its unique properties and consequent environmental issues, coals in northeast India draw special attention (Zamuda and Sharpe 2007; Saikia *et al.* 2014a).

The main objective of this paper is to summarize coal characteristics and associated environmental issues in northeast India. In addition, this paper also reviews the current state of research in the field of various treatment approaches to reduce environmental impacts of coal.

2. Coal Distribution and Its Characteristics in Northeast INDIA

As on April 2014, India's inventory of coal resource was 300 Billion Tons (BT) comprising of: Proven—125BT; Indicated—142BT and Inferred—32BT (Ministry of Coal 2014). Northeast India contributes 105 Million Tons (MT) of the Gondwana coal and 1492MT of tertiary coal reserves. Meghalaya and Assam in northeast India contain 73% of the total tertiary coal reserves. Nagaland and Arunachal Pradesh contribute 21% and 6% of the total tertiary coal reserves, respectively. Coal inventory of northeast India is given in **Table 1**.

Table 1. Northeast India's coal inventory of Gondwana and tertiary coal in million tones (Ministry of coal, India as on April 1, 2014).

Coalfield	State	Proved	Indicated	Inferred	Total
Gondwana	Assam	0	4	0	4
	Sikkim	0	58	43	101
	Total	0	62	43	105
Tertiary	Arunachal Pradesh	31	40	19	90
	Assam	465	43	3	511
	Meghalaya	89	17	471	577
	Nagaland	9	0	307	316
Total		594	99	799	1492

Source Ministry of coal (2014).

Sub-bituminous tertiary coal of northeast India was deposited under the influence of marine environment (Rajarathnam *et al.* 1996). These coals have high sulfur and low ash content, with high organic sulfur, due to the influence of marine sources during diagenesis (Chandra *et al.* 1983; Singh and Singh 2000; Ward *et al.* 2007; Widodo *et al.* 2010). Coal can be termed as low sulfur (<1% sulfur content), medium sulfur (1%–<3% sulfur content) and high sulfur coals ([3% sulfur content) based on their sulfur contents (Chou 2012). In addition to high sulfur content, northeastern coals have a high content of volatile matter and vitrinite contents, yielding double the amount of tar in comparison to other Indian coals. Mining of these coals generates a large volume of waste materials. These coals generally contain 2%–8% total sulfur, where 75%–90% is of organic sulfur, while the rest is in inorganic form viz. sulfate and pyritic sulfur (Baruah and Khare 2007a). Ledo and Baragolai coal of Makum coalfield, Assam, India contains 28.2% and 21.5% of inorganic and 71.7% and 78.5% of organic sulfur, respectively (Baruah *et al.* 2006). In a study, proximate and ultimate analysis of coal collected from Makum coalfield, Assam showed 3.31%, 2.95% and 2.16% of ash content, total sulfur and organic sulfur, respectively (Saikia *et al.* 2014a). The volatile matter and vitrinite content of the Makum coalfield, Assam were 42.3% and around 93%, respectively (Saikia *et al.* 2014a). Total sulfur of Namchik coalfield, Arunachal Pradesh ranged 1.23%–4.84%, with organic sulfur constituting ~41%–74% of total sulfur and volatile matter ranged 41.8%–46.6% (Chandra *et al.* 1984). Similarly, in Bapung coals of Jaintia Hills, Meghalaya, organic sulfur was more abundant among the different sulfur species constituting an average 62% of organic sulfur of the total sulfur content of 4.59% in it (Ahmed and Rahim 1996). Total sulfur sometimes exceeding 7 wt% out of which the organic sulfur content accounts for about 75% and the rest is inorganic sulfur. Tiru valley coals of Nagaland, India are sub-bituminous to bituminous-D in rank characterized by low to medium moisture (4%–7%), moderately high volatile matter content (22% and 42%) and high sulfur (5%–11%) content (Singh *et al.* 2012a, b). Ash, volatile matter and total sulfur content of Northern Mongchen and Moulong Kimong coalfields, Nagaland, India ranged 2.01%–19.5%, 34.9%–44.8% and 3.23%–5.21%, respectively (Das *et al.* 2015).

In addition to high sulfur and volatile matter, and low ash content of northeast coal, many environmental sensitive organic and mineral bound elements remain enriched in these coals that can cause air, water and land pollution. Sub-bituminous coals of Assam obtained from Makum coalfield showed that Fe, Co, Ni, Cu and Zn are significantly mineral bound, Mg, Ca and Mn are organic bound, while Cd is 50% bound to either organic or mineral matter (Baruah *et al.* 2003). The aqueous leaching of these coals showed their tendency to atmospheric weathering and highly acidic water formed during the leaching process enhanced the mobilization of associated trace and heavy metals (Fe, Mg, Bi, Al, V, Cu, Cd, Ni, Pb, and Mn) above the regulatory levels (Baruah *et al.* 2006). Element concentrations such as Cr, Mn, Ni, Cu, Zn, As and Pb in coals obtained from Makum coalfield, Assam were 5, 23, 5, 2, 27, 1 and 4mg/kg, respectively. Concentrations of these elements in coals from Moulong Kimong coalfield, Nagaland were 4, 2289, 3, 2, 49, 2 and 1mg/kg, respectively (Saikia *et al.* 2014a). Study further indicated that many of these elements were associated with hematite, magnetite, and goethite in the coals.

3. Environmental Issues Associated with Coal Mining and Its Utilization in the Region

Unscientific mining of minerals poses a serious threat to the environment, resulting in reduction of forest cover and loss of biodiversity, soil erosion and pollution of air, water and land. The primitive and unscientific 'rat-hole' method of mining adopted by private operators and related activities have caused large-scale environmental degradation and severe ecosystem destruction in Meghalaya (Swer and Singh 2003, 2004; Sarma 2005a, b). Large scale denudation of forest cover, scarcity of water, air and water pollution, degradation of soil and agricultural lands, land subsidence, haphazard dumping of coal and overburden are some of the conspicuous environmental implications of coal mining in north eastern coal mines of Meghalaya, India (Swer and Singh 2004). Based on a study in the Nokrek Biosphere Reserve in Meghalaya, India, it is revealed that coal mining has adversely affected the vegetation and the density of trees, shrubs and herbs in mined areas (Sarma and Barik 2011).

The mining and cleaning of coal at local processing sites creates large quantities of ambient particulate matter (Ghose and Banerjee 1995; Ghose and Majee 2000). Opencast mining operations contribute major air pollutants to the atmosphere and are responsible for environmental degradation by deteriorating the air quality in respect to dust, fine coal particles and other gaseous pollutants (Mukhopadhyay *et al.* 2010). The major sources of air pollution in coal mining area include: drilling and blasting, loading and unloading of coal and overburden, movements of heavy vehicles on haul roads, dragline operations, crushing of coal in feeder breakers, presence of fire, exposed pit faces, wind erosion and exhaust of heavy earthmover machinery (Nair and Sinha 1987; Ghose and Majee 2007; Huertas *et al.* 2011). According to Ghose and Banerjee (1997), air pollution caused by washeries is more acute than any other coal processing operations.

Based on a report on ambient air quality around northeastern coalmines in Margherita, Assam, the maximum daily average values of SPM (Suspended Particulate Matter), RPM (Respirable Particulate Matter), SO_2 and NO_x were found to be 214, 60, 25 and 52lg/m^3, respectively (Envirocon 2010). Except for SPM, all other values were within CPCB guidelines (CPCB 2009). Atmospheric concentration of gaseous NH_3, SO_2 and NO_2 released from the mining activities in open cast mine area of Tirap colliery, Margherita (Assam), ranged between 4.7–40.03, 1.47–6.14, and 1.92–2.40lg/m^3, respectively, and particulate NH_4^+ in PM_{10} and $PM_{2.5}$ ranged between 0.02–0.07 and 0.008–0.031g/m^3, respectively (Sarmah *et al.* 2012). The study further suggested that low emission and deposition of NO_x and SOx prevents the greater formation of acidic species due to neutralization with NH_4^+. Source apportionment of $PM_{2.5}$ levels at the suburban site of northeast India (Khare and Baruah 2010) showed that largest contribution to aerosol mass in $PM_{2.5}$ is from crustal sources (38%) followed by coal combustion (26%), industrial and vehicular emissions (19%), wood burning (9%) and secondary aerosol formation (8%). Among different elements, emissons of Te, Fe, Mn, Cd, Sn and Sb were related to coal burning (Khare and Baruah 2010).

Coke industry is one of the major coal utilization industries in northeast India. To assess the impact of coke oven burning high sulfur and volatile matter containing coal on ambient air quality, levels of SO_2, $PM_{2.5}$ and trace metals were investigated (Khare and Baruah 2011). The study showed that total emissions of $PM_{2.5}$, total carbon (TC), black carbon (BC) and organic carbon (OC) ranged between 72–306, 49–217, 0.71–2.9 and 48–214 t/year, respectively and the concentration of trace metals was in the decreasing order as: Te > Mn ~ V > Cr > Co > Mo > Cu > Zn ~ Sb > Sn > Cd ~ Ni > As > Se > Hg. The study further indicated that emission rates of metals were dependent on the volatility of the metals, condition of coke ovens and rank of coal (Khare and Baruah 2011).

In addition to air pollution, problems of AMD (Acid Mine Drainage) are intensely localized in the coalfields of northeast India, where ecology of the surrounding area is badly disrupted. The rejects and coals dumped near the pit entrance are exposed to the environment. Being highly enriched with sulfur, pyrite present in these materials is oxidized and hydrolysed and therefore is well known for the generation of AMD (Tiwary 2001; Baruah *et al.* 2005, 2006; Baruah 2009; Baruah and Khare 2007a). Metals concentrations in mine water in India and the world is shown in **Table 2**. As it is visible, metals such as Fe, Cu, Mn, Zn, Ni and Pb in mine water of northeast India (Jaintia and Makum) showed higher concentrations as compared to other mining sites in India. Zn and Pb showed the maximum concentrations in Jaintia coalfield of Meghalaya, northeast India and Ni showed the maximum level in Makum coalfield of Assam. Such high concentrations of metals in these sites can be attributed to higher leaching under acidic conditions in these coalfields. However, elemental contents in leachate water are controlled by three factors: the oxidation rate of pyrite, the acidity of the leachate water and the mineralogy of the rejects (Baruah and Khare 2010). Further, it depends on the element content in the coal. Concentration of toxic elements present in northeastern and other coals in India is shown in **Table 3**. Toxic metal such as Cd showed the maximum level in Jaintia coal of Meghalaya, northeast India. In a

Table 2. Metal contents in mine water (lg/L) in few coal mines in northeast India, India and the world.

Parameter	Fe	Cu	Mn	As	Zn	Ni	Pb	Cr	Cd	References
Jaintia coalfield (Meghalaya, India)	118,400	320	4070	-	4220	1080	430	60	30	Sahoo *et al.* (2012)
Jharia coalfield (Jharkhand, India)	423	32.3	136	3.4	106.1	17.6	14.9	8.1		Singh *et al.* (2009)
Raniganj (West Bengal, India)	329	18.8	39.4	10.06	60	45.6	22.6	44.6		Singh *et al.* (2009)
West Bokaro coalfields (Jharkhand, India)	652	46	1431	7.21	194	154	34.3	81.2		Singh *et al.* (2009)
Makum (Assam, India)	105,300	310	10,200		1530	3120	270	56	35	Equeenuddin *et al.* (2010)
Karnen (Iran)	192,500	350	30,900		2070	1060	180	850	18	Shahabpour *et al.* (2005)
Dogye coalmine (Korea)	176,300	430	8360		2120					Chon and Hwang (2000)

Table 3. Concentration of some toxic elements present in northeast India and other Indian coals/lignite (mg/kg).

Parameters	As	Cu	Mn	Zn	Ni	Pb	Cr	Cd	References
Jaintia coalfield (Meghalaya, India)	1–3	2.8–40	36.6–81.5	8.5–36.6	2–9.8	2.4–13.7	17.9–55.5	5	Baruah and Khare (2010)
Makum coalfield (Assam, India)	0.04–0.24	9.86–30.35	15.27–63.81	-	-	5.06–24.13	-	-	Mukherjee and Srivastava (2005)
Jammu and Kashmir, India	9.5	16.7	39	17.3	42.5	13.5	31.5	1.8	Banerjee *et al.* (2000)
Damodar Koel Valley coal	8.2	21.4	57.7	33.3	28	17.9	47.5	2.2	Banerjee *et al.* (2000)
Wardha Godavari Valley coal	2.1	29.5	58.6	29.2	25	4.5	54.5	2.8	Banerjee *et al.* (2000)
Pench Kanhan Tawa Valley coal	5.8	24.3	85	26	22.7	10.2	33.7	2.1	Banerjee *et al.* (2000)

study on elemental leaching of Meghalaya coals, elements such as Al, P, S, K, Ti, Cr, Co, Zn showed negative correlations with pH (Baruah and Khare 2010).The release of Al, Si, P, Cl, K, Ti, Mn, Co and Ni concentrations in the leachates depends on pyrite oxidation and dissolution (Yue and Zhao 2008), whereas Cd, Sn, Sb and Te contents in the leachates are mainly controlled by adsorption on Fe hydroxides, which is indirectly influenced by pH. The concentrations of trace and potentially harmful elements (Sb, As, Cd, Cr, Co, Cu, Pb, Mn, Ni, V, and Zn) in the Meghalaya coals mine rejects ranged (mg/kg): 11.1–12.6, 1.3–25.9, 5–5.1, 259–361, 20.9–22, 23.6–32.9, 98–149, 87–104, 36.4–58, 50–55 and 35.8–55, respectively, and among these Sb and Cd showed high enrichment factor showing build up in the environment (Baruah and Khare 2010).

Impact of AMD in the streams and groundwater at the vicinity of collieries is a growing problem in northeast India. The Meghalaya State Pollution Control Board, Shillong (MSPCB 2007) reported a case of massive fish death in Lukha River on the eastern border of Jaintia Hills district, which was attributed to AMD contaminating the stream water and sediments. Swer and Singh (2003, 2004) have reported the lack of commonly found aquatic life forms such as fish, frogs and benthic macroinvertibrate such as Plecoptera, Ephemeroptera and Tricoptera in water bodies of coal mining areas in Jaintia Hills, Meghalaya. Overall, socio-economic and ecological impacts in the area includes: severe scarcity of freshwater resources for domestic use and drinking purposes by the local community causing breach of basic human right; lack of aquatic life in many rivers and streams and reduced vegetation diversity; decreased agricultural productivity etc. (Swer and Singh 2004). Swer and Singh (2004) further reported that water quality in the Jaintia Hills of Meghalaya is highly affected as evidenced by low pH (in the range of 3−5), high conductivity, high concentration of sulphates, iron and other toxic metals, low dissolved oxygen (DO) and high biological oxygen demand (BOD). Such low pH, low DO, higher sulphate content and turbidity in water of coal mining areas are affecting the aquatic life.

Singh and Sinha (1992) reported variation of pH in northeastern coalfields, pH 2.8−4.1 in Churcha, pH 4.2−5.0 in West Chirimir, pH 5.2−5.6, pH 5.3−6.0 in Rakhikhol and pH 4.0−4.6 in Gorbi coalfields. Highly acidic mine water with high sulphate (up to 1500mg/L) and Fe (40mg/L) were reported in Margherita group of mines in Assam (Rawat and Singh 1982). Bhole (1994) reported pH of 3.9, 3.10 and 4.3 in Ledo, Tirap and Bargolia mines of Assam. Based on a similar study carried out in Makum coalfields in Assam by Equeenuddin *et al.* (2010), it was found that the mine discharges were highly acidic (up to pH 2.3) to alkaline (up to pH 7.6) in nature with high concentration of SO_4^{2-} and mine water was highly enriched with Fe, Al, Mn, Ni, Pb and Cd. In addition, ground water close to the collieries and AMD affected creeks were highly contaminated by Mn, Fe and Pb but major rivers were not much impacted by AMD due to their large volume of water. Different physico-chemical parameters of surface and groundwater near coalfields in northeast and other parts of India are shown in **Table 4**. As can be seen in **Table 4**, pH of surface water near Jaintia coalfield, Meghalaya, India is highly acidic as compared to surface water in other sites in India. The maximum concentrations of metals detected in groundwater near Makum coalfield, Assam, India was (mg/L): 0.018 for Cr, 0.2 for Ni, 0.108 for Zn, 2.18 for Mn, 3.9 for Fe, 1.1 for Al, 0.061 for Pb, and 0.009 for Cu, in river water, the maximum concentrations were (mg/L): 0.06 for Ni, 0.016 for Zn, 0.94 for Mn, 2.47 for Fe, 0.42 for Al, 0.017 for Cd, 0.056 for Pb and 0.021 for Cu (Equeenuddin *et al.* 2010). In a study

Table 4. Physico-chemical characteristics of surface water around coalmines in northeast and other parts of India (in mg/L except pH).

Item	Parameter	pH	TDS	Ca	Mg	Fe (μg/L)	Cu (μg/L)	Mn (μg/L)	Zn	Ni	Pb	Cr	Cd	References
Surface water	Jaintia coalfield (Meghalaya, India)	2.6–5.6	174–2078	0.33–108.2	0.72–26.27	1.5–66.1	bdl-0.09	0.01–3.25	0.011–2.05		bdl-0.46	bdl-0.05	bdl-0.06	Sahoo et al. (2012)
	Jharia coalfield(Jharkhand, India)	6.83–9.71	5–885	6.9–100.1	3.8–136.9	0.01–0.82		0.01–3.29						Sarkar et al. (2007)
	Singrauli, Madhya Pradesh	7.6–8.6	315–1425	17.6–38.5	1.95–12.7	0.072–2.27	0.004–0.013	0.005–0.153	0.015–0.058	0.012–.034		bdl-0.200		Khan et al. (2013)
	Makum (Assam, India)	6.1–7.4		6.2–31.4	2.5–16.5	0.41–2.47	bdl-0.021	0–0.94	bdl-0.016	bdl-0.06	0.008–0.056		bdl-0.017	Equeenuddin et al. (2010)
Groundwater	Jaintia coalfield (Meghalaya, India)	4.8–6.8	93–234	2.6–23.5	2.2–11.7	0.024–2.3	bdl-0.01		bdl-0.06		bdl-0.03	bdl-0.01		Sahoo et al. (2012)
	Jharia coalfield (Jharkhand, India)	4.6–7.68	5–320	16.4–175.6	4.4–174.5	0.001–11.94	0.002–0.21		0.01–3.95					Sarkar et al. (2007)
	Jharia coalfield (Jharkhand, India)					0.693	0.0282	0.74	0.153	0.0223	0.0125			Chandra and Jain (2013)
	Singrauli coal field (Madhya Pradesh, India)	7.83–8.7	176–1845	9.62–41.68	bdl-26.3	bdl-1.45	0.003–0.068	0.002–0.222	0.014–0.797	0.009–0.03		0.004–0.266		Khan et al. (2013)
	Makum (Assam, India)	4.2–7.8		4.8–27.52	2.3–12.5	0.15–3.76	bdl-0.009	0.01–2.18	bdl-0.108	bdl-0.2	0.0–0.061	bdl-0.018		Equeenuddin et al. (2010)

by Abhishek *et al.* (2006), water quality parameters in groundwater in Jharia coalfield ranged: pH (6.72−7.94), TDS (213−530mg/L), SO_4^{2-} (8.8−41.2mg/L), Cl^- (19.8−96mg/L), NO_3^- (3−77.7mg/L), Fe (0.13−2.18mg/L), Zn (0.02−0.04mg/L), Pb (0.01−0.04mg/L). The maximum TDS, NO_3^- and Fe concentrations exceeded the Bureau of Indian Standards (BIS) limit for drinking water quality. In surface water, water quality parameters varied between (Abhishek *et al.* 2006): pH (7.15−7.76), EC (250.6−470.6lS/cm), TDS (237−616mg/L), DO (2.5−5.8mg/L), BOD (3.8−13.7mg/L), Pb (0.01−0.03mg/L), Zn (0.03−0.09mg/L) and Fe (0.15−1.91).

Metals concentrations in stream sediments around Makum coal field of Assam ranged (mg/kg): 5.5−71.7, 100−386, 3.1−21.1, 0.48−2.1, 23.1−231, 101−9163 and 17.8−264 for Cu, Cr, Pb, Cd, Zn, Mn and Ni, respectively (Equeenuddin *et al.* 2013). The study further indicated that higher concentrations of all metals were available in exchangeable fraction under strongly acidic environment. Based on their mobility and potential bioavailability, metals were in the order of Cd > Pb> Mn > Ni ≥ Zn > Cu > Cr.

4. Management and Treatment Strategies to Reduce Environmental Impacts of Coals

Since coal mining and its utilization in coal based industries is associated with environmental issues, it is necessary to manage or mitigate its impact on environment or clean coal prior to its utilization. An attempt was made by Dowarah *et al.* (2009) to achieve eco-restoration of a high-sulfur containing coal mine overburden dumping site through primary and secondary ecological succession of native plant species in Tirap Collieries, Assam, India. The study revealed that planting of herbaceous monocots with fibrous root systems such as citronella, lemon grass, Saccharum spontaneum, lianes and shrub species accelerates the ecological processes in an adverse mine overburden environment of Tirap colliery and a secondary sere ecological succession was observed in the restored mine site. In addition, 80%−100% vegetation coverage was observed, the plant species density was more than 80%, and soil organic matter increased from 0.001%−0.005% to 0.5%−1.3%. Restoration refers to reinstatement of the pre-mining ecosystem in all its structural and functional aspects (Bradshaw 2000). Re-vegetation plays a crucial role in enhancing the soil fertility status in mine spoil and in the stabilization of dump slopes by creating mechanical reinforcement of dump material and enhancing shear strength of dump material (Singh 2011; Singh *et al.* 2012a, b). Soil structure development, nutrient cycling, and soil chemical and physical limitations to plant growth are mediated and mitigated by microorganisms and they play a very important role in eco-restoration (Singh and Singh 2006).

Mineral matter and sulfur exhibit harmful effects on utilization of coal. Desulfurization and de-ashing are essential for sustainable utilization of low rank high sulfur coals used in different industries (Baruah *et al.* 2006; Baruah and Khare 2007b; Saikia *et al.* 2013). Sequential solvent extraction was found to be an effective method of desulfurization of high sulfur containing Assam coal, especially for organic sulfur, which could be removed up to 89% (Das and Sharma 2001). Investigation on desulfurization of coal samples from Boragolai and Ledo collieries of Makum coal field, Assam, India using alkali treatment leads to over 70% removal of inorganic sulfur, and removal of sulfur increased with increase in alkali concentration and treatment time (Mukherjee and Borthakur 2003). In another study, for the same coal, solvent extraction and alkali treatment showed successful removal of organic and inorganic sulfur. Solvent extraction using dimethyl formamide (DMF) increased desulfurization of the oxidized Baragolai and Ledo coals up to 95% and 93% for inorganic sulfur and 31% and 23% organic sulfur, respectively, while the alkali treatment showed complete removal of inorganic sulfur and a maximum of 33% and 26.4% organic sulfur for these coals, respectively (Baruah and Khare 2007b). Alkali treatment of high sulfur Assam coal using mixtures (1:1) of 16% sodium hydroxide and potassium hydroxide solution followed by 10% hydrochloric acid could remove 50%–54% of the ash, total inorganic sulfur, and around 25% organic sulfur (Mukherjee 2003). 9.4% of the total organic sulfur was removed by electron transfer process (Borah and Baruah 1999). In another study, approximately 93% and 98% of the pyritic sulfur was removed in the case of the Baragolai and Ledo coal of Makum, Assam, respectively, using 15% (v/v) hydrogen peroxide ? 0.1 N sulfuric acid (Mukherjee and Srivastava 2004). An attempt was made to clean some low rank medium to high sulfur coal samples from northeast India using low ultrasonic energy (20 kHz) in the presence of H2O2 solutions and it showed removal of 31%, 48%, 51%, 48% and 32% of total sulfur, organic sulfur, pyritic, sulfate sulfur and ash, respectively (Saikia *et al.* 2014b). In a similar study, treatment using application of ultrasonic energy (20 kHz) in aqueous and mixed alkali media (1:1 KOH and NaOH) on coals collected from Assam and Nagaland, India showed that the maximum removal of ash, pyritic sulfur, sulphate sulfur and total sulfur were 87.52%, 83.92%, 12.50% and 18.80%, respectively (Saikia *et al.* 2014c). Ultrasound assisted coal de-sulfurization and de-ashing is partially green approach that has been recently studied by other researchers (Hoffmann *et al.* 1996; Ze *et al.* 2007; Wang and Yang, 2007; Mello *et al.* 2009; Shen *et al.* 2012).

In addition to several physico-chemical desulfurization methods, biodesulfurization using Thiobacillus ferrooxidans (ATCC 13984) was attempted for Assam coal (Dastidar *et al.* 2000). Results showed that the rate of pyritic sulfur removal was retarded at higher concentrations of ferrous and ferric ions that need to be controlled to maintain high rate of removal (Dastidar *et al.* 2000). In general, AMD can be remediated by two generic approaches i.e. active or passive treatment (Skousen *et al.* 1998; Wolkersdorfer 2008). Active treatment requires the use of alkaline materials (lime, limestone, hydrated lime, caustic soda, soda ash, etc.) or aeration to reduce acidity and precipitate metals, while passive (abiotic and biological) treatment allows chemical and biological processes to take place naturally in a controlled environment (Costello 2003; Johnson and Hallberg 2005; Sheoran and Sheoran 2006; Rios *et al.* 2008; Sheoran *et al.* 2010). A pilot plant consisting of sequential alkalinity producing (SAP) system coupled with biological processes was designed for treatment of AMD from coalmines of Meghalaya, northeast India (Baruah *et al.* 2010). The treatment system was found to be effective in reducing TDS, conductivity, sulphate and toxic elements.

In India, the Ministry of Environment and Forests (MoEF) plays a key role in regulating the environmental impacts of mining and in providing clearances for mining in forest lands. Some environmental protection measures include: prevention of pollution at source; ensuring polluters pay principle; protection of heavily polluted areas and river stretches; encouragement of development and application of best available technological solutions; and involving the public in decision making (Mehta 2002). Under Mineral Concession Rules, 1960, it is required to specify the area indicating impact of mining activity on forest, land and environment, scheme for restoration of the area by afforestation, adoption of pollution control devices. According to Article 23 of the Mineral Conservation and Development Rules (1988), conditions for the abandonment of any mine need to be laid down by the mining company and provision of a plan for dealing with the environment, and is liable to protect and control pollution during the mining and post mining operations. The law further lays guidelines to restore or protect the flora of the area under the mining lease and nearby areas, technically, economically and environmentally.

The main environmental acts that impact the mining industry in India are: The Wildlife (Protection) Act, 1972 (amended in 1991); The Water (Prevention and Control of Pollution) Act, 1974 (amended in 1988); The Forest (Conservation) Act, 1980 (amended in 1988); The Air (Prevention and Control of Pollution) Act, 1981 (amended in 1988); and The Environment (Protection) Act, 1986 (with rules 1986 and 1987). Separate pollution standards for air quality and coal mine effluents has been laid down by Central Pollution Control Boards for coal mining in India (**Table 5**, **Table 6**).

Table 5. Pollution standards for air quality in India.

Pollutant	Time-weighted averages	Concentration in ambient air (mg/L)		
		New coal mines (after December 1998)	Existing coalfields/mines	Old coal mines (Jharia, Raniganj, Bokaro)
SPM	Annual average	360	430	500
	24h	500	600	700
RPM	Annual average	180	215	250
	24h	250	300	300
SO_2	Annual average	80	80	80
	24h	120	120	120
NO_x	Annual average	80	80	80
	24h	120	120	120

Table 6. Pollution standards for coal mine effluents.

Parameter	Level
pH	5.5–9.0
TSS (mg/L)	100
Oil and grease (mg/L)	10
COD (mg/L)	250
BOD (mg/L)	30
Phenolics (mg/L)	1.0

In order to achieve sustainable utilization of coal resources integrated approach considering various aspects to reduce its environmental impacts is necessary. Proper implementation of regulatory rules and policies is as important as other management strategies to deal with environmental issues.

5. Conclusions

Demand for coal in India is projected to increase dramatically in short to medium term. This would result in increased coal mining in different parts of India including northeast region. Since, coals in northeast India is characterised by high sulfur and volatile matter contents that exhibits more potential harmful impacts, extra efforts are required to manage these coals to reduce its environmental impacts in the region. More studies need to be done in the field to assess the impact of coal mining on biodiversity, soil, air, surface and ground water in northeast India. Although several researches on desulfurization, de-ashing and demineralization techniques have been made, effort should be made to do further research on developing effective, low cost and environmental friendly technologies to clean coal and to use these techniques in the field. Further, it is essential to encourage and emphasize on alternative clean sources of energy to meet future energy demands.

Source: Chabukdhara M, Singh O P. Coal mining in northeast India: an overview of environmental issues and treatment approaches[J]. International Journal of Coal Science & Technology, 2016:1–10.

References

[1] Abhishek R, Tiwary K, Sinha SK (2006) Status of surface and groundwater quality in coal mining and industrial areas of Jharia coalfield. Ind J Environ Prot 26:905–910.

[2] Ahmed M, Rahim A (1996) Abundance of sulfur in Eocene coal beds from Bapung, Northeast India. Int J Coal Geolgy 30:315–318.

[3] Banerjee NN, Ghosh B, Das A (2000) Trace metals in Indian Coals; CFRI golden jubilee monograph. Allied Publishers Ltd, New Delhi, p 13.

[4] Baruah BP (2009) Environmental studies around makum coalfields. LAP Lambert Academic Publishing, India.

[5] Baruah BP, Khare P (2007a) Pyrolysis of high sulfur Indian coals. Energy Fuels 21:3346–3352.

[6] Baruah BP, Khare P (2007b) Desulfurization of oxidized Indian coals with solvent extraction and alkali treatment. Energy Fuels 21:2156–2164.

[7] Baruah BP, Khare P (2010) Mobility of trace and potentially harmful elements in the environment from high sulfur Indian coal mines. Appl Geochem 25:1621–1631.

[8] Baruah MK, Kotoky P, Borah GC (2003) Distribution and nature of organic/mineral bound elements in Assam coals, India. Fuel 82:1783–1791.

[9] Baruah BP, Kotoky P, Rao PG (2005) Genesis of acid mine drainage from coalfields of Assam, India. In: Proceedings of international seminar on coal science and technology—emerging global dimensions: global coal. Allied Publishers. ISBN:81-7764-818-7.

[10] Baruah BP, Saikia BK, Kotoky P, Rao PG (2006) Aqueous leaching of high sulfur sub-bituminous coals in Assam, India. Energy Fuels 20:1550-1555.

[11] Baruah BP, Khare P, Rao PG (2010) Management of acid mine drainage (AMD) in Indian coal mines In: Proceedings of international seminar on mineral processing technology (MPT-2010), pp 1163–1170.

[12] Bhole AG (1994) Acid mine drainage and its treatment. Proceedings of International Symposium on the impact of mining on the envirnment. Paithankar *et al.* (ed) Oxford & IBH Pub., Nagpur, Jan 11-16, pp 131–141.

[13] Borah D, Baruah MK (1999) Electron transfer process 1. Removal of organic sulphur from high sulphur Indian coal. Fuel 78:1083–1088.

[14] Bradshaw AD (2000) The use of natural processes in reclamation-advantages and difficulties. Landsc Urban Plan 51:89–100.

[15] Centre for Science and Environment, CSE (2012) Coal mining, pp 1–5. http://www.cseindia.org/userfiles/fsheet2.pdf. Accessed 5 Apr 2015.

[16] Chandra A, Jain MK (2013) Evaluation of heavy metals contamination due to overburden leachate in groundwater of coal mining area. J Chem Biol Phys Sci 3: 2317–2322.

[17] Chandra D, Mazumdar K, Basumallick S (1983) Distribution of sulfur in the tertiary coals of Meghalaya, India. Int J Coal Geol 3:63-75 Chandra D, Ghose S, Chaudhuri SG (1984) Abnormalities in the chemical properties of tertiary coals of Upper Assam and Arunachal Pradesh. Fuel 63:1318–1323.

[18] Chaulya SK, Chakraborty MK (1995) Perspective of new national mineral policy and environmental control for mineral sector. In: Proceedings of national seminar on status of mineral exploitation in India, New Delhi, India, pp 114–123.

[19] Chon HT, Hwang JH (2000) Geochemical characteristics of the acid mine drainage in the water system in the vicinity of the Dogye coalmine in Korea. Environ Geochem

Health 22:155–172.

[20] Chou CL (2012) Sulfur in coals: a review of geochemistry and origins. Int J Coal Geol 100:1–13.

[21] Costello C (2003) Acid mine drainage: innovative treatment technologies. US Environmental Protection Agency, Office of Solid Waste and Emergency Response. Technology Innovation Office, Washington, DC. http://www.cluin.org/download/studentpapers/costello_amd.pdf.

[22] CPCB (Central Pollution Control Board) (2009) National Ambient Air Quality Standards (NAAQS), under Gazette Notification B-29016/20/90/PCI-1, New Delhi.

[23] Das A, Sharma DK (2001) Organic desulfurization of assam coal and its sulfur-rich lithotypes by sequential solvent extraction to obtain cleaner fuel. Energy Source 23:687–697.

[24] Das T, Saikia BK, Baruah BP, Das D (2015) Characterizations of humic acid isolated from coals of two Nagaland coalfields of India in relation to their origin. J Geol Soc India 86:468–474.

[25] Dastidar MG, Malik A, Roychoudhury PK (2000) Biodesulfurization of Indian (Assam) coal using Thiobacillus ferrooxidans (ATCC 13984). Energy Conver Manag 41:375–388.

[26] Dowarah J, DekaBoruah HP, Gogoi J, Pathak N, Saikia N, Handique AK (2009) Eco-restoration of a high-sulphur coal mine over-burden dumping site in northeast India: a case study. J Earth Syst Sci 118:597–608.

[27] Envirocon (2010) A report on ambient air quality, water/waste water analysis & noise level measurement at north eastern coalfields. Coal India Ltd, Margherita, pp 1–20.

[28] Equeenuddin SM, Tripathy S, Sahoo PK, Panigrahi MK (2010) Hydrogeochemical characteristics of acid mine drainage and water pollution at Makum Coalfield, India. J Geochem Explor 105:75–82.

[29] Equeenuddin SM, Tripathy S, Sahoo PK, Panigrahi MK (2013) Metal behavior in sediment associated with acid mine drainage stream: role of pH. J Geochem Explor 124:230–237.

[30] Ghose MK, Banerjee SK (1995) Status of air pollution caused by coal washery projects in India. Environ Monit Assess 38:97–105.

[31] Ghose MK, Banerjee SK (1997) Physico-chemical characteristics of air-borne dust emitted by coal washery in India. Energy Environ Monit 13:11–17.

[32] Ghose MK, Majee SR (2000) Assessment of dust generation due to opencast coal mining—an Indian case study. Environ Monit Assess 61:257–265.

[33] Ghose MK, Majee SR (2007) Characteristics of hazardous airborne dust around an Indian surface coal mining area. Environ Monit Assess 130:17–25.

[34] Hoffmann MR, Hua I, Hochemer R (1996) Application of ultrasonic irradiation for the degradation of chemical contaminants in water. Ultrason Sonochem 3:163–172.

[35] Huertas J, Camacho D, Huertas M (2011) Standardized emissions inventory methodology for open pit mining areas. Environ Sci Pollut Res 19:2784–2794.

[36] Johnson DB, Hallberg KB (2005) Acid mine drainage remediation options: a review. Sci Tot Environ 338:3–14.

[37] Khare P, Baruah BP (2010) Elemental characterization and source identification of PM2.5 using multivariate analysis at the suburban site of North-East India. Atmos Res 98:148–162.

[38] Khare P, Baruah BP (2011) Estimation of emissions of SO2, PM2.5, and metals released from coke ovens using high sulfur coals. Environ Progr Sustain Energy 30:123–129.

[39] Khan I, Javed A, Khurshid S (2013) Physico-chemical analysis of surface and groundwater around Singrauli Coal Field, District Singrauli, Madhya Pradesh, India. Environ Earth Sci 68:1849–1861.

[40] Mehta PS (2002) The Indian mining sector: effect on the environment and FDI inflows. In: Conference on foreign direct investment and the environment, 7–8 February, Paris, France.

[41] Mello PA, Duarte FA, Nunez MAG, Alencar MS, Moreira EM, Korn M *et al* (2009) Ultrasound-assisted oxidative process for sulfur removal from petroleum product feedstock. Ultrason Sonochem 16:732–736.

[42] Ministry of Coal (2014) Government of India, provisional coal statistics 2013–14, coal reserves. http://coal.nic.in. Accessed 12 Aug 2015.

[43] Mukherjee S (2003) Demineralization and desulfurization of high-sulfur Assam coal with alkali treatment. Energy Fuels 17:559–564.

[44] Mukherjee S, Borthakur PC (2003) Effects of alkali treatment on ash and sulphur removal from Assam coal. Fuel Process Technol 85:93–101.

[45] Mukherjee S, Srivastava SK (2004) Kinetics and energetics of high-sulfur northeastern India coal desulfurization using acidic hydrogen peroxide. Energy Fuels 18:1764–1769.

[46] Mukherjee S, Srivastava SK (2005) Trace elements in high-sulfur Assam coals from the makum coalfield in the northeastern region of India. Energy Fuels 19:882–891.

[47] Mukhopadhyay S, Pal S, Mukherjee AK, Ghosh AR (2010) Ambient air quality in opencast coal mining areas of Bankola area (under Eastern coal field ltd.) of Asansol-Raniganj regions. Ecoscan 4:19–24.

[48] Nair PK, Sinha JK (1987) Dust control at deep hole drilling for open peak mines and development of a arrester. J Mines Met Fuel 35(8):360–364.

[49] Rajarathnam S, Chandra D, Handique GK (1996) An overview of chemical properties

of marine-influenced Oligocene coal from the northeastern part of the Assam-Arakan basin, India. Intern J Coal Geol 29:337–361.

[50] Rawat NS, Singh G (1982) The role of micro-organisms in the formation of acid mine drainage in the north eastern coal field of India. Int J Mine Wat 2:29–36.

[51] Rios CA, Williams CD, Roberts CL (2008) Removal of heavy metals from acid mine drainage (AMD) using coal fly ash, natural clinker and synthetic zeolites. J Hazard Mater 156:23–35.

[52] Sahoo PK, Tripathy S, Equeenuddin SM, Panigrahi MK (2012) Geochemical characteristics of coal mine discharge vis-à-vis behavior of rare earth elements at Jaintia Hills coalfield, northeastern India. J Geochem Explor 112:235–243.

[53] Saikia BK, Kakati N, Khound K, Baruah BP (2013) Chemical kinetics of oxidative desulfurization of Indian coals. Int J Oil Gas Coal Tech 6:720–727.

[54] Saikia BK, Ward CR, Oliveira MLS, Hower JC, Baruah BP, Braga M, Silva LF (2014a) Geochemistry and nano-mineralogy of two medium-sulfur Northeast Indian coals. Intern J Coal Geol 121:26–34.

[55] Saikia BK, Dutta AM, Baruah BP (2014b) Feasibility studies of desulfurization and de-ashing of low grade medium to high sulfur coals by low energy ultrasonication. Fuel 123:12–18.

[56] Saikia BK, Dutta AM, Saikia L, Ahmed S, Baruah BP (2014c) Ultrasonic assisted cleaning of high sulphur Indian coals in water and mixed alkali. Fuel Process Technol 123:107–113.

[57] Sarkar BC, Mahanta BN, Saikia K, Paul PR, Singh G (2007) Geo-environmental quality assessment in Jharia coalfield, India using multivariate statistics and GIS. Environ Geol 51:1177–1196.

[58] Sarma K (2005a) Impact of coal mining on vegetation: a case study in Jaintia Hills district of Meghalaya, India. M.Sc. diss., International Institute for Geoinformation Science and Earth Observation (ITC), Enschede, the Netherlands.

[59] Sarma K (2005b) Impact of Coal Mining on Vegetation: A Case Study in Jaintia Hills District of Meghalaya, India, M.Sc Thesis, International Institute for Geo-information Science and Earth Observation, the Netherlands and Indian Institute of Remote Sensing (IIRS), India.

[60] Sarma K, Barik SK (2011) Coal mining impact on vegetation of the Nokrek Biosphere Reserve, Meghalaya, India. Biodiversity 12:154–164.

[61] Sarmah M, Khare P, Baruah BP (2012) Gaseous emissions during the coal mining activity and neutralizing capacity of ammonium. Water Air Soil Pollut 223:4795–4800.

[62] Shahabpour J, Doorandish M, Abbasnejad A (2005) Mine-drainage water from coal mines of Kerman region, Iran. Environ Geol 47:915–925.

[63] Shen Y, Sun T, Liub X, Jiaa J (2012) Rapid desulfurisation of CWS via ultrasonic enhanced metal boron hydrides reduction under ambient conditions. RSC Adv 2:4189–4197.

[64] Sheoran AS, Sheoran V (2006) Heavy metal removal mechanism of acid mine drainage in wetlands: a critical review. Miner Eng 19:105–116.

[65] Sheoran AS, Sheoran V, Choudhary RP (2010) Bioremediation of acid-rock drainage by sulphate-reducing prokariotes: a review. Miner Eng 23:1073–1100.

[66] Singh TN (2011) Assessment of coal mine waste dump behavior using numerical modeling. In: Fuenkajorn K, Phien-wej N (eds) Rock mechanics. Proceedings of the third Thailand symposium. ISBN 978 974 533 636 0, pp 25–36.

[67] Singh MP, Singh AK (2000) Petrographic characteristics and depositional conditions of Eocene coals of platform basins, Meghalaya, India. Int J Coal Geol 42:315–356.

[68] Singh AN, Singh JS (2006) Experiments on ecological resto-ration of coalmine spoil using native trees in a dry tropical environment, India: a synthesis. New For 31:25–39.

[69] Singh G, Sinha DK (1992) The problem of acid mine drainage its occurrence and effects. In: Proceedings of Environment Management of Mining Operations, Department of Environment & Forest Government of India, pp 156–167.

[70] Singh AK, Mondal GC, Tewary BK, Sinha A (2009) Major ion chemistry, solute acquisition processes and quality assessment of mine water in Damodar valley coalfields, India. In: International mine water conference proceedings ISBN Number: 978-0-9802623-5-3, 19th–23rd October, Pretoria, South Africa, pp 267–276.

[71] Singh PK, Singh MP, Singh AK, Naik AS (2012a) Petrographic and geochemical characterization of coals from Tiru valley Nagaland, NE India. Energy Explor Exploit 30: 171–192.

[72] Singh RS, Tripathi N, Chaulya SK (2012b) Ecological study of revegetated coal mine spoil of an Indian dry tropical ecosystem along an age gradient. Biodegradation 23:837–849.

[73] Skousen J, Rose A, Geidel G, Foreman J, Evans R, Hellier W (1998) A handbook of technologies for avoidance and reclamation of acid mine drainage. National Mine Land Reclamation Center, West Virginia University, Morgantown, p 131.

[74] Swer S, Singh OP (2003) Coal mining impacting water quality and aquatic biodiversity inJ aintia Hills District of Meghalaya. ENVIS Bull Himal Ecol 11:26–33.

[75] Swer S, Singh OP (2004) Status of water quality in coal mining areas of Meghalaya, India. In: National seminar on environmental engineering with special emphasis on mining environment. NSEEME, 19–20 March, Dhanbad, India. pp 26–33.

[76] Tiwary RK (2001) Environmental impact of coal mining on water regime and its management. Water Air Soil Pollut 132:185–199.

[77] Wang Y, Yang RT (2007) Desulfurization of liquid fuels by adsorption on carbon-based sorbents and ultrasound-assisted sorbent regeneration. Langmuir 23:3825–3831.

[78] Ward CR, Li Z, Gurba LW (2007) Variations in elemental composition of macerals with vitrinite reflectance and organic sulphur in the Greta coal measures, New South Wales, Australia. Intern J Coal Geol 69:205–219.

[79] Widodo S, Oschmann W, Bechtel A, Sachsenhofer RF, Anggayana K, Puettmann W (2010) Distribution of sulfur and pyrite in coal seams from Kutai basin (east Kalimantan, Indonesia): implications for paleoenvironmental conditions. Intern J Coal Geol 81:151–162.

[80] Wolkersdorfer Ch (2008) Water management at abandoned flooded underground mines. Fundamentals, tracer tests, modelling, water treatment. Springer, New York, pp 235–275.

[81] Yue M, Zhao F (2008) Leaching experiments to study the release of trace elements from mineral separates from Chinese coals. Int J Coal Geol 73:43–51.

[82] Zamuda CD, Sharpe MA (2007) A case for enhanced use of clean coal in India: an essential step towards energy security and environmental protection. In: Workshop on Coal Beneficiation and Utilization of Rejects. Ranchi, India.

[83] Ze KW, Xin XH, Tao CJ (2007) Study of enhanced fine coal desulfurization and de-ashing by ultrasonic floatation. J Chin Univ Min Technol 17:358–362.

Chapter 13

Conceptualizing and Communicating Management Effects on Forest Water Quality

Martyn N. Futter[1], Lars Högbom[2], Salar Valinia[3], Ryan A. Sponseller[4], Hjalmar Laudon[5]

[1]Department of Aquatic Sciences and Assessment, Swedish University of Agricultural Sciences, 750 07 Uppsala, Sweden
[2]Skogforsk, Uppsala Science Park, 751 83 Uppsala, Sweden
[3]Norwegian Institute for Water Research, Gaustadalle'en 21, 0349 Oslo, Norway
[4]Department of Ecology and Environmental Science, Umeå University, 901 87 Umeå, Sweden
[5]Department of Forest Ecology and Management, SLU, Skogsmarksgränd, 901 83 Umeå, Sweden

Abstract: We present a framework for evaluating and communicating effects of human activity on water quality in managed forests. The framework is based on the following processes: atmospheric deposition, weathering, accumulation, recirculation and flux. Impairments to water quality are characterized in terms of their extent, longevity and frequency. Impacts are communicated using a "traffic lights" metaphor for characterizing severity of water quality impairments arising from forestry and other anthropogenic pressures. The most serious impairments to water quality in managed boreal forests include (i) forestry activities causing excessive sediment mobilization and extirpation of aquatic species and (ii) other anthropogenic pressures caused by long-range transport of mercury and acidifying pollutants. The framework and tool presented here can help evaluate, summarize and communicate the most important issues in circumstances where land management and other anthropogenic pressures combine to impair water quality and may also assist in implementing the "polluter pays" principle.

Keywords: Boreal, Environmental Communication, Forestry, Water Quality

1. Introduction

Forests cover approximately 2/3 of Sweden and forestry contributes 2% of GDP (Skogsstyrelsen 2014). Because they cover a relatively large proportion of the Baltic Sea drainage basin, runoff from Swedish forests has a major influence on water quality in the marine environment (Brandt *et al.* 2008). The vast majority of Swedish forests are managed for biomass production, and there are demands for further intensification to meet the goals of an emerging bioeconomy (Egnell *et al.* 2011). This near universal anthropogenic shaping of the forest landscape has been ongoing for several centuries, making it difficult to separate background or reference condition levels from the effects of present-day management activities (Renberg *et al.* 2009). Furthermore, Swedish forests have been subject to a range of non-forestry-related environmental stresses which have degraded water quality. Much of the forest area in southern Sweden is still recovering from the legacy of acid deposition (Akselsson *et al.* 2013; Moldan *et al.* 2013) which has led to ongoing surface water acidification (Futter *et al.* 2014) and slow biological recovery (Valinia *et al.* 2014). Most of the nitrogen (N) and mercury (Hg) deposited on Swedish forests is the result of emissions in other regions and long-range transport. Almost all of the organic micro pollutants (OMPs; including legacy and emerging persistent organic pollutants) are anthropogenic in origin. Forestry activities can, if carried out without proper consideration, exacerbate negative effects on water quality by altering rates of biogeochemical cycles, depleting element pools or mobilizing atmospherically deposited pollutants (Kreutzweiser *et al.* 2008; Lattimore *et al.* 2009; Laudon *et al.* 2011; Thiffault *et al.* 2011; Palviainen *et al.* 2015).

In 2000, member states in Europe adopted the Water Framework Directive (WFD) as an overall goal for water management (EC 2000). The WFD moved towards ecological integrity as a focal point of management instead of traditional sectoral strategies. This led to a comprehensive list of physical, biological and chemical parameters to be used when classifying surface waters in Europe (Hatton-Ellis 2008). The overall goal of the WFD is to reach Good Ecological Status (GES) which is defined as a state with minor influence from anthropogenic alterations, hence an undisturbed state (EC 2000, Annex V). The undisturbed state is determined by reference conditions, which are assumed to have existed before major industrialization, urbanization and intensification of agriculture (EC 2003a). The reference condition concept has been criticized for problems with interpretation and identification of the undisturbed state (Moss 2008; Hering *et al.* 2010; Valinia *et al.* 2012). In particular, it is important to recognize that reference conditions cannot and should not be equated with "natural conditions" (sensu Siipi 2008). Indeed, given the long history of human habitation and that almost all forests in Sweden are managed, reference conditions represent something of an idealization.

The WFD also enshrines the "polluter pays principle" (EC 2000) which embodies the concept that polluters are responsible for the pollution they have caused. While this principle appears simple, its implementation can be complicated, especially in situations where pollution is caused by more than one polluter (Lindhout and Van den Broek 2014). This is especially relevant in managed forests where water pollution may be the result of a combination of deposition of pollutants from long-range transport and their subsequent mobilization by forest management activities.

While water quality in managed Swedish forests is generally good when compared to agricultural and urban regions (Sponseller *et al.* 2014), as well as to other countries in Europe, there are valid concerns about the potential consequences of forestry activities for achieving Good Ecological Status. However, one of the main obstacles when using the WFD to communicate the effects of forestry on Swedish surface waters is that its complexity overwhelms foresters, decision makers, scientists and other actors (Futter *et al.* 2011; Berglund 2014; Keskitalo 2015) and that the results of status classifications can be counterintuitive. For example, the "one out, all out" principle under which the worst result from a series of metrics (e.g. phytobenthos, fish and insects) is used for ecological status classification leads to near-pristine forest streams failing to achieve good ecological status (Löfgren *et al.* 2009). However, this could be resolved with type-specific reference conditions since using individual classification of surface water bodies as the

WFD requires, where a naturally acidic system should be classified as naturally acidic without major anthropogenic influence. Compared to previous ecological quality criteria (EQC), where threshold values were used, naturally acidic systems would be wrongly classified.

Effects of forestry on boreal ecosystem status and surface water quality have been the subject of numerous reviews (Kreutzweiser *et al.* 2008; Bishop *et al.* 2009; Lattimore *et al.* 2009; Laudon *et al.* 2011; Thiffault *et al.* 2011; Palviainen *et al.* 2015). We have no intention of duplicating this material, but instead focus on frameworks for the conceptualization and communication of water quality issues related to forests and forestry.

We focus on eight surface water quality parameters which can be adversely affected by forestry or other anthropogenic activities. These include runoff volume, suspended sediments, N, phosphorus (P), dissolved organic carbon (DOC), base cations (BC; Calcium, Potassium, Sodium and Magnesium), Hg and OMPs. These parameters represent key physical and chemical attributes of streams, lakes and rivers; biological responses to anthropogenic disturbance are considered insofar as they are caused by the above eight issues. Hydromorphological alterations, while important, are not considered further.

Here, we propose a simple conceptual framework for evaluating biogeochemical cycles in the boreal forest and a tool for communicating the manner in which forestry operations may alter these cycles. We use the framework to explore controls on water quality in intact forests and to rank the impacts of forestry-related disturbances on water quality at local, landscape and national scales. Specifically, we pose three questions about water quality connected to forests and forest management. First, do forests or forestry affect the cycling of the chemical species in question, and if so, how strong is the effect?; second, what are the effects of present-day forestry on the water quality issue?; third, and most important, how certain is the science used to answer the first two questions?

2. Conceptual Frameworks

Biogeochemical cycles in forest stands or headwater catchments can be conceptualized using the mnemonic DWARF: Deposition, Weathering, Accumulation, Recirculation and Flux (**Figure 1**). Deposition is the wet or dry input of dissolved and particulate compounds and elements from the atmosphere to a forest stand. Deposition includes rain and snowfall. The "forest filter" effect and the waxy needles of conifers enhance the deposition of some classes of compounds, especially OMPs (Di Guardo *et al.* 2003; Nizzetto *et al.* 2006) and acidifying N and sulphur compounds (Helliwell *et al.* 2014). Typically, deposited pollutants are the product of long-range transport. Weathering is the physical, chemical or biological breakdown of geologic parent material. Weathering makes elements including phosphorus and base cations available for biological uptake and is the primary source of sediment. Accumulation is the process by which deposited and weathered materials are incorporated into the soil or biota. Accumulation also includes biological fixation of C and N from gaseous to organic form. Carbon fixation (*i.e.* photosynthesis) is the ultimate source of nearly all living and nonliving organic matter in forests, including DOC. Recirculation is a broad term which includes recycling, and redistribution of material within a stand. Examples of recycling processes include vertical transfers between plant and soil (as with litter fall and element uptake by roots), or the movement of base cations on and off ion exchange complexes. Redistributive processes include lateral redistribution of material within a stand including buildup of material in riparian zones and wetlands,

Figure 1. DWARF: a conceptual framework for forest biogeochemical cycles. Forest biogeochemical cycles are a combination of Deposition (D), Weathering (W), Accumulation (A) in soils or vegetation, Recirculation (R) between different stocks (*i.e.* vegetation, soils and litter) and Fluxes (F) to surface waters.

paludification, the slow movement of contaminants through the soil profile and vertical redistribution related to e.g. podsolisation. Fluxes out of the system include gravity driven processes such as surface water runoff and mass wasting as well as the return of material to the atmosphere (e.g. via trace gas production or evapotranspiration). Redistribution of elements can be extremely important in delaying the impact of atmospheric deposition on stream water fluxes. Klaminder *et al.* (2011) suggested that the flux of atmospherically deposited lead in stream water might be delayed more than a century due to slow movement through the soil profile. There is also some evidence to suggest that changing rates of inputs can change outputs at a more rapid time scale. For example, Kothawala *et al.* (2011) showed that declines in atmospheric N deposition led to contemporaneous declines in stream water flux.

There can be positive feedbacks between the various components of the DWARF framework. For example, accumulation of material in growing forest biomass can enhance the "forest filter" effect whereby deposition of acidifying substances is increased. This, in turn, alters rates of base cation cycling (Helliwell *et al.* 2014). There is also some evidence from soil experiments that more rapidly growing forests (with higher assimilation rates) will affect base cation cycling through increased weathering rates (Palviainen *et al.* 2012).

Anthropogenic stressors, including forestry, alter the rates of one or more of the DWARF processes. These alterations may lead to impaired water quality, either through direct or indirect mechanisms. For example, atmospheric deposition of OMPs has a direct effect on their accumulation in forest ecosystems. On the other hand, acid deposition has an indirect effect on base cations, leading to alterations in their rates of weathering, accumulation and flux.

The magnitude of potentially negative effects of forest management on forest biogeochemical cycles can be conceptualized using the mnemonic ELF: extent, longevity and frequency. ELF can be used to weight the severity of forestry-related impacts to the spatial extent of an effect, its longevity and the frequency with which it occurs. For example, if it is assumed that elevated N leaching occurs after final felling of a whole stand (E = 1), for 10 years (L = 10) and a forest rotation lasts 100 years [thus F = 1/(rotation length)] then, at a stand scale, the ELF score is $1 \times 10 \times (1/100)$, or 0.1. It should be noted that the same ELF score will be obtained over a landscape where 0.01 of the stands are harvested on an annual frequency and the effect of individual harvesting events lasts 10 years.

$$\text{ELF} = \text{Extent} \times \text{Longevity} \times \text{Frequency}$$

The downstream extent of negative effects is factored into the ELF score. For example, at a stand scale, E values greater than 1 will result if an impact is observed in both the stand and downstream watercourses. Quantifying the extent of downstream impacts can be somewhat subjective, especially when data are lacking. Forestry impacts that lead to significant negative effects downstream of harvested stands, including sediment pollution and Hg accumulation by fish in downstream lakes, will have higher ELF scores than impacts that are mostly observed at the stand level or immediately downstream (such as N leaching). The longer the duration of a negative effect, the higher the ELF score. If sediment pollution were to destroy the habitat of long-lived, slow-growing species such as freshwater pearl mussel (Margaritifera margaritifera) (Österling and Högberg 2014), it is possible that effect longevity would be greater than the length of a forest rotation. Such lags in the ecological effects of stream sedimentation have been observed elsewhere (Harding *et al.* 1998). Most negative effects are associated with final harvest. However, activities which occur more frequently, such as those caused by soil compaction associated with driving damage, will receive higher ELF scores.

ELF scores are closely related to the scaling of water quality problems. Forestry activities that have a bigger footprint in space or time will typically have higher ELF scores. Regional and national scale problems are typically associated with spatially extensive or long-lasting forestry impacts. Because of the uncertainties associated primarily with longevity and assessing the extent of downstream influence, we report qualitative high and low values corresponding to ELF scores above and below 0.01, respectively. When sufficient data are not available to estimate an ELF score, as is the case with OMP, a value of "unknown" is reported (**Figure 2**).

2.1. Spatial Scale

Effects of anthropogenic activities including forestry on water quality can be manifested at local, landscape and national scales (**Figure 3**). The local scale

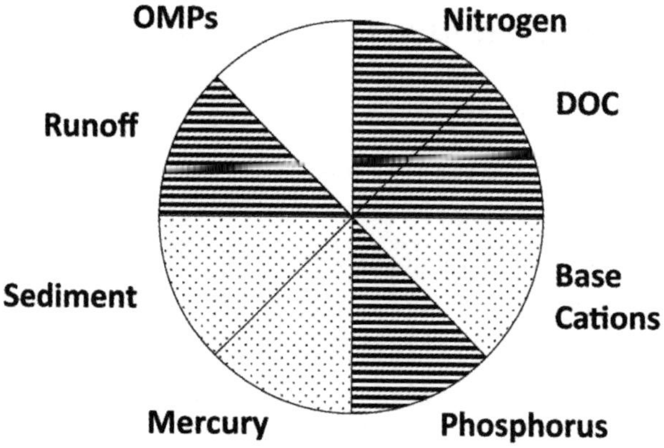

Figure 2. Pie chart showing ELF scores for forest water quality issues. Horizontal lines denote low ELF scores (*i.e.* <0.01) while dots are indicative of high scores (*i.e.* >0.01). There is insufficient information to assign values to blank cells.

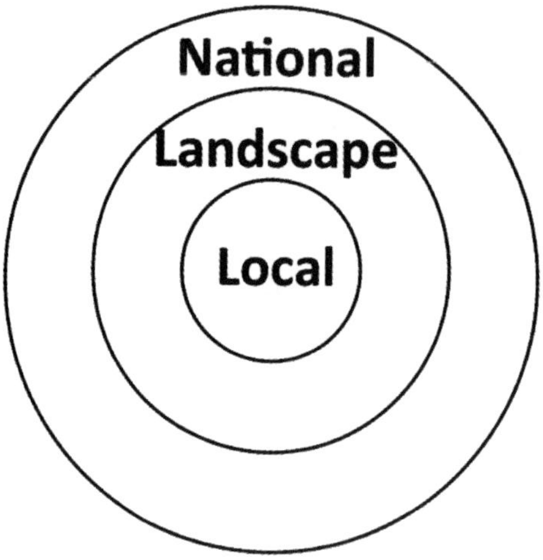

Figure 3. "Dart board" representation of spatial scales assessed here: local (headwater or stand scale effects) are presented in the innermost circle, landscape (10's–100's km^2) scale effects are shown in the middle circle and national (Baltic Sea drainage basin) scale effects in the outer circle. Scale-dependent water quality impacts are communicated by overlaying the pie chart structure in **Figure 2** with the scale representation in this figure.

corresponds to individual forest stands or headwater catchments with areas of a few hectares to a maximum of approximately 10 km^2. The landscape-scale is representative of tens to hundreds of km^2. The national scale in our analysis is synonymous with the Baltic Sea drainage basin. The severity of each water quality issue and forestry effect is assessed at all three spatial scales. Effects at a local scale can be more or less severe at the landscape and national scales.

2.2. Scale for Issue Severity, Effect Magnitude and Uncertainty

To visually summarize multiple water quality parameters, we use a "traffic light" coding of red, yellow and green to communicate severity and effect magnitude for different stressors (**Figures 4–6**). When insufficient data are available to make an assessment, the cell is left blank. The "traffic light" approach has received widespread use in healthcare (Peters *et al.* 2007) and marine environmental assessment (Hargrave 2002; Foden *et al.* 2008). Foden *et al.* (2008) note that the strengths of the approach are that it provides users with a general, easy to track overview of impacts and gives a simplified presentation of potentially complex quantitative data. They caution, however, that any characterization scheme may be subjective and fine detail lost.

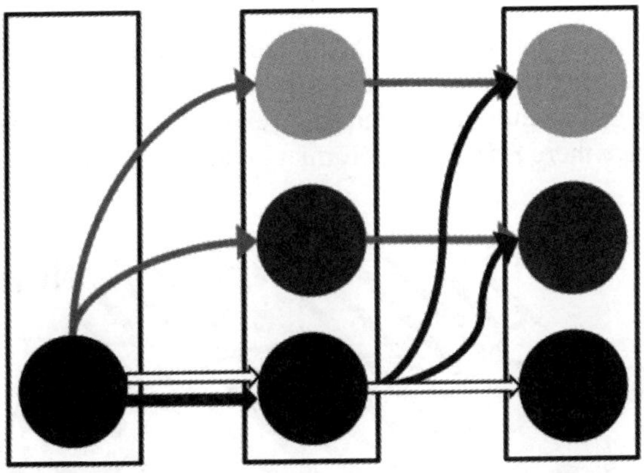

Figure 4. Possible trajectories in water quality as a result of impacts caused by other anthropogenic pressures or forestry. It is assumed that all surface waters are in reference conditions (green) when anthropogenic pressures are absent. Other anthropogenic pressures (e.g. long-range transport, climate change, etc.) may cause a range of deviations from reference conditions spanning from no (green) to moderate (yellow) severe impairments (red). Forestry may not lead to any further appreciable deviation in water quality above and beyond that caused by other anthropogenic pressures (horizontal arrow), or it may result in a further detectable deterioration of water quality. Type I trajectories are shown with white arrows; neither forestry nor other anthropogenic pressures lead to meaningful deviations from reference conditions. Grey arrows show Type II trajectories where other anthropogenic pressures lead to degraded water quality which is not further exacerbated by forestry. Black arrows show Type III trajectories where forestry is the cause of degraded water quality.

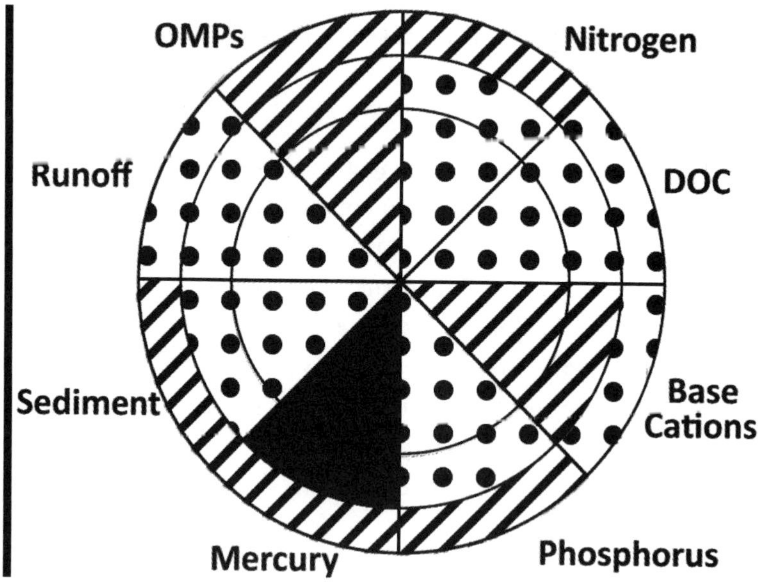

Figure 5. Water quality issues in the Swedish forest landscape at a local (inner), landscape (middle) and national (outer) scale caused by anthropogenic pressures other than forestry. Severity is coded as green (little or no impact), yellow (moderate impact) and red (severe impact) or white where there is too little information to make an assessment.

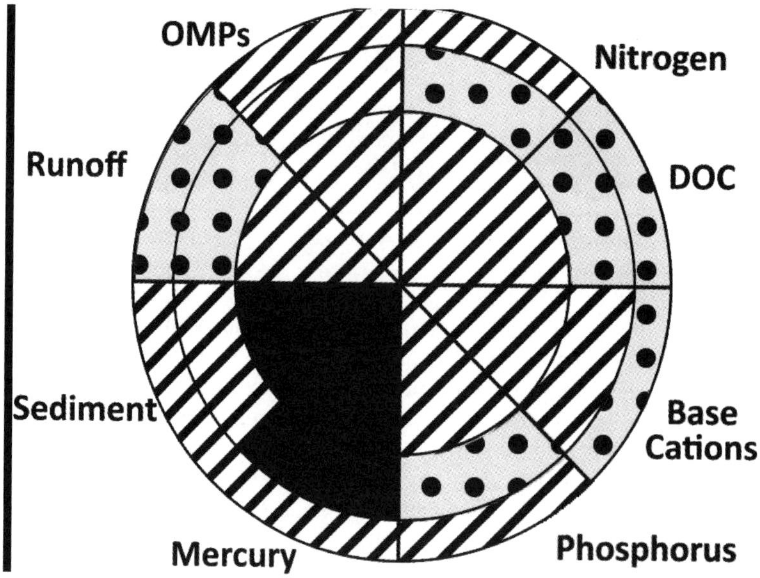

Figure 6. Net impacts of other anthropogenic pressures and forestry impacts on water quality in the forest landscape at local (inner), landscape (middle) and national (outer circle) scales. Severity is coded as green (little or no impact), yellow (moderate impact) and red (severe impact) or white where there is too little information to make an assessment.

In the assessment presented here, minor impairments of water quality are coded green. A minor impairment is a detectable deviation from reference conditions which, on the basis of present scientific knowledge, is not believed to cause unacceptable harm to ecosystem function (*i.e.* minor impairments are analogous to WFD good ecological status). For instance, in acidification assessments of Swedish surface waters, the accepted deviation from reference conditions is a decline of 0.4 pH units since it is assumed that changes smaller than this do not adversely affect aquatic biota (Fölster *et al.* 2007).

Significant impairments of water quality are coded yellow. An impairment is deemed to be significant if it leads to an undesirable deviation from reference conditions. Conceptually, this deviation is analogous to the WFD moderate status (EC 2000, Annex V, WFD). Significant water quality impairments in the forest landscape can occur as a direct result of forestry activities or when other anthropogenic stressors have already pushed ecosystems into a degraded state. Thus, the relatively small impact of forestry on Baltic Sea eutrophication is still considered to be a significant impairment of water quality since that ecosystem is already in a degraded state due to excessive nutrient inputs from agriculture and sewage discharge. In a similar manner, forestry can potentially have a significant impact on base cation concentrations in surface waters already affected by acidification (Aherne *et al.* 2008; Zetterberg *et al.* 2013).

Severe impairments to water quality are coded red, similar to the WFD poor or bad ecological status (EC 2000; Annex V, WFD). An impairment is deemed to be severe if it results in unacceptable negative effects including demonstrable effect on human health, or if it leads to local- or regional-scale species extirpation. Severe impairments can be caused by both forestry activities and other anthropogenic pressures, primarily long-range atmospheric transport.

We assume that the effects of other anthropogenic pressures and forestry are additive. In the absence of forestry effects, other anthropogenic pressures such as long-range transport of pollutants cause one of the following: no appreciable deviation from reference conditions, significant, or severe impairments to water quality. Forestry may cause no further deterioration in water quality, or it may exacerbate the problem. The trajectories in **Figure 4** can thus be classified into three types depending on the traffic light colour under reference conditions, and due to other anthropogenic pressures and forestry.

- Type I—Neither forestry nor other anthropogenic impacts lead to appreciable deviations from reference conditions (white arrows in **Figure 4**).

- Type II—Forestry does not appreciably worsen water quality above and beyond the effects of other anthropogenic pressures. That is to say, there is no increase in severity when moving from other anthropogenic pressures to forestry effects (grey arrows in **Figure 4**).

- Type III—Forestry activities result in water quality impairments, whereas other anthropogenic pressures do not result in appreciable deviations from reference conditions (black arrows in **Figure 4**).

It is possible that other anthropogenic pressures will cause significant impairments of water quality which are then exacerbated by forestry to cause severe degradation. However, none of the examples presented here appear to follow this trajectory.

While the framework presented here only accounts for negative effects of forestry on water quality, it should be noted that forestry can have positive effects, also. Globally, land use conversion through afforestation is widely used as a means of improving water quality (Neary *et al.* 2009) and can be an important contributor to sustainable flood management (Iacob *et al.* 2014). In Sweden, actively growing managed forests are strongly N retentive and thus may mitigate negative eutrophication and acidification effects in surface waters associated with excessive N deposition (Sponseller *et al.* 2016).

3. Forests and Forestry

Water quality in Swedish forests is directly and indirectly affected by a number of human activities. Emissions from fossil fuel burning in Sweden and elsewhere contribute to N pollution and exacerbate problems with base cations losses. In the past, forestry had a much greater impact on water quality than it does today. For example, alteration of river channels to facilitate log transport has had severe and long-lasting ecological consequences (Nilsson *et al.* 2005). While poorly planned forestry activities have the potential to negatively affect water quality, well-managed forests may have less water quality problems than some un-managed forests.

Here, we focus on stand-level forestry operations including site preparation, drainage, ash return, planting, thinning, fertilizing, fire prevention, final felling, harvesting and terrain transport. Thus, we do not consider water quality impairments associated with, inter alia, historical stream channel alteration for timber transport (Nilsson *et al.* 2005) or the negative effects of the forest products industry such as fibre banks associated with pulp mills on the Baltic coast (Assefa *et al.* 2014).

Severity of forestry impacts may differ depending on whether stem only (SOH) or whole tree (WTH) harvesting is practiced. Depending on site quality, the typical rotation time in a Swedish forest ranges between 60 and 120 years. Over that period, the following management activities are applied in the following, or slightly adjusted, sequence: final felling, biomass removal through harvesting, mechanical site preparation, optional ditch maintenance, planting, optional ash return, pre-commercial thinning, commercial thinning, optional fertilization and final felling. Throughout the rotation, road building and maintenance occurs. The majority of water quality impacts are associated with roads, harvesting (including thinning and final felling) and ditch maintenance.

The impacts of other anthropogenic stressors (**Figure 5**) and their combined effects with forestry on water quality (**Figure 6**) can be represented using the "traffic lights" colour coding from **Figure 4**, the spatial scale representation (**Figure 3**) and the water quality issue pie chart (**Figure 2**). The differences in colours between the other anthropogenic stressor effects (**Figure 5**) and combined impacts (**Figure 6**) are related to impact type trajectories (**Figure 4**). The impact types for different spatial scales and water quality issues are graphically summarized in **Figure 7**. It is notable that forestry is not responsible for any Baltic-scale effects and a minority of landscape-scale impacts (**Figure 7**).

3.1. Runoff

The hydrological cycle is the key driver of forest biogeo-chemical cycling. Globally, precipitation is a limiting factor for forest establishment in many regions.

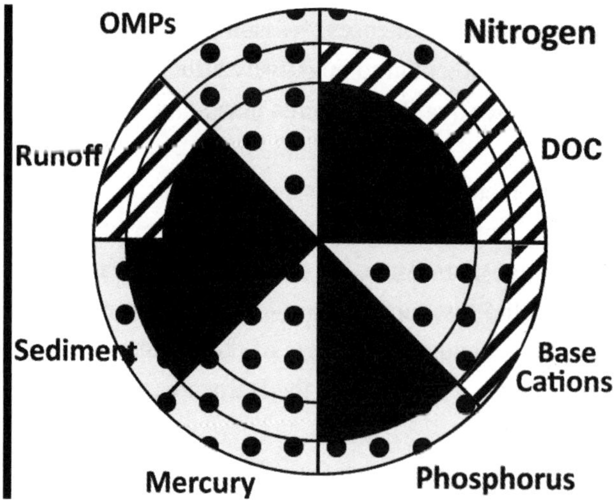

Figure 7. Impact type scores for water quality issues at local (inner), landscape (middle) and national (outer circle) scales. Type I impacts, shown in white, do not deviate significantly from reference conditions. Type II impacts, shown in grey, occur when other anthropogenic impacts are the primary reason for deterioration in water quality. Types III, shown in black, impacts occur when forestry is the primary cause of deterioration in water quality.

In the boreal ecozone, water is generally not the primary limiting factor for forest growth, and a significant fraction of annual precipitation falls as snow. The effects of forests and forestry on the hydrological cycle are strongly scale dependent (Ellison *et al.* 2012). Intact forests return a significant fraction of incoming precipitation to the atmosphere through evaporation and transpiration and afforestation can effectively reduce runoff (Iacob *et al.* 2014). Felling reduces transpiration and canopy interception, leading to wetter soils, a greater fraction of precipitation contributing to runoff, and increased lateral fluxes of water (for a recent review of the processes, see Launiainen *et al.* 2014). Wetter soils can contribute to increases in surface water DOC concentration (Schelker *et al.* 2013), mercury methylation rates (Lattimore *et al.* 2009) and potentially production of greenhouse gases (e.g. CH4 and NO2) from anoxic soils (Vor *et al.* 2003). Forest ditches which are established to dry out soils so as to improve forest growth in saturated areas increase water fluxes beyond reference condition levels and their maintenance can lead to elevated fluxes of sediments and nutrients (Manninen 1998). Forestry operations on wet soils are an underappreciated threat to water quality; Laudon *et al.* (2016) discuss some of the issues of soil wetness and consequences for forests and forestry. At a local scale, forestry has a Type III effect on runoff.

3.2. Nitrogen

Nitrogen is an essential plant nutrient (see Sponseller *et al.* 2016) that also can limit rates of biological processes in boreal streams (Burrows *et al.* 2015) and lakes (Bergström *et al.* 2008). Atmospheric N deposition has increased considerably over the past 100 years as a result of fossil fuel burning and increased fertilizer use. The health of the Baltic Sea ecosystem is under threat from excessive N inputs associated mostly with sewage and agriculture (Conley 2012). Tree growth in most Swedish forests is N-limited and N fertilizer is added into approximately 25,000ha annually in northern and central Sweden to increase yields. Over the course of a whole rotation, boreal forests tend to be net N sinks, in that they effectively take up the N deposition derived from fossil fuel burning in Sweden and elsewhere.

Forestry activities affect the accumulation, recirculation and fluxes of N from forest stands. SOH and WTH remove N from the stand, decreasing the size of the N pool and potentially slowing rates of recirculation (Lundborg 1997; Palviainen and Fine'r 2012). Effects are more pronounced with WTH due to the removal of large amounts of N in needles. While N leakage can occur following final felling, the total amount lost is small relative to total atmospheric deposition (Futter *et al.* 2010). The concentrations of N in groundwater following final felling are elevated when compared to undisturbed forests but are not high enough to cause problems of compliance with European legislation or human health issues. However, forest lands are the largest single-net source of N entering the Baltic Sea from Sweden (Brandt *et al.* 2008). Forestry clearly causes local increases in N fluxes but the legacy of greater deposition and other pollution sources in the Baltic Sea catchment mean that many of the negative impacts are caused by other anthropogenic pressures. Forestry has a Type III effect on N at the local scale; the legacy of atmospheric deposition contributes to the Type II effect at the landscape and national scales.

3.3. Phosphorus

Phosphorus (P) is also an essential plant nutrient that can further influence algal growth in lakes and rivers. This effect is most pronounced in southern Sweden, where significant atmospheric deposition means that systems are not N-limited. Very little P is lost from intact forests. However, significant amounts can be released when soils or sediments are disturbed during site preparation and ditch clearing. At the local scale, levels of P in surface waters can be high enough to cause significant changes in aquatic plant communities.

Following harvest, P is removed in biomass. Ditch maintenance may increase P fluxes out of the affected stands through the mobilization of sediments (Manninen 1998). Increases in both concentration and flux of particulate phosphorus may be seen even when soil disturbance is minimal as the increased runoff following clearfelling can flush fine sediments from ditches (Kaila *et al.* 2014). The local scale effects of forestry operations on P cycling must be balanced against the observed longterm decline in tree mineral nutrition status and the increasing likelihood that forests are P limited (Jonard *et al.* 2015). At a national scale, the situation for P is similar to that for N. Any additional inputs are problematic for the already eutrophied Baltic Sea ecosystem. At local and national scales, forestry results in Type III effects on surface water P.

3.4. Base Cations

Base cations (Calcium, Magnesium, Potassium and Sodium) are essential plant nutrients and some of the most important elements buffering soil and surface water acidification. Acid deposition increases the rate at which base cations are leached from the soil. Following reductions in acid deposition, surface water base cation concentrations may decline further due to lack of a mobile co-anion for transport. The acidification caused by long-range pollutant transport is largely an issue of the past. However, modelling studies have suggested that whole tree harvesting, which may remove base cations from forest soils (Zetterberg *et al.* 2013) faster than they can be replaced by mineral weathering. If this were to occur, it could possibly lead to further acidification of sensitive waters (Akselsson *et al.* 2007). Unfortunately, weathering rates are too uncertain to draw firm conclusions about the sustainability of forest harvesting (Klaminder *et al.* 2011; Futter *et al.* 2012). However, experiments suggest that more rapidly growing forests may increase weathering rates (Palviainen *et al.* 2012).

Water quality impairment associated with declining base cation concentrations probably follows a Type II trajectory. The regional legacy of acid deposition has depleted soil base cations, resulting in ongoing acidification of many soils

(Akselsson *et al.* 2013) and surface waters (Moldan *et al.* 2013; Futter *et al.* 2014) in southern Sweden. Biomass removal following forest harvest will reduce the BC pool in a stand, leading to reductions in the rates of recirculation and potentially lower fluxes to surface waters.

3.5. Dissolved Organic Carbon

Dissolved organic carbon (DOC) originates ultimately from plants fixing atmospheric carbon and is derived from the breakdown of plant material in soils and litter. Concentrations of DOC are increasing in many surface waters and it has been hypothesized that declines in acid deposition (Monteith *et al.* 2007; Valinia *et al.* 2015), historical land management practices (Meyer-Jacob *et al.* 2015) and a changing climate (Oni *et al.* 2014) are important drivers. This is a concern for a number of reasons. DOC is a naturally occurring acid that if elevated above its reference condition can contribute to a delay in acidification recovery (Futter *et al.* 2014) and acidity-related fish kills in some parts of Sweden (Serrano *et al.* 2008). Elevated DOC concentrations can lead to significant alterations of lake ecology including changing the light environment which inhibits gross primary productivity (Solomon *et al.* 2015), fuelling heterotrophic processes and altering the amount and bioavailability of contaminants (Rask *et al.* 2014). Finally, the flux of DOC from Swedish forests may contribute to acidification in the Baltic Sea (Omstedt *et al.* 2010).

Final harvesting almost always leads to increases in surface water DOC (e.g. Schelker *et al.* 2012; Palviainen *et al.* 2015). This effect is difficult to detect at all but the smallest spatial scales (Lepistö *et al.* 2014). Thus, forestry has a Type III effect on DOC at the local scale.

3.6. Mercury

Mercury (Hg) is a potent neurotoxin which is banned in Sweden. There is a high degree of concern about mercury in Swedish forest waters (Eklöf *et al.* 2016). In its methylated form (MeHg), it is able to bio-accumulate in food webs and cause neurological damage in humans, other mammals and birds. Concentrations of MeHg in fish from many Swedish lakes are high enough to constitute a possible human health risk (Åkerblom *et al.* 2014). The environmental behaviour of mercury is complicated: MeHg is produced in environments with low-ambient oxygen concentrations, including lake sediments and wetlands with high concentrations of DOC. Forestry activities (such as final felling) which result in wetter soils in some cases can lead to higher concentrations of MeHg (Porvari *et al.* 2003), which in turn can result in elevated Hg concentrations in fish (Garcia and Carignan 2005; Martin 2014). In a survey of intact and harvested Swedish sites, Skyllberg *et al.* (2009) observed significantly higher MeHg concentrations in streams draining areas with clearcuts than those draining intact forests. However, de Wit *et al.* (2014) reported no increase in MeHg concentrations following clearcutting in a Norwegian study.

The trajectories for Hg are assigned to Type II at all spatial scales. While it is clear that forestry activities can sometimes lead to increased Hg fluxes, the Hg concentrations and fluxes associated with the legacy of atmospheric deposition will continue to pose health threats for many years to come even in the absence of forestry.

3.7. Organic Micro Pollutants

Organic micro pollutants (OMPs) include a wide range of natural and anthropogenic compounds including dioxins, polychlorinated biphenyls (PCBs), polycyclic aromatic hydrocarbons (PAHs) and perfluorinated compounds (PFAs). Many of these compounds are found at toxic levels in Baltic Sea biota and sediments. With few exceptions, OMP accumulate in boreal forests as a result of wet and dry atmospheric deposition. While their concentrations are typically very low in Swedish forests, they are a potential concern because, if mobilized, they can be transported to the Baltic Sea. Very little is known about the behaviour of OMPs in Swedish forests. They appear to be co-transported with organic carbon (Bergknut *et al.* 2011a) and can be found at concentrations similar to those in contaminated sites (Bergknut *et al.* 2011b). There are large uncertainties in estimates of OMP fluxes and further research is needed to evaluate the importance of boreal forest waters as a source of OMPs to the Baltic Sea. It is not clear what effect forestry operations will have on OMP cycling, but it is likely that activities which contribute to increased fluxes of DOC and sediments will also increase flux of OMP. This is of special concern at the national scale as any extra inputs of OMPs to the

Baltic Sea are undesirable. Because of the uncertainty associated with forestry effects on OMP cycling and the clear link between long-range transport and subsequent deposition, this issue is coded as Type II at all spatial scales.

3.8. Sediments

Sediments can be mobilized as a result of increased runoff following final felling, ditch maintenance and site preparation. Excess suspended sediments can have serious negative effects on aquatic biota (Wood and Armitage 1997). The sediments produced by ditch clearing and poorly planned or constructed forest roads and stream crossings can be a serious water quality issue. Forestry activities can have both direct and indirect effects on water quality. There are direct negative effects of increased sediment loads on aquatic habitat (Stenberg *et al.* 2015) as well as indirect effects associated with co-transport of nutrients and contaminants. Specifically, sediments can transport and subsequently release large amounts of P (Kaila *et al.* 2014). More importantly, sediments can destroy aquatic habitats, smother spawning beds, cause the loss of fish populations, and severely alter the abundance and biodiversity of aquatic invertebrates (Burdon *et al.* 2013). While sediment pollution is often a local issue, the effects can be long-lasting as it can take many years for habitats to recover and be re-colonized (Harding *et al.* 1998). For example, if excess sediment results in extirpation of freshwater pearl mussels, it can take decades before recolonization occurs (Österling *et al.* 2010). Kreutzweiser *et al.* (2009) suggest that environmentally sensitive forestry practices, which take extra precautions when working near water, can potentially minimize sediment pollution. While most local-scale impacts are the result of too much sediment, too little sediment can also be problematic. Legacy hydromorphological alterations to river channels to facilitate log transport led to lowered sediment production and transport. At the national scale, hydroelectricity reservoir impoundments have resulted in declines in sediment transport with negative effects on Baltic Sea silica concentrations (Humborg *et al.* 2000). Thus, sediments are coded as Type III at the local and landscape-scale but Type II at the national scale.

3.9. Uncertainty

There is some degree of scientific uncertainty about all environmental issues and there are significant challenges in communicating this to decision makers (Beven 2010; Petersen 2012). The science behind forest water quality and forestry related impacts is more or less certain, depending on the particular issue. We have used sky colours to represent, and what, in our opinion, is the relevant degree of uncertainty with each issue (**Figure 8**). Under blue skies, it is possible to obtain a relatively good view of the surrounding landscape. When skies are grey as a result of low clouds or fog, features in the surrounding landscape are less certain. Thus, a high degree of certainty is coloured blue and a low degree of certainty is coloured grey.

Effects of forestry on N, P and DOC and runoff in the boreal forest are relatively well understood at all spatial scales. The potential local scale effects of forest harvesting on Hg cycling are also well documented (Bishop *et al.* 2009; Eklöf *et al.* 2016), but more work is needed to upscale these results to regional and Baltic Sea levels. The effects of forestry on base cation cycling are uncertain and need further investigation as modelling and measurement suggest contradictory results. The water quality impacts of OMP are not well established at either the stand or

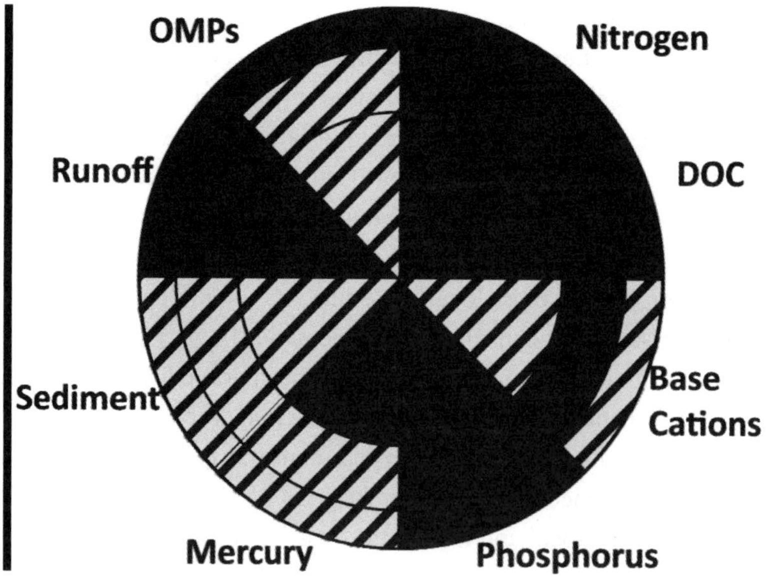

Figure 8. Uncertainty associated with water quality issues and forestry impacts at local (inner), landscape (middle) and national (outer circle) scales. Cells are coded blue when there is limited or no uncertainty and grey where there is significant uncertainty associated with forestry impacts on a water quality issue at the local (inner circle), landscape (middle circle) or national (outer circle) scales.

landscape level. However, it is clear that any additional loading of these compounds to the Baltic is undesirable. The most important knowledge gaps are related to sediment production and mobilization. It seems highly likely that excessive sediment mobilization is having widespread negative effects on aquatic biota dependent on well-oxygenated streambeds.

4. Discussion

Successful policy implementation is dependent on a dialogue between all relevant stakeholders, as their involvement leads to a diversity of experiences and views and knowledge (EC 2003b). The framework presented here can help this dialogue as it provides a set of tools for communicating the potential effects of forestry and other sources of impaired water quality to policy makers, regulators, land managers and other stakeholders. It provides a "dashboard" for the forestry sector and decision makers to quantify, assess and communicate water quality-related risks associated with forestry activities on a level that is understandable. This may be especially helpful in linking top-down and bottom-up initiatives to maintain or improve forest water quality.

The connection between WFD measures for achieving good ecological status and Swedish forestry is quite weak, without any real guidance about programs of measures to improve water quality (Futter *et al.* 2012; Berglund 2014). As it is today, the WFD mandates ecological status assessment on the basis of deviations from reference conditions. As shown in **Figure 8**, there are significant uncertainties associated with forestry effects on water quality. The uncertainty in reference condition estimates reduces the credibility of water management systems and complicates communication with stakeholders in the forest sector. Furthermore, the relatively short-5-year planning cycles in the WFD may be inappropriate for forest management based on a whole rotation. It has been suggested that 100-year planning cycles would be more appropriate in the WFD (Josefsson 2012). This would be more consistent with the 60–120-year rotation period used in forest planning.

The WFD enshrines the "polluter pays principle", the goal of which is to ensure that those who cause water pollution are held responsible for pollution monitoring and cleanup (Lindhout and Van den Broek 2014). Today, it is easy for relevant authorities to identify point source polluters, while sectors such as forestry and other recipients of long-range transported pollutants pose challenges in application of the polluter pays principle. While it is clear that the forestry sector should be held accountable for the direct impacts of forestry related water pollution, the responsibility of actors in the forestry sector for water quality impairments caused by other anthropogenic actions is less clear. While forestry operations should be as environmentally sensitive as possible, it does not seem entirely appropriate to hold the forest industry responsible for the legacy of impaired water quality caused by long-range pollutant transport. Forestry measures to maintain or improve water quality should focus on Type III issues where forest management is the main cause of water quality impairments. Remediation of Type II water quality issues caused primarily by other anthropogenic pressures cannot be the sole responsibility of the forestry sector.

Top-down, regulatory approaches to water quality management must be complemented by non-policy options such as forest certification (Lattimore *et al.* 2009) and bottom-up initiatives. For example, Nordlund *et al.* (2014) report on attitudes of forest machine operators to soil disturbance associated with driving damage. In general, machine operators were sensitive to and aware of the potential for driving damage and water quality impairment. The results of the framework analysis presented here, showing the potentially severe negative consequences of forestry activities on sediment mobilization could help to reinforce the sense of stewardship already felt by some actors in the forest sector. Specifically, forestry operations should be conducted in a manner which minimize sediment loads to surface waters. This could include hydro-mapping measures (Laudon *et al.* 2016) such as water sensitive driving, better road planning and use of brash to minimize soil compression.

Furthermore, separating the effects of forestry from other anthropogenic stressors could help to achieve more ethical forest management. Berglund (2014) notes that participatory approaches are needed in forest management. The conceptual framework presented here can be used as a simple tool to facilitate dialogue between the forestry sector, relevant authorities and other stakeholders so as to achieve a deliberative democracy and work towards consensually agreed upon goals as prescribed by the WFD. We believe that this framework could aid in the democratic process by allowing all stakeholders to rank and communicate the effects a management decision may have on forest surface waters. Newig *et al.* (2005) have stated that public participation is a key component for reducing uncertainties in the WFD planning and implementation process. This framework will encourage participation from local to national levels and present the effects of forestry while at the same time facilitating active involvement from stakeholders. Furthermore, the simplicity of this approach offers the possibility to use the conceptual framework outside Europe and for sectors other than forestry.

5. Conclusions

Water quality in Swedish forests is generally good, and the effects of modern forestry are often relatively minor when compared to other industries and to past forestry activities. This does not mean we can be complacent. Any forestry activity leading to increased sediment mobilization can have serious negative consequences and the legacy of OMP and Hg deposition is a persistent and pernicious threat to water quality, whether forestry occurs or not. Also, any activity which results in increased nutrient fluxes to the Baltic is a concern. Climate change and increasing demands for bio-energy may alter forest management strategies, leading to more N, P, Hg and sediment pollution. Lastly, overcoming the legacy of forest ditching may be difficult or impossible. However, the simple conceptual framework presented here creates an opportunity for relevant authorities, actors and other stakeholders to identify, rank and communicate potential effects of forestry at local, regional and national scales. It also gives the forestry sector the opportunity to measure its effects (direct and indirect) against long-range pollution. By identifying those responsible for impaired water quality, appropriate measures for enforcing the polluter pays principle can be developed and appropriate remediation measures can be taken.

Acknowledgements

We are grateful to our colleagues both within and outside Future Forests for many thought-provoking discussions over the years. The ideas presented here also benefited from the many meetings and workshops held by the Nordic Centre for Advanced Research in Ecosystem Services (CAR-ES). The quality of the final manuscript was improved by comments from the editor and reviewer. We thank S. Wennermark for preparing the DWARF illustration.

Source: Futter M N, Högbom L, Valinia S, *et al.* Conceptualizing and communicating management effects on forest water quality[J]. Ambio A Journal of the Human Environment, 2016, 45(2):188–202.

References

[1] Aherne, J., M. Posch, M. Forsius, J. Vuorenmaa, P. Tamminen, M. Holmberg, and M. Johansson. 2008. Modelling the hydrogeochemistry of acid-sensitive catchments in Finland under atmospheric deposition and biomass harvesting scenarios. Biogeochemistry 88: 233–256.

[2] Åkerblom, S., A. Bignert, M. Meili, L. Sonesten, and M. Sundbom. 2014. Half a century of changing mercury levels in Swedish freshwater fish. Ambio 43: 91–103.

[3] Akselsson, C., O. Westling, H. Sverdrup, J. Holmqvist, G. Thelin, E. Uggla, and G. Malm. 2007. Impact of harvest intensity on longterm base cation budgets in Swedish forest soils. Water, Air, & Soil Pollution: Focus 7: 201–210.

[4] Akselsson, C., H. Hultberg, P.E. Karlsson, G.P. Karlsson, and S. Hellsten. 2013. Acidification trends in south Swedish forest soils 1986–2008—Slow recovery and high sensitivity to sea-salt episodes. Science of the Total Environment 444: 271–287.

[5] Assefa, A.T., M. Tysklind, A. Sobek, K.L. Sundqvist, P. Geladi, and K. Wiberg. 2014. Assessment of PCDD/F source contributions in baltic sea sediment core records. Environmental Science and Technology 48: 9531–9539.

[6] Bergknut, M., K. Wiberg, and J. Klaminder. 2011a. Vertical and lateral redistribution of POPs in soils developed along a hydrological gradient. Environmental Science and Technology 45: 10378–10384.

[7] Bergknut, M., H. Laudon, S. Jansson, A. Larsson, T. Gocht, and K. Wiberg. 2011b. Atmospheric deposition, retention, and stream export of dioxins and PCBs in a pristine boreal catchment. Environmental Pollution 159: 1592–1598.

[8] Berglund, E. 2014. Forest and water governance in Sweden. SLU Masters Thesis.

[9] Bergström, A.K., A. Jonsson, and M. Jansson. 2008. Phytoplankton responses to nitrogen and phosphorus enrichment in unproductive Swedish lakes along a gradient of atmospheric nitrogen deposition. Aquatic Biology 4: 55–64.

[10] Beven, K. 2010. Environmental modelling: An uncertain future? Abindgon: Routledge.

[11] Bishop, K., C. Allan, L. Bringmark, E. Garcia, S. Hellsten, L. Högbom, K. Johansson, A. Lomander, *et al.* 2009. The effects of forestry on Hg bioaccumulation in nemoral/boreal waters and recommendations for good silvicultural practice. Ambio 38: 373–380.

[12] Brandt, M., H. Ejhed, and L. Rapp. 2008. Nutrient loading to the Baltic Sea and the Swedish West Coast 2006. Sweden's contribution to HELCOM. Pollution Load Compilation. Swedish Environment Protection Agency Report, 5815.

[13] Burdon, F.J., A.R. McIntosh, and J.S. Harding. 2013. Habitat loss drives threshold response of benthic invertebrate communities to deposited sediment in agricultural streams. Ecological Applications 23: 1036–1047.

[14] Burrows, R.M., E.R. Hotchkiss, M. Jonsson, H. Laudon, B.G. McKie, and R.A. Sponseller. 2015. Nitrogen limitation of heterotrophic biofilms in boreal streams. Freshwater Biology. doi:10.1111/ fwb.12549.

[15] Conley, D.J. 2012. Ecology: Save the Baltic Sea. Nature 486: 463–464.

[16] de Wit, H.A., A. Granhus, M. Lindholm, M.J. Kainz, Y. Lin, H.F.V. Braaten, and J. Blaszczak. 2014. Forest harvest effects on mercury in streams and biota in Norwegian boreal catchments. Forest Ecology and Management 324: 52–63.

[17] Di Guardo, A., S. Zaccara, B. Cerabolini, M. Acciarri, G. Terzaghi, and D. Calamari. 2003. Conifer needles as passive biomonitors of the spatial and temporal distribution of DDT from a point source. Chemosphere 52: 789–797.

[18] EC 2000. Directive 2000/60/EC of the European Parliament and on the Council of 23 October 2000 establishing a framework for Community action in the field of water policy. (Official Journal L, 327).

[19] EC 2003a. Common implementation strategy for the Water Framework Directive (2000/60/EC), Guidance Document No. 10, Rivers and Lakes—Typology, Reference Conditions and Classification Systems. Produced by Working Group 2.3— REF-COND.

[20] EC 2003b. Guidance on Public Participation in Relation to the Water Framework Directive. Active Involvement, Consultation, and Public Access to Information. Office for Official Publications of the European Communities: Luxembourg.

[21] Egnell, G., H. Laudon, and O. Rosvall. 2011. Perspectives on the potential contribution of Swedish forests to renewable energy targets in Europe. Forests 2: 578–589.

[22] Eklöf, K., R. Lidskog, and K. Bishop. 2016. Managing Swedish forestry's impact on mercury in fish: Defining the impact and mitigation measures. Ambio. doi:10.1007/s13280-015-0752-7.

[23] Ellison, D., M.N. Futter, and K. Bishop. 2012. On the forest cover-water yield debate: From demand-to supply-side thinking. Global Change Biology 18: 806–820.

[24] Foden, J., S.I. Rogers, and A.P. Jones. 2008. A critical review of approaches to aquatic environmental assessment. Marine Pollution Bulletin 56: 1825–1833.

[25] Fölster, J., C. Andre'n, K. Bishop, I. Buffam, N. Cory, W. Goedkoop, K. Holmgren, R.K. Johnson, *et al.* 2007. A novel environmental quality criterion for acidification in Swedish lakes—An application of studies on the relationship between biota and water chemistry. Water, Air, and Soil pollution 7: 331–338.

[26] Futter, M.N., E. Ring, L. Högbom, S. Entenmann, and K. Bishop. 2010. Consequences of nitrate leaching following stem-only harvesting of Swedish forests are dependent on spatial scale. Environmental Pollution 158: 3552–3559.

[27] Futter, M.N., E.C.H. Keskitalo, D. Ellison, M. Pettersson, A. Strom, E. Andersson, J. Nordin, S. Löfgren, *et al.* 2011. Forests, forestry and the Water Framework Directive in Sweden: A transdisciplinary commentary. Forests 2: 261–282.

[28] Futter, M.N., J. Klaminder, R.W. Lucas, H. Laudon, and S.J. Köhler. 2012. Uncertainty in silicate mineral weathering rate estimates: Source partitioning and policy implications. Environmental Research Letters 7(2): 024025.

[29] Futter, M.N., S. Valinia, S. Löfgren, S.J. Kö hler, and J. Fö lster. 2014. Long-term trends in water chemistry of acid-sensitive Swedish lakes show slow recovery from historic acidification. Ambio 43: 77–90.

[30] Garcia, E., and R. Carignan. 2005. Mercury concentrations in fish from forest harvesting and fire-impacted Canadian Boreal lakes compared using stable isotopes of nitrogen. Environmental Toxicology and Chemistry 24: 685–693.

[31] Harding, J.S., E.F. Benfield, P.V. Bolstad, G.S. Helfman, and E.B.D. Jones. 1998. Stream biodiversity: The ghost of land use past. Proceedings of the National Academy of Sciences 95: 14843–14847.

[32] Hargrave, B.T. 2002. A traffic light decision system for marine finfish aquaculture siting. Ocean and Coastal Management 45: 215–235.

[33] Hatton-Ellis, T. 2008. The Hitchhiker's Guide to the Water Frame-work Directive. Aquatic Conservation-Marine and Freshwater Ecosystems 18: 111–116.

[34] Helliwell, R.C., J. Aherne, T.R. Nisbet, G. MacDougall, S. Broad-meadow, J. Sample, L. Jackson-Blake, and R. Doughty. 2014. Modelling the long-term response of stream water chemistry to forestry in Galloway, Southwest Scotland. Ecological Indicators 37: 396–411.

[35] Hering, D., A. Borja, J. Carstensen, L. Carvalho, M. Elliott, C.K. Feld, A.-S. Heiskanen, R.K. Johnson, *et al.* 2010. The European Water Framework Directive at the age of 10: a critical review of the achievements with recommendations for the future. Science of the Total Environment 408: 4007–4019.

[36] Humborg, C., D.J. Conley, L. Rahm, F. Wulff, A. Cociasu, and V. Ittekkot. 2000. Silicon retention in river basins: Far-reaching effects on biogeochemistry and aquatic food webs in coastal marine environments. Ambio 29: 45–50.

[37] Iacob, O., J.S. Rowan, I. Brown, and C. Ellis. 2014. Evaluating wider benefits of natural flood management strategies: An ecosystem-based adaptation perspective. Hy-

drology Research 45: 774–787.

[38] Jonard, M., A. Fürst, A. Verstraeten, A. Thimonier, V. Timmermann, N. Potočič, P. Waldner, S. Benham, *et al.* 2015. Tree mineral nutrition is deteriorating in Europe. Global Change Biology 21: 418–430.

[39] Josefsson, H. 2012. Achieving ecological objectives. Laws 1: 39–63.

[40] Kaila, A., S. Sarkkola, A. Laure'n, L. Ukonmaanaho, H. Koivusalo, L. Xiao, C. O'Driscoll, Z.-U.-Z.-Asam, *et al.* 2014. Phosphorus export from drained Scots pine mires after clear-felling and bioenergy harvesting. Forest Ecology and Management 325: 99–107.

[41] Keskitalo, E.C.H. 2015. Actors' perceptions of issues in the implementation of the first round of the Water Framework Directive: Examples from the water management and forestry sectors in southern Sweden. Water 7: 2202–2213.

[42] Kothawala, D.N., S.A. Watmough, M.N. Futter, L. Zhang, and P.J. Dillon. 2011. Stream nitrate responds rapidly to decreasing nitrate deposition. Ecosystems 14: 274–286.

[43] Klaminder, J., R.W. Lucas, M.N. Futter, K. Bishop, S.J. Köhler, G. Egnell, and H. Laudon. 2011. Silicate mineral weathering rate estimates: Are they precise enough to be useful when predicting the recovery of nutrient pools after harvesting? Forest Ecology and Management 261: 1–9.

[44] Kreutzweiser, D.P., P.W. Hazlett, and J.M. Gunn. 2008. Logging impacts on the biogeochemistry of boreal forest soils and nutrient export to aquatic systems: A review. Environmental Reviews 16: 157–179.

[45] Kreutzweiser, D.P., S. Capell, K. Good, and S. Holmes. 2009. Sediment deposition in streams adjacent to upland clearcuts and partially harvested riparian buffers in boreal forest catchments. Forest Ecology and Management 258: 1578–1585.

[46] Lattimore, B., C.T. Smith, B.D. Titus, I. Stupak, and G. Egnell. 2009. Environmental factors in woodfuel production: Opportunities, risks, and criteria and indicators for sustainable practices. Biomass and Bioenergy 33: 1321–1342.

[47] Launiainen, S., M.N. Futter, D. Ellison, N. Clarke, L. Fine'r, L. Högbom, A. Laure'n, and E. Ring. 2014. Is the water footprint an appropriate tool for forestry and forest products: The Fennoscandian case. Ambio 43: 244–256.

[48] Laudon, H., L. Kuglerova', R.A. Sponseller, M. Futter, A. Nordin, K. Bishop, T. Lundmark, G. Egnell, and A.M. Ågren. 2016. The role of biogeochemical hotspots, landscape heterogeneity and hydrological connectivity for minimizing forestry effects on water quality. Ambio. doi:10.1007/s13280-015-0751-8.

[49] Laudon, H., R.A. Sponseller, R.W. Lucas, M.N. Futter, G. Egnell, K. Bishop, A. Ågren, E. Ring, *et al.* 2011. Consequences of more intensive forestry for the sustainable management of forest soils and waters. Forests 2: 243–260.

[50] Lepistö, A., M.N. Futter, and P. Kortelainen. 2014. Almost 50 years of monitoring

shows that climate, not forestry, controls longterm organic carbon fluxes in a large boreal watershed. Global Change Biology 20: 1225–1237.

[51] Lindhout, P.E., and B. Van den Broek. 2014. The polluter pays principle: Guidelines for cost recovery and burden sharing in the case law of the European court of justice. Utrecht Law Review 10: 46–59.

[52] Lö fgren, S., M. Kahlert, M. Johansson, and J. Bergengren. 2009. Classification of two Swedish forest streams in accordance with the European Union Water Framework Directive. Ambio 38: 394–400.

[53] Lundborg, A. 1997. Reducing the nitrogen load: whole-tree harvesting: A literature review. Ambio 26: 387–393.

[54] Manninen, P. 1998. Effects of forestry ditch cleaning and supplementary ditching on water quality. Boreal Environment Research 3: 23–32.

[55] Martin, J. 2014. Forest harvest effects on mercury in European perch. SLU Masters Thesis.

[56] Meyer-Jacob, C., J. Tolu, C. Bigler, H. Yang, and R. Bindler. 2015. Early land use and centennial scale changes in lake-water organic carbon prior to contemporary monitoring. Proceedings of the National Academy of Sciences, United States of America 112: 6579–6584.

[57] Moldan, F., B.J. Cosby, and R.F. Wright. 2013. Modeling past and future acidification of Swedish lakes. Ambio 42: 577–586.

[58] Monteith, D.T., J.L. Stoddard, C.D. Evans, H.A. de Wit, M. Forsius, T. Høgåsen, A. Wilander, B.L. Skjelkvåle, *et al.* 2007. Dissolved organic carbon trends resulting from changes in atmospheric deposition chemistry. Nature 450: 537–540.

[59] Moss, B. 2008. The Water Framework Directive: Total environment or political compromise? Science of the Total Environment 400: 32–41.

[60] Neary, D.G., G.G. Ice, and C.R. Jackson. 2009. Linkages between forest soils and water quality and quantity. Forest Ecology and Management 258: 2269–2281.

[61] Newig, J., C. Pahl-Wostl, and K. Sigel. 2005. The role of public participation in managing uncertainty in the implementation of the Water Framework Directive. European Environment 15: 333–343.

[62] Nilsson, C., F. Lepori, B. Malmqvist, E. Törnlund, N. Hjerdt, J.M. Helfield, D. Palm, J. Ö stergren, *et al.* 2005. Forecasting environmental responses to restoration of rivers used as log floatways: an interdisciplinary challenge. Ecosystems 8: 779–800.

[63] Nizzetto, L., C. Cassani, and A. Di Guardo. 2006. Deposition of PCBs in mountains: The forest filter effect of different forest ecosystem types. Ecotoxicology and Environmental Safety 63: 75–83.

[64] Nordlund, A., E. Ring, L. Högbom, and I. Bergkvist. 2014. Beliefs among formal actors in the Swedish forestry related to rutting caused by logging operations,

807–2013. Arbetsrapport nr: Skogforsk.

[65] Omstedt, A., M. Edman, L.G. Anderson, and H. Laudon. 2010. Factors influencing the acid-base (pH) balance in the Baltic Sea: A sensitivity analysis. Tellus series B-Chemical and Physical Meteorology 62: 280–295.

[66] Oni, S.K., M.N. Futter, C. Teutschbein, and H. Laudon. 2014. Crossscale ensemble projections of dissolved organic carbon dynamics in boreal forest streams. Climate Dynamics 42: 2305–2321.

[67] Österling, M., and J.O. Högberg. 2014. The impact of land use on the mussel Margaritifera margaritifera and its host fish Salmo trutta. Hydrobiologia 735: 213–220.

[68] Österling, M.E., B.L. Arvidsson, and L.A. Greenberg. 2010. Habitat degradation and the decline of the threatened mussel Margaritifera margaritifera: Influence of turbidity and sedimentation on the mussel and its host. Journal of Applied Ecology 47: 759–768.

[69] Palviainen, M., and L. Fine'r. 2012. Estimation of nutrient removals in stem-only and whole-tree harvesting of Scots pine, Norway spruce, and birch stands with generalized nutrient equations. European Journal of Forest Research 131: 945–964.

[70] Palviainen, M., M. Starr, and C.J. Westman. 2012. The effect of site fertility and climate on current weathering in Finnish forest soils: Results of a 10-16 year study using buried crushed test-rock material. Geoderma 183: 58–66.

[71] Palviainen, M., L. Fine'r, A. Laure'n, T. Mattsson, and L. Högbom. 2015. A method to estimate the impact of clear-cutting on nutrient concentrations in boreal headwater streams. Ambio 44: 521–531.

[72] Peters, E., N. Dieckmann, A. Dixon, J.H. Hibbard, and C.K. Mertz. 2007. Less is more in presenting quality information to consumers. Medical Care Research and Review 64: 169–190.

[73] Petersen, A.C. 2012. Simulating nature: A philosophical study of computer-simulation uncertainties and their role in climate science and policy advice, 2nd ed. Boca Raton, Florida: CRC Press, 208 pp.

[74] Porvari, P., M. Verta, J. Munthe, and M. Haapanen. 2003. Forestry practices increase mercury and methyl mercury output from boreal forest catchments. Environmental Science and Technology 37: 2389–2393.

[75] Rask, M., L. Arvola, M. Forsius, and J. Vuorenmaa. 2014. Preface to the special issue" Integrated monitoring in the Valkea-Kotinen catchment during 1990-2009: Abiotic and biotic responses to changes in air pollution and climate. Boreal Environment Research 19: 1–3.

[76] Renberg, I., C. Bigler, R. Bindler, M. Norberg, J. Rydberg, and U. Segerström. 2009. Environmental history: A piece in the puzzle for establishing plans for environmental management. Journal of Environmental Management 90: 2794–2800.

[77] Schelker, J., K. Eklöf, K. Bishop, and H. Laudon. 2012. Effects of forestry operations

on dissolved organic carbon concentrations and export in boreal first-order streams. Journal of Geophysical Research: Biogeosciences 117: G01011. doi:10.1029/2011JG001827.

[78] Schelker, J., L. Kuglerova, K. Eklöf, K. Bishop, and H. Laudon. 2013. Hydrological effects of clear-cutting in a boreal forest—Snowpack dynamics, snowmelt and streamflow responses. Journal of Hydrology 484: 105–114.

[79] Serrano, I., I. Buffam, D. Palm, E. Brännäs, and H. Laudon. 2008.

[80] Thresholds for survival of brown trout during the spring flood acid pulse in streams high in dissolved organic carbon. Transactions of the American Fisheries Society 137: 1363–1377.

[81] Siipi, H. 2008. Dimensions of naturalness. Ethics and the Environment 13: 71–103.

[82] Skogsstyrelsen, 2014. Swedish Statistical Yearbook of Forestry.

[83] Skyllberg, U., M.B. Westin, M. Meili, and E. Bjorn. 2009. Elevated concentrations of methyl mercury in streams after forest clearcut: A consequence of mobilization from soil or new methylation? Environmental Science and Technology 43: 8535–8541.

[84] Solomon, C.T., S.E. Jones, B.C. Weidel, I. Buffam, M.L. Fork, J. Karlsson, S. Larsen, J.T. Lennon, *et al.* 2015. Ecosystem consequences of changing inputs of terrestrial dissolved organic matter to lakes: Current knowledge and future challenges. Ecosystems 18: 376–389.

[85] Sponseller, R.A., M.J. Gundale, M. Futter, E. Ring, A. Nordin, T. Näsholm, and H. Laudon. 2016. Nitrogen dynamics in managed boreal forests: Recent advances and future research directions. Ambio. doi:10.1007/s13280-015-0755-4.

[86] Sponseller, R.A., J. Temnerud, K. Bishop, and H. Laudon. 2014. Patterns and drivers of riverine nitrogen (N) across alpine, subarctic, and boreal Sweden. Biogeochemistry 120: 105–120.

[87] Stenberg, L., L. Fine'r, M. Nieminen, S. Sarkkola, and H. Koivusalo. 2015. Quantification of ditch bank erosion in a drained forested catchment. Boreal Environment Research 20: 1–18.

[88] Thiffault, E., K.D. Hannam, D. Pare', B.D. Titus, P.W. Hazlett, D.G. Maynard, and S. Brais. 2011. Effects of forest biomass harvesting on soil productivity in boreal and temperate forests—A review. Environmental Reviews 19: 278–309.

[89] Valinia, S., M.N. Futter, B.J. Cosby, P. Rose'n, and J. Fölster. 2015. Simple models to estimate historical and recent changes of total organic carbon concentrations in lakes. Environmental Science and Technology 49: 386–394.

[90] Valinia, S., G. Englund, F. Moldan, M.N. Futter, S.J. Köhler, K. Bishop, and J. Fö lster. 2014. Assessing anthropogenic impact on boreal lakes with historical fish species distribution data and hydrogeochemical modeling. Global Change Biology 20: 2752–2764.

[91] Valinia, S., H.P. Hansen, M.N. Futter, K. Bishop, N. Sriskandarajah, and J. Fölster. 2012. Problems with the reconciliation of good ecological status and public participation in the Water Framework Directive. Science of the Total Environment 433: 482–490.

[92] Vor, T., J. Dyckmans, N. Loftfield, F. Beese, and H. Flessa. 2003. Aeration effects on CO2, N2O, and CH4 emission and leachate composition of a forest soil. Journal of Plant Nutrition and Soil Science 166: 39–45.

[93] Wood, P.J., and P.D. Armitage. 1997. Biological effects of fine sediment in the lotic environment. Environmental Management 21: 203–217.

[94] Zetterberg, T., B.A. Olsson, S. Löfgren, C. von Brömssen, and P.-O. Brandtberg. 2013. The effect of harvest intensity on long-term calcium dynamics in soil and soil solution at three coniferous sites in Sweden. Forest Ecology and Management 302: 280–294.

Chapter 14

Economic Benefits of Methylmercury Exposure Control in Europe: Monetary Value of Neurotoxicity Prevention

Martine Bellanger[1], Céline Pichery[1], Dominique Aerts[2], Marika Berglund[3], Argelia Castaño[4], Mája Čejchanová[5], Pierre Crettaz[6], Fred Davidson[7], Marta Esteban[4], Marc E. Fischer[8], Anca Elena Gurzau[9], Katarina Halzlova[10], Andromachi Katsonouri[11], Lisbeth E Knudsen[12], Marike Kolossa-Gehring[13], Gudrun Koppen[14], Danuta Ligocka[15], Ana Miklavčič[16], M. Fátima Reis[17], Peter Rudnai[18], Janja Snoj Tratnik[16], Pál Weihe[19], Esben Budtz-Jørgensen[12], Philippe Grandjean[20,21], DEMO/COPHES

[1]EHESP School of Public Health, Rennes Cedex, France
[2]FPS Health, Food Chain Safety and Environment, Brussels, Belgium
[3]Karolinska Institutet, Stockholm, Sweden
[4]Instituto de Salud Carlos III, Majadahonda, Madrid, Spain
[5]National Institute of Public Health, Prague, Czech Republic
[6]EDI, Bundesamt für Gesundheit, Liebefeld, Switzerland
[7]Health Service Executive South, Cork, Ireland
[8]Laboratoire National de Santé, Luxembourg, Luxembourg
[9]Environmental Health Center, Cluj-Napoca, Romania
[10]Public Health Authority of the Slovak Republic, Bratislava, Slovakia
[11]Cyprus State General Laboratory, Nicosia, Cyprus

[12]Department of Public Health, University of Copenhagen, Copenhagen, Denmark
[13]Umweltbundesamt, Berlin, Germany
[14]Flemish Institute for Technological Research, Mol, Belgium
[15]Nofer Institute of Occupational Medicine, Lodz, Poland
[16]Jožef Stefan Institute, Ljubljana, Slovenia
[17]Faculdade de Medicina de Lisboa, Lisboa, Portugal
[18]National Institute of Environmental Health, Budapest, Hungary
[19]Faroese Hospital System, Tórshavn, Faroe Islands
[20]Institute of Publicinterventions to minimize exposure to this hazardous pollutant. Health, University of Southern Denmark, Odense, Denmark
[21]Department of Environmental Health, Harvard School of Public Health, Boston, MA, USA

Abstract: Background: Due to global mercury pollution and the adverse health effects of prenatal exposure to methylmercury (MeHg), an assessment of the economic benefits of prevented developmental neurotoxicity is necessary for any cost-benefit analysis. Methods: Distributions of hair-Hg concentrations among women of reproductive age were obtained from the DEMOCOPHES project (1,875 subjects in 17 countries) and literature data (6,820 subjects from 8 countries). The exposures were assumed to comply with log-normal distributions. Neurotoxicity effects were estimated from a linear dose-response function with a slope of 0.465 Intelligence Quotient (IQ) point reduction per μg/g increase in the maternal hair-Hg concentration during pregnancy, assuming no deficits below a hair-Hg limit of 0.58μg/g thought to be safe. A logarithmic IQ response was used in sensitivity analyses. The estimated IQ benefit cost was based on lifetime income, adjusted for purchasing power parity. Results: The hair-mercury concentrations were the highest in Southern Europe and lowest in Eastern Europe. The results suggest that, within the EU, more than 1.8 million children are born every year with MeHg exposures above the limit of 0.58μg/g, and about 200,000 births exceed a higher limit of 2.5μg/g proposed by the World Health Organization (WHO). The total annual benefits of exposure prevention within the EU were estimated at more than 600,000 IQ points per year, corresponding to a total economic benefit between €8,000 million and €9,000 million per year. About four-fold higher values were obtained when using the logarithmic response function, while adjustment for productivity resulted in slightly lower total benefits. These calculations do not include the less tangible advantages of protecting brain development against neurotoxicity or any other adverse effects. Conclusions: These estimates document that efforts to combat mercury pollution and to reduce MeHg exposures will have very substantial economic benefits in Europe, mainly in southern countries. Some data may not be entirely representative, some countries were not covered, and anticipated changes in mercury pollution all suggest a need for extended biomonitoring of human MeHg exposure.

Keywords: Economic Evaluation, Methylmercury, Prenatal Exposure, Neuro Developmental Deficits

1. Background

Methylmercury (MeHg) is a well-documented neurotoxicant, and prenatal exposures are therefore of particular concern[1][2]. The main sources of exposure are seafood and freshwater fish[3]. Thus, MeHg exposures vary with dietary habits, contamination levels, and species availability. While the distribution of MeHg exposures has been studied in substantial detail in the United States[4], only scattered information is available on MeHg exposures in Europe.

Because the critical effect of MeHg exposure is developmental brain toxicity, exposures among women of reproductive age groups are of primary concern[5][6]. As has previously been determined in regard to lead exposure[7], developmental MeHg exposure is linked to a loss in Intelligence Quotient (IQ), with associated lower school performance and educational attainment, thereby leading to long-term impacts on societal benefits of pollution abatement[8]. These consequences may be expressed in terms of economic impacts, as has been demonstrated in United States[9][10]. However, few economic evaluations have been performed in Europe[8][11][12], primarily because of the lack of exposure data.

Based on harmonised protocols developed in COPHES[13], the DEMOCOPHES project has just completed a multi-country study of hair-mercury concentrations in women of reproductive age groups in 17 European countries. In conjunction with literature data, we now utilise the exposure data to generate estimates of economic impacts of MeHg exposures in Europe.

The economic assessment relies on several assumptions. The hair-Hg concentrations is used as the main exposure indicator in this study, and any blood-based measurements also considered are expressed in terms of hair-mercury using a conversion factor of 250[14][15]. In regard to the dose-response function (DRF), a linear model is usually the default[14], although it may not necessarily provide the best statistical fit to the data[16]. We therefore used the linear slope as the primary DRF and then conducted a sensitivity analysis using the log function, where each doubling of exposure above the background causes the same deficit of 1.5 IQ points[10].

With regard to background exposures and the possible existence of a threshold, the U.S. EPA's Reference Dose (RfD) of 0.1μg/kg body weight/day corresponds to a hair-Hg concentration of about 1μg/g hair[14]. Updated calculations[17] resulted in an adjusted biological limit about 50% below the recommended level, corresponding to 0.58μg/g hair. The validity of this lower cut-off point below the RfD is supported by recent studies of developmental neurotoxicity at exposure levels close to the background[18]-[21]. We assumed that, below the 0.58μg/g cut-off point, only negligible adverse effects would exist. As additional reference point, we use a tolerable limit proposed by the World Health Organization (WHO), which corresponds to a hair-Hg concentration of approximately 2.5μg/g[22]. This limit takes into account the possible compensation of MeHg toxicity by beneficial nutrients in seafood[22].

2. Methods

2.1. Exposure Information

DEMOCOPHES is a cross-sectional survey of European population exposure to environmental chemicals. The human exposure biomarkers included the hair-mercury concentration and was collected in 17 European countries based on children aged 6–11 years and their mothers. A common European protocol, developed by the COPHES project, was followed in each country. The main inclusion and exclusion criteria were (1) residence in the study area for at least five years, and (2) not having metabolic disturbances. The period of sampling was September 2011 to February 2012. A total of 1,875 child-mother pairs were recruited from urban and rural communities in the participating countries, while excluding exposure hot-spots. Major efforts were carried out to achieve high quality and comparability of data. Standard operational procedures for total mercury concentrations in hair were developed and validated by the Laboratory of Environmental Toxicology in Spain, to ensure comparable measurements, which included a strict quality assurance programme, in which seventeen European laboratories participated. Each DEMO-COPHES partner contributed information to allow estimation of the underlying distribution of exposures in the population, where rural and urban results were merged. In addition, each partner provided the frequencies of results above the cut-off levels of 0.58μg/g, 1.0μg/g, and 2.5μg/g. The latter corresponds to WHO's tolerable limit, which takes into account likely toxicity compensation by beneficial nutrients in seafood[22].

Additional information on MeHg exposures in Europe was obtained to complement the DEMOCOPHES data. Thus, information of similar quality was extracted from published articles (Miklavčič, unpublished data), and distribution information from comparable studies was obtained from Belgium, Denmark, France, Norway, Slovenia, and the United Kingdom. As explained below, missing information was calculated assuming a log-normal distribution of the exposures.

2.2. Exposure Distributions

Using the number of births in 2008 and the observed hair-Hg concentrations, we estimated the number of births exceeding the three exposure limits for each country and obtained the sum for all of the EU. For missing EU member states, MeHg exposures were assumed to be the same as a neighbouring country. The year 2008 was chosen as the closest to the time during which the exposure data had been collected, and it allowed complete information for the calculations envisaged. Due to the existence of sampling uncertainty, "smoothed" proportions exceeding the three limits were calculated assuming log-normal distributions. Because log-transformed concentrations would follow a normal distribution, the parameters in the log-normal distributions could be estimated by standard normal distribution methods. Each data set included probabilities (prob) for being below specific percentiles (perc). The parameters in the logarithmic distributions were therefore obtained as the intercept and slope when regressing log(perc) on $\Phi - 1$ (prob), where Φ is the cumulative distribution function of the standard normal distribution. Using the total numbers of births in 2008, numbers of births exceeding the three cut-off limits in each country were calculated from observed and smoothed distributions.

2.3. Calculation of IQ Benefits

A linear dose-response function was applied as the default model[14]. Thus, as a 1μg/L increase of the cord-blood mercury concentration is associated with an average adverse impact on IQ of 0.093 times the standard deviation (which is standardised to be 15), each increase in the maternal hair-mercury by 1μg/g is associated with an average loss of 0.465 IQ points[10]. This slope is based on a range of neuropsychological tests and subtests administered in the Faroe Islands study at age 7 years[23]. As some recent studies[18]-[21] suggest MeHg-associated deficits close to or below the cut-off level of 0.58μg/g hair, the calculations may represent an underestimate. In addition, the slope may be steeper at low exposure levels. Thus, a log model was used for sensitivity analyses. In this model, a doubling in prenatal MeHg exposures is associated with a delay in development of 1.5–2 months at age 7, which corresponds to about 10% of the standard deviation, *i.e.* 1.5 IQ points[1]. Again, we applied this slope for exposures above the 0.58μg/g the cut-off point.

To estimate the benefits at exposures above the cut-off point, we calculated the average hair-mercury concentration in women exceeding 0.58μg/g based on 1,000,000 simulations from the estimated log-normal distribution (as described above). After deduction of the 0.58μg/g and multiplication by the slope factor, an average IQ benefit was obtained. This amount was then multiplied by the annual number of births exceeding the cut-off level. A similar calculation was made in the logarithmic dose-response model except that here we calculated the average log-transformed mercury concentration in women exceeding 0.58μg/g, deducted log(0.58) and multiplied by the slope factor of the logarithmic dose-response model [1.5/log(2)].

2.4. Annual Benefits of Exposure Reduction

The major component of the social costs incurred by an IQ reduction is loss of productivity and thus a lower earning potential[9][24]. The economic consequence of prenatal exposure to MeHg is valued as the lifetime earning loss per person. We assumed singleton births only, so that the number of women was equal to the cohort size. We also assumed that IQ deficits present at age 7 years or preschool ages are permanent[25]. The estimated individual benefits are the avoided lifetime costs using 2008 data (slightly lower benefits are obtained if referring to more recent years, and benefits are only minimally affected by subsequent membership of the Euro zone). The benefit estimates originate from the 2008 figure of €17,363 per IQ point as recently calculated for France based on data from the United States[24]. For the various European countries involved, this value is adjusted for differences in purchasing power. While simple currency exchange conversion and Gross Domestic Product (GDP) per capita do not adjust for price differences, Purchasing Power Parity (PPP) conversion rates allow for comparison based on a common set of average international prices[26][27]. We also carried out the calculations after adjustment for productivity as the ratio of PPP-adjusted real GDP/capita in each country in relation to the US as a reference. The estimated value of an IQ point then takes into account the impact of labour costs and productivity (Additional file 1).

3. Results

Table 1 and Additional file 2 show summary information on MeHg exposures in the European countries covered by DEMOCOPHES or other exposure studies. There is a clear trend from north and east to southern countries, most likely due to differences in dietary habits and availability of large fish species from the Mediterranean (the sources of exposure were not considered in the present study). In **Table 1**, exposures in Austria were assumed to be similar to those in Germany, as suggested by available data[28]. Exposure information from the Flemish part of Belgium[29][30] do not differ much from the national data obtained in DEMOCOPHES, which were therefore used for the calculations. The Flemish data were used to represent exposures in The Netherlands. In the absence of exposure data from Estonia, Finland, Latvia, and Lithuania, the DEMOCOPHES exposure information from Sweden was applied. National data from France are available[31] and have been used in recent economic calculations[8]. Data for Croatia and Greece were obtained from a recent birth cohort study[32]. Two exposure studies had been carried out in Italy, one in the northeast[32] and one in Naples[33], and a joint distribution was therefore used to obtain national exposure distributions that would also apply to Malta. Thus, a log-normal distribution was first fitted to each Italian data

Table 1. Annual numbers of births and numbers exceeding three cut-off limits, as indicated by hair-mercury analyses (inμg/g) in population samples in European countries.

Countrya	Annual number of births (2008)	Number of samples[b]	Above 0.58μg/g		Above 1.0μg/g		Above 2.5μg/g	
			Proportion in sample (%)	Estimated Number of births	Proportion in sample (%)	Estimated Number of births	Proportion in sample (%)	Estimated number of births
Austria	77,800	NA	(6.7)	5,213	(0.8)	622	(0)	0
Belgium	127,200	129	28.7	36,506	9.3	11,830	0	0
		242[c]	23.2	29,510	7.2	9,158	0	
Bulgaria	77,700	NA	(4.2)	3,263	(1.2)	932	(0.8)	622
Croatia	43,800	234[d]	52.0	22,776	22.0	9,636	4.7	2,059
Cyprus	9,200	60	36.7	3,376	18.3	1,684	3.3	304
Czech Republic	119,600	120	5.0	5,980	0.8	957	0	0
Denmark	65,000	145	36.6	23,790	13.1	8515	0.7	455
Estonia	16,000	NA	(10.0)	1,600	(2.0)	320	(0)	0
Faroe Islands	675	505[e]	62.6	423	30.2	204	5.3	36
Finland	59,500	NA	(10.0)	5,950	(2.0)	1,190	(0)	0
France	829,300	126[f]	44.0	364,892	14.51	120,331	0.61	5,059
Germany	682,500	120	6.7	45,728	0.8	5,460	0	0
Greece	118,300	454[d]	78	92,274	57	67,431	14	16,562
Hungary	99,100	120	0.83	823	0	0	0	0
Ireland	74,000	120	10.8	7,992	2.5	1,850	0	0
Italy	576,700	891[d]+115[g]	(65.6)	378,315	(36.8)	212,226	(5.7)	32,872
Latvia	23,834	NA	(10.0)	2,383	(2.0)	477	(0)	0
Lithuania	35,100	NA	(10.0)	3,510	(2.0)	702	(0)	0
Luxembourg	5,600	55	32.7	1,831	18.2	1,019	0	0
Malta	4,100	NA	(65.6)	2,690	(36.8)	1,509	(5.7)	234
Netherlands	184,600	NA	(23.2)	42,827	(7.2)	13,291	(0)	0

Continued

Norway	60,500	119[h]	27.7	16,759	5.9	3,570	0	0
Poland	414,500	120	1.7	7047	0	0	0	0
Portugal	104,600	120	90.8	94,977	57.5	60,145	8.3	8,682
Romania	221,900	120	4.2	9,320	1.2	2,663	0.8	1,775
Slovakia	57,400	129	5.43	3,117	0.8	459	0	0
Slovenia	21,800	156	22.0	4,796	7.7	1,679	1.9	414
Spain	519,800	120	88.5	460,023	74.2	385,692	31.7	164,777
Sweden	109,300	100	10.0	10,930	2.0	2,186	0	0
Switzerland	76,700	120	5.0	3,835	2.1	1,611	0	0
United Kingdom	794,400	4134[h]	31.0	246,264	5.1	40,200	0	0
Total EU (27)	5,400,000			1,865,416		903,169		231,754

Exposures in EU countries without recent data are estimated from neighbouring countries (modelled results not based on observed distributions are given in parenthesis). [a]For countries without available exposure data (for number of samples, NA denotes not available), data from a neighbouring country have been applied to allow, EU-wide estimates, and frequencies are given in parenthesis. This applies to Austria (data from Germany were used), Bulgaria (Romania), Netherlands (Flanders[30]), and Estonia, Finland, Latvia, and Lithuania (Sweden); [b]All data are from DEMOCOPHES, unless otherwise noted; [c][30]; [d][32]; [e]Pal Weihe, unpublished data; [f][31]; [g][33]; [h]Jean Golding, pers. comm.

subset, and then the parameters of a joint log-normal distribution were determined as the mean of the parameters for the two distributions. Recent results from the Norwegian national birth cohort were used for this country[34]. As DEMOCOPHES data from the United Kingdom covered only a small rural sample, we relied on data on blood-mercury in pregnant women obtained from the ALSPAC birth cohort study in the 1990s[35]. Additional exposure data from Ukraine[36] supported the notion that MeHg exposures in Eastern Europe are low, with only small percentages exceeding the cut-off level, but this study was considered too small to be used for detailed calculations. The same applied to several other sources identified (Miklavčič, unpublished data).

The estimated number of annual births in the EU that exceed the 0.58μg/g cut-off is about 1.8 million (**Table 1**, Additional file 3). The EPA limit is exceeded in about 900,000 births, and the WHO limit in 200,000 births within the EU. As each study is subject to sampling uncertainty, log-normal distribution models showed similar, though sometimes slightly higher, proportions exceeding the 0.58 cut-off level (**Table 2**). The data from Eastern European countries and from Croatia, the Faroe Islands, Norway, and Switzerland suggest that, within Europe, the great majority of births exceeding the various limits occur in EU member states.

Table 2 presents the estimated IQ losses associated with the MeHg exposures using the linear model, along with the estimates of economic impacts. We used both the observed data and the modelled distributions, and only small differences were seen, thus supporting the notion that the log-normal exposure distribution has an appropriate fit. The greatest benefits accrue for the largest countries with the highest proportions of subjects with exposures above the cut-off level. The total benefit from control of MeHg exposure was the highest for Spain and the lowest for Hungary. On a per capita basis, the calculated benefits are the greatest in the Faroe Islands and the southern countries, Spain, Greece, Portugal, Italy, and Croatia. The total annual benefits in terms of IQ points within the EU were estimated to be in excess of 600,000 per year for the linear DRC. With an average benefit of €13,579 per IQ point, the total economic benefits are estimated to exceed €9,000 million per year. When adjustment for productivity is included, the benefits are somewhat lower for several countries, and the EU total is slightly less than €8,000 million per year (Additional file 3).

For comparison, **Table 3** shows the estimated IQ losses and economic benefits using the log transformed DRF. Due to the steeper curve shape at exposures close to the cut-off point of 0.58μg/g, the estimated benefits are about 4-fold greater, at about 2.7 million IQ points per year, which correspond to total benefits for the EU of approximately €39,000 million or, after productivity adjustment, €33,000 million.

4. Discussion

This study provides for the first time regional European data on economic benefits of controlling MeHg exposure in relation to prevention of developmental neurotoxicity. It relies on data from a multi-country study of hair-Hg concentrations with a high level of quality assurance and with similar population sampling

Table 2. Annual number of births with excess exposure, average hair-Hg concentration, IQ benefit from prevention of excess exposure, and the value of the IQ benefits.

Country	Number of births above 0.58μg/g		Average concentration above 0.58μg/g	Benefit in IQ points		Value of 1 IQ point (Euro)	Total benefit (million Euro)	
	Modelled	Observed		Modelled	Observed		Modelled	Observed
Austria	3,812	5,213	0.917	597	817	16,044	9.6	13.1
Belgium	39,686	36,506	0.939	6,625	6,094	16,458	109.0	100.3
Bulgaria	3,186	3,263	1.455	1,296	1,328	7,529	9.8	10.0
Croatia	21,769	22,776	1.355	7,845	8,208	11,320	88.8	92.9
Cyprus	3,514	3,376	1.311	1,195	1,148	13,747	16.4	15.8
Czech Republic	5,143	5,980	0.847	639	742	10,797	6.9	8.0
Denmark	22,815	23,790	1.027	4,742	4,945	20,220	95.9	100.0
Estonia	1,840	1,600	0.846	228	198	10,339	2.4	2.0
Faroe Islands	406	423	1.323	140	146	20,220	2.8	2.9
Finland	6,843	5,950	0.846	846	736	17,288	14.6	12.7
France	405,528	364,892	0.989	70,186	69,397	17,363	1,218.6	1,204.9
Germany	33,443	45,728	0.917	5,241	7,166	15,292	80.1	109.6
Greece	94,403	92,274	1.563	50,131	49,000	13,201	661.8	646.9
Hungary	892	823	0.884	126	116	9,691	1.2	1.1
Ireland	7,104	7,992	0.946	1,209	1,360	17,927	21.7	24.4
Italy	378,315	(378,315)	1.045	81,801	(81,801)	17,062	1,395.7	(1,395.7)
Latvia	2,741	2,383	0.846	339	295	11,568	3.9	3.4
Lithuania	4,037	3,510	0.846	499	434	9,661	4.8	4.2
Luxembourg	1,870	1,831	1.212	550	538	17,062	9.4	9.2
Malta	2,690	(2,690)	1.045	582	(582)	11,111	6.5	6.5
Netherlands	45,227	42,827	0.909	6,919	6,552	15,857	109.7	103.9
Norway	16,759	16,759	0.866	2,237	2,229	20,051	44.8	44.7

Continued

Poland	6,218	7,047	0.751	494	560	9,979	4.9	5.6
Portugal	94,349	94,977	1.482	39,573	39,836	12,221	483.6	486.8
Romania	9,098	9,320	1.455	3,702	3,797	8,187	30.3	31.1
Slovakia	2,468	3,117	0.899	366	462	10,037	3.7	4.6
Slovenia	4,840	4,796	1.194	1,382	1,369	11,939	16.5	16.3
Spain	479,775	460,023	2.136	347,137	332,845	13,558	4,706.5	4,512.7
Sweden	12,570	10,930	0.846	1,555	1,352	17,167	26.7	23.2
Switzerland	6,520	3.835	0.902	976	574	18,346	17.9	10.5
United Kingdom	248,647	246,200	0.81	26,593	26,338	15,324	407.5	403.5
EU Total	1,926,652	1,865,365		654,551	639,804		9,458	9,256

Data are for European countries with information on methylmercury exposure distributions. For countries without detailed observed data available, the modelled results are given in parenthesis. Sources of underlying data are as in **Table 1**.

criteria. In addition, available data from other studies have been taken into consideration to provide supplementary information, thereby allowing EU-wide estimates to be calculated. Given the low MeHg exposures in Eastern Europe and the relatively small contributions from Croatia, the Faroe Islands, Norway, and Switzerland, the results suggest that benefits for all of Europe will not be substantially above the benefits calculated for the EU.

Several assumptions and caveats must be acknowledged. The hair-Hg concentration is an established biomarker of human MeHg exposure and is generally considered reliable[14]. We used available data from DEMOCOPHES and other sources, with most studies including only about 120 subjects. The sampling size and strategy may have underestimated the occurrence of uncommon high-level exposures, which would weigh more in the calculation of IQ benefits. Adjustment for this bias is obtained in the modelled distributions, which tended to show slightly greater benefits. Although these calculations rely on an assumption of a log-normal distribution of the exposures, the concurrence of the two sets of estimates support the validity of this assumption.

Table 3. Annual number of births with excess exposure, the average log hair-Hg concentration, and IQ benefit and value from prevention of excess exposure (logarithmic dose-effect relationship).

Country	Number of births above 0.58μg/g	Average log concentration above 0.58μg/g	Benefit in IQ points	Value of 1 IQ point (Euro)	Total benefit (million Euro)
Austria	3,812	−0.157	3,199	16,044	51.3
Belgium	39,686	−0.128	35,790	16,458	589.0
Bulgaria	3,186	0.128	4,638	7,529	34.9
Croatia	21,769	0.142	32,350	11,320	366.2
Cyprus	3,514	0.109	4,972	13,747	68.3
Czech Republic	5,143	−0.216	3,658	10,797	39.5
Denmark	22,815	−0.060	23,932	20,220	483.9
Estonia	1,840	−0.214	1,317	10,339	13.6
Faroe Islands	406	0.139	600	20,220	12.1
Finland	6,843	−0.214	4,897	17,288	84.7
France	405,528	−0.053	368,742	17,363	6,402.5
Germany	33,443	−0.157	28,060	15,292	429.1
Greece	94,403	0.355	183,808	13,201	2,426.4
Hungary	892	−0.186	692	9,691	6.7
Ireland	7,104	−0.132	6,345	17,927	113.7
Italy	378,315	−0.036	416,490	17,062	7.106.2
Latvia	2,741	−0.214	1,962	11,568	22.7
Lithuania	4,037	−0.214	2,889	9,661	27.9
Luxembourg	1,870	0.053	2,419	17,062	41.3
Malta	2,690	−0.036	2,961	11,111	32.9
Netherlands	45,227	−0.155	38,144	15,857	604.8
Norway	16,759	−0.198	12,574	20,051	252.1
Poland	6,218	−0.312	3,131	9,979	31.2
Portugal	94,349	0.277	167,777	12,221	2,050.4
Romania	9,098	0.128	13,245	8,187	108.4
Slovakia	2,468	−0.173	1,986	10,037	19.9
Slovenia	4,840	0.034	6,061	11,939	72.4
Spain	479,775	0.561	1,148,026	13,558	15,564.9
Sweden	12,570	−0.214	8,996	17,167	154.4
Switzerland	6,520	−0.167	5,329	18,346	97.8
United Kingdom	248,647	−0.244	161,816	15,324	2,479.7
EU Total	1,884,563		2,645,953		39,061

Data from European countries, sources of underlying data are as in **Table 1**.

In calculating the IQ benefits, we used a linear dose-response function for the decrease in IQ at increased prenatal MeHg exposures, and this curve shape is an approximation of unknown validity. As has been documented for lead[37], a logarithmic DRF may be plausible, and a log curve shows a slightly better fit[16]. As the results for the log curve (**Table 3**) are about 4-fold higher than those obtained for the linear curve, the benefits calculated in **Table 2** must be considered likely underestimates. In recent calculations using French data using similar methods[8], the logarithmic curve shape also resulted in substantially higher estimates.

The cut-off level assumed to be 0.58μg/g hair may also result in underestimated benefits. Recent data from Poland[20], Japan[21] and the United States[18][19] suggest that a lower threshold is likely. If the threshold is indeed lower than we have assumed, the benefits of controlling MeHg exposures will likely be greater, although an additional effort may be required to achieve such lower exposures. Further, given that the much higher tolerable limit of 2.5μg/g is likely exceeded by 200,000 births in the EU per year, clear benefits will accrue already from controlling the very highest exposures.

The IQ benefits from controlling mercury pollution were translated into economic impacts based on the calculated current life-time income benefits from a higher IQ level. These benefits are mainly based on studies carried out in the United States[24][38], and it is possible that IQ-linked differences in life-time incomes may not be the same in Europe. Adjustment for differences in purchasing power has been included to take this issue into partial account. We used data from 2008 to secure complete data sources; the use of more recent records would change the estimates only slightly. An alternative approach might be to calculate benefits from prevention of specific diseases, e.g. for mental retardation or autism, associated with MeHg exposure. However, the attributable risks associated with increases in MeHg exposure are unknown, and such calculations are therefore uncertain[10][39].

Some sources of imprecision in exposure estimates must be emphasized. Thus, in several cases when exposure information was not available for an EU member state, data from a neighbouring country were used as a proxy. Further, the results reported in DEMOCOPHES and in published reports may not be representative for each country. Although high fish consumers may possibly have been oversampled, it is more likely that the avoidance of known exposure hotspots resulted in lowered exposure estimates. In addition, especially for small studies, an element of uncertainty exists with regard to the frequencies of the highest exposures, although this problem was addressed by modelling a log-normal distribution of exposures. Temporal variation and time trends may also play a role, especially in regard to older data. We have assumed stable diets, so that any seasonal or other time trends as well as the time dependence of MeHg sensitivity during brain development would not matter for the calculation of impacts.

Our focus on the loss in life-time earnings is similar to the avoidable costs previously calculated in relation to lead exposure[24]. Other costs were ignored, such as direct medical costs linked to treatment or interventions for children with neuro developmental disorders. We also neglected indirect costs, such as those related to special education or additional years of schooling for children as a consequence of these disorders, as well as intangible costs. In addition, our study did not consider other avoided direct health care costs in the longer term, such as those potentially related to the treatment of cardiovascular or neurodegenerative effects of MeHg exposure, which could be important for high fish consumers[2], but would be difficult to estimate. Any compensation of the IQ benefit due to special education and other remedies was not taken into account. Overall, the estimates presented in **Table 2** are likely underestimates of the total benefits of MeHg exposure abatement.

Clear differences are apparent between European countries. Seafood and freshwater fish constitute the main source of exposure, but countries with high fish consumption levels, such as Spain and Norway, clearly show great differences in MeHg exposure that are undoubtedly related to the choice of fish species consumed as well as the contamination level. The high exposure levels observed in Spain are in accordance with other studies[40][41]. The elevated exposures in the Faroes are likely related to the occasional consumption of pilot whale meat[23].

Calculations from the United States have resulted in several greatly varying estimates, depending on the DRF assumptions. One comparable estimate put the aggregate economic benefit for each annual birth cohort in the US at \$8.7 billion (range: \$0.7–\$13.9 billion for year 2000)[10]. We recently calculated the annual benefit for the US at about 264,000 IQ points, which would correspond to benefits of approximately \$5 billion[42]. The EU benefits of over 600,000 IQ points are much higher. However, in comparing the figures for the US and the EU, note should be taken that annual number of births in the EU (5.4 million) are 27% greater than the 4.2 million births in the US per year. In addition, MeHg exposures in parts of Europe are higher than in the US[4]. On a global scale, benefit estimates can be extended on the basis of GDP values adjusted for PPP and productivity, but the validity of such calculations is limited by the lack of exposure assessments[43]. However, the present study leaves little doubt that global benefits substantially exceed $20 billion.

The present study did not aim at calculating annual costs of investments in pollution abatement due to the paucity of available data. Relevant investment costs would consider mercury emissions controls in coal-fired power plants, reduction of mercury usage in the chlorine industry, measures taken in dentistry, plus expenses for recycling and treatment of mercury releases. Some information is available and suggests that one-time expenses may be quickly balanced by the cumulated annual benefits from exposure abatement[9]. However, mercury emissions control needs to be carried out on a global level due to the regional and hemispherical dispersion of mercury releases[43]. These costs would likely have additional socioeconomic yields from better control of mercury emissions, e.g. job creation and modernization of capital equipment.

The control of inorganic mercury emissions will only result in diminished MeHg exposure in the long term, and the benefits will therefore be delayed. As MeHg exposure mainly originates from seafood and freshwater fish, public health advice on dietary choices is an important element of the intervention[6][44]. Due to the essential nutrients present in seafood[3], a reduction in MeHg exposure should not be sought through a decrease or replacement of fish in the diet. A prudent advice would be to maintain fish consumption and minimise the MeHg exposure by consumption of fish known to have lower MeHg concentrations, e.g., smaller species, younger fish, and catches from less polluted waters. Such advice should be directed toward women during pregnancy as the most cost-effective preventive action. Restricted consumption of large, piscivorous fish species may also benefit overfished populations of pelagic fish, such as tuna[45].

The successful completion of the DEMOCOPHES project and the complements from other exposure studies in Europe illustrate the feasibility and usefulness of biological monitoring approaches, in particular when relying on hair samples that may be easily obtained, stored and transported. While such studies have become a routine function in the United States through the National Health And Nutrition Examination Survey[4], and the biomonitoring reports from the Centers for Disease Control and Prevention have become key resources for research on human exposures to environmental chemicals, Europe has lagged behind. Following international policy decisions to decrease global mercury pollution, such human biomonitoring studies will be crucial to monitor the effects of the interventions.

5. Conclusions

Annual benefits of removing Hg exposure can be estimated to be approximately €9 billion in Europe. While our results support enhanced public policies for the prevention of MeHg exposure, the economic estimates are highly influenced by uncertainties regarding the dose-response relationship. Thus, a logarithmic response curve results in 4-fold higher benefit estimates. In addition, benefits might be underestimated because costs linked to all aspects of neurotoxicity and long-term disease risks have not been considered. These European data and the calculated economic benefits support the need for interventions to minimize exposure to this hazardous pollutant.

Additional Files

Additional file 1: Conversion rates, 2008. http://www.biomedcentral.com/content/supplementary/1476-069X-12-3-S1.docx.

Additional file 2: Exposure distributions. http://www.biomedcentral.com/content/supplementary/1476-069X-12-3-S2.doc.

Additional file 3: IQ calculation spreadsheet._http://www.biomedcentral.com/content/supplementary/1476-069X-12-3-S3.xls.

Abbreviations

DRF: Dose-response Function; EPA: Environmental Protection Agency; EU: European Union; GDP: Gross Domestic Product; hair-Hg: Mercury concentration in hair; MeHg: Methylmercury; IQ: Intelligence Quotient; perc: Percentile; PPP: Purchasing Power Parity; prob: Probability; RfD: Reference Dose; US: United States; WHO: World Health Organization.

Competing Interests

PG is an editor of this journal but did not participate in the editorial handling of this manuscript. The authors declare that they have no competing interests.

Authors' Contributions

MB, CP, EBJ and PG planned the economic evaluation, carried out the calculations, and drafted the manuscript. AM reviewed published data on MeHg exposure. DA coordinated the contributions of the 17 DEMOCOPHES countries. AC and ME were responsible for the development and follow-up of the Standard Operating Procedures and Quality Assurance for hair sampling and mercury analyses in support to comparability of DEMOCOPHES measurements. DA, MB2, AC, MČ, PC, FD, MEF, AEG, KH, AK, LEK, MK-G, GK, DL, AM, MFR, PR, JST, and PW contributed unpublished exposure data from European countries and act as guarantors of the data applied. All authors commented on the draft manuscript, and all authors read and approved the final version.

Authors' Information

National guarantors of the DEMOCOPHES data are listed as co-authors. The DEMO/COPHES Consortium that established and tested harmonised human biomonitoring on a European scale (www.eu-hbm.info) also included Jürgen Angerer, Pierre Biot, Louis Bloemen, Ludwine Casteleyn, Milena Horvat, Anke Joas, Reinhard Joas, Greet Schoeters, and Karen Exley.

Acknowledgements

Exposure data were contributed from the DEMOCOPHES project (LIFE09

ENV/BE/000410) carried out thanks to joint financing of 50% from the European Commission programme LIFE + along with 50% from each participating country (see the national implementation websites accessible via http://www.eu-hbm.info/democophes/project-partners). Special thanks go to the national implementation teams. The COPHES project that provided the operational and scientific framework was funded by the European Community's Seventh Framework Programme-DG Research (Grant Agreement Number 244237). Additional exposure data were supported by the PHIME project (FOOD-CT-2006-016253) and ArcRisk (GA 226534). We are grateful to Yue Gao and colleagues for sharing Flanders exposure data from the Flemish Center of Expertise on Environment and Health, financed and steered by the Ministry of the Flemish Community. National exposure data from the 2006–2007 French national survey on nutrition and health (Etude Nationale Nutrition Santé) were made available by Nadine Fréry, French Institute for Public Health Surveillance. Data from the Norwegian Mother and Child Cohort Study (a validation sample) were kindly provided by Anne Lise Brantsæter, National Institute of Public Health, Oslo. The UK mercury data were obtained from the ALSPAC pregnancy blood analyses carried out at the Centers for Disease Control and Prevention with funding from NOAA (the US National Oceanographic and Atmospheric Administration). The studies in the Faroe Islands were supported by the US National Institutes of Health (ES009797 and ES012199). The contents of this paper are solely the responsibility of the authors and do not necessarily represent the official views of the funding agencies.

Source: Bellanger M, Pichery C, Aerts D, et al. Economic benefits of methylmercury exposure control in Europe: Monetary value of neurotoxicity prevention[J]. Environmental Health A Global Access Science Source, 2013, 12(1):1–10.

References

[1] Grandjean P, Herz KT: Methylmercury and brain development: imprecision and underestimation of developmental neurotoxicity in humans. Mt Sinai J Med 2011, 78(1):107–118.

[2] Mergler D, Anderson HA, Chan LH, Mahaffey KR, Murray M, Sakamoto M, Stern AH: Methylmercury exposure and health effects in humans: a worldwide concern. Ambio 2007, 36(1):3–11.

[3] Mahaffey KR, Sunderland EM, Chan HM, Choi AL, Grandjean P, Marien K, Oken E,

Sakamoto M, Schoeny R, Weihe P, et al: Balancing the benefits of n-3 polyunsaturated fatty acids and the risks of methylmercury exposure from fish consumption. Nutr Rev 2011, 69(9):493–508.

[4] Mahaffey KR, Clickner RP, Jeffries RA: Adult women's blood mercury concentrations vary regionally in the United States: association with patterns of fish consumption (NHANES 1999–2004). Environ Health Perspect 2009, 117(1):47–53.

[5] Shimshack JP, Ward MB: Mercury advisories and household health tradeoffs. J Health Econ 2010, 29(5):674–685.

[6] Pouzaud F, Ibbou A, Blanchemanche S, Grandjean P, Krempf M, Philippe HJ, Verger P: Use of advanced cluster analysis to characterize fish consumption patterns and methylmercury dietary exposures from fish and other sea foods among pregnant women. J Expo Sci Environ Epidemiol 2010, 20(1):54–68.

[7] Landrigan PJ, Schechter CB, Lipton JM, Fahs MC, Schwartz J: Environmental pollutants and disease in American children: estimates of morbidity, mortality, and costs for lead poisoning, asthma, cancer, and developmental disabilities. Environ Health Perspect 2002, 110(7):721–728.

[8] Pichery C, Bellanger M, Zmirou-Navier D, Frery N, Cordier S, Roue-Legall A, Hartemann P, Grandjean P: Economic evaluation of health consequences of prenatal methylmercury exposure in France. Environ Health 2012, 11(1):53.

[9] Rice G, Hammitt JK: Economic Valuation of Human Health Benefits of Controlling Mercury Emissions from U.S. Coal-fired Power Plants. Boston: Northeast States for Coordinated Air Use Management; 2005.

[10] Trasande L, Schechter C, Haynes KA, Landrigan PJ: Applying cost analyses to drive policy that protects children: mercury as a case study. Ann N Y Acad Sci 2006, 1076:911–923.

[11] Pacyna JM, Sundseth K, Pacyna EG, Jozewicz W, Munthe J, Belhaj M, Astrom S: An assessment of costs and benefits associated with mercury emission reductions from major anthropogenic sources. J Air Waste Manag Assoc 2010, 60(3):302–315.

[12] Swain EB, Jakus PM, Rice G, Lupi F, Maxson PA, Pacyna JM, Penn A, Spiegel SJ, Veiga MM: Socioeconomic consequences of mercury use and pollution. Ambio 2007, 36(1):45–61.

[13] Budtz-Jorgensen E, Bellinger D, Lanphear B, Grandjean P: An international pooled analysis for obtaining a benchmark dose for environmental lead exposure in children. Risk Anal 2012, in press: http://www.ncbi.nlm.nih.gov/ pubmed/22924487.

[14] National Research Council: Toxicological effects of methylmercury. Washington, DC: National Academy Press; 2000.

[15] Budtz-Jorgensen E, Grandjean P, Jorgensen PJ, Weihe P, Keiding N: Association between mercury concentrations in blood and hair in methylmercury-exposed subjects at different ages. Environ Res 2004, 95(3):385–393.

[16] Budtz-Jorgensen E: Estimation of the benchmark dose by structural equation models. Biostatistics 2007, 8(4):675–688.

[17] Grandjean P, Budtz-Jorgensen E: Total imprecision of exposure biomarkers: implications for calculating exposure limits. Am J Ind Med 2007, 50(10):712–719.

[18] Lederman SA, Jones RL, Caldwell KL, Rauh V, Sheets SE, Tang D, Viswanathan S, Becker M, Stein JL, Wang RY, et al: Relation between cord blood mercury levels and early child development in a World Trade Center cohort. Environ Health Perspect 2008, 116(8):1085–1091.

[19] Oken E, Osterdal ML, Gillman MW, Knudsen VK, Halldorsson TI, Strom M, Bellinger DC, Hadders-Algra M, Michaelsen KF, Olsen SF: Associations of maternal fish intake during pregnancy and breastfeeding duration with attainment of developmental milestones in early childhood: a study from the Danish National Birth Cohort. Am J Clin Nutr 2008, 88(3):789–796.

[20] Jedrychowski W, Perera F, Jankowski J, Rauh V, Flak E, Caldwell KL, Jones RL, Pac A, Lisowska-Miszczyk I: Fish consumption in pregnancy, cord blood mercury level and cognitive and psychomotor development of infants followed over the first three years of life: Krakow epidemiologic study. Environ Int 2007, 33(8):1057–1062.

[21] Suzuki K, Nakai K, Sugawara T, Nakamura T, Ohba T, Shimada M, Hosokawa T, Okamura K, Sakai T, Kurokawa N, et al: Neurobehavioral effects of prenatal exposure to methylmercury and PCBs, and seafood intake: neonatal behavioral assessment scale results of Tohoku study of child development. Environ Res 2010, 110(7): 699–704.

[22] Joint Expert Committee on Food Additives: Sixty-first meeting of the Joint FAO/ WHO Expert Committee on Food Additives held in Rome, 10–19 June 2003, World Health Organ Techn Rep Ser 922. Geneva: World Health Organization; 2004. http:// whqlibdoc.who.int/trs/WHO_TRS_922.pdf.

[23] Grandjean P, Weihe P, White RF, Debes F, Araki S, Yokoyama K, Murata K, Sorensen N, Dahl R, Jorgensen PJ: Cognitive deficit in 7-year-old children with prenatal exposure to methylmercury. Neurotoxicol Teratol 1997, 19(6):417–428.

[24] Pichery C, Bellanger M, Zmirou-Navier D, Glorennec P, Hartemann P, Grandjean P: Childhood lead exposure in France: benefit estimation and partial cost-benefit analysis of lead hazard control. Environ Health 2011, 10:44.

[25] Debes F, Budtz-Jorgensen E, Weihe P, White RF, Grandjean P: Impact of prenatal methylmercury exposure on neurobehavioral function at age 14 years. Neurotoxicol Teratol 2006, 28(5):536–547.

[26] EUROSTAT: Methodological manual on Purchasing Power Parities. Paris: OECD; 2006.

[27] Schreyogg J, Tiemann O, Stargardt T, Busse R: Cross-country comparisons of costs: the use of episode-specific transitive purchasing power parities with standardised cost categories. Health Econ 2008, 17:S95–S103.

[28] Gundacker C, Frohlich S, Graf-Rohrmeister K, Eibenberger B, Jessenig V, Gicic D, Prinz S, Wittmann KJ, Zeisler H, Vallant B, et al: Perinatal lead and mercury exposure in Austria. Sci Total Environ 2010, 408(23):5744–5749.

[29] Schoeters G, Den Hond E, Colles A, Loots I, Morrens B, Keune H, Bruckers L, Nawrot T, Sioen I, De Coster S, et al: Concept of the Flemish human biomonitoring programme. Int J Hyg Environ Health 2012, 215(2):102–108.

[30] Croes KDCS, De Galan S, Morrens B, Loots I, Van de Mieroop E, Nelen V, Sioen I, Bruckers L, Nawrot T, Colles A, Den Hond E, Schoeters G, van Larebeke N, Baeyens W, Gao Y: Health effects in the Flemish population in relation to low levels of mercury exposure: from organ to transcriptome level.; 2013 (Submitted).

[31] Fréry N, Saoudi A, Garnier R, Zeghnoun A, Falq G: Exposition de la population française aux substances chimiques de l'environnement. Tome 1. Présentation de l'étude. Métaux et métalloïdes. Paris: Institut de Veille Sanitaire (InVS); 2011.

[32] Miklavcic A, Casetta A, Snoj Tratnik J, Mazej D, Krsnik M, Mariuz M, Sofianou K, Spiric Z, Barbone F, Horvat M: Mercury, arsenic and selenium exposure levels in relation to fish consumption in the Mediterranean area. Environ Res 2012, in press: http://www.ncbi.nlm.nih.gov/pubmed/22999706.

[33] Diez S, Montuori P, Pagano A, Sarnacchiaro P, Bayona JM, Triassi M: Hair mercury levels in an urban population from southern Italy: fish consumption as a determinant of exposure. Environ Int 2008, 34(2):162–167.

[34] Brantsaeter AL, Haugen M, Thomassen Y, Ellingsen DG, Ydersbond TA, Hagve TA, Alexander J, Meltzer HM: Exploration of biomarkers for total fish intake in pregnant Norwegian women. Public Health Nutr 2010, 13(1):54–62.

[35] Boyd A, Golding J, Macleod J, Lawlor DA, Fraser A, Henderson J, Molloy L, Ness A, Ring S, Davey Smith G: Cohort Profile: The 'Children of the 90s'-the index offspring of the Avon Longitudinal Study of Parents and Children. Int J Epidemiol 2012, in press: http://www.ncbi.nlm.nih.gov/pubmed/22507743.

[36] Gibb H, Haver C, Kozlov K, Centeno JA, Jurgenson V, Kolker A, Conko KM, Landa ER, Xu H: Biomarkers of mercury exposure in two eastern Ukraine cities. J Occup Environ Hyg 2011, 8(4):187–193.

[37] Lanphear BP, Hornung R, Khoury J, Yolton K, Baghurst P, Bellinger DC, Canfield RL, Dietrich KN, Bornschein R, Greene T, et al: Low-level environmental lead exposure and children's intellectual function: an international pooled analysis. Environ Health Perspect 2005, 113(7):894–899.

[38] Gould E: Childhood lead poisoning: conservative estimates of the social and economic benefits of lead hazard control. Environ Health Perspect 2009, 117(7): 1162–1167.

[39] Landrigan PJ: What causes autism? Exploring the environmental contribution. Curr Opin Pediatr 2010, 22(2):219–225.

[40] Castaño ANC, Cañas A, Díaz G, García JP, Esteban M, Lucena MA, Arribas M,

Jiménez JA: A biomonitoring study of mercury in hair and urine of 267 adults living in Madrid (Spain). Toxicol Lett 2008, 180:S79–S80.

[41] Ramon R, Murcia M, Aguinagalde X, Amurrio A, Llop S, Ibarluzea J, Lertxundi A, Alvarez-Pedrerol M, Casas M, Vioque J, et al: Prenatal mercury exposure in a multicenter cohort study in Spain. Environ Int 2011, 37(3):597–604.

[42] Grandjean P, Pichery C, Bellanger M, Budtz-Jorgensen E: Calculation of Mercury's effects on Neurodevelopment. Environ Health Perspect 2012, 120(12):a452.

[43] Spadaro JV, Rabl A: Global health impacts and costs due to mercury emissions. Risk Anal 2008, 28(3):603–613.

[44] Weihe P, Grandjean P, Jorgensen PJ: Application of hair-mercury analysis to determine the impact of a seafood advisory. Environ Res 2005, 97(2): 200–207. le Clezio P: Les indicateurs du développement durable et l'empreinte écologique. Paris: Conseil économique, social et environnemental; 2009.

Chapter 15

Economics of Carbon Dioxide Capture and Utilization—A Supply and Demand Perspective

Henriette Naims

IASS—Institute for Advanced Sustainability Studies e.V., Berliner Strasse 130, 14467 Potsdam, Germany

Abstract: Lately, the technical research on carbon dioxide capture and utilization (CCU) has achieved important breakthroughs. While single CO_2-based innovations are entering the markets, the possible economic effects of a large-scale CO_2 utilization still remain unclear to policy makers and the public. Hence, this paper reviews the literature on CCU and provides insights on the motivations and potential of making use of recovered CO_2 emissions as a commodity in the industrial production of materials and fuels. By analyzing data on current global CO_2 supply from industrial sources, best practice benchmark capture costs and the demand potential of CO_2 utilization and storage scenarios with comparative statics, conclusions can be drawn on the role of different CO_2 sources. For near-term scenarios the demand for the commodity CO_2 can be covered from industrial processes, that emit CO_2 at a high purity and low benchmark capture cost of approximately 33€/t. In the long-term, with synthetic fuel production and large-scale CO_2 utilization, CO_2 is likely to be available from a variety of processes at benchmark costs of approx. 65€/t. Even if fossil-fired power generation is phased out, the CO_2 emissions of current industrial processes would suffice for ambitious CCU demand scenarios. At current economic conditions, the business case for CO_2 utilization is technology specific and depends on whether efficiency gains or substitution of volatile priced raw materials can be achieved. Overall, it is argued that CCU should be advanced complementary to mitigation technologies and can unfold its potential in creating local circular economy solutions.

Keywords: Carbon Capture and Utilization, Supply and Demand Scenarios, Commodity CO_2, Costs of CO_2 Capture, Circular Economy

1. Motivations for Using CO_2

In the context of the global climate change debate, the motivation behind the research on CO_2 utilization seems obvious: If there are possibilities to make use of the industrial CO_2 emissions that are a major cause of global warming they should be harvested. If recovered CO_2 emissions can be used as feedstock for industrial production processes the existing resource base could be broadened. Especially in the context of circular economy thinking as promoted by the World Economic Forum (2014) making use of waste emissions offers a promising new perspective. However, the largest part of worldwide industrial emissions is still unregulated and unpriced. Currently, only approx. 6Gt of the estimated annual 37Gt of global anthropogenic CO_2 emissions are regulated by some form of carbon pricing instrument (Le Quéré *et al.* 2014; World Bank 2014). Meanwhile, in the few existing schemes the emission allowance price is rather low, for example around 8€/t in the EU Emission Trading Scheme (EEX 2015) and approx. 13US$/t in the California Cap and Trade Program (California Carbon Dashboard 2015). Consequently, the economic incentives to tackle the CO_2 problem are largely insufficient. The development of technically, environmentally, and economically viable ways of utilizing CO_2 as a feedstock for industrial production can imply a complementary route to existing mitigation strategies such as the deployment of renewable energy and other green technologies.

As almost all materials that surround us in our everyday lives are carbon-based, the option to use recovered CO_2 to substitute fossil carbon sources remains an attractive possibility that could be worth further investigation. In the last years, several public funding programs, e.g., by the US Department of Energy or the German Federal Ministry of Education and Research have encouraged research in this field and already the first technological breakthroughs and advances to a demonstration scale can be observed (Federal Ministry of Education and Research 2014; US DOE n.d.). Even though the first CO_2-based products are just entering global markets in the near future their number and scale is expected to grow (Aresta *et al.* 2013). Consequently, based on a literature review, the potential supply and demand of the commodity CO_2 is presented in this paper to discuss the fundamentals of the commodity CO_2 from an economic perspective of comparative statics.

Since the utilized CO_2 in most cases is reemitted at a later point in time a simple aggregation of the used volumes of CO_2 is not an indicator of ecologic performance (von der Assen *et al.* 2013). Instead, a detailed environmental analysis is necessary to calculate the real carbon footprint of a certain CCU technology compared to a conventional technology (von der Assen *et al.* 2015). Indeed, the same principle applies to the business case of CO_2 utilization. In some cases, using comparatively cheap CO_2 as a feedstock and replacing more costly and volatile priced fossil-based raw materials can lead to a cost reduction which sets the business case for CCU. However, for those production processes that use CO_2 still inefficiently or are not competitive to conventional fossil-based production, there is no business case until further research and development or political incentives prove otherwise. While CO_2 can generally be used in many processes, this paper focuses on potential commodity CO_2 from industrial capture and does not include biological fixation and conversion via the cultivation of crops or algae for example for making biofuels.

2. Supply Side: Potential Sources and Cost of CO_2

The potential sources of waste CO_2 emissions are numerous. Industrial plants emit CO_2 in different quantities and at diverse qualities. Several capture technologies can be applied, for example adsorption, absorption, cryogenic separation, or membranes (de Coninck and Benson 2014). The costs of capturing CO_2 at a certain source depend on the technological efforts that must be undertaken to collect the CO_2 in the required quality from the industrial exhaust gas. Thus, the costs are largely influenced by the concentration of CO_2 in the exhaust gas. Moreover, the CO_2 needs to be purified and any toxic or hazardous chemicals removed (Aresta and Dibenedetto 2010). Furthermore, a larger plant size can lower the investment and operating costs per captured tonne of CO_2 through economies of scale (Faulstich *et al.* 2009; Möllersten *et al.* 2003). Consequently, despite technical feasibility, not all emitting sources represent economically viable options at current conditions.

Table 1 summarizes current data on respective global emission volumes, concentrations, estimated capture rates, and benchmark capture costs per type of source for the largest point sources of CO_2 based on a literature review. The presented data are based on several selected sources with heterogeneous technological

Table 1. Potential sources of waste CO_2 (most recent available estimates).

CO_2 emitting source	Global emissions[a] (Mt CO_2/year)	CO_2 content[a] (vol%)	Estimated capture rate[b] (%)	Capturable emissions (Mt CO_2/year)	Benchmark capture cost[b] (€ 2014/t CO_2) [rank]	Groups of emitters
Coal to power	9031[c]	12–15	85	7676	34 [6]	Fossil-based power generation
Natural gas to power	2288c	3–10[d]	85	1944	63 [9]	Fossil-based power generation
Cement production	2000	14–33	85	1700	68 [10]	Industry large emitters
Iron and steel production	1000	15	50	500	40 [7]	Industry large emitters
Refineries[e]	850	3–13	40	340	99 [12]	Industry large emitters
Petroleum to power	765[c]	3–8	Not available	Not available	Not available	Fossil-based power generation
Ethylene production	260	12	90	234	63 [8]	Industry large emitters
Ammonia production	150	100	85	128	33 [5]	Industry high purity
Bioenergy[f]	73[d]	3–8[d]	90	66	26 [2]	High purity/ power generation
Hydrogen production[f]	54[g]	70–90[h]	85	46	30 [4]	Industry high purity

Continued

Natural gas production	50	5–70	85	43	30 [3]	Industry high purity
Waste combustion	60[i]	20	Not available	Not available	Not available	Industry large emitters
Fermentation of biomass[f]	18[d]	100[d]	100	18	10 [1]	Industry high purity
Aluminum production	8	<1[j]	85	7	75 [11]	Industry large emitters

[a]Data from Wilcox (2012) if not indicated otherwise; [b]See **Table 2** for literature reference, assumptions, and calculation methods; [c]Data from IEA (2014) based on the largest point sources suitable for capture and not including the emissions of the large amount of emissions that are caused by small decentral point sources in the mobility and residential sector; [d]Data from Metz *et al.* (2005); [e]Refineries could include ammonia and hydrogen production. A separate listing is nevertheless interesting to differentiate these two high purity from general refinery CO_2 streams. The capturable emission data based on the estimated capture rates should ensure that emissions are not included twice; [f]Undisclosed technological assumptions for emissions volumes and CO_2 content, if not indicated otherwise. For technological assumptions for cost data see **Table 2**. For bioenergy and fermentation, emission estimates are only for North America and Brazil; [g]Data from Mueller-Langer *et al.* (2007); [h]Data for hydrogen from steam methane reformer from Kurokawa *et al.* (2011); [i]Data from Bogner *et al.* (2007); [j]Data from Jilvero *et al.* (2014), Jordal *et al.* (2014).

and financial assumptions, calculation methods, and reference years. Therefore, the collected data do not allow for a detailed cost comparison or technological discussion. Nevertheless, it is useful to provide general insights on potential large-scale supply of CO_2 as a commodity. The presented emission volumes represent recent global direct CO_2 emissions. The capturable emissions are calculated by applying the estimated lowerbound capture rate as a benchmark to the global emission volume per type of source. The benchmark costs represent minimum cost of CO_2 captured per tonne that are possible to achieve with a certain benchmark technology, a so-called best practice process. The purity of the recovered CO_2 depends on the respective process but can be assumed between 95% and 99.9%. The detailed underlying assumptions are further explained in the following and in **Table 2**.

2.1. Costs of CO_2 Capture

Today, CO_2 capture is technologically feasible and industrial practice on a

Table 2. Overview of capture cost assumptions of CO_2 point sources.

Literature reference	CO_2 emitting source	Capture cost (€/t CO_2) and benchmark case	Reference year	Region and currency	Technology assumptions	Economic assumptions
Finkenrath (2011) *Cost and performance of carbon dioxide capture from power generation*, IEA Working Paper.	**Coal to power**	• Average 34–46 € (43–58 $)/t CO_2 • *Benchmark case:* pre-combustion integrated gasification combined cycle (IGCC) technology at a 20 % net efficiency decrease • ***CEPCI adjusted 2014 benchmark cost: 34 €***[a]	Aligned data for 2010 (original studies from 2006 to 2010)	Summarized OECD data (especially from US and EU), in $	• Average normalized performance data from several studies • Data for generic, new-build, early commercial power plants (no pilots or retrofits) • Most advantageous **capture rates between 85 and 100 %**	• **Costs of CO_2 avoided** • Average normalized cost data from several studies • Including levelised cost of electricity (LCOE) • Including capital costs as overnight costs without
	Natural gas to power	• Average 64 € (80 $)/t CO_2 • *Benchmark case:* post-combustion capture from natural gas combined cycle at a 15 % net efficiency decrease • ***CEPCI adjusted 2014 benchmark cost: 63 €***[a]			• Undisclosed CO_2 compression but for transport at supercritical level (>74 bar) • **CO_2 purity above 99.9 %**	interest costs • Discount rate of 10 % • Not including emission price • not including R&D costs
US EIA (2014) *Assumptions to the Annual Energy Outlook 2014*, Independent Statistics and Analysis.	**Cement production** **Ammonia production** **Hydrogen production** **Natural gas production**	• **66 €** (82 $)/t CO_2 • **32 €** (40 $)/t CO_2 • **30–37 €** (37–47 $)/t CO_2 • **30 €** (37 $)/t CO_2	2014 estimates	US six regions, in $	• Undisclosed technological assumptions of data • Background: CO_2 for EOR/EGR from industrial sources, • The CO_2 is compressed for pipeline transport, thus 90-120 bar and a **CO_2 purity of ≥95 % can be assumed** (Metz et al. 2005) • source of cost estimates: US EIA Office of Energy Analysis • ***undisclosed capture rate, thus 85-100 % are assumed***[a]	• **Costs of capture and compression** • Including regional transport costs (for EOR/EGR purposes) • not including interregional transport costs • unclarity about inclusion of capital costs
Kuramochi et al. (2011) *Techno-economic assessment and comparison of CO_2 capture technologies for industrial processes: preliminary results for the iron and steel sector*, Energy Procedia.	**Iron and steel production**	• 40–50 €/t CO_2 with a large sensitivity to energy prices • *Benchmark case:* vacuum pressure swing adsorption (VPSA) from top gas recycling blast furnace (TGRBF) with a net efficiency decrease due to an increased power consumption • ***CEPCI adjusted 2014 benchmark cost: 40€***[a]	Aligned data for 2008	Undisclosed region but reference to several international studies, in €	• Normalized performance data from several studies • Normalized technical parameters: pressure, emission factors and plant scale • **Capture rate of 50 %** • CO_2 compression to 110 bar • The CO_2 is compressed for transport, thus a **CO_2 purity of ≥95 % can be assumed** (Metz et al. 2005)	• **Costs of CO_2 avoided** • Average normalized cost data from several studies: fuel, electricity and capital cost • Including capital costs as total capital requirement (TCR) • Capture costs are very sensitive to energy prices
van Straelen et al. (2010) *CO_2 capture for refineries, a practical approach*, International Journal of Greenhouse Gas Control.	**Oil refineries**	• 90–120 €/t CO_2 for a number of large flue gas sources in oil refining • Above 160 €/t CO_2 from a large number of scattered, small concentration sources • *Benchmark case:* capture with an amine-based solvent from a combined stack • ***CEPCI adjusted 2014 benchmark cost: 99 €***[a]	2007	Western Europe, in €	• Techno-economic case study: post-combustion capture at one large-scale complex oil refinery • **Capture rate of 40–50 %** • Excluding emissions from hydrogen production (approx. 5–20 %) and a large number of small concentration sources (approx. 50 %) • The CO_2 is compressed for transport and storage, thus 90-120 bar and a **CO_2 purity of ≥95 % can be assumed** (Metz et al. 2005)	• **Costs of CO_2 avoided** • Including costs of capture and compression • Including capital costs based on a Shell-internal cost estimation tool • Discount rate of 7 % • Excluding transport cost

Continued

Weikl and Schmidt (2010) *Carbon capture in cracking furnaces,* AIChe 2010 Spring Meeting and 6th Global Congress on Process Safety.	**Ethylene production**	• 60 € (85 $)/t CO_2 with a net efficiency decrease due to additional power and steam consumption • ***CEPCI adjusted 2014 benchmark cost: 63 €***[a]	Not specified, assumed 2010	No specified region, in $ and €	• Techno-economic case study: post-combustion and oxyfuel capture at one ethylene plant • **Capture rate of 90–99 %** • 20–30 % higher overall emissions • CO_2 compression to 100 bar • **CO_2purity: >98 % with post-combustion; 85 % with oxyfuel purposes**	• **Costs of capture and compression** • Including fuel, steam and electricity costs • Including capital costs but with high uncertainty • Excluding transport costs • Annuity method with a discount rate of 8 %
Möllersten et al. (2003) *Potential market niches for biomass energy with CO_2capture and storage—Opportunities for energy supply with negative CO_2emissions,* Biomass and Bioenergy.	**Bioenergy**	• 18–42 € (23–53 $)/t CO_2 • *Benchmark case:* pre-combustion CO_2 capture in a pulp mill with black liquor integrated gasification combined cycle (BLGCC) technology • ***CEPCI adjusted 2014 benchmark cost: 26 €***[a]	Not specified, assumed 2003	No specified region, in $	• Techno-economic case study: post-combustion and IGCC capture at a pulp mill and an ethanol plant • **Capture rate of 90 %** • CO_2 compression to 100 bar • The CO_2 is compressed for pipeline or tanker transport, thus a **CO_2 purity of ≥95 % can be assumed** (Metz et al. 2005)	• **Costs of capture and compression** • Including lost electricity production • Including capital cost • Excluding transport cost
	Fermentation	• 7 € (9 $)/t CO_2 • *Benchmark case*: only CO_2 compression necessary • ***CEPCI adjusted 2014 benchmark cost: 10 €***[a]			• Techno-economic case study: capture from sucrose fermentation in a sugar cane-based ethanol production plant • For non-pressurized fermentation vessels except compression no processing of CO_2 is needed • **Capture rate of 100 %** • CO_2 compression to 100 bar • The CO_2 is compressed for pipeline or tanker transport, thus a **CO_2 purity of ≥95 % can be assumed** (Metz et al. 2005)	• **Capture cost only correspond to compression cost** • Excluding capital costs
Jilvero et al. (2014) *Techno-economic analysis of carbon capture at an aluminum production plant—Comparison of post-combustion capture using MEA and ammonia,* Energy Procedia.	**Aluminum production**	• 74–97 €/t CO_2 • *Benchmark case*: post-combustion CO_2 capture with ammonia as a solvent at a new aluminum plant (with an increased CO_2 concentration in the flue gas of 10 %) • ***CEPCI adjusted 2014 benchmark cost: 75 €***[a]	2013	Norway, in €	• Techno-economic case study: reference aluminum plant in Norway with post-combustion capture by amines and ammonia • **Capture rate of 85 %** • Additionally emitted CO_2 from capture is not included • CO_2 compression for transport to 70–100 bar • **CO_2purity of 99.5 %**	• **Costs of capture and compression** • Net present value (NPV) method combined with Aspen process and cost software • Including capital costs with 25 % contingency fund • But excluding retrofitting costs • Discount rate of 7.5 %

[a] Adjusted/estimated by the author, not included in the cited original literature

small scale around the world. However, due to a lack of incentives, large-scale capture is currently not economically viable. Hence, the costs of capture are essential when considering potential sources and technologies for recovering CO_2 emissions. Capture costs are generally defined as the costs of CO_2 separation and compression at a single facility (e.g., an industrial plant), disregarding any costs of transport, storage, or further conversion steps (Metz *et al.* 2005). They are usually derived from comparing a system with CO_2 capture to a reference system without capture. In the literature, two main measures for CO_2 capture costs exist: costs of capture and costs of avoidance of CO_2. According to the IPCC (Metz *et al.* 2005), the two measures are clearly defined as follows:

(a) Cost of CO_2 captured represents the rather straightforward cost of capturing per amount of captured CO_2. They can be determined through formula (1):

$$\text{cost of CO}_2\text{ captured}\left(\frac{€}{\text{tCO}_2}\right)=\frac{\text{additional costsof CO}_2\text{ capture}(€)}{\text{amount of CO}_2\text{ captured }(\text{tCO}_2)}=\frac{\text{costs}_{\text{capture plant}}(€)-\text{costs}_{\text{reference plant}}(€)}{\text{CO}_2\text{ captured }(\text{tCO}_2)} \tag{1}$$

These capture cost can reveal Bthe viability of a CO_2 capture system given a market price for CO_2 (as an industrial commodity) (Bogner *et al.* 2007). Consequently, if these costs can be reimbursed, e.g., through CO_2 utilization options or political incentives such as a carbon tax, then carbon capture could make economic sense. Due to this information value, costs of capture are the preferred measure in this study.

(b) Cost of CO_2 avoided gauges the effect of the overall emission reduction by calculating the cost of capturing per amount of CO_2 reduced compared to a reference process. The respective formula is (2):

$$\text{cost of CO}_2\text{ avoided}\left(\frac{€}{\text{tCO}_2}\right)=\frac{\text{additional costs of CO}_2\text{ capture}(€)}{\text{amount of CO}_2\text{ reduction }(\text{tCO}_2)}=\frac{\text{costs}_{\text{capture plant}}(€)-\text{costs}_{\text{reference plant}}(€)}{\text{CO}_2\text{ emitted}_{\text{reference plant}}(\text{tCO}_2)-\text{CO}_2\text{ emitted}_{\text{captuprleant}}(\text{tCO}_2)} \tag{2}$$

The cost of CO_2 avoided considers actual emission reductions and thus contains more ecologically relevant information. As the process of capturing usually requires additional energy and decreases the plant's efficiency, the capturing often produces additional CO_2 emissions. Therefore, the amount of CO_2 reduced or avoided compared to a reference system will largely be smaller than the amount of CO_2 captured. Consequently, the cost of CO_2 avoided will generally be higher than the cost of capture (Metz *et al.* 2005). In an optimal capturing case, where additional energy and emissions can be avoided, both measures will be equal.[1] The IPCC recommends the cost of avoidance especially for complete carbon capture and storage (CCS) systems and less for capture only analysis (Metz *et al.* 2005). For the market perspective of this paper, the value of the commodity CO_2 is however more important. Moreover, it is recommended, that any utilization technology should be accompanied by an LCA that measures the full ecologic impact of all production steps.

Unfortunately, the two measures are often commingled in the literature, so that a clear differentiation is not always possible. The preferred cost estimate of this study is the cost of CO_2 captured. Deviating cost measures were included where necessary. As described above, the cost of CO_2 avoided are potentially higher than the capture costs since the reduced emissions in the denominator are smaller than the captured emissions for most capture systems. The difference depends on the additional amounts of CO_2 emissions caused by the capturing efforts and how they are accounted for. This cost difference is expected to increase more or less proportionally with the efforts that are necessary for capture at the respective sources. Thus, for sources with higher capture costs, the difference in costs of capture and avoidance should be larger. Moreover, even if a measure is clearly and consistently selected in an assessment, varying assumptions and system boundaries limit the possibility of comparing cost data across studies (Metz *et al.* 2005). Instead, a comparison would only make sense, if the calculation methods and underlying assumptions were fully transparent and either consistent or could be aligned. To improve the measurement quality and comparability of capture costs—within a single study as well as across studies—the reference systems with and without capture should ideally be based on the same assumptions. Firstly, general

[1]This constraint only applies to the capturing system and its reference plant, as well as for CO_2 storage. If further process steps of CO_2 utilization are included, the relationship of the two measures can vary in all directions. For example, through raw material substitution or process efficiency improvements the amount of CO_2 avoided can be larger than the CO_2 captured. An LCA permits a thorough environmental assessment of utilization options.

conditions such as reference year, region, and type of data (real or hypothetical) should be identical at best. Secondly, the technical and economical parameters such as production process, plant size, fuel type, energy cost, and efficiency standards should be comparable. Especially the measuring of capital costs which can strongly influence the capture costs should be consistent.

In a cross-technology comparison, meeting these standards is often not possible. Especially, when emerging technologies and future scenarios are evaluated, reliable and consistent data can be scarce. Instead, a more heuristic approach must be adopted and the best available data analyzed. Thus, this paper summarizes the recent techno-economic literature on carbon capture in order to establish a large-scale picture of CO_2 supply in the near-term. A secondary database for the largest industrial CO_2 emitting sources is established in **Table 2** that gathers the most recent and reliable cost data available. The presented measures and assumptions are heterogeneous and the data should be considered as estimates and benchmark values for best practice processes. To maximize cost data quality, recent peer-reviewed as well as broader government studies were preferably selected. Other studies were included to fill data gaps. The origin and relative assumptions of the capture cost studies are detailed in **Table 2** as far as they were disclosed. Since capture from coal- and natural gas-fired power plants has been discussed the most extensively in the literature, the summarized average costs from the IEA study seem a reliable data source. Moreover, the capture costs for the higher concentrated sources of ammonia, hydrogen, and natural gas as well as cement production derived from the annually updated assumptions of US Energy Information Agency (EIA) seem a reliable data source for the purpose of analysis even though the data regionally cover only the USA. For the other potential sources of CO_2, less research has been performed and average cost data are not available. Thus, recent peer-reviewed techno-economic studies have been included for capture from iron and steel, refineries, bioenergy, fermentation, and aluminum production. For capture from ethylene production a non-governmental, non-peer reviewed data source was included.

The original data have been adjusted slightly to allow for a common depiction in € per tonne on a 2014 basis. Firstly, cost data in US$ have been converted to € based on Oanda exchange rates[2] from the end of the respective year of data

[2]The Oanda currency converter is available at http://www.oanda.com/currency/converter.

reference. Then, data that were older than 2014 were adjusted with the annual Chemical Engineering Plant Cost Index (CEPCI). The CEPCI is a composite index that reflects the development of equipment, construction labor, buildings and engineering and supervision costs over time since the 1960s (Chem. Eng. 2008; Chem. Eng. 2015a; Jenkins 2015; Mignard 2014). Thus, it helps to correct for changing economic conditions for chemical plants over time while slightly adjusting the original cost data from older reference years.

2.2. Evaluation of Potential CO_2 Supply

Based on the presented data in **Table 1** and **Table 2**, a benchmark CO_2 supply base can be established as a merit order and insights on the feasibility of selected utilization scenarios can be derived. The benchmark CO_2 supply is based on the best practice minimum capture costs. Upper limit or maximum costs are not presented since it is assumed that a variety of processes exist that could lead to even higher capture costs than those presented in **Table 2**. The supply cost function thus can be described by formula (3) that sorts and aggregates the capturable quantities of the potential sources (q_i) according to their benchmark capture costs (p_i):

$$p(q) = p_i \forall q \in]q_{i-1}; q_i] \tag{3}$$

For the space U as the union of all intervals

$$\cup_{i=1}^{n} (q_{i-1}; q_i]$$

with

i = rank of CO_2 emitting source

n = number of ranked sources

p_i = benchmark capture cost of source ranked in ith position; in € per tonne;

$p_1 \leq p_2 \leq \cdots \leq p_n$

q_i = aggregated capturable emissions of sources ranked $\leq -i$; in Mt/year

For the following considerations, the scenarios will be split in a near-term view of up to 10 years and a long-term horizon of more than 10 years. This differentiation is considered useful for technology development since it is a common assumption that new industrial technologies can grow from lab to commercial scale within 10 years. For near-term scenarios, the presented current volumes and costs of CO_2 capture seem an adequate estimate. For future volumes of CO_2 emissions, numerous scenarios exist for different policy scenarios and time horizons, most notably those of the IPCC and the IEA. Future capture costs will vary depending on changing overall economic conditions and energy prices. Improved technological efficiency and performance usually decreases costs over time when the technologies are deployed (Finkenrath 2011). Assuming that capture technologies are advanced further in the future, at stable economic conditions current benchmark costs can likely be lowered or at least maintained.

As assigned in **Table 1**, the CO_2 point sources can be divided into four major groups of emitters:

1) High purity sources

For certain industrial processes such as ammonia production, the CO_2 emitted is very pure and capture requires only small additional efforts (IEA 2011 and UNIDO 2011). Therefore, these processes yield relatively cheap CO_2 as an output. These high-concentration sources represent only approx. 2% of the 12.7Gt capturable point source emissions (see **Figure 1**). Today, capture of CO_2 is an established process predominantly in hydrogen, ammonia, and natural gas purification plants as they allow for comparatively cost efficient CO_2 separation (Wilcox 2012). While raw natural gas can contain CO_2 in different concentrations depending on the respective source, the processing of the gas to achieve pipeline quality often includes carbon dioxide separation (Baker and Lokhandwala 2008).

2) Fossil-based power generation

The largest CO_2 emitting group—the combustion of coal and gas for power generation—currently is responsible for approx. 76% of the 12.7Gt capturable emissions from point sources (see **Figure 1**). However, CO_2 capture at power plants is often connected to significant efficiency losses of approx. 10%–30% of

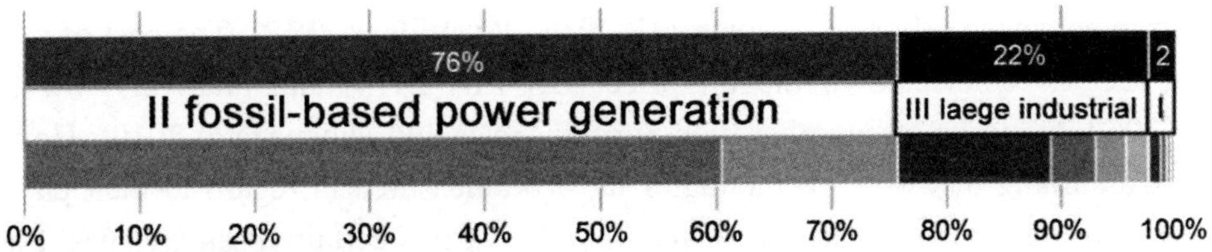

Figure 1. Groups of capturable CO_2 emissions from large industrial point sources (based on **Table 1**, 100% correspond to 12.7Gt CO_2).

the output energy (de Coninck and Benson 2014; Finkenrath 2011). Consequently, coal and natural gas power plants currently lack business incentives for large-scale capture. If CCS was incentivized by political regulation (e.g. via emission performance standards) power companies could start to implement capture technologies for new plants and possibly retrofits. For power plants, economies of scale can play an important role for lowering capital and operating capture costs per tonne.

3) Large industrial emitters

Large industrial CO_2 emitting processes together currently make up for approx. 22% of the 12.7Gt capturable emissions from point sources (see **Figure 1**). They include the production of industrial materials such as iron and steel, cement, aluminum as well as refineries. As these processes emit CO_2 in different quantities and qualities, CO_2 capture at such plants is also connected to varying efficiency penalties and benchmark costs. Moreover, a large number of other industrial manufacturing plants are potential candidates for CO_2 capture. Often, they are comparatively smaller than power plants (Bennaceur *et al.* 2008; Faulstich *et al.* 2009; Weikl and Schmidt 2010). Thus, economies of scale can be more difficult to achieve. For example, waste incineration so far has barely been analyzed in regard to CO_2 capture although reusing such CO_2 would conceptually close resource cycles. The comparatively small size of the incinerators however entails higher capture costs per tonne than those of other CO_2 sources (Faulstich *et al.* 2009).

4) Natural wells

It must be noted, that part of the current market is covered by CO_2 from natural wells instead of recovered CO_2 emissions (Aresta and Dibenedetto 2010). For example, in the USA, approx. 45Mt of CO_2 from natural reservoirs are currently used in enhanced oil or gas recovery (EOR/EGR) (Wilcox 2012). The cost of natural CO_2 is connected to the oil price (US EIA 2014) and relatively low at 15–20€/t due to its often rather high purity (Aresta and Dibenedetto 2010). However, assuming that CCU technologies must be measured in regard to their environmental performance, the use of natural CO_2 carries certain disadvantages. Extracting CO_2 that is naturally stored underground for the purpose of using it in the production of fuels and materials will result in higher total emissions than when using CO_2 that is emitted anyway, e.g., by an industrial plant. Thus, it is recommended to replace CO_2 from natural wells currently in use with recovered CO_2 to achieve a net emission reduction (Aresta and Dibenedetto 2010; Metz *et al.* 2005). Since this paper focuses on recovered CO_2 emissions, CO_2 from natural sources is consequently not further included in the presented data.

Based on the presented emission and cost data a potential supply curve for the commodity CO_2 is established in **Figure 2** and **Figure 3**. The low cost sources represent largely the high purity emitters (group I) and are detailed in **Figure 2**. These can collectively provide approx. 300Mt of CO_2 at a benchmark capture cost of approx. 33€/t or less. The aggregated capturable CO_2 supply of 12.7Gt CO_2 from all listed point sources is displayed in **Figure 3**. Coal power plants can provide large amounts of CO_2 at relatively low benchmark capture costs of approx. 34€/t.

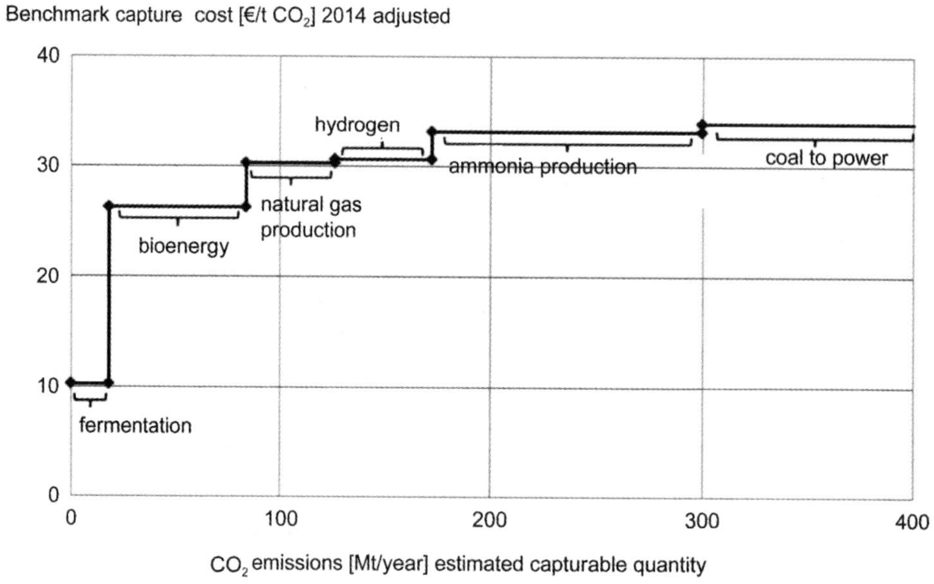

Figure 2. CO_2 supply curve: high purity and low capture cost sources.

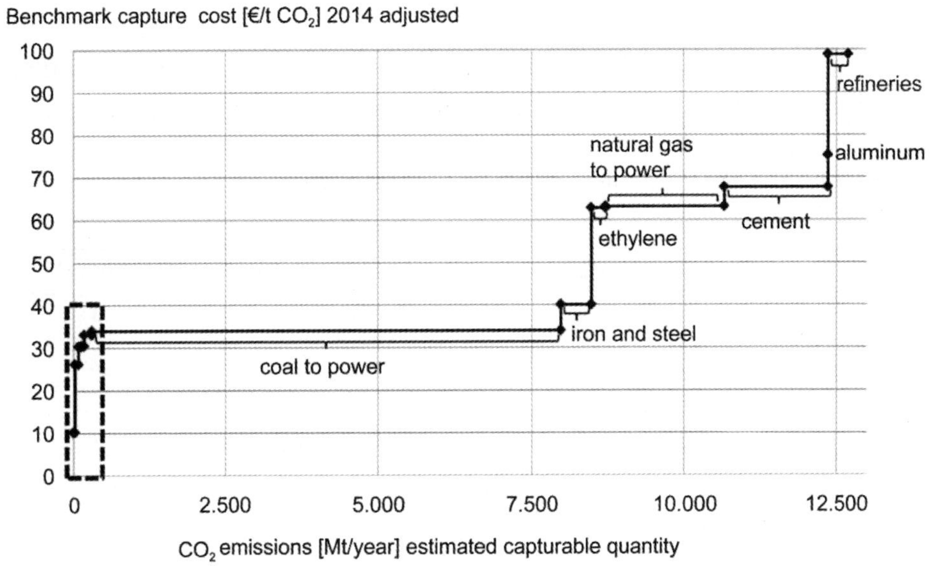

Figure 3. CO_2 supply curve: fossil power and large industrial sources.

Nevertheless, even if coal power was phased out in the future, several other industrial processes would cause large amounts of CO_2 emissions that can be captured at benchmark costs below 100€/t. Such a supply curve can also be designed for a single plant where different processes emit CO_2 at various capture costs, as exemplified in van Straelen *et al.* (2010).

3. Demand Side: Carbon Dioxide Capture and Utilization

The idea of using CO_2 as feedstock is as old as the chemical industry (Aresta and Dibenedetto 2010), but so far very few applications have been realized. The conversion of CO_2 with a catalyst evolved in the 1970s, when chemical engineers first succeeded in developing catalysis processes inspired by nature's CO_2 conversion cycles (Aresta and Dibenedetto 2010; Aresta *et al.* 2013). Due to the oil crises at that time, the discovery of alternative feedstock to lower the dependency of fossil resources was economically very attractive. With rising political and public awareness on climate change a large field of research has developed around possibilities to reduce industrial CO_2 emissions. Accordingly, technological research on CCU technologies slowly but surely has gained momentum in the last decades.

CO_2 can either be used directly or as feedstock for a variety of products. Overall, approx. 222Mt of the commodity are used in industrial applications worldwide (see current est. volumes in **Table 3**). Firstly, direct utilization of liquid

Table 3. Current and near-term markets of CO_2 utilization (based on Aresta *et al.* (2013) if not indicated otherwise).

Product/application	Current est. volumes[a]		Near-term est. volumes[b]	
In kt p.a.	CO_2	Product	CO_2	Product
Direct utilization	42,400		42,400	
Beverage carbonation[c]	2900	2900	2900[d]	2900d
Food packaging[c]	8200	8200	8200[d]	8200d
Industrial gas[c]	6300	6300	6300[d]	6300d
Oil and gas recovery (EOR/EGR)[e]	25,000	7%–23% of oil reserve, <5 % of gas reserve[f]	25.000[d]	7%–23% of oil reserve, <5% of gas reserve[f]
Materials	167,515		212,400	
Urea	114,000	155,000	132,000	180,000
Inorganic carbonates	50,000	200,000	70,000	250,000
Formaldehyde	3500	21,000	5000	25,000
PC (polycarbonates)	10	4000	1000	5000
Carbonates	5	200	500	2000
Acrylates	0	2500	1500	3000
Carbamates	0	5300	1000	6000
Formic acid	0	600	900	1000
PUR (polyurethanes)	0	8000	500	10,000
Fuels	12,510		20,000	
Methanol	8000	50,000	10,000	60,000
DME (dimethyl ether)	3000	11,400	>5000	>20,000
TBME (tertiary butyl methyl ether)	1500	30,000	3000	40,000
Algae to biodiesel	10	5	2000	1000
Total	222,425		274,800	

[a]Current data is based on the 2013 estimates from Aresta *et al.* (2013); [b]Near-term data is based on the former 2016 estimates from Aresta *et al.* (2013) and includes CCU technologies that could be implemented within the next 10 years; [c]Data from IHS (2013), worldwide data without Latin America and Asia except Japan; [d]Estimated as constant by the author, not included in the cited original literature; [e]Data from Global CCS Institute (2014); [f]Estimate from (Metz *et al.* 2005).

or gaseous carbon dioxide usually requires a very high purity especially in the food and beverage industry which currently consumes approx. 11Mt CO_2 per year. Furthermore, around 6Mt CO_2 are used as process gas in various industrial applications (IHS 2013). The largest direct use of 25Mt of CO_2 can be found in EOR/EGR which represent a borderline case, as they combine a utilization and storage function (Global CCS Institute 2014). Largely, they are attributed to CCS rather than CCU since after the extraction of additional fuels through CO_2, the CO_2 can potentially be stored permanently in the depleted oil and gas fields. As EOR/EGR is a potential market for recovered CO_2, it needs to be included when analyzing market volumes of CO_2 (see **Table 3**). Secondly, the conversion of CO_2 to materials still is limited to few applications at a smaller scale, except for urea synthesis which globally currently consumes approx. 130Mt CO_2 per year. Indeed, urea and ammonia production are often combined, so that an estimated half of the high purity CO_2 from ammonia production is used for urea synthesis while the rest is often vented (IEA 2013; Metz *et al.* 2005). Apart from that, a marginal amount of CO_2 is used for the production of several specialty chemicals, e.g., of salicylic acid used for making aspirin pills. Commercial plants producing CO_2-based fuels currently can be found only at demonstration scale of several thousand tonnes, e.g., by the companies Carbon Recycling International (CRI) in Iceland and Audi and Sunfire in Germany (CRI 2016; Strohbach 2013; Sunfire 2014). As R & D on CCU technologies continues and some important breakthroughs have been observed further CO_2-based products are expected to enter global markets soon as depicted in the near-term (up to 10 years) estimates in **Table 3**. Thus, the demand for CO_2 as a commodity might increase in the future.

3.1. CO_2 Utilization and Emission Reductions

CCS aims to store large amounts of CO_2 underground for long periods of time—approx. 1000 years (Metz *et al.* 2005). By contrast, when CO_2 is used directly or as feedstock for materials and fuels it will be reemitted to the atmosphere depending on the durability of the product, ranging from days to several years. As described earlier, the amounts of CO_2 used thus do not correspond to the amount of CO_2 avoided. Each CO_2 utilization process has a different environmental impact which needs to be determined in a life cycle assessment (von der Assen *et al.* 2015). The crux lies in the efficiency gains connected to the process: If the fossil raw material consumption of a production process can be reduced by the introduction of a CO_2-based process the environmental balance can be positive. A recent example illustrates how for polyols used for the production of foams up to 3 t of CO_2 emissions can be avoided per tonne of CO_2 used compared to a conventional production process (von der Assen and Bardow 2014). Hence, despite the short durability of CO_2 utilization compared to storage there is an unknown overall mitigation potential that can possibly be significantly larger than the volumes of CO_2 utilized. However, for a market perspective, the volumes of the commodity CO_2 that can be captured and used need to be matched. A judgment in regard to mitigation potential is not possible on that basis. While CCS is a recognized emission reduction instrument and commonly accounted for in existing carbon management schemes CO_2 utilization per se is not accounted as direct emission reduction. If CCU can lead to reductions in fossil raw material use it is possible that CCU indirectly affects emission accounting just as other efficiency measures.

3.2. Evaluation of Potential CO_2 Demand

The presented status quo of CO_2 utilization has demonstrated the limited demand for the commodity CO_2. However, ongoing worldwide CCU-related research covers a diverse array of utilization options. In the best case "recycling of CO_2 from anthropogenic sources provides a renewable, inexhaustible carbon source and could allow the continued use of derived carbon fuels in an environmentally friendly, carbon neutral way" (Mikkelsen *et al.* 2010). Even in direct utilization innovations are possible that go beyond the substitution of fossil feedstock. As currently shown by the CO_2-based dry cleaning innovation of the US company CO_2 Nexus large-scale dry cleaning with CO_2 could potentially lead to economic and environmental benefits by replacing and reducing the consumption of a combination of valuable raw materials such as water, natural gas, and energy (Madsen *et al.* 2014). At the same time, replacing existing refrigerants with CO_2 as a standard coolant in automobile air conditioning systems is currently considered at the European Commission (EurActiv.com 2013; Malvicino 2011). Consequently, even when the CO_2 is directly used it can potentially substitute various substances that are hazardous or have a higher climate impact (Aresta and Dibenedetto 2010).

Overall, future estimates for the potential of CO_2 utilization in the literature vary, but all range around the same maximum potential. For chemical materials, the estimated large-scale potential is around 200Mt CO_2 p.a. [212Mt est. by Aresta *et al.* (2013), 200Mt by Mikkelsen *et al.* (2010), 180Mt est. by VCI (2009), and 115Mt est. by Metz *et al.* (2005)]. By contrast, the estimated large-scale potential for fuel production with CO_2 is much bigger with approx. 2Gt CO_2 p.a. (VCI 2009). Altogether, large-scale CCU can hence potentially require a maximum of 5%–6% of the estimated 37Gt of anthropogenic CO_2 emissions (Le Quéré *et al.* 2014). In order to build a potential demand for the commodity CO_2, next to the utilization potential the storage potential must also be taken into account. Thus, CCS targets of the IEA will also be included in the scenarios (IEA 2013). Since EOR/EGR is usually considered as storage, it will be attributed to the CCS and not the CCU scenarios in this study. For direct utilization no long-term estimates exist, but due to the limited volumes currently required even significant demand changes are expected to have a minor impact on overall CO_2 demand. Consequently, the following five demand scenarios were identified:

A. CCU current

This scenario represents the current (2013 est.) CO_2 demand presented in **Table 3**. The aggregate CO_2 demand amounts to approx. 200Mt CO_2 excluding EOR/EGR.

B. CCU near-term

This scenario includes the expected near-term development of CO_2 utilization over the next 10 years based on the near-term scenario of **Table 3**. Next to a continued direct utilization of approx. 17Mt p.a. (excluding EOR/EGR) the production of some CCU-based fuels consumes an estimated 20Mt of CO_2 and the aggregated demand for CO_2 amounts to 250Mt p.a. Similar demand scenarios of mixed CCU material and fuel activities have also been projected in the range of 250–350Mt by Quadrelli and Centi (2011) and 300–400Mt by Aresta and Dibenedetto (2010) for the medium-term.

C. CCU and CCS near-term

This scenario combines the projected shorter-term development of 250Mt CO_2 used with 40Mt CO_2 stored (Carbon Visuals 2014). A combined 290Mt CO_2 thus depict a realistic demand for CO_2 in the near-term of up to 10 years.

D. CCU fuels long-term

This scenario illustrates a large-scale potential of CO_2 utilization including a transformation of the fuels sector to synthetic CO_2-based fuels. Due to the relatively low current prices of fossil energy, the transition away from fossil fuels is a larger endeavor however that will take much more than 10 years to implement. Implementing this scenario would be connected to severe infrastructure investments in particular in regard to renewable energy and fuel refineries and thus requires strong political efforts. Consequently, this scenario serves rather as an optimistic long-term vision of CCU with 2300Mt of CO_2 (250Mt for materials and direct use and 2050Mt for fuel production (VCI 2009).

E. CCU fuels and CCS long-term

The combination of the 2300Mt CO_2 of scenario D and the 2050 target for CCS deployment according to the IEA (2013) of 7870Mt CO_2 provides a long-term overall potential demand of combined use and storage of more than 10Gt CO_2.

In summary, scenarios A, B, and C represent realistic, near-term scenarios that show a potential demand of 200−300Mt CO_2 and represent less than 1% of anthropogenic emissions while scenarios D and E with 2−10Gt show an optimistic, long-term potential of materials, fuels, and storage and consume potentially 5%−27% of the current estimated anthropogenic emissions of 37Gt (Le Quéré *et al.* 2014).

4. Supply and Demand Scenarios

As a next step, the merit order of supply of CO_2 from industrial point sources will be matched with the described demand developments for CCU and CCS. For this, the discussed current and potential demand volumes (d) for scenarios A to E are combined with the supply of CO_2 in order to determine the equilibrium best practice benchmark capture prices (p^*) for each scenario. These are determined by applying formula (4) to the presented data.

$$p^*(d) = p(d) = p_i \, for \, d \in]q_{i-1}; q_i]$$

with

d = current and potential demand volumes for scenarios A to E

p^* = equilibrium best practice benchmark capture prices

For simplification purposes, it is assumed that the cost of capture correspond to the price of carbon dioxide paid by the consumer which implies that no profits are made on the commodity CO_2. The presented scenarios are useful to understand the potential allocation of captured emissions in global supply and demand of CO_2. The analysis of comparative statics is again grouped into short-term and long-term visions.

4.1. Short-Term Scenarios

The current and near-term potential is depicted in **Figure 4**. It becomes evident that there is enough CO_2 from high purity sources at a comparatively low equilibrium capture cost of approx. 33€/t to cover the CCU demand in the short-term. From a global perspective, the volumes captured from high purity sources

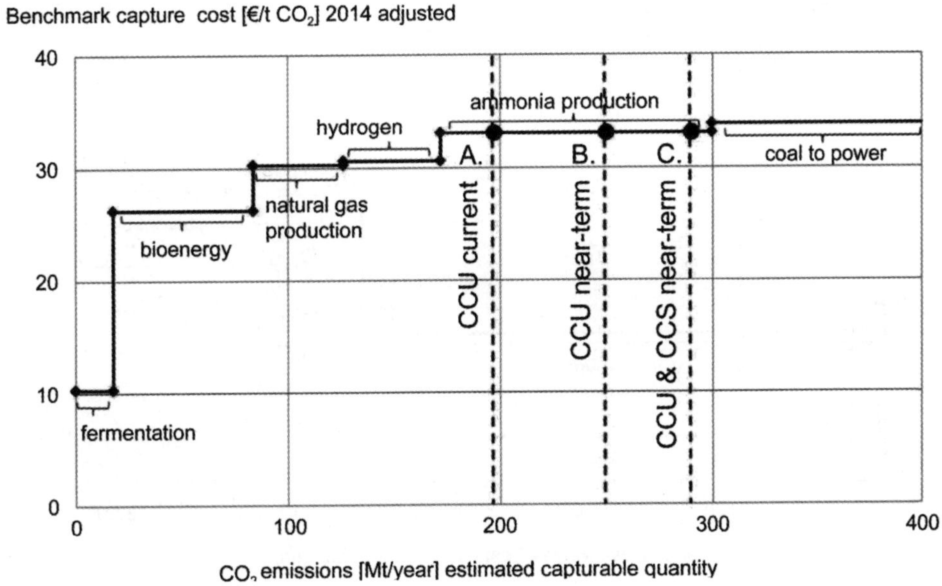

Figure 4. Short-term supply and demand scenarios.

such as fermentation, bioenergy, natural gas production, hydrogen, and ammonia would suffice to cover the current and upcoming CO_2 demand. Consequently, for the small total volumes required in each specific case the source is usually chosen based on local availability, respective quality, and cost of available CO_2. A trend towards using a plant's own or a nearby partner's waste emissions can be observed at several demonstration plants (Chem. Eng. 2015b; German Embassy Pretoria 2013; Tieman 2013).

4.2. Long-Term Scenarios

Considering the long-term development of CCU and CCS technologies, an optimistic full implementation of described target scenarios D and E is illustrated in **Figure 5**. Next to the high purity sources, other industrial plants will become relevant to capture the demanded CO_2 emissions. Currently, CO_2 from coal can be captured in large amounts (>7.5Gt p.a.) at benchmark costs of approx. 33€/t. Beyond coal and gas, other industrial processes such as iron and steel, ethylene, and cement production can provide CO_2 at benchmark capture costs of less than 70€/t. Considering the laid out business as usual supply base the long-term scenario D for CCU including fuels could be achieved with CO_2 at an equilibrium CO_2 price

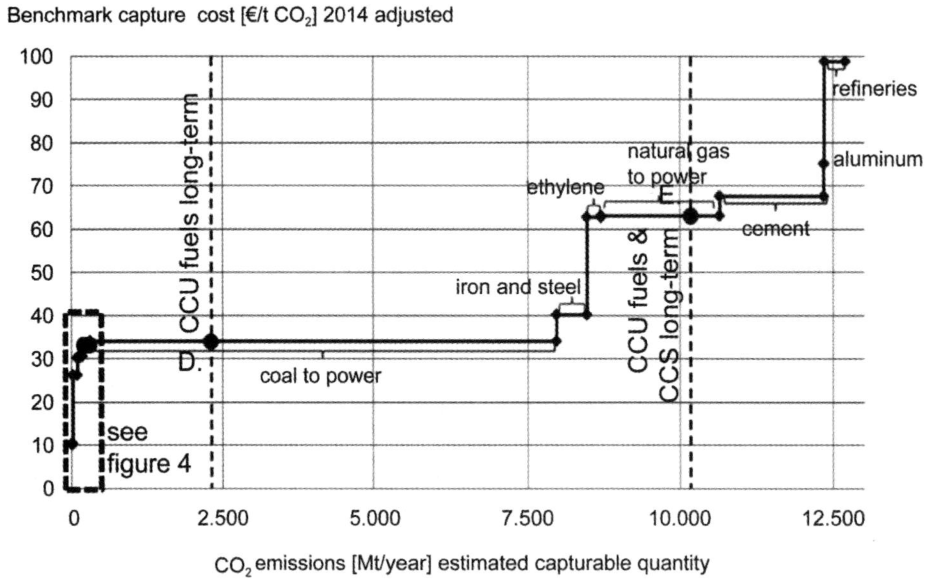

Figure 5. Long-term supply and demand scenarios.

of approx. 33€/t. For a combined long-term target for CCU and CCS, the equilibrium price per tonne based on current conditions would be approx. 63€/t neglecting possible price changes and inflation.

4.3. Scenarios without Fossil-Fired Power Generation

As stated earlier, the future development of CO_2 emission volumes and prices is unclear due to a variety of political scenarios. The static equilibrium of the current CO_2 supply with long-term demand scenarios illustrated in **Figure 5** can only provide insights for a business as usual scenario. If however major technological or political changes occur and significantly larger volumes of CO_2 will continually be mitigated or removed the investments into CCS and potentially also CCU can turn into a dead end and lead to sunk costs in the long run. For future scenarios assuming a larger share of renewable energy, remaining fossil power plants might have to work at lower load factors and efficiencies (Finkenrath 2011). Thus, capture costs at these plants could potentially be higher than the current data suggest. Moreover, for other industrial plants, new technologies can become available in the long run that emit significantly less CO_2. Then, the relevancy and cost of carbon capture could change—in both directions. For example in the iron and steel production the recently demonstrated Hisarna process can on the one hand reduce approx. 20% of the conventional CO_2 emissions. On the other hand, the process allows for a very efficient combination with CO_2 capture (Pfeifer 2015). Consequently, advancing green technologies across industries will impact available volumes and costs of recovering CO_2 emissions.

Since CCS technologies are largely considered as an instrument to improve the carbon footprint of continued fossil-based power generation (Metz *et al.* 2005), their implementation becomes less relevant once the energy sector would be based on renewables. To understand the effects on CCU, a modified merit order for CO_2 supply excluding power generation based on coal and natural gas has been developed in **Figure 6** and matched with a demand scenario excluding CCS. It becomes evident that in the near-term (scenario B) CO_2 supply for CCU would be provided from industrial high purity sources at an equilibrium price of approx. 33€/t, in line with scenario B depicted in **Figure 4**. Meanwhile the long-term, large-scale potential of CCU including fuels (scenario D) would be served from recovered emissions of various industrial production processes such as iron and steel, ethylene,

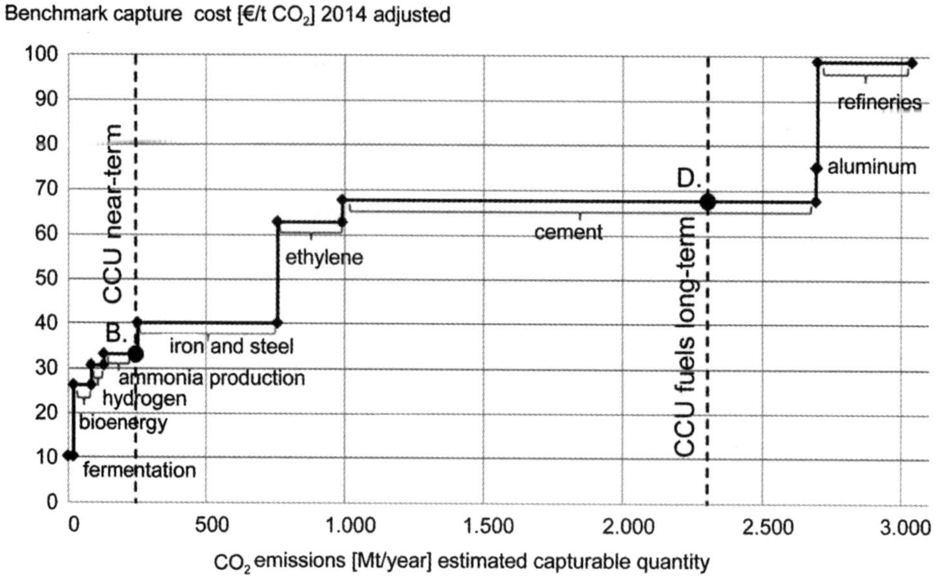

Figure 6. Supply and demand scenario without fossil-fired power generation.

and cement at equilibrium cost of approx. 68€/t. This modified supply scenario shows that the future development of CCU technologies is independent from the fossil power industry. By contrast, the amounts of CO_2 required even for the visionary potential of CCU can possibly be recovered from various industrial sources at estimated costs of less than 70€/t. CCU technologies thus do not conceptually contradict renewable energy or energy efficiency technologies. Instead, they can be seen as a complementary route.

4. Conclusions and Perspectives

With carbon capture and utilization, recovered CO_2 emissions could be turned into a valuable feedstock for the production of consumer goods. CCS in contrast aims to sequester recovered CO_2 permanently underground. For both technology fields CO_2 represents a commodity good that is potentially supplied from capture at industrial plants. The comparative statics of CO_2 demand and supply show that depending on the targeted scenario for CCU and CCS different industrial sources of CO_2 emissions will play a role: For the near future, smaller scale CCU scenarios, industrial plants with higher CO_2 concentration and lower benchmark capture costs of approx. 33€/t will be relevant. Meanwhile, for the long-term large-scale scenarios including the fuel sector and CCS coal-fired power plants or other larger emitters must also be included and CO_2 captured for benchmark costs of up to 65€/t will be needed. Even if fossil-fired power generation is phased out, capture at industrial production processes can provide more than enough CO_2 for large-scale CCU visions. Consequently, reusing recovered industrial CO_2 emissions can unfold its environmental and economic potential in creating regional or local circular economy solutions. For example, half of the total CO_2 emissions from waste incineration in Germany would sufficiently supply CO_2 as a carbon source for the domestic polymer consumption (Bringezu 2014). Moreover, capture of CO_2 from ambient air remains a potential alternative for recovering CO_2 emissions. However, technological development is at present still at an early stage and the costs of air capture are highly uncertain and depend on a variety of factors (Lackner *et al.* 2012). In any case, further research on environmental and economic impacts should accompany the technological advancement of carbon capture and utilization technologies. Harmonizing methods and measures in environmental and techno-economic assessments and establishing best practices would improve the interpretability of the results enormously.

Since the expected near-term status of CO_2 utilization does not involve large volumes of CO_2 smaller regional solutions gain importance. When new plants are set up that reuse CO_2 emissions these can be planned next to a convenient source of CO_2 at sufficient quality and a competitive cost. Emissions from fossil-fired power plants are not required to meet the potential CO_2 utilization demand. Even large-scale visions for CCU can therefore not serve as an argument to prolong fossil-fired power generation. When implementing large-scale CO_2 utilization scenarios involving synthetic fuels based on power-to-liquid or -gas technologies a broader infrastructure especially for renewable energy but also for CO_2 supply will be needed. Until then, from a mitigation perspective, differentiating recovered CO_2 by source can even be misleading as in sum it does not play a role to the environment where the reused CO_2 comes from. Instead, market mechanisms will balance supply and demand. Nevertheless, sustainability aspects always need to be considered when further deploying CCU technologies for example by conducting life-cycle analysis and considering alternative technologies based on renewable energy and raw materials.

Moreover, if energy efficient CCU technologies can be developed, the presented CO_2 supply benchmark cost range of 10−100€/t CO_2 can prove to be a relatively cheap alternative feedstock compared to more expensive or volatile priced chemicals based on fossil raw materials such as crude oil. The search for such technologies that use CO_2 to replace fossil raw materials and their derivatives consequently remains attractive even in times of relatively low or decreasing energy prices. Currently, in existing emission trading schemes, the CO_2 is largely underpriced. At current conditions, investments into CCU or CCS hence can only allow for future profits if substitution of expensive raw materials is possible, increased process and production efficiencies can be achieved or government subsidies compensate potential losses. In the future, however, the combined total of approx. 6Gt of global CO_2 emissions that are currently regulated by some form of carbon pricing instrument (World Bank 2014) could increase in amount and price. A sufficiently high carbon tax or emission trading price could then incentivize further CCU and CCS activities in certain regions. If the combined costs of capture, transport, and storage of a certain emitting source are lower than the CO_2 tax or certificate price CCS will have a business case. For example, the relatively old as well as high Norwegian carbon tax, especially for offshore petroleum businesses of up to 69€/t CO_2 in 2014 (World Bank 2014) has led to several investments by the affected players into CCS (de Coninck and Benson 2014) and energy efficiency (Bruvoll and Larsen 2004). In contrast, CCU technologies—as efficiency measures generally—are indirectly impacted by the carbon price. Depending on respective energy requirements and prices, certain technologies are profitable even at current conditions. A higher carbon price can be expected to stimulate the deployment of more CCU technologies. Best practice benchmark capture cost can give an indication but do not include potential substitution or efficiency effects connected to CCU processes. An equilibrium price of carbon dioxide for enabling the implementation of CCU from an economic perspective is thus technology specific.

Eventually, while policy makers and businesses must continue to work on mitigating global emissions, pathways for using waste emissions should be pursued complementary. For any desired future, reliable political targets and regulations will be important to permit optimal investment decisions when technologies are implemented and scaled up. Overall, moving ahead, CCU should be considered as a means for improving regional resource security and as enabler of smaller circular economy solutions. Making use of recovered CO_2 seems to be one option for imitating nature's no waste philosophy into industrial design and consumption.

Acknowledgements

The author thanks her colleagues Barbara Olfe-Kräutlein, Thomas Bruhn, Ana Maria Lorente Lafuente and Patrick Matschoss at the IASS, Christoph Gürtler from Covestro, and the reviewers for their valuable feedback on the manuscript. Moreover, she thanks the various helpful colleagues at the Gordon Research Conference on CCUS as well as at the ICCDU Singapore whose comments greatly improved the research.

Compliance with Ethical Standards

Conflict of Interest

The author declares that she has no conflict of interest.

Source: Naims H. Economics of carbon dioxide capture and utilization—a supply and demand perspective[J]. Environmental Science & Pollution Research, 2016: 1–16.

References

[1] Aresta M, Dibenedetto A (2010) Industrial utilization of carbon dioxide (CO_2). In: Maroto-Valer MM (ed) Developments and innovation in carbon dioxide (CO_2) capture and storage technology: Volume 2: Carbon dioxide (CO2) storage and utilisation, vol 2. Woodhead Publishing, Great Abington, pp 377–410.

[2] Aresta M, Dibenedetto A, Angelini A (2013) The changing paradigm in CO_2 utilization. Journal of CO_2 Utilization 3:65–73. doi:10.1016/j. jcou.2013.08.001.

[3] Baker RW, Lokhandwala K (2008) Natural gas processing with membranes: an overview. Ind Eng Chem Res 47:2109–2121. doi:10. 1021/ie071083w.

[4] Bennaceur K, Gielen D, Kerr T, Tam C (2008) CO_2 capture and storage: a key carbon abatement option, Energy Technology Analysis. OECD/ IEA, Paris.

[5] Bogner J *et al* (2007) Waste management. In: Metz B, Davidson OR, Bosch PR, Dave R, Meyer LA (eds) Climate Change 2007: Mitigation, Contribution of working group III to the fourth assessment report of the Intergovernmental Panel on Climate Change. Cambridge University Press, Cambridge, pp 585–618.

[6] Bringezu S (2014) Carbon recycling for renewable materials and energy supply: recent trends, long-term options, and challenges for research and development. J Ind Ecol 18:327–340. doi:10.1111/jiec.12099.

[7] Bruvoll A, Larsen BM (2004) Greenhouse gas emissions in Norway: do carbon taxes work? Energy Policy 32:493–505. doi:10.1016/S0301-4215(03)00151-4.

[8] California Carbon Dashboard (2015) Carbon price. California Carbon Dashboard., http://calcarbondash.org/. Accessed August 27, 2015.

[9] Carbon Visuals (2014) CCS: a 2 degree solution: a film by carbon visuals—methodology, WBCSD. http://wbcsdservers.org/web/wbcsdfiles/ClimateEnergy/WBCSD_CCS_film_methodology.pdf. Accessed May 27, 2015.

[10] Chem. Eng. (2008) Chemical Engineering Plant Cost Index (CEPCI).

[11] Chem Eng January 2008:60

[12] Chem. Eng. (2015a) Chemical Engineering Plant Cost Index (CEPCI).

[13] Chem Eng January 2015:64. http://www.chemengonline.com/ issues/2015-01.

[14] Chem. Eng. (2015b) Polymer produced from CO2 waste gas makes commercial debut. http://www.chemengonline.com/polymer-produced-co2-waste-gas-makes-commercial-debut/. Accessed 27 Aug 2015.

[15] CRI (2016) World's largest CO_2 methanol plant. http://carbonrecycling. is/projects-1/. Accessed 21 May 2016.

[16] de Coninck H, Benson SM (2014) Carbon dioxide capture and storage: issues and prospects. Annu Rev Env Resour 39:243–270. doi:10.1146/annurev-environ-032112-095222.

[17] EEX (2015) Market data: EEX Primary auction phase 3. European Energy Exchange (EEX). http://www.eex.com/en/market-data#/market-data. Accessed August 27, 2015.

[18] EurActiv.com (2013) Volkswagen follows Daimler in opting for CO_2 refrigerant. EurActiv.com/Reuters. http://www.euractiv.com/energy-efficiency/volkswagen-rules-honeywell-dupon-news-518389. Accessed August 28, 2015.

[19] Faulstich M, Leipprand A, Eggenstein U (2009) Relevance of CCS technology for waste incineration. In: Bilitewski B, Urban AI, Faulstich M (eds) Thermic waste treatment, vol 14. Kassel University Press, Kassel, pp 73–86 (in German).

[20] Federal Ministry of Education and Research (2014) Technologies for sustainability and climate protection—Chemical processes and use of CO2: Federal Ministry of Education and Research funding programme information brochure. Federal Ministry of Education and Research. http://www.chemieundco2.de/_media/technologies_for_

sustainability_climate_protection.pdf. Accessed August 25, 2015

[21] Finkenrath M (2011) Cost and performance of carbon dioxide capture from power generation. OECD/ IEA. http://www.iea.org/publications/freepublications/publication /costperf_ccs_powergen.pdf. Accessed January 27, 2015.

[22] German Embassy Pretoria (2013) LANXESS commissions CO2 concentration unit at Newcastle site in South Africa. German Missions in South Africa, Lesotho and Swaziland. http://www.southafrica.diplo.de/Vertretung/suedafrika/en/__pr/2__Embassy/ 2013/4thQ/10-Lanxess.html. Accessed January 21, 2015.

[23] Global CCS Institute (2014) Status of CCS project database. http://www.globalcc sinstitute.com/projects/status-ccs-project-database. Accessed 22 May 2015.

[24] IEA (2013) Technology roadmap: Carbon capture and storage—2013 edition. OECD/ IEA. http://www.iea.org/publications/freepublications/publication/technologyroadmap carboncaptureandstorage.pdf. Accessed May 27, 2015.

[25] IEA (2014) CO2 emissions from fuel combustion: highlights. OECD/IEA. https:// www.iea.org/publications/freepublications/publication/CO2EmissionsFromFuelComb ustionHighlights2014.pdf. Accessed August 20, 2015.

[26] IEA, UNIDO (2011) Technology roadmap: carbon capture and storage in industrial applications. OECD/IEA/UNIDO. http://www.iea.org/publications/freepublications /publication/ccs_industry.pdf. Accessed December 15, 2014.

[27] IHS (2013) Carbon dioxide. Chemical Economics Handbook. IHS, Englewood.

[28] Jenkins S (2015) Economic indicators: CEPCI. http://www.chemengonline.com/ economic-indicators-cepci/?printmode=1. Accessed August 14, 2015.

[29] Jilvero H, Mathisen A, Eldrup N-H, Normann F, Johnsson F, Müller GI, Melaaen MC (2014) Techno-economic analysis of carbon capture at an aluminum production plant—comparison of post-combustion capture using MEA and ammonia. Energy Procedia 63:6590−6601. doi:10.1016/j.egypro.2014.11.695.

[30] Jordal K *et al* (2014) Feeding a gas turbine with aluminum plant exhaust for increased CO2 concentration in capture plant. Energy Procedia 51:411−420. doi:10.1016/j.egypro.2015.03.055.

[31] Kuramochi T, Ramírez A, Turkenburg W, Faaij A (2011) Technoeconomic assessment and comparison of CO2 capture technologies for industrial processes: preliminary results for the iron and steel sector. Energy Procedia 4:1981−1988. doi:10.1016/j.egypro.2011.02.079.

[32] Kurokawa H, Shirasaki Y, Yasuda I (2011) Energy-efficient distributed carbon capture in hydrogen production from natural gas. Energy Procedia 4:674−680. doi:10.1016/j.egypro.2011.01.104.

[33] Lackner KS, Brennan S, Matter JM, Park A-HA, Wright A, van der Zwaan B (2012) The urgency of the development of CO2 capture from ambient air. Proceedings of the National Academy of Sciences 109:13156−13162. doi:10.1073/pnas.1108765109.

[34] Le Quéré C *et al* (2014) Global carbon budget 2014. Earth System Science Data Discussions 7:521–610. doi:10.5194/essdd-7-521-2014.

[35] Madsen S, Normile-Elzinga E, Kinsman R (2014) CO2 - based cleaning of commercial textiles: The world's first CO2 solution for cleanroom texti l es . C O 2 No xu s I n c. http://www.energy.ca.gov/2014publications/CEC-500-2014-083/CEC-500-2014-083.pdf. Accessed August 26, 2015.

[36] Malvicino C (2011) Final report summary—B-COOL (Low cost and high efficiency CO2 mobile air conditioning system for lower segment cars). European Commission. http://cordis.europa.eu/result/rcn/ 46813_en.html. Accessed August 28, 2015.

[37] Metz B, Davidson O, De Coninck H, Loos M, Meyer L (2005) IPCC special report on carbon dioxide capture and storage. Cambridge University Press. http://www.ipcc.ch/pdf/special-reports/srccs/srccs_wholereport.pdf. Accessed February 11, 2015.

[38] Mignard D (2014) Correlating the chemical engineering plant cost index with macro-economic indicators. Chemical Engineering Research and Design 92:285–294. doi:10.1016/j.cherd.2013.07.022.

[39] Mikkelsen M, Jorgensen M, Krebs FC (2010) The teraton challenge. A review of fixation and transformation of carbon dioxide. Energy & Environmental Science 3:43–81. doi:10.1039/B912904A.

[40] Möllersten K, Yan J, R Moreira J (2003) Potential market niches for biomass energy with CO2 capture and storage—opportunities for energy supply with negative CO2 emissions. Biomass Bioenerg 25:273–285. doi:10.1016/S0961-9534(03)00013-8.

[41] Mueller-Langer F, Tzimas E, Kaltschmitt M, Peteves S (2007) Technoeconomic assessment of hydrogen production processes for the hydrogen economy for the short and medium term. Int J Hydrogen Energ 32:3797–3810. doi:10.1016/j.ijhydene.2007.05.027.

[42] Pfeifer H (2015) Iron and steel industry. In: Fischedick M, Görner K, Thomeczek M (eds) CO2: Capture, storage, utilization: Holistic evaluation for the energy sector and industry. Springer, Berlin, pp 375–384 (in German).

[43] Quadrelli EA, Centi G (2011) Green carbon dioxide. ChemSusChem 4: 1179–1181. doi:10.1002/cssc.201100518.

[44] Strohbach O (2013) World premiere: Audi opened power-to-gas facility. Audi. http://www.volkswagenag.com/content/vwcorp/info_center/en/themes/2013/06/Audi_opens_power_to_gas_facility.html. Accessed August 24, 2015.

[45] Sunfire (2014) Sunfire presents power-to-liquids. http://www.sunfire.de/wp-content/ uploads/Sunfire-PM-2014-05-Einweihung-PtL_final_EN.pdf. Accessed August 24, 2015.

[46] Tieman R (2013) Chemicals: science views waste in role as raw material of the future. http://www.ft.com/cms/s/0/2ef57b06-2c26-11e3-acf4-00144feab7de.html#axzz3PS8ZsD8B. Accessed January 21, 2015.

[47] US DOE (n.d.) Innovative concepts for beneficial reuse of carbon dioxide. U.S. Department of Energy. http://energy.gov/fe/innovative-concepts-beneficial-reuse-carbon-dioxide-0. Accessed January 20, 2015.

[48] US EIA (2014) Assumptions to the annual energy outlook 2014. U.S. Department of Energy. http://www.eia.gov/forecasts/aeo/assumptions/pdf/0554(2014).pdf. Accessed November 19, 2014.

[49] van Straelen J, Geuzebroek F, Goodchild N, Protopapas G, Mahony L (2010) CO2 capture for refineries, a practical approach. Int J Greenh Gas Con 4:316–320. doi:10.1016/j.ijggc.2009.09.022.

[50] VCI, DECHEMA (2009) Position paper utilisation and storage of CO2. DECHEMA. http://www.dechema.de/dechema_media/Positionspapier_co2_englisch-p-2965.pdf. Accessed December 12, 2014.

[51] von der Assen N, Bardow A (2014) Life cycle assessment of polyols for polyurethane production using CO2 as feedstock: insights from an industrial case study. Green Chem 16:3272-3280. doi:10.1039/ c4gc00513a.

[52] von der Assen N, Jung J, Bardow A (2013) Life-cycle assessment of carbon dioxide capture and utilization: avoiding the pitfalls. Energ Environ Sci 6:2721–2734. doi:10.1039/c3ee41151f.

[53] von der Assen N, Lorente Lafuente AM, Peters M, Bardow A (2015) Environmental assessment of CO2 capture and utilisation. In: Armstrong K, Styring P, Quadrelli EA (eds) Carbon dioxide utilisation. Elsevier, Amsterdam, pp 45–56.

[54] Weikl MC, Schmidt G (2010) Carbon capture in cracking furnaces. In: AIChE 2010 Spring Meeting & 6th Global Congress on Process Safety, San Antonio, 2010.

[55] Wilcox J (2012) Carbon capture. Springer, New York.

[56] World Bank (2014) State and trends of carbon pricing 2014. World Bank Group. http://www-wds.worldbank.org/external/default/WDSContentServer/WDSP/IB/2015/06/01/090224b0828bcd9e/1_0/Rendered/PDF/State0and0trends0of0carbon0pricing0 2014.pdf. Accessed August 28, 2015.

[57] World Economic Forum (2014) Towards the circular economy: accelerating the scale-up across global supply chains. World Economic Forum. http://www3.weforum.org/docs/WEF_ENV_TowardsCircularEconomy_Report_2014.pdf. Accessed November 28, 2014.

Chapter 16

Energy and Sustainable Development in Nigeria: The Way Forward

Sunday Olayinka Oyedepo

Mechanical Engineering Department, Covenant University, Ota 2023, Nigeria

Abstract: Access to clean modern energy services is an enormous challenge facing the African continent because energy is fundamental for socioeconomic development and poverty eradication. Today, 60% to 70% of the Nigerian population does not have access to electricity. There is no doubt that the present power crisis afflicting Nigeria will persist unless the government diversifies the energy sources in domestic, commercial, and industrial sectors and adopts new available technologies to reduce energy wastages and to save cost. This review examines a set of energy policy interventions, which can make a major contribution to the sustainable economic, environmental, and social development of Africa's most populated country, Nigeria. Energy efficiency leads to important social benefits, such as reducing the energy bills for poor households. From an economic point of view, implementing the country's renewable energy target will have significant costs, but these can partly be offset by selling carbon credits according to the rules of the 'Clean Development Mechanism' agreed some 10 years ago, which will result in indirect health benefits. Nigeria could benefit from the targeted interventions that would reduce the local air pollution and help the country to tackle greenhouse gas emissions. Many factors that need to be considered and appropriately addressed in the shift to its sustainable energy future are examined in this article. These include a full exploitation and promotion of renewable energy resources, energy efficiency practices, as well as the application of energy conservation measures in various sectors such as in the construction of industrial, residential, and office buildings, in transportation, etc.

Keywords: Sustainable Energy, Renewable Energy, Energy Efficiency, Energy Conservation

1. Background

Energy plays the most vital role in the economic growth, progress, and development, as well as poverty eradication and security of any nation. Uninterrupted energy supply is a vital issue for all countries today. Future economic growth crucially depends on the long-term availability of energy from sources that are affordable, accessible, and environmentally friendly. Security, climate change, and public health are closely interrelated with energy[1]. Energy is an important factor in all the sectors of any country's economy. The standard of living of a given country can be directly related to the per capita energy consumption. The recent world's energy crisis is due to two reasons: the rapid population growth and the increase in the living standard of whole societies. The per capita energy consumption is a measure of the per capita income as well as a measure of the prosperity of a nation[2].

Energy supports the provision of basic needs such as cooked food, a comfortable living temperature, lighting, the use of appliances, piped water or sewerage, essential health care (refrigerated vaccines, emergency, and intensive care), educational aids, communication (radio, television, electronic mail, the World Wide Web), and transport. Energy also fuels productive activities including agriculture, commerce, manufacturing, industry, and mining. Conversely, a lack of access to energy contributes to poverty and deprivation and can contribute to the economic decline. Energy and poverty reduction are not only closely connected with each other, but also with the socioeconomic development, which involves productivity, income growth, education, and health[3].

The energy crisis, which has engulfed Nigeria for almost two decades, has been enormous and has largely contributed to the incidence of poverty by paralyzing industrial and commercial activities during this period. The Council for Renewable Energy of Nigeria estimates that power outages brought about a loss of 126 billion naira (US$ 984.38 million) annually[4]. Apart from the huge income loss, it has also resulted in health hazards due to the exposure to carbon emissions caused by constant use of 'backyard generators' in different households and business enterprises, unemployment, and high cost of living leading to a deterioration of living conditions.

Moreover, according to the Central Bank estimate in 1985, Nigeria consumed 8,771,863 tonnes of oil equivalent[5]. This is equal to about 180,000 barrels of oil per day. Since then, oil consumption in Nigeria has drastically increased. The effect of this increase on the economy relying solely on revenue from oil is tremendous. Also, the Department for Petroleum Resources[6] reported an amount of petroleum of more than 78% of the total energy consumption in Nigeria. In the present predicament as a nation, it is obvious that depending mainly on fossil fuel (petroleum) is not enough to meet the energy needs of the country. Since Nigeria is blessed with abundant renewable energy resources such as hydroelectric, solar, wind, tidal, and biomass, there is a need to harness these resources and chart a new energy future for Nigeria. In this regard, the government has a responsibility to make renewable energy available and affordable to all.

Many indigenous researchers have looked into the availability of renewable energy resources in Nigeria with a view to establishing their viability in the country. Onyebuchi[7] estimated the technical potential of solar energy in Nigeria with a 5% device conversion efficiency put at 15.0 × 1014kJ of useful energy annually. This equates to about 258.62 million barrels of oil equivalent annually, which corresponds to the current national annual fossil fuel production in the country. This will also amount to about 4.2 × 105GW/h of electricity production annually, which is about 26 times the recent annual electricity production of 16,000 GW/h in the country. In their work, Chineke and Igwiro[8] show that Nigeria receives abundant solar energy that can be usefully harnessed with an annual average daily solar radiation of about 5.25kW h/m^2/day. This varies between 3.5kW h/m^2/day at the coastal areas and 7kW h/m^2/day at the northern boundary. The average amount of sunshine hours all over the country is estimated to be about 6.5h. This gives an average annual solar energy intensity of 1,934.5kW h/m^2/year; thus, over the course of a year, an average of 6,372,613PJ/year (approximately 1,770TW h/year) of solar energy falls on the entire land area of Nigeria. This is about 120,000 times the total annual average electrical energy generated by the Power Holding Company of Nigeria (PHCN). With a 10% conservative conversion efficiency, the available solar energy resource is about 23 times the Energy Commission of Nigeria's (ECN) projection of the total final energy demand for Nigeria in the year 2030[9]. To enhance the developmental trend in the country, there is every need to support the existing unreliable energy sector with a sustainable source of power supply through solar energy.

Moreover, many indigenous researchers have also explored the availability of wind energy sources in Nigeria with a view of implementing them if there is a likelihood for their usage. Adekoya and Adewale[10] analyzed the wind speed data of 30 stations in Nigeria, determining the annual mean wind speeds and power flux densities, which vary from 1.5 to 4.1m/s to 5.7 to 22.5W/m^2, respectively. Fagbenle and Karayiannis[11] carried out a 10-year wind data analysis from 1979 to 1988, considering the surface and upper winds as well as the maximum gusts, whereas Ngala *et al.*[12] performed a statistical analysis of the wind energy potential in Maiduguri, Borno State, using the Weibull distribution and 10-year (1995 to 2004) wind data. A cost benefit analysis was also performed using the wind energy conversion systems for electric power generation and supply in the State. Each of these reports point to the fact that the nation is blessed with a vast opportunity for harvesting wind for electricity production, particularly at the core northern states, the mountainous parts of the central and eastern states, and also the offshore areas, where wind is abundantly available throughout the year. The issue then is for the country to look at ways of harnessing resources towards establishing wind farms in various regions and zones that have been identified as possessing abilities for the harvesting of wind energy.

Akinbami[13] reported that the total hydroelectric power potential of the country was estimated to be about 8,824MW with an annual electricity generation potential in excess of 36,000GW h. This consists of 8,000MW of large hydropower technology, while the remaining 824MW is still small-scale hydropower technology. Presently, 24% and 4% of both large and small hydropower potentials, respectively, in the country have been exploited.

Akinbami *et al.*'s assessment[14] indicated that the identified feedstock substrate for an economically feasible biogas program in Nigeria includes water lettuce, water hyacinth, dung, cassava leaves, urban refuse, solid (including industrial) waste, agricultural residues, and sewage. The authors' views include the following: Nigeria produces about 227,500 tonnes of fresh animal wastes daily. Since 1 kg of fresh animal wastes produces about 0.03m^3 gas, then Nigeria could produce about 6.8 million m^3 of biogas every day. In addition to all these, 20kg of municipal solid wastes per capital has been estimated to be generated in the country annually.

The prime objectives of this paper are (1) to review the current status of the energy resources, the energy demand, and supply in Nigeria and (2) to explore the prospects of utilizing renewable energy resources and to increase the energy efficiency as a possible means of sustainable development in Nigeria.

2. Energy Situation in Nigeria

Nigeria is Africa's energy giant. It is the continent's most prolific oil-producing country, which, along with Libya, accounts for two-thirds of Africa's crude oil reserves. It ranks second to Algeria in natural gas[15]. Most of Africa's bitumen and lignite reserves are found in Nigeria. In its mix of conventional energy reserves, Nigeria is simply unmatched by any other country on the African continent. It is not surprising therefore that energy export is the mainstay of the Nigerian economy. Also, primary energy resources dominate the nation's industrial raw material endowment.

Several energy resources are available in Nigeria in abundant proportions. The country possesses the world's sixth largest reserve of crude oil. Nigeria has an estimated oil reserve of 36.2 billion barrels. It is increasingly an important gas province with proven reserves of nearly 5,000 billion m^3. The oil and gas reserves are mainly found and located along the Niger Delta, Gulf of Guinea, and Bight of Bonny. Most of the exploration activities are focused in deep and ultra-deep offshore areas with planned activities in the Chad basin, in the northeast. Coal and lignite reserves are estimated to be 2.7 billion tons, while tar sand reserves represent 31 billion barrels of oil equivalent. The identified hydroelectricity sites have an estimated capacity of about 14,250MW. Nigeria has significant biomass resources to meet both traditional and modern energy uses, including electricity generation[16]. **Table 1** shows Nigeria's energy reserves/potentials. There has been a supply and demand gap as a result of the inadequate development and inefficient management of the energy sector. The supply of electricity, the country's most used energy resource, has been erratic[17].

The situation in the rural areas of the country is that most end users depend on fuel wood. Fuel wood is used by over 70% of Nigerians living in the rural areas. Nigeria consumes over 50 million tonnes of fuel wood annually, a rate which exceeds the replenishment rate through various afforestation programs. Sourcing fuel wood for domestic and commercial uses is a major cause of desertification in the arid-zone states and erosion in the southern part of the country. The rate of deforestation is about 350,000ha/year, which is equivalent to 3.6% of the present area of forests and woodlands, whereas reforestation is only at about 10% of the deforestation rate[19].

Table 1. Nigeria's energy reserves/capacity as in December 2005.

Resource type	Reserves	Reserves (BTOE)[c]	Reserves ($\times 10^7$)TJ
Crude oil	36.2 billion barrels	4.896	20.499
Natural gas	166 trillion SCF[a]	4.465	18.694
Coal and lignite	2.7 billion tonnes	1.882	7.879
Tar sands	31 billion barrels of oil equivalent	4.216	17.652
Subtotal Fossil		15.459	64.724
Hydropower, large Scale	11,000MW		0.0341/year
Hydropower, small Scale	3,250MW		0.0101/year
Fuel wood	13,071,464ha[b]		
Animal waste	61 million tonnes/year		
Crop residue	83 million tonnes/year		
Solar radiation	3.5 to 7.0kW h/m^2/day		
Wind	2 to 4m/s (annual average) at 10m in height		

[a]SCF, standard cubic feet; [b]forest land estimate for 1981; [c]BTOE, billion tonnes of oil equivalent. Adapted from ECN[18].

The rural areas, which are generally inaccessible due to the absence of good road networks, have little access to conventional energy such as electricity and petroleum products. Petroleum products such as kerosene and gasoline are purchased in the rural areas at prices 150% in excess of their official pump prices. The daily needs of the rural populace for heat energy are therefore met almost entirely from fuel wood. The sale of fuel wood and charcoal is mostly uncontrolled in the unorganized private sector. The sale of kerosene, electricity and cooking gas is essentially influenced and controlled by the Federal Government or its agencies—the Nigerian National Petroleum Corporation (NNPC) in the case of kerosene and cooking gas, and the PHCN in the case of electricity. The policy of the Federal Government had been to subsidize the pricing of locally consumed petroleum products, including electricity. In a bid to make the petroleum downstream sector more efficient and in an attempt to stem petroleum product consumption as a policy focus, the government has reduced and removed subsidies on various energy resources in Nigeria. The various policy options have always engendered price increases of the products[20].

With the restructuring of the power sector and the imminent privatization of the electricity industry, it is obvious that for logistic and economic reasons especially in the privatized power sector, rural areas that are remote from the grid and/or have low consumption or low power purchase potential will not be attractive to private power investors. Such areas may remain unserved into the distant future[21].

Meanwhile, electricity is required for such basic developmental services as pipe borne water, health care, telecommunications, and quality education. The poverty eradication and Universal Basic Education programs require energy for success. The absence of reliable energy supply has not only left the rural populace socially backward, but has also left their economic potentials untapped. Fortunately, Nigeria is blessed with abundant renewable energy resources such as solar, wind, biomass, and small hydropower potentials. The logical solution is increased penetration of renewables into the energy supply mix[15].

3. Energy Consumption Pattern in Nigeria

Energy consumption patterns in the world today shows that Nigeria and indeed African countries have the lowest rates of consumption. Nevertheless, Nigeria suffers from an inadequate supply of usable energy due to the rapidly increasing demand, which is typical of a developing economy. Paradoxically, the country is potentially endowed with sustainable energy resources. Nigeria is rich in conventional energy resources, which include oil, national gas, lignite, and coal. It is also well endowed with renewable energy sources such as wood, solar, hydropower, and wind[17].

The patterns of energy usage in Nigeria's economy can be divided into industrial, transport, commercial, agricultural, and household sectors[22]. The household sector accounts for the largest share of energy usage in the country—about 65%. This is largely due to the low level of development in all the other sectors.

The major energy-consuming activities in Nigeria's households are cooking, lighting, and use of electrical appliances. Cooking accounts for a staggering 91% of household energy consumption, lighting uses up to 6%, and the remaining 3% can be attributed to the use of basic electrical appliances such as televisions and pressing irons[9].

The predominant energy resources for domestic and commercial uses in Nigeria are fuel wood, charcoal, kerosene, cooking gas and electricity[20]. Other sources, though less common, are sawdust, agricultural crop residues of corn stalk, cassava sticks, and, in extreme cases, cow dung. In Nigeria, among the urban dwellers, kerosene and gas are the major cooking fuels. The majority of the people rely on kerosene stoves for domestic cooking, while only a few use gas and electric cookers[23].

The rural areas have little access to conventional energy such as electricity and petroleum products due to the absence of good road networks. Petroleum products such as kerosene and gasoline are purchased in the rural areas at prices very high in excess of their official pump prices. The rural population, whose needs are often basic, therefore depends to a large extent on fuel wood as a major traditional source of fuel. It has been estimated that about 86% of rural households in Nigeria depend on fuel wood as their source of energy[24]. A fuel wood supply/demand imbalance in some parts of the country is now a real threat to the energy security of the rural communities[22].

The energy consumption per capita in Nigeria is very small—about one-sixth of the energy consumed in developed countries. This is directly linked to the level of poverty in the country. Gross domestic product (GDP) and per capita income are indices that are used to measure the economic well-being of a country and its people[25]. GDP is defined as the total market value of all final goods and services produced within a given country in a given period of time (usually a calendar year). The per capita income refers to how much each individual receives, in monetary terms, of the yearly income that is generated in his/her country through productive activities. That is what each citizen would receive if the yearly income generated by a country from its productive activities were divided equally between everyone.

4. Current Electricity Situation in Nigeria

The electricity system in Nigeria centers on PHCN, which accounts for about 98% of the total electricity generation[26]. Power generation by other agencies such as the Nigerian Electricity Supply Company relies on thermal power for electricity generation unlike PHCN, which relies on both hydro- and thermal power. However, electricity is also a consumer of fuel and energy such as fuel oil, natural gas, and diesel oil. The importance of these sources of energy and fuel for generating electricity has been decreasing in recent years. However, hydropower that is relatively cheaper than these sources has grown to be more important than other sources[27]. However, more recently, the Power Authority has generated electricity through a mix of both thermal and hydro systems. All the power, distribution, and substations are specially interlinked by a transmission network popularly known as the national grid. The entire electricity generated nationwide is pooled into the National Control Centre, Osogbo, from where electricity is distributed to all parts of Nigeria.

The national electricity grid presently consists of 14 generating stations (3 hydro and 11 thermal) with a total installed capacity of about 8,039MW as shown in **Table 2**. The transmission network is made up of 5,000km of 330-kV lines, 6,000km of 132-kV lines, 23 of 330/132-kV substations, with a combined capacity of 6,000 or 4,600MVA at a utilization factor of 80%. In turn, the 91 of 132/33-kV substations have a combined capacity of 7,800 or 5,800MVA at a utilization factor of 75%. The distribution sector is comprised of 23,753km of 33-kV lines,

Table 2. Summary of generation capabilities of PHCN power stations as operated in 2008 (January to December).

Plant	Operator	Age (year)	Type	Installed capacity (MW)	Average availability (MW)	Availability factor	Number of units installed	Current number available
Kainji	PHCN	38 to 40	Hydro	760	438.86	0.58	8	6
Jebba	PHCN	25	Hydro	578.4	529.40	0.92	6	4
Shiroro	PHCN	22	Hydro	600	488.82	0.81	4	4
Egbin	PHCN	23	ST	1320	694.97	0.53	6	5
AES	AES	7	GT	315	233.91	0.77	9	9
Ajaokuta	STS	NA	GT	110	24.88	0.23	2	2
Sapele	PHCN	26 to 30	ST/GT	1020	156.60	0.15	10	1
Okpai	AGIP	3	GT/ST	480	394.56	0.88	3	3
Afam	PHCN	8 to 45	GT	709.6	82.12	0.09	20	3
Delta	PHCN	18	GT	912	211.67	0.24	18	12
Geregu	PHCN	NA	GT	414	305.14	0.74	3	3
Omoku	RS	3	GT	150	87.27	0.87	6	4
Omotosho	PHCN	1	GT	335	256.58	0.77	8	2
Olorunsogo	PHCN	1	GT	335	271.46	0.81	8	2
Total				8,039	4176.24	0.50	93	45

Adapted from PHCN[30].

19,226km of 11-kV lines, and 679 of 33/11-kV substations. There are also 1,790 distribution transformers and 680 injection substations[28]. **Table 2** shows a summary of the generation capabilities of PHCN power stations as operated in the year 2008 (January to December)[29].

As it can be seen in **Table 2**, the existing plants operate at far below their installed capacity as many of them have units that need to be rehabilitated, retrofitted, and upgraded[31]. The percentage of generation capability from hydro turbines is 34.89%; from gas turbine, 35.27%; and from steam turbines, 29.84%. The relative contribution of the hydropower stations to the total electricity generation (megawatt per hour) is greater than that of the thermal power stations.

In terms of the consumption of electricity, a classification into three groups has been proposed (industrial, residential, and street light consumption). In 1970, the total electricity consumption stood at 145.3MW/h; this increased to about 536.9MW/h in 1980. However, in 2005, the total electricity consumption had increased to 1,873.1MW/h[32]. On the generation side, these values of 176.6MW/h in 1970 increased to 815.1MW/h in 1980. By the end of 2005, the achieved total electricity generation was 2,997.3MW/h[32]. Comparing the per capita power generation to that of other countries, Nigeria has the lowest among the countries, as shown in **Table 3**, while the USA has the highest per capita electricity generation.

In spite of the contribution of electricity to the total gross domestic product, it is evident that Nigeria is facing several problems. The incapacity of the electricity subsector to efficiently meet the demand for electricity in the country has been caused by a number of problems, which have been detrimental to economic growth. The Central Bank of Nigeria[26] has identified nine problems associated with the National Electric Power Authority (NEPA) (now PHCN):

Table 3. Country statistics of electricity generation and per capita consumption.

Continent	Country	Population (million)	Generation capacity (MW)	Per capita consumption (kW)
North America	USA	250	813,000	3.2
South America	Cuba	10.54	4,000	0.38
Europe (central)	UK	57.5	76,000	1.1
Europe (eastern)	Ukraine	49	54,000	1.33
Middle East	Iraq	23.6	10,000	0.42
Far East	Republic of Korea	47	52,000	1.10
Africa	Nigeria	140	<4,000	0.03
	Egypt	67.9	18,000	0.27
	South Africa	44.3	45,000	1.02

Adapted from Okafor and Joe-Uzuegbu[17].

1) Lack of preventive and routine maintenance of NEPA's facilities, resulting in huge energy losses.

2) Frequent major breakdowns, arising from the use of outdated and heavily overloaded equipment.

3) Lack of coordination between town planning authorities and PHCN, resulting in poor overall power system planning and overloading of PHCN equipment.

4) Inadequate generation due to operational/technical problems arising from machine breakdown, low gas pressure, and low water levels.

5) Poor funding of the organization.

6) Inadequate budgetary provision and undue delay in release of funds to PHCN.

7) PHCN's inefficient billing and collection system.

8) High indebtedness to PHCN by both public and private consumers who are reluctant to pay for electricity consumed when due.

9) Vandalizing and pilfering of PHCN equipment.

In addition to these, most of the existing electricity plants in Nigeria are underutilized or not functioning at all. Numerous reasons could be sighted as responsible for the underutilization of these plants. Some of which are (1) scarcity of relevant manpower for adequate maintenance and general consumer indiscipline; (2) lack of essential spare parts for maintenance of the plants; (3) absence of local manufacturing capabilities; (4) lack of systematic studies of distribution networks to reduce the extraordinary losses that usually accompany haphazard system expansion; and (5) inability to convert gas flares to a source of electricity[33].

The inefficiency as well as the inadequate facilities to boost electricity supply also have been major causes of the increasing gap between the demand and the supply of electricity. This could be due to the fact that there are only 14 generating stations in Nigeria (3 hydro and 11 thermal stations). Out of the approximated 8,039MW of installed capacity in Nigeria, not more than 4,500MW is ever produced. This is due to poor maintenance, fluctuation in water levels powering the hydro plants, and the loss of electricity in transmission. It could also be due to the 80-MW export of electricity each to the republic of Niger and Benin. 'Apart from serving as a pillar of wealth creation in Nigeria, electricity is also the nucleus of operations and subsequently the engine of growth for all sectors of the economy'[34]. It has been indirectly re-echoed that electricity consumption is positively related to economic growth and that the former is a causal factor of the latter. This means that electricity consumption has diverse impacts on a range of socioeconomic activities and consequentially the living standards of Nigerians.

Notwithstanding the above pitfalls that had rendered public electricity supply in Nigeria unreliable and inefficient, the trend of its utilization has grown significantly over the past years. **Figure 1** shows the total electricity consumption in megawatts per hour and the various sectorial decompositions. Electricity utilization by the industrial sector has been fairly static because of the unreliable nature of the public electricity supply system in the country. Thus, many companies have resolved to provide their own power-generating sets as sources of electricity, leading to huge transfer costs on their products and services.

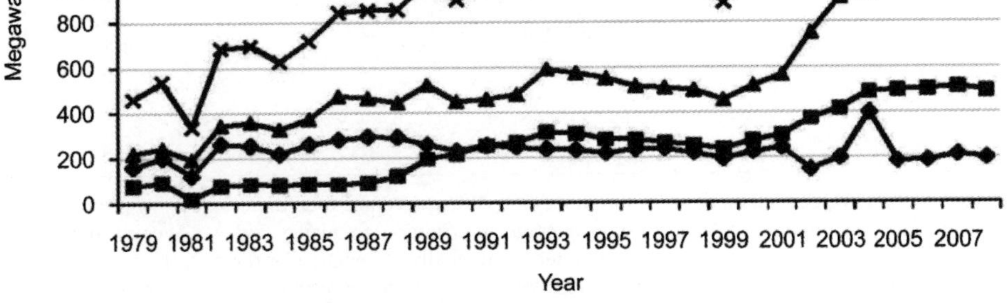

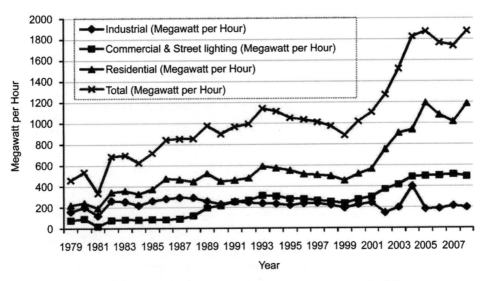

Figure 1. Electricity consumption pattern in Nigeria. Adapted from CBN[35].

Studies and experiences have shown that power generation in the country has been dismal and unable to compare with what has been obtained in smaller African countries. Manufacturers Association of Nigeria (MAN) gave the following performance indicators in **Table 4** for Nigeria's electricity sector compared with those of some other countries[28]. The data for some Southern Africa Development Community (SADC) countries such as Botswana and South Africa are comparable to those of the USA and France. The performance of the Nigerian power sector on the International Best Practices comparative rating is disgraceful. Perhaps, no other sector feels it as much as the manufacturing industrial sector wherein some notable international companies and organizations are on self-generated electricity 24 h/day for the 365 days of each year, as confirmed by the United Nations Industrial Development Organization in 2009[36]. The survey showed that, on average, manufacturers generated about 72% of the total power required to run their factories.

5. The Nigerian Energy Challenge

Nigeria's energy need is on the increase, and its increasing population is not adequately considered in the energy development program. The present urban-centered energy policy is deplorable, as cases of rural and sub-rural energy demand and supply do not reach the center stage of the country's energy development policy. People in rural areas depend on burning wood and traditional biomass for their energy needs, causing great deforestation, emitting greenhouse gases, and

Table 4. Power supply reliability indices (international best practices).

Index	USA	Singapore	France	Nigeria (NEPA data)	Nigeria (MAN study)
SAIDI[a](min)	88	1.5	52	900	≥60,000
SAIFI[b](number/year)	1.5	NA	NA	5	≥600
CAIDI[c](h)	0	NA	0	9	15
ASAI[d]	1	1	1	NA	≤0.4

[a]SAIDI, System average international duration index—Annual average total duration of power interruption to a consumer, in minutes; [b]SAIFI, System average interruption frequency index—Average number of interruptions of supply that a consumer experiences annually; [c]CAIDI, Consumer average interruption duration index—Average duration of an interruption of supply for a consumer who experiences the interruption on an annual basis, in hours; [d]ASAI, Average service availability index-Ratio of (Consumer hours service availability)/(Consumer hours service demanded). Adapted from Fagbenle *et al.*[28].

polluting the environment, thus creating global warming and environmental concerns. The main task has been to supply energy to the cities and various places of industrialization, thereby creating an energy imbalance within the country's socioeconomic and political landscapes. Comparing the present and ever increasing population with the total capacity of the available power stations reveals that Nigeria is not able to meet the energy needs of the people. The rural dwellers still lack electric power[37].

The nature of Nigeria's energy crises can be characterized by two key factors. The first concerns the recurrent severe shortages of the petroleum product market of which kerosene and diesel are the most prominent. Nigeria has five domestic refineries owned by the government with a capacity to process 450,000 barrels of oil per day, yet imports constitute more than 75% of petroleum product requirements. The state-owned refineries have hardly operated above a 40% capacity utilization rate for any extended period of time in the past two decades. The gasoline market is much better supplied than kerosene and diesel because of its higher political profile. This factor explains why the government has embarked on large import volumes to remedy domestic shortages of the product. According to the Minister for Energy, the subsidy to support the imports of gasoline alone will be in the range of 700 to 800 billion naira in 2008[38]. The weaker political pressures exerted by the consumers of kerosene (the poor and low middle class) and diesel (industrial sector) on the government and the constraints on public financing of large-scale imports of these products, as in the case of gasoline, largely explain their more severe and persistent market shortages[39].

The second dimension of Nigeria's energy crises is exemplified by such indicators as electricity blackouts, brownouts, and pervasive reliance on self-generated electricity. This development has occurred despite abundant energy resources in Nigeria. The electricity market, dominated on the supply side by the state-owned PHCN, formerly called NEPA, has been incapable of providing minimum acceptable international standards of electricity service reliability, accessibility, and availability for the past three decades[40]. The nature of this poor record in electricity supply is apparent in the trend in transmission and distribution losses shown in **Figure 2**. The double-digit transmission and distribution losses are extremely large by international standards and are among the highest in the world. The system losses are five to six times higher than those in well-run power systems. The high

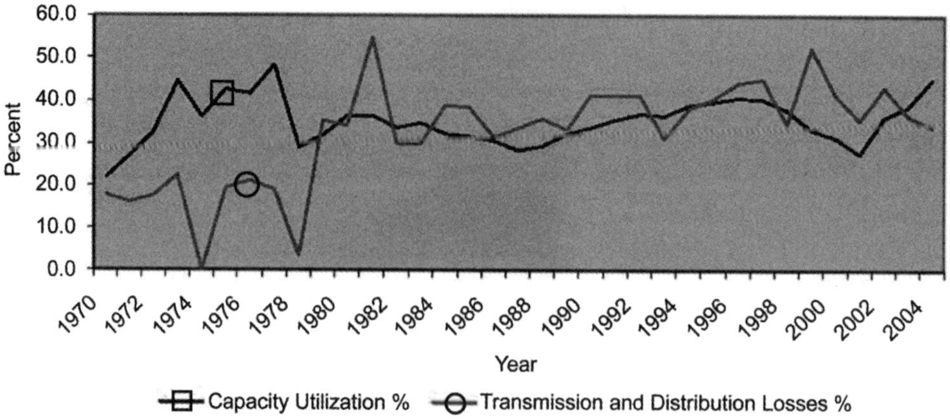

Figure 2. Indicators of the electricity crisis in Nigeria from 1970 to 2004. Adapted from Iwayemi[38].

level of power losses and the significant illegal access to the public power supply are indicative of the crisis in the industry.

Though the peak electricity demand has been less than half of the installed capacity in the past decade, load shedding occurs regularly. Power outages in the manufacturing sector provide another dimension to the crisis. In 2004, the major manufacturing firms experienced 316 outages. This increased by 26% in 2005, followed by an explosive 43% increase between 2006 and 2007. Though no published data exist, the near collapse of the generating system to far below 2,000MW for prolonged periods of time suggests a reason for the number of outages in 2008 to be very high. This poor service delivery has rendered public supply a standby source as many consumers who cannot afford irregular and poor quality service substitute more expensive captive supply alternatives to minimize the negative consequences of power supply interruptions on their production activities and profitability. An estimated 20% of the investment into industrial projects is allocated to alternative sources of electricity supply[3].

In summary, the causal factors in Nigeria's energy crisis include the following:

- Prevalence of a regime of price control.
- Weak concern for cost recovery and lack of adequate economic incentives to

induce the state-owned companies (NNPC and PHCN) to engage in efficient production and investment behavior. This seems apparent in the existence of large input and output subsidies.

- Multiplicity of economic and noneconomic objectives without proper identification of the trade-offs among these different objectives. This is implicit in its pricing policies in both electricity and petroleum products markets.

- Institutional and governance failures which induced gross distortions and inefficiency in production, investment choices and high costs of operation, low return on investment, and expensive delays along with cost overruns in the state energy enterprises.

6. Energy Demand Projection

There is an increasing demand for fuel energy due to the increase in economic development and civilization all over the world. Industry is one of the most important energy-consuming sectors in the world. According to Mitchel[41], energy is essential to our way of life. It provides us with comfort, transportation, and the ability to produce food and material goods. Historically, energy consumption has been directly related to the gross national product, which is a measure of the market value of the total national output of goods and services[42].

According to Sambo *et al.*[43], population is a major driver of energy demand, while its most important determinant is the level of economic activity and its structure measured by the total gross domestic product (GDP) alongside the various sectors and sub-sectors of the economy. Population projection of Nigeria was expected to grow from 115.22 million in 2000 to 281.81 million by 2030 at an average annual rate of 2.86% between 2000 and 2030.

Based on the models developed by the ECN, the country's energy demand was analyzed for the period from 2000 to 2030 with the use of the Model for the Analysis of the Energy Demand (MAED) and the Wien Automatic System Planning (WASP) package (**Table 5**). It can be said that the energy demand of Nigeria will be approximately 2.5-, 3-, 3.5-, and 4.5-fold between the years 2000 and 2015 and approximately 8-, 13-, 17-, and 22.5-fold between the years 2000 and 2030

Table 5. Total projected energy demand (MTOE).

Scenario	2000	2010	2015	2020	2025	2030
Reference (7%)	32.01	51.40	79.36	118.14	169.18	245.19
High growth (10%)	32.01	56.18	94.18	190.73	259.19	414.52
Optimistic (11.5%)	32.01	56.18	108.57	245.97	331.32	553.26
Optimistic (13%)	32.01	72.81	148.97	312.61	429.11	715.70

Adapted from ECN[44].

based on a 7% (reference), 10% (high growth), 11.5% (optimistic), and 13% (optimistic) GDP growth rate per annum, respectively. This increase in the energy demand is due to the high level of economic activities expected in Nigeria as measured by the total GDP.

The trends of the projected energy demand are shown in **Figure 3**. In 2005, the total energy demand based on a 10% GDP growth rate revealed that the household segment had the largest share of all the sectors. The sectorial energy demands in the 2030 plan period, however, showed the highest growth rates for the industrial, followed by the services, household, and transport sectors in that order (**Table 6**). The electricity demand (extracted from the total energy demand) shows an increasing trend from the base year 2005 to 2030 in the four adopted growth scenarios, respectively, as shown in **Figure 4**, indicating a high economic growth rate leading to a substantial increase in the electricity demand. The energy consumed over the years shows a decreasing trend with an increasing population, necessitating a corresponding increase in the energy output. Hence, the country's large energy efficiency potential needs to be exploited (**Table 7**). In 2007, the total primary energy consumed was 11.4 million tons of oil equivalent (MTOE) with petroleum products having the largest share of 67.3% of the total consumption, amounting to a total average consumption of 78.7% between 2002 and 2007. This level of consumption was followed by that of hydropower at 23.9%, natural gas at 8.7%, and coal at 0.05% with their respective total average consumption standing at 16.08%, 5.17%, and 0.04% for the period from 2002 to 2007 as shown in **Table 8**. Flaring adversely reduced the maximum contribution of natural gas to the total energy consumption mix in spite of its abundance in the country as most of the oil fields lack appropriate infrastructure for gas production. The general Niger Delta

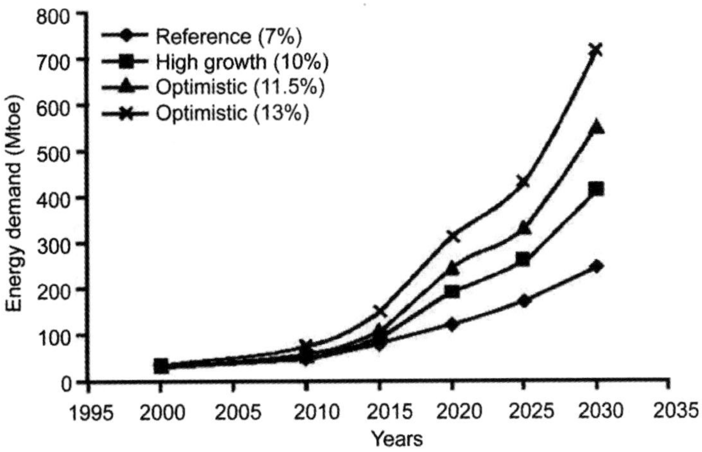

Figure 3. Graph showing the projected electricity demand between 2000 and 2030.

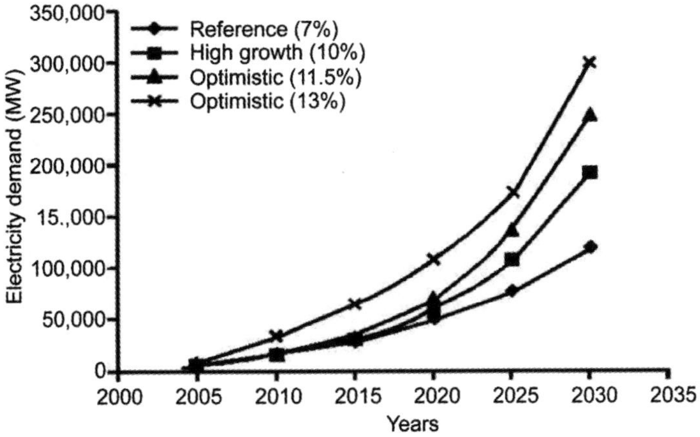

Figure 4. Graph showing the projected electricity demand between 2005 and 2030.

Table 6. Total energy demand based on a 10% GDP growth rate (MTOE).

Item	2005	2010	2015	2020	2025	2030	Average growth rate (%)
Industry	8.08	12.59	26.03	39.47	92.34	145.21	16.2
Transport	11.70	13.48	16.59	19.70	26.53	33.36	4.7
Household	18.82	22.42	28.01	33.60	33.94	34.27	2.6
Services	6.43	8.38	12.14	15.89	26.95	38.00	8.7
Total	45.01	56.87	82.77	108.66	179.75	250.84	8.3

Adapted from ECN[46].

Table 7. Per capita primary energy consumption in Nigeria.

Year	Energy consumed (MTOE)	Population (million)	Per capita energy consumption (TOE/capita)
2002	18.783	122.365	0.153
2003	19.106	126.153	0.151
2004	16.267	129.927	0.125
2005	17.707	133.702	0.132
2006	12.421	140.003	0.089
2007	11.387	144.203	0.097

Adapted from CBN and NBS[32][47][48].

Table 8. Commercial primary energy consumption by type (average percentage of total).

Type	2002	2003	2004	2005	2006	2007	Average
Coal	0.03	0.03	0.03	0.03	0.05	0.05	0.04
Hydro	11.93	14.20	17.39	12.04	17.03	23.90	16.08
Natural gas	2.84	1.9	4.54	5.5	7.52	8.73	5.17
Petroleum products	85.20	83.87	78.04	82.45	75.44	67.32	78.71

Adapted from CBN[32][49].

security issue (bunkering, sabotage, etc.) have also weakened most of the oil and gas projects[45].

Throughout the world, electricity is the most widely used and desirable form of energy. It is a basic requirement for economic development, national development, meeting the Millennium Development Goals (MDGs), and for an adequate standard of living. As a country's population grows and its economy expands, its demand for electrical energy multiplies. If this demand is not met adequately, a shortage in supply occurs. This shortage can assume crisis proportions and possibly affect achieving sustainable energy development.

The electric power capacity demand by projection in Nigeria would be approximately 3.5-fold between 2010 and 2020 and 7.5-fold between 2010 and 2030, respectively, at a growth rate of 7%, while the projected supply by fuel mix shows a similar trend with the demand at both growth rates of 7% and 13% (**Table 9**). There is a wide disparity in the energy demand to the supply ratio in Nigeria both in the present and the future. This necessitates an urgent need for alternative energy sources and efficient energy usage in order to avert looming energy crises.

These projections for continued rapid energy growth imply some severe problems for the future resource depletion, energy degradation, associated environmental problems, fuel shortage, etc. Indeed, many of these problems are already happening; thus, energy conservation is concerned with ways to reduce energy demand, but yet achieve the same objective as before.

To achieve its objective of sustainable development, Nigeria needs to substantially increase the supply of modern affordable energy services to all its citizens while, at the same time, maintaining environmental integrity and social cohesion. In addition, a robust mix of energy sources (fossil and renewable), combined

Table 9. Electric power capacity in Nigeria.

	Electric power demand											
	2010				2020				2030			
	Demand (MW%)		Supply (MW%)		Demand (MW%)		Supply (MW%)		Demand (MW%)		Supply (MW %)	
Fuel type	7	13	7	13	7	13	7	13	7	13	7	13
Coal			0	0			6.515	16.913			15.815	63.896
Gas			13.555	31.935			37.733	78.717			85.585	192.895
Hydro			3.702	3.902			6.479	6.749			11.479	11.479
Nuclear			0	0			3.530	11.005			11.872	36.891
Small-hydro			40	208			140	1.000			701	2.353
Solar			5	30			34	750			302	4.610
Wind			0	500			1.471	3.791			5.369	15.567
Total	15.730	33.250	17.303	36.576	50.820	107.600	55.903	118.836	119.200	297.900	131.122	327.690

with an improved end-use efficiency, will almost certainly be required to meet the growing demand for energy services in the country. Technological development, decentralized non-grid networks, diversity of energy-supply systems, and affordable energy services are imperative to meeting the future demand.

7. The Role of Renewable Energy Technologies in Sustainable Development

Renewable energy has an important role to play in meeting the future energy needs in both rural and urban areas[50]. The development and utilization of renewable energy should be given a high priority, especially in the light of increased awareness of the adverse environmental impacts of fossil-based generation. The need for sustainable energy is rapidly increasing in the world. A widespread use of renewable energy is important for achieving sustainability in the energy sectors in both developing and industrialized countries.

Nigeria is blessed with a large amount of renewable natural resources (**Table 1**), which, when fully developed and utilized, will lead to poverty reduction and sustainable development.

Renewable energy resources and technologies are a key component of sustainable development for the following primary reasons:

- They generally cause less environmental impact than other energy sources. The implementation of renewable energy technologies will help to address the environmental concerns that emerged due to greenhouse gas emissions such as carbon dioxide (CO_2), oxides of nitrogen (NO_x), oxides of sulfur (SO_x), and particulate matters as a result of power generation from oil, natural gas, and coal. A variety of renewable energy resources provide a flexible array of options for their use.

- They cannot be depleted. If used carefully in appropriate applications, renewable energy resources can provide a reliable and sustainable supply of energy almost indefinitely. In contrast, fossil fuel resources are diminished by extraction and consumption.

- They favor system decentralization and local solutions that are somewhat independent of the national network, thus enhancing the flexibility of the system and providing economic benefits to small isolated populations.

To seize the opportunities presented by renewable energy resources in sustainable development, Nigeria needs to establish renewable energy markets and gradually develop experience with renewable energy technologies. The barriers and constraints to the diffusion of renewable energy should be overcome. A legal, administrative, and financing infrastructure should be established to facilitate planning and application of renewable energy projects. Government must play a useful role in promoting renewable energy technologies by initiating surveys and studies to establish their potential in both urban and rural areas.

Because renewable energies are constantly being replenished from natural resources, they have security of supply, unlike fossil fuels, which are negotiated on the international market and subject to international competition, sometimes even resulting in wars and shortages. They have important advantages, which could be stated as follows:

- Their rate of use does not affect their availability in the future; thus, they are inexhaustible.

- The resources are generally well distributed all over the world, even though wide spatial and temporal variations occur. Thus, all regions of the world have reasonable access to one or more forms of renewable energy supply.

- They are clean and pollution-free and are therefore a sustainable natural form of energy.

- They can be cheaply and continuously harvested and are therefore a sustainable source of energy.

Unlike the nuclear and fossil fuels plants which belong to big companies, governments, or state-owned enterprises, renewable energy can be set up in small units and is therefore suitable for community management and ownership. In this way, the returns from renewable energy projects can be kept in the community. In

Nigeria, this has particular relevance since the electricity grid does not extend to remote areas, and it is prohibitively expensive to do so. This presents a unique opportunity to construct power plants closer to where they are actually needed. In this way, much needed income, skill transfer, and manufacturing opportunities for small businesses would be injected into rural communities.

8. Energy and Sustainable Development in Nigeria

Sustainable energy involves the provision of energy services in a sustainable manner, which in turn necessitates that energy services be provided for all people in ways that, now and in the future, are sufficient to provide the basic necessities, affordable, not detrimental to the environment, and acceptable to communities and people[51]-[53]. Linkages between sustainable energy and factors such as efficiency and economic growth have been investigated[54].

The energy sector plays a pivotal role in attempts to achieve sustainable development, balancing economic and social developments with environmental protection (encapsulated in the 'strap line' for the 2002 Johannes-burg World Summit on Sustainable Development of 'people, planet, and prosperity'). Energy is central to practically all aspects of sustainable development. Energy is central to the economy because it drives all economic activities. This characterization of energy directs our attention to its sources in nature, to activities that convert and reconvert this energy, and finally to activities that use the energy to produce goods and services and household consumption. Traditionally, energy is treated as an intermediate input in the production process. This treatment of energy's role understates its importance and contribution to development. All economic activities and processes require some form of energy. This effectively makes energy a critical primary factor of production. Given the state of technological advancement in the economy, capital and labor perform supporting roles in converting, directing, and amplifying energy to produce goods and services needed for growth and poverty reduction[3].

Energy services are essential ingredients of all three pillars of sustainable development-economic, social, and environmental. Economies that have replaced human and animal labor with more convenient and efficient sources of energy and technology are also the ones that have grown fastest. No country in modern times has succeeded in substantially reducing poverty without adequately increasing the provision and use of energy to make material progress[55]. Indeed, by not ensuring a minimum access to energy services for a broad segment of the population, economic development of developing countries such as Nigeria beyond the level of subsistence has proven to be a real challenge.

At the national level, energy propels economic development by serving as the launch pad for industrial growth and, via transport and communications, providing access to international markets and trade. Reliable, efficient, and competitively priced energy supplies also attract foreign investment—a very important factor in boosting economic growth in recent times. At the local level, energy facilitates economic development by improving productivity and enabling local income generation through improved agricultural development (irrigation, crop processing, storage, and transport to market) and through non-farm employment, including micro-enterprise development. As an indicator of local recognition of the importance of energy for businesses, Nigerian manufacturers, who were asked to rank the constraints on their firms' activities, identified power breakdowns, and voltage fluctuations as their top two problems[46]. Recent developments in Ghana's energy sector support this point[56].

Energy has also strong and important links to the environment. Many energy sources are drawn directly from the environment, requiring a sound management for these sources to be sustainable. Furthermore, energy use affects the environment. Emissions from fossil fuels, for example, reach beyond the local and national levels to affect the global environment and contribute to climate change. The poorest people often live in the most ecologically sensitive and vulnerable physical locations. These areas may be the most affected by the predictable effects of climate change such as an increased frequency of extreme events, for example floods, drought, rising sea levels, and melting ice caps. The risks facing poor people are often increased by the unsustainable use of biomass resources[3].

The connection between energy, the environment, and sustainable development is worth highlighting. Energy supply and use are related to climate change as well as such environmental concerns as air pollution, ozone depletion, forest destruction, and emissions of radio-active substances. These issues must be addressed if society is to develop while maintaining a healthy and clean environment. Ideally, a society seeking sustainable development should use only energy resources which have no environmental impact. However, since all energy resources lead to some environmental impact, an improved efficiency and environmental stewardship can help overcome many of the concerns regarding the limitations imposed on sustainable development by environmental emissions and their negative impacts[55].

Energy is directly linked to the broader concept of sustainability and affects most of civilization. That is particularly evident since energy resources drive much if not most of the world's economic activity, in virtually all economic sectors. Also, energy resources, whether carbon-based or renewable, are obtained from the environment, and wastes from energy processes (production, transport, storage, utilization) are typically released to the environment. Given the intimate ties between energy and the key components of sustainable development, the attainment of energy sustainability is being increasingly recognized as a critical aspect of achieving sustainable development[55].

Use of renewable natural resources, combined with efficient supply and use of fossil fuels with cleaner technologies, can help reduce the environmental effects of energy use and help Nigeria replacing the existing, inefficient fossil fuel technologies that pollute the environment.

As a complementary measure, careful management of energy resources is important to promote economic growth, protect ecosystems and provide sustainable natural resources.

Thus, energy sustainability is considered to involve the sustainable use of energy in the overall energy system. This system includes processes and technologies for the harvesting of energy sources, their conversion to useful energy forms, to provide energy services such as operating communications systems, lighting buildings, and cooking[57].

The reform of the energy sector is critical to sustainable development in Nigeria. This includes reviewing and reforming subsidies, establishing credible regulatory frameworks, developing policy environments through regulatory interventions, and creating market-based approaches such as emission trading[58]. Globally, countries are developing strategies and policies to enable a sustainable development of their energy resources, thus contributing to fuel economic and social developments, while reducing air pollution and greenhouse gas emissions.

The energy sector is very strategic to the development of the Nigerian economy. In addition to its macroeconomic importance, it has major roles to play in reducing poverty, improving productivity, and enhancing the general quality of life. If Nigeria is to take the path of sustainable energy, it is important to accurately and technically model the energy demand and supply scenarios and their impacts on the economy, resources, and society along with the environment, for both medium and long terms. From such analyses, we can derive information that is vital for policy construction and investment[59].

9. Energy Efficiency and Energy Conservation in Sustainable Development

Energy efficiency means an improvement in practices and products that reduce the energy necessary to provide services. Energy efficiency products essentially help to do more work with less energy[60]. Energy efficiency is also defined as essentially using less energy to provide the same service[55]. In this sense, energy efficiency can also be thought of as a supply of resource—often considered an important, cost-effective supply option. Investment into energy efficiency can provide additional economic value by preserving the resource base (especially combined with pollution prevention technologies) and mitigating environmental problems.

Energy efficiency (EE) improvements have multiple advantages, such as the efficient exploitation of natural resources, the reduction in air pollution levels, and lower spending by the consumers on energy-related expenditure. Investments in EE result in long-term benefits, such as reduced energy consumption, local environmental enhancement, and overall economic development. Energy use has environmental impacts, regardless of the source or mechanism. For example, hydroelectric projects affect their local ecological systems and displace long-standing social systems. Fossil fuel power creates pollution in the extraction, transportation, and combustion of its raw materials. The long-term storage of waste products of the nuclear power industry is an issue to be resolved. Cost-effective energy efficiency is the ultimate multiple pollutant reduction strategy[61].

In Nigeria, a lot of energy is wasted because house-holds, public and private offices, as well as industries use more energy than is actually necessary to fulfill their needs. One of the reasons is that they use outdated and inefficient equipment and production processes. Unwholesome practices also lead to energy wastage.

In Nigeria, the need for energy is exceeding its supply. In view of these circumstances, primary energy conservation, rationalization, and efficient use are immediate needs. Getting all the possible energy from the fuel into the working fluid is the goal of efficient equipment operations. This leads to a higher productivity and saves not only money, but also influences the safety and life of the equipment and reduces pollution[62]. Steps taken to minimize energy consumption, or to use the energy more effectively, are steps in the right direction to preserve the global environment. Energy conservation measures or recommendations are often referred to more positively as opportunities. Two primary criteria for applying energy conservation are that it is easy to implement and that its payback is brief. Ease of implementation and duration of payback period have been used to classify Energy conservation opportunities into three general categories for use: in maintenance and operation measures, in process improvement projects, and in large capital projects[61].

Energy conservation and energy efficiency are separate but related concepts. Energy efficiency is achieved when energy intensity in a specific product, process, or area of production or consumption is reduced without affecting output, consumption, or comfort levels. Promotion of energy efficiency will contribute to energy conservation and is therefore an integral part of energy conservation promotional policies[63].

Energy efficiency encompasses conserving a scarce resource; improving the technical efficiency of energy conversion, generation, transmission and end-use devices; substituting more expensive fuels with cheaper ones; and reducing or reversing the negative impact of energy production and consumption activities on the environment. Energy conservation is a tangible resource by itself that competes economically with contemporary energy supply options. In addition to this, it offers a practical means of achieving four goals that should be of high priority in any nation that desires quick and sustainable economic growth and development. These are economic competitiveness, utilization of scarce capital for development, environmental quality, and energy security. It enhances the international competitiveness of the industries in the world markets by reducing the cost of production. It optimizes the use of capital resources by directing lesser amounts of money in conservation investment as compared with capital-intensive energy supply options. It protects the environment in the short run by reducing pollution and in the long run by reducing the scope of global climate change. It strengthens the security of supply through a lesser demand and a lesser dependence on petroleum product imports. No energy supply option may be able to provide all these benefits. Energy conservation is a decentralized issue and is largely dependent on individual, distinct decisions of energy supply, which are highly centralized. The housewife, the car driver, the housing developer, the house owner, the boiler operator in industry, and every other individual who consumes energy in some form or another are required to participate in energy-saving measures. It calls for a collective endeavor and is dependent upon the actions of people in diverse fields although the people involved may not be sufficiently informed or motivated to conserve energy[64].

10. Renewable Energy and Energy Efficiency as Climate Change Mitigation Strategies

The Inter-government Panel on Climate Change (IPCC), a body set up in 1988 by the World Meteorological Organization and the United Nations Environmental Programme to provide authoritative information about the climate change phenomenon, asserts that the warming of the last 100 years was unusual and unlikely to be natural in origin[58]. The IPCC has attributed the warming of at least the second half of the century to an increase in the emission of greenhouse gases into the atmosphere. Human activity is largely responsible for the emission of these gases into the atmosphere: CO_2 is produced by the burning of fossil fuels (coals, oil, gas) as well as by land-use activities such as deforestation; methane is produced by cattle, rice agriculture, fossil fuel use, and landfills; and nitrous oxide is produced by the chemical industry, cattle feed lots, and agricultural soils. As humans have increased their levels of production and consumption, greenhouse gas emissions have also increased; since 1750, at the time of the industrial revolution, CO_2 emission has increased by 31%, methane by 15%, and nitrous oxide by 17%. Moreover, the emissions of these gases continue to rise steadily[65].

The Clean Development Mechanism (CDM) was integrated to the Kyoto Protocol as the United Nations Framework Convention on Climate Change[66]. CDM projects allow investment by entities from industrialized countries into projects in developing countries. In return for this investment, carbon credits (in this case, certified emission reductions) are received by the investor in the industrialized country. This enables the industrialized country to meet its emission reduction targets given by the Kyoto Protocol more cost-effectively, while promoting sustainable development in developing countries. CDM projects may also be unilateral, i.e., they take place in the developing country without a project partner from an industrialized nation.

Investment into clean energy facilities is recognized as the best way to increase the participation of Nigerian proponents in the CDM process and hence the global carbon market. Clean energy investment is defined as follows: investment into an energy supply and utilization system that provides the required energy with minimal negative environmental and social consequences[67]. Investment into clean energy systems can also be viewed as an investment into energy sources and technologies that are significantly less environmentally damaging than in the status quo case. Investment into clean energy systems provides the most effective and optimally efficient path to an increased CDM participation in Nigeria and hence an effective participation in the global carbon market.

The salient characteristics of clean energy investment are as follows:

- The resulting system results in little or no emissions of obnoxious gases and particulates;
- The clean energy technologies have a carbon footprint that is much lower than the baseline emission scenario;
- The technology is accessible, and the required investment is available for adoption in developing countries like Nigeria;

- The implementation of the clean energy technology will contribute to sustainability.

Energy efficiency and renewable energy technologies are prominent in most sustainable development programs, for example, the Agenda 21[68]. According to the Intergovernmental Panel on Climate Change (IPCC), the second assessment report, the stabilization of atmospheric greenhouse gas concentrations at levels that will prevent serious interference with the climate system can only be achieved by dramatically increasing the implementation of renewable energy. In one IPCC scenario, in which greenhouse gases are stabilized by the year 2050, the share of renewable energy in the global energy balance must increase tenfold from the current level. In developing countries, the required increase is even more dramatic, estimated at 20-fold between 1990 and 2050. Further improvements in energy efficiency and energy conservation can reduce emissions in the shorter term, thus 'buying time' for the required changes in energy production[69].

Nigeria is one of the highest emitters of greenhouse gases in Africa. The practice of flaring gas by the oil companies operating in Nigeria has been a major means through which greenhouse gases are released into the atmosphere. Carbon dioxide emissions in this area are among the highest in the world[70]. Some 45.8 billion kW of heat are discharged into the atmosphere of the Niger Delta from flaring 1.8 billion ft3 of gas every day[58]. Gas flaring has raised temperatures and rendered large areas uninhabitable. Between 1970 and 1986, a total of about 125.5 million m^3 of gas was produced in the Niger Delta region, about 102.3 (81.7%) million m^3 were flared, while only 2.6 million m^3 were used as fuel by oil-producing companies and about 14.6 million m^3 were sold to other consumers[71][72]. The use of renewable energy sources will reduce the over dependence on the burning of fossil fuel. Moreover, instead of flaring gas in Nigeria, the gases can be converted to methanol and used as a fuel for both domestic and industrial use. With good energy efficiency practices and products, the burning of fossil fuel for energy will be greatly minimized.

11. Conclusions

From the energy outlook of Nigeria, it is very clear that the energy demand is very high and is increasing geometrically while the supply remains inadequate, insecure, and irregular and is decreasing with time; the mix has hitherto been dominated by fossil resources which are fast being depleted apart from being environmentally non-friendly. The energy supply mix must thus be diversified through installing an appropriate infrastructure and creating full awareness to promote and develop the abundant renewable energy resources present in the country as well as to enhance the security of supply.

There is clear evidence that Nigeria is blessed with abundant resources of fossil fuels as well as renewable energy resources. The major challenge is an inefficient usage of energy in the country. As a result, there is an urgent need to encourage the evolvement of an energy mix that will emphasize the conservation of petroleum resources in such a manner enabling their continued exportation for foreign earnings for as many years as possible.

The opportunities for conserving energy in our various sectors—office building and residential areas, manufacturing industries, transportation, electricity generation and distribution, and electricity equipment and appliances—were presented in this work. The various areas where savings in energy can be made have also been identified. Several guidelines and measures have been suggested to conserve energy in these areas, and if the guidelines and measures are strictly adhered to, then substantive savings in energy will be carried out.

In this study, four economic growth scenarios were dealt with in the review of the energy requirements. These are the reference scenarios of a 7% total GDP growth rate that will ensure the MDGs of reducing poverty by 50% of the 2,000 value by 2015. The high growth scenario of a 10% GDP growth rate in the attempt to eradicate poverty by 2030 and the optimistic scenarios of 11.5% and 13% GDP growth rates that will further increase the rate of economic development.

In order to ensure the sustainability of energy supply and subsequently the sustainable economic development of the country, the government has to intensify the further implementation of renewable energy and energy efficiency programs. As observed in quite a number of successful countries promoting renewable energy, such as Germany, Denmark, and Japan, a strong and long-term commitment from the government is crucial in implementing any kind of policies which will lead to the development of renewable energies, in particular, and a sustainable development, in general.

12. Recommendations

In this study, it is established that renewable energy and energy efficiency are two components that should go together to achieve sustainable development in Nigeria. The need to conserve the present energy generated in the country using energy-efficient products and the appropriate practices is essential for sustainable development. Therefore, it is recommended that the country should do the following:

- Develop policies on energy efficiency and integrate them into the current energy policies. A comprehensive and coherent energy policy is essential in guiding the citizens towards an efficient usage of its energy resources.

- Promote energy-efficient products and appropriate practices at the side of the end users and energy generation.

- Create awareness on renewable energy and energy efficiency.

- Establish an agency to promote the use of energy-efficient products and ensure the appropriate practices.

- Develop and imbibe energy efficiency technologies.

- Carry out a resource survey and assessment to determine the total renewable energy potential in the country as well as identify the local conditions and local priorities in various ecological zones.

- Establish a testing and standards laboratory for renewable energy technologies similar to that in South Africa.

- Take advantage of global partnerships, such as the Residential Energy Efficiency Project initiative of UK, to assist the country in a creative integration of renewable energy systems.

- Establish a renewable energy funding/financing agency such as India's Indian Renewable Energy Agency.

- Develop appropriate drivers for the implementation of energy efficiency policies.

- Clean energy facilities should be embraced in the different sectors of the Nigerian economy.

In the following, a partial list of potential clean energy opportunities in Nigeria is presented:

- More efficient passive and full usage of solar technologies in the residential, commercial, and industrial sectors.

- Biogas from wastes as a source of cooking fuel in homes.

- Use of energy-efficient lighting.

- Implementation of renewable biomass as a fuel in highly efficient cook stoves.

- Efficient production of charcoal as a fuel in homes and small and medium enterprises.

- Use of biofuels in efficient cooking stoves and lamps in homes.

- Energy-efficient lighting.

- Use of compressed natural gas (CNG) as a transport fuel.

- Use of biofuels as a transport fuel.

- Introduction of a bus rapid transit system to other cities and expansion of the Lagos system.

- Shift from high carbon intensive fuels to natural gas for energy generation in industries.

- Development of a CNG infrastructure to distribute natural gas to industries located at sites remote from the existing pipelines.
- Implementation of combined heat and power (CHP) facilities in industries.
- Implementation of energy efficiency improvements in manufacturing industries.
- Implementation of CHP facilities in commercial facilities.
- Use of solar and wind energy for irrigation water pumping and farm electricity supply.
- Utilization of agricultural residues for electricity generation.
- Generation of biogas from wastes produced by the livestock and animal husbandry.

In addition to these, the existing research and development centers and technology development institutions should be adequately strengthened to support the shift towards an increased use of renewable energy. Human resource development, critical knowledge, and knowhow transfer should be the focus for project development, project management, monitoring, and evaluation. The preparation of standards and codes of practices, maintenance manuals, life cycle costing, and cost-benefit analysis tools should be undertaken on urgent priority.

Competing Interests

The author declares that there are no competing interests.

Source: Oyedepo S O. Energy and sustainable development in Nigeria: the way forward[J]. Energy Sustainability & Society, 2012, 2(1):1−17.

References

[1] Ramchandra P, Boucar D (2011) Green Energy and Technology. Springer, London Dordrecht Heidelberg New York.

[2] Rai GD (2004) NonConventional Energy Sources. Khanna Publishers, Delhi.

[3] Nnaji C *et al* (2010) In: Nnaji CE, Uzoma CC (eds) CIA World Factbook., Nigeria. http://www.cia.gov/library/publications/the-world-factbook/geos/ni. html.

[4] Council for Renewable Energy, Nigeria (CREN) (2009) Nigeria Electricity Crunch. available at www.renewablenigeria.org.

[5] CBN (1985) Central Bank of Nigeria Annual Reports and Statement of Account.

[6] Department of Petroleum Resources (DPR) (2007) Nigeria, Nigeria, available at http: //www.DPR.gov.ng.

[7] Onyebuchi EI (1989) Alternative energy strategies for the developing world's domestic use: A case study of Nigerian household's final use patterns and preferences. The Energy Journal 10(3):121–138.

[8] Chineke TC, Igwiro EC (2008) Urban and rural electrification: enhancing the energy sector in Nigeria using photovoltaic technology. African Journal Science and Tech 9(1):102–108.

[9] Energy Commission of Nigeria (ECN) (2005) Renewable Energy Master Plan.

[10] Adekoya LO, Adewale AA (1992) Wind energy potential of Nigeria. Renewable Energy 2(1):35–39.

[11] Fagbenle RO, Karayiannis TG (1994) On the wind energy resources of Nigeria. International Journal of Energy research 18(5):493–508.

[12] Ngala GM, Alkali B, Aji MA (2007) Viability of wind energy as a power generation source in Maiduguri, Borno state, Nigeria. Renewable energy 32 (13):2242–2246.

[13] Akinbami JFK (2001) Renewable Energy Resources and Technologies in Nigeria: Present Situation, Future Prospects and Policy Framework'. Mitigation and Adaptation Strategies for Global Change 6:155–181. Kluwer Academic Publishers, Netherlands.

[14] Akinbami JFK, Ilori MO, Oyebisi TO, Akinwumi IO, Adeoti O (2001) Biogas energy use in Nigeria: Current status, future prospects and policy implications. Renewable and Sustainable Energy Review 5:97–112.

[15] Sambo AS (2008) Matching Electricity Supply with Demand in Nigeria. International Association of Energy Economics 4:32–36.

[16] Ighodaro CAU (2010) Co-Integration And Causality Relationship Between Energy Consumption And Economic Growth: Further Empirical Evidence For Nigeria.

Journal of Business Economics and Management 11(1):97–111.

[17] Okafor ECN, Joe-Uzuegbu CKA (2010) Challenges To Development Of Renewable Energy For Electric Power Sector In Nigeria. International Journal Of Academic Research 2(2):211–216.

[18] Energy Commission of Nigeria (ECN) (2007) Draft National Energy Masterplan.

[19] (August 2000) Report of the Inter-Ministerial Committee on Combating Deforestation and Desertification.

[20] Famuyide OO, Anamayi SE, Usman JM (2011) Energy Resources' Pricing Policy And Its Implications On Forestry And Environmental Policy Implementation In Nigeria. Continental J Sustainable Development 2:1–7.

[21] Sambo AS (2009) Strategic Developments in Renewable Energy in Nigeria. International Association of Energy Economics 4:15–19.

[22] Energy Commission of Nigeria (ECN) (2003) National Energy Policy. Federal Republic of Nigeria, Abuja.

[23] Abiodun R (2003) Fuel price Hike Spells Doom for Nigeria's Forest. http://www.islamonline.net/English/index.shtml.

[24] Williams CE (1998) 'Reaching the African Female Farmers with Innovative Extension Approaches: Success and Challenges for the Future'. Paper Presented To The International Workshop on Women Agricultural Intensification and Household Food Security At University of Cape Coast, Ghana, 25th-28th June.

[25] Karekezi R (1997) Renewable Energy Technologies in Africa, Zed Books Limited with African Energy Policy Research Network (AFREPREN) and Stockholm Environment Institute (SEI).

[26] Central Bank of Nigeria (CBN) (2000) 'The changing structure of the Nigerian economy and implications for development'. Research Department, Central Bank of Nigeria; Realm Communications Ltd, Lagos, August.

[27] Famuyide OO, Adu AO, Ojo MO (2004) Socio-Economic Impacts of Deforestation in the Sudano-Sahelian Belt of Nigeria. Journal of Forestry Research and Management 1(1&2):94–106.

[28] Fagbenle RO, Adenikinju A, Ibitoye FI, Yusuf AO, Alayande O (2006) Draft Final Report on Nigeria's Electricity Sector Executive Report.

[29] PHCN (2008) Generation and Transmission Grid Operations. Annual Technical Report National Control Center (NCC), Osogbo.

[30] PHCN (2009) Generation and Transmission Grid Operations. National Control Center (NCC), Osogbo.

[31] Imo EE (2008) 'Challenges of Hydropower Development in Nigeria'. Hydrovision.

[32] CBN (Central Bank of Nigeria) (2007), Vol. 18th edn., December.

[33] Emeka EE (2010) 'Causality Analysis of Nigerian Electricity Consumption and Economic Growth'. Journal of Economics and Engineering 4:80–85. ISSN: 2078-0346.

[34] Odularu GO, Okonkwo C (2009) 'Does Energy Consumption Contribute To Economic Performance? Empirical Evidence From Nigeria', Journal of Economics and Business Vol. XII, No 2:43–47.

[35] CBN (2009) Central Bank of Nigeria Statistical Bulletin. CBN Press, Abuja.

[36] UNIDO (United Nations Industrial Development Organization) (2001) UNIDO (United Nations Industrial Development Organization). UNIDO (United Nations Industrial Development Organization), UNIDO (United Nations Industrial Development Organization).

[37] Ajayi OO, Ajanaku KO (2007) Nigeria's Energy Challenge and Power Development: The Way Forward. Bulletin of Science Association of Nigeria 28:1–3.

[38] Iwayemi A (2008) Nigeria's Dual Energy Problems: Policy Issues and Challenges. International Association for Energy Economics, pp 17–21, Fourth Quarters.

[39] Ibitoye F, Adenikinju A (2007) (2007) Future Demand for Electricity in Nigeria. Applied Energy 84:492–504.

[40] Adenikinju A (2005) ' Analysis of the cost of infrastructure failures in a developing economy the case of electricity sector in Nigeria'. African Economic Research Consortium AERC Research Paper 148, Nairobi, February 2005.

[41] Mitchel JW (1983) Energy engineering. John Wiley and Sons, New York.

[42] Adeyemo SB (2001) 'Energy potentials of organic wastes'. In: Proceedings of the first national conference, pp 55–61. ISBN 978-35533-0-5.

[43] Sambo AS, Iloeje OCJ, Ojosu OJ, Olayande S, Yusuf AO (2006) Nigeria's Experience on the Application of IAEA"s Energy Models (MAED & WASP) for National Energy Planning, Paper presented during the Training Meeting/Workshop on Exchange of Experience in Using IAEA's Energy Models and Assessment of Further Training Needs. held at the Korea Atomic Energy Research Institute, Daejon, Republic of Korea, pp 24–28, April.

[44] Energy Commission of Nigeria (ECN) (2006) National Energy Policy. Federal Republic of Nigeria, Abuja.

[45] Ohunakin OS (2010) Energy utilization and renewable energy sources in Nigeria. Journal of Engineering and Applied Sciences 5(2):171–7.

[46] Energy Commission of Nigeria (ECN) (2008) National Energy Policy. Federal Republic of Nigeria, Abuja.

[47] CBN (2006) Central Bank of Nigeria Statistical Bulletin. CBN Press, Abuja.

[48] National Bureau of Statistics (NBS) (2007) National Core Welfare Indicator Ques-

tionnaire (CWIQ) Survey. National Bureau of Statistics (NBS), Abuja, p 2, Summary Sheet.

[49] CBN (2005) Central Bank of Nigeria Statistical Bulletin. CBN Press, Abuja.

[50] Hui SCM (1997) From renewable energy to sustainability: the challenge for Honk Kong'. Hong Kong Institution of Engineers :351–358.

[51] Lior N (2008) Energy resources and use: The present situation and possible paths to the future. Energy 33:842–857.

[52] Haberl H (2006) The global socioeconomic energetic metabolism as a sustainability problem. Energy 31:87–99.

[53] Rosen MA (2002) Energy efficiency and sustainable development. Int J Global Energy Issues 17:23–34.

[54] Ayres RU, Turton H, Casten T (2007) Energy efficiency, sustainability and economic growth. Energy 32:634–648.

[55] Rosen MA (2009) Energy Sustainability: A Pragmatic Approach and illustrations. Sustainability 1:55–80.

[56] Ashong SN (2001) Macroeconomic Framework for Poverty Reduction Within the Context of Debt Relief:The Case of Ghana. Paper Presented at WIDER Development Conference on Debt Relief Helsinki, Finland.

[57] Hammond GP (1998) Alternative energy strategies for the United Kingdom revisited; market competition and sustainability. Technological Forecasting and Social Change 59:131–151.

[58] Sims REH, Schock RN, Adegbululgbe A, Fenhann J, Konstantinaviciute I, Moomaw W, Uyigue E (2007) CREDC Conference on Promoting Renewable Energy and Energy Efficiency in Nigeria. held at University of Calabar Hotel and Conference Centre, 21st November.

[59] Winkler H (ed) (2006) Energy policies for sustainable development in South Africa Options for the future. Energy Research Centre University of Cape Town, Private Bag Rondebosch 7701 South Africa, Website: www.erc.uct.ac.za.

[60] SECCP (2002) 'Getting to grips with sustainable energy'.

[61] Adeyemo SB, Odukwe AO (2008) 'Energy Conservation as a Viable Pathway towards Energy Stability. Journal of Engineering and Applied Sciences 3(3):233–238.

[62] Habib MA, Said SAM, Igbal MO, El-Mahallawy FM, Mahdi EA (1999) Energy Conservation and Early Failure Prediction in Boilers and Industrial Furnaces. Symposium on Management of Energy Consumption in Industry, Chamber of Commerce, Dammam, Kingdom of Saudi Arabia, October.

[63] Dubin FS, Long CG (1980) 'Energy conservation standards for Building design, construction and operation. McGraw-Hill Book Company.

[64] Etiosa U (ed) (2009) Energy Efficiency Survey in Nigeria: A Guide for Developing Policy and Legislation. International, Rivers, pp 1–37.

[65] Uyigue E (2007) 'Renewable energy and energy efficiency and sustainable development in Nigeria'. In: CREDC Conference on Promoting Renewable Energy and Energy Efficiency in Nigeria.

[66] Sarah La P (2002) 'Climate Change and Poverty'. A Publication of Tearfund.

[67] Winkler H, Van Es D (2007) Energy Efficiency and CDM in South Africa: Constraints and Opportunities. Journal of Energy in South Africa 18:29–37.

[68] Dayo FB (2008) Clean Energy Investment in Nigeria The domestic context. International Institute for Sustainable Development (IISD).

[69] The United Nations (UN) (1993) Report on the National Fuel wood Substitution Programme. United Nations Publications, New York, Energy Statistics Yearbook.

[70] Martinot E, McDom O (2002) Promoting Energy Efficiency and Renewable Energy GEF Climate Change Projects and Impacts; Global Environmental Facility. Washington, DC.

[71] Iyayi F (2004) 'An integrated approach to development in the Niger Delta'.A paper prepared for the Centre for Democracy and Development (CDD).

[72] Awosika LF (1995) Impacts of global climate change and sea level rise on coastal resources and energy development in Nigeria. In: Umolu JC (ed) Global Climate Change:Impact on Energy Development. DAMTECH Nigeria Limited, Nigeria.

Chapter 17

Environmental Contamination by Canine Geohelminths

Donato Traversa[1*], Antonio Frangipane di Regalbono[2], Angela Di Cesare[1], Francesco La Torre[3], Jason Drake[4], Mario Pietrobelli[2]

[1]Faculty of Veterinary Medicine, University of Teramo, Teramo, Italy
[2]Department MAPS, University of Padua, Padua, Italy
[3]Novartis Animal Health, Origgio, VA, Italy
[4]Novartis Animal Health, Greensboro, NC, USA

Abstract: Intestinal nematodes affecting dogs, *i.e.* roundworms, hookworms and whipworms, have a relevant health-risk impact for animals and, for most of them, for human beings. Both dogs and humans are typically infected by ingesting infective stages, (*i.e.* larvated eggs or larvae) present in the environment. The existence of a high rate of soil and grass contamination with infective parasitic elements has been demonstrated worldwide in leisure, recreational, public and urban areas, *i.e.* parks, green areas, bicycle paths, city squares, playgrounds, sandpits, beaches. This review discusses the epidemiological and sanitary importance of faecal pollution with canine intestinal parasites in urban environments and the integrated approaches useful to minimize the risk of infection in different settings.

Keywords: Toxocara Canis, Ancylostoma Caninum, Trichuris Vulpis, Faeces, Dog, Urban Areas

1. Introduction

Soil-transmitted helminthoses affects more than 2 billion people worldwide[1]. Other than human-specific parasites, intestinal nematodes affecting dogs have a relevant health-risk impact for both animals and human beings. The importance of these pathogens is often minimized by veterinarians and the general public, although Toxocara canis, hookworms (*i.e.* Ancylostoma spp.) and whipworms (*i.e.* Trichuris vulpis) are the most relevant canine helminths in terms of geographic distribution and clinical importance[2][3].

The presence of infective eggs or larvae in the environment has a crucial role among the different routes of transmission of dog intestinal nematodes in both humans and animals. In fact, human beings become infected by canine Toxocara spp. and Ancylostoma spp. most frequently via contaminated soil[4]–[7].

Studies from various countries have demonstrated a high rate of soil and grass contamination with infective parasitic elements in leisure, recreational, public and urban areas, *i.e.* parks, green areas, bicycle paths, city squares, playgrounds, sandpits, beaches.

When using these areas, people often take their pets with them. Owned dogs and stray animals may defecate in public streets and areas, thus contaminating the environment with parasites and favoring zoonotic transmission and (re-) infection for other animals.

While readers interested in biology, pathology and general control of canine intestinal nematodes are referred to[2][3][7]–[9], the present article reviews the epidemiological importance of faecal pollution in urban environments with canine intestinal parasites in terms of veterinary and human health and discusses the integrated approaches useful to minimize the risk of infection.

2. The Environment Is Incessantly Contaminated

Toxocara canis and Ancylostoma caninum are, respectively, the primary species of roundworms and hookworms infecting dogs worldwide. Other species of ascarids and ancylos-tomatids may be present in particular areas, e.g. Toxascaris leonina in Europe and USA, Uncinaria stenocephala in colder areas of temperate and subarctic regions, and Ancylostoma braziliense in the southern hemisphere. Additionally, the whipworm T. vulpis is the ubiquitous whipworm inhabiting the large intestine of dogs[2][3].

Parasitic burdens and egg output are higher in puppies but patent intestinal infections may occur in dogs of all ages and categories[10]–[19], even when under regular control programs[15][20]. Bitches are a relevant source of infection for other animals and environmental contamination because they often harbor somatic larvae, which mobilize during pregnancies and infect subsequent litters even when re-infections do not occur. Puppies become infected in utero and via the milk, but a proportion of mobilized larvae reach adulthood in the intestine of the dam and cause a patent infection with a long-lasting high egg shedding[21][22]. The patent infection in the bitch can be re-enforced when suckling puppies defecate immature ascarids, which are ingested by the dam and become adults in her intestine[23]. Altogether these biological features make nursing bitches and puppies a very important source of environmental contamination by T. canis.

Remarkably, pre-vaccination confinement of puppies would often imply that eggs are shed into the home or private gardens and backyards, thus posing a potential health risk for the owners[24]. This is of great importance considering that virtually 100% of puppies acquire toxocarosis by transmammary and/or transplacental route/s and that they pass thousands of T. canis eggs per gram of feces every day (**Figure 1**).

Hookworm filariform larvae present in the soil infect a suitable host by actively penetrating the skin (especially for Ancylostoma spp.) and/or via the oral route (*i.e.* Ancylostoma spp., Uncinaria spp.)[3][23][25]–[27]. As with T. canis, hypobiotic larvae may survive for years in the tissues of adult dogs and when reactivated during oestrus and in the last 2–3 weeks of pregnancy, they are passed via the milk to the litter[27]–[30]. Adult dogs may suffer patent ancylostomosis when they become infected with environmental larvae or when hypobiotic stages are re-activated by drivers of stress[3]. Remarkably, dogs infected by A. caninum may shed millions of hookworm eggs for weeks[7].

The absence of a vertical transmission in T. vulpis, its long pre-patent period

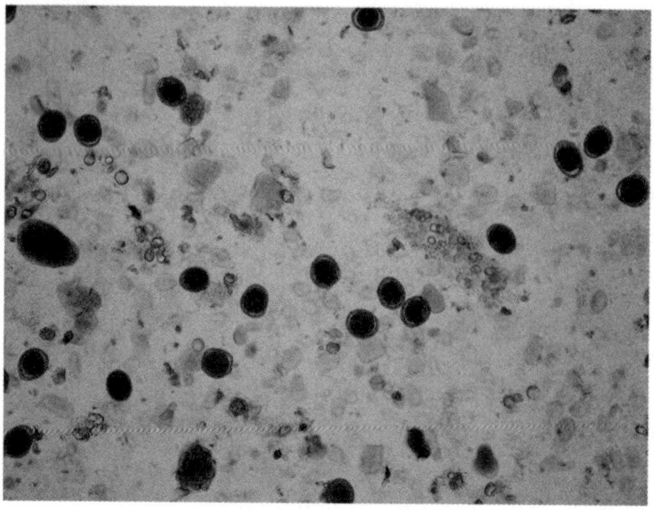

Figure 1. Copromicroscopic examination of a puppy: microscopic field (10×) showing a high shedding of Toxocara canis eggs.

and a partial ability to stimulate a protective immune response[31][32], explain the high degree of intestinal trichurosis in adult dogs rather than in puppies. Hence, it could be erroneously argued that this parasite is not spread as easily as roundworms and that the environment is not as contaminated by whipworm eggs.

It is estimated that the contamination of soil with Toxocara eggs may be more than the 90% of the investigated areas worldwide[33]. This is explained by the fact that mature eggs of ascarids (and T. vulpis as well) can survive in contaminated soil even in harsh conditions (e.g. they may resist to chemicals, broad temperature ranges and several degrees of moisture), thus are available for ingestion at any time by susceptible hosts[8][9][34]. Also, viability and infectivity of environmental larvated eggs persist for years, thus explaining the high number of chances that dogs have of becoming infected and the difficulties in controlling these intestinal parasitoses. As an example, eggs of T. vulpis survive from cold winter to hot summer, especially in wet and shady areas, which are widely distributed in green areas of metropolitan cities[9].

Larvated eggs of T. canis and larval ancylostomatids are an efficient environmental source of infection for various animals, which act as paratenic hosts. These animals greatly contribute to maintaining the biological cycle of toxocarosis and ancylostomosis everywhere. In fact, dogs can become infected by Toxocara by ingesting tissues of invertebrates (e.g. earthworms), ruminants (e.g. sheep), rodents, birds (e.g. chicken)[3][7][31].

The role of wildlife is another exogenous factor contributing to the environmental contamination. In fact, movements of wildlife to sub-urban and urban environments due to destruction or reduction of their habitat is another source of soil contamination by T. canis[35]. The key example is represented by synantropic fox populations, which reinforce environmental contamination and risk of infection for humans and stray and domestic dogs[36].

Thus, a combination of these factors is the basis for an extremely high environmental contamination and a life-long risk of infection for dogs living in contaminated areas.

The analysis of datasets from field investigations has recently described general principles and approaches useful to quantify levels of contamination with ascarid eggs and to prioritize control measures. In particular, the relative role of dogs, cats and foxes in disseminating parasite eggs in a given environment (*i.e.* the city of Bristol, UK) was investigated. This study, carried out in an urban setting in the absence of stray animals, showed that pet dogs are the source of most of the eggs that contaminate the environment[24]. Obviously, this study example would differ in terms of results and conclusions upon different localities, but in general it demonstrated that an estimation of egg density in urban settings is possible and provides local epidemiological models of egg outputs and sources of contamination. Also, this study illustrated that education of pet owners is crucial to minimize the risk of disease transmission to animals and humans and that stray dogs are not the culprits of faecal urban pollution in every city. It is obvious that the number of eggs contaminating the environment is dependent on the amount of faeces eliminated by owned and stray dogs and on the extent of feces removal by the owners. However, there is a lack of information on rates of deposition and removal of dog faeces from public spaces in several areas[24]. In this regard, recent field studies conducted during summer 2012 by operators observing dogs and their owners in parks and green public areas located in the cities of Rome and Padua (Italy), showed that 15.6% pet-owners did not remove dog faecal deposits from the ground, with a few differences between the investigated cities (13.5% and 16.9%, respectively) (unpublished data).

3. Risk for Humans

Human beings become infected by T. canis most commonly by ingesting embryonated eggs from the soil. Other sources of transmission with dog intestinal nematodes include ingestion of larvae resting in tissues of paratenic hosts, or hookworm larvae in contaminated soil, which can penetrate the skin of humans walking barefoot.

The presence of eggs on the ground is not only implicated with the direct infection for humans but could represent a source of contamination for pets' coats. Indeed, the role of embryonated ascarid eggs present on the fur of dogs has been evocated as a source of human infections via hand-to-mouth contact[6][37][38].

Indeed, infective eggs have been found on the coat of dogs in different studies suggesting that direct contact with these animals could bc a potential risk for humans. Eggs of T. canis may be present on the hair of both stray and privately owned dogs, with the latter considered as a more important risk for human infection due to the frequent contact with people[39]. On the other hand, close-contact with a pet has been considered an unlikely risk of infection with intestinal parasites for humans because the strong adherence of eggs on the animal's fur, the relatively high number of eggs which should be ingested to establish an infection and the long time for the embryonation (*i.e.* minimum 2 weeks)[7][40]. Rather than a self-contamination (e.g. with self-grooming transmitting eggs from the peri-anal region to other parts of the body), dogs may pick up Toxocara eggs on their hair by the scent-rolling[6]. In any case, regarding the actual risk for human infection via touching or petting a pet, scent-rolling can be a relevant cause of contamination for the animals coat when a pet is taken out in contaminated areas. Interestingly, the presence of non-canine parasite eggs on the fur of dogs indicates that the contact with a contaminated environment plays a key role in the acquisition of eggs by the animals[41]. The presence of embryonated eggs on the fur of owned dogs in some studies[37][40][42][43] may account for a lack of care in terms of anthelmintic treatment programmes. Surveys in Ireland and in the Netherlands have shown the presence of eggs on the coat of owned dogs with a percentage of 8.8%[42] and 12.2%[40] respectively. However, eggs in both studies were not infective. Relatively old private dogs have been found with a higher percentage of eggs on their coats than puppies[37][40][42]. Additionally, the absence of a correlation between intestinal worm burden and intensity of coat contamination suggests that pick-up from a contaminate soil is the main reason for the presence of parasite eggs on the coat of a dog[6][40].

Dogs with patent toxocarosis do not represent an immediate risk for human infection for a variety of reasons[44]–[46] and direct contact with an infected dog is considered of minor importance in the zoonotic transmission of intestinal nematodes[47][48].

Canid ascarids can cause different syndromes (e.g. visceral, neural or ocular larva migrans, covert toxocarosis) in human beings, especially children and toddlers.

In fact, children are the subjects at highest risk of infection, due to exposure to areas (e.g. sandpits, green areas, gardens, playgrounds) potentially contaminated by T. canis eggs[44]. Children suffering by geophagic pica caused by mineral deficiency or behavior disorders are also at high risk[44][49]. For example, the impact of human infection by larval Toxocara in childhood is demonstrated by the hundreds of cases of blindness and eye damage calculated to occur yearly in the USA, which in the past has often led to eye enucleations due to misdiagnosis with retinoblastomas[3][48][50]. However, the role of migrating larvae of the feline ascarid Toxocara cati has been repeatedly also evocated in causing human syndromes[5]. Thus, the importance of environmental contamination by T. cati should not be neglected considering the likely absence of differences in terms of zoonotic potential between dog and cat roundworms[51]. People with a soil-related job (e.g. mechanics, gardeners, farmers, street cleaners) may be at more risk of infection with toxocarosis, as shown by their higher seroprevalence compared with values found in people with non-soil related occupations[52].

A survey from Ireland showed that garden soil contamination is not associated with the household presence of pets[53]. In general, ownership of companion animals is not definitively associated with seropositivity and seroprevalence for toxocarosis[52]–[54]. Contrariwise, human seropositivity to Toxocara spp. has been put in relation with the contamination of soil with parasite eggs in some US areas[55], although actual risk factors for human infections may change according to different geographical and epidemiological settings[56][57]. A study carried out in a city of Brazil showed that almost all seropositive children had the behavior disorder of geophagy and that they played nearly every day of the week in public squares with a minimum contamination of 1 Toxocara egg/gram of sand[58]. Additionally, it was also shown that contamination in the neighborhood of domiciles in the same areas was again positively correlated with seropositivity in children in the presence of infected animals. Interestingly, seronegative children played infrequently in public squares[58].

Zoonotic hookworms may cause different pictures of skin, enteric and pulmonary diseases, being the cutaneous larva migrans the most important. Interested readers are referred to[7][8][59]. A relationship between the presence of Ancylostoma spp. larvae in soil of public squares and occurrence of cutaneous larva migrans in children has been demonstrated in Brazil[60].

It is obvious that tourists sunbathing on beaches in risky areas where zoonotic hookworms are endemic are at risk of infection with larval hookworms.

The dog whipworm T. vulpis is not included in zoonotic intestinal nematodes of pets[48] and its zoonotic potential is questioned although presumed cases of visceral larva migrans and of patent intestinal infections have been described in people. At the moment T. vulpis cannot be ultimately considered as a zoonotic canine parasite and readers interested may find more details in[9].

Despite its high zoonotic potential, few references are available on the presence of Strongyloides stercoralis in public areas. For instance, S. stercoralis-like larvae have been found in soil samples from Iran[61] and Nigeria[62].

4. Contamination and Geography

Eggs of Toxocara spp., eggs and larvae of Ancylostoma spp. and eggs of T. vulpis have been found from soil and faecal samples in public areas from Europe, the Americas, Africa and Asia.

Table 1 reports key examples of surveys carried out in different countries to evaluate the frequency of canine parasites due to faecal pollution in various human settings.

Table 1. Key examples of studies that evaluated the frequency (%) of soil contamination of public areas by roundworm, hookworm and whipworm eggs in different continents.

Geographical area	Site	Frequency (%)			Reference
		Roundworms	Hookworms	Whipworms	
Africa					
Niger	Kaduna		9.0		[63]
Americas					
USA	Connecticut	14.4			[64]
Argentina	Buenos Aires	13.2			[65]
	Buenos Aires	1.7	20.5	2.6	[66]
Brazil	Fernandopolis	79.4	6.9		[67]
	Itabuna		47.9		[68]
	São Paulo	29.7			[69]
	Guarulhos, São Paulo	68.1	64.8		[70]
Chile	Santiago	66.7			[71]
Venezuela	Ciudad Bolívar		61.1		[72]
Asia					
Japan	Tokushima	63.3			[73]
Thailand	Bangkok	5.7			[74]
Turkey	Ankara	45.0			[75]
Europe					
Ireland	Dublin	15.0			[76]
Spain	Madrid	16.4	3.0		[77]
Italy	Marche region	33.6			[78]
	Milan	7.0	3.0	5.0	[79]
	Bari	2.5	1.6	2.5	[80]
	Naples	0.7−1.4	2.4	10.1	[81]
	Messina	3.6	2.6	1.3	[82]
	Alghero	0.5−8.0	4.0	1.9	[83]
Poland	Wrocław	3.2	4.9	4.9	[84]
	Warsaw	26.1			[85]
	Kraków	15.6−19.8			[86]
Turkey	Erzurum	64.3			[87]
Czech Republic	Prague	20.4			[88]
Hungary	Eastern and northern areas	24.3−30.1	8.1−13.1	20.4−23.3	[89]
Slovak Republic	Bratislava	18.7			[90]

In a recent survey, canine faecal deposits were collected from June 2012 to January 2013 in public green areas (e.g., historic gardens, children's playgrounds or green places for physical activities or fitness) in three different municipalities of Italy (*i.e.*, Padua, Rome and Teramo). Out of a total of 677 collected samples, 38 (5.6%) scored positive upon copromicroscopical examination for at least one canine geo-helminth, *i.e.* 22/209 (10.6%) from Rome, 13/198 (6.6%) from Teramo, and 4/270 (1.5%) from Padua. Overall, the highest prevalence was detected for T. vulpis (30; 4.4%), followed by T. canis (13; 1.9%), and A. caninum (3; 0.4%), distinguished from Uncinaria based on the egg size differences reported in literature[91][92]. More specifically, prevalence values for T. vulpis and T. canis showed a similar trend in each municipality (7.7% and 1.9% in Rome, 5.1% and 3.6% in Teramo, 1.5% and 0.7% in Padua, respectively), whereas A. caninum-positive samples (1.4%) were observed solely in Rome (unpublished data).

Although parasite eggs may be found in several urban and industrialised settings, the risk of environmental contamination is particularly relevant in resource poor communities due to the fact that extensive worm control programs are limited by financial constraints. Also, in those poor settings the public health system is deficient, there is usually a high number of stray and feral animals and people lack awareness of health risks[93]. In these settings the bond between physicians, veterinarians and the whole community should be re-enforced to minimize as much as possible the risk of public hazard.

5. What Can We Do to Reduce Environmental Contamination?

A reduction of the contamination of public areas by dog helminths can be achieved only with a combination of approaches, e.g. reliable worm control programs, awareness of veterinarian and behavior of pet owners and the general public.

No reliable methods exist to realistically eliminate eggs or larvae of intestinal nematodes of pets present on the ground. Therefore, preventing the initial contamination of the environment is of paramount importance. The individualized treatment of parasitized animals is mandatory to control infection in pets and environmental pollution. Unfortunately, negligence in performing diagnostic copromicroscopy in veterinary practices is frequent, due to the fallacy in considering an antiparasitic treatment powerful enough to "generically clear parasites".

Contrariwise, copromicroscopic examinations should be regular for pets, given that virtually all dogs are at risk of becoming infected by intestinal nematodes for all their life. The role of veterinarians is crucial, because pet owners should be convinced of the importance of periodic faecal examinations. Veterinarians have a plethora of parasiticides, which can be administered according to each individual possible scenario and both owner and animal compliance to treat infected animals[7][9]. Thorough indications for worm control programs have been released by the US Companion Animal Parasite Council (CAPC) and the European Scientific Counsel Companion Animal Parasites (ESCCAP)[7][94][95].

A key point for controlling pet parasites is the lifelong chemopreventative program. Using year-round treatment is of importance where there is the necessity to perform the annual chemoprophylaxis for other severe parasites and not only for intestinal nematodes, e.g. for the prevention of cardio-pulmonary nematodes, *i.e.* Dirofilaria immitis and Angiostrongylus vasorum. In addition, several formulations containing compounds effective against intestinal nematodes also contain cestocides which are powerful for controlling infections caused by tapeworms distributed worldwide (e.g. Dipylidium caninum) or hazardous for humans (e.g. Echinococcus spp.).

Broad-spectrum formulations with an easy mode of administration (e.g. chewy tablets, spot-on) fit particularly with year-long worm control programs. Faecal examinations should be performed whether or not a monthly-based treatment program is used, even when the dog appears healthy, as there are parasites that may not be covered by the treatment program or there may be poor compliance with the program. In fact, owners may be not interested in paying for faecal examination if the animals are asymptomatic, because they are commonly considered parasite-free. While puppies and their thousands of eggs shed daily are the major source of contamination for the environment, a US study has shown that after young dogs, the most parasitized category of pets are >10 years old[96]. This high degree of parasitism in old animals could reside in a lack of willingness of owners in chemopreventative and/or worm control programs in old pets[96]. Indeed, there is no reason to consider an old animal a less effective source of infection for pets, human beings and the environment.

Unfortunately, public risk perception and awareness may be poor in veterinarians, the general public and pet owners of several countries[97]–[100]. Interview-based studies have been conducted to understand how the risk perception is present in the human population and to implement awareness of the general public and of pet owners. For example, a British survey has unveiled that less than the half of the participants (*i.e.* pet and non-pet owners) were aware of the potential for transmission of parasites via animal faeces with no differences between who had a pet and who did not[55].

Similarly, a recent Italian interview-based study carried out during summer 2012 in the cities of Rome and Padua illustrated that out of 469 participants, 246 (52.5%) were aware of the health risk associated with canine faecal pollution in urban settings, with no differences between pet and non-pet owners. In the same study, the awareness of the health risks was higher in Padua (205/339, corresponding to 60.4%) than in Rome (41/130; 31.5%), again with no differences between pet and non-pet owners (unpublished data).

Veterinarians should routinely inform clients about source of infections for both pets and humans and on reliable measures to prevent transmission to other animals and people. Regrettably, this is not a frequent behavior. As a key example, less than the half of interviewed veterinarians in a Canadian survey discussed the zoonotic risk of pet ownership with clients, while the remainder did this only in particular cases or not at all[100].

Given that public squares, sandpits, playgrounds, beaches are always at a high risk for heavy contamination by pet faeces and public parks and green areas are always contaminated by parasites of dogs[4][8][101]–[103], avoiding animal defecation in public areas or immediate collection of stool by the pet owner is crucial (**Figure 2**). Veterinarians should educate owners on regular removal and disposal of faeces, which is at the basis to minimize environmental contamination and risk of transmission[44][48]. When walking their pets in public areas, all owners should respect local indications and keep their animals in reserved areas, if present (**Figure 3**).

Figure 2. Indication for dog-owners in New York City, USA.

Figure 3. Beach area reserved for dogs in Broome, Australia.

A very "creative" measure was recently adopted by the Municipality of Brunete (Spain), in which undercover volunteers were recruited to patrol the streets, and to confront dog-owners who did not remove the faeces of their pet. Approaching the guilty owner for a friendly conversation, volunteers swindled some useful information to identify his/her domicile by the preexisting pet-registration database. At the end of the conversation, when the dog-owner was out of sight, the volunteers picked up the dog's faeces, the excrements were boxed, and hand-delivered to the pet owner along with an official fine and warning[104].

Other than constant municipal cleaning and maintenance, controlled access of green areas and public parks by fences is an effective way of prevention of faecal contamination. A study in Japan has shown that placing vinyl plastic covers over sandboxes at night is able to discourage animals from defecating there[105]. An extreme measure chosen by some municipalities is the elimination of sandboxes from parks and playgrounds[8]. It is important to note, however, that while T. canis eggs are most prevalent in public parks, sandboxes are mainly contaminated with eggs of T. cati due to the common behavior of cats during defecation[103].

Surveillance of the presence of parasite eggs in public soil is also important in this integrated approach to control intestinal parasites. In general, microscopic examination of soil samples is performed to identify Toxocara eggs, although this method may have low sensitivity and specificity[106][107]. DNA-based approaches have been developed to discriminate eggs of ascarids in soil samples[107][108], although some pitfalls may impair a routine use, e.g. low throughput analysis and risk of carry-on contamination. A duplex Real-Time PCR has been recently validated for the detection and discrimination of T. canis and T. cati eggs in different samples, including soil. This assay is promising for the implementation of standardized methods able to evaluate the presence of roundworm eggs in contaminated soil on a large scale. In particular, this novel molecular tool can be used to investigate, with a high throughput, the occurrence and the level of contamination of eggs of T. canis (and T. cati) in urban parks, green areas, playgrounds and sandpits[109].

This is of importance because different investigations have shown that some urban environments may be heavily contaminated by T. cati rather than by T. canis[101].

6. Conclusion

Canine faeces in cities are an important source of pathogens for the pet population, for dog owners and for the community in general. Prevention of initial contamination is the most important way to avoid human and animal infections, given that no practical methods are available to actually minimize environmental egg contamination. The non-polite habit of dog owners of not removing feces of their pet from streets and green areas (**Figure 4**) represents a concern for hygiene and health of both animals and humans. Hence, polluted public environments represent the principle risk for human health with zoonotic intestinal nematodes of dogs[38]. Other than social responsibility in eliminating dog faeces from streets, parks and squares, appropriate worm control programs, especially in young dogs, are crucial to control faecal contamination and minimize the risk of infection for humans and other animals.

Unfortunately, public education in reducing the risk of exposure for both humans and companion animals is poor. In recent years sociological changes have influenced the relationships between physicians and veterinarians, towards the concept of the "One Health Program" (*i.e.* "the collaborative work of multiple disciplines to help attain optimal health of people, animals, and our environment")[110]. Thus, there is the necessity for physicians, veterinarians and the general

Figure 4. Dog faeces in a green park of Dublin, Ireland (left) and in a public square of Padua, Italy (right).

public to foster interest and efforts in appropriate control programs towards a reduction of pollution of the cities and of the risk of infection for both animals and people.

Competing Interests

Studies whose unpublished results are reported in the review were financed by Novartis Animal Health, of which FLT and JD are employees.

Authors' Contributions

DT, AFdR and MP conceived the article and all authors contributed to its drafting, preparation and intellectual content. AFdR and MP were scientifically responsible for the studies whose unpublished results are reported in the review. All authors read and approved the final manuscript.

Acknowledgements

The authors are grateful to all the co-workers who contributed in the studies whose unpublished results are presented in the review.

Source: Traversa D, Regalbono A F D, Cesare A D, et al. Environmental contamination by canine geohelminths[J]. Parasites & Vectors, 2014, 7(3):1–9.

References

[1] World Health Organization and partners unveil new coordinated approach to treat millions suffering from neglected tropical disease. 2006. http://whqlibdoc.who.int/press_release/2006/PR.

[2] Soulsby EJL: Helminths, Arthropods and Protozoa of Domesticated Animals. 7th edition. London, UK: Bailliere Tindall; 1982.

[3] Bowman DD: Georgi's Parasitology for Veterinarians. 9th edition. Philadelphia, USA: Saunders Company; 2009.

[4] Holland CV, Smith HV: Toxocara: The Enigmatic Parasite. Wallingford, UK: CABI Publishing; 2006.

[5] Fisher M: Toxocara cati: an underestimated zoonotic agent. Trends Parasitol 2003, 19(4):167–170.

[6] Roddie G, Stafford P, Holland C, Wolfe A: Contamination of dog hair with eggs of Toxocara canis. Vet Parasitol 2008, 152(1–2):85–93.

[7] Traversa D: Pet roundworms and hookworms: a continuing need for globalworming. Parasit Vectors 2012, 5:91.

[8] Despommier D: Toxocariasis: clinical aspects, epidemiology, medical ecology, and molecular aspects. Clin Microbiol Rev 2003, 16(2):265–272.

[9] Traversa D: Are we paying too much attention to cardio-pulmonary nematodes and neglecting old-fashioned worms like Trichuris vulpis? Parasit Vectors 2011, 4:32.

[10] Visco RJ, Corwin RM, Selby LA: Effect of age and sex on the prevalence ofintestinal parasitism in dogs. J Am Vet Med Assoc 1977, 170:835–837.

[11] Visco RJ, Corwin RM, Selby LA: Effect of age and sex on the prevalence of intestinal parasitism in cats. J Am Vet Med Assoc 1978, 1978(172):797–800.

[12] Lloyd S: Toxocarosis. In Zoonoses. Biology, Clinical Practice and Public Health Control. Edited by Palmer SR, Soulsby EJL, Simpson DIH. Oxford: Oxford University Press; 1998:841–854.

[13] Malloy WF, Embil JA: Prevalence of Toxocara spp. and other parasites in dogs and cats in Halifax, Nova Scotia. Can J Comp Med 1978, 42:29–31.

[14] Martínez-Barbabosa I, Vázquez Tsuji O, Cabello RR, Cárdenas EM, Chasin OA: The prevalence of Toxocara cati in domestic cats in Mexico City. Vet Parasitol 2003, 114:43–49.

[15] Fahrion AS, Staebler S, Deplazes P: Patent Toxocara canis infections in previously exposed and in helminth-free dogs after infection with low numbers of embryonated eggs. Vet Parasitol 2008, 152:108–115.

[16] Little SE, Johnson EM, Lewis D, Jaklitsch RP, Payton ME, Blagburn BL, Bowman DD, Moroff S, Tams T, Rich L, Aucoin D: Prevalence of intestinal parasites in pet dogs in the United States. Vet Parasitol 2009, 166:144–152.

[17] Scorza AV, Duncan C, Miles L, Lappin MR: Prevalence of selected zoonotic and vector-borne agents in dogs and cats in Costa Rica. Vet Parasitol 2011, 183:178–183.

[18] Savilla TM, Joy JE, May JD, Somerville CC: Prevalence of dog intestinal nematode parasites in south central West Virginia, USA. Vet Parasitol 2011, 178:115–120.

[19] Barutzki D, Schaper R: Results of parasitological examinations of faecal samples from cats and dogs in Germany between 2003 and 2010. Parasitol Res 2011, 109 (Suppl 1):S45–S60.

[20] Sager H, Moret CS, Grimm F, Deplazes P, Doherr MG, Gottstein B: Coprological study on intestinal helminths in Swiss dogs: temporal aspects of anthelminthic treatment. Parasitol Res 2006, 98:333–338.

[21] Lloyd S, Amersinghe PH, Soulsby EJL: Peripaturient immunosuppression in the bitch and its influence on infection with Toxocara canis. J Small Anim Pract 1983, 24:237–247.

[22] Lloyd S: Toxocara canis: the dog. In Toxocara and Toxocariasis Clinical, Epidemiological and Molecular Perspectives. Edited by Lewis JW, Maizels RM. London: British Society for Parasitology. Institute of Biology; 1993:11–24.

[23] Epe C: Intestinal nematodes: biology and control. Vet Clin North Am Small Anim Pract 2009, 39:1091-1107. vi–vii.

[24] Morgan ER, Azam D, Pegler K: Quantifying sources of environmental contamination with Toxocara spp. eggs. Vet Parasitol 2013, 193(4):390–397.

[25] Anderson RC: Nematode Parasites of Vertebrates. In Their development and transmission. 2nd edition. Guilford: CABI; 2000.

[26] Prociv P: Zoonotic hookworm infections (Ancylostomosis). In In Zoonoses. 1st edition. Edited by Palmer SR, Soulsby EJL, Simpson DIH. Oxford, UK: Oxford Medical Publications; 1998.

[27] Bowman DD, Montgomery SP, Zajac AM, Eberhard ML, Kazacos KR: Hookworms of dogs and cats as agents of cutaneous larva migrans. Trends Parasitol 2010, 26: 162–167.

[28] Stoye M: Galactogenic and prenatal Toxocara canis infections in dogs (Beagle). Dtsch Tierarztl Woch 1976, 83:107–108. in German.

[29] Stoye M: Biology, pathogenicity, diagnosis and control of Ancylostoma caninum. Dtsch Tierarztl Woch 1992, 99:315–321. in German.

[30] Bosse M, Manhardt J, Stoye M: Epidemiology and Control of neonatal Helminth Infections of the dog. Fortschr Vet Med 1980, 30:247–256. in German.

[31] Taylor MA, Coop RL, Wall RL: Veterinary Parasitology. 3rthth edition. Oxford, UK: Blackwell Publishing; 2007.

[32] Fontanarrosa MF, Vezzani D, Basabe J, Eiras DF: An epidemiological study of gastrointestinal parasites of dogs from Southern Greater Buenos Aires (Argentina): age, gender, breed, mixed infections, and seasonal and spatial patterns. Vet Parasitol 2006, 136:283–295.

[33] Kirchheimer R, Jacobs DE: Toxocara species egg contamination of soil from children's play areas in southern England. Vet Rec 2008, 163(13):394–395.

[34] Parsons JC: Ascarid infections of cats and dogs. Vet Clin North Am Small Anim Pract 1987, 17:1307–1339.

[35] Brochier B, De Blander H, Hanosset R, Berkvens D, Losson B, Saegerman C: Echinococcus multilocularis and Toxocara canis in urban red foxes (Vulpes vulpes) in Brussels. Belgium. Prev Vet Med 2007, 80(1):65–73.

[36] Antolová D, Reiterová K, Miterpáková M, Stanko M, Dubinský P: Circulation of Toxocara spp. in suburban and rural ecosystems in the Slovak Republic. Vet Parasitol 2004, 126:317–324.

[37] Wolfe A, Wright IP: Human toxocariasis and direct contact with dogs. Vet Rec 2003, 152:419–422.

[38] Deplazes P, van Knapen F, Schweiger A, Overgaauw PA: Role of pet dogs and cats in the transmission of helminthic zoonoses in Europe, with a focus on echinococcosis and toxocarosis. Vet Parasitol 2011, 182(1):41–53.

[39] El-Tras WF, Holt HR, Tayel AA: Risk of Toxocara canis eggs in stray and domestic dog hair in Egypt. Vet Parasitol 2011, 178(3–4):319–323.

[40] Overgaauw PA, van Zutphen L, Hoek D, Yaya FO, Roelfsema J, Pinelli E, van Knapen F, Kortbeek LM: Zoonotic parasites in fecal samples and fur from dogs and cats in The Netherlands. Vet Parasitol 2009, 163:115–122.

[41] Holland C, O'Connor P, Taylor MR, Hughes G, Girdwood RW, Smith H: Families, parks, gardens and toxocariasis. Scand J Infect Dis 1991, 23(2):225–231.

[42] Keegan JD, Holland CV: Contamination of the hair of owned dogs with the eggs of Toxocara spp. Vet Parasitol 2010, 173:161–164.

[43] Aydenizöz-Ozkayhan M, Yağci BB, Erat S: The investigation of Toxocara canis eggs in coats of different dog breeds as a potential transmission route in human toxocariasis. Vet Parasitol 2008, 152(1–2):94–100.

[44] Overgaauw PAM: Aspects of Toxocara epidemiology: human toxocarosis. Crit Rev Microbiol 1997, 23:215–231.

[45] Overgaauw PAM, Van Knapen F: Dogs and nematodes zoonoses. In Dogs Zoonoses and Public Haelth. 1st edition. Edited by MacPhersom CNL, Melsin FX, Wandeler A. New York: CABI Publishing Oxon; 2000:213–245.

[46] Overgaauw PAM, Van Knapen F: Negligible risk of visceral or ocular larva migrans from petting a dog. Ned Tijdschr Geneeskd 2004, 148:1600–1603.

[47] Overgaauw PAM: Aspects of Toxocara epidemiology: toxocarosis in dogs and cats. Crit Rev Microbiol 1997, 23:233–251.

[48] Robertson ID, Thompson RC: Enteric parasitic zoonoses of domesticated dogs and cats. Microbes Infect 2002, 4:867–873.

[49] Schantz PM: Toxocara larva migrans now. Am J Trop Med Hyg 1989, 41:21–34.

[50] Glickman LT, Schantz PM: Epidemiology and pathogenesis of zoonotic toxocariasis. Epidemiol Rev 1981, 3:230–250.

[51] Overgaauw PA, van Knapen F: Veterinary and public health aspects of Toxocara spp. Vet Parasitol 2013, 193(4):398–403.

[52] Raschka C, Haupt W, Ribbeck R: Studies on endoparasitization of stray cats. Mon Vet 1994, 49:307–315.

[53] Jenkins DJ: Hydatidosis a zoonosis of unrecognised increasing importance? J Med Microbiol 1998, 47:1–3.

[54] Tenter AM, Heckeroth AR, Weiss LM: Toxoplasma gondii: from animals to humans. Int J Parasit 2000, 30:1217–1258.

[55] Won KY, Kruszon-Moran D, Schantz PM, Jones JL: National seroprevalence and risk factors for Zoonotic Toxocara spp. infection. Am J Trop Med Hyg 2008, 79(4): 552–557.

[56] Andrade C, Alava T, De Palacio IA, Del Poggio P, Jamoletti C, Gulletta M, Montresor A: Prevalence and intensity of soil-transmitted helminthiasis in the city of Portoviejo (Ecuador). Mem Inst Oswaldo Cruz 2001, 96(8):1075–1079.

[57] Matsuo J, Nakashio S: Prevalence of fecal contamination in sandpits in public parks in Sapporo City, Japan. Vet Parasitol 2005, 128(1–2):115–119.

[58] Manini MP, Marchioro AA, Colli CM, Nishi L: Falavigna-Guilherme AL: Association between contamination of public squares and seropositivity for Toxocara spp. in children. Vet Parasitol 2012, 188(1–2):48–52.

[59] Hotez PJ, Wilkins PP: Toxocariasis: America's most common neglected infection of poverty and a helminthiasis of global importance? PLoS Negl Trop Dis 2009, 3(3):e400.

[60] Santarém VA, Giuffrida R, Zanin GA: Larva migrans cutânea: ocorrência de casos humanos e identificação de larvas de Ancylostoma spp. em parque público do município de Taciba, São Paulo. Rev Soc Bras Med Trop 2004, 37:179–181.

[61] Motazedian H, Mehrabani D, Tabatabaee SH, Pakniat A, Tavalali M: Prevalence of helminth ova in soil samples from public places in Shiraz. East Mediterr Health J 2006, 12:562–565.

[62] Ogbolu DO, Alli OA, Amoo AO, Olaosun II, Ilozavbie GW, Olusoga-Ogbolu FF: High-level parasitic contamination of soil sampled in Ibadan metropolis. Afr J Med Med Sci 2011, 40:321–325.

[63] Maikai BV, Umoh JU, Ajanusi OJ, Ajogi I: Public health implications of soil contaminated with helminth eggs in the metropolis of Kaduna, Nigeria. J Helminthol 2008, 82(2):113–118.

[64] Chorazy ML, Richardson DJ: A survey of environmental contamination with ascarid ova, Wallingford, Connecticut. Vector Borne Zoonotic Dis 2005, 5(1):33–39.

[65] Fonrouge R, Guardis MV, Radman NE, Archelli SM: Soil contamination with Toxocara sp. eggs in squares and public places from the city of La Plata. Buenos Aires,

Argentina. Bol Chil Parasitol 2000, 55(3–4):83–85.

[66] Rubel D, Wisnivesky C: Dog fouling and helminth contamination in parks and sidewalks of Buenos Aires City, 1991–2006. Medicina (B Aires) 2010, 70(4):355–363.

[67] Cassenote AJ, Pinto Neto JM, Lima-Catelani AR, Ferreira AW: Soil contamination by eggs of soil-transmitted helminths with zoonotic potential in the town of Fernandópolis, State of São Paulo, Brazil, between 2007 and 2008. Rev Soc Bras Med Trop 2011, 44(3):371–374.

[68] Campos Filho PC, Barros LM, Campos JO, Braga VB, Cazorla IM, Albuquerque GR, Carvalho SM: Zoonotic parasites in dog feces at public squares in the municipality of Itabuna, Bahia, Brazil. Rev Bras Parasitol Vet 2008, 17(4):206–209.

[69] Muradian V, Gennari SM, Glickman LT, Pinheiro SR: Epidemiological aspects of Visceral Larva Migrans in children living at São Remo Community, São Paulo (SP), Brazil. Vet Parasitol 2005, 134(1–2):93–97.

[70] Marques JP, Guimarães Cde R, Boas AV, Carnaúba PU, Moraes J: Contamination of public parks and squares from Guarulhos (São Paulo State, Brazil) by Toxocara spp. and Ancylostoma spp. Rev Inst Med Trop Sao Paulo 2012, 54(5):267–271.

[71] Castillo D, Paredes C, Zañartu C, Castillo G, Mercado R, Muñoz V, Schenone H: Environmental contamination with Toxocara sp. eggs in public squares and parks from Santiago, Chile, 1999. Bol Chil Parasitol 2000, 55(3–4):86–91.

[72] Devera R, Blanco Y, Hernández H, Simoes D: Toxocara spp. and other helminths in squares and parks of Ciudad Bolívar, Bolivar State (Venezuela). Enferm Infecc Microbiol Clin 2008, 26(1):23–26.

[73] Shimizu T: Prevalence of Toxocara eggs in sandpits in Tokushima city and its outskirts. J Vet Med Sci 1993, 55(5):807–811.

[74] Wiwanitkit V, Waenlor W: The frequency rate of Toxocara species contamination in soil samples from public yards in a urban area "Payathai", Bangkok, Thailand. Rev Inst Med Trop Sao Paulo 2004, 46(2):113–114.

[75] Avcioglu H, Burgu A: Seasonal prevalence of Toxocara ova in soil samples from public parks in Ankara, Turkey. Vector Borne Zoonotic Dis 2008, 8(3):345–350.

[76] O'Lorcain P: Prevalence of Toxocara canis ova in public playgrounds in the Dublin area of Ireland. J Helminthol 1994, 68(3):237–241.

[77] Dado D, Izquierdo F, Vera O, Montoya A, Mateo M, Fenoy S, Galván AL, García S, García A, Aránguez E, López L, del Águila C, Miró G: Detection of zoonotic intestinal parasites in public parks of Spain. Potential epidemiological role of microsporidia. Zoonoses Public Health 2012, 59(1):23–28.

[78] Habluetzel A, Traldi G, Ruggieri S, Attili AR, Scuppa P, Marchetti R, Menghini G, Esposito F: An estimation of Toxocara canis prevalence in dogs, environmental egg contamination and risk of human infection in the Marche region of Italy. Vet Parasitol 2003, 113(3–4):243–252.

[79] Genchi M, Ferroglio E, Traldi G, Passera S, Mezzano G, Genchi C: Fecalizzazione ambientale e rischio parassitario nelle città di Milano e Torino. Professione Veterinaria 2007, 41:15–17.

[80] Lia R, La Montanara C, Leone N, Pantone N, Llazari A, Puccini V: Canine helminthic fauna and environmental faecalization in the town of Bari (Apulia region, Southern Italy). Parassitologia 2002, 44(1):92.

[81] Rinaldi L, Biggeri A, Carbone S, Musella V, Catelan D, Veneziano V, Cringoli G: Canine faecal contamination and parasitic risk in the city of Naples (southern Italy). BMC Vet Res 2006, 2:29.

[82] Risitano AL, Brianti E, Gaglio G, Ferlazzo M, Giannetto S: Environmental contamination by canine feces in the city of Messina: parasitological aspects and zoonotic hazards. In Proceedings of LXI Congress of the Italian Society for Veterinary Science (S.I.S.Vet.). Salsomaggiore Terme, Italy; 2007:135–136.

[83] Scala A, Garippa G, Pintus D: Environmental contamination by canine feces in the city of Alghero (SS): parasitological aspects and zoonotic hazards. In Proceedings of LXIII Congress of the Italian Society for Veterinary Science (S.I.S.Vet.). Udine, Italy; 2009:180–182.

[84] Perec-Matysiak A, Hildebrand J, Zaleśny G, Okulewicz A, Fatuła A: The evaluation of soil contamination with geohelminth eggs in the area of Wrocław. Poland. Wiad Parazytol 2008, 54(4):319–323.

[85] Borecka A, Gawor J: Prevalence of Toxocara canis infection in dogs in the Warszawa area. Wiad Parazytol 2000, 46(4):459–462.

[86] Mizgajska H: Soil contamination with Toxocara spp. eggs in the Kraków areaand two nearby villages. Wiad Parazytol 2000, 46(1):105–110.

[87] Avcioglu H, Balkaya I: The relationship of public park accessibility to dogs to the presence of Toxocara species ova in the soil. Vector Borne Zoonotic Dis 2011, 11(2):177–180.

[88] Dubná S, Langrová I, Nápravník J, Jankovská I, Vadlejch J, Pekár S, Fechtner J: The prevalence of intestinal parasites in dogs from Prague, rural areas, and shelters of the Czech Republic. Vet Parasitol 2007, 145(1–2):120–128.

[89] Fok E, Szatmári V, Busák K, Rozgonyi F: Prevalence of intestinal parasites in dogs in some urban and rural areas of Hungary. Vet Q 2001, 23(2):96–98.

[90] Totková A, Klobusický M, Holková R, Friedová L: Current prevalence of toxocariasis and other intestinal parasitoses among dogs in Bratislava. Epidemiol Mikrobiol Imunol 2006, 55(1):17–22.

[91] Ehrenford FA: Differentiation of the ova of Ancylostoma caninum and Uncinaria stenocephala in dogs. Am J Vet Res 1953, 14(53):578–580.

[92] Sloss MW, Kemp RL, Zajac AM: Veterinary Clinical Parasitology. Veterinary Clinical Parasitology: Iowa State University Press; 1994.

[93] Heukelbach J, Mencke N, Feldmeier H: Editorial: cutaneous larva migrans and tungiasis: the challenge to control zoonotic ectoparasitoses associated with poverty. Trop Med Int Health 2002, 7(11):907–910.

[94] Companion Animal Parasite Council. http://www.capcvet.org.

[95] European Scientific Counsel Companion Animal Parasites. http://www.esccap.org.

[96] Gates MC, Nolan TJ: Endoparasite prevalence and recurrence across different age groups of dogs and cats. Vet Parasitol 2009, 166:153–158.

[97] Rubinstensky-Elefant G, Hirata CE, Yamamoto JH, Ferreira MU: Human toxocariasis: diagnosis, worldwide seroprevalences and clinical expression of the systemic and ocular forms. Ann Trop Med Parasitol 2010, 104:3–23.

[98] Harvey JB, Roberts JM, Schantz PM: Survey of veterinarians' recommendations for treatment and control of intestinal parasites in dogs: public health implications. J Am Vet Med Assoc 1991, 199:702–707.

[99] Overgaauw PAM: Effect of a government educational campaign in the Netherlands on awareness of Toxocara and toxocarosis. Prev Vet Med 1996, 28:165–174.

[100] Stull JW, Carr AP, Chomel BB, Berghaus RD, Hird DW: Small animal deworming protocols, client education, and veterinarian perception of zoonotic parasites in western Canada. Can Vet J 2007, 48:269–276.

[101] Lee CY, Schantz PM, Kazacos KR, Montgomery SP, Bowman DD: Epidemiologic and zoonotic aspects of ascarid infections in dogs and cats. Trends Parasitol 2010, 26:155–161.

[102] Tharaldsen J: Parasitic organisms from dogs and cats in sandpits from nursery schools in Oslo. Norsk Veterinaertidsskrift 1982, 94:251–254.

[103] Jansen J, Van Knapen F: Toxocara eggs in public parks and sandboxes in Utrecht. Tijdschr Diergeneeskd 1993, 118:611–614.

[104] Brunete entrega a domicilio las cacas de perro 'extraviadas' por sus dueños. http://www.elmundo.es/elmundo/2013/06/03/madrid/1370257901.html.

[105] Uga S, Kataoka N: Measures to control Toxocara egg contamination in sandpits of public parks. Am J Trop Med Hyg 1995, 52:21–24.

[106] Uga S, Matsuo J, Kimura D, Rai SK, Koshino Y, Igarashi K: Differentiation of Toxocara canis and T. cati eggs by light and scanning electron microscopy. Vet Parasitol 2000, 92(4):287–294.

[107] Borecka A, Gawor J: Modification of gDNA extraction from soil for PCR designed for the routine examination of soil samples contaminated with Toxocara spp. eggs. J Helminthol 2008, 82(2):119–122.

[108] Fogt-Wyrwas R, Jarosz W, Mizgajska-Wiktor H: Utilizing a polymerase chain reaction method for the detection of Toxocara canis and T. cati eggs in soil. J Helminthol

2007, 81(1):75–78.

[109] Durant JF, Irenge LM, Fogt-Wyrwas R, Dumont C, Doucet JP, Mignon B, Losson B, Gala JL: Duplex quantitative real-time PCR assay for the detection and discrimination of the eggs of Toxocara canis and Toxocara cati (Nematoda, Ascaridoidea) in soil and fecal samples. Parasit Vectors 2012, 7(5):288.

[110] Paul M, King L, Carlin EP: Zoonoses of people and their pets: a US perspective on significant pet-associated parasitic diseases. Trends Parasitol 2010, 26:153–154.

Chapter 18

Envirotyping for Deciphering Environmental Impacts on Crop Plants

Yunbi Xu[1,2]

[1]Institute of Crop Science, Chinese Academy of Agricultural Sciences, Beijing, China
[2]International Maize and Wheat Improvement Center (CIMMYT), El Batan, Texcoco CP 56130, Mexico

Abstract: Global climate change imposes increasing impacts on our environments and crop production. To decipher environmental impacts on crop plants, the concept "envirotyping" is proposed, as a third "typing" technology, complementing with genotyping and phenotyping. Environmental factors can be collected through multiple environmental trials, geographic and soil information systems, measurement of soil and canopy properties, and evaluation of companion organisms. Envirotyping contributes to crop modeling and phenotype prediction through its functional components, including genotype-by-environment interaction (GEI), genes responsive to environmental signals, biotic and abiotic stresses, and integrative phenotyping. Envirotyping, driven by information and support systems, has a wide range of applications, including environmental characterization, GEI analysis, phenotype prediction, near-iso-environment construction, agronomic genomics, precision agriculture and breeding, and development of a four-dimensional profile of crop science involving genotype (*G*), phenotype (*P*), envirotype (*E*) and time (*T*) (developmental stage). In the future, envirotyping needs to zoom into specific experimental plots and individual plants, along with the development of high-throughput and precision envirotyping platforms, to integrate genotypic, phenotypic and envirotypic information for establishing a high-efficient precision breeding and sustainable crop production system based on deciphered environmental impacts.

1. Introduction

Climate change has resulted in significant changes in weather pattern, precipitation distribution, temperature and moisture fluctuation, soil erosion, and desertification (FAO 2008), although the average annual environmental measurements may not change significantly. Extreme conditions caused by these changes bring about many unexpected and more frequent biotic and abiotic stresses (Bebber *et al.* 2013; Trenberth *et al.* 2014). To feed increasing world population, total crop production will need to be significantly increased with less arable land under much severe environmental conditions (Tilman *et al.* 2002). For the past 50 years, such demanding has been met by continuous yield improvement. Unfortunately, yield growth has been slowing, rather than increasing as required by global population increase. For example, annual yield growth for three major cereals, rice, wheat and maize, has been decreased to 0.79%–1.74% for 1990-2010 from 2.19%–2.95% for 1960–1990. Moreover, if such reduction tendency continues, yield growth for 2010–2050 will decrease to 0.62%–1.33% (FAO 2014; Pardey *et al.* 2014). For the next 50 years, we will have more people, but less water on the planet, and have to develop two times better crops for a world free of poor, poverty, and environment degradation. To meet the challenges, we need to keep enhancing yield potential while filling the yield gap created by various abiotic and biotic stresses largely caused by climate change. Therefore, environmental factors that affect plant growth and yield should be understood and managed better for less degradation and input but more output.

Crop production has been largely affected by environmental factors that affect all the processes from metabolism to gene expression during plant growth and development. Increasing yield and filling yield gap largely depend on the management, control and improvement of the environments where crop plants grow. The genotypes (G) that determine the yield potential and their responses to environmental factors can be now investigated and measured through molecular and genomic approaches using chip or microarray (Hoheisel 2006) and sequencing technologies (Koboldt *et al.* 2013). The phenotypes (*P*) can be also measured precisely with the development of high-throughput phenotyping tools and methodologies (Araus and Cairns 2014). Compared with genotyping and phenotyping, determination and measurement of environmental factors (*E*) has fallen behind, largely due to three reasons. First, environmental factors have been largely considered as a whole and treated as a blackbox that interacts with genotypes to affect plant growth and yield, without dissection for individual plants. Second, only major environment factors have been considered and measured at the level of the whole experiment station or trial. Third, most environmental factors are dynamic and constantly changing throughout the plant growing period. Dissection of quantitative traits into individual Mendelian factors using molecular markers allows quantitative genetics walk out of the multiple-gene circle taking all the relevant genes as a whole (Paterson *et al.* 1988; Lander and Botstein 1989). A similar significant impact would be made if complex environments could be partitioned into individual factors and measured for individual plants and every developmental stage.

Understanding better the environment where plants live is critical to our future crop science for several reasons. Firstly, gene expression is largely dependent on the environment where the crop grows. Secondly, genetic mapping and gene cloning depends on the environment where the phenotyping is performed. Thirdly, many phenotyping procedures, including abiotic and biotic stress evaluation, is conducted under managed environmental conditions. Lastly, environmental assay becomes increasingly important for many procedures of crop production. Precise dissection of complex environmental factors for both target environments and specific genotypes provides us a novel opportunity for management, control and optimization of environmental factors for enhanced genetic improvement and more efficient crop production. All environmental factors that affect plant growth and yield can be defined as envirotypes (environment + types). The process for determination and measurement of all the environmental factors is called envirotyping (Xu 2015). The concept was first proposed at two international conferences as "etyping" (Xu 2011, 2012), followed by journal articles with more details (Xu *et al.* 2012; Xu 2015). The term "envirotyping" has also been used recently by other researchers to refer to the collective body of methodologies that are applied to characterize environments within multiple environmental trials and the frequent repeatable environment types within the target population of environments (Cooper *et al.* 2014, 2016). Envirotyping is different from conventional environmental assay in three aspects. First, envirotyping will measure all environmental factors that affect plant growth and production instead of only for the major ones. Second, envirotyping will zoom into specific field plots and individual plants so that the envirotypic data will be collected to match up with the corresponding genotypic and phenotypic data. Third, crop management and companion organisms will be included as a part of environmental factors so that their effects on crop plants can be investigated. As a new concept, envirotyping will be fully discussed in this article, including its conception, implementation, and application in crop science, by which environmental impacts on crop plants can be deciphered.

2. Environmental Variables and Envirotyping

2.1. Environmental Variables

Environmental factors can be micro- or macro-, non-organic or organic, and internal or external. Plant growth and yield are coordinated by both intercellular and external environments. Intercellular environments in plants are largely dominated by what are essentially enclosed in vacuoles, which consist of water (inorganic and organic molecules), waste products and small molecules with internal hydrostatic pressure or turgor, temperature, and an acidic pH maintained. The internal environments are largely affected due to the changes of pH, osmotic pressure and temperature, etc., caused by material exchange and signal transduction with external environments. Through a series of receptors, signal transductions and responses, plants make full responses to specific external environmental factors, resulting in ion transmembrane transport, metabolic pathway regulation, cytoskeleton modification and gene expression regulation (Nicotra *et al.* 2010).

External environmental factors can be classified into four categories, climate, soil factors, biotic factors, and crop management or cropping system (**Table 1**). Climate factors, such as temperature, radiation, precipitation or water availability and wind, determine where a plant can grow, while other factors determine how a plant grows. Some companion organisms, such as pathogens, pests and weeds,

Table 1. External environmental factors affecting plant growth and yield.

Category	Description	Effects and associated stresses
Climate factors		
Light	Solar radiation, light intensity (elevation, latitude, and season; clouds, dust, smoke, fog and smog), day length (photoperiod)	Most crucial factor for plant growth and development; shading stress
Temperature	Effective accumulated temperature; average, minimum and maximum daily temperatures	Photosynthesis, water and nutrient absorption, transpiration, respiration and enzyme activity, germination, flowering, pollen viability, fruit/seed set, rates of maturation and senescence, yield, quality, harvest duration and shelf life; cold, frost, and heat stresses
Water	Precipitation (rainfall, snow, hail, fog and dew)	Crop productivity and quality; drought, flooding and waterlogging stresses
	Atmospheric humidity (relative humidity)	Soil evaporation and plant transpiration; dry air stress
Air	Wind velocity	Supply of moisture, heat, and fresh CO_2; strong wind stress
	Atmospheric gases (CO_2, O_2, N); pollutants (SO_2, CO, CH_4)	Air pollution and shading stresses
Soil factors		
Soil type	Soil type (clay, clayey loam, loam, sandy loam, and sand)	Soil's capacity to store water and nutrients, aeration, drain-age, and ease of field operations; soil-related stresses
Soil structure	Soil structure (texture, soil sealing, erosion, contamination, compaction, hydro-geological risks)	Crop productivity and quality contributed by soil fertility, organic matter and soil biodiversity; soil-borne stresses
Soil components	Soil moisture	Crop productivity and quality; drought, flooding and waterlogging stresses
	Soil air	Water absorption, respiration of roots and micro organisms, nutrient availability, decomposition of organic matter; soil air stresses including O_2 limitation
	Soil temperature	Soil physical and chemical processes, absorption of water and nutrients, germination of seeds and growth rate, microbial activity and processes in the nutrient availability; cold and heat soil stresses

Continued

	Soil pH	Nutrient availability and microorganism activities; acidic, saline and alkaline soil stresses
	Soil fertility (N, P, K, micronutrients/mineral and soil organic matters)	Plant nutrients and their balance for plant growth; nutrient deficiency stresses and nutrient use efficiency
	Soil salinity (electrical conductivity)	Osmotic tension and water takeup; salinity stress
Biotic factors		
Companion animals	Soil fauna (protozoa, nematode, snails, and insects)	Decomposition of raw organic matter, fixation of atmospheric nitrogen; damages to plant roots and other parts
	Animals around plants (pest insects, parasites, fungi, bacteria, viruses, predators, honey bees, wasps, human)	Cross-pollination and increasing yield, damage to crop yield; various abiotic stresses
Companion plants	Weeds, epiphytic and allelopathic plants	Competition forspace, water, light and nutrients, mutual benefit (synergistic effect), interference with crop plants, releasing compounds, volatilization or decomposition of plant residues, inhibition or prevention of plant growth; various biotic stresses
Cropping system		
Intercropping	Companion crop(s)	Competition for space, water, light and nutrients, buffering and mutual benefit (synergistic effect); various biotic stresses
Rotating cropping	Fore-rotating crop(s)	Residual effects of agronomic practices from the forerotating crop; various biotic stresses

cause damages or stresses to the plants, while others, such as azotobacteria, are beneficial. Crop management, as a unique environment component, involves inter-cropping, rotating and agronomic practices. Environmental factors that affect plant growth and yield can be modified or dramatically changed by human activities. Controlled or artificial environments can be created using growth chambers, phytotrons, hydroponic or other manmade facilities. Crop production activities per se have contributed to some significant environmental changes such as fertility depletion, air pollution, acid rain, water contamination (toxic element accumulation), noise (dynamic disturbance), salinity, land weathering, and desertification. Climate change and globe warming may result in extreme environments and wild fluctuation of environmental factors, which impose severe stresses on plants, including biotic stresses caused by companion organisms and abiotic stresses associated with climate and soil factors. For survival and sustainable production, crop plants must cope with all the challenges from climate change catastrophes, including stressful water regimes, extreme temperatures, elevated CO_2 and salinity, which impact on all aspects of plant architecture individually or in combination (Ahuja *et al.* 2010). Expanding of human population needs to increasingly explore less-farmable land with significant abiotic stresses, particularly for the pool soils with abnormal pH, low fertility, and salinity stress (Masuka *et al.* 2012). As a major constraint to crop yield in tropical regions, poor and depleted soil fertility force farmers into marginal lands and non-farming areas (Pingali and Pandey 2000). To stabilize crop production, stresses caused by pathogens, pests and other companion or symbiontic organisms should be paid more attention, as they are less predictable than soil or climate stresses. On the other hand, multiple abiotic stresses become increasingly prevalent. For example, heat stress often happens with water deficiency, while drought is accompanied by salinity (Ahuja *et al.* 2010). Abiotic and biotic stresses may occur simultaneously (Bostock *et al.* 2014; Kissoudis *et al.* 2014; Ramegowda and Senthil-Kumar 2015; Prasch and Sonnewald 2015), and one stress may show positive or negative impact over the other.

2.2. Multiple Environmental Trial Data

Envirotyping can be implemented with a large amount of environmental information accumulated in crop science and production (**Figure 1**). Multiple environmental trials (METs), involving a large number of genotypes tested in multiple locations, each with multiple replications, for multiple years (Johnson *et al.* 1955), can be considered a basic type of envirotyping for systematic collection of environment-related data. To identify the environments best suitable for commercialization of the on-trial varieties (genotypes), weather, climate and soil data have been collected systematically along with records of crop management including fertilization and control of diseases, pests, and weeds. In developed countries, such

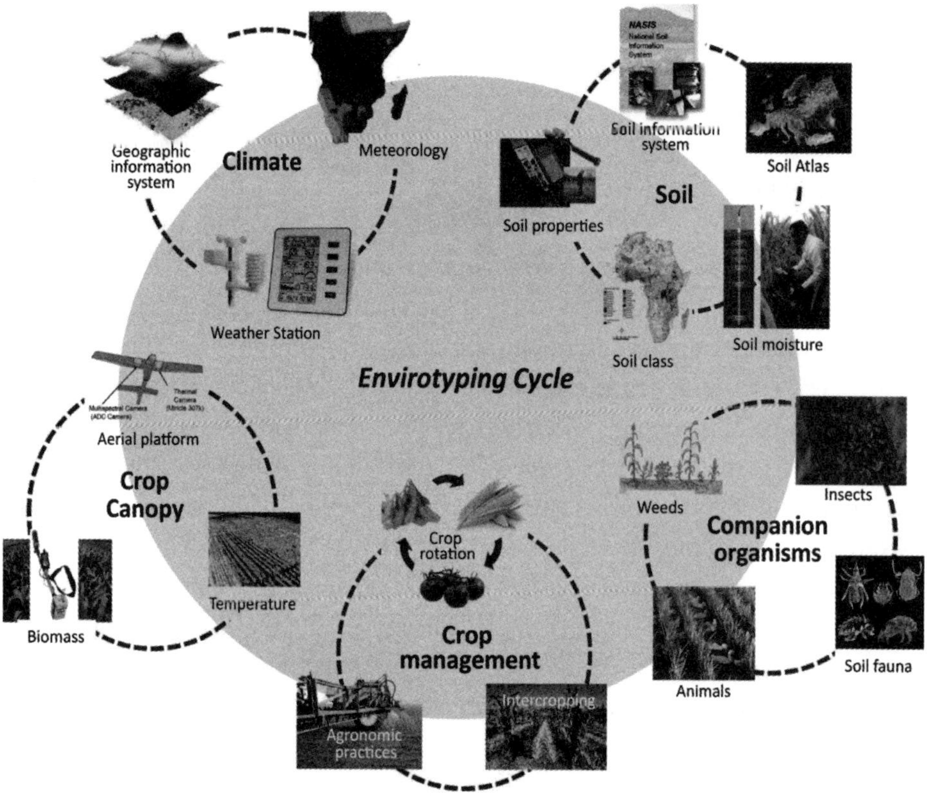

Figure 1. Envirotyping cycle. Environmental factors can be classified into five major groups, climate, soil, crop canopy, crop management and companion organisms, each containing several subgroups that describe important environmental factors affecting plant growth and development. Photos used for illustration were selected from public websites.

practices have been for decades. As an average effect of the environment, the empirical mean response for the *i*th genotype can be measured in the *j*th environment with *r* replications. Under Consultative Group on International Agricultural Research (CGIAR), International Maize and Wheat Improvement Center (CIMMYT) and International Rice Research Institutes (IRRI), for example, have implemented international breeding programs yield testing for many years, with a large amount of environmental data collected for three major crops, rice, wheat and maize. Such effort has been expanded to more countries and trial sites in recent years. However, envirotyping has not been well conducted in METs for three reasons: locations of MET sites are not precisely determined as needed; daily climate data linked to the trial sites are not available or difficult to collect; and data collection and completeness vary a lot across the sites.

2.3. Geographic Data

Geographic information system (GIS) has been established with the merging of cartography, statistical analysis and database technology, which is designed for collecting, storing, integrating, analyzing, and managing all types of geographical data (**Figure 1**). The data for any location in Earth space-time can be collected as dates/times of occurrence, with longitude, latitude, and elevation determined by *x*, *y*, and *z* coordinates, respectively. GIS integrates various data sources with existing maps and up-to-date records from climate satellites. To capture climate data, various types of weather observatory stations have been established worldwide, including ground, radiosonde, wind, rocket, radiation, agrometeorological, and automatic weather stations. These stations document climate data for numerous locations and sites, which are transferred in international or national central databases and become a part of GIS data.

2.4. Data from Soil Information Systems

Soil data have been accumulating in worldwide soil information systems (**Figure 1**). International Soil Reference and Information Centre (ISRIC) provides the international community with the world soil information. With a worldwide collaboration in soil data, soil mapping and their applications in global development issues, a centralized and user-focused World Soil Database is being developed, by which users can extract all validated and authorized data, including soil profiles and area-class soil maps (http://www.isric.org). To help bridge the soil information gap on the African continent, ISRIC has produced predicted information for various soil properties for the whole African continent at 250m spatial resolution with multiple standard soil depths (http://www.isric.org). European Soil Portal (http://eusoils.jrc.ec.europa.eu/) is the focal point for soil data, contributing to a thematic data infrastructure with data and information regarding soils at European level, including maps and Atlases. At the national level, National Soil Information System (NASIS), USA, is one of the most comprehensive national systems, providing a dynamic resource of soil information for a wide range of needs. Soil Information System of China (SISChina) has been established to include soil spatial and attribute data, and China 1:1,000,000 soil database (Shi *et al.* 2007).

Lack of consistent soil classification systems across countries or organizations has hindered the communication and organizational functions. To bridge the gap, translations between systems should be developed. As a soil classification system for naming soils and creating soil map legends, the World Reference Base for Soil Resources (WRB) with its third edition has been released (IUSS Working Group WRB 2014), as an adopted system for soil correlation and international communication. It allocates every soil into one of the 32 Reference Soil Groups and then characterizes further each soil by a set of qualifiers. Information on the named soil, such as its genesis, ecological function and properties, will be provided through the system. To provide comprehensive spatial information, the same system can be refined slightly, and used to name the units of soil map legends. By accommodating national soil classification systems, WRB facilitates the worldwide correlation of soil information.

2.5. Soil Properties

Agricultural soils can be classified based on their physical texture, the size of the particles that make up the soil. Based on the particle-size distribution, soil texture can be further clarified into sand, silt and clay. Then crop suitability can be determined for each soil class, and the soil responses to environmental stresses and agronomic practices, such as drought or nutrient requirements, can be explored.

Currently available and potentially useful techniques for proximal, on-the-go monitoring of important soil physical properties (Whelan and Taylor 2013) can be used to measure soil texture/type, soil water storage capacity, soil water in season and waterlogging. In general, soil conductivity can be measured as a product of soil composition and formation. A typical "spread" of soil types gives a certain range of conductivity and resistivity, which match up with different soil types from sand to saline (Bevan 1998). Apparent soil electrical conductivity (ECa) measurement has been improved and becomes a widely accepted means to determine several soil physicochemical properties (Corwin and Lesch 2005). Ground-based remote sensing technology has brought out a series of instruments for measurement of soil properties. For example, EM38, designed particularly for agricultural surveys and soil salinity measurement, provides a quick survey over large areas at depths of 1.5 and 0.75 m with its vertical and horizontal dipole modes, respectively. In a recent report on maize, measuring soil water content to 300 cm depth identified significant water extraction to a depth of 240–300cm (Reyes *et al.* 2016), indicating that in-depth measurement of soil properties is required to capture a full profile of available resources for crops with large plant and root sizes such as maize.

Time domain reflectometry (TDR) systems, designed to detect cable breaks, are now widely used to determine soil water content, bulk electrical conductivity, and rock mass deformation. To monitor soil water profiles, PR2 soil moisture probe measures soil moisture at 4–6 depths down to 40–100cm (http://www.delta-t.co.uk). As a portable and robust device, Diviner 2000 can be used to measure soil water over multiple depths (at 10cm intervals) in the profile. With its probe and hand-held data logging display unit, onsite management decisions can be made at up to 99 sites (http://www.sentek.com.au).

Currently available Soil Atlases provide all information about soil sealing, erosion, organic matter loss, biodiversity decline, contamination, compaction, hydro-geological risks and salinization (http://eusoils.jrc.ec.europa.eu). GIS can be used to sample, test and localize precisely to evaluate soil fertility and nutrients. Soil fertility can be assessed accurately with GIS-assisted sampling, testing, and mapping. With grid or zone soil sampling, many soil properties, including macro- and micro-nutrients, pH, and salinity carbon content, can be tested.

2.6. Crop Canopy

Remote sensing techniques, such as spectroradiometrical reflectance, digital imagery, thermal images, near Infrared reflectance spectroscopy and infrared photography, provide tools for characterization of crop canopy. These tools can be used with airborne remote sensing platform to collect data for temperature, humidity, light, air, biomass and overage of the crop canopy. Robotic imaging platforms and computer vision-assisted analytical tools developed for high-throughput plant phenotyping (Fahlgren *et al.* 2015) can be used for measurement of the crop canopy. Automated recovery of three-dimensional models of plant shoots can be used for multiple color images (Pound *et al.* 2014). The 3-D structure can be also determined directly using laser scanning (Paulus *et al.* 2013) and deep time-flight sensor (Chéné *et al.* 2012).

2.7. Companion Organisms

Companion organisms are those surrounding crop plants, including bacteria, fungi, viruses, insects, weeds and even other intercropping plants (**Figure 1**), which should be considered an important component of the environments. A series of methods and protocols have been developed to measure or determine companion organisms for different crops through multidisciplinary collaborations. For example, rhizospheric microorganisms can be extracted from bulked soil samples followed by comprehensive analysis and evaluation. Bulked sample analysis combined with metagenomics and DNA or RNA seq can be used to determine precisely the species, quantity, and mutual relationships of the organisms in bulked soil samples (Myrold *et al.* 2014). Using bulked samples collected from leaves or crop canopy, the organisms on the plant surface can be analyzed for their species, quantity, origin, distribution, developmental stages, and possible symbiontic relationships.

3. Environmental Characterization

Environments can be favorable and adverse to crop plants. The favorable environments are crop-friendly and resource-use efficient, while the adverse ones involve the pollutions and stresses of air, water and soil, and unfavorable climate changes. Environmental characterization is essential to experimental error control, data interpretation, data meta-analysis, and, in case of abiotic stresses, understanding patterns of resource availability (**Figure 2**; Masuka *et al.* 2012; Trenberth *et al.* 2014). Envirotypic information can be used to reveal a series of important features for experimental and crop production environments (Xu 2015).

3.1. Determination of Field Properties and Within-Site Variability

Potential variables in a trial site can be largely reduced or eliminated while its historical features can be evaluated by environmental characterization. Soil mechanical impedance and depth can be measured by proximal sensors such as cone penetrometers. Due to the close relationship between ECa and clay, water, and ionic

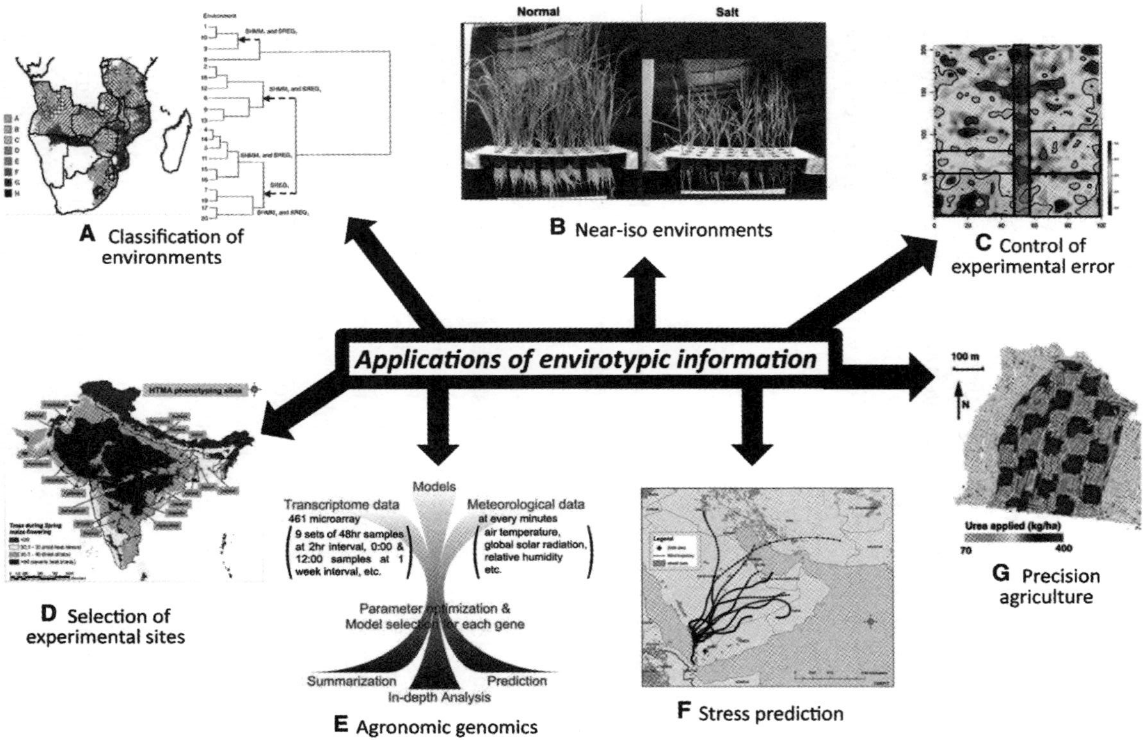

Figure 2. Applications of envirotypic information. Major applications include characterization of environments (**a** Bänziger *et al.* 2006; Crossa and Cornelius 2012), development of near-iso-environments (**b** http://www.google.com), control of experimental errors (**c** Prasanna *et al.* 2013), selection of experimental sites (**d** P. H. Zaidi, CIMMYT-India, personal comm.), agronomic genomics, studying, the effects of crop management on gene expression (**e** Nagano *et al.* 2012), prediction of disease epidemics (**f** Singh *et al.* 2006), and precision crop production (**g** McBratney and Whelan 2001). Photos used for application illustration were selected from public websites or provided by CIMMYT colleagues, except for those indicated otherwise.

content, field gradients can be determined with electromagnetic surveys (Rebetzke *et al.* 2013; Gebbers and Adamchuk 2010). As the best indicator of field variability, crop performance, when combined with imaging techniques, wireless sensor networks and GIS, can be used to map and monitor spatial variability precisely (Lee *et al.* 2010). Linked with GPS, aerial high-throughput phenotyping platform enables fast non-destructive measurements of biomass. By conversion of biomass into normalized difference vegetation index (NDVI), such platforms provide a tool to measure field and within-experiment variability, which can be used to develop performance maps to guide next planting (Araus and Cairns 2014).

3.2. Classification of Environments

In mega-environment analysis, target environments can be classified into three types (Yan *et al.* 2007): single, simple mega-environments, which show no crossover genotype-by-environment interactions (GEI) with phenotypic performance repeatable across years; multiple mega-environments, which show crossover GEI that is repeatable across years; and single but complex mega-environments, which show crossover GEI that is not repeatable across years. Based on all available environmental information, potential trial sits can be accurately evaluated and thus the best trial sites can be selected. Such effort has been done in Africa for selection of the trial sites best suitable for stress tests of drought, low nitrogen, low pH, stemborer, and striga. The need for environmental data is particularly important in screening for drought tolerance where availability of soil moisture should be checked to ensure that the field condition and the drought stress imposed represent the target environment well (Römer *et al.* 2011). Three major criteria, maximum temperature, season precipitation and subsoil pH, have been used in sequential retrospective pattern analysis of environment similarity, by which eight maize mega-environments could be identified in southern Africa [**Figure 2(a)**; Bänziger *et al.* 2006]. Similarly, typical temporal modes of environmental variation for the soil-plant water balance have been identified for the US corn-belt target population of environments (Cooper *et al.* 2014).

3.3. Construction of Near-Iso-Environments

In many cases, experiments need to be done in two contrasting environments. The concept of near-iso-environments (NIEs), which is conceptually similar to neariso-genic lines, was proposed to represent two contrasting environments that are significantly different in one major factor (**Figure 2**; Xu 2002, 2010). One environment imposes much less stress on plants than the other. The less stress or normal environment can be used as control to measure the effect of the stress environment. A relative trait is then obtained from two direct traits phenotyped in the two environments to measure the plant sensitivity to the stress. If different plants show a similar phenotype under the less stress, the sensitivity can be measured using the direct trait value obtained in the more-stress environment. When both environments impose little stress the sensitivity should be measured instead using the relative trait value, that is, the difference of trait values measured in the NIEs, divided by the trait value measured in one of the NIEs or in the normal environment (Xu 2002). Traits suitable for measurement under NIEs include all abiotic/biotic stresses and plant responses to different environmental factors or crop management practices. NIEs can be used to facilitate identifying genes with environment-specific effects. In rice, days-to-heading (flowering) and photo-thermo-sensitivity were studied under conditions of field (Xu 2002) and greenhouse (Maheswaran *et al.* 2000), with different sets of quantitative trait loci (QTL) identified. Such QTL showing environment-specific effect have also been revealed for abiotic stresses under contrasting environments by integrative analysis of a large number of QTL reports (Des Marais *et al.* 2013).

3.4. Control of Environmental Errors

Experimental errors are mainly contributed by micro-climate variability, non-uniformity of crop management and soil fertility, unpredictable influences of insects, diseases, weeds, companion microorganisms/plants/animals, winds, rainstorms and hails, and differences contributed by observation, measurement and production methods, tools, instruments, experimenters and farmers. Envriotyping can play a vital role in reducing environmental errors (**Figure 2**). Experimental errors can be controlled through various approaches by reducing "signal-to-noise" ratio (Xu *et al.* 2012, 2013).

Trial sites selected should have uniform soil texture and fertility, relatively large in size, and fitting-in an appropriate rotation system, with good record of land utilization, representative of soil texture, climate, natural and economic conditions of the target environment. (2) Experimental materials under testing should be genetically homogenous with uniform individual samples (in terms of, e.g., seed quality and seedling age/size). (3) The common standards should be taken for all experiment managements and tests, with uniform application of resources and consistent control of weeds, pests and diseases. (4) Multiple replications, random treatments and controls should be included along with an appropriate experimental design. (5) Border effects should be minimized using boarder protect planting and selecting the trial sites far from villages, trees and highways. (6) New field-based techniques, such as precision resource application and remote sensing technology, should be used to measure secondary traits, by correctly selecting, calibrating and applying instruments, such as neutron probes, radiation sensors, and chlorophyll and photosynthesis meters (Xu *et al.* 2013). Climatic and soil moisture conditions can be characterized by wireless sensor networks, by which environmental conditions can be brought under real-time monitoring so that the environmental errors can be controlled (Araus and Cairns 2014).

Well-controlled or managed environments are often favored by molecular biologists because unwanted environmental variation can be minimized using pots, soil-filled pipes, hydroponics, growth chambers and greenhouses. However, it is still needed to achieve a better understanding of the environmental stresses prevailing under the nature conditions. Facilities for environment management become increasingly important, as they enable selection under controlled stresses (Rebetzke *et al.* 2013), improve crop performance measurement, and dissect phenotypic effects and their underlying genetic makeup (Blum 2011). However, controlled or managed environments could be very different from the nature or target environments (Masuka *et al.* 2012; Basu *et al.* 2015). Therefore, the results from controlled or managed environments could be far removed from what plants will experience in the field and will thus limit their application in germplasm development (Masuka *et al.* 2012; Araus and Cairns 2014). Compared to the field condition, for example, a pot is considerably smaller with a limited volume of soil available to roots, and the amount of water and nutrients will be limited to plants (Poorter *et al.* 2012; Reynolds *et al.* 2012; Basu *et al.* 2015).

4. Crop Modeling and Phenotype Prediction

4.1. Models

As a third “typing” technology, envirotyping can be applied in many fields of crop science (**Figure 2**). One of the applications is in crop modeling and phenotype prediction. With envirotypic effect as a new component, phenotype (*P*) can be partitioned into those contributed by genotypic effects (*G*), envirotypic effects (*E*), GEI (GE) and experimental error:

$$P = G + E + \mathrm{GE} + \mathrm{error}$$

With genotypic and envirotypic information available, *G* and *E* and thus the phenotype can be further partitioned. *E* can be partitioned into major components each consisting of several key environmental factors. In addition to crop management (*M*), both process and social economics can be also included as a part of *E*. For hybrid crops, phenotypic prediction can be performed for both inbreds and hybrids. Interactions of general and special combining abilities with *E*, GCA × *E* and SCA × *E*, can be estimated based on the responses of inbreds and hybrids to environmental factors.

Phenotype prediction can be performed with three modalities, *i.e.*, new genotypes with known environments, known genotypes with new environments, and new genotypes with new environments (Bustos-Korts *et al.* 2016). Phenotype prediction can be simplified with consideration of co-variances and conditional variances. For example, prediction can be optimized using known soil properties and texture, etc., as co-variances, performed crop management as conditional variances, and improved simulation and modeling methodologies as facilitators. The prediction will be improved with managed conditions and precision envirotyping. If the contribution from *E* and GEI to phenotype is relatively small, the phenotype can be predicted largely by the genotype alone. Under well-managed environments, uniformity or good control of major environmental factors can be achieved and used to eliminate a large part of *E* and GEI, and thus they can be largely taken out of the prediction equation. On the other hand, when identical, homogeneous genotypes are phenotyped under natural conditions, the relevant *E* and GEI can be estimated. With some major environmental effects fixed, the rest *E* effects and thus the phenotype can be predicted. Near-iso-environments can be used to estimate the relevant major *E* and GEI. With the availability of known information about several major environmental factors for a target environment, a certain level of reliability can be achieved for prediction of important traits. For example, using over ten decades of actual data for maize yield and seasonal precipitation in Africa, a highly positive correlation between them has been established (CIMMYT, internal comm.), and thus the maize yield in Africa can be predicted largely based on the seasonal precipitation.

Phenotypic prediction needs to be conducted for individual traits and their conceptual models. In wheat, a conceptual model for yield and heat-adaptive traits has been developed (Cossani and Reynolds 2012):

$$\text{YIELD} = \text{LI} \times \text{RUE} \times \text{HI}$$

where LI is light interception which involves rapid ground cover and functional stay-green, RUE is radiation use efficiency, and HI is partitioning of total assimilates. RUE is dominated by three major components, photo-protection, efficient metabolism and water use. Photo-protection involves leaf morphology, down-regulation, pigment composition and antioxidants. Efficient metabolism is determined by CO_2 fixation, canopy photosynthesis, spike photosynthesis and respiration. Water use efficiency involves the roots that match evaporative demand and regulation of transpiration. Partitioning (HI) consists of spike fertility, stress signaling, regulating, grain filling, and stem carbohydrate storage and remobilization (Reynolds *et al.* 2011). A similar model has also been constructed for yield under drought.

As one of the great efforts in phenotype prediction, the Genomes to Fields Initiative was established to predict traits from genotype and environment, thereby leading to improved maize production (http://www.genomes2fields.org). One of the subprojects, Genome by Environment, aims to assess environmental effects, using 31 inbreds and nearly 1000 hybrids tested in 22 environments across 14 states in the US and Canada. Similarly, Next-Generation Crop Breeding Platform for Predicting Germplasm Performance in Target Environments, proposed by University of Florida in collaboration with CG centers, is to develop a breeding platform for integrating and harmonizing genotype, phenotype, environment, and management data, and build next-generation crop-based models for predicting performance of genotypes in different environments (M. P. Reynolds 2015, presented at CIMMYT Science Week). However, greater efforts will be needed to establish a comprehensive predictive model to reveal the diverse biological networks, by which plants respond to combined climate change catastrophes. Such a model could be utilized to improve plant adaptation to changing climates (Ahuja *et al.* 2010).

4.2. Genotype-by-Environment Interaction

GEI has been investigated recently through QTL mapping and gene cloning. QTL mapping using phenotypic data collected from multiple locations can be explored for understanding of the mechanisms involving GEI and their relative importance. The QTL with additive effects have four main GEI patterns (Des Marais *et al.* 2013): (a) antagonistic pleiotropy, with sign or direction changing of additive effects; (b) conditional neutrality with environment-dependent additive effects, which are limited to specific environmental conditions; (c) differential sensitivity, where the magnitude of additive effects are environment dependent; and (d) no GEI with no detectable change in additive effects across environments. One of the earliest GEI evaluations is to simply compare QTL identified across three locations in tomato (Paterson *et al.* 1991). QTL mapping under contrasting environments with one significantly different factor can be explored for understanding of actual GEI. Such studies are largely performed for abiotic stresses. From over 700 research reports, 37 of them with QTL mapped for abiotic stresses and complete QTL information available were selected for an integrative analysis (Des Marais *et al.* 2013), revealing that nearly 60% of QTL exhibited GEI caused by antagonistic pleiotropy or environment-specific effects. These two GEI types showed strong influence on QTL effect plasticity as measured by the absolute difference in the standardized additive effects across environments.

Two types of GEI analyses have been done based on expression QTL (eQTL). One is based on the gene that shows different environment-dependent expression patterns in two genotypes. Another is based on genome-wide association study using expression levels of many genotypes across multiple environments. However, only few typical eQTL mapping reports are available with mapping populations tested under controlled abiotic conditions and quantified phenotypic expressions. One of the early formal eQTL studies involving an abiotic manipulation focused on *Brassica rapa* leaf tissue from plants grown under two levels of phosphorus (P) availability and identified over 3226 transcripts and several notable hot spots responsive to P, without formal tests of GEI (Hammond *et al.* 2011). Using an RIL population tested across soil drying treatments, thousands of genes that responded to soil drying and hundreds of main-effect eQTL were identified in Arabidopsis by eQTL mapping. However, very few eQTL were identified with significant interaction with the soil drying treatment (Lowry *et al.* 2013). There is accumulating evidence in model organisms such as yeast and flies that GEI are ubiquitous, accounting perhaps for the greater part of the phenotypic variation (Grishkevich and Yanai 2013). Such interactions appear to be caused by the changes to upstream regulators rather than local changes to promoters. Moreover, genes show different levels of GEI, and many factors, including promoter architecture, expression level, regulatory complexity and essentiality, are associated with the environment-induced differential gene regulation. For example, significant differences in response to drought or cold were found between the genes with consistent expression and the genes with variable expression under abiotic stresses. On average, the consistently expressed genes tended to share relatively more pairwise haplotypes, with lower promoter diversity and fewer nonsynonymous poly-morphisms (Lasky *et al.* 2014).

To understand the origin, spread, and evolutionary processes of GEI, the specific genes that control GEI phenotypes and the mutational variants that define functionally distinct alleles should be identified (Des Marais *et al.* 2013). We need to determine which of the following factors are more often GEI driver in plant abiotic stresses: gene type (e.g., environmental sensors, biosynthetic enzymes, or regulatory proteins), gene characteristics (e.g., paralogs, or complex cis-regulatory control), mutation type (coding/noncoding, or transposable elements, etc.), and molecular mechanisms (condition-dependent epistasis, gene expression, enzymatic activity). Genes exhibiting GEI can be identified and used in comparative analyses across lineages and tests for parallel and convergent evolution in responses to the environment. The complex regulatory systems plants have evolved to control phenology are driven largely by environmental cues such as photoperiod, temperature, and circadian signals (Kim *et al.* 2009). As summarized for cloned GEI genes using flowering time and soil and water availability as examples, a variety of natural variants and mechanisms have been revealed at the molecular level, including nonsynonymous changes in receptor proteins, loss-of-function mutations in transcriptional repressors, splicing variants in biosynthetic enzymes, and gene duplication in transcription factors (Des Marais *et al.* 2013).

To predict GEI, environmental covariates and crop modeling have been integrated recently into the genomic selection framework through factorial regression model (Heslot *et al.* 2014). Stress covariates for predicted crop development stages were derived from daily weather data, with model tested by a large wheat dataset. For unobserved environments with available weather data, the accuracy of genotype performance (phenotype) prediction increased by 11.1% on average. With insight into the genetic architecture of GEI provided by this model, genotype performance could be predicted based on available environment data such as past and future weather scenarios. In another report, using covariance functions GEI was modeled through interactions between high-dimensional sets of markers and environmental covariates (Jarquín *et al.* 2014). Using data from 139 wheat lines genotyped by 2395 SNPs and phenotyped for grain yield over 8 years across locations in northern France, prediction accuracy substantially increased (17%–34%) by including interaction terms in the models compared to models with main effects only. Similarly, GEI was modeled in genomic selection using a marker × environment interaction (Lopez-Cruz *et al.* 2015), which was used to analyze three CIMMYT wheat datasets with over 1000 lines genotyped by GBS and phenotyped at CIM-MYT. The model had substantially greater prediction accuracy, compared to an across-environment analysis with GEI excluded.

So far, environmental information has been used, collectively in almost all cases, as a component in the model to investigate GEI and the phenotypic performance across different environments, without partitioning into individual environmental factors. Such evaluation usually does not involve any envirotypic information, by assuming that different locations have different environmental effects and thus difference revealed in genetic and molecular analysis can be attributed to GEI. With more thorough understanding of genotypic information, we can now precisely describe genes, alleles, haplotypes and their integrative contribution to a phenotype. As a result of envirotyping, we should be able to dissect the *E* component into individual factors. By incorporating precise measurements of *G* and *E* with precision phenotyping, therefore, GEI can be evaluated precisely, and predicted based on the theoretical model established with *G* and *E* information.

4.3. Environment-Responsive Genes

Plastic responses to environmental signals can occur at the molecular level. To initiate a signaling cascade, a receptor at the cell surface must first perceive an external stimulus. The environment-responsive genes can be classified into two major categories: one responding to neutral environments such as photoperiods, regular temperatures and normal nutrient levels, and the other to the environments with abiotic stresses such as drought, waterlogging, extreme temperatures and deficiency of essential nutrients. Under half of the studies Alvarez *et al.* (2015) reviewed (41%) addressed how gene expression can be affected by environmental stimuli, such as abiotic stress, environmental heterogeneity in time or space, host-parasite interactions and potentially selective biotic and abiotic interactions. Ten years of transcriptomics in natural environmental fluctuations have shown that stress responses can have significant impacts on many categories of genes, and transcription may be affected by even small environmental changes. Responses to the environmental change can involve the post-translational modifications of the components of signaling pathways (Nühse *et al.* 2007). Alternatively, regulatory gene transcription can respond to a wide range of external stimuli. Gene expression alternation and thereby plasticity generation can be created by epigenetic processes, such as DNA methylation, histone modification and transposable element activation (Chinnusamy and Zhu 2009). Variation in small RNA populations can lead to post-transcriptional control (RNAi) as well as changes in chromatin modification. Lastly, gene expression can be also affected by the expansion of short repeat sequences (Nicotra *et al.* 2010). With large-scale epigenomic analysis involving large numbers of genetic samples, spatial and temporal effects on families or inbreds can be partitioned to optimize the genetic variation to identify frequent epialleles.

Under well-managed or near-iso-environments, genes responsive to climate change catastrophes can be dissected. It is important to decipher and predict plant dynamics under field or natural conditions (Izawa 2015). Many genes have been cloned with function analyses for their responses to major neutral or extreme stressful environments. Identification of specific genetic determinants of stress adaptation to waterlogging, drought, low temperature, Al toxicity and salinity has revealed that the genetic loci are often associated with distinct regulation or function, duplication and/or neofunctionalization of genes that maintain plant homeostasis (Mickelbart *et al.* 2015). At the same time, a large number of genes have been cloned for biotic stress tolerance, including over ten genes for rice blast resistance (as summarized in Chen *et al.* 2015). In addition, a series of genes responsive to neutral or normal environmental factors, such as normal light and temperature (for photo- and thermo-sensitivity), have been cloned (as summarized in Matsubara *et al.* 2014).

Due to selection of dramatic fluctuation of diverse environmental factors, wild plants have evolved with their genetic networks responsive to such complex nature conditions. Plant adaptation to environmental stresses are coordinated and fine-tuned by adjusting growth, development and cellular and molecular activities. Responses to stresses are usually accompanied by major changes in the levels of transcriptome, proteome and metabolome. The metabolic adaptations to environmental stress factors involve increase, decrease or accumulation of various metabolites in leaves, shoots, roots, flowers, seedlings, grains, and nodules of plants (Ahuja *et al.* 2010). As an important contributor to abiotic stresses, miR156 isoforms are highly induced by heat shock, and the miR156-SPL module mediates the response to recurring heat shock in *Arabidopsis thaliana* and thus may function to integrate stress responses with development (Stief *et al.* 2014). Plant adaptation to environmental stresses is modulated by a myriad of genes, proteins and metabolites, and their corresponding metabolic pathways or biological networks. Phenotypic expression over both space and time is inconsistently affected by environmental variability, which should be accounted for any statistical models for estimation of parameters of interest (Cobb *et al.* 2013; Araus and Cairns 2014). Further identification of the genome architecture associated with responses to particular stimuli might help us predict plastic responses to adverse environments imposed by climate change (Nicotra *et al.* 2010).

4.4. Prediction of Biotic and Abiotic Stresses

Based on the environmental factors and their predominant changes that affect disease and pest epidemics, epidemic time, place and distribution of biotic stresses can be predicted, in combination with other relevant social factors (**Figure 2**). Such prediction can be also done to forecast new abiotic stresses caused by significant weather variation and climate change. In general, biotic and abiotic stresses can be predicted based on Gb (genotype of biofactors in case of biotic stresses), Gh (genotypes of the host crop), *E* and their interactions. *E* should include all the environmental factors that affect crop growth and yield, dominate the boom, bust and epidemics of diseases and pests, and impose abiotic stresses to crop plants.

Prediction of biotic stresses involves examining new diseases and races due to weather variation and climate change, establishing prediction model to forecast new diseases and races, and their spread paths and speeds. The prediction should be done for major environments, major experimental stations and production zones. With detailed envirotypic information available, prediction may be done for experimental blocks, plots and even individual plants. To predict abiotic stresses, environments should be well characterized, particularly for the prevalent factors.

Biotic stresses can be predicted for both short and long terms. The former is important for farmers to take actions to reduce stress-related losses. Such prediction should be done largely for the coming season or year based on currently available and forecasting weather data, in combination with historical data on diseases, pests and climate. With the prediction, forecasting before the season, boom or bust cycle is highly preferred. The long-term prediction is important for scientists to develop techniques and varieties to be prepared for the diseases and pests to change, evolve, or move. It is largely done for the coming years and for their movement and spread, based on historical and current climate data and foreseeable climate change with updating knowledge on the epidemic diseases/pests, host-pathogen interaction (GbXGh), GEIs (GhXE, GbXE, and GhXGbXE), and the genes and genotypes of the host plants against diseases and pests. Based on a large number of pests and pathogens examined, an average poleward shift of 2.7 km per year since 1960 has been demonstrated, while a significant variation in trends was detected among taxonomic groups (Bebber *et al.* 2013). The positive latitudinal trends observed in many taxa provide evidence to support the hypothesis that global warming has driven the pest movement.

Wheat rust Ug99, detected first in East Africa, is one of the best examples for disease prediction. Relevant factors that determine movement of Ug99 in wheat have been integrated to predict the rust epidemics (Singh *et al.* 2006). The factors include the current status and distribution of the rust, prevailing winds, climatic factors favoring survival and sporulation, wheat production zones (geographical distribution and associated human populations), historical migration patterns for the rust races with East African origin, and responses of existing cultivars to the rust. The rust monitoring can be established and optimized by developing standardized data collection, building up lab capacity for rapid diagnostics, centralizing data management and information dissemination, monitoring both pathogen and host in an integrative way, and establishing early warning/forecasting system.

Prediction for abiotic stresses should be more straight-forward as the relevant environments are the direct causal factors. Stressful factors may come from climate (radiation, temperature, precipitation, air, etc.), soil (nutrients, moisture, pH, salinity) and water (Deinlein *et al.* 2014; Lobell *et al.* 2014; López-Arredondo *et al.* 2014; Hu and Xiong 2014). Abiotic stresses caused by soil factors, modified largely by production activities, are more predictable and measurable than biotic stresses. Climate factors can be largely predicted based on latitude, longitude and elevation. Compared to climate changes that cause long-term variation, weather changes cause short-term, less-predictable fluctuations. Both climate and weather changes may cause significant variation of abiotic factors and thus stresses on crop plants.

It has been predicted that global climate change will have significant impacts on crop productivity by creating significant abiotic stresses on crop plants such as ozone and heat. Depending on production zones, some crops show primary sensitivity to single stresses such as ozone (e.g., wheat) and heat (e.g., maize) (Tai *et al.* 2014). High temperatures contribute to reduced crop yields, and predicted global warming has raised growing concern regarding future crop productivity and food security. Without adaptation, losses in crop production are expected for three major cereals (wheat, rice and maize) in both temperate and tropical production zones by 2°C of local warming (Challinor *et al.* 2014). Yield gains in most wheat-growing regions have been slowing down by global warming, and global wheat production would become more variable over locations and time, with production decrease by 6% for each °C of further temperature increase (Asseng *et al.* 2015). With climate change, more frequent adverse weather conditions will happen in European wheat zones. For example, adverse conditions for 14 representative European wheat sites might substantially increase by 2060 compared to 1981–2010, with more frequent crop failure expected (Trnka *et al.* 2014). In maize, when adaptation is accounted for, average yield losses in the US from a 2°C warming would be reduced from 14% to only 6% (Butler and Huybers 2013). Rainfed maize yields in the US (Schlenker and Roberts 2009; Lobell *et al.* 2013) and elsewhere (Lobell *et al.* 2011) have indicated a strong negative yield response to accumulation of temperatures above 30°C. Maize also shows increased yield sensitivity to drought stress caused by high vapor pressure deficits. Translation of improved drought tolerance into higher average yields becomes more acceptable agronomic changes than decreasing yield sensitivity to drought at the field scale (Lobell *et al.* 2014).

As abiotic and biotic stress combination becomes very common, it is essential to predict all relevant stresses simultaneously. Responses to combined stresses are genetically controlled to a great extent by different, even functionally opposing, signaling pathways that may interact and inhibit each other, and therefore, it would be impossible to extrapolate directly the response to simultaneous multiple stresses from those to single stresses (Suzuki *et al.* 2014; Prasch and Sonnewald 2015; Ramegowda and Senthil-Kumar 2015). By omics and functional analyses of individual genes, a convergence of signaling pathways has been revealed for abiotic and biotic stress adaptation. Complicated potential effects of abiotic stress have been expected on resistance components, for example, extra-cellular receptor proteins, R-genes and systemic acquired resistance. Such elaborated stress resistance crosstalk would also happen at the levels of hormone, reactive oxygen species, and redox signaling (Kissoudis *et al.* 2014). The prediction for possible combined stresses would help develop breeding strategies for manipulation of individual common regulators and pyramiding of non-interacting components.

4.5. Integration of Phenotyping with Envirotyping

Crop modeling and phenotype prediction depend on precision phenotyping as a feedback correct for the best model-fitting. A new trend in crop science will be to combine high-throughput precision phenotyping with large-scale envirotyping to collect, mine and utilize the two sets of information comprehensively. Currently, many robotic, high-throughput phenotyping systems are built under controlled or well-managed environments. Automatic phenotyping platform has been established for plants in growth chambers with major environmental factors controlled (Jansen *et al.* 2009; Massonnet *et al.* 2010). Expensive sensors used in phenotying can be either in fixed location with moving plants, in mobile device with fixed plants, or colocalized with plants. However, it is very challenging to combine the high-throughput phenotyping with large-scale envirotyping for several reasons. First, there are a large number of genotypes to be tested under a wide range of environments. Second, precision phenotyping needs to be done across most, if not all, of the developmental stages and spaces. Third, it is very expensive. Due to the cost and capability limitation, early phenotyping efforts are seldom close to meeting the requirements of complete phenomics (Houle *et al.* 2010). Probably, more challenges come from establishing such high-throughput phenoltyping platforms for the crops with big plants and long life cycle such as maize and sorghum, because it is very difficult to build up large enough controlled environments for a large number of big plants to grow for a long time.

Precision phenotyping needs to be coupled with precision envirotyping, as envirotyping plays a vital role in generating phenotypic data of high quality, consequently, improving crop research. Field variation in soil, moisture and fertility contributes to error variances, thereby masking major genetic variation for important traits and reducing repeatability, regardless of the cost and precision of available phenotyping platforms (Masuka *et al.* 2012). High-throughput platforms allow phenotyping a huge number of genotypes growing in a larger field, thereby increasing soil variability. In general, the larger the land is required for an experiment, the harder it becomes to identify a land with minimum soil variability (Araus and Cairns 2014).

Phenotypic and envirotypic information collected for the same set of genotypes will greatly contribute to crop modeling and phenotype prediction by complementary and comparison analyses. Firstly, phenotypes collected for homogeneous, identical genotypes under large-scale experiments can be used to reveal within-site variation and environmental variability, because significant phenotypic variation can be attributed to significant environmental variation and GEI. Secondly, under well-managed environments with appropriate experimental error control, phenotypic variation in genetically different materials can be largely attributed to genetic contribution. Thirdly, under near-iso-environments, or controlled vs uncontrolled environments, phenotypic differences are the genotypes' response to the major environment factor. Fourthly, envirotyping of all environmental factors along with phenotyping will reveal integrative phenotypes for different genotypes, by which we can determine the contribution of genotypes and their interaction to the phenotype. Fifthly, phenotyping overtime using homogenous genotypes will reveal dynamic environmental variation. A recent integrated phenotyping with soil content in maize rooting zones indicates that under water-limited conditions grain yield increased significantly without significant increase in total water extraction (Reyes *et al.* 2016). Therefore, the measured long-term genetic gain for yield must have been achieved through improved maize adaptation to water stress conditions by either increased water use efficiency or increased carbon partitioning to the grain.

4.5. Evirotyping for Single Plots and Individual Plants

For some major environment factors, such as soil moisture, nutrients and pH, we can now implement envirotyping at the level of individual plots, single rows, or even single plants. For most environmental factors, however, it is not possible for a real envirotyping until some great technical innovations have been achieved. In one hand, variation for some environmental factors might be too small to be detectable, and thus facilities currently available need to be improved significantly in terms of sensitivity, precision and resolution. On the other hand, envirotyping process with well-equipped facilities may significantly disturb or interfere with plant growth and development, resulting in significant micro-environment changes around the crop plants. It can be expected that technical innovations with increased precision and minimized disturbance on crop plants will allow our current environmental data collection move from the level of experimental station to the whole block and then to individual plots and single plants, as a zooming-in process of photographing to focus on specific details with high resolution (**Figure 3**). By the end, we can generate plot- or plant-based envirotypic data to match up with genotypic and phenotypic data for each genotype.

The bottleneck for a plot- or plant-based envirotyping could be the high cost involved along with a data tsunami. Significant advances in development of platforms and facilities are essential for envirotyping to match the current scale and resolution of genotyping and phenotyping, since crop improvement and production involves a large number of varieties (genotypes) each involving a huge number of plants. With such a numbers game, significant cost reduction will be needed, including much cheaper high-throughout and precision envirotyping facilities, less labor-intensive processes, and highly effective information management platform.

In additional to the technical difficulties and potential high cost, dynamic envirotyping across development stages and individual plots/plants will result in a data tsunami that might be more difficult to manage compared to genotypic and phenotypic data, as envirotyping data formats vary greatly with huge numbers of images and videos generated across developmental stages. Moreover, understanding

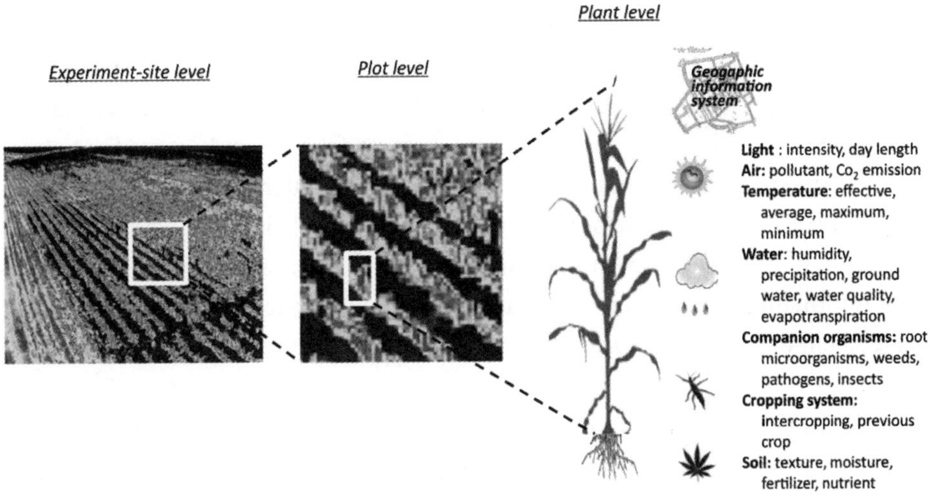

Figure 3. A zooming-in process of envirotyping. It is vital to move envirotyping from the levels of experimental stations and the whole field blocks to individual experimental plots and plants by a zooming in process so that envirotypic information collected can be matched up with genotypic and phenotypic data for each entry or target plant. Revised from Xu (2015).

the effect of any single environmental factor often means that we have to control the rest, which might be difficult for some environmental factors. On the other hand, standardization of environmental designs for single environmental factors, which is required for cross-research and cross-lab comparisons, needs to consider the masking effects of one major environmental factor over the others.

5. Envirotyping-Driven Precision Agriculture and Breeding

5.1. Precision Agriculture

Agricultural activities have accumulated many types of data, which associate with climate, soil, disease and pest epidemics, relationship between yield and plant density, and market prices of agricultural products. Data-based integration, modeling and simulation help make decisions on large-scale agricultural production. In industrial countries, modern agriculture is moving from mechanization to informatics, particularly, envirotyping driven, allowing high precision and efficient breeding and crop production (**Figure 2**). As a farming management concept, precision agriculture is developed based on observing, measuring and responding to field variability, weather conditions and other external environmental factors. More generally, it should include climate-smart agriculture and intensification involving irrigation, fertilization, cropping system and intercropping. Crop performance varies typically with both space and time which involve statistical treatments. The holy grail of research in precision agriculture will be the ability to develop information management and decision support systems for functional farm management to optimize returns on inputs while preserve resources.

Multi-national corporations have been investing heavily to establish a research and development system with decisions supported by big data (Cooper *et al.* 2016). To do that the seed business giant Monsanto recently purchased two companies, Precision Planting and Climate Corporation. Under guidance of these two companies, samples are collected for soil tests from each point of every 4-acre of land, providing detailed soil information for famers to make decisions on fertilization, irrigation and also to predict the yield that can be achieved for different crops. Through its climate software, Climate Corporation provides farmers with farm-wide real-time weather data, including temperature, humidity, wind, precipitation, which can be used to determine when planting, harvesting and crop management should be done for a given block of the farm land.

Using mobile systems supported by big data, real-time soil moisture, temperature and crop growth status can be obtained, so that farming precision can be significantly improved. Precision planting manufactured precision agriculture equipment that can be fixed in powerful tractors, parallel running planters or other machines. With built-in application software, decisions can be made on when the crop field should be checked and when pesticides and fertilizers should be applied. Real-time soil moisture can be collected and used to make irrigation decisions. Such real-time monitoring has become increasingly important with increased climatic variability, particularly during the off-season and in managed drought phenotyping (Araus and Cairns 2014). For precision farming supported by big data, all agricultural inputs can be brought under precise control so that energy, fertilizers, water and pesticides can be significantly saved. Through GPS, auto-driving system, computer facilities and essential sensors, the information provided by big data software can be accessed, manipulated and transferred to realize intelligent agricultural mechanization. According to soil properties, such intelligent systems can adjust their planting to make the seed sowed at the same depth. They can also improve the operation quality, for example, by increasing the ratio of single seed planting to 99%.

With historic climate data, Climate Corporation can provide more accurate weather prediction for a small area of land. The basic model is to develop an informative farm map using global germplasm information, historic yield data and Climate Corporation's climate data, and then put all possible climate information onto the map. With all the data support, the company can provide farmers with crop insurance service, and the farmers can access to the map to search for specific information and to determine which crops and crop varieties should be planted and under what conditions they have a good harvest.

In the public sector, landscape-scale crop assessment tool (LCAT), a cloud-based tool developed by CIMMYT scientists in collaboration with Oak Ridge National Laboratory and GEOGLAM (University of Maryland), can be used for crop land identification, crop classification, phenology test, crop status measurement, and in season-forecasting input. Data sources include those from satellites (Landsat, Aster, MODIS, VIR, etc.), satellite-derived soil moisture and other products, and weather (Urs Schulthess, CIMMYT, personal comm.). As an example of applications, N-rate can be calculated by GreenSat for a crop field using NDVI maps derived from SPOT satellite with N-rich strips identified. Within-field (or rather treatment) variability can be detected, as ground cover is measured directly by the amount of sun light that a crop captures for photosynthesis (Ortiz-Monasterio *et al.* 2013). Actionable advice on crop management, planning, decisions and field operations can reach farmers through mobile phones to guide nutrient management (sources, rates, and times for field size, and N timing adjusted for water availability), crop establishment (seed source and seed rate), crop protection (weed, disease and insect management), irrigation (amount, time and frequency), and yield estimation (before and at harvest) (Urs Schulthess, CIMMYT, personal comm.).

5.2. Agronomic Genomics and Improvement of Companion Organisms

To better understand the effects of agronomic practices on crop growth and yield, the genomic approaches widely used in genetics and crop improvement should be explored for agronomy. Agronomic genomics aims to integrate agronomy with omics to develop a high-efficient, cost-effective, and environment-friendly crop management system to optimize the gene expression and thus the crop production (**Figure 2**). Effects of agronomic practices, such as fertilization, irrigation, pest/ disease management, weed control, etc., can be revealed by the changes of the level and pattern of gene expression as revealed by DNA-, RNA- and protein-sequencing technologies. Dynamics of gene expression patterns may be directly determined under complex and changing environments. Currently, environmental factors that shape transcriptomics are complex, vulnerable and multiplexed, making it difficult to integrate all the data collected for crop management for meaningful data mining. Nevertheless, agronomic genomics will facilitate breeding crop varieties with improved responses to crop management.

Compared to the natural conditions where crop plants experience, complicate environmental changes across their growth and developmental stages, *i.e.*, the dynamics of gene expression changes, can be determined more directly and accurately under managed or controlled environments. Using transcriptome data collected from rice leaves along with the meteorological data in the field, including wind intensity, air temperature, relative humidity, atmospheric pressure, global solar radiation and precipitation, statistical models were developed for the endogenous and external influences on gene expression (Nagano *et al.* 2012; **Figure 2**). The transcriptome dynamics is revealed to be predominantly governed by several key factors such as endogenous diurnal rhythms, ambient temperature, plant age, and solar radiation. Diurnal gates for environmental stimuli affected transcription and pointed to relative influences on different metabolic genes exerted by circadian and environmental factors. In Arabidopsis, DNA microarrays were used to reveal how gene expression changes with development and responds to environmental conditions (Richards *et al.* 2012). Differences in accession and developmental status could equally explain the variation in gene expression in two accessions of *A. thaliana* grown in field conditions, and gene expression was significantly predicted with temperature and precipitation. Using a relatively simple design and several environmental factors, these two studies identified the molecular basis of response to environmental changes and teased apart the influences of development and complex environmental variables (Nagano *et al.* 2012; Richards *et al.* 2012), and should be applicable to other crops for deciphering the impacts of complex environments on transcriptome fluctuations.

Companion organisms exist on or around the plants, particularly in the rhizosphere. Crop yield, quality and stress tolerance are largely affected by soil environments, particularly the rhizosphere microorganism community. Therefore, simultaneous improvement of crop and its rhizosphere microorganisms has significant implications in genetics, ecology and agronomy. Improvement of the rhizosphere microorganisms can help establish better environmental conditions for crop plants. The improvement involves inorganic and organic conditions. The former includes upgrading of water and fertility maintainability of the soil, transfer of ineffective inorganic nutrients to effective ones, and degrading of soil toxic matters. The latter involves inhibiting unfavorable soil organisms but enhancing favorable ones. Improving rhizosphere microorganisms will play roles in increasing crop yield similar to improving abiotic and biotic stresses of crop plants per se. Therefore, companion organisms should be improved under the guidance of envirotyping. Molecular biology can be used to modify and optimize the environments surrounding the crop plants, and it is important to transform crop improvement from the crop oriented to crop community oriented.

5.3. Four-Dimensional Profile of Crop Breeding

Environmental information has not been well exploited for improving our understanding of plant adaptation. It is being complemented with environmental characterization through large-scale envirotyping via information systems such as GIS. It can be exploited for plant breeding and crop production in various ways, including but not limited to, precision measurement of environmental factors affecting specific developmental processes, selection of target environments for specific experiments, designation of environmental factors associated with phenotypic variation, dissection of GEI factors into specific components, and breeding for improved response or adaptation to specific environments, environmental factors and their combinations. Ultimately, an optimized precision breeding and crop production system can be built up with a four-dimensional (4D) profile, the first three (3D) being spatial determined by genotype, phenotype and envirotype, while the fourth being temporal involving developmental stages (**Figure 4**), and thus a cultivar or genotype architecture can be designed with an optimized phenotype and best adaptation to a target environment or crop management system. With genome-

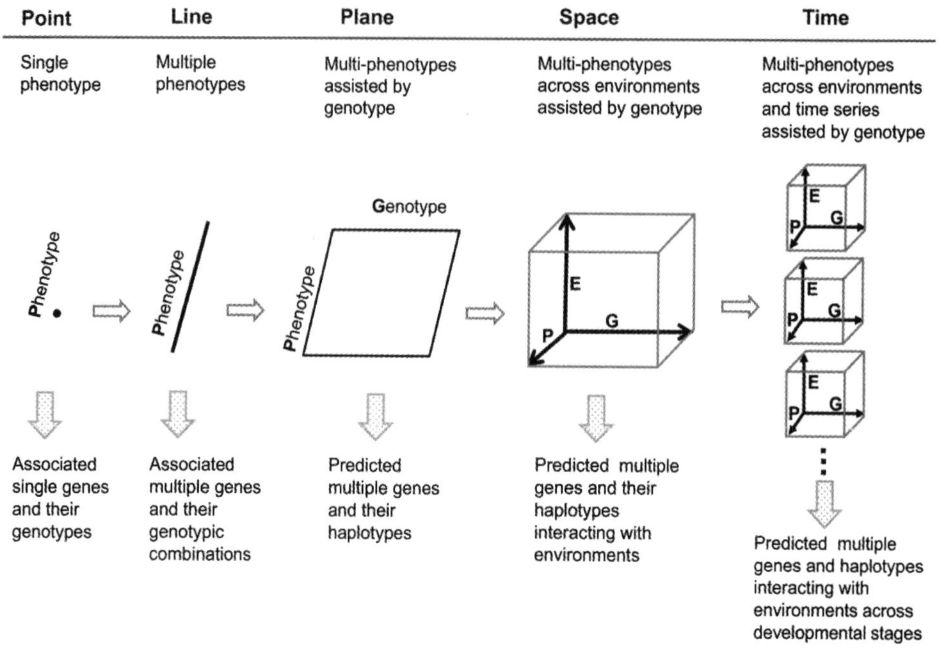

Figure 4. A four-dimensional profile of precision breeding and crop production system with the concept evolving from point to line, plane and space. Selection in early plant breeding was performed based on single desirable phenotypes one at a time ("point"). Conventional breeding has been based on selection of multiple phenotypes ("line"). Marker-assisted breeding uses selection criteria determined by both multiple phenotypes and genotypes ("plane"). Our future breeding and crop production system will be built upon the knowledge generated by genotyping, phenotyping and envirotyping, which forms the three spatial dimensions ("space"). Considering the temporal variation across different growth and developmental stages, a fourth dimension (time) should be also included. Green arrows represent the evolutionary steps of plant breeding; gray-dotted arrows represent the selection targets that can be inferred from selection strategies. Revised from Xu *et al.* (2012).

wide understanding of the environmental impacts on crop plants—enviromics and long-term trial data and outputs from general circulation models to form climate change-oriented breeding—breeding by design can be driven by incorporating information from genotyping, phenotyping, and envirotyping across developmental stages. With the support of the 3D (*G-P-E*) information, for example, management of wheat rust race Ug99 can facilitate establishing more effective *G-P-E* models for biotic stresses (Dave Hudson, CIMMYT, personal comm.). Monitoring systems for tackling biotic stresses that have been established with *G-P-E* informatics are well in the position to predict temporal change of the relevant stresses for a 4D informatics-driven precision breeding.

"Sustainable Intensification" aims at increasing crop production through more efficient use of all resources, while minimizing pressure on the environment and developing resilience, natural capital and environmental services. Breeding under conservation agricultural system can help understand if full selection under conservation agriculture conditions results in genotypes with better performance under such conditions, better emergence vigor/crop establishment, and better performance under water-limited conditions. It may also help us understand if such a full selection would result in genotypes with excessive height and tendency to lodge, and less earliness. However, little has been done through breeding to fully realize the yield potential of new germplasm, for example, under no-tillage. On the other hand, low GEI (15% of variability) suggests that no separate breeding program would be required for conservation agriculture (Hlatywayo *et al.*, Cairns and Thierfelder; CIMMYT, personal comm.). As envirotypic information becomes available for all the factors under conservation agriculture conditions, a precision breeding program can be designed and optimized based on the 4D breeding profile so that the breeders can deal with complexity without being lost in complexity (**Figure 4**).

The concept of 4D plant breeding profile can be used to identify the best environments for a specific crop and to adjust breeding strategies for the changing target environments. As one of the examples for movement of crop belts, U.S maize production increase from 1.8 to 12.7 billion bushels during 1879–2007 accompanied a substantial change of the footprint of production (Beddow and Pardey 2015). During this period, the US corn growing areas moved 279 km north and 342 km west. The new spatial output indices developed showed that such spatial movement contributed to 16%–21% of the yield increase in US maize production over the 128 years. This long-run perspective provides historical precedent for how much crop production might adjust to future climate change and technology innovation.

5.4. Information and Support Systems

The phenotype we observed is the outcome of the interactions between constantly changing environmental factors and a certain genotype. A full understanding of this process posts a great challenge and depends on high-throughput envirotyping platforms and the power of computation to analyze big datasets. In addition to currently available large datasets created by genotyping and phenotyping, envirotyping will generate a huge amount of information including images and videos, causing a data tsunami in the near future. Therefore, the *G-P-E* data space should be expanded to include a fourth dimension contributed with the time (*T*) that lasts through the whole developmental stages of crop plants. There are 148,000 wheat accessions stored in CIMMYT Genebank and 113,000 rice accessions in IRRI Genebank. These accessions can be genotyped using SNP chips to produce thousands to millions of genotypic data points per accession, or sequenced to produce Tb levels of DNA, RNA and protein data. To meet the challenges of increasing envirotypic information and to prevent agronomists and biologists from devouring by the data tsunami, standardized data generation and collection procedures, data collecting tools, sampling technologies, controlled vocabularies and ontologies, and interoperable query systems should be established. An integrative information system will need a general database with ontology control to bring genotypic, phenotypic and envirotypic information together, which is particularly important for comparison and integration across crop species. In general, we need to develop a generic database and information system for all the data from different *G-P-E-T* sources. As significant efforts have been made in development of controlled vocabularies and ontologies for genotyping and phenotyping (Xu 2010), we need to follow the scenarios to develop the similar system for envirotyping. Unique identifiers that are associated with each concept in envirotyping should be developed and used for linking and querying databases.

High-performance computing cluster, hardware and software are needed for data analysis. Considering that *G-P-E-T* data have been generated and maintained largely through different research programs and groups, it is vital to establish a one-step-shopping system to allow all the data accessible to all relevant scientists and users. Systems biology approaches driven by information could prove beneficial and biological models could be generated finally to show the contribution of different signaling pathways through building plant '-omic' architectural responses to climate change catastrophes (Ahuja *et al.* 2010).

To meet the data tsunami challenge, a successful crop science research program needs to back up with appropriate decision support tools. Integrative Breeding Platform (IBP) provides tools to help breeders in designing and managing experiments, collecting and storing data, and conducting analyses (https://www.integratedbreeding.net). Its interconnected software, Breeding Management System (BMS), is designed for breeders to manage their daily activities throughout their breeding programs. However, the databases and tools currently available or under development are largely for genotypic and phenotypic information, with little consideration of environmental data. Apparently, incorporation of all environmental information into databases and tools deserves a special attention. Future changes for long-term breeding program orientation can be predicted by utilizing downscaled general circulation model outputs in combination with mega-environment definitions, crop production areas, specific areas or locations, and socioeconomic data. Learning from the best and the past through data mining will help us to develop an integrated approach to optimize crop production.

6. Future Prospects

In the future, envirotyping will face more challenges than what genotyping and phenotyping have ever. Human resources and investments are needed to be dispatched among genotyping, phenotyping and envirotyping for a balanced development. As high-throughput and sequence-based genotyping becomes routine and high-throughput and precision phenotyping becomes achievable, a full set of next-generation high-throughput precision envirotyping technologies will be also showing up at the corner. As in many cases, where biologists present challenges while computational scientists present solutions, envirotyping tools and methodologies need to be developed through multidisciplinary collaborations by standing on the shoulder of history. Companion organisms have more complicated interactions with host plants than other environmental factors. Predictive phenomics would become possible with more and more envirotypic information available. Understanding the effects of crop management on gene expression will help us design more cost-effective, sustainable and multifunctional crop production systems. Envirotyping will help identify the best sets of crop management practices to optimize yield and yield components such as plant density. A great contribution to breeding and crop production by envirotyping would be the precision agriculture largely driven by envirotypic information and reducing drudgery. Due to increasing soil and water pollutions, demanding for healthier and nutritious food, which is becoming increasingly important in developing countries, will drive to establish soil-human health relationship and to develop new tools and methodologies for dynamic and continuous envirotyping to monitor the entire environmental profile. Cutting childhood mortality in half (FAO 2008) through biofortification by 2050 means that we need to develop new varieties with improved nutrition by combining varietal improvement for the changing soil nutrients. Envirotyping-guided crop science will help us to face all the challenges through genetic studies, genotype reconstruction, variety development, abiotic/biotic stress prediction, and precision agriculture and breeding. Envirotyping, as a third "typing" technology for crop science, should be used to decipher all environmental impacts on growth, reproduction and survival of crop plants. With large-scale envirotyping, in combination with known genotypic information, environments for crops can be optimized and phenotypic performance under specific environments can be largely predicted, thus enhanced cost-benefit efficiency for precision breeding and crop production.

Acknowledgements

The research activities at CIMMYT-China and Institute of Crop Science, CAAS have been supported by National Natural Science Foundation of China (31271736), China National 973 Project (2014CB138206), National International Science and Technology Collaboration Program of China (2012DFA32290), Agricultural Science and Technology Innovation Program (ASTIP) of CAAS, Bill and Melinda Gates Foundation, and CGIAR Research Program MAIZE.

Source: Xu Y. Envirotyping for deciphering environmental impacts on crop plants[J]. Theoretical & Applied Genetics, 2016, 129(4):1–21.

References

[1] Ahuja I, de Vos RCH, Bones AM, Hall RD (2010) Plant molecular stress responses face climate change. Trends Plant Sci 15:664–674.

[2] Alvarez M, Schrey AW, Richards CL (2015) Ten years of transcriptomics in wild populations: what have we learned about their ecology and evolution? Mol Ecol 24:710–725.

[3] Araus JL, Cairns JE (2014) Field high-throughput phenotyping: the new crop breeding frontier. Trends Plant Sci 19:52–61.

[4] Asseng S, Ewert F, Martre P, Rötter RP, Lobell DB, Cammarano D, Kimball BA, Ottman MJ, Wall GW, White JW, Reynolds MP, AldermanPD Prasad PVV, Aggarwal PK, Anothai J, Basso B, Biernath C, Challinor AJ, De Sanctis G, Doltra J, Fereres E, Garcia-Vila M, Gayler S, Hoogenboom G, Hunt LA, Izaurralde RC, Jabloun M, Jones CD, Kersebaum KC, Koehler AK, Müller C, Naresh Kumar S, Nendel C, O'Leary G, Olesen JE, Palosuo T, Priesack E, Eyshi Rezaei E, Ruane AC, Semenov MA, Shcherbak I, Stöckle C, Stratonovitch P, Streck T, Supit I, Tao F, Thorburn PJ, Waha K, Wang E, Wallach D, Wolf J, Zha Z, Zhu Y (2015) Rising temperatures reduce global wheat production. Nat Clim Change 5:143–147.

[5] Bänziger M, Setimela PS, Hodson D, Vivek B (2006) Breeding for improved abiotic stress tolerance in maize adapted to southern Africa. Agric Water Manag 80:212–224.

[6] Basu PS, Srivastava M, Singh P, Porwal P, Kant R, Singh J (2015) High-precision phenotyping under controlled versus natural environments. In: Kumar J, Pratap A, Kumar S (eds) Phenomics in crop plants: trends options and limitations. Springer, India.

[7] Bebber DP, Ramotowski MAT, Gurr SJ (2013) Crop pests and pathogens move polewards in a warming world. Nat Clim Change 3:985–988.

[8] Beddow JM, Pardey PG (2015) Moving matters: the effect of location on crop production. J Eco Hist 7:219–249.

[9] Bevan B (1998) Geophysical exploration for archaeology: an introduction to geophysical exploration. U.S. National Park Service Publications and Papers. Paper 91. http://digitalcommons.unl.edu/natlpark/91.

[10] Blum A (2011) Drought resistance—is it really a complex trait? Funct Plant Biol 38:753–757.

[11] Bostock RM, Pye MF, Roubtsova TV (2014) Predisposition in plant disease: exploiting the nexus in abiotic and biotic stress perception and response. Annu Rev Phytopathol 52:517–549.

[12] Bustos-Korts D, Malosetti M, Chapman S, van Eeuwijk F (2016) Modelling of genotype by environment interaction and prediction of complex traits across multiple environments as a synthesis of crop growth modelling, genetics and statistics. In: Yin X, Struik PC (eds) Crop systems biology, narrowing the gaps between crop modelling and genetics. Springer, Switzerland, pp 55–82.

[13] Butler EE, Huybers P (2013) Adaptation of US maize to temperature variations. Nat Clim Change 3:68–72.

[14] Challinor AJ, Watson J, Lobell DB, Howden SM, Smith DR, Chhetri N (2014) A meta-analysis of crop yield under climate change and adaptation. Nat Clim Change 4:287–291.

[15] Chen J, Peng P, Tian J, He Y, Zhang L, Liu Z, Yin D, Zhang Z (2015) *Pike*, a rice blast resistance allele consisting of two adjacent NBS-LRR genes, was identified as a novel allele at the *Pik* Locus. Mol Breed 35:117 (online first).

[16] Chéné Y, Rousseau D, Lucidarme P, Bertheloot J, Caffier V, Morel P, Belin E, Chapeau-Blondeau F (2012) On the use of depth camera for 3D phenotyping of entire plants. Computers Electron Agric 82:122–127.

[17] Chinnusamy V, Zhu JK (2009) Epigenetic regulation of stress responses in plants. Curr Opin Plant Biol 12:133–139.

[18] Cobb JN, DeClerck G, Greenberg A, Clark R, McCouch S (2013) Next-generation phenotyping: requirements and strategies for enhancing our understanding of genotype-phenotype relationships and its relevance to crop improvement. Theor Appl Genet 126:867–887.

[19] Cooper M, Messina CD, Podlich D, Totir LR, Baumgarten A, Hausmann NJ, Wright D, Graham G (2014) Predicting the future of plant breeding: complementing empirical evaluation with genetic prediction. Crop Pasture Sci 65:311–336.

[20] Cooper M, Technow F, Messina C, Gho C, Totir LR (2016) Use of crop growth models (CGM) with whole genome prediction (WGP): application 1 of CGM-WGP to a maize multi-environment trial. Crop Sci. doi:10.2135/cropsci2015.08.0512.

[21] Corwin DL, Lesch SM (2005) Apparent soil electrical conductivity measurements in agriculture. Computers Electron Agric 46:11–43.

[22] Cossani CM, Reynolds MP (2012) Physiological traits for improving heat tolerance in wheat. Plant Physiol 160:1710–1718.

[23] Crossa J, Cornelius PL (2012) Linear-bilinear models for the analysis of genotype-environment interaction. In Kang (ed) Quantitative genetics, genomics and plant breeding. CABI, pp 305–322.

[24] Deinlein U, Stephan AB, Horie T, Luo W, Xu G, Schroeder JI (2014) Plant salt-tolerance mechanisms. Trends Plant Sci 19:371–379.

[25] DeMers MN (2005) Fundamentals of geographic information systems, 3rd edn. Wiley, New York.

[26] Des Marais DL, Hernandez KM, Juenger TE (2013) Genotype-by-environment interaction and plasticity: exploring genomic responses of plants to the abiotic environment. Annu Rev Ecol Evol Syst 44:5–29.

[27] Fahlgren N, Gehan MA, Baxter I (2015) Lights, camera, action: high-throughput plant phenotyping is ready for a close-up. Curr Opin Plant Biol 24:93–99.

[28] FAO (Food and Agriculture Organization of the United Nations) (2008) Climate change and food security: a framework document. Food and Agriculture Organization of the United Nations, Rome.

[29] FAO (Food and Agriculture Organization of the United Nations) (2014) FAOSTAT online data. Food and Agriculture Organization of the United Nations, Rome.

[30] Gebbers R, Adamchuk VI (2010) Precision agriculture and food security. Science 327:828–831.

[31] Grishkevich V, Yanai I (2013) The genomic determinants of genotype X environment interactions in gene expression. Trends Genet 29:479–487.

[32] Hammond JP, Mayes S, Bowen HC, Graham NS, Hayden RM, Love CG, Spracklen WP, Wang J, Welham SJ, White PJ, King GJ, Broadley MR (2011) Regulatory hotspots are associated with plant gene expression under varying soil phosphorus supply in *Brassica rapa*. Plant Physiol 156:1230–1241.

[33] Heslot N, Akdemir D, Sorrells ME, Jannink JL (2014) Integrating environmental covariates and crop modeling into the genomic selection framework to predict genotype by environment interactions. Theor Appl Genet 127:463–480.

[34] Hoheisel JD (2006) Microarray technology: beyond transcript profiling and genotype analysis. Nat Rev Genet 7:200–210.

[35] Houle D, Govindaraju DR, Omholt S (2010) Phenomics: the next challenge. Nat Rev Genet 11:855–866.

[36] Hu H, Xiong L (2014) Genetic engineering and breeding of drought-resistant crops. Annu Rev Plant Biol 65:715–741.

[37] IUSS Working Group WRB (2014) World reference base for soil resources 2014. International soil classification system for naming soils and creating legends for soil maps. World Soil Resources Reports No. 106. FAO, Rome.

[38] Izawa T (2015) Deciphering and prediction of plant dynamics under field conditions. Curr Opin Plant Biol 24:87–92.

[39] Jansen M, Gilmer F, Biskup B, Nagel KA, Rascher U, Fischbach A, Briem S, Dreissen G, Tittmann S, Braun S, De Jaeger I, Metzlaff M, Schurr U, Scharr H, Walter A (2009) Simultaneous phenotyping of leaf growth and chlorophyll fluorescence via GROWSCREEN FLUORO allows detection of stress tolerance in *Arabidopsis thaliana* and other rosette plants. Funct Plant Biol 11:902–914.

[40] Jarquín D, Crossa J, Lacaze X, Du Cheyron P, Daucourt J, Lorgeou J, Piraux F, Guerreiro L, Pérez P, Calus M, Burgueño J, de los Campos G (2014) A reaction norm model for genomic selection using high-dimensional genomic and environmental data. Theor Appl Genet 127:595–607.

[41] Johnson HW, Robinson HF, Comstock RE (1955) Estimates of genetic and environmental variability in soybeans. Agron J 47:314–318.

[42] Kim DH, Doyle MR, Sung S, Amasino RM (2009) Vernalization: winter and the timing of flowering in plants. Annu Rev Cell Dev Biol 25:277–299.

[43] Kissoudis C, van de Wiel C, Visser RGF, Van Der Linden G (2014) Enhancing crop resilience to combined abiotic and biotic stress through the dissection of physiological and molecular crosstalk. Front Plant Sci 5:207.

[44] Koboldt DC, Steinberg KM, Larson DE, Wilson RK, Mardis ER (2013) The next-generation sequencing revolution and its impact on genomics. Cell 155:27–38.

[45] Lander ES, Botstein D (1989) Mapping Mendelian factors underlying quantitative traits using RFLP linkage maps. Genetics 121:185–199.

[46] Lasky JR, Des Marais DL, Lowr DB, Povolotskaya I, McKay JK, Richards JH, Keitt TH, Juenger TE (2014) Natural variation in abiotic stress responsive gene expression and local adaptation to climate in *Arabidopsis thaliana*. Mol Biol Evol 31:2283–2296.

[47] Lee WS, Alchanatis V, Yang C, Hirafuji M, Moshou D, Li C (2010) Sensing technologies for precision specialty crop production. Computers Electron Agric 74:2–33.

[48] Lobell DB, Banziger M, Magorokosho C, Vivek B (2011) Nonlinear heat effects on African maize as evidenced by historical yield trials. Nat Clim Change 1:42–45.

[49] Lobell DB, Hammer GL, McLean G, Messina C, Roberts MJ, Schlenker W (2013) The critical role of extreme heat for maize production in the United States. Nat Clim Change 3:497–501.

[50] Lobell DB, Roberts MJ, Schlenker W, Braun N, Little BB, Rejesus RM, Hammer GL (2014) Greater sensitivity to drought accompanies maize yield increase in the U.S. Midwest. Science 344:516–519.

[51] López-Arredondo DL, Leyva-González MA, González-Morales SI, López-Bucio J, Herrera-Estrella L (2014) Phosphate nutrition: improving low-phosphate tolerance in crops. Annu Rev Plant Biol 65:95–123.

[52] Lopez-Cruz M, Crossa J, Bonnett D, Dreisigacker S, Poland J, Jannink JL, Singh RP, Autrique E, de los Campos G (2015) Increased prediction accuracy in wheat breeding trials using a marker × environment interaction genomic selection model. G3: Genes|Genomes|Genetics 5:569–582.

[53] Lowry DB, Logan TL, Santuari L, Hardtke CS, Richards JH, DeRose-Wilson L, McKay JK, Sen S, Juenger TE (2013) Expression quantitative trait locus mapping across water availability environments reveals contrasting associations with genomic features in Arabidopsis. Plant Cell 25:3266–3279.

[54] Maheswaran M, Huang N, Sreerangasamy SR, McCouch SR (2000) Mapping quantitative trait loci associated with days to flowering and photoperiod sensitivity in rice (*Oryza sativa* L.). Mol Breed 6:145–155.

[55] Massonnet C, Vile D, Fabre J, Hannah MA, Caldana C, Lisec J, Beemster GTS, Meyer RC, Messerli G, Gronlund JT, Perkovic J, Wigmore E, May S, Bevan MW, Meyer C, Rubio-Díaz S, Weigel D, Micol JL, Buchanan-Wollaston V, Fiorani F, Walsh S, Rinn B, Gruissem W, Hilson P, Hennig L, Willmitzer L, Granier C (2010) Probing the reproducibility of leaf growth and molecular phenotypes: a comparison of three Arabidopsis accessions cultivated in ten laboratories. Plant Physiol 152: 2142–2157.

[56] Masuka B, Araus JL, Sonder K, Das B, Cairns JE (2012) Deciphering the code: successful abiotic stress phenotyping for molecular breeding. J Integr Plant Biol 54:238–249.

[57] Matsubara K, Hori K, Ogiso-Tanaka E, Yano M (2014) Cloning of quantitative trait genes from rice reveals conservation and divergence of photoperiod flowering pathways in Arabidopsis and rice. Front Plant Sci 5:193.

[58] McBratney A, Whelan B (2001) Precision AG-OZ style. NSW Agriculture, GIA2001, pp 274–281.

[59] Mickelbart MV, Hasegawa PM, Bailey-Serres J (2015) Genetic mechanisms of abiotic stress tolerance that translate to crop yield stability. Nat Rev Genet 16:237–251.

[60] Myrold DD, Zeglin LH, Jansson JK (2014) The potential of metagenomic approaches for understanding soil microbial processes. Soil Sci Soc Am J 78:3–10.

[61] Nagano AJ, Sato Y, Mihara M, Antonio BA, Motoyama R, Itoh H, Nagamura Y, Izawa T (2012) Deciphering and prediction of transcriptome dynamics under fluctuating field conditions. Cell 151:1358–1369.

[62] Nicotra AB, Atkin OK, Bonser SP, Davidson AM, Finnegan EJ, Mathesius U, Poot P, Purugganan MD, Richards CL, Valladares F, vanKleunen M (2010) Plant phenotypic plasticity in a changing climate. Trends Plant Sci 15:684–692.

[63] Nühse TS, Bottrill AR, Jones AME, Peck SC (2007) Quantitative phosphoproteomic analysis of plasma membrane proteins reveals regulatory mechanisms of plant innate immune responses. Plant J 51:931–940.

[64] Ortiz-Monasterio I, Schulthless U, Govaerts B, Dobler C (2013) From GreenSeeker to GreenSat in irrigated wheat in Mexico. Mexico. In: "Remote sensing—beyond images" Workshop, 14–15 December 2013. International Maize and Wheat Improvement Center (CIMMYT).

[65] Pardey PG, Beddow JM, Hurley TM, Beatty TKM, Eidman VR (2014) The International agricultural prospects model: assessing consumption and production futures through 2050 (version 2.1). Department of Applied Economics Staff Paper P14–09. University of Minnesota, St. Paul.

[66] Paterson AH, Lander ES, Hewitt JD, Peterson S, Lincoln SE, Tanksley SD (1988) Resolution of quantitative traits into Mendelian factors, using a complete linkage map of restriction fragment length polymorphisms. Nature 335:721–726.

[67] Paterson AH, Damon S, Hewitt JD, Zamir D, Rabinowitch HD, Lincoln SE, Lander ES, Tanksley SD (1991) Mendelian factors underlying quantitative traits in tomato: comparison across species, generations, and environments. Genetics 127:181–197.

[68] Paulus S, Dupuis J, Mahlein AK, Kuhlmann H (2013) Surface feature based classification of plant organs from 3D laser scanned point clouds for plant phenotyping. BMC Bioinform 14:238.

[69] Pingali P, Pandey S (2000) Meeting world maize needs: technological opportunities and priorities for the public sector. Part 1 of CIM-MYT World maize facts and figures. International Maize and Wheat Improvement Center (CIMMYT), Mexico D.F.

[70] Poorter H, Bühler J, van Dusschoten D, Climent J, Postma JA (2012) Pot size matters:

a meta-analysis of the effects of rooting volume on plant growth. Funct Plant Biol 39:839–850.

[71] Pound MP, French AP, Murchie EH, Pridmore TP (2014) Automated recovery of three-dimensional models of plant shoots from multiple color images. Plant Physiol 166:1688–1698.

[72] Prasanna BM, Cairns J, Xu Y (2013) Genomic tools and strategies for breeding climate resilient cereals. In: Kole C (ed) Genomics and breeding for climate-resilient crops, vol 1. Springer, Berlin Heidelberg, pp 213–239.

[73] Prasch CM, Sonnewald U (2015) Signaling events in plants: stress factors in combination change the picture. Environ Exp Bot 114:4–14.

[74] Ramegowda V, Senthil-Kumar M (2015) The interactive effects of simultaneous biotic and abiotic stresses on plants: mechanistic understanding from drought and pathogen combination. J Plant Physiol 176:47–54.

[75] Rebetzke GJ, Chenu K, Biddulph B, Moeller C, Deery DM, Rattey AR, Bennett D, Barrett-Lennard EG, Mayer JE (2013) A multisite managed environment facility for targeted trait and germplasm phenotyping. Funct Plant Biol 40:1–13.

[76] Reyes A, Messina CD, Hammer GL, Liu L, van Oosterom E, Lafitte R, Cooper M (2016) Soil water capture trends over 50 years of single-cross maize (*Zea mays* L.) breeding in the US corn-belt. J Exp Bot. doi:10.1093/jxb/erv430.

[77] Reynolds M, Bonnett D, Chapman SC, Furbank RT, Mane Y, Mather DE, Parry MAJ (2011) Raising yield potential of wheat. I. Overview of a consortium approach and breeding strategies. J Exp Bot 62:439-452.

[78] Reynolds MP, Pask AJD, Mullan DM (eds) (2012) Physiological breeding i: interdisciplinary approaches to improve crop adaptation. CIMMYT.

[79] Richards CL, Rosas U, Banta J, Bhambhra N, Purugganan MD (2012) Genome-wide patterns of Arabidopsis gene expression in nature. PLoS Genet 8:e1002662.

[80] Römer C, Bürling K, Hunsche M, Rumpf T, Noga G, Plümer L (2011) Robust fitting of fluorescence spectra for presymptomatic wheat leaf rust detection with support vector machines. Computers Electron Agric 79:180–188.

[81] Schlenker W, Roberts MJ (2009) Nonlinear temperature effects indicate severe damages to U.S. crop yields under climate change. Proc Natl Acad Sci USA 106:15594–15598.

[82] Shi XZ, Yu DS, Gao P, Wang HJ, Sun WX, Zhao YC, Gong ZT (2007) Soil Information System of China (SISChina) and its application. Soils 39:329–333.

[83] Singh RP, Hodson DP, Jin Y, Huerta-Espino J, Kinyua MG, Wanyera R (2006) Current status, likely migration and strategies to mitigate the threat to wheat production from race Ug99 (TTKS) of stem rust pathogen. In: CAB reviews: perspectives in agriculture, veterinary science, nutrition and natural resources 1, No. 054.

[84] Stief CWA, Altmann S, Hoffmann K, Pant BD, Scheible WR, Bäurle I (2014) Arabidopsis miR156 regulates tolerance to recurring environmental stress through SPL transcription factors. Plant Cell 26:1792–1807.

[85] Suzuki N, Rivero RM, Shulaev V, Blumwald E, Mittler R (2014) Abiotic and biotic stress combinations. New Phytol 203:32–43.

[86] Tai APK, Val Martin M, Heald CL (2014) Threat to future global food security from climate change and ozone air pollution. Nat Clim Change 4:817–821.

[87] Tilman D, Cassman KG, Matson PA, Naylor R, Polasky S (2002) Agricultural sustainability and intensive production practices. Nature 418:671–677.

[88] Trenberth KE, Dai A, van der Schrier G, Jones PD, Barichivich J, Briffa KR, Sheffield J (2014) Global warming and changes in drought. Nat Clim Change 4:17–22.

[89] Trnka M, Rötter RP, Ruiz-Ramos M, Kersebaum KC, Olesen JE, Žalud Z, Semenov MA (2014) Adverse weather conditions for European wheat production will become more frequent with climate change. Nat Clim Change 4:637–643.

[90] Whelan B, Taylor J (2013) Precision agriculture for grain production systems. CSIRO Publishing, Collingwood.

[91] Xu Y (2002) Global view of QTL: rice as a model. In: Kang MS (ed) Quantitative genetics, genomics and plant breeding. CABI Publishing, Wallingford, pp 109–134.

[92] Xu Y (2010) Molecular plant breeding. CAB International, Wallingford.

[93] Xu Y (2011) From line to space: a 3-D profile of molecular plant breeding. In: The first congress of cereal biotechnology and breeding, May 23–27, 2011, Szeged, Hungary.

[94] Xu Y (2012) Environmental assaying or e-typing as a key component for integrated plant breeding platform. In: Marker-assisted selection workshop, 6th international crop science congress, August 6–10, 2012, Bento Goncalves, RS, Brazil.

[95] Xu Y (2015) Envirotyping and its applications in crop science. Scientia Agricultura Sinica 48:3354–3371.

[96] Xu Y, Lu Y, Xie C, Gao S, Wan J, Prasanna BM (2012) Wholegenome strategies for marker-assisted plant breeding. Mol Breed 29:833–854.

[97] Xu Y, Xie C, Wan J, He Z, Prasanna PM (2013) Marker-assisted selection in cereals: platforms, strategies and examples. In: Gupta PK, Varshney RK (eds) Cereal Genomics II. Springer, Dordrecht, pp 375–411.

[98] Yan W, Kang MS, Ma B, Woods S, Cornelius PL (2007) GGE biplot vs. AMMI analysis of genotype-by-environment data. Crop Sci 47:643–655.

Chapter 19

Explaining the High PM_{10} Concentrations Observed in Polish Urban Areas

Magdalena Reizer, Katarzyna Juda-Rezler

Faculty of Environmental Engineering, Warsaw University of Technology, Nowowiejska 20, 00-653 Warsaw, Poland

Abstract: The main goal of this paper is to identify the drivers responsible for the high particulate matter concentrations observed in recent years in several urban areas in Poland. The problem was investigated using air quality and meteorological data from routine monitoring network, air mass back trajectories and multivariate statistical modelling. Air pollution in central and southern part of the country was analysed and compared with this in northern-eastern "The Green Lungs of Poland" region. The analysis showed that in all investigated locations, there is a clear annual cycle of observed concentrations, closely following temperature-heating cycles, with the highest concentrations noted in January. However, the main drivers differ along the country, being either connected with regional background pollution (in the central part of the country) or with local emission sources (in the southern part). The occurrence of high PM_{10} concentrations is most commonly associated with the influence of high-pressure systems that brought extremely cold and stable air masses form East or South of Europe. During analysed episodes, industrial point sources had the biggest (up to 70%–80%) share in PM_{10} levels on the days with maximum PM pollution, while remote and residential/traffic sources determined the air quality in the early stages of the episodes.

Principal component analysis (PCA) shows that secondary inorganic aerosols account for long-range transported pollution, As, Cd, Pb and Zn for industrial point sources, while Cr and Cu for residential and traffic sources of PM_{10}, respectively.

Keywords: PM_{10} Episode, Poland, Coal Combustion, Source Apportionment, PCA-MLRA, Backward Trajectories

1. Introduction

Particulate matter (PM) is a complex mixture of solid and liquid particles with varying physical, mineralogical and chemical characteristics dependent on categories identified according to their aerodynamic diameter, as either ultrafine $PM_{0.1}$ (particles with an aerodynamic diameter, d_a < 100nm), fine $PM_{2.5}$ (d_a < 2.5μm), coarse $PM_{10-2.5}$ (d_a > 2.5 and d_a < 10μm) or PM_{10} (d_a <10μm). Common chemical constituents of PM include crustal material, sea salt, organic carbon (OC), elemental carbon (EC), secondary inorganic aerosols (SIA including sulphates, nitrates, ammonia) and trace elements (TEs), as well as particle-bound water and unspecified compounds. Particles can either be directly emitted from natural sources and virtually all kinds of anthropogenic activities, or be formed in the atmosphere from other pollutants by gas-to-particle conversion processes. So far, numerous studies have shown the association of adverse health effects with exposure to PM over both short (days) and long (a year or more) periods, ranging from modest temporary changes, through increased risk of symptoms requiring hospital admission and increased risk of death from cardiovascular and respiratory diseases or lung cancer (e.g. Pope *et al.* 2002; Brunekreef and Forsberg 2005; Rückerl *et al.* 2011). Most recently, the specialized cancer agency of the World Health Organization (WHO), the International Agency for Research on Cancer (IARC) concluded that there is strong evidence that exposure to outdoor air pollution, and PM specifically, is associated with increases in genetic damage, including cytogenetic abnormalities, mutations in both somatic and germ cells and altered gene expression, which have been linked to increased cancer risk in humans and thus classified outdoor air pollution and particulate matter from this pollution as carcinogenic to humans—IARC group 1 (Loomis *et al.* 2013). According to the latest data, atmospheric PM pollution constitutes the 6th leading risk factor (among 43 ranked), which corresponds to over 3 million deaths worldwide every year (Lim *et al.*

2012). If no new policies are implemented, by 2050, outdoor air pollution is projected to become the top cause of environmentally related deaths worldwide (OECD 2012).

Since 1990, global emissions of primary PM and main precursors of secondary PM have either declined (PM_{10}, SO_2, NMVOCs) or slightly increased (NO_x, NH_3), but there is strong spatial variability, with Europe and North America continuing emission reduction and increasing role of Asia and Africa (based on data from the Emissions Database for Global Atmospheric Research—EDGAR v. 4.2). However, poor urban air quality due to high concentrations of PM remains a major public health problem worldwide. Severe PM episodes are occurring repeatedly in many cities all over the world, e.g. in Cracow and Warsaw (January 2006), New Delhi (November 2011), Athens (December 2012), Beijing (January 2013), Salt Lake City (January 2013), London (February 2014) and Paris (March 2014). According to the latest report of the European Environmental Agency, up to a third of the European Union's (EU) urban population is exposed to levels of PM_{10} exceeding daily air quality limit value ($50 \mu g \cdot m^{-3}$), while the exposure to PM_{10} levels that do not meet the annual air quality guideline ($20 \mu g \cdot m^{-3}$) set by the WHO is significantly higher, comprising over 80% of the EU urban population (EEA 2014). Moreover, only 2% of the global urban population is living in areas where PM_{10} concentrations are lower than WHO air quality guideline (OECD 2012).

Therefore, it is crucial to gain knowledge about the types of emission sources responsible for severe PM events. To address this issue, different source apportionment (SA) methods are widely used in worldwide studies. Two main types of approaches are employed: (1) bottom-up methods based on numerical dispersion modelling and (2) top-down methods applying receptor modelling (RM) techniques. The usage of dispersion models, which simulate aerosol emission, formation, transport and deposition, basing on detailed emission inventories as well as meteorological and topographical data, is limited in SA analyses by the accuracy of the input data, especially emission data (Juda-Rezler 2010; Kiesewetter *et al.* 2015). However, these models are frequently preferred when natural sources of PM are of special interest (Fragkou *et al.* 2012). The most regularly used bottom-up models are the Eulerian chemical transport models, followed by the Lagrangian ones. On the contrary, RM models based on multidimensional statistical analysis of ambient PM concentrations and its chemical composition are independent from emission inventories and meteorological data. The number of RM tools, ranging from·simple techniques applying elementary mathematical calculations and basic physical assumptions, up to complex models requiring pre- and post-processing of data, are currently available. Until 2005, principal component analysis (PCA) was the most frequently used RM model (Viana *et al.* 2008), while in the recent years, shifting towards more advanced methods, which require also more input data—such as positive matrix factorization (PMF) and chemical mass balance (CMB)—is observed (Belis *et al.* 2013). PMF method requires uncertainties of ambient concentrations, while CMB requires emission profiles of relevant sources as well as uncertainties of both ambient concentrations and emission profiles. Meta-analysis of 108 European SA studies conducted by Belis *et al.* (2013) defined six major source categories for PM in Europe: atmospheric formation of secondary inorganic aerosol, traffic, resuspension of crustal/mineral dust, biomass burning, (industrial) point sources and sea/road salt. A review of SA techniques conducted by Johnson *et al.* (2011) identified 11 common PM source categories in 18 developing countries of Asia, Africa and Latin America, grouped into 4 main types: (1) dust emissions, including road dust, soil dust, resuspension, fugitive dust and construction; (2) transport (gasoline, diesel); (3) industrial activities, including coal and oil burning, brick kilns and power plants; as well as (4) non-urban, including biomass burning, long-range transport and marine sources.

In Poland, as in the majority of central-eastern European countries, the dependence of economy on coal is still higher than in western European countries. Hard coal and lignite amount to approximately 50%, 80% and 77% in the structures of primary energy, electricity and heat consumption, respectively. It is well recognized that coal combustion is one of the main source of primary PM and precursor gases emissions. However, according to our knowledge, there has no previously been a study differentiating the source profile of coal combustion in different utilities, *i.e.* domestic (residential) stoves/boilers and industrial high-efficiency boilers.

Therefore, the main goal of this paper is to distinguish different types of PM emission sources as well as to discern coal combustion in industrial and domestic sources that determine high PM_{10} concentrations recorded in Polish urban areas. In recent years, a number of PM episodes varied in spatial range and strength, which occurred in Polish cities (Reizer 2013). In general, they are occurring in different synoptic situations; however, the largest frequency is observed in anti-cyclonic conditions with the air masses coming from eastern, south-eastern and southern directions (Reizer 2013; GIOŚ 2013). For the present study, two such wintertime episodes of January 2009 and January 2010, connected with eastern and southern inflow of air masses, have been selected for investigation.

2. Materials and Methods

2.1. Selection of Episodes

In general, there is no definite rule for selecting episode value from a measurement database and three different approaches are usually applied. The most commonly used method is to assume the daily air quality limit value for a selected pollutant as a threshold value (e.g. Kukkonen *et al.* 2005; Muir *et al.* 2006; Aarnio *et al.* 2008; Im *et al.* 2010). In some studies, either the maximum value is classified as an episode (e.g. Bessagnet *et al.* 2005; Amodio *et al.* 2008) or a given threshold based on personal experience is assumed (e.g. Niemi *et al.* 2009). Finally, the concentration exceeding the 75th percentile (e.g. Karaca *et al.* 2009) or the 95th percentile (e.g. Chu 2004) is marked as an episode. In this study, the severe episode is defined as a situation with the daily PM_{10} concentration at the urban background site exceeding $100 \mu g \cdot m^{-3}$ (200% of the EU limit value) for at least three subsequent days. According to the above definition, 77 PM_{10} episodes that occurred in analysed cities in 2005–2012 period were recognized. Two wintertime episodes of 7–16th January 2009 and 22–28th January 2010, connected with eastern and southern inflow of air masses, were investigated and hereinafter are referred to as episodes 1 and 2, respectively.

2.2. Characteristics of Investigated Cities

The air quality during the episodes was considered in four Polish cities, situated in central (Warsaw) and southern (Cracow, Zabrze, Jelenia Góra) part of the country (**Figure 1**).

The selected cities are characterized by different conditions with respect to climate, topography and emission sources.

Figure 1. Location of urban background (white), traffic (grey) and regional/rural background (marked by stars) air quality monitoring sites.

Warsaw, the capital city of Poland, is located in the Mazovian Lowland in central Poland. With 1.7 million inhabitants, it is the largest agglomeration in the country and 11th largest city in the EU-28 (EUROSTAT 2013). Nonetheless, it is characterized by relatively good air quality regarding the main pollutants. The majority of the city's households are connected to a central heating supply system which greatly limits the pollution that originates from private use of fossil fuels for domestic heating. The most important PM local emission sources remain road transport and cogeneration plants (AQP for Warsaw 2013). Due to the lack of a real bypass road, Warsaw is one of the most congested cities in Poland, where most of the traffic is routed through the city streets, which are quite narrow in many areas.

Cracow, with a population of approximately 760,000 inhabitants, is the second largest city in Poland, situated in the Lesser Poland close to the most extensively industrialized and highly polluted Upper Silesian region. The city is also influenced by surrounding industrial sources that include steelworks (the second largest steel plant in Poland) as well as coking, electric power and cogeneration plants. The most important local emission sources are residential burning of coal/ wood and road transport (AQP for Lesser Poland Voivodeship 2013). Moreover, Cracow, located in the wide valley of the Vistula river, is enclosed by shallow hills that aggravates the dispersion of pollutants.

The city of Zabrze together with other ten cities belongs to the Silesian Metropolis which population counts over 2.7 million inhabitants. It is located in the Upper Silesia region of Southern Poland, where coal mining and metallurgy, as well as the coke factories, are the major industrial activities. The largest steel plant in Poland is also located in the neighbourhood of the city. However, the most important local emission source in Zabrze is burning of coal/wood in small residential boilers that are used in large part of the city (AQP for Silesian Voivodeship 2010).

Jelenia Góra which constitutes a health resort area is situated in the Lower Silesian region of Southern Poland, in the area belonging to the former "Black Triangle" extremely polluted region of Europe. The most important local emission source of PM is again the residential sector (AQP for Lower Silesian Voivodeship 2010). Air pollution dispersion in the city is aggravated due to its location in the valley surrounded by the mountains on all sides, with the highest range—Karkonosze—in the south.

2.3. Air Quality Monitoring Sites

For each city, air quality monitoring data from urban background (UB), traffic (TR) and the nearest regional background (RB) sites were considered. Due to its character, rural background site in Diabla Góra (PL0005R) belonging to the European Monitoring and Evaluation Programme (EMEP) network was chosen as representative for continental background (**Figure 1**). The nearest town (>10,000 inhabitants) and road (>50 cars per day) are situated at 20 and 16 km from this site, respectively. The characteristics of investigated air quality monitoring sites are given in **Table 1**.

The data of PM_{10} mass concentrations for all sites and the concentrations of PM_{10} and its constituents for Diabla Góra site: TEs (As, Cd, Cr, Cu, Ni, Pb, Zn) and

Table 1. Characteristics of air quality monitoring sites.

City-site name	Code	Label	Method of PM_{10} analysis	Longitude (° ′ ″ E)	Latitude (° ′ ″ N)	Altitude (m a.s.l)
Warszawa-Targówek	PL0143A	W-UB	(2)	21°02′33″	52°17′36″	85
Warszawa-Komunikacyjna	PL0140A	W-TR	(2)	21°00′17″	52°13′00″	103
Belsk IGPAN	PL0014A	W-RB	(2)	20°47′30″	51°50′00″	176
Kraków-Nowa Huta	PL0039A	C-UB	(3)	20°03′07″	50°04′00″	195
Kraków-Krasińskiego	PL0012A	C-TR	(3)	19°55′20″	50°03′00″	175
Zabrze-Skłodowskiej	PL0242A	Z-UB	(1)	18°46′19″	50°19′12″	255
Chorzów	PL0235A	Z-TR	(2)	18°56′13″	50°15′00″	285
Złoty Potok	PL0243A	Z-RB (C-RB)	(3)	19°27′36″	50°42′36″	291
Jelenia Góra-Cieplice	PL0189A	J-UB	(3)	15°44′06″	50°51′36″	341
Czerniawa	PL0028A	J-RB	(3)	15°18′51″	50°54′36″	645
Diabla Góra	PL0005R	–	(1)	22°04′01″	54°08′00″	157

(1) A reference gravimetric sampler (Standard EN 12341); (2) Tapered Element Oscillating·microbalance (TEOM®); (3) beta ray attenuation method (MP101M).

SIA (SO_4^{2-}, $NH_3 + NH_4^+$, $HNO_3 + NO_3^-$) were extracted from the Voivodeship air pollution networks which belong to the Polish Voivodeship Environment Protection Inspectorates. Data was also extracted from EMEP (http://www.emep.int) and Air-Base, European Air quality (http://acm.eionet.europa.eu/databases/airbase) databases. The data series fulfilling the criterion of 75% completeness of the data sets within analysed periods were retained for the analysis. Depending on the monitoring site, concentrations of PM_{10} were determined by either (1) a reference gravimetric method (Standard EN 12341), (2) Tapered Element Oscillating·microbalance (TEOM®) or (3) beta ray attenuation method (MP101M) (**Table 1**). None of the two latter methods of PM_{10} measurements are considered to be the reference methods regarding the standard EN 12341, hence Poland, as the EU member state, is required to demonstrate that their methods yield results equivalent to the reference (EC 2010). However, up to now, the procedures for equivalency demonstration have not yet been accustomed. Therefore, herein, we have used the official data as they were reported to the EMEP and AirBase repositories.

The analysis of TEs content was performed using inductively coupled plasma-atomic emission spectroscopy (ICP-AES), while sulphate and sum of nitric acid and nitrate, as well as sum of ammonia and ammonium were determined by capillary electrophoresis (CE) and spectrophotometry, respectively.

2.4. Methodology

In order to identify the emission sources determining PM_{10} concentrations during the most severe episode noted in the history of air quality measurements in Poland (January 2006), the original method of PM source identification was applied in the study of Juda-Rezler *et al.* (2011). In the present paper, this method was further developed and extended. The influence of specific anthropogenic sources on PM_{10} levels during the episodes was investigated by the combination of different apportionment techniques applied to routinely available air quality and meteorological monitoring data, *i.e.* (1) analysis of the PM_{10} patterns at both urban and regional background monitoring sites, following the Lenschow approach (Lenschow *et al.* 2001); (2) analysis of the synoptic situation and variability of the local meteorological parameters; (3) analysis of the air mass back trajectories; and (4) receptor modelling using principal component analysis with multivariate linear regression analysis (PCA-MLRA).

The Hybrid Single-Particle Lagrangian Integrated Trajectory (HYSPLIT) model of the Air Resources Laboratory (ARL) of the National Oceanic and Atmospheric Administration (NOAA) was used, in order to calculate transport of air masses to each city (Draxler and Rolph 2013). The model was forced by the global reanalysis meteorological data produced by the National Weather Service's National Service for Environmental Prediction (NCEP) to compute advection and dispersion of air parcels. Three-day backward trajectories arriving at the studied locations at 50, 100, 200, 500 and 700m were generated. In Poland, winter is characterised by long-lasting ground-based inversions (also during daytime) and a weak convective mixing·mainly around the midday (GIOŚ 2013; Godłowska *et al.* 2015). Such stagnant conditions were present also during elaborated episodes. Therefore, it was supposed that near-surface air quality (measured at receptor site) could be affected by sources along trajectories only when backward trajectory starts within mixing layer or ground inversion layer. Hence, the heights of trajectories starting points were situated in the bottom part of PBL.

In this study, the air pollution concentrations were obtained from routine air quality monitoring networks, which do not provide the data uncertainties. Thus, the use of more advanced receptor models such as PMF or CMB was precluded. Therefore, receptor modelling technique based on PCA with Varimax rotation was applied to TEs and SIA concentrations in order to identify the main sources determining the aerosol composition at the rural background site, while the Lenschow approach was employed to ascertain PM·sources in urban areas. In order to facilitate comparison between the different seasons (spring, summer, autumn, winter), PCA was performed for each period separately, resulting in input data matrices of sizes between 10×70 and 10×90. The decision of selecting the number of principal components (PC) was based on the eigenvalue of the factors. Following the Kaiser criterion, only factors having an eigenvalue greater than 1.0 were retained for further analysis. The interpretation of the principal components was based on the variables with factor loadings with absolute values greater than 0.6. The contribution of each source to the PM burden was further quantitatively assessed by the means of MLRA procedure proposed by Thurston and Spengler (1985).

The statistical significance of the differences between mean PM_{10} concentrations in heating and non-heating season was assessed using Student's t test. All statistical analyses were performed with STATISTICA software.

3. Results and Discussion

3.1. PM_{10} Levels

3.1.1. Seasonal Variation

The seasonal variation of PM_{10} concentrations for the years 2009–2010 presents a common pattern for both urban and regional background sites (**Figure 2**), which is defined by emission cycles of primary PM and precursors of secondary PM. The seasonal trend is characterized by two annual maxima: January-March and October-December, which corresponds with heating season in Poland. Student's t test revealed that statistically significant differences between average urban

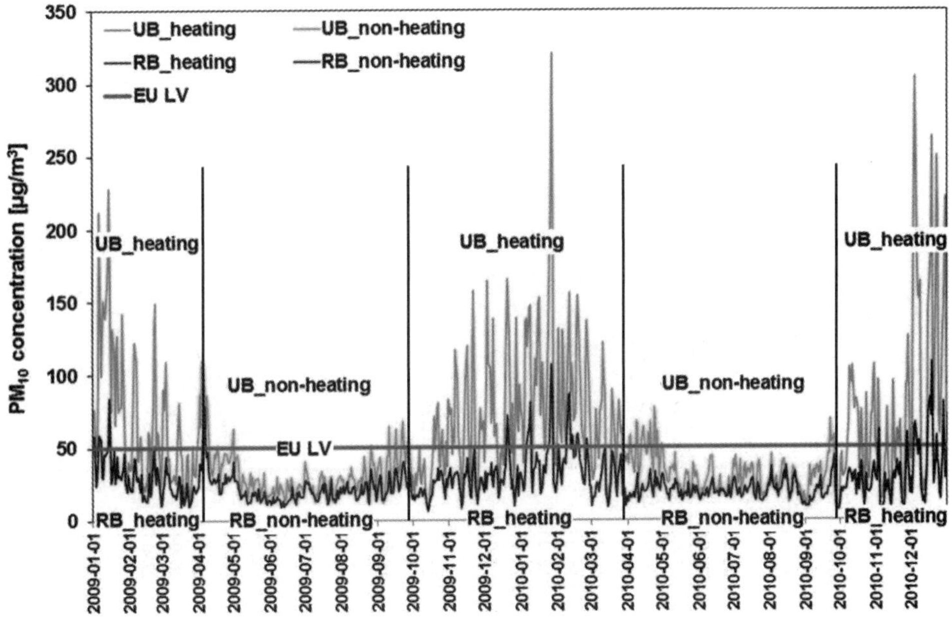

Figure 2. Patterns of daily PM_{10} concentrations ($\mu g \cdot m^{-3}$) during the 2009–2010 period averaged for urban (light lines) and regional (dark lines) background monitoring sites. Blue and red lines indicate heating (October-March) and non-heating (April-September) seasons, respectively, while grey solid line—EU daily limit value for PM_{10} ($50\mu g \cdot m^{-3}$).

background PM_{10} concentrations recorded during heating (65.0 and $83.6\mu g \cdot m^{-3}$ in 2009 and 2010, respectively) and non-heating (32.4 and $31.9\mu g \cdot m^{-3}$ in 2009 and 2010, respectively) season can be observed ($p < 0.00001$). The seasonal differences of the average PM_{10} concentrations for regional background sites for both analysed years are also statistically significant ($p < 0.00001$). In addition, the average PM_{10} concentrations for the urban background sites are significantly higher than for the regional ones ($p < 0.00001$). This could be explained as due to the larger PM emissions of urban origin.

3.1.2. Characteristics of Air Pollution Episodes

Daily PM_{10} concentrations observed in Januaries 2009 [**Figure 3(a)**] and 2010 [**Figure 3(b)**] at both urban and regional background air quality monitoring sites indicate substantially increased pollutant concentrations during all investigated PM_{10} events being·more pronounced in UB sites than in RB ones. Maximum PM_{10} levels in episode periods were found to exceed the daily EU air quality limit

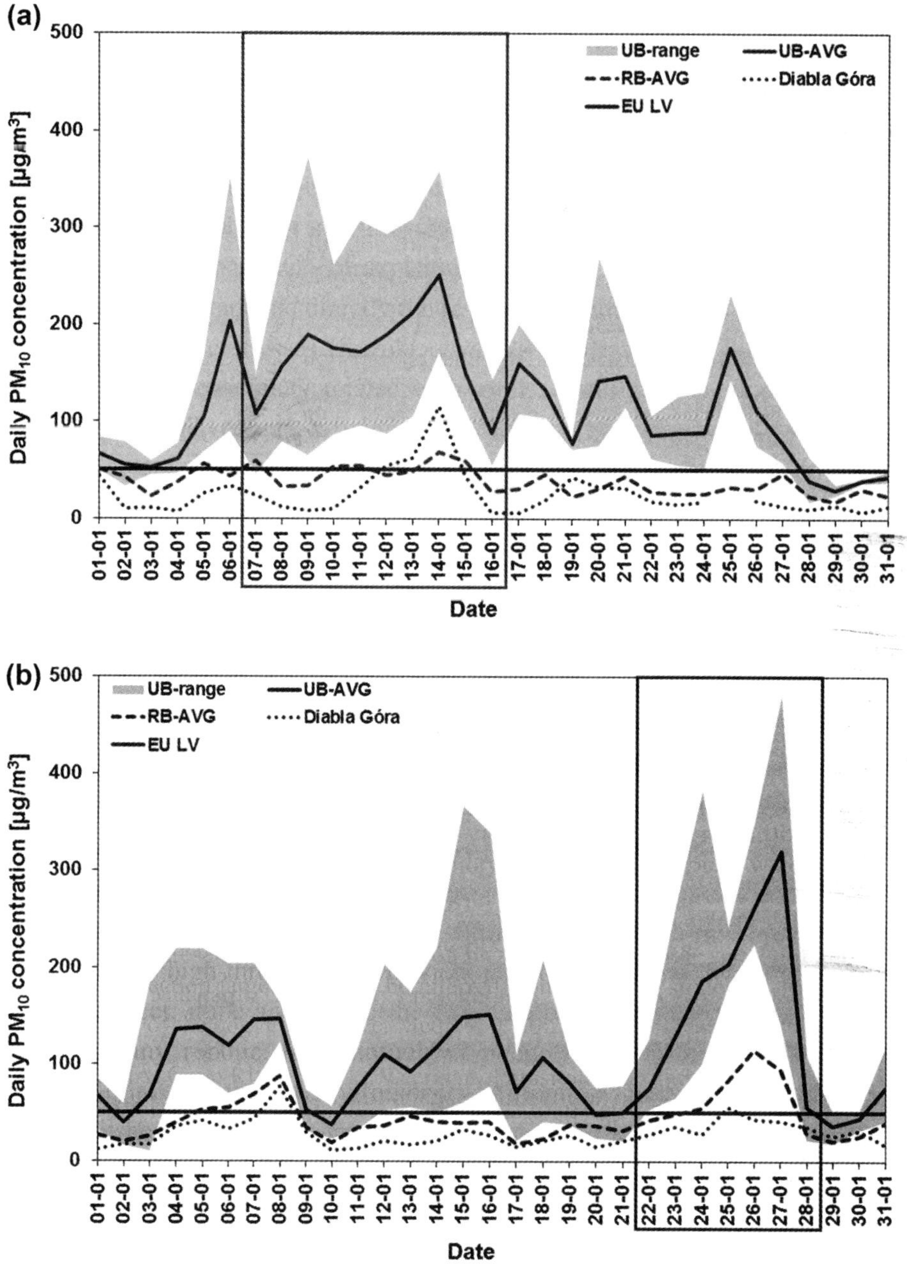

Figure 3. Patterns of daily PM_{10} concentrations (µg·m^{-3}) during (a) January 2009 (green) and (b) January 2010 (blue) averaged for urban (solid lines) and regional (dashed lines) background monitoring sites and at rural background EMEP site in Diabla Góra (black dashed lines). Colour light backgrounds indicate the range of PM_{10} concentrations at urban background sites, colour frames—the period of PM episodes, while red solid lines-EU daily limit value for PM_{10} (50µg·m^{-3}).

value for PM_{10} (50μg·m^{-3}) from·sevenfold in January 2009 (372μg·m^{-3}) up to tenfold in January 2010 (481μg·m^{-3}).

Episode 1 of 7–16th January 2009 occurred only in southern cities and can be regarded as the regional one. In northern and central Poland, the daily PM_{10} limit value was in general not exceeded. In Warsaw, the maximum daily PM_{10} concentration (160μg·m^{-3}) was recorded in the same date as for the rest of cities (14th January); however, the exceedance of the limit value persisted shorter than 3 days and therefore PM_{10} pollution recorded in this city cannot be classified as an episode according to the definition set in the present study. In Zabrze and Jelenia Góra, the episode was characterized by two peaks of PM_{10} concentrations with the first peak being slightly higher [**Figure 3(a)**]. The maximum PM_{10} levels, up to 372 and 181μg·m^{-3} in Jelenia Góra and Zabrze, respectively, were observed on 10th and 14th January. In Cracow, three maxima of PM_{10} levels on 7th, 9th and 14th January were recorded with the third peak being the highest (225μg·m^{-3}). During the episode, the patterns of daily PM_{10} concentrations at RB sites were analogous to those registered at UB sites with much lower values up to 25 and 90μg·m^{-3} at J-RB and Z-RB sites, respectively. At the rural background site in Diabla Góra, the maximum PM_{10} level was recorded on 14th January (115μg·m^{-3}). It was the highest PM_{10} concentration recorded at this site during the period of 2005-2012 (Reizer 2013).

Episode 2 of 22–28th January 2010 was preceded by two episodes with lower PM concentrations: the 2–10th and the 11–20th January [**Figure 3(b)**]. During episode period, high PM_{10} levels were observed at monitoring sites across the whole country. In three out of four cities, the episode was characterized by only one peak of high PM_{10} concentrations observed on 27th January. At rural background site in Diabla Góra, one peak concentration (56μg·m^{-3}) was also recorded; however, it occurred 2 days before the maximum PM_{10} levels registered at the rest of the sites. Two peaks of PM_{10} levels were observed on 24th and 27th January with the highest value of 480μg·m^{-3} in Jelenia Góra. During the episode, the patterns of daily PM_{10} concentrations at RB sites were similar to those recorded at UB ones with the maximum values up to 70μg·m^{-3} at J-RB site and around 140μg·m^{-3} at the rest of RB sites.

The analysis of daily PM_{10} concentration patterns at UB and RB sites indicates that during the episodes, urban sources were significantly involved in the build-up of pollution events in all cities. However, the regional background was also quite high with the ratios of RB/UB increasing from the southern (densely industrialized) to the central part of Poland (**Table 2**). The highest contribution of regional pollution was found in both Warsaw and Zabrze, amounting to more than a half of the urban episodes (RB/UB = 0.57 and 0.54–0.55, respectively). In contrast, for the southern city of Jelenia Góra, the regional background pollution was the lowest (RB/UB = 0.06–0.15), indicating that during all the episodes air quality in this city was strongly determined by the local "hot-spot" emission sources. At the same time, comparably high TR/UB ratios were obtained for Cracow during episodes 1 (1.26) and 2 (1.21), indicating typical for urban areas traffic component to the overall urban pollution.

3.2. Synoptic Situation and Local Meteorological Conditions

The patterns of the measured PM_{10} concentrations during the episodes can be explained by the variation of the local meteorological parameters (**Figure 4**). During episode 1, southern part of Poland was under the influence of very stable high-pressure system with its centre over the Czech Republic in the early stage of the episode and over Hungary in the rest of the episode period, while the rest of the country was under the influence of other high-pressure systems. The maximum values of air pressure were observed on 11th January reaching 1030–1035 hPa. At

Table 2. Average (avg) PM_{10} concentrations measured during analysed PM episodes at urban (UB) or regional (RB) background and traffic (TR) monitoring sites.

Episode	City	UB avg (μg·m^{-3})	TR avg (μg·m^{-3})	RB avg (μg·m^{-3})	RB/UB	TR/UB
Episode 1	Cracow	132.1	165.9	64.1	0.49	1.26
	Zabrze	115.9	95.3	64.1	0.55	0.82
	Jelenia Góra	259.1	–	16.3	0.06	–
Episode 2	Warsaw	134.6	95.7	76.9	0.57	0.71
	Cracow	162.7	197.0	75.9	0.47	1.21
	Zabrze	141.6	117.2	75.9	0.54	0.83
	Jelenia Góra	266.7	–	39.6	0.15	–

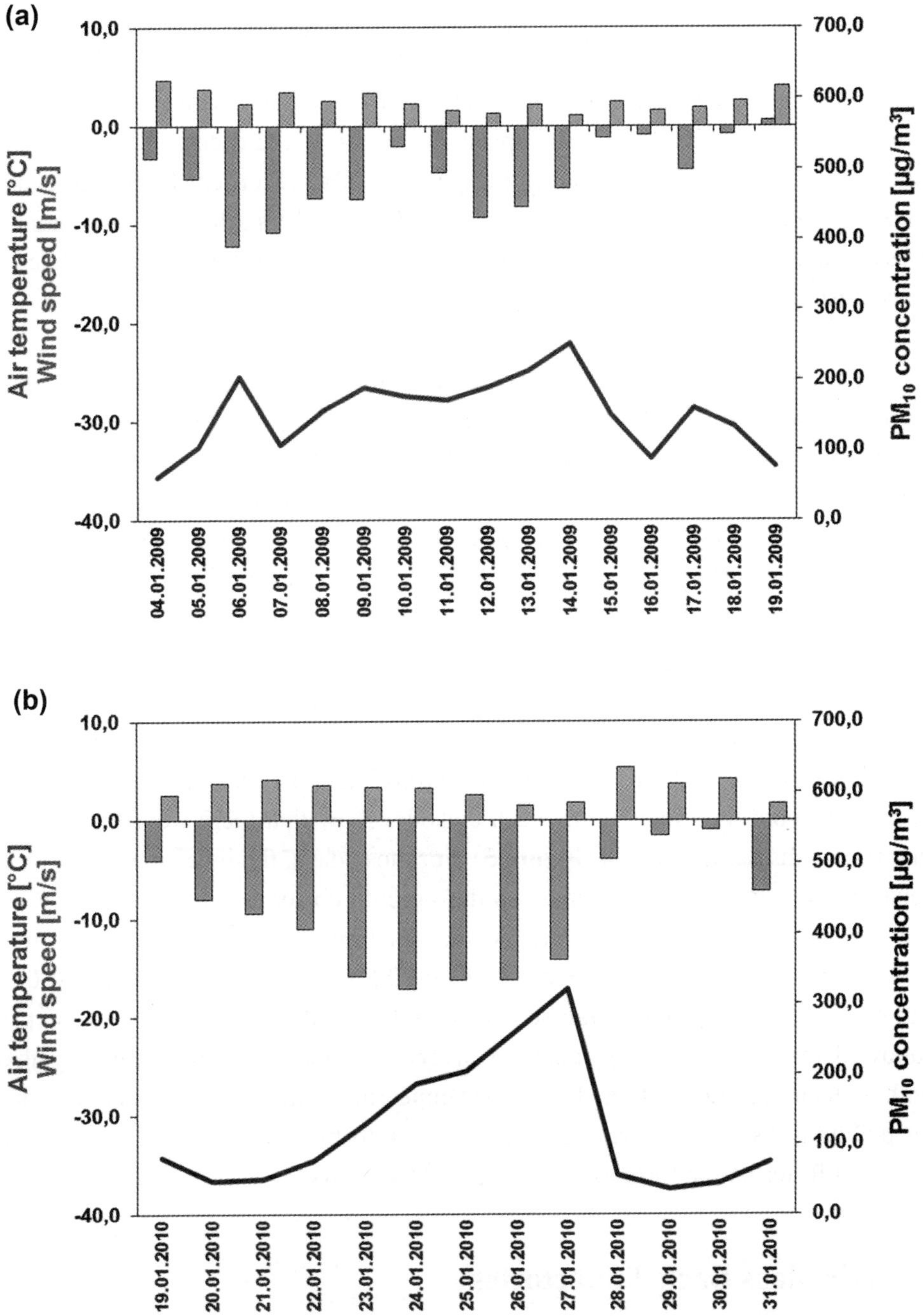

Figure 4. Patterns of daily PM_{10} concentrations ($\mu g \cdot m^{-3}$) (colour solid lines) during the period of a 4–19th January 2009 and b 19–31th January 2010 averaged for urban background monitoring sites in comparison with daily mean of meteorological parameters: air temperature (°C) (blue bars) and wind speed ($m \cdot s^{-1}$) (beige bars).

the beginning of the episode (6–9th January), average daily temperature varied from −10°C to −15°C, falling down below −20°C during night. Second, a smaller decrease of air temperature was noted on 12–14th January with minimum daily values close to −10°C. Low wind speeds up to $5 m \cdot s^{-1}$, and below $2 m \cdot s^{-1}$ in the days with the highest PM_{10} concentrations were also recorded. During this episode, the peak PM_{10} concentrations occurred 2 and 3 days after the maximum air pressure and minimum air temperature in Zabrze and in both Cracow and Jelenia Góra, respectively.

During episode 2, Poland was under the influence of the Siberian High ridge with maximum value of 1040 hPa. Low air temperatures were observed with daily minimum below −17°C in the southern cities (on 24th January) and −19°C in the central city of Warsaw (on 25th January). Weak wind up to $5 m \cdot s^{-1}$ was also recorded, whereas in Zabrze, the wind speed decreased below $1 m \cdot s^{-1}$ during the days with the highest PM_{10} concentrations. Episode 2 was characterized by the occurrence of the peak PM_{10} concentrations up to 3 days after the maximum air pressure and minimum air temperature.

The vertical soundings (skewT-logP diagrams), from the atmospheric sounding taken at Legionowo station in central Poland at 00Z on days with maximum PM_{10} concentrations during each episode, show that there were strong thermal inversions below 900hPa (**Figure 5**). For those days, the inversion layers were found to be approximately 700m deep during both episodes. In the early stage of the episodes, the **Figure 5** Vertical soundings from Legionowo station (central Poland) measured on a 14th January 2009 at 00Z and b 27th January 2010 at 00Z (http://weather.uwyo.edu/upperair/sounding.html) subsidence inversions were noted, followed then by the strong radiation inversions. The inversion layers together with high air pressure, extremely low air temperature and low wind speed impeded air pollution dispersion during episodes leading to the accumulation of air pollutants and fostering the formation of severe PM_{10} events.

3.3. Air Mass Back Trajectories

Transport of air masses to pairs of UB and RB sites of each city was examined by means of atmospheric 3-days back trajectories. Due to the fact that the air mass trajectories calculated for each day of the episode demonstrate similar

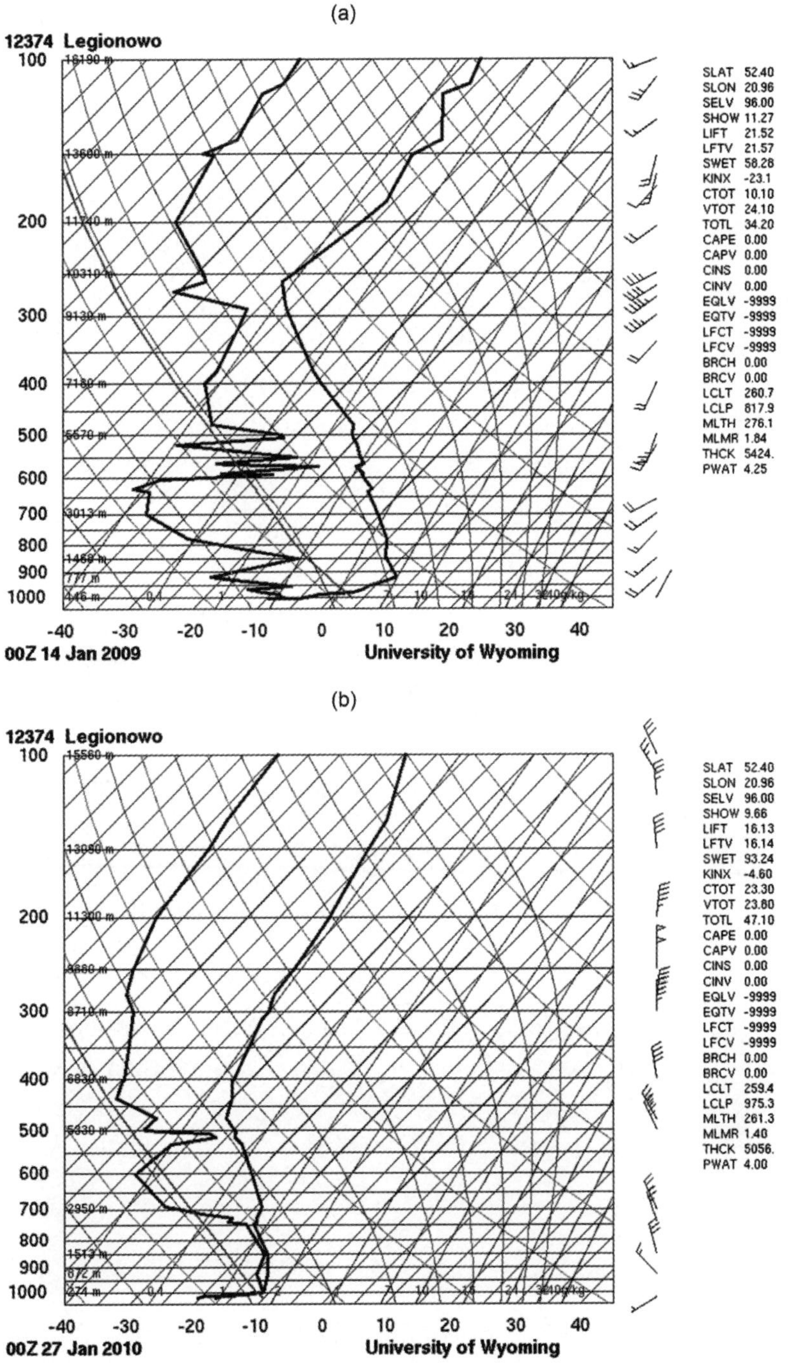

Figure 5. Vertical soundings from Legionowo station (central Poland) measured on (a) 14th January 2009 at 00Z and (b) 27th January 2010 at 00Z (http://weather.uwyo.edu/upperair/sounding.html).

pattern and that the trajectories determined at all altitudes were almost identical for each city, only back trajectories calculated at 50 m of altitude are presented in **Figure 6**.

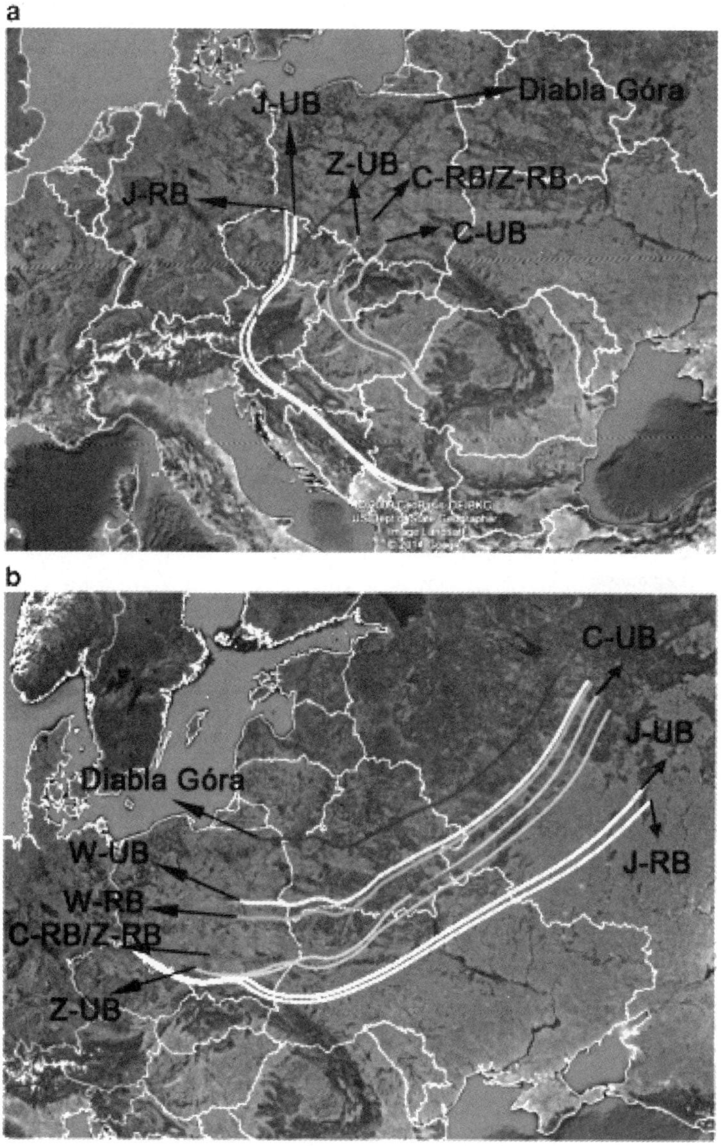

Figure 6. Three-day air mass back trajectories at the altitude of 50 m a.g.l. at urban (light colours) and regional (shaded colours) background monitoring sites of the four Polish cities: Warsaw (green), Cracow (blue), Zabrze (purple), Jelenia Góra (white) and at rural background EMEP site in Diabla Góra (red), calculated by HYSPLIT model at 00 UTC on a 14th January 2009 and b 25th January 2010.

Analysis of air mass back trajectories shows that during the analysed episodes, all pairs of UB and RB sites were under the influence of the same air masses moving along the high-pressure system edges. During episode 1, the air masses arriving to Jelenia Góra and Diabla Góra were coming from serbia and passing through Bosnia and Herzegovina, Croatia, Slovenia and Czech Republic and then from·southern to northern Poland in case of Diabla Góra. On the contrary, the air masses arriving to Zabrze and Cracow were transported from the western part of Romania through Hungary and Slovakia. During episode 2, the air masses, coming·mainly from the European part of Russia, were transported through Ukraine, Belarus and further over Poland.

The analysis of air mass back trajectories indicates that during episode 2, both UB and RB sites were under the influence of the same air masses of north-eastern origin that brought polar air masses from a Siberian high ridges coming mainly from central parts of European Russia and transported through Belarus and Ukraine. Thus, all investigated Polish cities appear to be under the influence of the same long-range transported air pollutants from remote sources in Eastern European countries. On the contrary, during episode 1, all types of sites were under the influence of a high-pressure system bringing air masses from southern European countries to both types of sites. Furthermore, regional transport of air pollutants determining high PM_{10} concentrations during episodes is associated with the transport of air masses over Poland. The air masses arriving to the so-called "Green Lungs of Poland", delimited in the north-eastern, highly forested and most sparsely populated region of the country, where the rural background site in Diabla Góra is located, brings there pollutants from both remote and regional industrial sources.

The results of air mass back trajectories analysis were confirmed by the air pollution roses calculated for Diabla Góra site (not shown here) which indicate that the highest concentrations of TEs occur at the southern or eastern direction of wind.

3.4. PCA-MLRA Analysis

PCA with Varimax rotation applied to TEs and SIA concentrations measured at Diabla Góra site allows to extract three principal components (PC) explaining up to 88% and 81% of the total variance for the winters (January-March) of 2009 and 2010, respectively (**Table 3**). A major group of TEs that includes As, Cd, Pb and Zn was collocated in the PC1 which explain around 40% of the total variance for both 2009 and 2010. The second set (PC2 with explained variance between 25 and 30%) is characterized by the group of SIA components, however, lacking sum of ammonia and ammonium in winter 2009 and sulphate in winter 2010. The third group (PC3 with almost 16% of explained variance) is represented mostly by Cr and Cu.

As SIA are commonly related with an "external" source of pollution (e.g. Querol *et al.* 2007; Aarnio *et al.* 2008), PC2 was identified as remote sources

Table 3. Factor loadings determined in the PCA analysis with Varimax rotation of trace elements and SIA measured at rural background EMEP site in Diabla Góra during January-March 2009 and January-March 2010.

PM_{10} component	January-March 2009			January-March 2010		
	PC1	PC2	PC3	PC1	PC2	PC3
As (ng·m^{-3})	0.92	0.24	0.21	0.82	0.37	0.05
Cd (ng·m^{-3})	0.93	0.17	0.12	0.86	0.09	0.37
Cr (ng·m^{-3})	0.30	0.05	0.92	0.44	0.27	0.62
Cu (ng·m^{-3})	0.04	0.72	0.63	0.30	0.30	0.75
Ni (ng·m^{-3})	0.65	0.63	0.13	0.53	0.68	0.09
Pb (ng·m^{-3})	0.84	0.41	0.26	0.91	0.28	0.21
Zn (ng·m^{-3})	0.71	0.53	0.32	0.91	0.20	0.27
SO_4^{2-} (µg(S) m^{-3})	0.53	0.73	0.13	0.49	0.32	0.52
$NH_3 + NH_4^+$ (µg(N) m^{-3})	0.35	0.89	0.01	0.31	0.85	0.17
$HNO_3 + NO_3^-$ (µg(N) m^{-3})	0.63	0.44	0.30	0.11	0.91	0.25
%Variance	42.4	29.6	15.8	39.7	25.5	15.7
PM source	LPS	Remote sources	Residential/traffic	LPS	Remote sources	Residential/traffic

Factor loadings >0.6 are represented in italics.

involved in building of background pollution in Poland. PC1 representing by majority of TEs was identified as originating from industrial large point sources (LPS), mainly metallurgy and power generation, concentrated in southern part of the country, as well as close to its southern and eastern boundary, while PC3 with Cr and Cu was attributed to local residential/traffic sources.

TEs are emitted to the atmosphere with fly ash particles created during combustion of solid fuels. Volatile TEs (such as Se and Hg) are completely vaporized at combustion temperatures, while other including As, Pb, Cu, Cd and Zn are partially vaporized at flame temperatures and subsequently condensed on the surfaces of fly ash (Helble 2000; Nelson 2007). Studies of the properties of TEs present in the coal as well as of processes that they undergo during and after coal combustion (e.g. Helble 2000; Senior *et al.* 2000; Riley *et al.* 2012) show presence of As, Cd, Co, Pb, Sb and Se in the finest emitted particles—able to be transported over 100s km—independently of combustion utility and the type of burned coal. This is not, however, a case of Cr and Ni, as their portioning depend on the coal origin and type. Besides, As, Cd, Pb and Zn are typically bonded to the mineral part of coal, while Cr, Cu and Ni are present in both mineral and organic fractions. TEs present in organic parts are more volatile than those bonded with minerals (as As in pyrite), even when coal is burned at low temperatures, typical for domestic stoves. From all these reasons, As, Cd, Pb and Zn can be regarded as markers of regionally transported pollution from LPS, while Cr can be attributed to coal combustion in small-scale installations, as domestic stoves/local boiler houses. Finally, Cu is commonly used as a marker of brake abrasion from road transport (e.g. Schauer *et al.* 2006; Querol *et al.* 2007). It can also come from industrial emissions, e.g. Cu, Zn or pyrite smelters (e.g. Bruinen de Bruin *et al.* 2006); however, the factor loadings of Cu for PC1, characterized by other industrial TEs, were very low (0.04 and 0.30 for 2009 and 2010, respectively). Therefore, both Cr and Cu were considered as originating most likely from local emission sources.

The contribution of each source type to the PM burden was further quantitatively assessed by means of MLRA. **Table 4** evidences the good correspondence between the modelled by PCA-MLRA and the measured PM_{10} concentrations, with R^2 values >0.78 for winter and spring 2010 and R^2 values > 0.80 for the rest of seasons. Only for autumn 2009, R^2 value is very low ($R^2 = 0.20$). **Figure 7** presents contributions of identified PM_{10} emission sources for each day of episodes as

Table 4. Correlation between measured PM_{10} concentration (x) and PCA-MLRA results (y) in different seasons of the 2009 and 2010 years.

Slope		R_2	Intercept
January-March2009	y = 0.87x	0.90	5.45
April-June 2009	y = 0.88x	0.89	3.19
July-September 2009	y = 0.22x	0.20	10.11
October-December 2009	y = 0.95x	0.95	5.20
January-March2010	y = 0.82x	0.79	5.55
April-June 2010	y = 0.78x	0.78	5.07
July-September 2010	y = 0.82x	0.82	3.30
October-December 2010	y = 0.85x	0.86	10.09

well as for different seasons and for the whole years 2009 and 2010. During both episodes, LPS had the biggest share in PM_{10} levels during the days with the highest PM concentrations, up to 81 and 73% in episodes 1 and 2, respectively. They were 2–4 times higher than the whole year averages (16%–40%). In the case of episode 1, the contribution of LPS was around two times higher than the averages for all seasons (35%–46%), while for episode 2, their share was almost three and six times higher than the averages for winter and spring (25%–27%) and for summer (12%), respectively. The contribution of the remote sources ranged during episode 1 from 7% to almost 25%. During episode 2, the share of remote sources (10%–21%) was 2–4 times lower in comparison to winter average (37%) and to the rest of seasons and the whole year averages (50%–68%). The share of the residential/traffic sources in PM_{10} (2%–43%) at the beginning of the episodes was almost at the same level (episode 2) and 2–3 times lower (episode 1) than the averages for all seasons and the whole years. During the last days of the episodes, five times lower (episode 2) or even no contributions of local sources were noted (episode 1). The share of sources unidentified by PCA-MLRA was relatively small for episode 2 accounting for 6%–12%, while in the early stages of episode 1, the unidentified part was as high as 80%. Such large share of unidentified sources in episode 1 is surprising; however, it can be expected that it could possibly be related to the various sources which markers are not measured routinely at the Diabla Góra site. According to Chow *et al.* (2015), the major PM components measured

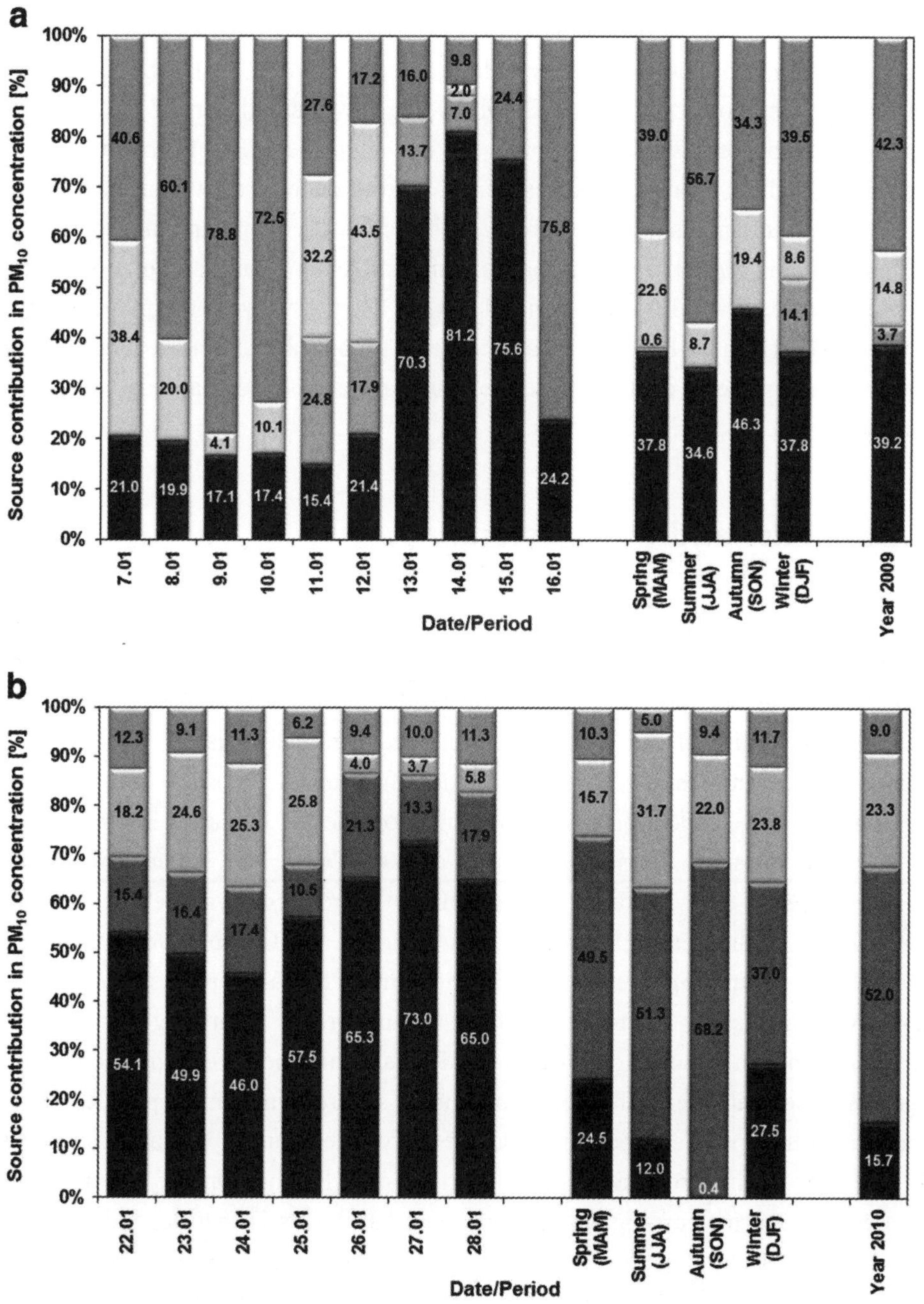

Figure 7. Contributions of PM_{10} emission sources: LPS (dark colours), remote (normal colours), local pollution (light colours), unknown (grey), determined in the PCA-MLRA analysis for each day of PM episodes, springs (March-May), summers (June-August), autumns (September-November), winters (December-February) and for the whole years: a 2009 and b 2010.

to explain gravimetric mass include (1) anions and cations (2) elements, including·metals [up to 51 elements from·sodium (Na) to uranium (U)], and (3) OC and EC and their carbon fractions. In Diabla Góra, only a small part of these components are measured; there is a lack of, e.g. chloride (Cl^-) from anions, water-soluble sodium (Na^+) and potassium (K^+), from cations, as well as carbon species. From 51 elements, only 7 were available. Measurement campaign performed at this station in the 2010 shown that in winter, SIA (44%) and carbon compounds (organic matter, 28%; EC, 16%) and Na+ with Cl^- (11%) were the main constituents of $PM_{2.5}$ (Rogula-Kozłowska *et al.* 2014). It could be therefore supposed that contaminated mineral dust, resuspended dust from unpaved roads as well as sea salt are among unidentified sources.

4. Summary and Conclusions

The combination of two source apportionment techniques: principal component analysis with multivariate linear regression analysis (PCA-MLRA) and the so-called Lenschow approach, complemented by the analysis of the synoptic situation and variability of the local meteorological parameters as well as of the air mass back trajectories was applied for identification of emission sources determining high PM_{10} concentrations during two episodes of Januaries 2009 and 2010 in Polish urban areas. During the period of 2009-2010, both urban and regional background sites show a statistically significant increase of average PM_{10} concentrations during the heating season (January-March and October-December), presumably due to anthropogenic emissions of primary PM and precursors of secondary PM. The results demonstrate that a number of anthropogenic emission processes and unfavourable synoptic- and local-scale meteoro-logical conditions play an important role in the determination of severe wintertime episodes. It was shown that the episode 2, occurring across the whole country, was associated with the presence of the strong Siberian High ridge over Poland. One of the main findings concerning meteorological influences on PM levels is that depending on the city, the peak PM concentrations occur 2–3 days after the maximum atmospheric pressure and minimum air temperature observed, which are related to the strongest expression of the high-pressure system impact. Three sources of PM_{10} events were identified: remote sources, LPS and residential/traffic sources. At the rural background site, representative for continental background pollution, PM_{10} levels were dominated by LPS situated south-east from it (up to 70%–80% on the days with the highest air pollution), followed by remote and local sources. The Lenschow approach indicate that in central Poland, ambient air pollution was predominantly influenced by LPS, while in the southern cities local residential/ traffic sources dominate during the episodes. Although limited number of elements were available in PM, results from analyses suggest that different trace elements characterize coal combustion in large industrial (As, Cd, Pb and Zn) and in small-scale domestic (Cr) furnaces. This study shows that if only routine measurements (as in EMEP network) are available, identification of main PM·sources and apportion PM to those sources is possible by the proposed method; however, to provide efficient air quality management tool, more data and more advanced methods are to be used.

Acknowledgements

This work was supported by the Polish National Science Centre under OPUS funding scheme 2nd edition, Project no. UMO-2011/03/B/ST10/04624. We wish to acknowledge the inspiring discussions and suggestions made by Dr. Jean-Paul Oudinet from INSERM UMR 788-University Paris-Sud, France. The NOAA Air Resources Laboratory (ARL) is gratefully acknowledged for the provision of the HYSPLIT transport and dispersion model and READY website (http://www.arl.noaa.gov/ready.php) used in this publication. The authors would like to thank anonymous reviewers for their constructive comments, which helped to improve the paper.

Source: Reizer M, Juda-Rezler K. Explaining the high PM_{10}, concentrations observed in Polish urban areas[J]. Air Quality Atmosphere & Health, 2015:1–15.

References

[1] Aarnio P, Martikainen J, Hussein T, Valkama I, Vehkamäki H, Sogacheva L, Härkönen J, Karppinen A, Koskentalo T, Kukkonen J, Kulmala M (2008) Analysis and evaluation of selected PM_{10} pollution episodes in the Helsinki Metropolitan Area in 2002. Atmos Environ 42: 3992–4005.

[2] Amodio M, Bruno P, Caselli M, de Gennaro G, Dambruoso PR, Daresta BE, Ielpo P, Gungolo F, Placentino CM, Paolillo V, Tutino M (2008) Chemical characterization of fine particulate matter during peak PM_{10} episodes in Apulia (South Italy). Atmos Res

90:313–325.

[3] AQP for Lesser Poland Voivodeship (2013) Lesser Poland Voivodeship Parliament Resolution XLII/662/13 on the Air Quality Plan for the Lesser Poland Voivodeship. Cracow [In Polish].

[4] AQP for Lower Silesian Voivodeship (2010) Lower Silesian Voivodeship Parliament Resolution III/44/10 on the Air Quality Plan for the areas of the Lower Silesian Voivodeship in which air quality standards have been exceeded. Part B—The city of Jelenia Góra. Wrocław [In Polish].

[5] AQP for Silesian Voivodeship (2010) Silesian Voivodeship Parliament Resolution III/52/15/2010 on the Air Quality Plan for the areas of the Silesian Voivodeship in which air quality standards have been exceeded. Katowice [In Polish].

[6] AQP for Warsaw (2013) Masovian Voivodeship Parliament Resolution 186/13 on the Air Quality Plan for Warsaw area, in which PM_{10} and NOx air quality standards have been exceeded. Warsaw [In Polish].

[7] Belis CA, Karagulian F, Larsen BR, Hopke PK (2013) Critical review and meta-analysis of ambient particulate matter source apportionment using receptor models in Europe. Atmos Environ 69:94–108.

[8] Bessagnet GB, Hodzic A, Blanchard O, Lattuati M, Le Bihan O, Marfaing H, Rouïl L (2005) Origin of particulate matter pollution episodes in wintertime over the Paris Basin. Atmos Environ 39: 6159–6174.

[9] Bruinen de Bruin Y, Koistinen K, Yli-Tuomi T, Kephalopoulos S, Jantunen M (2006) A review of source apportionment techniques and marker substances available for identification of personal exposure, indoor and outdoor sources of chemicals. Report EUR 22349 EN. European Comission Joint Research Centre, Ispra.

[10] Brunekreef B, Forsberg B (2005) Epidemiological evidence of effects of coarse airborne particles on health. Eur Respir J 26:309–318.

[11] Chow JC, Lowenthal DH, Chen L-WA, Wang X, Watson JG (2015) Mass reconstruction methods for PM2.5: a review. Air Qual Atmos Health 8:243–263.

[12] Chu S-H (2004) PM2.5 episodes as observed in the speciation trends network. Atmos Environ 38:5237–5246.

[13] Draxler RR, Rolph GD (2013) HYSPLIT (HYbrid Single-Particle Lagrangian Integrated Trajectory) Model access via NOAA ARL READY Website. www.arl.noaa.gov/HYSPLIT.php.

[14] EC (2010) Guide to the demonstration of equivalence of ambient air monitoring·methods. Report by an EC Working Group on Guidance for the Demonstration of Equivalence.

[15] EEA (2014) Air Quality in Europe-2014 report, European Environment Agency. Publications Office of the European Union, Luxembourg.

[16] EUROSTAT (2013) Eurostat regional yearbook 2013. Publications Office of the European Union, Luxembourg.

[17] Fragkou E, Douros I, Moussiopoulos N, Belis CA (2012) Current trends in the use of models for source apportionment of air pollutants in Europe. Int J Environ Pollut 50:363–375.

[18] GIOŚ (2013) Analysis of selected episodes of high PM_{10} concentrations, based on measurements, meteorological data and trajectory analysis. Bureau of Studies and Proecological Measurements EKOMETRIA, Gdańsk [In Polish].

[19] Godłowska J, Hajto MJ, Tomaszewska AM (2015) Spatial analysis of air masses backward trajectories in order to identify distant sources of fi ne particulate matter emission. Arch Environ Prot 41:28–35.

[20] Helble JJ (2000) A model for the air emissions of trace metallic elements from coal combustors equipped with electrostatic precipitators. Fuel Process Technol 63:125–147.

[21] Im U, Markakis K, Unal A, Kindap T, Poupkou A, Incecik S, Yenigun O, Melas D, Theodosi C, Mihalopoulos N (2010) Study of a winter PM episode in Istanbul using the high resolution WRF/CMAQ modeling system. Atmos Environ 44:3085–3094.

[22] Johnson TM, Guttikunda S, Wells GJ, Artaxo P, Bond TC, Russell AG, Watson JG, West J (2011) Tools for improving air quality management: A review of top-down source apportionment techniques and their application in developing countries. Report 339/11, Energy Sector Management Assistance Program. World Bank Group, Washington.

[23] Juda-Rezler K (2010) New challenges in air quality and climate modelling. Arch Environ Prot 36:3–28.

[24] Juda-Rezler K, Reizer M, Oudinet JP (2011) Determination and analysis of PM_{10} source apportionment during episodes of air pollution in Central Eastern European urban areas: the case of wintertime 2006. Atmos Environ 45:6557–6566.

[25] Karaca F, Anil I, Alagha O (2009) Long-range potential source contributions of episodic aerosol events to PM_{10} profile of a megacity. Atmos Environ 43:5713–5722.

[26] Kiesewetter G, Borken-Kleefeld J, Schöpp W, Heyes C, Thunis P, Bessagnet B, Terrenoire E, Fagerli H, Nyiri A, Amann M (2015) Modelling street level PM_{10} concentrations across Europe: source apportionment and possible futures. Atmos Chem Phys 15: 1539–1553.

[27] Kukkonen J, Pohjola M, Sokhi RS, Luhana L, Kitwiroon N, Fragkou L, Rantamaki M, Berge E, Ødegaard V, Slørdal LH, Denby B, Finardi S (2005) Analysis and evaluation of selected local-scale PM_{10} air pollution episodes in four European cities: Helsinki, London, Milan and Oslo. Atmos Environ 39:2759–2773.

[28] Lenschow P, Abraham HJ, Kutzner K, Lutz M, Preuss JD, Reichenbächer W (2001) Some ideas about the sources of PM_{10}. Atmos Environ 35:123–133.

[29] Lim·sS, Vos T, Flaxman AD, Danaei G *et al* (2012) A comparative risk assessment of burden of disease and injury attributable to 67 risk factors and risk factor clusters in 21 regions, 1990–2010: a systematic analysis for the Global Burden of Disease Study 2010. Lancet 380:2224–2260.

[30] Loomis D, Grosse Y, Lauby-Secretan B, El Ghissassi F, Bouvard V, Benbrahim-Tallaa L, Guha N, Baan R, Mattock H, Straif K (2013) The carcinogenicity of outdoor air pollution. Lancet Oncol 14: 1262–1263.

[31] Muir D, Longhurst JWS, Tubb A (2006) Characterisation and quantification of the sources of PM_{10} during air pollution episodes in the UK. Sci Total Environ 358:188–205.

[32] Nelson PF (2007) Trace metal emissions in fine particles from coal combustion. Energy Fuel 21:477–484.

[33] Niemi JV, Saarikoski S, Aurela M, Tervahattu H, Hillamo R, Westphal DL, Aarnio P, Koskentalo T, Makkonen U, Vehkamäki H, Kulmala M (2009) Long-range transport episodes of fine particles in southern Finland during 1999-2007. Atmos Environ 43:1255–1264.

[34] OECD (2012) Environmental Outlook to 2050, The Consequences of Inaction. Organisation for Economic Co-operation and Development Publishing, Paris.

[35] Pope CA III, Burnett RT, Thun MJ, Calle EE, Krewski D, Ito K, Thurston GD (2002) Lung cancer, cardiopulmonary mortality and long-term exposure to fine particulate air pollution. J Am Med Assoc 287: 1132–1141.

[36] Querol X, Alastuey A, Moreno T, Viana M, Castillo S, Pey J, Escudero M, Rodríguez S, Cristóbal A, González A, Jiménez S, Pallarés M, de la Rosa J, Artíñano B, Salvador P, García Dos Santos S, Fernández-Patier R, Cuevas E (2007) Atmospheric particulate matter in Spain: levels, composition and source origin. In: EMEP Particulate Matter Assessment Report, EMEP/CCC-Report 8/2007. Norwegian Institute for Air Research, Kjeller.

[37] Reizer M (2013) Methodology for identification of the causes of particulate matter episodes in Polish conditions. PhD dissertation. Publishing House of Warsaw University of Technology, Warsaw [In Polish with English summary].

[38] Riley KW, French DH, Farrell OP, Wood RA, Huggins FE (2012) Modes of occurrence of trace and minor elements in some Australian coals. Int J Coal Geol 94:214–224.

[39] Rogula-Kozłowska W, Klejnowski K, Rogula-Kopiec P, Ośródka L, Krajny E, Błaszczak B, Mathews B (2014) Spatial and seasonal variability of the mass concentration and chemical composition of PM2.5 in Poland. Air Qual Atmos Health 7:41–58.

[40] Rückerl R, Schneider A, Breitner S, Cyrys J, Peters A (2011) Health effects of particulate air pollution: a review of epidemiological evidence. Inhal Toxicol 23:555–592.

[41] Schauer JJ, Lough GC, Shafer MM, Christensen WF, Arndt MF, DeMinter JT, Park J-S (2006) Characterization of metals emitted from motor vehicles. Research Report 133. Health Effects Institute, Boston.

[42] Senior CL, Zeng T, Che J, Ames MR, Sarofim AF, Olmez I, Huggins FE, Shah N, Huffman GP, Kolker A, Mroczkowski S, Palmer C, Finkelman R (2000) Distribution of trace elements in selected pulverized coals as a function of particle size and density. Fuel Process Technol 63:215–241.

[43] Thurston GD, Spengler JD (1985) A quantitative assessment of source contributions to inhalable particulate matter pollution in metropolitan Boston. Atmos Environ 19:9–25.

[44] Viana M, Kuhlbusch TAJ, Querol X, Alastuey A, Harrison RM, Hopke PK, Winiwarter W, Vallius M, Szidat S, Prévôt ASH, Hueglin C, Bloemen H, Wåhlin P, Vecchi R, Miranda AI, Kasper-Giebl A, Maenhaut W, Hitzenberger R (2008) Source apportionment of particulate matter in Europe: a review of methods and results. J Aerosol Sci 39:827–849.

Chapter 20

Coupling Socioeconomic and Lake Systems for Sustainability: A Conceptual Analysis Using Lake St. Clair Region as a Case Study

Georgia Mavrommati, Melissa M. Baustian, Erin A. Dreelin

Center for Water Sciences, Michigan State University, 301 Manly Miles Building, 1405 S. Harrison Rd, East Lansing, MI 48823, USA

Abstract: Applying sustainability at an operational level requires understanding the linkages between socioeconomic and natural systems. We identified linkages in a case study of the Lake St. Clair (LSC) region, part of the Laurentian Great Lakes system. Our research phases included: (1) investigating and revising existing coupled human and natural systems frameworks to develop a framework for this case study; (2) testing and refining the framework by hosting a 1-day stakeholder workshop and (3) creating a causal loop diagram (CLD) to illustrate the relationships among the systems' key components. With stakeholder assistance, we identified four interrelated pathways that include water use and discharge, land use, tourism and shipping that impact the ecological condition of LSC. The interrelationships between the pathways of water use and tourism are further illustrated by a CLD with several feedback loops. We suggest that this holistic approach can be applied to other case studies and inspire the development of dynamic models capable of informing decision making for sustainability.

Keywords: Coupled Human and Natural Systems, Lake St. Clair, Clinton River Watershed, Ecosystem Services, Stakeholders, Sustainability

1. Introduction

The concept of sustainability includes various assumptions about the relationship between human-made and natural capital as well as the needs and preferences of current and future generations. Divergence in these assumptions has led to two opposing schools of thought: strong and weak sustainability (Neumayer 2010). The main difference between the two paradigms is that strong sustainability rejects the major assumption of weak sustainability in that natural capital can be substituted by human-made capital (Daly 1996). With regard to the preferences of future generations, strong sustainability assumes preferences are unknown and unknowable, whereas weak sustainability anticipates that they will be similar to those of the present generation (Bithas 2008). Another fundamental difference between the two schools of thought is the way the benefits of future generations are taken into account. Low and zero discount rates are proposed by strong sustainability approaches, while positive discount rates are used by weak sustainability approaches, thereby reflecting different ethical considerations (Sumaila 2004; Howarth 2009). Implicitly, weak sustainability views human and natural systems as independent, while strong sustainability views them as interdependent systems (Mavrommati and Richardson 2012).

Recent evidence suggests that further deterioration of natural systems may provoke tremendous impacts to humanity (Rockstrom *et al.* 2009; Folke *et al.* 2011), thus the sustainability of human and natural systems cannot be examined separately. Under the current model of economic development and increasing human population size, the dependence of socioeconomic activities on natural resources is becoming so high that we cannot ignore the 'limits to growth' stemming from natural systems (Burger *et al.* 2012). Therefore, applying sustainability at an operational level requires deep understanding of the structure, processes, and interactions of human and natural systems, the so-called coupled human and natural systems (CHANS) approach (Liu *et al.* 2007).

A CHANS approach can enhance our understanding regarding the dependence of socioeconomic systems on natural systems and enable decision makers to design more effective policies for managing ecosystem services (ES) (Pickett *et al.* 2005; Liu *et al.* 2007; Alberti *et al.* 2011). CHANS frameworks constitute the basis for building integrated models that specify factors and processes for applying the principles of sustainability (Pickett *et al.* 2005; Carpenter *et al.* 2009). Conceptual frameworks are essential starting points to illustrate components, pathways, and hypothesized responses among the subsystems in a CHANS framework (Alberti *et al.* 2011). Delineating the couplings taking place among human and natural systems is a complicated process that requires both extensive scientific and applied knowledge from various disciplines (Ostrom 2009). Ostrom (2009) and Ostrom and Cox (2010) suggest a general framework for studying sustainability of socio-ecological systems and identifying variables that might affect resource users to "avert the tragedy of the commons" by designing and implementing costly governance systems. Interacting with diverse stakeholders is necessary to incorporate expert knowledge, especially at various stages of the research, planning, and building of frameworks (Maxwell 1996; Schmolke *et al.* 2010; Cumming 2011). The use of conceptual frameworks also assists and encourages discussion across scientific disciplines and stakeholder sectors and is therefore a critical prerequisite for building complex models (Heemskerk *et al.* 2003; Alberti *et al.* 2011).

The most widely known CHANS framework is the Millennium Ecosystem Assessment (MEA) framework (2003) that includes four subsystems: ES, direct and indirect drivers of change, and human well-being (HWB). The innovative element of the MEA framework is the linkage between HWB and ES. Based on this framework, a revised CHANS framework has been proposed by emphasizing the fundamental role that changes in HWB have in the formulation of environmental policies (Stevenson 2011). The revised MEA framework highlights the importance of environmental policy for regulating the impacts of human activities on ES and sustaining HWB. Our goal was to build a framework that identifies the key parameters and pathways affecting the function of both natural and human systems and incorporates the knowledge of stakeholders. We began by further refining the MEA and revised MEA frameworks (Stevenson 2011) by explicitly including ecosystem condition responses to the stressors produced by human activities and how those responses may affect ES. We applied our conceptual framework to a real-world, dynamic case study in order to provide an example of the necessary steps for developing a CHANS approach. This approach is based on the premise that the maintenance of minimally-disturbed ecological condition of western Lake St. Clair (LSC) is necessary for ensuring the satisfaction of human needs now and in the future (Holden and Linnerud 2007). This is the first time a framework that couples the socioeconomic and natural systems has been developed for the LSC region (Clinton River watershed and western shore of LSC). Previous work in this watershed reported findings on the conditions of human or natural systems without providing insights for the couplings and feedbacks between them (Bricker *et al.* 1976; Selegean *et al.* 2001; Francis and Haas 2006; van Hees *et al.* 2010).

Developing a framework as a tool to understand the interactions between humans and the environment is needed in this area for several reasons. First, even with serious threats (*i.e.*, beach closures) to the ES in the region, there are still local initiatives that promote outdoor recreational activities (*i.e.*, Macomb County Blue Economy Initiative, LSC Tourism Initiative) and encourage their increased use. With respect to sustainability, there is a need to understand how ES that support and are affected by recreational activities can be maintained in the long-term. Second, the Clinton River, which flows into and influences the western shore of LSC, is one of the most ecologically impaired rivers in the state of Michigan based on fish and macro-invertebrate communities (Riseng *et al.* 2010) implying the need for better understanding of pollutant sources and impacts. Third, LSC is surrounded by counties with high human population densities that have an increased demand on the ES provided by the LSC region and to a great extent is linked historically with the socioeconomic evolution of the city of Detroit (SEMCOG 2002; Baustian *et al.* unpublished).This multidisciplinary approach is not intended for only this case study, but can be applied in other systems to assist scientists and stakeholders in building dynamic models that incorporate interactions between systems and ES and apply the principles of sustainability at an operational level.

2. Methods

2.1. Case Study

LSC is considered to be the "heart of the Great Lakes" because it provides an important connecting channel in the Laurentian Great Lakes system (**Figure 1**). It links Lake Huron to Lake Erie via the St. Clair River to the north and Detroit River to the south; these waters are known as the Huron-Erie corridor. The prevailing winds and the St. Clair River water that enters the lake from the north produce a

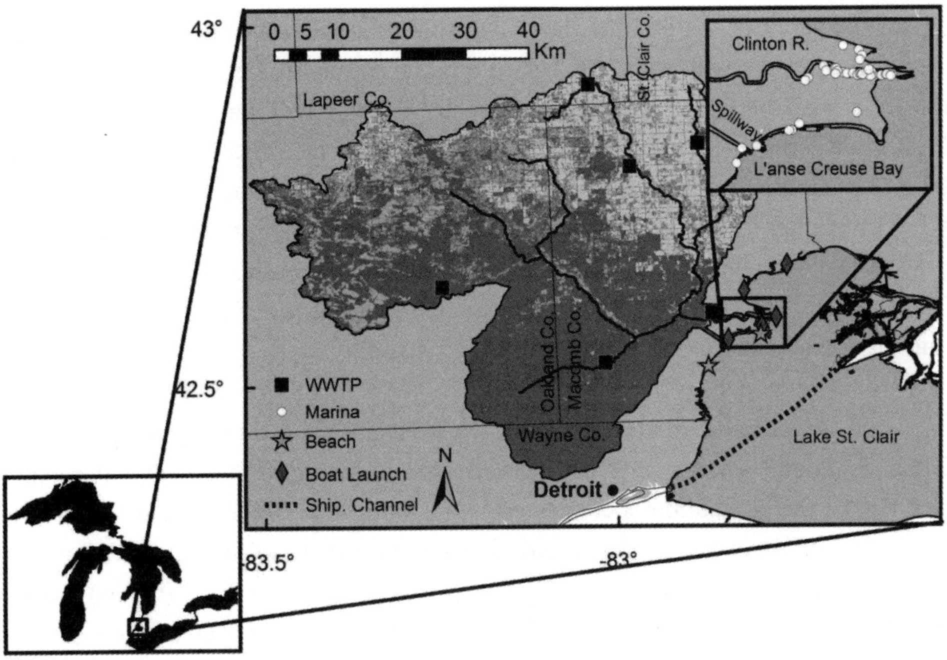

Figure 1. Map of the land use (developed dark gray, open water and wetlands blue, forest green, and agriculture yellow) of Clinton River watershed based on the 2006 NLCD (Fry *et al.* 2011) and the western shore of LSC, a connecting channel in the Laurentian Great Lakes system. Upper map is a close-up of the mouth of the Clinton River. WWTP wastewater treatment plants. Marinas, beaches (nearest to river), and boat launches indicate the recreational use of the area.

coastal current that tends to direct the Clinton River plume to the south (Schwab *et al.* 1989; Anderson and Schwab 2011) along the coastline, including the sandy beaches. LSC provides essential ES to the region, such as drinking water and recreational activities. These ES are also popular, as demonstrated by the total number of visitors at Metro Beach (Lake St. Clair Metropark Beach) which is estimated >200,000 people per month during the summer months.

The Clinton River is approximately 128km long and is located in eastern Michigan, USA (**Figure 1**). It drains an area of about 1980km^2 that includes the northern suburbs of the Detroit Metropolitan area and the majority of Oakland and Macomb Counties and smaller areas of St. Clair, Lapeer, and Wayne Counties (Jweda and Baskaran 2011). The Clinton River watershed encompasses approximately 60 cities, towns, and villages and is home to more than 1.5 million people (van Hees *et al.* 2010) who are mainly employed in automotive manufacturing and associated services. As of 2001, land use within the watershed was approximately 31% forest, 26% developed, 23% agricultural land, and 14% grasslands/open areas (Healy *et al.* 2008). The counties within the Clinton River watershed have one of highest densities of permitted point-source pollution facilities (>75 National Pollutant Discharge Elimination System permits per county) in the Great Lakes Basin (US Government Accountability Office 2005) and according to the last census the growth rate of population and number of households is positive (2.6% and 8.7%). The river and spillway, which was constructed in 1952 to relieve flooding, empty into the western shore of LSC near L'anse Creuse Bay.

Six wastewater treatment plants (WWTPs) with a total carrying capacity 264,675$m^3 \cdot day^{-1}$ operate in the watershed and serve more than 500,000 people. The rest of the population is connected to the Detroit's' WWTP or has septic systems. An important issue in the region is the replacement of the aging wastewater infrastructure as inflow, infiltration and combined sewer overflows (CSOs) impact water quality and human health (SEMCOG 2001). Financial constraints pose a major challenge on developing funding schemes for replacing wastewater infrastructure and sustain water quality in the area.

The Clinton River watershed and the western lake shoreline, downstream of the river mouth have been designated as an Area of Concern since 1987 under the Great Lakes Water Quality Agreement (International Joint Commission 2006). There are currently 41 Areas of Concern listed in the Great Lakes region due to one or more of 14 beneficial use impairments (BUIs) (http://www.epa.gov/glnpo/aoc/clintriv.html). Examples of the BUIs in the Clinton River watershed and western shore of LSC (8 of 14 BUIs) are: eutrophication or undesirable algae, degradation of fish and wildlife populations, and beach closures.

2.2. Research Phases

2.2.1. Phase I: Developing a Revised Conceptual Framework and Applying It to a Case Study

The construction of the conceptual framework was based on the principles and assumptions underlying the MEA and revised frameworks (Millennium Ecosystem Assessment 2005; Stevenson 2011) as well as existing theory and research concerning the coupling of human and natural systems that considers the possible linkages, interdependencies, and feedbacks between the natural and socioeconomic systems (Liu *et al.* 2007; Walsh and McGinnis 2008; Alberti *et al.* 2011). A conceptual framework was developed reflecting our understanding of how the socioeconomic and lake system interact with respect to the maintenance of the ecological condition for ES and HWB and then modified based on the published literature for the Clinton River and western shore of LSC (Bricker *et al.* 1976; Healy *et al.* 2008; van Hees *et al.* 2010).

2.2.2. Phase II: Testing and Refining the Framework by Hosting a 1-Day Stakeholder Workshop

A 1-day workshop was held to elicit the expert knowledge of professionals working in the LSC region. Fifteen stakeholders from a pool of 30 individuals with various expertise (e.g., ecology, community planning, engineering, economics, public health) and organizations (e.g., public utilities, universities, county, state and federal agencies), brought many years of experience in working in the area to discuss CHANS. Invited stakeholders were not aware of our hypothesized parameters and linkages in the conceptual framework because we did not want to influence their opinions. Before the workshop, we met frequently with a professional facilitator to help plan, organize and think about expected outcomes. We also emailed the stakeholders a short assignment beforehand that consisted of filling out their own conceptual diagram (the boxes of the two systems with no parameters or arrows) (see Appendix S1). We asked them to think about the indicators of ecosystem health, the socioeconomic activities that impact them and how ecological condition affects HWB. The main objective of workshop was to test and refine the parameters and linkages in the conceptual framework. The workshop goals were achieved by three exercises in which the stakeholders worked in small groups. First, each group listed key parameters of concern in the socioeconomic and lake (*i.e.*, natural) systems (**Figure 2**). Second, as a team they added arrows to show how the parameters affected each other (**Figure 2**). Lastly, a discussion followed about potential future management options. Based on this stakeholder input, we refined the conceptual framework by compiling the parameters and arrows from the groups into one diagram. We compared and contrasted our framework to the results of the workshop and focused on parameters and linkages

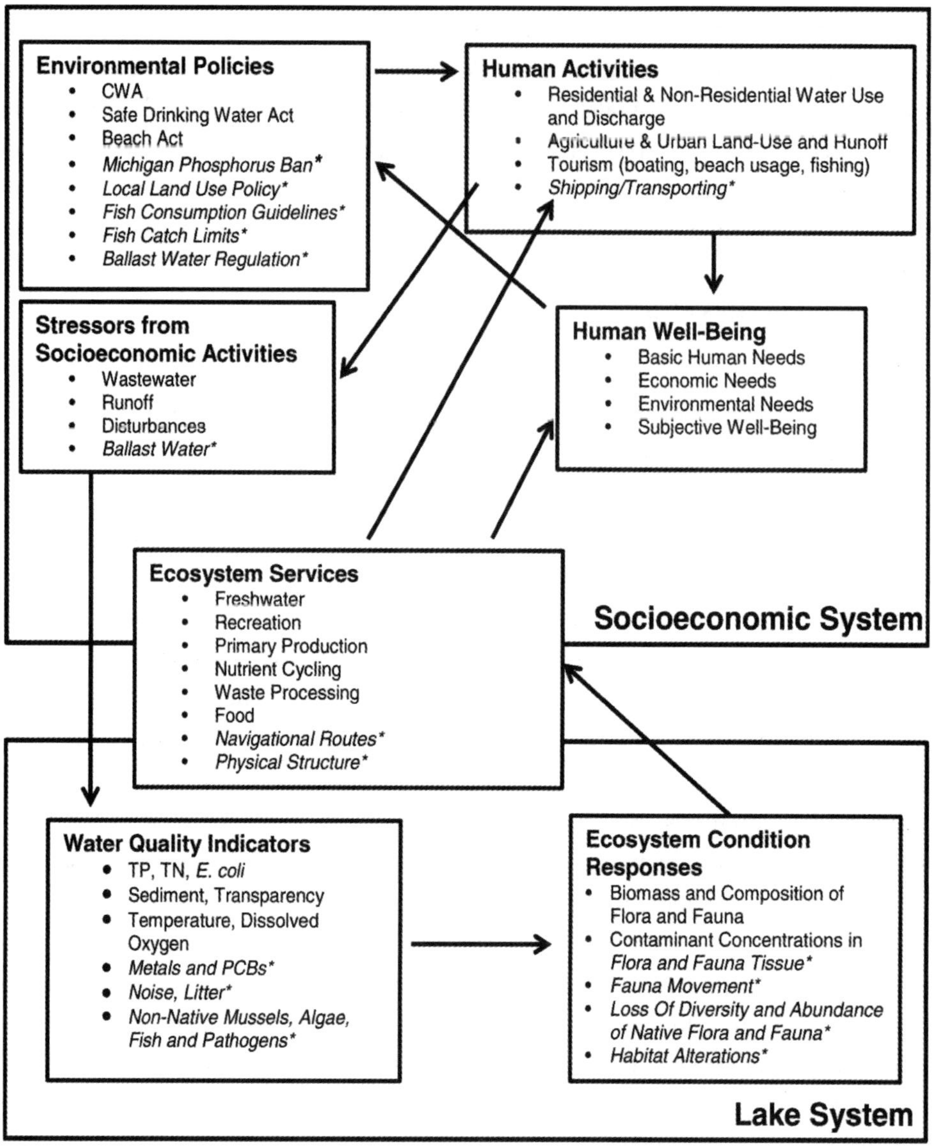

Figure 2. A CHANS framework for the Clinton River watershed and western shore and LSC. *indicates stakeholders' input. TP total phosphorus, TN total nitrogen.

identified as important for LSC region by both the stakeholders (see **Figure 2** with *) and our team. Most of our stakeholders commented that they were not familiar or accustomed to working with CHANS conceptual frameworks and they expressed that the workshop gave them the opportunity to think about CHANS and discuss more broadly about their own discipline and how it connects to others.

2.2.3. Phase III: Creating a Causal Loop Diagram Based on System Dynamics

A causal loop diagram (CLD) was developed based on system dynamics methodology to better understand the complexity and feedback loops among some of the parameters indicated in the conceptual framework. We used Stella Software (Version 10.02) to create the CLDs. One of most useful features provided by the software is the sensitivity analysis that occurs after running the model to learn if basic patterns of results are sensitive to changes in the uncertain parameters (Ford 1999).

3. Results

3.1. Phase I and Phase II: A CHANS Framework for Lake St. Clair

The proposed framework has two main systems: the socioeconomic and the lake (**Figure 2**). The socioeconomic system is composed of four interrelated sub-systems: human activities, stressors from socioeconomic activities, HWB, and environmental policies (**Figure 2**). The lake system is described through two subsystems: water quality indicators and the ecosystem condition responses.

ES link both systems (**Figure 2**). The lake system provides ES that directly benefit the socioeconomic system in two ways: first, by providing inputs to economic activities (e.g., commercial water use) and second, by influencing HWB (e.g., provisioning drinking water to meet basic biological human needs). Human activities taking place in the socioeconomic system indirectly affect ES by the production of stressors that decrease water quality and disturb the ecosystem condition, which in turn affects the provision of ES.

Using the CHANS framework and its application to LSC, we identified four main pathways and seven essential categories that describe key relationships between the socioeconomic and lake systems (**Table 1**). All of these pathways describe how human activities degrade the ecological condition and ES provided by the lake, but the pathways also indicate how humans depend on these natural systems for their well-being. Human activities might affect different ES than the one

Table 1. Four pathways that connect the socioeconomic system, ecosystem services and lake system in the Clinton River watershed and western shore of LSC.

Socioeconomic system (SS)				Ecosystem services (ES)		Lake system (LS)	
Human activities (HA) that produce stressors	Stressors from HA	Human well being components from HA and ES	Environmental policies regulating stressors	ES as inputs to SS	ES impacted by LS responses	Water quality indicators that are impacted by stressors	Ecosystem condition response to water quality indicators
Pathway 1: residential and non-residential, water use and discharge	Wastewater: BOD, TSS, nutrients, waterborne pathogens, temperature, metals, PCBs	Basic human needs, economic needs, environmental needs, subjective well-being	Clean Water Act, Safe Drinking Water Act, MI Phosphorus Ban	Freshwater (D), waste processing (ID), nutrient cycling (ID)	Freshwater, primary production, nutrient cycling, waste processing, recreation, biodiversity, food	TP, TN, *E. coli*, sediment, transparency, temperature, dissolved oxygen, mercury, PCBs	Biomass and composition of flora and fauna; concentrations in flora and fauna tissue
Pathway 2: land use (agr and urban runoff)	Runoff: nutrients, TSS, waterborne pathogens, altered hydrology	Basic human needs, economic needs, subjective well-being	Clean Water Act, local land use policy	Freshwater (D), recreation, aesthetics, navigational routes, physical structure	Same as Pathway 1	TP, TN, *E. coli*, sediment, transparency, temperature, dissolved oxygen, metals, PCBs	Same as Pathway 1
Pathway 3: tourism (boating, beach usage and fishing)	Disturbances: physical, noise, sediment, submerg. veg., litter; discharges from vessels	Economic needs, subjective well-being	Beach Act, Fish catch limits, fish consumption guidelines	Recreation (D), primary production (ID), nutrient cycling (ID), waste processing (ID), food (D), aesthetics, navigational routes (D)	Freshwater, recreation, primary production, biodiversity, food	Noise, transparency, litter, concentrations of anti-fouling paints	Flora and fauna movement and distribution, habitat alterations, flora and fauna tissue; fish communities composition and abundance
Pathway 4: shipping/ transporting	Ballast water: non-native mussels, algae, fish, and pathogens; dredging	Economic needs, environmental needs, subjective well-being	Ballast Water Regulation, Clean Water Act	Physical structure (D), navigational routes (D)	Recreation, food, biodiversity	Non-native organisms; sediment	Diversity and abundance of native flora and fauna; habitat alterations

D direct use of ES, *ID* indirect usages, *BOD* biological oxygen demand, *TSS* total suspended solids, *TP* total phosphorus and *TN* total nitrogen.

they depend on and for this reason two columns of ES are presented in **Table 1**.

3.2. Description of the Pathways

The first pathway is the human use of water and, consequently, the production of pollutant loads through residential sewage and non-residential discharge (**Table 1**). In this respect, water is an input to the socioeconomic system which is transformed into a pollutant load (wastewater) output to the Clinton River.

The second pathway is non-point source pollution related to land use, which is a significant component of contaminant loads in the runoff from the Clinton River watershed (**Table 1**). Approximately one-fourth of the Clinton River watershed is agricultural (**Figure 1**). The type of products (e.g., fertilizers and pesticides) and cultural practices used in agricultural watersheds drives pollutant loads (Jung *et al.* 2008). Another one-fourth of land use in the watershed is developed and thus stormwater runoff from impervious surfaces is also a key stressor and source of pollution to the river and lake (Environmental Consulting and Technology 2007). It is important to note that mixtures of pollutants derived from various sources can interact in the environment and have the potential to produce adverse effects to ecological and human health (Ravichandran 2004; Sumpter *et al.* 2006; Barber *et al.* 2013).

Tourism is the third pathway where humans and the environment significantly interact (**Table 1**). The importance of this activity to HWB was reiterated by the stakeholders. Beach usage, boating and fishing can pollute and disturb the natural environment by causing wave action, visual disturbances, noise pollution, re-suspension of sediment and submerged vegetation, and increase litter (fishing line, tackle, food wrappers, etc.) into the river and lake (Mosisch and Arthington 1998; Graham and Cooke 2008) (**Table 1**). Conversely, tourism is an important part of the economy of LSC region and it is estimated that beach closures in LSC results in a welfare loss \$13.89 per person per trip (Song *et al.* 2010). Stakeholders discussed the potential tradeoffs among the human activities and the derived ES. For example, pollutant sources, such as nutrients could positively impact the primary production that influences the recreational fishery through food web and habitat alterations. However, algal production has negative impacts on beach aesthetics and may harbor waterborne pathogens (Whitman *et al.* 2003) (**Table 1**).

The last pathway (**Table 1**) illustrates how shipping and transporting goods and the associated HWB heavily relies on the ecosystem for its physical structure, navigational routes, and freshwater (lake levels) (Rothlisberger *et al.* 2012). Shipping contributes significantly to the local economy by providing jobs and transporting goods (e.g., coal) (Martin Associates 2011); however, shipping activities can introduce aquatic invasive species. Ships release ballast water for stabilization purposes, which may contain non-native mussels, algae, fish, and pathogens (Mills *et al.* 1993). The best known example is the invasion of zebra (Dreissena polymorpha) and quagga (D. rostriformis bugenis) mussels in the mid 1980s and 1990s, which has been significantly impacting the lake's ecological structure and function (Nalepa and Gauvin 1988; Nalepa *et al.* 1996; David *et al.* 2009). Federal ballast water regulations have since been put in place to prevent introduction of aquatic invasive species.

3.3. Phase III: Coupling Water Use, Tourism, and Ecological Indicators

A snapshot of the interactions among the systems' components was developed to show the complexity in developing integrative models (**Figure 2**; **Table 1**). To illustrate how our conceptual framework can be transformed into a system dynamics model, we describe in detail the first pathway (human residential water use) and how it relates to the third pathway (tourism) by developing a CLD (**Figure 3**). The complexities and feedbacks between the systems in CLD help reinforce the notion that conceptual frameworks are an essential prerequisite when building system dynamics models that can help operationalize sustainability.

HWB is defined through four components based on a recently proposed classification scheme that captures physical, mental, and social aspects of well-being (Summers *et al.* 2012): basic human needs, economic needs, environmental needs, and subjective well-being. In our simplified example, the two ES, freshwater—drinking water (basic human need) and recreation (subjective well-being and income), contribute to HWB.

Figure 4 shows nine loops controlling the interplay between human activities and the ecological condition of freshwater ecosystems. The positive (+) arrows

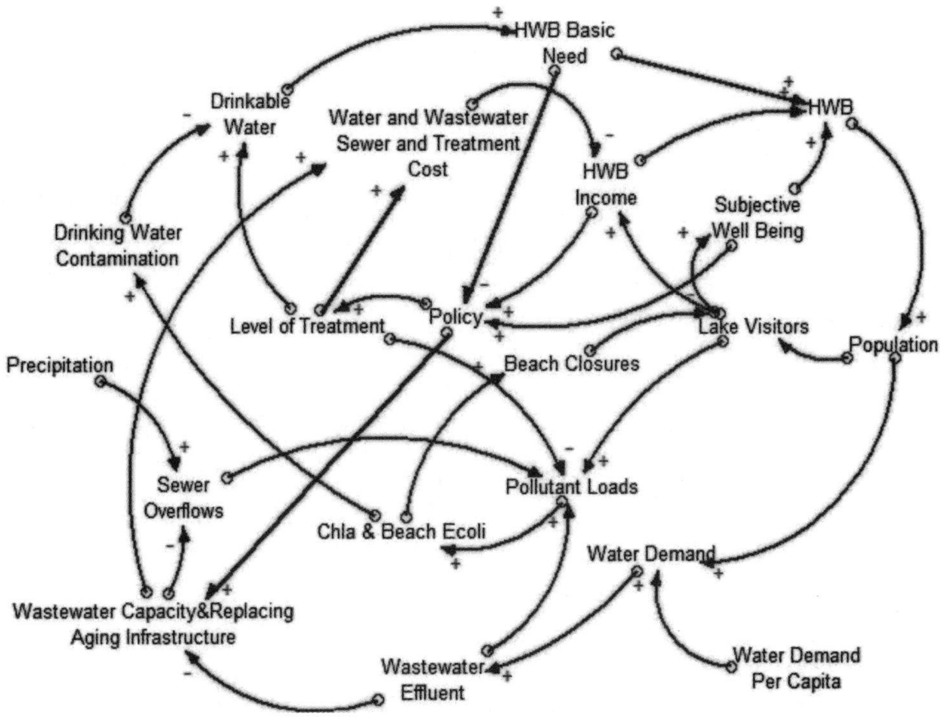

Figure 3. A causal loop diagram that represents Pathways 1 and 3 (see **Table 1**), which are residential water demand and tourism, and their impacts on the lake's ecological condition.

represent a cause and effect relationship in which the two parameters change in the same direction, while the negative (−) arrows represent two parameters that change in the opposite direction (Ford 1999).

Explanation of the Feedback Loops

Human population drives the water use that produces point source pollutant loads in the Clinton River watershed [**Figure 4(a)**]. Indicative determinants of the volume of waste-water and the pollutant loads entering the Clinton River and LSC include: residential water consumption per capita or household, number of served people or households, water use per type of industrial product and other activities, and available treatment technology. For simplicity reasons, the CLD only contains residential water demand per capita as an exogenous parameter. Discharged pollutants (e.g., nutrients and waterborne pathogens) enter the lake through the Clinton River and the spillway (**Table 1**). The impact from these pollutants can be measured

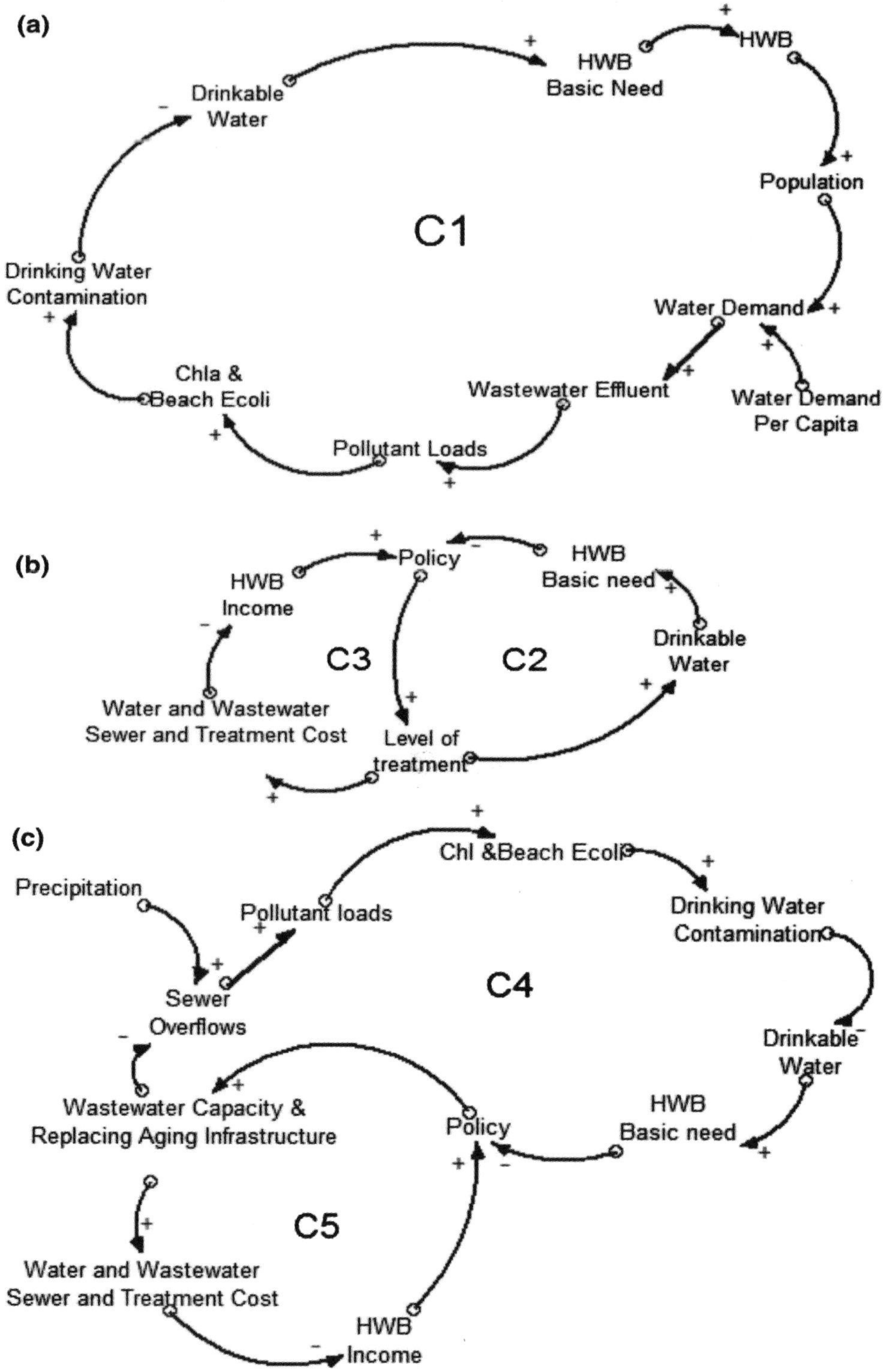
(a)
HWB
HWB
Basic Need
Drinkable
Water
Population
C1
Drinking Water
Contamination
Water Demand
Chla &
Beach Ecoli
Wastewater Effluent
Water Demand
Per Capita
Pollutant Loads
(b)
Policy
HWB
Basic need
HWB
Income
Drinkable
Water
C3
C2
Water and Wastewater
Sewer and Treatment Cost
Level of
treatment
(c)
Chl &Beach Ecoli
Precipitation
Pollutant loads
Drinking Water
Contamination
Sewer
Overflows
C4
Drinkable
Water
Wastewater Capacity &
Replacing Aging Infrastructure
Policy
HWB
Basic need
C5
Water and Wastewater
Sewer and Treatment Cost
HWB
Income

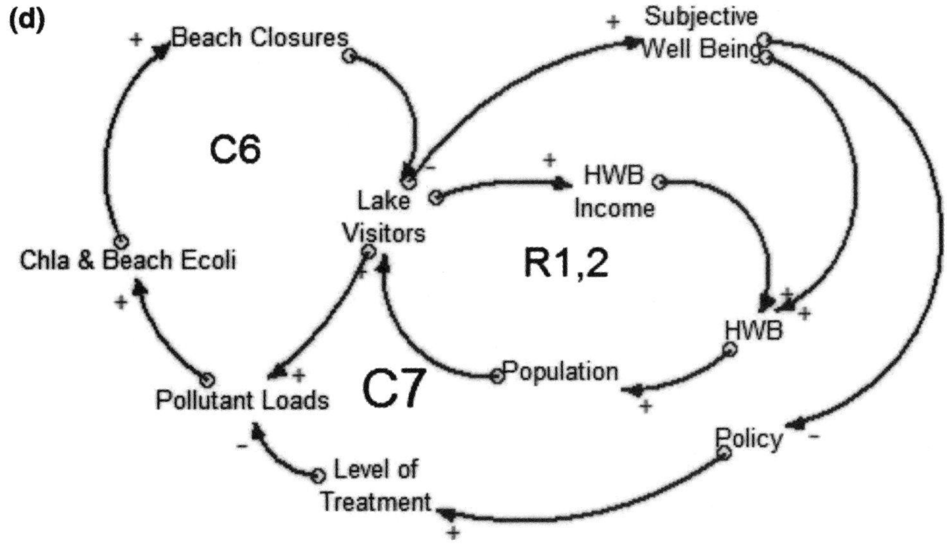

Figure 4. Details of the loop diagrams that represent Pathways 1 and 3 (see **Table 1**), the human activities of residential water demand and tourism. The panels include: (a) C1; (b) C2, C3; (c) C4, C5; (d) C6, C7, R1,2. Reinforcing or positive feedback loops, which are symbolized with "R", represent a feedback loop that reinforces the original change. Counteracting or negative feedback loops, "C", counteract the original change (Sterman 2000).

by ecological responses of primary production (macrophyte distribution and phytoplankton biomass indicated by chlorophyll a) and concentrations of the fecal indicator bacteria Escherichia coli (*E. coli*). High concentrations of chlorophyll a and *E. coli* impact the flow of ES such as drinking water or recreation, which in turn both affect HWB that drives the population size [**Figure 4(a)**, C1]. At the same time, if the ES of drinking water is not addressed under the current conditions, policymaking is activated to protect public health. In this case, two opposing dimensions of HWB affect the policymaking process, the satisfaction of basic human needs such as drinking water and economic needs (utility) such as income. Policymaking can set the legislative framework for employing the appropriate water and/or wastewater treatment infrastructure to mitigate the effects of human activities on water quality and protect human health [**Figure 4(b)**, C2]. In this case, the most relevant policy action is revision of National Pollutant Discharge Elimination System permits to require greater treatment of effluent. If state water quality standards are violated, development of a Total Maximum Daily Load (TMDL) is required; the TMDL allocates pollution loads among point and nonpoint sources with the goal of achieving overall pollutant load reductions. However, the process of policymaking that might lead to the adoption of new technologies (higher levels of water and wastewater treatment) or replacing aging infrastructure is subject to the induced cost for the residents. More advanced methods of treatment lead to higher water and wastewater treatment costs (negative arrow), and the human population is left with lower income (and well-being), which would encourage a policy response to ease costs [**Figure 4(b)**, C3].

Sewer overflows arising from CSOs along with aging infrastructure are considered by stakeholders to be the most serious sources of pollution in the study area [**Figure 4(c)**, C4]. Sewer overflows can increase pollutant loads which in turn can decrease the lakes' ecological condition (increased concentrations of chlorophyll a and beach *E. coli*) resulting in drinking water contamination and beach closures due to violations of beach quality regulations. Drinking water contamination reduces the availability of drinkable water, which is a major human need, and results in the decrease of HWB. When basic human needs such as drinking water cannot be satisfied, policymaking for controlling sewer overflows is activated [**Figure 4(c)**, C4]. However, the control of sewer overflows increases the sewer cost implying that residents are left with a lower income (HWB income), which also triggers policymaking [**Figure 4(c)**, C5].

Irrespective of the source of pollutant loads, violation of beach quality regulations leads to beach closures and decreases lake visitors and in turn the pollutants generated from direct beach usage [**Figure 4(d)**, C6]. At the same time, two reinforcing feedback loops related to subjective well-being and income are positively connected to beach usage (lake visitors), and increase the overall HWB, population in the area and the number of lake visitors [**Figure 4(d)**, R1,2]. Lastly, as lake visitors decrease and subjective well-being decays, new policy measures are needed to increase the level of treatment, to decrease pollutant loads that impact ecological condition (chlorophyll a and beach *E. coli*), and to decrease the beach closures for increasing lake visitors [**Figure 4(d)**, C7].

The parameters used in our example represent specific aspects of socio-ecological systems. Defining numerical values for those parameters is a challenge as some of them are uncertain. However, sensitivity and policy analysis can help determine if the estimated values for uncertain parameters reproduce the same pattern of results and if a policy creates the desired outcome (Ford 1999). This example represents the policy response only through the water and wastewater infrastructure (e.g., treatment technology or replacing aging infrastructure) ensuring safe drinking water and clean beaches for the local communities.

The cost for employing the appropriate infrastructure can be considered as an indirect payment for ES (e.g., drinking water). Other policy instruments might be capable of mitigating human effects on water quality such as government incentive payments, voluntary payments, and institutional changes (Brauman *et al.* 2007; Daily *et al.* 2009; Molnar and Kubiszewski 2012). The ability of those instruments to foster the maintenance of ES in the long run needs further investigation.

4. Discussion

Our research reveals that the four proposed pathways are highly dependent on each other, implying that deriving the desired ES from the natural environment requires holistic, integrated management of the pathways in the systems. For example, developing and sustaining tourist activities (Pathway 3) will depend on management of point and non-point sources of pollution (Pathways 1 and 2). Another example is the impact of shipping through the discharge of ballast water which can introduce non-indigenous species (Pathway 4) that impact local ecological condition and subsequently can impact tourism and increase costs of maintaining water infrastructure. In our case study, water quality issues such as *E. coli* and algal blooms, create constraints on the tourist activities that directly depend on ecosystem condition. Conversely, there are human activities such as water use, boating and fishing that disturb the natural environment. Evaluating the potential tradeoffs between the benefits and costs among various human activities with respect to ecological condition and services can enable decision makers to manage valuable aquatic resources (Kremen and Ostfeld 2005). However, when evaluating the tradeoffs among ES, an important criterion is to include uncertainty for two main reasons. First, it remains a challenge to assess the full value of ecosystems in providing services (Brauman *et al.* 2007). Second, even if we can estimate the value that current users place in ES, it not possible to do so for future generations, implying the need to maintain the full range of services provided by the ecosystems (Bithas 2011).

Building CHANS frameworks for specific areas and defining the key parameters and data needs by eliciting stakeholder knowledge enhances our ability to develop dynamic models that capture real world systems (Carpenter *et al.* 2009). From the stakeholder workshop, we found our proposed framework can be used as a basis for further discussion and collaboration among interested parties as it ultimately suggests a holistic approach for watershed management and sustainability. Stakeholders' knowledge helped our team to identify more pathways (e.g., Pathway 4), define essential feedbacks among systems' components (e.g., arrows in **Figure 4**) and explore potential policy management options. In terms of system dynamics, stake-holders provided us with the "casual knowledge" on how systems function and are interrelated (Jones *et al.* 2011).

We developed a CLD to present the complexity and the various feedback loops underlying water use and some aspects of recreation (beach usage defined through lake visitors). As long as the human population increases, the socioeconomic activities to serve this population will increase and as a result there will be increasing pressures on the natural systems. Maintaining aggregate HWB requires policy actions. In our example, the determinant parameter to mitigate the effects of human activities on water quality and maintain the ES of drinking water and recreation is the establishment of relevant pollution-prevention policies and construction and maintenance of appropriate infrastructure. Investment in infrastructure is critical for addressing sewer overflows, inadequate treatment, leaking pipes, and stormwater discharges, which are the major stressors in the study area. SEMCOG (2001) estimates that southeast Michigan will need to invest $14–26 × 10^9$ to maintain and improve the current sewer system and to remediate overflows. Although water and WWTPs can provide a substitute for some ES (*i.e.*, waste processing) that maintain water quality for human consumption and recreation, pollution prevention strategies are still required to improve other measures of water quality to protect other ES. The quality of source water entering treatment plants also affects treatment and the drinking water quality that local communities enjoy (Levy *et al.* 2012). Although residents do not directly pay for ES, they pay indirectly for their loss in terms of substitutes (Summers *et al.* 2012) such as treatment technologies and the operation of water and WWTPs. The costs of substitutes may be large and some ES have no substitutes or technological fixes. This implies that people who mostly suffer the impacts of ecosystems' deterioration are those who cannot employ instruments like technology to mitigate water quality issues. In some cases, technologies may have "lower resilience, cost effectiveness, suitability and life span than the ES they replace" (Brauman *et al.* 2007, p. 15). For example, the cities of New York and Boston found that watershed protection was more beneficial than constructing and maintaining filtration plants. Multiple groups in Michigan have been promoting green infrastructure as a cost-effective means of reducing pollutant loads.

Although we can conceptually and qualitatively define the key components of HWB, it remains a challenge to quantify the impacts of ES on HWB. The inability to quantitatively link ES and HWB can be viewed as a limitation as it constrains what can be represented in typical models. Approaches like creating CLDs by using system dynamics methodology allows to focus on critical linkages within CHANS. This can be useful for policy and for targeting research to get the necessary data for HWB quantification. These complex systems or case studies and data limitations should not prevent us from thinking about and fostering solutions to achieve sustainability. Our next step is to develop a system dynamics model based on our conceptual framework to move beyond conceptual linkages between socioeconomic and natural systems and to evaluate causal relationships (Sterman 2012). This dynamic model can provide decision makers useful tools for attaining sustainability under alternative scenarios.

5. Conclusions

Designing for and achieving sustainability demands interdisciplinary approaches (Kremen and Ostfeld 2005) that fully integrate the knowledge of socioeconomic and natural sciences (Mavrommati and Richardson 2012). We suggest that evaluating the systems as a whole can enhance understanding of the importance for maintaining the functions and processes of natural systems for the healthy function of the socioeconomic system. Applying the concept of sustainability at an operational level remains a challenge given that increasing the current generations' HWB arises from the degradation of ES (Raudsepp-Hearne *et al.* 2010). Maintaining certain components of the natural system by managing the socioeconomic systems' activities is necessary for sustaining key ES that contribute to the well-being of current and future generations.

Acknowledgments

This work is supported in part by the National Science Foundation under Grant No. EAR-1039122. The expert workshop was approved by MSU IRB# x12-477e and we appreciate the expertise and involvement of fifteen stakeholders. We thank J. Urban-Lurain for facilitating, P. Esselman, and R. McNinch for note-taking, and S. Carver for assisting in the preparations of the workshop. Thank you to S. Schultze and P. Esselman for assistance in developing the map and J. Rose, R.J. Stevenson, and two anonymous reviewers for providing comments on earlier drafts.

Source: Mavrommati G, Baustian M M, Dreelin E A. Coupling Socioeconomic and Lake Systems for Sustainability: A Conceptual Analysis Using Lake St. Clair Region as a Case Study[J]. Ambio, 2014, 43(3):275–287.

References

[1] Alberti, M., H. Asbjornsen, L.A. Baker, N. Brozovic, L.E. Drinkwater, S.A. Drzyzga, C.A. Jantz, J. Fragoso, *et al.* 2011. Research on coupled human and natural systems (CHANS): Approach, challenges, and strategies. Bulletin of the Ecological Society of America 92: 218–228.

[2] Anderson, E.J., and D.J. Schwab. 2011. Relationships between wind-driven and hydraulic flow in Lake St. Clair and the St. Clair River Delta. Journal of Great Lakes Research 37: 147–158.

[3] Barber, L.B., S.H. Keefe, G.K. Brown, E.T. Furlong, J.L. Gray, D.W. Kolpin, M.T. Meyer, M.W. Sandstrom, *et al.* 2013. Persistence and potential effects of complex organic contaminant mixtures in wastewater-impacted streams. Environmental Science and Technology 47: 2177–2188.

[4] Bithas, K. 2008. Tracing operational conditions for the ecologically sustainable economic development: The Pareto optimality and the preservation of the biological crucial levels. Environment, Development and Sustainability 10: 373–390.

[5] Bithas, K. 2011. Sustainability and externalities: Is the internalization of externalities a sufficient condition for sustainability? Ecological Economics 70: 1703–1706.

[6] Brauman, K.A., G.C. Daily, T.K.E. Duarte, and H.A. Mooney. 2007. The Nature and value of ecosystem services: An overview highlighting hydrologic services. Annual Review of Environment and Resources 32: 67–98.

[7] Bricker, K.S., F.J. Bricker, and J.E. Gannon. 1976. Distribution and abundance of zooplankton in the U.S. Waters of Lake St. Clair, 1973. Journal of Great Lakes Re-

search 2: 256–271.

[8] Burger, J.R., C.D. Allen, J.H. Brown, W.R. Burnside, A.D. Davidson, T.S. Fristoe, M.J. Hamilton, N. Mercado-Silva, *et al.* 2012. The macroecology of sustainability. PLoS Biology 10: e1001345.

[9] Carpenter, S.R., H.A. Mooney, J. Agard, D. Capistrano, R.S. DeFries, S. D′ıaz, T. Dietz, A.K. Duraiappah, *et al.* 2009. Science for managing ecosystem services: Beyond the millennium ecosystem assessment. Proceedings of the National Academy of Sciences of the United States of America 106: 1305–1312.

[10] Cumming, G. 2011. Spatial resilience in social–ecological systems, 247 pp. London: Springer.

[11] Daily, G.C., S. Polasky, J. Goldstein, P.M. Kareiva, H.A. Mooney, L. Pejchar, T.H. Ricketts, J. Salzman, *et al.* 2009. Ecosystem services in decision making: Time to deliver. Frontiers in Ecology and the Environment 7: 21–28.

[12] Daly, H. 1996. Beyond growth: The economics of sustainable development, 253 pp. Boston: Beacon Press.

[13] David, K.A., B.M. Davis, and R.D. Hunter. 2009. Lake St. Clair Zooplankton: Evidence for Post-Dreissena changes. Journal of Freshwater Ecology 24: 199–209.

[14] Environmental Consulting and Technology, Inc. 2007. Water quality sampling & analysis: Final Report, 129 pp. Clinton Township, MI Environmental Consulting and Technology, Inc.

[15] Folke, C., Å. Jansson, J. Rockström, P. Olsson, S.R. Carpenter, F.S. Chapin, A.-S. Cre′pin, G. Daily, *et al.* 2011. Reconnecting to the biosphere. AMBIO 40: 719–738.

[16] Ford, A. 1999. Modeling the environment: An introduction to system dynamics modeling of environmental systems, 401 pp. Washington, DC: Island Press.

[17] Francis, J.T., and R.C. Haas. 2006. Clinton River Assessment (Trans. F. Division), 97 pp. Lansing, MI: Department of Natural Resources.

[18] Fry, J.A., G. Xian, S. Jin, J.A. Dewitz, C.G. Homer, Y. Limin, C.A. Barnes, N.D. Herold, et al. 2011. Completion of the 2006 National Land Cover Database for the Conterminous United States. Photogrammetric Engineering and Remote Sensing 77: 858–864.

[19] Graham, A.L., and S.J. Cooke. 2008. The effects of noise disturbance from various recreational boating activities common to inland waters on the cardiac physiology of a freshwater fish, the Largemouth Bass (Micropterus salmoides). Aquatic Conservation: Marine and Freshwater Ecosystems 18: 1315–1324.

[20] Healy, D.F., D.B. Chambers, C.M. Rachol, and R.S. Jodoin. 2008. Water quality of the St. Clair River, Lake St. Clair, and their U.S. tributaries, 1946–2005, 92 pp. Reston: U.S. Geological Survey.

[21] Heemskerk, M., K. Wilson, and M. Pavao-Zuckerman. 2003. Conceptual models as tools for communication across disciplines. Conservation Ecology 7: 9.

[22] Holden, E., and K. Linnerud. 2007. The sustainable development area: Satisfying basic needs and safeguarding ecological sustainability. Sustainable Development 15: 174–187.

[23] Howarth, R.B. 2009. Discounting, uncertainty, and revealed time preference. Land Economics 85: 24–40.

[24] International Joint Commission. 2006. A guide to the Great Lakes Water Quality Agreement: Background for the 2006 Governmental Review, 32 pp.

[25] Jones, N.A., H. Ross, T. Lynam, P. Perez, and A. Leitch. 2011. Mental models: An interdisciplinary synthesis of theory and methods. Ecology and Society 16: 46.

[26] Jung, K.-W., S.-W. Lee, H.-S. Hwang, and J.-H. Jang. 2008. The effects of spatial variability of land use on stream water quality in a coastal watershed. Paddy and Water Environment 6: 275–284.

[27] Jweda, J., and M. Baskaran. 2011. Interconnected riverine–lacustrine systems as sedimentary repositories: Case study in Southeast Michigan using (210)Pb and (137)Cs-based sediment accumulation and mixing models. Journal of Great Lakes Research 37: 432–446.

[28] Kremen, C., and R.S. Ostfeld. 2005. A call to ecologists: Measuring, analyzing, and managing ecosystem services. Frontiers in Ecology and the Environment 3: 540–548.

[29] Levy, K., G. Daily, and S.S. Myers. 2012. Human health as an ecosystem service: A conceptual framework. In Integrating ecology and poverty reduction, ed. J.C. Ingram, F. DeClerck, and C. Rumbaitis Del Rio, 231–251. New York: Springer.

[30] Liu, J., T. Dietz, S.R. Carpenter, C. Folke, M. Alberti, C.L. Redman, S.H. Schneider, E. Ostrom, *et al.* 2007. Coupled human and natural systems. AMBIO 36: 639–649.

[31] Martin Associates. 2011. The economic impacts of the Great Lakes-St. Lawrence Seaway System, 98 pp. Lancaseter: Martin Associates.

[32] Mavrommati, G., and C. Richardson. 2012. Experts' evaluation of concepts of ecologically sustainable development applied to coastal ecosystems. Ocean and Coastal Management 69: 27–34.

[33] Maxwell, J.A. 1996. Qualitative research design. An interactive approach, 174 pp. Thousand Oaks: Sage.

[34] Millennium Ecosystem Assessment. 2003. Ecosystems and human well-being: A framework for assessment, 266 pp. Washington: World Resources Institute.

[35] Millennium Ecosystem Assessment. 2005. Ecosystems and human well-being: Current state and trends, vol. 1, 901 pp. Washington, DC: World Resources Institute.

[36] Mills, E.L., J.H. Leach, J.T. Carlton, and C.L. Secor. 1993. Exotic species in the Great Lakes: A history of biotic crises and anthropogenic introductions. Journal of Great Lakes Research 19: 1–54.

[37] Molnar, J.L., and I. Kubiszewski. 2012. Managing natural wealth: Research and implementation of ecosystem services in the United States and Canada. Ecosystem Services 2: 45–55.

[38] Mosisch, T.D., and A.H. Arthington. 1998. The impacts of power boating and water skiing on lakes and reservoirs. Lakes & Reservoirs: Research & Management 3: 1–17.

[39] Nalepa, T.F., and J.M. Gauvin. 1988. Distribution, abundance, and biomass of freshwater mussels (Bivalvia: Unionidae) in Lake St. Clair. Journal of Great Lakes Research 14: 411–419.

[40] Nalepa, T.F., D.J. Hartson, G.W. Gostenik, D.L. Fanslow, and G.A. Lang. 1996. Changes in the freshwater mussel community of Lake St Clair: From Unionidae to Dreissena polymorpha in eight years. Journal of Great Lakes Research 22: 354–369.

[41] Neumayer, E. 2010. Weak versus strong sustainability. Cheltenham: Edward Elgar.

[42] Ostrom, E. 2009. A general framework for analyzing sustainability of social–ecological systems. Science 325: 419–422.

[43] Ostrom, E., and M. Cox. 2010. Moving beyond Panaceas: A multitiered diagnostic approach for social–ecological analysis. Environmental Conservation 37: 451–463.

[44] Pickett, S.T.A., M.L. Cadenasso, and J.M. Grove. 2005. Biocomplexity in coupled natural–human systems: A multidimensional framework. Ecosystems 8: 225–232.

[45] Raudsepp-Hearne, C., G.D. Peterson, M. Tengö, E.M. Bennett, T. Holland, K. Benessaiah, G.K. MacDonald, and L. Pfeifer. 2010. Untangling the environmentalist's paradox: Why is human well-being increasing as ecosystem services degrade? BioScience 60: 576–589.

[46] Ravichandran, M. 2004. Interactions between mercury and dissolved organic matter—A review. Chemosphere 55: 319–331.

[47] Riseng, C.M., M.J. Wiley, P.W. Seelbach, and R.J. Stevenson. 2010. An ecological assessment of Great Lakes Tributaries in the Michigan Peninsulas. Journal of Great Lakes Research 36: 505–519.

[48] Rockstrom, J., W. Steffen, K. Noone, A. Persson, F.S. Chapin, E.F. Lambin, T.M. Lenton, M. Scheffer, *et al.* 2009. A safe operating space for humanity. Nature 461: 472–475.

[49] Rothlisberger, J., D. Finnoff, R. Cooke, and D. Lodge. 2012. Shipborne nonindigenous species diminish great lakes ecosystem services. Ecosystems 15: 1–15.

[50] Schmolke, A., P. Thorbek, D.L. DeAngelis, and V. Grimm. 2010. Ecological models supporting environmental decision making: A strategy for the future. Trends in Ecology & Evolution 25: 479–486.

[51] Schwab, D.J., A.H. Clites, C.R. Murthy, J.E. Sandall, L.A. Meadows, and G.A. Meadows. 1989. The effect of wind on transport and circulation in Lake St. Clair. Journal of Geophysical Research 94: 4947–4958.

[52] Selegean, J.P.W., R. Kusserow, R. Patel, T.M. Heidtke, and J.L. Ram. 2001. Using zebra mussels to monitor in environmental waters. Journal of Environmental Quality 30: 171–179.

[53] SEMCOG. 2001. Investing in Southeast Michigan's quality of life: Sewer infrastruc-

ture needs, 84 pp. Detroit: Southeast Michigan Council of Governments.

[54] SEMCOG. 2002. Historical population and employment by Minor Civil Division, Southeast Michigan, 29 pp.

[55] Song, F., F. Lupi, and M. Kaplowitz. 2010. Valuing Great Lakes Beaches. Paper read at Agricultural & Applied Economics Association 2010 AAEA, CAES, &WAEA Joint Annual Meeting, July 25–27, Denver, Colorado.

[56] Sterman, J.D. 2000. Business dynamics. Systems thinking and modeling for a complex world, ed. J.J. Shelstad, 1008 pp. Irwin: McGraw-Hill.

[57] Sterman, J.D. 2012. Sustaining sustainability: Creating a systems science in a fragmented academy and polarized world. In Sustainability science, ed. M.P. Weinstein, and R.E. Turner, 21–58. New York: Springer.

[58] Stevenson, J.R. 2011. A revised framework for coupled human and natural systems, propagating thresholds, and managing environmental problems. Physics and Chemistry of the Earth, Parts A/B/ C 36: 342–351.

[59] Sumaila, U.R. 2004. Intergenerational cost–benefit analysis and marine ecosystem restoration. Fish and Fisheries 5: 329–343.

[60] Summers, J.K., L.M. Smith, J.L. Case, and R.A. Linthurst. 2012. A review of the elements of human well-being with an emphasis on the contribution of ecosystem services. AMBIO 41: 327–340.

[61] Sumpter, J.P., A.C. Johnson, R.J. Williams, A. Kortenkamp, and M. Scholze. 2006. Modeling effects of mixtures of endocrine disrupting chemicals at the river catchment scale. Environmental Science and Technology 40: 5478–5489.

[62] US Government Accountability Office. 2005. Great Lakes Initiative EPA needs to better ensure the complete and consistent implementation of water quality standards, 52 pp.

[63] van Hees, E.H.P., E.I.B. Chopin, T.M. Sebastian, G.D. Washington, L.M. Germer, P. Domanski, D. Martz, and L. Schweitzer. 2010. Distribution, sources, and behavior of trace elements in the Clinton River Watershed, Michigan. Journal of Great Lakes Research 36: 606–617.

[64] Walsh, S.J., and D. McGinnis. 2008. Biocomplexity in coupled human–natural systems: The study of population and environment interactions. Geoforum 39: 773–775.

[65] Whitman, R.L., D.A. Shively, H. Pawlik, M.B. Nevers, and M.N. Byappanahalli. 2003. Occurrence of Escherichia coli and Enterococci in Cladophora (Chlorophyta) in Nearshore Water and Beach Sand of Lake Michigan. Applied and Environmental Microbiology 69: 4714–4719.

Chapter 21

Focus on Potential Environmental Issues on Plastic World towards a Sustainable Plastic Recycling in Developing Countries

Onwughara Innocent Nkwachukwu[1,2], Chukwu Henry Chima[3], Alaekwe Obiora Ikenna[2], Lackson Albert[2,4]

[1]Reliable Research Laboratory Service, D30 Orji Kalu Housing Estate, Umuahia, Abia State, Nigeria
[2]Department of Pure and Industrial Chemistry, Nnamdi Azikiwe University Awka, P.M.B. 5052, Awka, Anambra State, Nigeria
[3]Department of Chemistry, Abia State Polytechnic, Aba, Abia State, Nigeria
[4]Yagai Academy, P.O. Box 1180, Jalingo, Taraba State, Nigeria

Abstract: Due to the tremendous growth of plastics in the world, it has brought about environmental concerns for over the past two or three decades. Most of these plastics, due to poor management, are currently disposed of in unauthorized dumping sites or burned uncontrollably in the fields. The paper outlines environmental concerns of so many applications of plastics. The most important mechanisms of degradation of plastics, environmental impacts and recommendations for sustainable development are fully discoursed, with recycling option being overviewed as the route under most intense development at this time because of its broad public appeal and obvious environmental advantages.

Keywords: Plastic Waste, Plastics' End-of-Life, Degradation, Environmental Impacts, Health Effects, Recycling and Recommendation

1. Introduction

Plastics are organic polymeric materials consisting of giant organic molecules. Plastic materials can be formed into shapes by one of a variety of processes, such as extrusion, moulding, casting or spinning. Modern plastics (or polymers) possess a number of extremely desirable characteristics: high strength-to-weight ratio, excellent thermal properties, electrical insulation, and resistance to acids, alkalis and solvents, to name but a few. Some have unique electrical insulating properties, such as their strength, stress resistance, flexibility and durability, which make them important materials for use in electronics. These polymers are made of a series of repeating units known as monomers. The structure and degree of polymerisation of a given polymer determine its characteristics. Linear polymers (a single linear chain of monomers) and branched polymers (linear with side chains) are thermoplastic; they soften when heated. Cross-linked polymers (polymers with bond formed between polymer chains, either between different chains or between different parts of the same chain.) are thermosetting, that is, they harden when heated.

Development of synthetic polymers, used to make plastics such as polyethylene, polypropylenes, polyesters and polyamides (including nylon), has revolutionized the types of containers for products, the types of materials for packaging and the materials used for carry bags. However, most of these polymers are not biodegradable and, once used and discarded, become major waste management challenges[1]. However, plastic waste can also impose negative externalities such as greenhouse gas emissions or ecological damage. It is usually non-biodegradable and therefore can remain as waste in the environment for a very long time; it may pose risks to human health and the environment. In some cases, it can be difficult to reuse and/or recycle. There is a mounting body of evidence which indicates that substantial quantities of plastic waste are now polluting marine and other habitats[2]. The widespread presence of these materials has resulted in numerous accounts of wildlife becoming entangled in plastic, leading to injury or impaired movement and, in some cases, resulting in death. Concerns have been raised regarding the effects of plastic ingestion as there is some evidence to indicate that toxic chemicals from plastics can accumulate in living organisms and throughout nutrient chains. There are also some public health concerns arising from the use of plastics treated with chemicals[2].

As with most materials, global plastic production is estimated to decrease from 245 million tonnes (Mt) in 2008 to around 230 million tonnes in 2009 as a result of the economic crisis. Over the past 50 years, however, there has been a very steep rise in plastic production, especially in Asia (**Figure 1**). The European Union (EU) accounts for around 25% of world production. The total consumption of plastics in Western Europe was approximately 39.7 million tonnes in 2003, which means about 98kg/person, and the amount has been increasing[3]. China produces more plastic than any other country, at 15% of global production. Germany

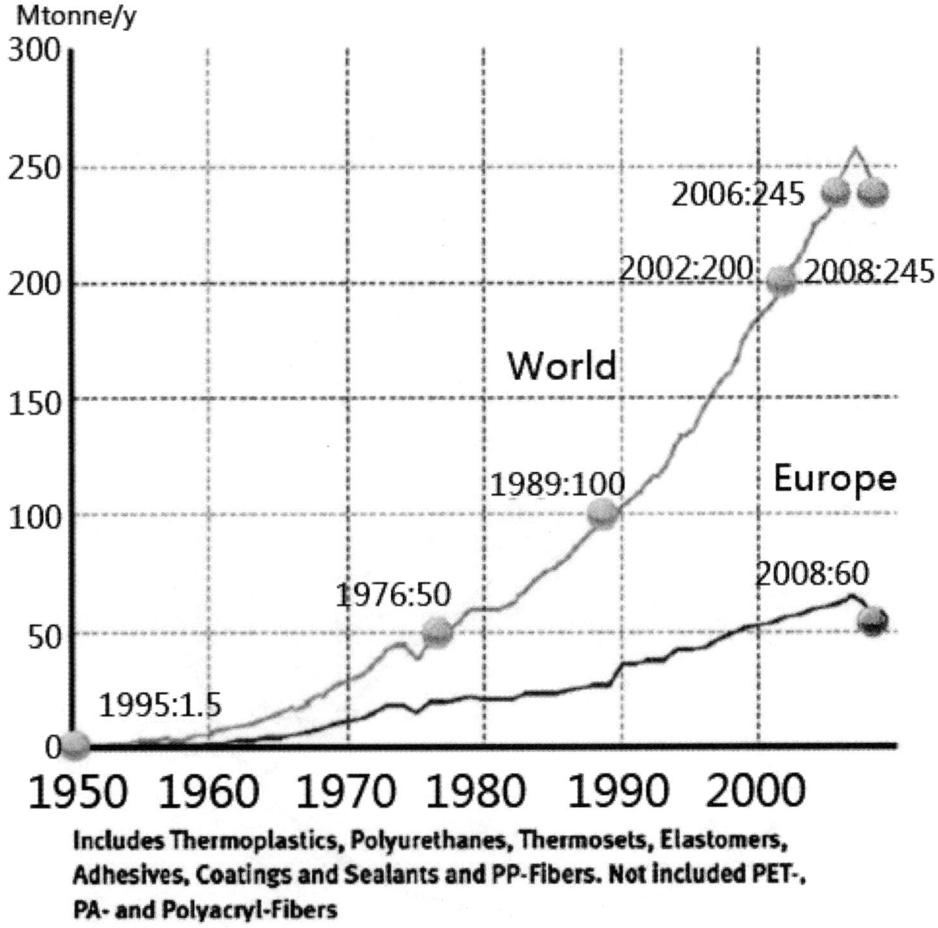

Figure 1. World plastic production, 1950 to 2008 (Mt) (adapted from[5]).

produces the greatest amount of any EU country, accounting for about 8% of global production as shown in **Figure 2**[4]. Societies are increasingly reliant on plastics, which are already a ubiquitous part of everyday life. As the development of new materials is ongoing, limiting their detrimental effects poses new challenges for policy makers. Regulatory instruments designed to mitigate the effects of plastics on human health and the environment must evolve in line with trends in production, use and disposal[2].

Polyethylene has the highest share of production of any polymer type, while four sectors: polyethylene terephthalate (PET), which accounts for 20% of thermoplastic resin capacity, followed by polypropylene (PP), which accounts for 18%, polyvinyl chloride (PVC) and polystyrene/expanded polystyrene (PS/EPS), represent 72% of plastic demand: packaging, construction, automotive and electrical and electronic equipment as shown in **Figure 3**. The rest includes sectors such as

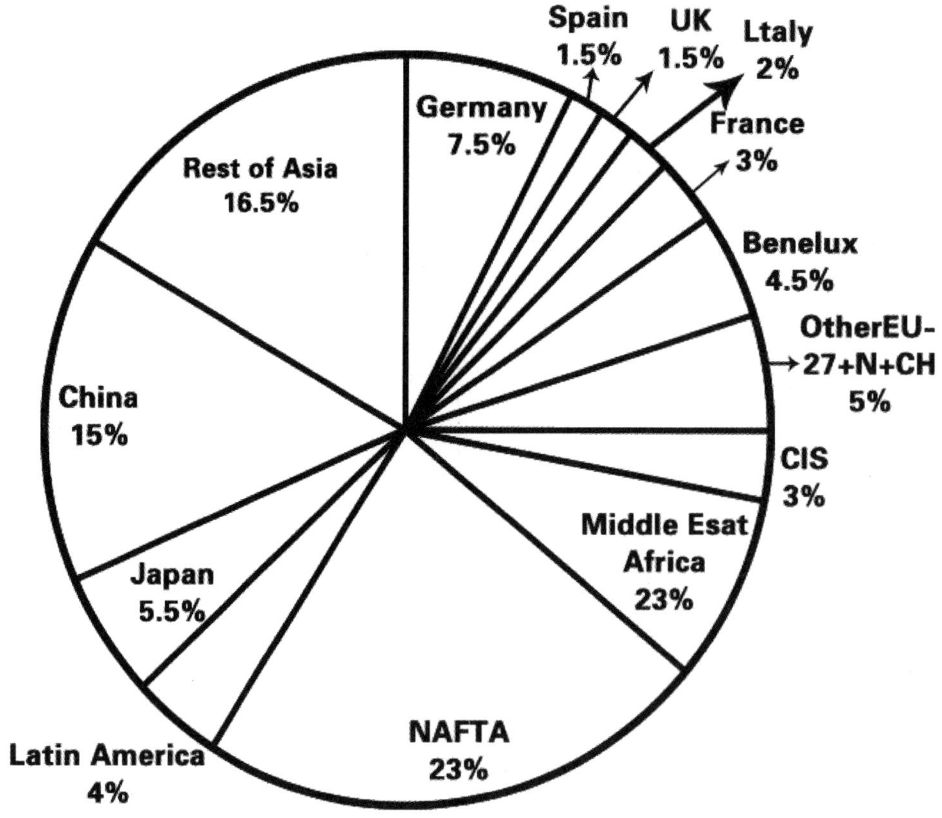

Figure 2. Distribution of world plastic production (adapted from[4]).

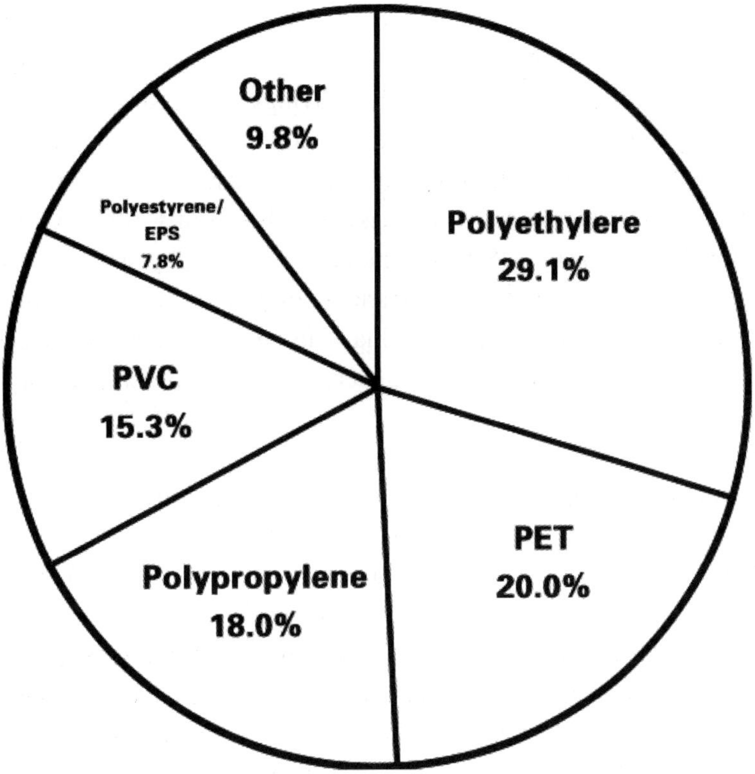

Figure 3. World thermoplastic resin capacity, 2008 (adapted from[10]).

household, furniture, agriculture and medical devices[4]. Plastic packaging accounts for the largest share of plastic production in the world level. About 50% of plastic is used for single-use disposable applications, such as packaging, agricultural films and disposable consumer items[6]. Plastics were the second largest component in waste from electrical and electronic equipment (WEEE), and approximately 30% of the mass electronic scrap consists of plastics[7]-[9]. Plastics consume approximately 8% of world oil production: 4% as raw material for plastics and 3% to 4% as energy for manufacture[1][6].

The plastic industry is in constant development, with technology evolving in response to the ever-changing demand. Some trends that emerge clearly are continued innovation and improvements such as weight reduction of individual items, increasing use of plastics (and bioplastics) in vehicle manufacturing, a shift in primary plastic production to transition and emerging economies and continued growth in the market share of bioplastics (despite some sorting and price barriers).

Bioplastics make up only 0.1% to 0.2% of total EU plastics[11]. It is estimated that plastics save 600 to 1,300 million tonnes of CO_2 through the replacement of less efficient materials, fuel savings in transport, contribution to insulation, prevention of food losses and use in wind power rotors and solar panels[12]. In 2000, the consumption of polymers for plastic applications in Western Europe was 36,769,000 tonnes, an increase of 3.4% from 1999[13]. Of the generated municipal solid waste (MSW) in Thailand, 14% were plastics[14]. According to Onwughara[15], the percentage components of plastics and nylon of different categories of solid generated in Umuahia, capital of Abia State, Nigeria were 1.5% and 10.2%, respectively. Of the generated wastes in Kathmandu Valley in Nepal, 22.65% were plastics[16].

The chemicals produced known as dioxins and furan from plastic, especially incinerating plastics, have been implicated in birth defects and several kinds of cancer. The slag and fly ash were found to be environmentally beneficial in cement production and for off-gases for power production[17]. Thermoplastics make up 80% of the plastics produced today[17]. Examples of thermoplastics include high-density polyethylene (HDPE) used in piping, automotive fuel tanks, bottles and toys; low-density polyethylene (LDPE) used in plastic bags, cling film and flexible containers; PET used in bottles, carpets and food packaging; PP used in food containers, battery cases, bottle crates, automotive parts and fibres; PS used in dairy product containers, tape cassettes, cups and plates; and PVC used in window frames, flooring, bottles, packaging film, cable insulation, credit cards and medical products.

There are hundreds of types of thermoplastic polymer, and new variations are regularly being developed. In developing countries, the number of plastics in common use, however, tends to be much lower. Thermosets make up the remaining 20% of plastics produced. They are hardened by curing and cannot be re-melted or re-moulded and are therefore difficult to recycle. They are sometimes ground and used as a filler material. They include polyurethane (PU)-coatings, finishes, gears, diaphragms, cushions, mattresses and car seats; epoxy-adhesives, sports equipment and electrical and automotive equipment; and phenolics-ovens, handles for cutlery, automotive parts and circuit boards. The global demand for plastic composites has grown significantly over the past few years (**Figure 4**).

Nowadays, the raw materials for plastics come mainly from petrochemicals,

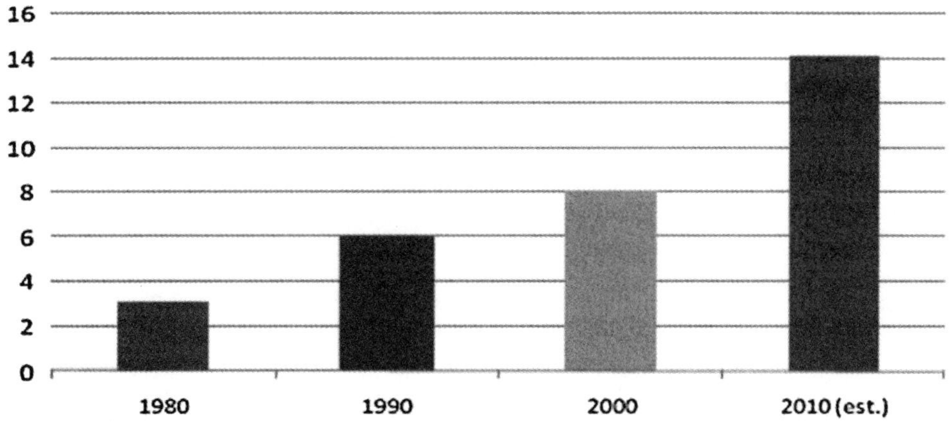

Figure 4. Global demand in the composite industry (Mt) (adapted from[18]).

although originally plastics were derived from cellulose, the basic material of all plant life. The materials used in electronics have several important characteristics. In PC monitors and televisions (TVs), acrylonitrile butadiene styrene (ABS) and high-impact polystyrene (HIPS) are used for cathode ray tube protection. Also, polyphenylene oxide (PPO) has good properties such as high temperature resistance, rigidity, impact strength and creep resistance.

Table 1 shows the summary of typical resins used in different electrical and electronics equipment[7], and **Table 2** shows the weight percentage of manufactured plastic from organic compounds[7][15]. Polymer types used in various construction applications are described in **Table 3**.

A more recent projection (**Figure 5**) shows slightly slower growth to just over 1.4Mt in 2013, but the trend is still strongly positive. The SRI study projects total consumption of biodegradable polymers worldwide at an average annual growth rate of 13% from 2009 to 2014[19].

According to Kurudufu[21], it is estimated that 100 million tonnes of plastics are produced each year. The average European throws away 36kg of plastics each year. Four percent of oil consumption in Europe is used for the manufacture of plastic products. Some plastic waste sacks are made from 64% recycled plastic. Plastic packaging totals 42% of the total consumption, and very little of this is recycled. In 2008, total generation of post-consumer plastic waste in EU-27, Norway and Switzerland was 24.9Mt. Packaging is by far the largest contributor to plastic

Table 1. Resins used in electronic products.

EEE	Resins
Computers	ABS, HIPS, PPO, PPE, PVC, PC/ABS
TVs	HIPS, PC, ABS, PPE, PVC
Miscellaneous	HIPS, PVC, ABS, PC/ABS, PPE, PC

ABS acrylonitrile butadiene styrene, HIPS high-impact polystyrene; PPO polyphenylene oxide; PPE polyphenylene ether; PVC polyvinylchloride; PC polycarbonate. Miscellaneous: fax, telephone, refrigerator etc.

Table 2. Manufactured plastic from organic compounds.

Manufactured plastic	Weight percentage (wt.%)
HIPS	59
ABS	20
PPO	16
PP or PE	2
Other	3

HIPS high-impact polystyrene, ABS acrylonitrile butadiene styrene; PPO polyphenylene oxide, PP polypropylene, PE polyethylene.

Table 3. Main polymers used for applications.

Application	Most common polymers used
Pipes and ducts	PVC, PP, HDPE, LDPE, ABS
Insulation	PU, EPS, XPS
Window profiles	PVC
Other profiles	
Floor and wall coverings	
Lining	PE, PVC
Fitted furniture	PS, PMMA, PC, POM, PA, UP, amino

PVC polyvinylchloride, PP polypropylene, HDPE high-density polyethylene; LDPE low-density polyethylene, ABS acrylonitrile butadiene styrene; PU polyurethane, EPS expanded polystyrene, XPS extruded polystyrene.

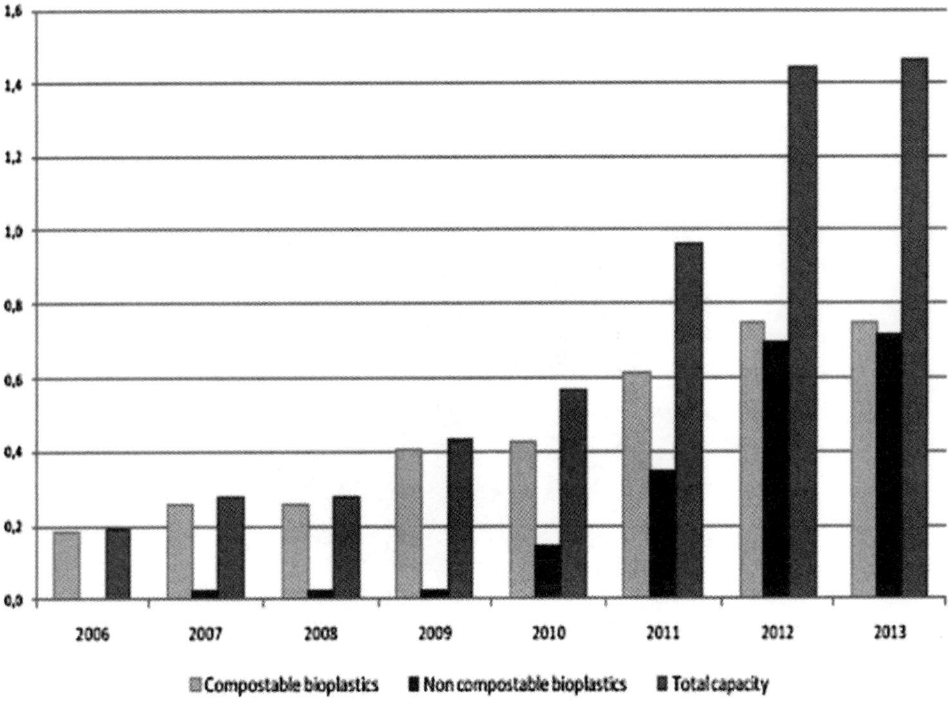

Figure 5. Global production capacity of bioplastics (Mt) (adapted from[20]).

waste at 63%. Average EU-27 per capita generation of plastic packaging waste was 30.6kg in 2007[4]. There are lots of different plastics, and they will give off lots of different vapours when they decompose. **Figure 6** shows various areas where plastics are used[22].

It could be just a simple hydrocarbon or it could contain cyanides, polychlorinated biphenyls (PCBs) or lots of other substances. Without knowing what the plastic was (including what additives might have been incorporated), it would be difficult to know what likely volatiles it would create; volatiles given off from plastics in house fires are a major cause of death. Halogenated plastics, those that are made from chlorine or fluorine, are problematic. This work will review environmental issue ascertained from the development of plastics.

Sources of Waste Plastics

Industrial waste (or primary waste) can often be obtained from the large

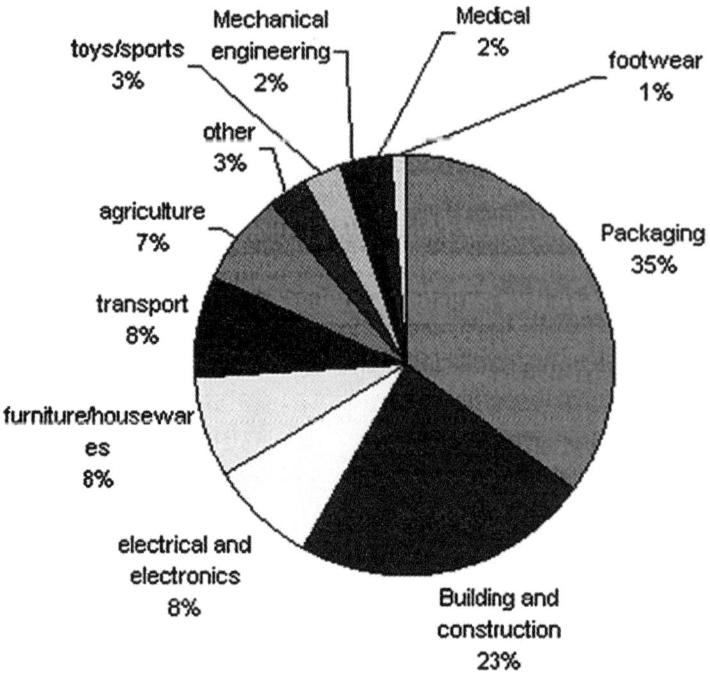

Figure 6. Utilization of plastic in various fields.

plastic processing, manufacturing and packaging industries. Rejected or waste material usually has good characteristics for recycling and thus will be clean. Although the quantity of material available is sometimes small, the quantities tend to be growing as consumption, and therefore production, increases. Commercial waste is often available from workshops, craftsmen, shops, supermarkets and wholesalers. A lot of the plastics available from these sources will be PE, often contaminated. Agricultural waste can be obtained from farms and nursery gardens outside the urban areas. This is usually in the form of packaging (plastic containers or sheets) or construction materials (irrigation or hosepipes). Municipal waste can be collected from residential areas (domestic or household waste), streets, parks, collection depots and waste dumps. In Asian cities, this type of waste is common and can either be collected from the streets or from households by arrangement with the householders[21].

2. Plastics' End-of-Life

Several end-of-life options exist to deal with plastic waste, including recycling, disposal and incineration with or without energy recovery. The plastic recycling rate was 21.3% in 2008, helping to drive total recovery (energy recovery and recycling) to 51.3%. The highest rate of recycling is seen in Germany at 34%[4]. As plastic packaging has the longest established system for the recovery and recycling of plastic waste, it is natural that its recycling rates are higher than those of other streams. It is followed by agricultural waste plastic, which, although not under direct legislative obligation to increase recovery, is subject to economic incentives linked to the availability of homogenous materials. Although WEEE and construction plastic waste sources have relatively low rates of recycling overall, the rate of energy recovery is relatively high. Overall, total recovery is the highest for plastic packaging at 59.8% and the lowest for ELV plastics at 19.2%[4].

The final stage in the life cycle of plastics is disposal. In India, there are three common ways of getting rid of plastics: by dumping them in landfills, by burning them in incinerators or by littering them. In the case of littering, plastic wastes fail to reach landfills or incinerators. It is the improper way of disposing plastics and is identified as the cause of manifold ecological problems. Incineration is a process in which plastic and other wastes are burnt, and the energy produced, as a result, is tapped. In combination with halogens in the plastic fraction, they can form volatile metal halides, but they also have a catalytic effect on the formation of dioxins and furans[9]. WEEE should not be combined with unsorted municipal waste destined for landfills or open burning of garbage because electronic waste can contain more than 100 different substances (toxic chemicals), many of which are toxic such as lead, mercury, hexavalent chromium, selenium, cadmium and arsenic[23]. Additional harmful substances in WEEE can include arsenic, PCBs, chlorofluorocarbons (CFCs) and hydrochlorofluorocarbons (HCFCs) and nickel. Some of these toxic chemicals, even when present in small amounts, can be potent pollutants and contribute to toxic landfill leachate and vapours, such as vaporization of metallic and dimethylene mercury[15].

During burning of WEEE, toxic chemicals such as dioxins and furans may be release to the environment; furthermore, runoff water carries leachate from acidic ash into the sea affecting the aquatic life. Also, the ash leached into the soil which causes groundwater contamination. The municipal solid wastes in Nigeria contain all sources of unsorted wastes, such as commercial refuse, construction and demolition debris, garbage, electronic wastes etc., which are dumped indiscriminately on roadsides and any available open pits irrespective of the health implication on people[15]. Most plastics (thermosets) are from electronic wastes[15]. With the rapid improvements in the electronic industry, electronic waste (e.waste), including all obsolete electronic products, has become the fastest growing component in the solid waste stream. This phenomenon has been a source of hazardous wastes such as PCs and TVs, which contain heavy metals and organic compounds that pose risk to the environment if not properly managed. Balakrishnan shows that 19% of plastic are found in WEEE[8].

More than 20,000 plastic bottles are needed to obtain 1 tonne of plastic[24]. The durability of the most widely used polyethylene plastic films used for protected cultivation varies from a minimum of one cultivating season to a maximum of 2 to 3 years, and at the end of their useful life, they are classified as waste. Most of this waste is currently disposed of in unauthorized dumping sites or burned uncontrollably in the fields. Management of the huge quantities of waste produced in this way represents a problem with great environmental implications. In order to minimize the environmental impact of this plastic waste stream, it is desirable that the films used in protected cultivation have an increased life, thus producing less waste per annum. However, sustainability requires that a degradable material breaks down completely by natural processes so that the basic building blocks can be used again by nature to make a new life form. Plastics made from petrochemicals are not a product of nature and cannot be broken down by natural processes. It is assumed that the breakdown products will eventually biodegrade. In the meanwhile, these degraded, hydrophobic, high surface area plastic residues migrate into the water table and other compartments of the ecosystem causing irreparable harm to the environment[25].

Mechanisms of Degradation

Degradation is a complicated non-linear time-dependent process which affects directly or indirectly several properties of the material related to its functional characteristics. In its final stage of degradation, a material does not meet its functional requirements and is easily prone to mechanical failure. As a practical rule, the useful life of a material is considered to be reached when its initial mechanical strength is reduced by 50%. There are several factors to monitor and criteria to define the degree of degradation. Not all properties are affected by degradation in the same way though. Thus, the elongation at break (expressed as a percentage) appears to be a more sensitive 'index' of degradation than the tensile strength, the stress at yield or the modulus of elasticity. In fact, the material becomes more brittle with degradation, so it cannot retain its initial elongation at break[7].

Degradation of polymers is induced by different external factors and mechanisms. Briefly, the various degradation types for polymers are the following:

1) Thermal degradation occurs due to use or processing of polymers at high temperatures.

2) Photo-induced degradation occurs when, on exposure to the energetic part of the sunlight, *i.e.* the ultraviolet (UV) radiation, or other high-energy radiation, the polymer or impurities within the polymer absorb the radiation and induce chemical reactions.

3) Mechanical degradation occurs due to the influence of mechanical stress-strain.

4) Ultrasonic degradation is the application of sound at certain frequencies which may induce vibration and eventually breaking of the chains.

5) Hydrolytic degradation occurs in polymers containing functional groups which are sensitive to the effects of water.

6) Chemical degradation occurs when corrosive chemicals, such as ozone or the sulphur in agrochemicals, attack the polymer chain causing bond breaking or oxidation.

7) Biological degradation is specific to polymer with functional groups that can be attacked by microorganisms.

3. Environmental Management Issues

3.1. Landfill Option

Landfill not only takes up large areas of land but can also generate bio-aerosols, odours and visual disturbance and may lead to the release of hazardous chemicals through the escape of leachate from landfill sites. Organic breakdown following landfill disposal of biodegradable waste, including bioplastics, causes the release of greenhouse gases. Landfill of waste usually implies an irrecoverable loss of resources and land (since landfill sites can normally not be used post-closure for engineering and/or health risk reasons), and in the medium to long term, it is not considered a sustainable waste management solution[26].

3.2. Incineration Option

The environmental impacts of incinerating plastic waste (as for most solid wastes or fuels) can include some airborne particulates and greenhouse gas emissions. Plants that are compliant with the Waste Incineration Directive are not thought to have any significant environmental impact. However, in some circumstances, energy recovery of plastic waste in MSW incinerators can result in a net increase in CO2 emissions due to substituted electricity and heat production[27]. Therefore, all incineration activities should be associated with suitable filter system trap for released toxic substances, where the incinerators operate in a way not to pollute the atmosphere, soil and groundwater. There will also be an environmental burden due to the disposal of ashes and slag. For example, flue gas cleaning residues often have to be disposed of as hazardous waste due to the toxicity of the compounds they absorb. The net societal cost or benefit would of course depend on the alternatives, e.g. the existing power generation mix and the risk of open-air burning or landfill fires.

3.3. Recycling Option

In western countries, plastic consumption has grown at a tremendous rate over the past two or three decades. In the consumer societies of Europe and America, scarce petroleum resources are used for producing an enormous variety of plastics for an even wider variety of products. Many of the applications are for products with a life cycle of less than 1 year and then the vast majority of these plastics are then discarded. In most instances, reclamation of this plastic waste is simply not economically viable. In the industry (the automotive industry for example), there is a growing move towards reuse and reprocessing of plastics for economic as well as environmental reasons with many praiseworthy examples of companies developing technologies and strategies for recycling of plastics. Plastic recycling needs to be carried out in a sustainable manner. However, it is attractive due to the potential environmental and economic benefits it can provide. There is a wide variety of recycled plastic applications, and the market is growing.

However, the demand depends on the price of virgin material as well as the quality of the recycled resin itself. The use of recycled plastics is marginal compared to virgin plastics across all plastic types due to a range of technological and market factors. Recycled plastics are not commonly used in food packaging (one of the biggest single markets for plastics) because of concerns about food safety and hygiene standards, though this is beginning to change. Another constraint on the use of recycled plastics is that plastic processors require large quantities of recycled plastics, manufactured to strictly controlled specifications at a competitive price in comparison to virgin plastic. Such constraints are challenging, in particular, because of the diversity sources and types of plastic waste and the high potential for contamination. Not only is plastic made from a non-renewable resource but it is also generally non-biodegradable (or the biodegradation process is very slow). This means that plastic litter is often the most objectionable kind of litter and will be visible for weeks or months, and waste will sit in landfill sites for years without degrading.

Although there is also a rapid growth in plastic consumption in the developing world, plastic consumption per capita in developing countries is much lower than in industrialised countries. These plastics are, however, often produced from expensive imported raw materials. Not all plastics are recyclable. There are four types of plastics which are commonly recycled:

1) Polyethylene—both high density and low-density polyethylene

2) Polypropylene

3) Polystyrene

4) Polyvinyl chloride

A common problem with recycling plastics is that plastics are made up from parts of more than one kind of polymer or there may be some sort of fibre added to the plastic (a composite) to give added strength. This can make recovery difficult. When thinking about setting up a small-scale recycling enterprise, it is advisable to first carry out a survey to ascertain the types of plastics available for collection, the type of plastics used by manufacturers (who will be willing to buy the reclaimed material) and the economic viability of collection. The method of collection can vary. The following gives some ideas:

- House-to-house collection of plastics and other materials (e.g. paper)
- House-to-house collection of plastics only (but all types of polymer)
- House-to-house collection of certain objects only
- Collection at a central point, e.g. market or church
- Collection from street boys in return for payment
- Regular collection from shops, hotels, factories etc.
- Purchase from scavengers on the municipal dump
- Scavenging or collecting oneself

The method will depend upon the scale of the operation, the capital available for the set-up, transport availability etc. It should be noted that the ideas above assume an expansion in recycling capacity, which will require associated expansion in collection activities, use of secondary plastic materials and, associated with the latter, better methods for separating the different types of plastic to reduce contamination levels. These will allow the delivery of higher quality plastic waste streams to facilitate higher levels of recycling and to ensure quality markets for the secondary raw materials that result. **Figure 7** is an example of the life cycle of recycled waste.

Recycled PET Post-consumer PET is often an attractive material for recycling. Unlike other polymers, nowadays, recycled PET can be produced and then

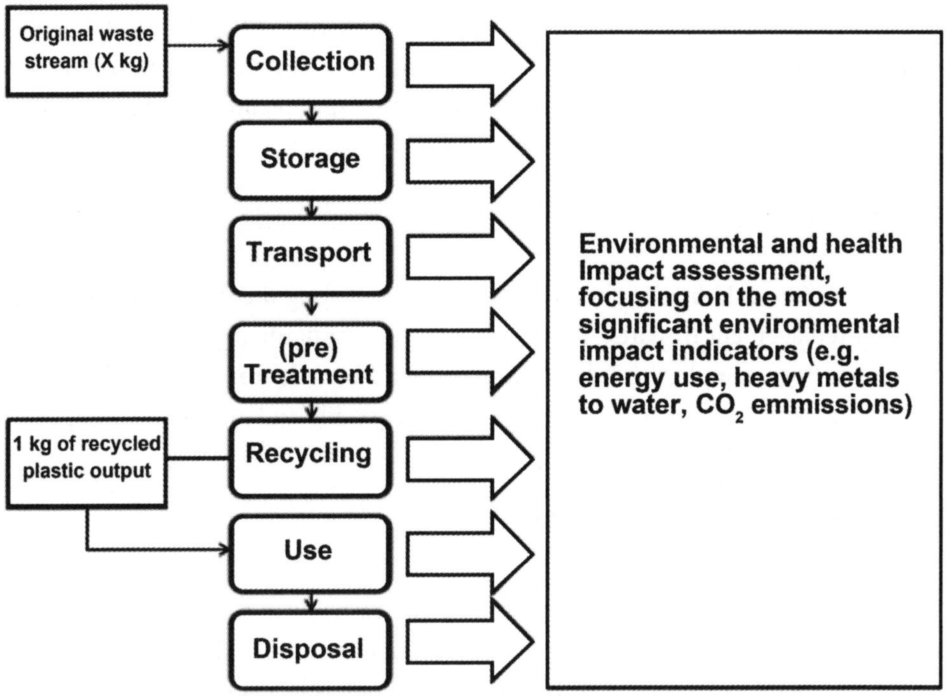

Figure 7. Life cycle approach for analysing the environmental impacts of plastic recycling.

directly suitable for contact with food if it is submitted to further decontamination steps such as super clean processes, which are able to decontaminate post-consumer contaminants to concentration levels of virgin PET materials[28]. PET can also be used in applications such as carpet fibres, geotextiles, packaging and fibre fill. PET can be converted into polybutylene terephthalate (PBT) resin, which can be a valuable material for injection and blow-moulding applications. PBT is created through chemical polymerisation which converts the PET molecular chain into 'small repeating units', and through additional catalyst-assisted processes, PBT is produced. The polymerised PBT contains approximately 60% of the original mass of PET and can reduce solid waste by up to 900kg for each tonne of PBT produced. Making PBT from recycled PET is often less energy consuming than producing the resin directly from oil stock (at 50 to 20GJ/tonne, respectively)[29].

The main trends of interest in terms of economic impacts are anticipated to be the relative expansion of the recycling sector and questions regarding the economic impact of potentially lower economic growth on plastic waste treatment and secondary raw material use. The main social impacts are anticipated to be associated with health, and in particular, the epidemiological impacts are associated with the treatment of waste in third countries and the social perceptions around the continued use and increasing levels of plastic consumption and waste production.

3.3. Re-Use Option

Products could be designed for re-use by facilitating the dismantling of products and replacement of parts. This could involve standardising parts across manufacturers. For example, LED lamp designs could benefit from standardisation of parts (many of which are plastic) to facilitate disassembly and remanufacturing[30].

4. Environmental Impacts

Pollutants released from burning plastic waste in a burn barrel are transported through the air either short or long distances and are then deposited onto land or into bodies of water. A few of these pollutants such as mercury, PCBs, dioxins and furans persist for long periods of time in the environment and have a tendency to bioaccumulate, which means they build up in predators at the top of the food web. Bioaccumulation of pollutants usually occurs indirectly through contaminated water and food rather than breathing the contaminated air directly. In wildlife, the range of effects associated with these pollutants includes cancer, deformed offspring, reproductive failure, immune diseases and subtle neurobehavioral effects. Humans can be exposed indirectly just like wildlife, especially through consumption of contaminated fish, meat and dairy products.

Environmental pollution can also be harmful to the structural integrity of the polyethylene due to chemical attack of the polymer bonds. Atmospheric pollutants such as nitrogen oxides, sulphur oxides, hydrocarbons and particulate can enhance the degradation of the polymers especially when combined with applied stress and must also be taken into account[31]. For instance, infrared studies have revealed that polyethylene reacts with NO_2 at elevated temperatures and that chemical attack is observed even at 25°C, probably due to the presence of olefinic bond impurities which react readily with NO_2. Similarly, SO_2 is rather reactive, especially in the presence of UV irradiation, which it readily absorbs and forms triplet excited sulphur dioxide ($3SO_2$*). This species is capable of abstracting hydrogen from the polymer chains leading to the formation of macroradicals in the polymer structure, which in turn can undergo further depolymerisation.

Overall, the level of environmental impact associated with plastic waste is anticipated to increase over the period until 2015 due to continued growth in plastic waste production (associated with continued rises in plastic waste consumption). More specifically, greenhouse gas emissions associated with the plastic life cycle are anticipated to increase, albeit on a lower trajectory than in the past. Negative consequences in terms of littering and plastic pollution in marine waters would also be anticipated to increase in the absence of any additional curbs[4]. Due to many factors, not the least of which is their ready availability, 96% of all plastic grocery bags as an example are thrown into landfills[32]. However, plastic bags decompose very slowly, if at all. In fact, a bag can last many years, inhibiting the breakdown of biodegradable materials around or in it[32].

Lightweight plastic grocery bags are additionally harmful due to their propensity to be carried away on a breeze and become attached to tree branches, fill roadside ditches or end up in public waterways, rivers or oceans. In one instance, Cape Town, South Africa had more than 3,000 plastic grocery bags that covered each kilometre of the road[32]. In this century, an estimated 46,000 pieces of plastic are floating in every square kilometre of the ocean worldwide[32].

What is most distressing is that over a billion seabirds and mammals die annually from ingestion of plastics[32]. According to UNEP, plastic waste in the ocean causes the deaths of up to one million seabirds, 100,000 marine mammals and countless fish every year[33]. Big animals (e.g. turtles, whales, seals, and sea lions) can be trapped by nets and films and eat the small particles of plastics which may block their digestive systems. Entanglement and ingestion of plastic fragments can even lead to death by drowning, suffocation, strangulation or starvation through reduced feeding efficiency. At least 267 different species are known to have suffered from entanglement or ingestion of marine debris, including seabirds, turtles, seals, sea lions, whales and fish[34]. According to Brown, in Newfoundland, 100,000 marine mammals are killed each year by ingesting plastic[35]. However, the impact of plastic bags does not end with the death of one animal; when a bird or mammal dies in such a manner and subsequently decomposes, the plastic bag will again be released into the environment to be ingested by another animal.

Another environmental issue involves blowing agents used to make foamed plastics. All blowing agents eventually escape to the atmosphere, and among them, there is a particular concern with the CFC stratospheric ozone layer. An international treaty was signed in 1990 on CFCs which was completely implemented in 2000. In some more restricted geographical area, the smoke-forming potential of hydrocarbon blowing agents is also considered an issue.

Because CFCs have been widely employed in foamed plastics, including polystyrene, rigid and flexible poly-urethanes and polyolefins, for example, there has been intense activity to develop satisfactory substitutes. Among the most promising of these are the HCFCs, which have only 2% to 10% of the ozone depletion potential of CFCs, and hydrofluorocarbons, which have zero ozone depletion potential. The current consensus is that the HCFCs represent a transitional solution to the problem. There has been promising development work with CO_2 as a blowing agent for polystyrene foam sheet. In this application, CO_2 is considered environmentally benign.

Dioxins

Dioxins are unintentionally but unavoidably produced during the manufacture of materials containing chlorine, including PVC and other chlorinated plastic feed-stocks. Burning these plastics can release dioxins. Polychlorinated dibenzo-*p*-dioxins and polychlorinated dibenzofurans, hexachlorobenzene and PCBs are unintentional persistent organic pollutants (U-POPs), formed and released from thermal processes involving organic matter and chlorine as a result of incomplete combustion or chemical reactions. These U-POPs are commonly known as dioxins because of their similar structure and health effects[36].

Dioxin is a known human carcinogen and the most synthetic carcinogen ever tested in laboratory animals. A characterization by the National Institute of Standards and Technology of cancer-causing potential evaluated dioxin as over 10,000 times more potent than the next highest chemical (diethanol amine), half a million times more than arsenic and a million or more times greater than all others. Also, open burning of plastic or incineration involves air emissions of sulphur dioxide, nitrogen dioxide, chlorine, dioxin and fine particulates and emissions of greenhouse gases of CO_2 and nitrous oxide (N_2O); ash which remains after incineration needs to be disposed of and can be toxic.

The pyrolysis or combustion of even a simple synthetic polymer produces mix grill of gases. Most of these gases are self-toxic, *i.e.* interfering with the normal biochemical processes of the body or exclude O_2 from the victim. It must be understood that the type and the concentration of these gases in any fire situation vary from material to material as well as the environment. A few of them together with their physiological effects are shown in **Table 4**.

5. Health Effects

Because of the persistent and bio-accumulative nature of dioxins and furans, these chemicals exist throughout the environment. Human exposure is mainly through consumption of fatty foods, such as milk. IPEP[36] notes that 90% to 95% of human exposure to dioxins is from food, particularly meat and dairy products. Their health effects depend on a variety of factors, including the level of exposure, duration of exposure and stage of life during exposure. Some of the probable health effects of dioxins and furans include the development of cancer, immune system suppression, reproductive and developmental complications and endocrine disruption[36]. The International Agency for Research on Cancer has identified 2, 3, 7, 8-tetrachlorodibenzodioxin as the most toxic of all dioxin compounds.

High exposure to PDBEs, which accumulate in the human body, has been linked to thyroid hormone disruption, permanent learning and memory impairment, behavioural changes, hearing deficits, delayed puberty onset, impaired infant neurodevelopment, decreased sperm count, fetal malformations and possibly cancer. These activities lead to severe pollution of soils by POPs and heavy metals in the countries concerned, which may also affect the surrounding environment such as rice fields and rivers via atmospheric movement and deposition[37]–[39].

Toxic emissions produced during the extraction of materials for the production of plastic grocery bags, their manufacturing and their transportation contribute to acid rain, smog and numerous other harmful effects associated with the use of petroleum, coal and natural gas, such as health conditions of coal miners and

Table 4. Physiological effects of some gases involved in combustion.

Gas concentration	Effect in all fire conditions
Oxygen (O_2 (%))	
21	Normal concentration in air
2 to 15	Shortness of breath, headache, dizziness, quickened pulse, fatigue on exertion, loss of muscular coordination for skilled movement
10 to 12	Nausea and vomiting, exertion impossible, paralysis of motion
6 to 8	Collapse and unconsciousness, but rapid treatment can prevent death
6	Death in 6 to 8min
2 to 3	Dcath in 45s
Carbon monoxide (CO (ppm))	
400	Nausea after 1 or 2h, collapse after 2h and death after 3 to 4h
1,000	Difficulty in ambulation, death after 2h
2,000	Death after 45min
3,000	Death after 30min
5,000	Rapid collapse, unconsciousness and death within a few minutes
Carbon dioxide (CO_2 (ppm))	
250 to 350	Normal concentration in air
25,000	Ventilation increased by 100%
50,000	Symptoms of poisoning after 30min, headache, dizziness and sweating
120,000	Immediate unconsciousness, death in a few minutes
Hydrogen cyanide (HCN (ppm))	
45 to 54	Tolerable for 1/2 to 1h[a]
110 to 135	Fatal after 1/2 to 1h[a]
181	Fatal after 10min[a]
280	Immediately fatal[a]
Hydrogen chloride (HCl (ppm))	
5 to 10	Mild irritation of mucus membranes[b]
50 to 100	Barely tolerable[b]
1,000	Danger of lung oedema after a short exposure[b]
1,000 to 2,000	Immediately hazardous to life[b]
Acrolein (CH_2CHCHO (ppm)) (PVC)	
1	Immediately detectable irritation[c]
5.5	Intense irritation[c]
<10	Lethal in short time[c]
24	Unbearable[c]

[a]Effect from nitrogen-containing polymeric materials, e.g. acrylics, wool, urethane foam etc.; [b]effect from chloride-containing polymers e.g. PVC; [c]effect from many polymeric materials, e.g. wool and polypropylene.

environmental impacts associated with natural gas and petroleum retrieval[40]. Heavy metals may be released into the environment from metal smelting and refining industries, scrap metal, plastic and rubber industries, various consumer products and from burning of waste containing these elements. On release to the air, the elements travel for large distances and are deposited onto the soil, vegetation and water depending on their density. Once deposited, these metals are not degraded and persist in the environment for many years poisoning humans through inhalation, ingestion and skin absorption. Acute exposure leads to nausea, anorexia, vomiting, gastrointestinal abnormalities and dermatitis.

Impacts on human health are perhaps the most serious of the effects associated with plastic grocery bags, ranging from health problems associated with emissions to death. In the year 2005, the city of Mumbai, India experienced massive monsoon flooding, resulting in at least 1,000 deaths, with additional people suffering injuries[32]. City officials blamed the destructive floods on plastic bags which clogged gutters and drains, preventing the rainwater from leaving the city through underground systems. Similar flooding happened in 1988 and 1998 in Bangladesh, which led to the banning of plastic bags in 2002[40]. By clogging sewer pipes, plastic grocery bags also create stagnant water; stagnant water produces the ideal habitat for mosquitoes and other parasites which have the potential to spread a large number of diseases, such as encephalitis and dengue fever, but most notably malaria.

6. Conclusion

Overall, the level of environmental impact associated with plastic waste is anticipated to increase over the period until 2015 due to continued growth in plastic waste production (associated with continued rises in plastic waste consumption). Over this period, the rise in environmental impacts is anticipated to be comparatively slower than in the past as much of this increase in production is dealt with by recycling and energy recovery expansion. However, disposal levels are only anticipated to remain static or drop in a limited way, maintaining the overall picture of the environmental footprint.

In terms of environmental impacts, the following trends are considered to be of most significance:

Rising use of plastics. The primary plastic feedstock will remain fossil fuels, despite the anticipated rapid rise in the production of bioplastics. This implies continued reliance on carbon-intensive production methods, with relatively high levels of embodied carbon and energy in the products. While traditional refineries might be driven to be more efficient over the projection period due to changes in rules surrounding for example the Fuel Quality Directive (which requires life cycle reductions in transport fuels), such efficiencies are likely to be offset by the increasing the level of production and demand.

Rising levels of plastic waste generation. This implies the need for an expanded waste management system simply to remain capable of dealing with the anticipated increase waste production.

Increasing levels of recycling. Recycling rates are anticipated to increase over the outlook period, and end markets are developing. However, the proportion of disposal is expected to remain significant. This implies a significant expansion in the overall Mt amount of waste recycled, *i.e.* a similar proportion of a greater quantity of waste will be recycled. This in turn implies three key evolutions in the plastic waste recycling business. Firstly, an expansion in the collection of plastic waste, secondly an expansion in processing capacity and thirdly an expansion in the use of secondary plastic materials. Legislation has already been proposed or passed in the USA of the federal, state and local levels restricting or banning the use of some plastics in applications where there is perceived problem. This trend is sure to continue. The technical community has not lagged in developing responses to these challenges expect in the developing nations. The principal routes being followed are degradation, incineration and recycling. Another important approach which involves both consumers and manufacturers is source reduction. The activity in this area is focused largely on warning consumers from the throwaway habits that have developed over recent decades, particularly with packaging waste stream. These efforts are evolving so rapidly that it is difficult to predict how each one will impact the problem over the longer term.

European approaches. In Europe, the principal measures implemented to deal with plastics are the producer responsibility mechanisms—these do not target plastic bags specifically but aim to encourage the recycling and recovery of plastics. Different member states use different approaches, but in most countries, the packaging industry makes payments to designated bodies that are responsible for arranging the collection, separation, recycling and recovery of a predetermined amount of packaging. A notable feature is that these fees paid by the packaging industry are not necessarily passed on to consumers in a transparent manner.

Therefore, the collection of taxes and public awareness can reduce indiscriminate use of plastic bags.

Recommendation

The redesign of plastic products, both at the scale of the individual polymer and in terms of the product's structure, could help alleviate some of the problems associated with plastic waste. With thoughtful development, redesign could have an impact at all levels of the hierarchy established by the European Waste Framework Directive: prevention, re-use, recycle, recovery and disposal[1]. Infrastructure for the safe disposal and recycling of hazardous materials and municipal solid waste should be developed. Approximately 50% of all waste is organic and can therefore be composted. Another large segment of the remainder can be recycled, leaving only a small portion to be disposed. The remaining portion can then be disposed through sanitary landfills, sewage treatment plants and other technologies.

In general, disposal via modern, sanitary landfills is not very effective with biodegradable plastic materials because the limited availability of oxygen and moisture retards most biodegradation processes. As yet, bioplastics cannot replace all types of plastic, particularly certain types of food packaging that require gas permeability. Biodegradable plastics are most effectively treated in compositing systems where aerobic processes predominate. Biodegradation may also influence the types and concentrations of soil microflora in disposal areas. Enrichment of soil with certain microflora could have unanticipated risks, such as an outbreak of a new microbial disease[41].

At present, there is a growing consensus that the concept of degradable plastics has been oversold as a solution to the waste disposal problem, primarily because a large portion of degradable plastics will end up in landfills where breakdown tends to be very slow. The most promising applications of degradable plastics probably are for more limited problems, such as litter, where sunlight, air, moisture and microorganism are generally available to accelerate polymer breakdown[42].

High-temperature incineration of waste plastics has at least two potential advantages. First, increasingly scarce landfill space is conserved because the volume reduction from feed to ash with a well-operated incinerator is in 90% range or higher; second, the high generation of stream, electricity or both. Incinerators have drawbacks, however. They can emit corrosives such as HCl and traces of highly toxic dioxins and furans if chlorine-containing materials such as PVC or bleached paper are in the feed stream. These emissions could pose hazards, especially to people living close to the incinerator; compounds of toxic heavy metals such as lead, chromium and cadmium are present in some plastic products as stabilizers and pigments, and these elements tend to end up concentrated in the ash. If the compounds of these metals in the ash are leachable, soil and groundwater contamination is possible. Advocates of incineration are convinced; however, that current technology will prevent stack discharge of most toxic and corrosive combustion products. They further claim that only a very small fraction of the heavy metals in the ash is leachable, and this should pose no problem if proper disposal procedures are followed. Moreover, heavy metal-based pigments and stabilizers gradually are being phased out of many applications in favour of organic substitutes. Current trends suggest increasing reliance on the incineration approach despite the claimed drawbacks.

Recycling has obvious environmental advantages and is not opposed by any strong voting blocs or economic interests. Recycling of plastics should be carried in a manner to minimize pollution during the process and enhance efficiency and conserve the energy. Much of the future success of plastic waste recycling will depend on the development of effective collection and separation systems, which, along with appropriate incentives, will ensure the broad and willing participation of industry and consumers. In this case, involving pre-consumer waste or scrap, where the material identifies and uniformity can be reasonably maintained, it is often possible to recycle the plastic back to the same product. In other instances, including those where carefully targeted post-consumer collection methods are possible. A secondary product can be made of a single recycled plastic, for example, PET beverage bottle scrap can be recycled to fabricate bottles for non-beverage applications[7].

Clearer certification and labelling schemes are needed to ensure that the public understands what is meant by biodegradable, compostable or eco-friendly. DG Environment's report on Plastic waste in the environment[11] proposed that any targets on bioplastics should be combined with a labelling system and initiatives to increase public awareness and education. Labelling of plastic parts with the type of polymer they contain could also help in sorting for recycling and re-use. Along with the plastic waste issue, significant new developments can be anticipated as the industry continues its aggressive search for solutions to these important environmental problems. The redesign of plastics and bioplastics has the potential to reduce the use of fossil fuels, decrease CO_2 emissions and decrease plastic waste. There is a need to increase public awareness through litter education as an important supporting element and other initiatives that may be undertaken to reduce plastic waste and their impacts.

Competing Interests

The authors declare that they have no competing interests.

Authors' Contributions

This work was finished through the collaboration of all authors. OIN conceived the study and drafted the manuscript together with AOI. CHC and LA carried out the computations in the manuscript. CHC, AOI and LA participated in the coordination and in revising the manuscript. All authors read and approved the final manuscript.

Acknowledgements

The authors would like to thank all the reviewers who read this paper carefully and provided valuable suggestions and comments. The editorial assistance of our colleagues Mr. Kanno Okechukwu Charles and Mr. Chukwuma Royal are very much appreciated.

Source: Nkwachukwu O I, Chima C H, Ikenna A O, *et al.* Focus on potential environmental issues on plastic world towards a sustainable plastic recycling in developing countries[J]. International Journal of Industrial Chemistry, 2013, 4(1): 1−13.

References

[1] Science for Environmental Policy (2011) Plastic waste: redesign and biodegradability. European Commission, Brussels, pp 1–8.

[2] Thompson RC, Swan SH, Moore CJ, Vom Saal FS (2009) Our plastic age. Philosophical Transactions of the Royal Society 364:1973 1976.

[3] Plastics Europe (2007) The compelling facts about plastics—an analysis of plastics production, demand and recovery for 2005 in Europe. Plastics Europe, Brussels. http: //www.plasticseurope.org/Content/Default.asp?PageID=517#.

[4] Mudgal S, Lyons L, Bain J, Débora D, Thibault F, Linda J (2011) Plastic waste in the environment: revised final report. European Commission DG Environment. Bio Intelligence Service, France. http://www.ec.europa.eu/environment/waste/studies/pdf/plastics.pdf. Accessed April 2011.

[5] Europe P (2009) The compelling facts about plastics—an analysis of European plastics production, demand and recovery for 2008. Plastics Europe, Brussels.

[6] Hopewell J, Dvorak R, Kosior E (2009) Plastics recycling: challenges and opportunities. Philosophical Transactions of the Royal Society 364:2115–2126.

[7] Onwughara IN, Nnorom IC, Kanno OC, Chukwuma RC (2010) Disposal methods and heavy metals released from certain electrical and electronic equipment wastes in Nigeria: adoption of environmental sound recycling system. International Journal of Environmental Science and Development 1(4):290–296.

[8] Balakrishnan RB, Anand KP, Chiya AB (2007) Electrical and electronic waste: a global environmental problem. Journal of Waste Management and Research 25: 307–317.

[9] Antrekowitsch H, Potesser M, Spruzina W, Prior F (2006) Metallurgical recycling of electronic scrap. Proceedings of the 135th The Minerals, Metals and Materials Society (TMS) annual meeting and exhibition. San Antonio, pp 899–904.

[10] Plasticsnews (http://plasticsnews.com/fyi-charts/index.html?id=17731. Accessed 13 October 2008) Paying more for less. http://plasticsnews.com/fyi-charts/index.html?id =17731. Accessed 13 October 2008.

[11] Mudgal S, Lyons L (2010) Plastic waste in the environment: final report.European Commission DG Environment. Bio Intelligence Service, France.

[12] Plastics Europe (2010) Plastics—the facts. An analysis of European plastics production, demand and recovery for 2009.

http://www.plasticseurope.org/document/plastics—the-facts-2010.aspx?FolID=2. Accessed 27 October 2010.

[13] Association of Plastics Manufacturers in Europe (APME) (2000) An analysis of plastics consumption and recovery in Europe. APME, Brussels.

[14] Thaniya K (2009) Sustainable solutions for municipal solids waste management in Thailand. World Academy of Science, Engineering and Technology 36:666–671.

[15] Onwughara IN, Nnorom IC, Kanno OC (2010) Issues of roadside disposal habit of municipal solid waste, environmental impacts and implementation of sound management practices in developing country: "Nigeria". International Journal of Environmental Science and Development 1(5):409–417.

[16] Luitel KP, Khanal SN (2010) Study of scrap waste in Kathmandu Valley. Kathmandu University Journal of Science, Engineering and Technology 6(1):116–122.

[17] Chi JO, Sung OL, Hyung SY, Tae JH, Myong JK (2003) Selective leaching of valuable metals from waste printed circuit boards. J Air Waste Manage Assoc 53:897–898.

[18] Witten E (2009) The composites market in Europe. AVK, Germany.

[19] Marcos NC (2010) Renewable chemicals and polymers. What's next? SRI Consulting 2010. http://www.apla.com.ar/img/conferencias/84_conf.pdf. Accessed November 2010.

[20] European Bioplastics (2011) European Bioplastics, driving the evolution of plastics. http://european-bioplastics.org/index.php?id=141.

[21] Kurudufu P (2009) Recycling plastic. Practical action Eastern Africa. http://practicalaction.org/docs/technical_information_service/recycling_plastics.pdf. Accessed 2 October 2009.

[22] Zereena BI (2010) Plastics and environment. Dissemination Paper-12. Centre of Excellence in Environmental Economics. http://coe.mse.ac.in/dp/ Paper12.pdf. Accessed 20 April 2010.

[23] Tippayawong NN, Khongkrapan P (2009) Development of a laboratory scale air plasma torch and its application to electronic waste treatment. International Journal for Environmental Science and Technology 6(3):407–411.

[24] Lardinois I, Van de K (1995) A plastic waste, option for small-scale resource recovery. TOOL, Amsterdam.

[25] Gautam SP (2009) Bio-degradable plastics impact on environment. Central Pollution Control Board Ministry of Environment and Forests Government of India, New Delhi.

[26] Commission E (2008) Green Paper on the management of bio-waste in the European Union. COM, European Commission, Brussels.

[27] Pilz H, Brandt B, Fehringer R (2010) The impact of plastics on life cycle energy consumption and greenhouse gas emissions in Europe. Denkstatt summary report, Plastics Europe, Brussels.

consumption and greenhouse gas emissions in Europe. Denkstatt summary report, Plastics Europe, Brussels.

[28] Welle F (2011) Twenty years of PET bottle to bottle recycling—an overview.Resources, Conservation and Recycling 55(11):865–875.

[29] Plastemart (2003) Green method of manufacturing virgin PET/PBT from recycled products offers energy saving. http://plastemart.com/upload/Literature/Green-method-manufacture-virgin%20PET-PBT-recycled-products-energy%20saving-Valox%20iQ-Xenoy%20iQ.asp. Accessed 15 May 2009.

[30] Hendrickson CT, Matthews DH, Ashe M, Jaramillo P, McMichael FC (2010) Reducing environmental burdens of solid-state lighting through end-of-life design. Environ Res Lett. doi:10.1088/1748-9326/5/1/014016.

[31] Dilara PA, Briassoulis D (2000) Degradation and stabilization of low-density polyethylene films used as greenhouse covering materials. Journal of Agricultural Engineering Resource 76:309–321. doi:10.1006/jaer.1999.0513. http://www.idealibrary.com.

[32] Ellis S, Kantner S, Saab A, Watson M (2005) Plastic grocery bags: the ecological footprint. VIPIRG, Victoria, pp 1–19.

[33] UNEP (2006) Ecosystems and biodiversity in deep waters and high seas. UNEP Regional Seas Reports and Studies No. 178. UNEP/IUCN, Switzerland. http://unep.org /pdf/EcosystemBiodiversity_DeepWaters_20060616.pdf.

[34] Derraik JGB (2002) The pollution of the marine environment by plastic debris: a review. Mar Pollut Bull 44:842–852.

[35] Brown S (2003) Seven billion bags a year. Habitat Australia 31(5):P28.

[36] The International POPs Elimination Project (IPEP) (2005) A study on waste incineration activities in Nairobi that release dioxin and furan into the environment. Environmental Liaison, Education and Action for Development (ENVILEAD), Kenya.

[37] Wong MH, Wu SC, Deng WJ, Yu XZ, Luo Q, Leung AOW, Wong CSC, Luksemburg WJ, Wong AS (2007) Export of toxic chemicals—a review of the case of uncontrolled electronic-waste recycling. Environ Pollut 149:131–140.

[38] Environmental Working Group (2003) Mother's milk—toxic fire retardants (PBDEs) in human breast milk. Environmental Working Group, Washington, DC.

[39] Herbstman JB, Andreas S, Matthew K, Sally AL, Richard SJ, Virginia R, Larry LN, Deliang T, Megan N, Richard YW, Frederica P (2010) Prenatal exposure to PBDEs and neurodevelopment. Environ Health Perspect 118(5):712–719.

[40] Environmental Literacy Council (2005) Paper or plastic? http://www.enviroliteracy.org/article.php/1268.html. Accessed 20 November 2005.

[41] Sudesh K, Iwata T (2008) Sustainability of biobased and biodegradable plastics.

Clean 36(5–6):433–442.

[42] Taylor DA (2010) Principles into practice: setting the bar for green chemistry. Environ Health Perspect 118(6):254–257.

Chapter 22

"We Are Used to This": A Qualitative Assessment of the Perceptions of and Attitudes towards Air Pollution amongst Slum Residents in Nairobi

Kanyiva Muindi[1,2], Thaddaeus Egondi[1,2], Elizabeth Kimani-Murage[1], Joacim Rocklov[2], Nawi Ng[2]

[1]African Population and Health Research Center (APHRC), Nairobi, Kenya
[2]Epidemiology and Global Health Unit, Umeå University, SE-901 85 Umeå, Sweden

Abstract: Background: People's perceptions of and attitudes towards pollution are critical for reducing exposure among people and can also influence the response to interventions that are aimed at encouraging behaviour change. This study assessed the perceptions and attitudes of residents in two slums in Nairobi regarding air pollution. Methods: We conducted focus group discussions with residents aged 18 years and above using an emergent design in the formulation of the study guide. A thematic approach was used in data analysis. Results: The discussions revealed that the two communities experience air pollution arising mainly from industries and dump sites. There was an apparent disconnect between knowledge and practice, with individuals engaging in practices that placed them at high risk of exposure to air pollution. Residents appear to have rationalized the situation in which they live in and were resigned to these conditions. Consequently, they expressed lack of agency in addressing prevalent air pollution within their communities. Conclusions: Community-wide education on air pollution and related health effects together with the measures needed to reduce exposure to air pollution are necessary towards reducing air pollution impacts. A similar city-wide study is recommended to enable comparison of perceptions along socio-economic groups and neighbourhoods.

Keywords: Air Pollution, Perceptions, Attitudes, Nairobi, Slums

1. Background

Urban air pollution remains a major health risk to millions of urban residents worldwide as it is estimated that about 1.3 million deaths annually are attributable to urban air pollution[1]. The problem is intensifying in many cities and towns in the developing world where growing urban populations and the attendant increase in activities which have led to a rise in air polluting emissions. This is coupled with industrial growth amidst weak or non-existent environmental protection laws, leading to levels of air pollution that often exceed emissions standards set by the World Health Organization[2]-[5]. In addition, urban poor populations are more disadvantaged in terms of exposure to air pollution because of poor housing structures, close proximity to air pollution sources and the types of fuel used for cooking. However, little is known about how people view air pollution in urban areas.

Studies on air pollution in Africa and other low- and middle-income countries (LMICs) are few[5]. As LMICs work towards gaining industrialised status, economic growth is emphasized at the expense of the environment and the health of the people. In these contexts, there is greater need for environmental stakeholders to address air [environmental] pollution in order to set the stage for inclusive and well-informed interventions aimed at reducing air[environmental] pollution effect. A starting point for such enquiry would be looking at individual perceptions and attitudes towards air pollution. Gaining knowledge of peoples' perceptions of air pollution is important, as it reflects the social dimensions and circumstances under which people understand pollution[6]. This knowledge helps to ensure that policy and communication frameworks achieve desired change in public attitudes and behaviour.

To involve lay opinions in the process of policy making, it would be important to understand public perceptions as they are important factors in the successful implementation of environmental policies[6][7]. First, understanding public perceptions of air pollution and associated risk will ensure there is a consultative process in the formulation of policies as opposed to the existing top-down approach where the public is largely a recipient of policy actions without giving any input in their formulation[7]. Secondly, perceptions of pollution and the associated consequences will give implementing agencies the opportunity to assess public knowledge regarding air pollution as well as any misconceptions that might exist. Addressing these misconceptions could increase involvement of the public in policy implementation[5]. Thirdly, understanding people's perceptions would inform the entry point of actions/interventions aimed at mitigating pollution-related risks[8].

In order to address the prevailing pollution, policies need to be formulated and implemented to curb emissions and encourage air friendly practices among the populace. The acceptance of formulated policies requires the cooperation of the populace as some of the policy actions would impact them directly or may require individual behaviour change. Indeed, the 1992 Earth Summit recommended the inclusion of the public in the environment policy process to ensure the effective implementation of formulated policies[9]. However, there has been little progress on inclusion of public globally as most environmental policies tend to lean more on science than on public opinions[7]. The top-down approach in policy formulation and implementation has been faced with serious challenges as the public view of risk does not overlay the scientific opinion of that risk[10][11]. This disconnect has led to failure of sections of the public adhering to existing policies. For example, a study in a province in China found that the population was largely ignorant of existing environmental protection policies and were not adhering to expected actions[12].

Perception of risk or vulnerability is a central component of health promotion theories and has been found to be shaped by several factors. This study is anchored on the general protection motivation theory[13] that posits that the intention to protect one-self is dependent on: 1) the perceived severity of a threatening event, 2) the perceived probability of the occurrence, or vulnerability, 3) the efficacy of the recommended preventive behaviour, and 4) the perceived self-efficacy to undertake the recommended preventive behaviour. In this study, the perceptions about air pollution levels and related health risks form the first and second components of the theory. The third component refers to the common practices among the residents that are expected to reduce air pollution and exposure levels. Lastly, the perceptions of residents regarding their role in addressing the prevailing air pollution represent the fourth component of the theory. Studies have found that perceptions on levels of air pollution are shaped by the presence of suspected sources of pollution such as industries or busy roads[8][14]. Further, social interactions that ensure the diffusion of knowledge have been shown to shape perceptions[15].

Studies in developed countries show that people in poor neighbourhoods are less likely to report air pollution as an issue of top concern[14]. However, there exist contrary findings in similar contexts[16]. Research evidence suggests that people's level of attachment to the place of residence and social capital determines how they responded about the levels of air pollution and their willingness to take action against pollution. People with high attachment to their place of residence and those with high social capital are less likely to report air pollution as an issue to avoid stigma, and are also likely to take action to address air pollution[17][18].

In slum areas around the world, environmental degradation ranks among the key challenges residents face[19][20]. This is compounded by the fact that many slums are located near industrial districts or close to busy highways. In addition, crowding, a characteristic of many slums, renders the adoption of measures to reduce pollution at individual household level ineffective due to a 'neighbourhood effect' in which adopting households may continue to suffer due to exposure from non-compliant households. Lastly, the political, social and economic exclusion of slum areas[21] puts them in a vulnerable position as they lack systems to manage such things as waste collection or find a collective voice to bargain for services and protection against external polluters such as industries.

There is limited evidence on people's perceptions and attitudes towards air pollution in Kenya. Therefore, the objective of the study was to assess the perceptions and attitudes of slum residents about air pollution. This study is expected to provide insights into people's perceptions on air pollution and what they consider to be their role in addressing air pollution. The results emerging from this study will be important in informing other larger studies in similar and/or different contexts as well as informing the design of quantitative studies on air pollution. The results will also be crucial to informing acceptable entry points for interventions to mitigate pollution.

2. Methods

2.1. Context

The study was conducted in two slums in Nairobi city; Korogocho and Viwandani. These sites were selected because of an on-going health and demographic surveillance initiated in 2002 which is a reliable sampling frame[22]. A detailed description of the Health and Demographic Surveillance System carried out in the two areas is given elsewhere[22]. One striking difference between the two slums is the proximity of Viwandani to the industrial area where diverse manufacturing activities occur and where traffic flow is constant as trucks deliver materials and pick up finished goods. Alternatively, Korogocho is near the city's municipal dumpsite and faces several environmental issues arising from proximity to this location. The two slums are also characterized by activities believed to raise the levels of both indoor and outdoor air pollution.

The two sites have some differences regarding the population structure, education and income generating activities. Viwandani residents are more educated and dependent on economic activities that are more stable, while in Korogocho, unstable economic activities dominate. Other differences can be seen in the physical structures of the houses with Korogocho having mostly mud-walled houses with zinc sheet roofing and mud flooring, while in Viwandani, the walls and roofs are made of zinc sheets and floors are cemented. Thus, Viwandani has a better household socio-economic outlook as compared to Korogocho.

In spite of these differences, the two sites are similar in that they are both slum communities located in close proximity to major pollution sources and face similar environmental challenges.

2.2. Design of Study and Data

We designed a qualitative study on the perceptions and attitudes towards air pollution of people living in Korogocho and Viwandani. This study was designed as an exploratory study in a context of non-existent data on people's perceptions and attitudes towards air pollution, nor data on levels of air pollutants. The study sought to inform an on-going qualitative study in the same communities as well as inform the monitoring of pollutant levels both at the community level and in individual households.

A total of eight focus group discussions (FGDs) were held with adult residents of the two communities, four in each community. The discussions were separately held with younger adults (18–29 years; n = 39) and older adults (30 and above; n = 42) as it was felt that the younger adults might be intimidated by the presence of older participants, affecting their contribution to the discussions. In addition, youths in slum settings use a colloquial language called "sheng" that is not widely understood by older people. The participants were of mixed gender as there was no anticipated personal information that would be withheld in the presence of the opposite sex. Groups had between nine and 11 participants with roughly equal representation of the sexes. Participants differed in their ethnic background as well as level of education, employment status and duration of stay in the community. This was preferred to ensure a diversity of opinions regarding air pollution. **Table 1** summarizes some of the background characteristics of the participants (see **Table 1**).

We employed an emergent design in which we analysed the data and revised the study guide based on results from the first set of discussions. The discussions were therefore conducted in two waves to allow the researchers, time to conduct some analysis of the collected data in order to revise the guide as necessary. In the first round of the FGDs, two groups were convened at each site during the month of November 2012. The second round of discussions was conducted in January 2013. With the help of community mobilizers, the researchers purposively selected the participants. The participants were selected as much as possible from all villages within each slum to ensure that different areas of the slums were represented.

The discussions were conducted in Kiswahili, which is the national language widely spoken in Kenya, and particularly, in the urban slums and well understood by all participants. However, participants were allowed to express themselves in Kiswahili, English or "Sheng" (a mix of Kiswahili, English and local languages).

Table 1. Distribution of FGD participants by select background characteristics.

	Percent
Sex	
Female	51.9
Male	48.2
Slum of residence	
Korogocho	49.4
Viwandani	50.6
Age	
19–29 years	48.2
30+ years	51.9
Duration of stay in slum	
<10 years	28.4
10–19 years	22.2
>20 years	49.4
Level of education	
None	12.4
Primary	50.6
Secondary or Higher	37.0
Occupation Type	
None	27.2
Business (petty or established)	44.4
Employee (casual or long term)	28.4
N = 81	

The FGDs were moderated by the second author while the first author took notes. In Korogocho, we conducted the first discussions in an office within a health facility. The second round of discussions was held within an office block a distance away from the health facility where there were fewer disturbances from those visiting or working at the facility. In Viwandani, all discussions were held in a community hall.

On average, the FGDs took 50 minutes and were recorded and later transcribed and translated into English by the note taker. Upon analysis of the first round of discussions (four FGDs), the FGD guide was revised, dropping some of the questions that were not eliciting any new information from the groups and adding new questions that occurred during the earlier discussions. The initial FGD guide had six questions ranging from what comes to the mind of participants when they heard about environmental pollution to issues about the sources of outdoor and indoor air pollution. Furthermore, participants were asked about their thoughts on the government's and residents' responsibilities in addressing air pollution in their communities. After revision, the questions on people's understanding of the environment and risk were dropped as the responses showed no variation in the first four groups. In addition, new questions on participants' concerns about air pollution and a question to compare perceived levels of indoor and outdoor air pollution were included.

The transcribed discussions were transferred to NVivo 9 to help organise the data for analysis and interpretation. Coding was done based on recurring themes that were identified through reading of transcripts or observations of recurrent issues raised by participants. Other themes were identified based on diversity of views and contradiction. Some broad themes identified through review of literature formed the basis of the interview guide prior to the discussions. Some of these themes were retained in the final analysis while others were refined or new ones formed based on participants' responses. This was achieved during the transcription process as well as through further reading of the transcribed discussions and analysis of the coded data which done by the lead author. The transcripts were shared with two of the co-authors and a qualitative researcher in Nairobi who is not a co-author but who gave insights into the analysis of the data. However, triangulation of these analyses was not done. Thematic analysis was used because of its appropriateness in selecting the most recurrent perceptions.

2.3. Ethical Considerations

The study was reviewed and granted ethical clearance by the African Medical Research Foundation's (AMREF) ethics review committee. Informed consent was sought in two stages. First, during recruitment, the researchers provided the potential participants with full disclosure regarding the study, which included information on the purpose of the study and procedures. Second, consent was obtained before recording the discussions. Written informed consent was sought for participation and for audio recording just before the discussions started. All participants wore numbered tags which were used as identifiers during the discussions. Participants were informed of the use of number tags as opposed to their actual names in the discussions. In addition they were informed that their identity would not be disclosed in any reports or publications arising from the data.

3. Results

The results from the narratives of adult participants from both communities formed seven thematic areas. These included mixed knowledge about sources of air pollution; sensing air pollution; who is to blame?; poor housing and neighbourhood effects; desperate practices; resistance and ignorance; and fatalism and helplessness. The results are structured into sections according to the seven thematic areas identified.

3.1. Knowledge on Sources of Air Pollution

3.1.1. Outdoor Air Pollution

Mixed knowledge about sources of air pollution We sought to learn about participants' thoughts regarding the sources of outdoor and indoor air pollution in their communities. It emerged from the discussions that residents had mixed knowledge about the sources of air pollution as well as some of the consequences of exposure to this pollution. While participants generally correctly identified sources of outdoor air pollution, there were occasions when it was evident that the knowledge prevailing in these communities was flawed. For example, the view that smelly drainage channels and toilets were a source of air pollution was frequently expressed by both the young and older participants from both communities. Similarly, participants from both communities raised the issue of drainage channels as important sources of air pollution. They were said to emit foul smells due to stagnant water and people's habits of dumping waste including faecal matter into the channels. Other opinions expressed by the participants was that lack of toilets also led to air pollution as open defecation and the use of 'flying' toilets were used as alternatives, as was the common method of emptying pit latrines using uncovered drums—exposing residents to fouls smells. In addition, poverty and preference for cheap fuel materials were mentioned as factors contributing to the use of alternative fuels such as plastic bags, gunny bags, and cloth rags, especially among roadside food vendors in Korogocho, and as such, contributing to air pollution.

Sensing air pollution During the discussions, it was apparent that participants relied more on their senses to assess their exposure to pollution and in identifying sources of pollution. Sensory perception of pollution sources was stressed throughout the discussions with participants using terms such as seeing a cloud of smoke covering the area, pungent smells from factories and dumpsites and soot falling on people and buildings. This brings to the fore the apparent reliance on the senses to inform perceptions on sources and individual exposure.

Who is to blame? There were mixed opinions towards who was responsible for air pollution in the community with some attributing the state of outdoor air to residents while others felt it was primarily due to sources external to the communities. Several participants from Korogocho mentioned the municipal garbage dumpsite as the biggest polluter. The participants mentioned that the dumpsite was always on fire, covering the entire community of Korogocho in smoke that was 'corrosive' as the following quote indicates:

That dumpsite is the biggest air pollutant, because when that dumpsite is lit, there is a lot of smoke coming here, it is dark, people cough and there are schools down there [near the dumpsite] sometimes teachers tell the pupils 'today there has been a lot of smoke so when you go home tell your parents to give you milk'... it is bad smoke, like acid; there are medicines and many chemicals burning in the dumpsite (older female, Korogocho).

In addition, the burning of medical waste in the open by local clinics, cigarette smoking, dust and motorcycle fumes were also mentioned as major sources of air pollution. Industrial emissions were also mentioned as sources of pollution but these were not seen as important polluters in Korogocho.

Conversely, discussions in Viwandani revealed that industries were perceived as the biggest sources of air pollution. Both the young and older adults were very emotive when discussing the industries' contribution to air pollution. Sometimes the emissions from the industries were said to be so bad that people had to step far away from the community in order to get fresh air. The following excerpts capture Viwandani residents' views on the role of industries in air pollution.

...these industries when night comes they start working, they emit smoke [expression of annoyance] even when you are sleeping you just have to wake up... to open the door and stand outside; they emit smoke! [Expression of annoyance] (young male, Viwandani).

There is an industry here eh; like yesterday they were releasing some chemical eh yesterday, we couldn't stay up here, those of us who stay up there we couldn't stay there we had to go elsewhere; you would see people moving away (young male, Viwandani).

In addition, an emerging unofficial dumpsite and people's habits of burning trash were indicated as other major sources of air pollution in Viwandani. During one of the discussion sessions, participants shared that their community is sandwiched between two major sources of pollution, the industries on one side and the illegal dumping site on the other. Further, participants shared that most of the smoke from the dumpsite was experienced at night when the garbage was burned. Other important sources of air pollution identified were burning of tyres to extract metal and cigarette smoking. It was surprising that as much as people in Viwandani were exposed to vehicular emissions from busy traffic coming in and out of the industrial area, this was not mentioned in any of the discussions.

3.1.2. Indoor Air Pollution

Poor housing and neighbourhood effect With regard to people's knowledge of indoor air pollution, participants were aware of the sources contributing to the poor quality of indoor air. These ranged from the type of cooking fuel, cooking stoves and smoking of cigarettes indoors. Other sources include the poorly constructed and congested houses in the two communities that were said to encourage cross-pollution from the outdoors and neighbours. Congestion was seen as limiting the use of corridors as cooking points instead of cooking in the same room where the family slept. It emerged that the kerosene used by many households was considered too smoky, and therefore, dangerous to the residents' health. In addition, poverty was said to promote the use of poor quality second-hand cooking stoves (*i.e.*, stoves that had previously been used) that were seen to emit more smoke as compared with newer ones. In both communities, participants shared that many people used plastic or rubber materials as well as old foam mattresses to light their charcoal stoves, emitting foul smelling smoke. In addition, the small rooms that households occupied were said to magnify the problem of indoor air pollution due to crowding and lack of vents to release bad indoor air; as the following excerpt indicates:

...because houses are not well ventilated and one room serves as the kitchen, bedroom and sitting room and you use a charcoal stove. I have a child who has asthma and I took him to the hospital to get oxygen... they advised me not to cook while the child is in the house. But if there is only one room, it is a must the child is present when cooking is going on... if the houses were proper it would reduce the smoke (Older female, Korogocho).

The discussions revealed the existence of a 'neighbourhood effect' that was acknowledged to contribute to indoor air pollution in people's homes. For example, participants mentioned that people took advantage of the poor structures they lived in and opted to not ventilate their houses when a stove was being used. Instead, they chose to create venting holes on the side of the neighbour's house, who then bears the brunt of their emissions. Outdoor influences, especially industry emissions and roadside cooking spots were mentioned as important contributors to indoor air pollution.

Desperate practices Other drivers of indoor air pollution that emerged from the discussions were the prevailing practices with regard to ventilation. When asked about use of ventilation when cooking, it was clear that there was a disparity between knowledge and practice. For example, respondents shared that one needs to open the windows and doors when cooking with a kerosene or charcoal stove to vent the emissions, however, they reported that this was indeed not practised because of the poor nature of houses and the levels of insecurity in the communities. In addition, some houses were said to lack windows and people only opened the door during the day and had to endure the emissions at night as insecurity forces them to keep the door closed. Fear of the cold outdoor air was also mentioned as a reason why people chose not to open the windows or doors. It also emerged that space constraints prevented people from opening windows to let in fresh air.

In Korogocho, it emerged that fear of illnesses such as malaria and of the cold outdoor air drove people to negatively modify the eaves of their houses. When asked whether houses had eaves, several respondents answered in unison that "many people fear malaria," and so to ward off mosquitoes and the cold, they chose to block the eaves with plastic bags and other materials available to them as indicated in these opinions:

There is a space between the roof and wall but to prevent the cold, we have blocked it with gunny bags (these are bags made of sisal or plastic fibre) and curtains so there is no way air from in can get out or from outside get in"(Older female, Korogocho).

...the ceiling is made of gunny bags, on the sides there are gunny bags so there are no spaces, so when smoke fills the room it just stays (older male, Korogocho).

3.1.3. Reducing Outdoor and Indoor Air Pollution

Resistance and ignorance When asked what residents could do to reduce the levels of air pollution in the community, there were mixed reactions with some people feeling that there was nothing they could do. Fear of fights was cited as prohibiting people from asking neighbours or other community members to stop practices that were contributing to air pollution. On the other hand, there were those who felt that residents had a responsibility of ensuring the environment was kept clean, instead of waiting for outsiders to come and clean their backyard. Discussions about possible relocation of the dumpsite from Korogocho emerged to be unpopular among residents. Many residents rely on the dumpsite as a source of livelihood through scavenging; therefore, even though they were aware of the hazards the dumpsite posed, they were opposed to its relocation. As one participant put it:

See that dumpsite, there are groups [involved in scavenging for recyclable materials] and together with some priests we formed one environmental group, we started recycling. That dumpsite has created employment for youth; it has created employment for 30% and caused 5% deaths (Older male, Korogocho).

Similar opinions emerged in Viwandani where it was felt that relocation of industries to a less habited place would be resisted as many people were employed in the industries. It was also felt that such a move would be pointless as people would 'follow' the industries and continue living in close proximity to these industries. Also discussions around the residents' opinions on temporary closure of the industries to allow installation of emission control systems seem to be an unpopular move due to its impact on jobs. However, participants agreed that there was need for the government to enforce strict emissions control measures in the industries.

Participants from both communities voiced the need to create awareness about air pollution and measures needed to reduce exposure as the following opinion indicates:

I think it would be good if people can be sensitized about the environment so that people know how their environment should be, the things they should do to avoid polluting or the effects of pollution; this will be achieved for example through discussions like this which has made us know where we were perhaps going wrong and we can correct that… If we had organizations that would sensitize people … So it is my appeal that for those who don't know we have organizations that educate them on the environment so that people have that awareness to help reduce pollution (young male, Viwandani).

Fatalism and helplessness There were sentiments of fatalism when asked whether residents were concerned about air pollution. The participants felt they could not do much to address the issue of air pollution. They appeared to have rationalized the state of the environment in which they lived and were resigned to it. Expressions indicating residents viewed their polluted space as normal were heard from both the young and older adults in both communities. One participant best summed it by saying, "We are used… we have got used to this [air pollution] we forgot it is a problem" (Older female, Korogocho).

In the discussions, there were those who felt helpless and they relied on religion to find solace as one participant indicated: "It's God who protects us, the kind of dirt we have seen in this place is a lot" (Young female, Korogocho). This sentiment also revealed a flawed state of knowledge on the health impacts of pollution and the lack of appreciation of the central role individuals ought to play in ensuring the environment is safe for them to live in.

There was a sense of frustration among respondents as they raised issues about the lack of a voice to petition leaders to address issues about the pollution occurring in their communities, especially, for those sources that were external to the slums. The Viwandani participants reported being threatened with eviction if they raised these issues. The following discussions reveal what residents faced:

We usually tell them and when we tell them, they mostly say we should know we are living under electric lines [high voltage lines] we will be moved. You hear a lot of things; we feel we are troubled; sewage line passes here, because we are not known by the government so we just decide to keep quiet so we can live here longer (Older male, Viwandani).

We tell them [government representatives] but we are told we should know that the government does not know there are people living here. We are forced to keep quiet but we know we are being oppressed (Older male, Viwandani).

On the other hand, Korogocho residents pointed fingers at the local leadership for worsening the pollution; for instance, allocating building space without leaving space for functions such as waste collection.

4. Discussion and Conclusion

This study explored the general perception and understanding among residents in two slums in Nairobi about air pollution and associated health risks in their community. We acknowledge that slum residents face various environmental challenges that are not limited to air pollution alone. However, since this study sought to assess the perceptions on air pollution, the discussion was limited to this objective. The main highlights indicate that residents of both slums relied on sensory perceptions to assess air pollution and its sources. There was an apparent disconnect between knowledge and practice as far as the use of ventilation was concerned. In addition, the participants expressed a lack of agency to address the current state of air quality in the two communities. These findings are similar to those from a recent quantitative study conducted in the two slums[23].

Outdoor air pollution in the two communities was mainly attributed to garbage dumpsites located in close proximity to both communities. Combustion of the garbage was cited to be a major contributor of smoke and soot in both locations, while industrial emissions were in addition, of great importance to Viwandani residents. Similar findings were observed by Howel and colleagues[24] on the important role of place in forming public perceptions. People's perceptions on pollution were informed by their sensory experience, for instance, the visible clouds of smoke and soot from the dumpsites and factories as well as the odorous emissions from industries and drainage channels. Sensory perceptions have been reported as important in informing perceptions of sources and exposure[8][18][25] and in subsequent response to the exposure[25]. The reliance on sensory perceptions in these communities also raises a red-flag for pollutants that are not odorous. For example, indoor air pollution from carbon monoxide might be ignored with fatal consequences.

There was a feeling that many of the activities carried out by residents, such as burning trash, were of a scale that was smaller in importance as compared to the major polluters, namely the industries and dump sites. This shift of responsibility to other entities has been reported in other studies as a mechanism adopted by individuals to distance themselves from any direct contribution to the problem at hand[6]. We found no evidence of a 'neighbourhood halo effect' that has been reported in other studies, where people are unwilling to attribute pollution to their place of residence as compared to other areas[8][16][24][25]. This can be attributed to the loose attachment many slum residents have to their place of residence given the informal nature of the settlements and the obvious disadvantage in terms of access to services compared to the neighbouring middle and upper class residential areas. The absence of a 'halo effect' would be an important attribute in the event programs aimed at reducing air pollution or exposure, are introduced. This is because residents already identify their communities as polluted spaces; a fact that might make them more accepting of programs/interventions to address the issue. In addition, the findings of this study are contrary to the review by Saksena[5] that pointed to studies indicating that among people of low socio-economic status, air pollution was out-ranked by other more urgent issues. In fact, this study finds that as much as residents had other issues to think about, pollution was on the forefront as an issue of concern.

The feeling of helplessness can be attributed to the lack of voice among the residents to approach their leaders and to demand action. Given their informal residential status, poverty and lack of alternatives, many felt entrapped in this polluted space. It is also a consequence of poor social capital among residents, which inhibits collective action against pollution[16][26]. Residents raised the issue of counter threats whenever they went to the local leaders to petition them to take action against polluters. This helplessness was exacerbated by the lack of security of the tenure on the land on which residents lived and the informal status of the slums. Studies elsewhere have shown that lack of attachment to a place limits people's investment in the place and can lead to inaction against pollution[8][16][18].

Further, there was an apparent lack of agency as people felt there was nothing they could do to reduce the levels of pollution in their communities; lack of agency has been attributed to lower socio-economic status[16][26]. This lack of agency could derail any efforts the government puts in place to address air pollution in these and similar communities. Therefore, the government must first address these barriers to ensure they effectively implement pollution control and other environmental protection policies.

People's ignorance about the 'true' effects of air pollution was evident in their daily practices such as blocking the eaves to prevent cold, dust and mosquitoes. This finding could be explained by the lack of public education on issues concerning air pollution and indeed, environmental issues in general. In addition, people's actions might be justified given the poor housing available in their communities and their lack of resources to facilitate residential moves to better houses. As such, residents had the hard choice of either letting in the cold, polluted outdoor air and mosquitoes or living with indoor air pollution.

This study has revealed the perceptions of slum residents regarding both indoor and outdoor air pollution. We find that only the first two pillars of the protection motivation theory informing this study are fulfilled in the study settings. However, the reported behaviour is not protective as it puts residents at higher risk of exposure to air pollution while lack of agency makes it difficult for residents to address air pollution. From the findings we conclude there is an urgent need to create awareness among residents on the effects of air pollution and the need for each individual to take part in reducing the levels of air and general environmental pollution.

However, the study faces some limitations on coverage and lack of inclusion of all stakeholders in the survey. The study covered only two slums and it would be useful to also understand the perceptions of residents in both formal and informal settlements. The current study interviewed people aged 18 years and above, however, it would be worthwhile to also to get the views of school going children particularly those aged 10–17 years. This age group represents a special group with high risk because of exposure to outdoor air pollution during outdoor activities. We also didn't conduct key informant interviews which would have provided information on the ways to support the residents in reducing the health burden from air pollution. Despite these limitations the study still provides useful insights on the perceptions about air pollution among the urban poor residents.

Future Research and Action

The following should be considered or included in future research as they were not considered in this study. First a similar study should be conducted in different parts of the city to enable a comparison of perceptions along socio-economic classes and across different neighbourhoods. Second, conduct a study to identify sustainable solutions to air pollution that can work not only in the study communities but city-wide and nationally by including key stakeholders. Third, conduct studies that include school-going children as participants.

Competing Interests

The authors declare there are no competing interests.

Authors' Contributions

All authors conceptualized the study. KM and TE designed the study guide and conducted the FGDs. TE moderated the discussions while KM took notes, transcribed the discussions, analysed the data and drafted the manuscript. TE, NN, EK, NN reviewed all earlier drafts of the manuscript and all authors signed off the final version of the manuscript. All authors read and approved the final manuscript.

Acknowledgements

We are greatly indebted to the participants from the two communities who took their time from their busy schedules to share their views with us. We are grateful to Dr. Netsayi Mudege and Dr. Benta Abuya for their input in various stages in the design of the study and data analysis, and in reading the manuscript. The work was undertaken within the Umea Centre for Global Health Research, with the support for graduate studies and data collection from the FAS, the Swedish Council for Working Life, and Social Research (Grant No. 2006-1512). Analysis and writing time was funded by the African Population and Health Research Center through a grant by the Bill and Melinda Gates Foundation (Grant No. OPP 1021893).

Source: Muindi K, Egondi T, Kimani-Murage E, *et al.* "We are used to this": a qualitative assessment of the perceptions of and attitudes towards air pollution amongst slum residents in Nairobi[J]. Bmc Public Health, 2014, 14(1):1–9.

References

[1] World Health Organization: Air quality and health. 2011. Fact sheet N°313. http://www.who.int/mediacentre/factsheets/fs313/en/index.html.

[2] Bruce N, Perez-Padilla R, Albalak R: The health effects of indoor air pollutionexposure in developing countries, in Protection of the Human Environment.Geneva: World Health Organization; 2002.

[3] Ezzati M, Saleh H, Kammen DM: The contributions of emissions and spatial micro-

environments to exposure to indoor air pollution from biomass combustion in Kenya. Environ Health Perspect 2000, 108(9):833–839.

[4] Kinney PL, Gichuru MG, Volavka-Close N, Ngo NK, Ndiba P, Law A, Gachanja A, Gaita SM, Chillrud SN, Sclar E; Traffic impacts on PM2.5 air quality in Nairobi, Kenya. Environ Sci Policy 2011, 14(4):369–378.

[5] Saksena S: Public perceptions of urban air pollution with a focus on developing countries. Environ Change Vulnerability Governance Series 2007, 65.

[6] Bickerstaff K, Walker G: The place(s) of matter: matter out of place—public understandings of air pollution. Prog Hum Geogr 2003, 27(1):45–67.

[7] Eden S: Public participation in environmental policy: considering scientific, counter-scientific and non-scientific contributions. Public Underst Sci 1996, 5:183–204.

[8] Brody SD, Mitchell Peck B, Highfield WE: Examining localised patterns of air quality perception in Texas: a spatial and statistical analysis. Risk Anal 2004, 24:6.

[9] Keating M: Making Decisions for Sustainable Development, in Agenda for Change: A Plain Language Version of Agenda 21 and Other Rio Agreements. Geneva, Switzerland: Centre for Our Common Future; 1993:70.

[10] Slovic P: Perception of risk. Sci New Series 1987, 236(4799):280–285.

[11] Slovic P, Weber EU: Perception of risk posed by extreme events. In Paper presented at the Risk Management strategies in an Uncertain World Palisades, New York 2002.

[12] Alford WP, Weller RP, Hall L, Polenske KR, Shen Y, Zweig D: The humandimensions of pollution policy implementation: air quality in rural China.J Contem China 2002, 32.

[13] University of Twente: Protection Motivation Theory. 2013. http://www.utwente.nl/cw/theorieenoverzicht/Theory%20Clusters/Health%20Communication/Protection_Motivation_Theory.doc/.

[14] Saksena S: Public perceptions of urban air pollution risks. Risk Hazard Crisis Public Policy 2011, 2:1.

[15] Howel D, Moffatt S, Bush J, Dunn CE, Prince H: Public views on the linksbetween air pollution and health in Northeast England. Environ Res 2003,91:163–171.

[16] Bickerstaff K: Risk perception research: socio-cultural perspectives on the public experience of air pollution. Environ Int 2004, 30:827–840.

[17] Bickerstaff K, Walker G: Public understandings of air pollution: the 'localisation' of environmental risk. Global Environ Change 2001, 11:133–145.

[18] Wakefield SEL, Elliott SJ, Cole DC, Eyles JD: Environmental risk and (re)action: air quality, health, and civic involvement in an urban industrial neighbourhood. Health Place 2001, 7(3):163–177.

[19] UN-HABITAT: The challenge of slums: global report on human settlements. London

and Sterling: United Nations Human Settlements Program; 2003.

[20] UN-HABITAT: Slums of the World: The face of urban poverty in the new millennium. Nairobi: United Nations Human Settlements Program; 2003.

[21] Arimah BC: Slums as Expressions of Social Exclusion: Explaining the Prevalenceof Slums in African Countries. Nairobi: UN-HABITAT.

[22] Emina J, Beguy D, Zulu E, Ezeh A, Muindi K, Elung'ata P, Otsola J, Yé Y: Monitoring of health and demographic outcomes in poor urban settlements: evidence from the Nairobi urban health and demographic surveillance system. J Urban Health 2011, 88:S200–S218.

[23] Egondi T, Kyobutungi C, Ng N, Muindi K, Oti S, van de Vijver S, Ettarh R,Rocklöv J: Community perceptions of air pollution and related health risks in Nairobi Slums. Int J Environ Res Public Health 2013, 10(10):4851–4868.

[24] Howel D, Moffatt S, Prince H, Bush J, Dunn CE: Urban air quality in North-East England: exploring the influences on local views and perceptions. Risk Anal 2002, 22:1.

[25] Claeson AS, Lide NE, Nordin M, Nordin S: The role of perceived pollution and health risk perception in annoyance and health symptoms: a population-based study of odorous air pollution. Int Arch Occup Environ Health 2013, 86(3):367–374.

[26] Wakefield SEL, Elliott SJ, Cole D: Social capital, environmental health and collective action: a Hamilton, Ontario case study. Can Geographer 2007, 51(4):428–443.

Chapter 23

Air and Water Pollution over Time and Industries with Stochastic Dominance

Elettra Agliardi[1], Mehmet Pinar[2], Thanasis Stengos[3]

[1]CIRI Energia e Ambiente and Department of Economics, University of Bologna, Piazza Scaravilli 2, 40126 Bologna, Italy
[2]Business School, Edge Hill University, St Helens Road, Ormskirk L39 4QP, Lancashire, UK
[3]Department of Economics, University of Guelph, Guelph, On N1G 2W1, Canada

Abstract: We employ a stochastic dominance (SD) approach to analyze the components that contribute to environmental degradation over time. The variables include countries' greenhouse gas (GHG) emissions and water pollution. Our approach is based on pair-wise SD tests. First, we study the dynamic progress of each separate variable over time, from 1990 to 2005, within 5-year horizons. Then, pair-wise SD tests are used to study the major industry contributors to the overall GHG emissions and water pollution at any given time, to uncover the industry which contributes the most to total emissions and water pollution. While CO_2 emissions increased in the first-order SD sense over 15 years, water pollution increased in a second-order SD sense. Electricity and heat production were the major contributors to the CO_2 emissions, while the food industry gradually became the major water polluting industry over time.

Keywords: Environmental Degradation, GHG Emissions, Stochastic Dominance,

Water Pollution

1. Introduction

There are various indicators and assessment methodologies for evaluating in practice the performance of industries, cities and countries, at global, national and regional level, related to economic and environmental sustainability (see e.g. Singh *et al.* 2012, providing a recent overview of a great number of indicators that are already common practice for policy-making; Blanchet and Fleurbaey 2013, which favor a dimension by dimension dashboard approach; Xepapadeas and Vouvaki 2008; Agliardi 2011; Pinar *et al.* 2014; Agliardi *et al.* 2015, for detailed discussions of environmental sustainability). In this paper we propose a novel methodology which allows us to assess temporal trends and industry contributions to air and water pollution and to identify the cases where externalities affect the overall pollution. Our methodology is sufficiently general and data-driven, so it can be employed to alternative units and at different levels.

We examine air and water pollution that have been extensively analyzed through their linkages to economic development (Dasgupta 2000; Persson *et al.* 2006; Tamazian *et al.* 2009; Orda's Criado *et al.* 2011; Sivakumar and Christakos 2011; Xepapadeas 2011; Li *et al.* 2014b; Paruolo *et al.* 2015). Air pollution is a major concern for various environmental policies and is perceived as one of the biggest threats to human health and global warming. CO_2 emissions, and also other greenhouse gases (GHG), affect air quality and have been identified as prime contributors. At the same time, water pollution is another major aspect of environmental degradation. Some preliminary information about these forms of environmental degradation can be obtained by pollution flow accounts. They track the generation of pollution by each industry and final demand sector. They also give data about the changes of pollution over time, to monitor the interaction between the environment and the economy and the progress toward meeting environmental protection goals.

In this paper we employ a stochastic dominance (SD) approach, which is a pretty general method allowing us to have a full picture of the environmental degradation over time and the major industry contributions to each polluting factor. It relies on pair-wise SD tests. Pair-wise SD tests are based on comparisons of cumulative distribution functions (CDFs), providing robust orderings in terms of welfare levels (e.g., Davidson and Duclos 2000; Barrett and Donald 2003; Anderson 2004). Stochastic orderings are defined on classes of probability distributions and represent intuitively, in case of welfare improvements, why one population's welfare is increased more than another, irrespective of the poverty lines (Davidson and Duclos 2000) or for all income levels (Anderson 2004). Pair-wise SD comparisons among populations allow one to ascertain whether there is an improvement, say, in the income levels of a given population over another one, for all income groups (*i.e.*, in all parts of the income distribution). For example, pair-wise SD is used to assess whether social programs and tax reforms improve social welfare, by analyzing the empirical distribution of income levels after and before tax reforms (see e.g., Duclos *et al.* 2005, 2008). In this respect, one evaluates the income distribution across the population before and after tax reforms by looking at its CDFs (and integral of CDFs), and if the income distribution after tax reforms dominates the income distribution before tax reforms, then one could suggest that there is always a higher proportion of population with higher income levels in all parts of the income distribution. More recently, pairwise SD tests have been used to compare male and female earnings in a competitive environment to ascertain whether one group has higher earnings at all earnings levels (Ors *et al.* 2013). Hence, SD tests compare the entire probability density function, rather than a finite number of moments, so SD approach can be considered less restrictive and more robust in comparisons across populations.

Although pair-wise SD comparisons are used extensively in well-being and poverty (see, e.g., Davidson and Duclos 2000; Pinar *et al.* 2013), to our knowledge, only Makdissi and Wodon (2004) apply SD analysis to compare CO_2 emissions between 1985 and 1998, and find that there has been first-order dominance up to a level, however not for all levels of CO_2 emissions. Furthermore, they find that there has been an overall increase in emissions over a 13-year period. In this paper, we extend the SD applications, both at first-order and second-order, to different types of emissions, water pollution and different polluting industries.

Our methodology is particularly well-suited to answer questions like these: Given that GHG emissions or water pollution not only vary over time but also across industries, is there a general increase (decrease) in GHG emissions or water pollution over time? If so, which industry has been the major contributor to those increases (decreases) in GHG emissions or water pollution? One could argue that an increase (or decrease) in GHG emissions over-time could be directly ascertained by counting the average GHG emissions. However, as discussed above, SD is more informative, considering the entire CDF rather than the average only. Indeed, this increase (or decrease) might be driven by a relatively larger increase (or decrease) of emissions of some countries (yielding a reallocation of emissions from central masses towards the tails of the distribution). For the purpose of distinguishing whether the changes have to be attributed to individual units (countries, industries, etc.) or there has been an overall change affecting all units, we adopt first-order and second-order SD. First-order SD (SD1 hereafter) would reveal information whether there has been a point-wise deterioration (improvement) over time. In this respect, SD1 analyzes the marginal CDFs of the environmental degradation at all levels of GHG emissions (or water pollution) and suggests whether there has been a proportional increase (decrease) in environmental degradation in all parts of the distribution, or not. For example, if emissions from industry A first-order dominate the emissions from industry B, this would suggest that there are always higher emission levels in industry A compared to B at all levels of emissions (*i.e.*, the proportion of countries that emit above a given emission level is always higher in industry A than B). In other words, the higher emissions in one industry are not driven by some specific countries, but they are higher at all emission levels (or, alternatively, the probability of having higher emissions above a given level in industry A is higher than in B, at all levels of emissions). Similarly, SD1 over time would suggest that there is always a higher proportion of countries that emit more above a given level over time. On the other hand, second-order SD (SD2 hereafter) would suggest that there is no point-wise deterioration (improvement), but an overall deterioration (improvement) over-time. In fact, SD2 does not analyze the CDFs, but the integrals of the CDFs (*i.e.*, sum of environmental degradation up to a level of environmental degradation). In this case, there might not be a higher proportion of countries that emit more above a given level over time, but a higher sum of the emissions above a given level by emitters over time. In other words, some countries' pollution levels might decrease and some others might increase over time, but if the sum of the pollution above a given level is higher over time, this would suggest that there has been an overall increase in air and/or water pollution for all given levels, even though not all countries experienced an increase in their pollution levels.

SD2 is particularly important when analyzing the possible negative externalities and free-riding issues in water pollution and overall GHG emissions. Negative externalities are defined as the social costs of the market activity (e.g., consumption and production) not covered by the private cost of the activity (e.g., Dahlman 1979). Producers make decisions based on the direct cost of production and revenues, but do not take into account the social costs of pollution (see Baumol 1972 for detailed discussion), such as acid precipitation and global warming (Arrow *et al.* 2004; Rezai *et al.* 2012). Tol (2009) suggests that low-income countries, which contribute the least to climate change because of their low production and consumption levels, are most vulnerable to its effects, as their adaptation to climate change is limited, due to the shortcomings in resources and institutions (e.g., Smit and Wandel 2006). Thus, even though the gains from economic activities linked with emissions are private, the costs associated with emissions are global. Therefore, it is not straightforward to identify which countries are responsible for the negative externalities of environmental degradation. In particular, CO_2 emissions have been mainly flowing to other partner countries through international trade (Peters and Hertwich 2008). For example, China's CO_2 emissions have been increasing over time due to its exports to other countries (Yunfeng and Laike 2010). Similarly, Dominguez-Faus *et al.* (2009) point out that water pollution increased over time due to major transportation biofuel needs across countries. Bernauer and Kuhn (2010) examine water pollution within Europe and analyze whether democracies that trade and are bound by international treaties are less likely to harm one another environmentally. They find that free-riding incentives are in place. Free-riding occurs when some users of the public good use these services without paying for them (see e.g., Gans *et al.* 2012). In this case, free-riding occurs when the cost of water pollution is not paid by some countries, even though they are responsible for it. Sigman (2002) found that free riding may substantially increase pollution in international rivers, whereas there is less free riding within the European Union, suggesting that international institutions might work as mitigating factors (see Sullivan 2011 which provides a multivariate model that assesses water vulnerability).

When there is no straightforward identification of contributors to water pollution and/or GHG emissions, we can employ SD2 to account for aggregate global contribution. Some countries' direct contribution to the environmental degradation might decrease over time (e.g., due to lower production), yet their indirect contribution to the aggregate level of environmental degradation might increase due to their consumption, as their imports would lead to higher levels of GHG emissions in their trading partners. In this case, even though one cannot find an absolute increase in environmental degradation for all countries at all levels, one can evaluate the aggregate environmental degradation levels at different levels (*i.e.*, sum of environmental degradation levels up to a given level) through SD2.

Here we implement two complementary SD approaches. Firstly, we employ consistent SD tests from Barrett and Donald (2003) to examine the dynamic progress of each separate GHG emissions (*i.e.*, CO_2, methane, nitrous and other greenhouse gas emissions) and water pollution over time from 1990 to 2005 within 5-year horizons. In other words, we examine whether there has been a general deterioration or improvement in each component. In that regard we will be able to obtain information on those environmental quality dimensions that are fast-moving (*i.e.*, fast deteriorating or fast improving dimensions) or slow-moving (*i.e.*, dimensions that remain at steady levels) for all countries over the period we analyze. Secondly, pair-wise SD tests allow us to examine the major industry contributors to the GHG emissions and water pollution at any given time. In order words, at a given time, we compare each industry contribution to GHG emissions and water pollution with all possible other industries to uncover the industry which contributes the most to total emissions and water pollution. The use of statistical tests allows us to obtain the level of statistical significance of environmental degradation (or improvement) over time.

Therefore, SD analysis provides a robust comparison of environmental degradation over time and industries, disentangles the effects of externalities, and determines the statistical significance level for such degradation. As such, it can be a useful guideline for the direction of environmental protection and public policy intervention. Fast-moving variables (in the components of GHG emissions and water pollution) provide an indication for pollution prevention, calling for the redesign of industrial processes and new technologies to reduce pollution. At the same time, they offer directions for policy instruments in the form of official restrictions and positive incentives designed to control activities that may be harmful to the quality of the environment.

This paper is organized as follows. Section 2 compares the SD method with other methods employed in the literature to evaluate spatio-temporal trends. Section 3 describes the methods and data and Section 4 discusses our results. Finally, Section 5 contains the main conclusions.

2. Comparison between Bayesian Approaches and SD

In his section we discuss the advantages of the SD method over alternative Bayesian approaches which have been employed to extract the spatio-temporal trends. Bayesian approaches have been employed to analyze different types of risk assessments, such as health, environmental and burglary risks—by allowing different levels of space-time dependence (Besag *et al.* 1991; Waller *et al.* 1997; Wikle *et al.* 1998). Bayesian methods consider specific spatial effects, time effects, and an interaction of these two effects (with prior assumptions about their interaction) to analyze the evolution of risk over time and to estimate the posterior risk levels. In particular, Bayesian approaches have been employed to analyze the environmental risk (Wikle 2003), where the spatio-temporal dependence is present, such as increase in PM10 pollution (Cocchi *et al.* 2007), rural ozone levels in the Ohio state (Sahu *et al.* 2007), risk of earthquake (Natvig and Tvete 2007), extreme precipitation (Sang and Gelfand 2009) and extreme waves (Scotto and Guedes Soares 2007; Vanem 2011), among other fields. Bayesian approaches are helpful in identifying the posterior risk by taking into account the spatial dependence; however, not only they classify risk relatively (prior choice of extreme events or risk categorization), but also they seem not to be suitable to analyse the environmental risk when there is no clear spatial dependence. In fact, Bayesian methods allocate spatial dependence a priori, estimating risk differently if space units share a common border or not. However, when dealing with environmental degradation, externalities in GHG emissions have global effects. Hence, our view is that the SD approach can be a more suitable method than the Bayesian ones, when there is no clear-cut spatial dependence. **Table 1** provides a comparison between BHM and SD approach, and gives details why SD approach is more suitable in analyzing the environmental degradation data than BHM.

3. Methods and Data

3.1. Pair-Wise SD Tests

Let us define SD pair-wise comparisons of a given variable over two points in time. In particular, we examine SD of the GHG emissions and water pollution in a 15-year and 10-year period, respectively (from 1990 to 2005 for GHG emissions,

Table 1. Comparison between stochastic dominance (SD) and Bayesian hierarchical methods (BHM).

Bayesian hierarchical methods (BHM)	Stochastic dominance (SD)
Takes into account the spatial dependence, but is not suitable when there is no clear spatial dependence	Captures global dependence when a priori spatial dependence is not a reasonable assumption. SD is more suitable if environmental degradation has global consequences rather than spatial
Takes into account the time-dependence (see, e.g., Law *et al.* 2014a), but time-effect is usually driven by the first two moments (mean and standard deviation of risk) only	Takes into account the time-effect, but analyses the empirical distribution of risk (*i.e.*, all moments), and hence provides a more robust comparison over-time and across industries
Provides posterior risk estimations; however, comparisons are usually relative to the distribution of risk in spatial units (see, e.g., Li *et al.* 2014a; Law *et al.* 2014a, b)	Suitable to analyse both absolute and relative risk over-time and space
It is based on prior probabilistic assumptions on the dependent variable for posterior risk estimations (Vanem 2011).	It is nonparametric as it does not impose any restrictions on the functional forms of probability distributions

and from 1995 to 2005 for water pollution) and determine whether there has been a deterioration or improvement in each environmental quality indicator over time above a given pollution level. Additionally, SD pairwise tests are employed for the sub-industry comparisons for GHG emissions and water pollution. In other words, we find major contributing industries to emissions and water pollution at a given time, comparing the CDFs of the pollution levels of the various industries. If there is SD1, this would suggest that degradation in one industry is clearly higher than in another at all levels of pollution. If there is no SD1, then we move to SD2 and analyze whether the sum of the pollution levels above a given pollution level is relatively higher in one industry than in another one at all levels of pollution. In particular, we apply the consistent SD tests provided by Barrett and Donald (2003).

Let us consider the pair-wise SD tests for water pollution comparisons over time. Denote by Z1 and Z2 the water pollution levels from two samples of countries at either two different points in time or different sub-industries at a given time. Suppose that Z1 and Z2 have associated cumulative distribution functions (CDFs) given by F1 and F2 respectively. In this context, Z1 stochastically dominates Z2 at the first-order if $F_1(z) \le F_2(z)$ for all z level, where z is the environmental degradation level (e.g., water pollution level). When this occurs, the water pollution level in sample Z1 is at least as large as that in sample Z2, for any utility function U that is a decreasing monotonic function of z—*i.e.*, $U'(z) \leq 0$ since the higher z (environmental degradation), the lower the utility.

How do we interpret SD1 of Z1 (e.g., water pollution levels of countries due to activities in industry A), over Z2 (e.g., water pollution levels of countries due to activities in industry B), *i.e.*, $F_1(z) \leq F_2(z)$? If the CDF of pollution levels due to activities in industry A is always below the CDF of pollution levels due to activities in industry B, then the proportion of countries that pollute due to activities in industry A is always greater than that of industry B at all levels of pollution, *i.e.*, z. Therefore, industry A stochastically dominates industry B in the first-order sense (see **Figure 2** as an example of SD1). In this respect, there is a clear ordering of industries in terms of environmental risk they impose.

If the CDF of pollution levels from one sample does not lie below the CDF of water pollution levels from the other sample at all z levels (*i.e.*, when the two CDF curves intersect), then there is no SD1 of one industry over another, and the ordering of industries in terms of environmental risk is ambiguous. This leads to an ambiguous situation which makes it necessary to test for SD2. SD2 of Z1 (water pollution levels due to activities in industry A) over Z2 (water pollution levels due to activities in industry B) corresponds to $\int_0^z F_1(p)\,\mathrm{d}p \leq \int_0^z F_2(p)\,\mathrm{d}p$ for all z level, where p is the pollution level that takes values between 0 and z. It holds for any utility function U that is a monotonically decreasing and concave, that is, $U'(z) \leq 0$ and $U''(z) \leq 0$. The utility function is monotonically decreasing, as pollution reduces welfare, and concave, as it is expected that most policy makers would be averse to an increased dispersion of pollution. SD2 of one sample over another is tested not by comparing the CDFs themselves, but comparing the integrals below them. If the area beneath the F1(z) distribution is less than the area beneath F2(z) at all levels of z, then F1(z) stochastically dominates F2(z) in the second-order. Thus, the sum of the pollution by countries that pollute above z is always higher in industry A than in industry B. In other words, SD2 of industry A over industry B implies that even though the proportion of countries that emit above a given pollution level is not higher in one industry than in another one, the sum of pollution is always greater in industry A than in B at all degradation levels.

We can also present the orders of SD using the integral operator, $\zeta_j(.; F)$, as a function of F defining SD of order $j-1$. Thus:

$$\zeta_1(z;F):=F(z),$$

$$\zeta_2(z;F):=\int_0^z F(p)dp=\int_0^z \zeta_1(p;F)dp,$$

where $\zeta_1(z;F)$ is the CDF of the population Z and $\zeta_2(z;F)$ is the integral counterpart of the CDF of the population Z. The general hypotheses for testing SD1 of Z^1 over Z^2 (e.g., pollution levels over-time or pollution levels from different industries) with respective CDFs of $F_1(z)$ and $F_2(z)$ can be written as:

$$H_0: F_1(z) \le F_2(z) \text{ for all } Z \in [0, \bar{Z}],$$

$$H_1: F_1(z) > F_2(z) \text{ for some } Z \in [0, \bar{Z}],$$

where the environmental degradation level, z, ranges between 0 and a finite upper level $\bar{Z}$. If one fails to reject the null hypothesis, then CDF, say in industry A, is always less than in industry B, that is, the proportion of countries that pollute due to activities in industry A is always greater than the proportion in industry B at all levels of emission. If there is some degradation level z at which the dominance relation between two samples change (*i.e.*, alternative hypothesis), then there is no clear ordering of samples compared (*i.e.*, two CDF curves intersect at some degradation levels of z), and therefore this is no SD1 of one sample over another. Similarly, we can write the general hypotheses for testing SD2 of Z1 over Z2. In this case the areas under the CDF curves of two samples are compared (see Section 2 of Barrett and Donald 2003 for asymptotic properties of the tests).

Let us assume that Z_i^1 and Z_j^2 are two samples with CDFs F1 and F2 respectively and the sample sizes might be different for each sample where $i = 1, 2, \cdots, N$ and $j = 1, 2, \cdots, M$. The empirical counterparts of the distributions to construct tests are, respectively:

$$\hat{F}_1(z)=\frac{1}{N}\sum_{i=1}^{N}1\left(Z_i^1\le z\right),\hat{F}_2(z)=\frac{1}{M}\sum_{j=1}^{M}1\left(Z_j^2\le z\right),$$

where $1\left(Z_i^1\le z\right)$ is an indicator function taking value of 1 if pollution level of spatial unit is less than or equal to z, and zero otherwise (Davidson and Duclos 2000). In other words, the empirical counterparts of the distributions calculate the proportion of spatial units in each sample that has a degradation level that is less than or equal to z.

The test statistics for testing the hypotheses can be written compactly using the integration operator as follows:

$$\hat{S}_j = \left(\frac{NM}{N+M}\right)^{1/2} \sup_z\left(\zeta_j\left(z;\hat{F}_1\right)-\zeta_j\left(z;\hat{F}_2\right)\right)$$

for first-order (second-order) of SD when $j = 1$ ($j = 2$) where sup operator denotes supremum difference between CDFs (integrals of CDFs) of samples Z_i^1 and Z_j^2 at a given degradation level of z, respectively.

We finally consider tests based on the decision rule: reject H_0^j if $\hat{S}_j > c_j$ where H_0^j is the null hypothesis for first-order (second-order) dominance of Z_i^1 over Z_j^2 when $j = 1$ ($j = 2$) and c_j are suitably chosen critical values to be obtained by simulation methods.

To make the result operational, one needs to find an appropriate critical value cj that satisfies $P\left(\overline{S}_j^{F_1} > c_j\right) \equiv \alpha$ or $P\left(\overline{S}_j^{F_1,F_2} > c_j\right) \equiv \alpha$ (some desired probability level such as 0.05 or 0.01). Since the distribution of the test statistic depends on the underlying distribution, we rely on bootstrap methods to simulate the p-values (see Section 3 of Barrett and Donald 2003 for the related bootstrapping to obtain test statistics for the hypotheses; SD tests are conducted with the use of GAUSS codes available on http://garrybarrett.com/research/).

3.2. Data

The dataset consists of different types of GHG emissions (CO_2 emissions, methane emissions, nitrous oxide emissions, and other GHG emissions) and water pollution, and their sub-industry contributions for several countries in various years, between 1990 and 2005. Although some types of pollutants have annual data and for longer periods, to keep the analysis the same for all variables, we only consider the periods where we have information for all variables. GHG emissions consist of total CO_2, methane, nitrous oxide and other GHG emissions (*i.e.*, perfluoro-carbon, hydrofluorocarbon, and sulfur hexafluoride) at a given year for a given country and the latter three emission types are measured in terms of CO_2 equivalent levels, which allow us to conduct pair-wise comparisons over time. Annual national estimates for the total fossil-fuel CO_2 emissions and respective fossil-fuel CO_2 emissions from solid (coal), liquid (oil) and gas (natural gas) consumption come from the Carbon Dioxide Information Analysis Center (CDIAC) of the U.S. Oak Ridge National Laboratory (see Boden *et al.* 2013). Data on carbon dioxide emissions by sector are from International Energy Agency (IEA) electronic files which are also reported in the World Bank's World Development Indicators (World Bank 2012). Methane, nitrous oxide and other GHG emissions and their sub-industry contributions are obtained from the European Commission, Joint Research Centre (JRC)/ Netherlands Environmental Assessment Agency (PBL). Emission Database for Global Atmospheric Research (EDGAR): http://edgar.jrc.ec.europa.eu/. Finally, water pollution is measured by biochemical oxygen demand (BOD) which is the amount of oxygen that bacteria in water will consume in breaking down waste. These data are initially obtained with the methodology of Hettige *et al.* (2000) where end of pipe discharge of organic emissions are measured using different sector information, and updated by the World Bank's Development Research Group using the same methodology. All the data sets are categorized and taken from the World Bank's World Development Indicators (World Bank 2012). **Table A1** in Appendix provides the list of countries used in our analysis for water pollution, total CO_2, methane, nitrous oxide and other GHG emissions. Sub-industry contributions to the water pollution and different type of emissions also cover the same countries listed under general categories. **Table A2** in Appendix offers the detailed variable definitions and sources, and provides electronic links to the data sources.

4. Results and Discussion

4.1. SD Comparisons in Air Pollution

4.1.1. CO_2 Emissions

First, we present our findings from the pair-wise SD1 and SD2 comparisons of CO_2 emissions from 1990 to 2005, based on the bootstrap methods from Barrett and Donald (2003), for total, sub-industry and sub-fuel CO_2 emissions. We first perform consecutive tests, comparing total CO_2 emissions, and then CO_2 emissions from each individual sector (e.g., emissions from the electricity and heat production), for each pair of 5-year horizons between 1990 and 2005. Furthermore, we also test CO_2 emissions from different sub-fuel consumptions for each pair of 5-year horizons between 1990 and 2005. These consecutive tests allow us to analyze whether over time deteriorations (or improvements) have occurred in CO_2 emissions and, additionally, which sector and/or sub-fuel consumption is mainly responsible for such deteriorations (or improvements).

Table 2 suggests that there has been no clear SD1 and SD2 (*i.e.*, no proportional increase or sum of aggregated environmental degradation at all risk levels) from 1990 to 2000 (*i.e.*, SD1 and SD2 are rejected in all cases). In other words, there has been an increase in some countries' emissions and decrease in some others at some risk levels (*i.e.*, CDF curves and their integrals for CO_2 emissions from 1990 to 2010 intersect at some risk level). However, there has been an increase in the total CO_2 emissions from 1990 to 2005, since there is dominance at first-order at the 10% significant level. Therefore, there has been a clear degradation in CO_2 emissions within 15 years by all type of emitters. Clearly, degradation here means that the proportion of countries that emits above a given emission level increased

Table 2. Pair-wise SD comparisons of total CO_2 emissions over time.

	1990	1995	2000
1995			
SD1	ND	–	–
SD2	ND	–	–
2000			
SD1	ND	ND	–
SD2	ND	ND	–
2005			
SD1	10%	ND	ND
SD2	10%	ND	ND

Notes: The vertical columns represent the years 1995 to 2005 that are tested for SD against years from 1990 to 2000. Percentage levels represent the significance level of SD. ND suggests no dominance at that order.

over the 15-year period of time at all emission levels, suggesting that distribution of CO_2 emissions shifted to the right at all levels. In other words, CO_2 emissions by low, medium and high emitters have increased significantly. On the other side, there has been no dominance in each sub-sector (*i.e.*, electricity and heat production; manufacturing industries and construction; other sectors, excluding residential buildings and commercial and public services; residential buildings, commercial and public services; and the transport sector) over the whole period, suggesting that emissions in each sub-sector have been increasing for some countries, and have been decreasing for some others between 1990 and 2005. We also performed the analysis for CO_2 emissions from different subfuel consumptions (*i.e.*, gaseous, solid and liquid fuel consumption). Given the space limitation, we do not present the tables, but results are available from the authors.

We find that there has been an increase in the CO_2 emissions from gaseous fuel consumption within a 15-year period (from 1990 to 2005) at all emission levels, since there is SD1 at the 5% significance level, suggesting that the emissions from gaseous fuel consumption increased for all type of emitters. Finally, we find no dominance over time from solid and liquid fuel consumption, suggesting that there is no corresponding decrease or increase in CO_2 emissions from solid and liquid fuel consumption throughout the distribution of emissions. Overall, there has been an increase in the total CO_2 emissions from 1990 to 2005 at all degradation levels, which was mostly driven by a corresponding increase in CO_2 emissions from the gaseous fuel consumption at all levels between the same periods.

Then, we study pair-wise SD comparisons by looking at CO_2 emissions from different sub-sectors in 1990, 1995, 2000 and 2005. Overall, electricity and heat production have been the most dominant sectors over the whole period for CO_2 emissions, since emissions in these industries have always been dominating all other sectors at the first-order. In other words, for given CO_2 emission level, there is always a higher proportion of countries that emits CO_2 above this level due to electricity and heat production than the proportion from other industries. This relationship holds at all CO_2 emission levels suggesting that emissions from electricity and heat production have been higher for all type of emitters. The transport sector has been the second contributor to total CO_2 emissions, since this sector significantly dominated all other sectors, except the electricity and heat production sector at the first-order. The contributions of other sectors to the CO_2

emissions are: manufacturing industries and construction; residential buildings and commercial and public services; and other sectors, excluding residential buildings and commercial and public services respectively from the highest to the lowest contributor. The significance level of the dominance of each sector on the other has been different at different periods, showing a robust ranking of sectors. (Results are available upon request from the authors).

Finally, a comparison among CO_2 emissions from different types of fuel consumption from 1990 to 2005 (see **Table 3**) suggests that over the whole period

Table 3. Pair-wise SD comparisons of CO_2 emissions from sub-fuel consumption.

(a) Sub-fuel comparisons in 1990			
Industry comparisons	Dominance outcome	SD1	SD2
GAS versus LIQUID	LIQUID dominates	1%	1%
GAS versus SOLID	SOLID dominates	ND	10%
LIQUID versus SOLID	LIQUID dominates	1%	1%
(b) Sub-fuel comparisons in 1995			
Industry comparisons	Dominance outcome	SD1	SD2
GAS versus LIQUID	LIQUID dominates	1%	1%
GAS versus SOLID	ND	ND	ND
LIQUID versus SOLID	LIQUID dominates	1%	1%
(c) Sub-fuel comparisons in 2000			
Industry comparisons	Dominance outcome	SD1	SD2
GAS versus LIQUID	LIQUID dominates	1%	1%
GAS versus SOLID	ND	ND	ND
LIQUID versus SOLID	LIQUID dominates	1%	1%
(d) Sub-fuel comparisons in 2005			
Industry comparisons	Dominance outcome	SD1	SD2
GAS versus LIQUID	LIQUID dominates	1%	1%
GAS versus SOLID	SOLID dominates	ND	10%
LIQUID versus SOLID	LIQUID dominates	1%	1%

Notes: Industry comparisons columns represent all possible sub-industry comparisons at a given year. Dominance Outcome column offers the dominating sub-fuel as a result of comparisons between different sub-fuels. SD1 and SD2 represent the significance levels for the first- and second-order dominance. ND suggests no dominance at that order.

the liquid fuel consumption has always been the major contributor to CO_2 emissions since CO_2 emissions from this type dominate the emissions from the gaseous and solid fuel consumption at first-order, at 1% significance level. On the other hand, CO_2 emissions from the solid fuel consumption dominate the emissions from the gaseous fuel consumption at the second-order, at 10% significance level in 1990 and 2005, but the relationship between these two types of fuel consumption is ambiguous in 1995 and 2000.

4.1.2. Methane Emissions

We then investigate the evolution of total methane emissions, methane emissions from agriculture and the energy sector, respectively, between 1990 and 2005. The findings suggest that there has been no general increase or decrease in total methane emissions over the whole period. Similarly, no general progress of methane emissions from different sub-sectors is found between the same periods. **Figure 1** presents the CDF of methane emissions for 1990, 1995, 2000 and 2005. Clearly, the CDF curves of methane emissions for different years overlap at almost all emission levels and there is no clear dominance at any order. **Figure 2** depicts the CDFs of methane emissions released by countries due to the activities in agriculture and energy sectors in 2005. Since the CDF of the methane emissions released due to the activities in agriculture sector is always below the CDF of the methane emissions released due to the activities in the energy sector, this suggests a clear SD1 of the agriculture sector over the energy sector. In other words, there is always a higher proportion of countries that emit methane gasses to the atmosphere due to the activities taking place in the agriculture sector than in the energy sector at all emission levels. Since there is a clear ordering of industries that contribute to the methane gas emissions, one could suggest a global action plan to reduce methane emissions released by the agriculture sector. It is not that different countries emit higher levels of methane emissions in different sectors (hence country-specific actions are required), but agriculture sectors' contribution is always higher than that of energy sector and therefore a global action targeting ways to eliminate methane emissions by agriculture sectors would be a more effective strategy.

We also conduct the pair-wise comparisons of methane emissions from the agriculture and the energy sectors in 1990, 1995, 2000 and 2005 (Results are

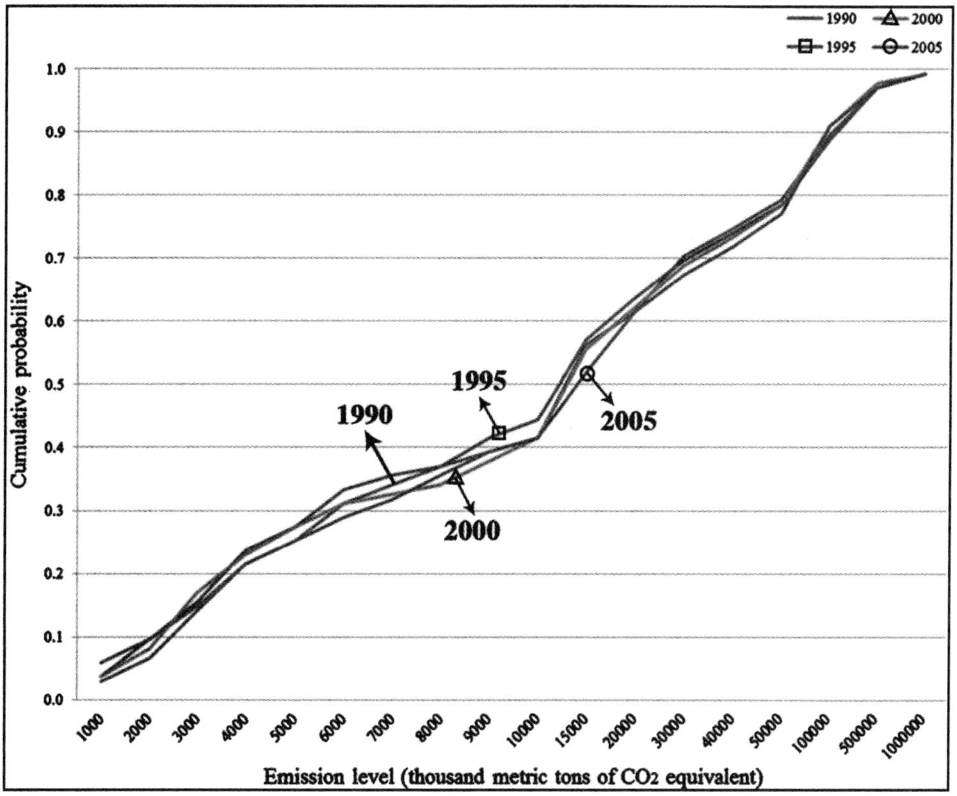

Figure 1. Cumulative distribution functions of methane emissions in 1990, 1995, 2000 and 2005.

available upon request from the authors). For the whole period, methane emissions from the agriculture sector have always been higher than from the energy sector. Methane emissions from agriculture dominate the energy sector at the first-order at 1% significance level. Thus, for any given methane emission level, there have been always more countries emitting above that level in the agriculture sector than the energy sector. Therefore, there has been a clear robust ranking of sectors (from the highest methane emitting sector to the lowest one) over the period 1990–2005.

4.1.3. Nitrous Oxide Emissions

We further analyze the progress of total nitrous oxide emissions, nitrous oxide emissions from the agriculture, the industrial and the energy sectors between 1990 and 2005 (Results are available from the authors). The findings suggest that

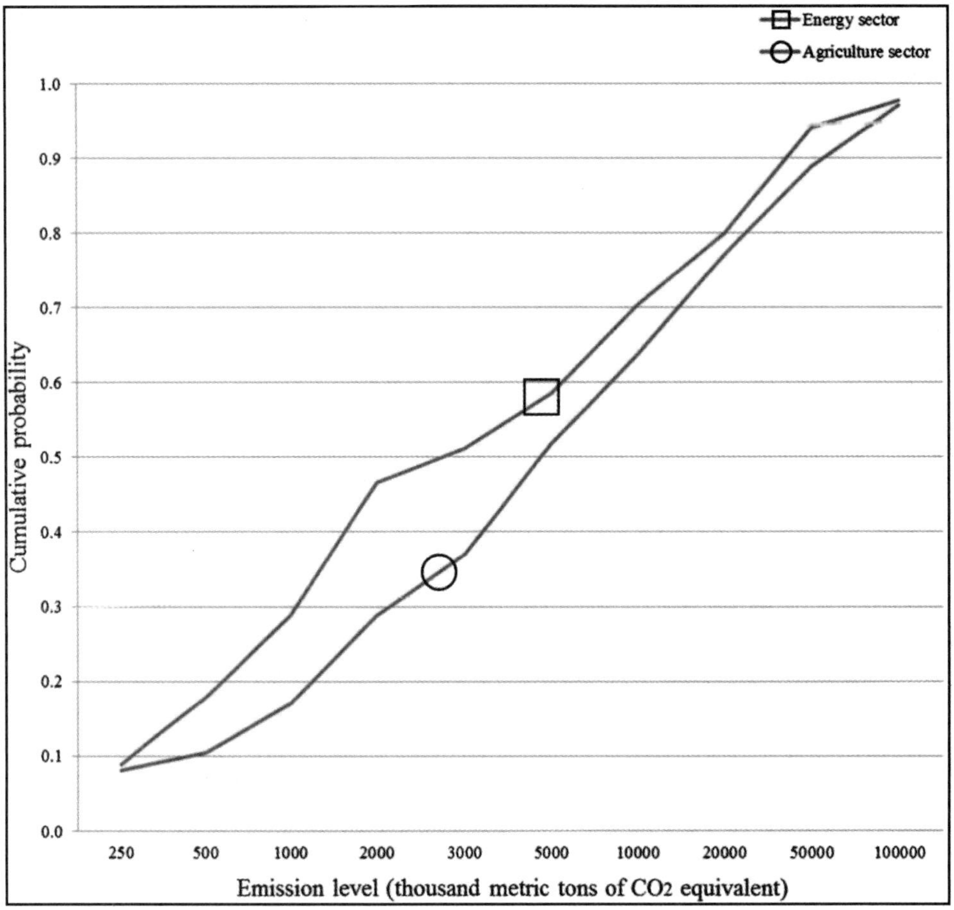

Figure 2. Cumulative distribution functions of methane emissions from agriculture and energy sector for 2005.

there has been neither a general increase or decrease in total nitrous oxide emissions nor the nitrous oxide emissions from different sub-sectors over time. This suggests that some countries' nitrous oxide emission levels increased and some other countries' emissions were decreased. Furthermore, increase in nitrous oxide emission levels for some countries was offset by the decrease in emissions by other countries (*i.e.*, there was no second-order SD). In other words, country-specific (or group of country-specific) policies will be more suitable to decrease the nitrous oxide emission levels as there is no clear increase in emissions for all type of emitters.

Similarly to the analyses above, we employ the pairwise comparisons between three sub-sectors (*i.e.*, agricultural, industrial and energy sectors) to find the major industry which releases the highest nitrous oxide emissions over time. For the whole period, nitrous oxide emissions from the agriculture sector has always been higher than the other two sectors, while nitrous oxide emissions from the energy sector have always been higher than the industrial sector for the whole period. Nitrous oxide emissions from agriculture dominate the energy and the industrial sectors at first-order at 1% significance level and, similarly, emissions from the energy sector dominate those of the industrial sector at first-order at a significance level of 1% over the whole period. In other words, for any given nitrous oxide emission level, there have been always more countries emitting above that level in agriculture sector than the energy and industrial sector. Overall, there has been a clear robust ranking of sectors (from the highest nitrous emitting sector to the lowest one) over the period 1990–2005.

4.1.4. Other GHG Emissions

Although the other GHG emissions have always been contributing less to the total, we still conduct pair-wise SD comparisons for the other GHG emissions and its subcomponents from 1990 to 2005. The four panels of **Table 4** present the results for the evolution of the total other GHG emissions, perfluorocarbon (PFC), hydrofluorocarbon (HFC), and sulfur hexafluoride (SF6) emissions respectively between 1990 and 2005. HFC emissions are mostly due to use of refrigeration, air-conditioning, and insulating foam products (see e.g., Velders *et al.* 2009). PFC emissions are mainly due to aluminum production (see e.g., Marks *et al.* 2013), whereas SF6 emissions are due to leakage and venting from the electricity sector, magnesium production, and other minor contributions (see e.g., Olivier *et al.* 2005).

We conduct our analysis for each type of emission and find that there has been a general increase in the total GHG emissions in 5-year horizons between 1990 and 2000 suggesting that there is always a higher proportion of countries that emit above a given level in 2000 than in 1990 for all emission levels, yet no clear indication was detected between 2000 and 2005 suggesting that increase in other GHG emission by some countries was offset by a decrease in other GHG emissions by some other countries. On the other hand, HFC emissions have been increasing in 5-year horizons over the whole period as the later 5-year HFC emissions dominate the earlier ones at first-order at the 1% significance level supporting the

Table 4. Pair-wise SD comparisons other GHG, HFC, PFC and SF6 emissions over time.

(a) Total other GHG emissions				(b) HFC emissions			
	1990	1995	2000		1990	1995	2000
1995				1995			
SD1	1%	–	–	SD1	1%	–	–
SD2	1%	–	–	SD2	1%	–	–
2000				2000			
SD1	1%	5%	–	SD1	1%	1%	–
SD2	1%	5%	–	SD2	1%	1%	–
2005				2005			
SD1	1%	1%	ND	SD1	1%	1%	1%
SD2	1%	1%	ND	SD2	1%	1%	1%

(c) PFC emissions				(d) SF6 emissions			
	1995	2000	2005		1990	1995	2000
	1990				1995		
SD1	5%	ND	1%	SD1	ND	–	–
SD2	5%	ND	1%	SD2	ND	–	–
	1995				2000		
SD1	–	ND	ND	SD1	ND	ND	–
SD2	–	ND	ND	SD2	ND	ND	–
	2000				2005		
SD1	–	–	ND	SD1	ND	ND	ND
SD2	–	–	ND	SD2	ND	ND	ND

Notes: The vertical columns represent the years 1995 to 2005 that are tested for SD against years from 1990 to 2000. Percentage levels give the significance level of SD. The vertical and horizontal axes are reversed for PFC emissions to represent the improvement over time. ND suggests no dominance at that order.

fact that increased demand for refrigeration, air-conditioning, and insulating foam products (*i.e.*, main contributors of the HFC emissions) and this has been the case for all type of emitters as there is always a higher proportion of countries that emit above a given HFC emission level in the following period than the previous one. On the other hand, we find no clear result for the SF6 emissions, since SD tests provide no dominance in the period as a whole. More interestingly, we find that there has been a general decrease of the PFC emissions from 1990 to 1995 and from 1990 to 2005. In other words, PFC emissions in 1990 dominate the PFC emissions in 1995 and 2005 at first-order at the 5 and 1% significance levels respectively. For PFC emissions, years on the vertical axis are tested against the horizontal but the years 1990 to 2000 are tested against the years 1995 and 2005 respectively. Since there has been a proportional decrease in PFC emissions at all emission levels over time, the testing horizon is reversed. Hence, for any given PFC emission level, there have been always more countries emitting above that level in 1990 when compared with 1995 and 2005. This confirms that there have been good adaptation strategies across the globe in reducing PFC emissions over time.

4.1.5. Comparison among GHG Emissions

Finally, we performed the pair-wise SD comparisons among CO_2, methane, nitrous oxide and other GHG emissions in 1990, 1995, 2000 and 2005 (Results are available upon request from the authors). Our findings suggest a clear difference between the types of emissions. CO_2 has always been the main component that has been releasing emissions when compared with the other type of greenhouse gases. As a result, for any given CO_2 equivalent emission level, there have been always more countries emitting CO_2 above that level when compared with methane, nitrous oxide and other GHG emissions. Furthermore, methane emissions dominate the nitrous and other GHG emissions between 1990 and 2005 at first order at the 1% significance level making it the second major GHG emissions contributor. Similarly, for any given CO_2 equivalent emission level, there have always been more countries emitting methane above that level when compared with nitrous oxide and other GHG emissions. Finally, other GHG emissions (*i.e.*, sum of the HFC, PFC and SF6 emissions), have been contributing the least, when compared with the other type of greenhouse gases. This result can help identify policies for achieving improvements in environmental quality. The implication here is that policies aiming to reduce CO_2 emissions need to be given priority when compared with the other types of emissions.

4.2. SD Comparisons in Water Pollution

For water pollution the sample period consists only of a 10-year horizon (from 1995 to 2005). There has been information on water pollution in 1990 for only 12 countries, which makes the application impossible before 1995 since the power of tests would not be reliable. The eight panels of **Table 5** give the pair-wise SD

Table 5. Pair-wise SD comparisons of total and sub-industry water pollution over time.

(a) Total water pollution			(b) Water pollution from chemistry industry		
	1995	2000		1995	2000
2000			2000		
SD1	ND	ND	SD1	ND	ND
SD2	ND	ND	SD2	10%	ND
2005			2005		
SD1	ND	ND	SD1	ND	ND
SD2	10%	ND	SD2	10%	ND
(c) Water pollution from clay and glass industry			(d) Water pollution from food industry		
	1995	2000		1995	2000
2000			2000		
SD1	ND	ND	SD1	ND	ND
SD2	ND	ND	SD2	10%	ND
	2005			2005	
SD1	ND	ND	SD1	ND	ND
SD2	10%	ND	SD2	5%	ND
(e) Water pollution from metal industry			(f) Water pollution from paper and pulp industry		
	1995	2000		1995	2000
2000			2000		
SD1	ND	ND	SD1	ND	ND
SD2	ND	ND	SD2	ND	ND
2005			2005		
SD1	ND	ND	SD1	ND	ND
SD2	10%	ND	SD2	ND	ND
(g) Water pollution from textile industry			(h) Water pollution from wood industry		
	1995	2000		1995	2000
2000			2000		
SD1	ND	ND	SD1	ND	ND
SD2	ND	ND	SD2	10%	ND
2005			2005		
SD1	ND	ND	SD1	ND	ND
SD2	ND	ND	SD2	5%	ND

Notes: The vertical columns represent the years 2000 and 2005 that are tested for SD against years from 1995 and 2000. Percentage levels represent the significance level of SD. ND suggests no dominance at that order.

test results for the evolution of total water pollution and its sub-industries' contributors over time. The first panel of **Table 5** suggests that there was no general increase in water pollution over the whole period. However, there has been an increase in water pollution in the 10-year horizon in a second-order sense, suggesting that sum of water pollution above a given level is higher in 2005 than in 1995 for all levels of pollution. Hence the sum of water pollution up to a given pollution level has always been higher in 2005 than in 1995 (*i.e.*, some countries' water pollution decreased, but some others experienced an increase in their water pollution, and the sum of the increases in water pollution has been higher than the sum of the decreases for a given level of pollution). **Figure 3** depicts the CDFs of the water pollutant emissions (measured as BOD levels per day) for 1995, 2000 and 2005. As the CDF curves of each year intersect with each other, the tests did not yield any SD1. However, when CDFs intersect, one could test whether there is any clear ordering over time when the integrals of water pollution at each respective year (*i.e.*, sum of the total water pollution up to a water pollution level) are compared. In this case, water pollution in 2005 dominates the water pollution in 1995 in the second-order sense at the 10% significance level. The CDFs of water pollution in 1995 and 2005 do intersect at some point (*i.e.*, no SD1), and yet one can discover that the sum of the water pollution up to a given level is always lower in

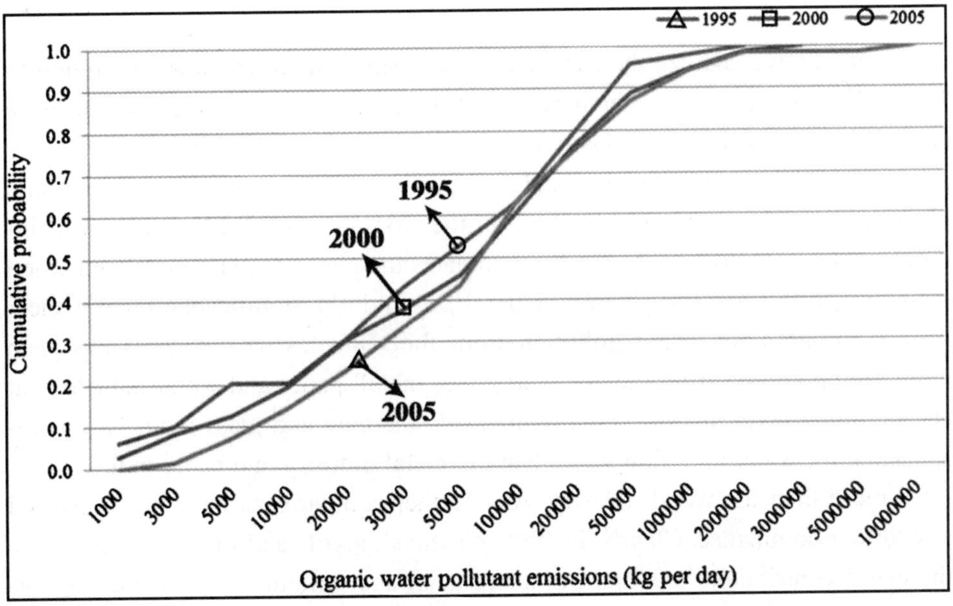

Figure 3. Cumulative distribution functions of water pollutant emissions for 1995, 2000 and 2005.

2005 than in 1995, suggesting SD2, where the sum of water pollution above a given level is always higher in 2005 than in 1995 for all emission levels.

Similarly to total water pollution, there has been no improvement or deterioration in sub-industry water pollution over the whole period at all emission levels, since there has been no dominance in the first-order sense for all industries. However, water pollution levels from different industries have shown different progress over time. The sum of water pollution from chemical, food and wood industries above a given level is always higher in 2000 than in 1995 suggesting that even though some countries' water pollution in these industries decreased, increase in water pollution by some other countries were relatively more than the decrease in those countries. Furthermore, water pollution from the chemical, food, wood, metal, and clay and glass industries increased between 1995 and 2005 in the second-order sense suggesting a similar trend as above but within 10-year horizon. Finally, no dominance of any order is found for textile and paper and pulp industries. Therefore, one can conclude that the increase in water pollution over time is mostly driven by the chemical, food and wood industries as those industries experienced an overall increase of water pollution in shorter horizons (*i.e.*, an overall increase within 5-year horizons) suggesting that the global action to reduce water pollution in these industries should be prioritized.

Finally, we analyze the sub-industry contributions to the water pollution in 1995, 2000 and 2005. The three panels of **Table 6** present all possible pair-wise comparisons between sub-industry water pollutions in 1995, 2000 and 2005 respectively. In 1995 the chemical industry pollutes water more than the clay and glass, metal and wood industries (*i.e.*, in the first panel of **Table 5**, chemical industry water pollution stochastically dominates the clay and glass metal and wood industries in the first-order sense at the 10, 5 and 1% significance level respectively). Furthermore, water pollution from food and textile industries has been more than pollution from the clay and glass, metal, paper and wood industries at any pollution level in 1995. Finally, in 1995, the clay and glass industry was responsible for water pollution more than the metal industry and paper industry polluted more than the wood industry. Any further comparisons have not suggested any further dominance. Clearly, in 1995, chemical, textile and food industries were the major contributors to water pollution, as at any pollution level there have always been more countries in those industries polluting water than remaining industries above that any given pollution level.

Table 6. Pair-wise SD comparison of water pollution from industries.

Industry comparisons	Water pollution industry comparisons in 1995			Water pollution industry comparisons in 2000			Water pollution industry comparisons in 2005		
	Dominating industry	SD1	SD2	Dominating industry	SD1	SD2	Dominating industry	SD1	SD2
Chemical vs. Clay	Chemical	10%	5%	Chemical	10%	5%	Chemical	5%	5%
Chemical vs. Food	ND	ND	ND	Food	5%	5%	Food	5%	5%
Chemical vs. Metal	Chemical	10%	5%	Chemical	5%	5%	Chemical	1%	1%
Chemical vs. Paper	ND	ND	ND	ND	ND	ND	Chemical	10%	10%
Chemical vs. Textile	ND	ND	ND	ND	ND	ND	ND	ND	ND
Chemical vs. Wood	Chemical	1%	1%	Chemical	1%	1%	Chemical	1%	1%
Clay versus Food	Food	1%	1%	Food	1%	1%	Food	1%	1%
Clay versus Metal	Clay	10%	10%	Clay	10%	10%	Clay	10%	10%
Clay versus Paper	ND	ND	ND	ND	ND	ND	ND	ND	ND
Clay versus Textile	Textile	10%	1%	Textile	5%	1%	Textile	5%	5%
Clay versus Wood	ND	ND	ND	ND	ND	ND	ND	ND	ND
Food versus Metal	Food	1%	1%	Food	1%	1%	Food	1%	1%
Food versus Paper	Food	10%	5%	Food	1%	1%	Food	1%	1%
Food versus Textile	ND	ND	ND	ND	ND	ND	Food	10%	10%
Food versus Wood	Food	1%	1%	Food	1%	1%	Food	1%	1%
Metal versus Paper	ND			Paper	10%	10%	Paper	10%	10%
Metal versus Textile	Textile	1%	1%	Textile	1%	1%	Textile	1%	1%
Metal versus Wood	ND	ND	ND	ND	ND	ND	ND	ND	ND
Paper versus Textile	Textile	10%	5%	Textile	5%	5%	Textile	10%	10%
Paper versus Wood	Paper	5%	5%	Paper	ND	10%	Paper	10%	10%
Textile versus Wood	Textile	1%	1%	Textile	1%	1%	Textile	1%	1%

Notes: First column represents all possible sub-industry water pollution comparisons. Second to fourth panels present the dominance outcomes between sub-industry comparisons for each respective case for the years 1995, 2000 and 2005 respectively. SD1 and SD2 represent the significance levels for the first- and second-order dominance. ND suggests no dominance at that order.

In 2000 the majority of the dominance relations between industries remain the same, with some differences with respect to 1995. Water pollution from the food industry dominates pollution from the chemical industry in the first-order sense at the 5% significance level. In 2000 the major contributors to water pollution are the food and textile industries. However, there is no clear SD ordering among food and textile industries, when water pollution is considered. Finally, in 2005, water pollution from the food industry contributes more than any other industry (*i.e.*, water pollution from the food industry dominates such pollution from any industry in the first-order sense). Therefore, a global action tackling the increase in water pollution due to activities in the food industry should be prioritized.

5. Conclusions

Our methodology based on consistent pair-wise SD tests can provide useful information to policy makers in their efforts to design policies that compare the risks from environmental degradation. Reducing CO_2 emissions needs to be given a priority, with special attention to those industrial sectors which are mainly responsible for these emissions. As the agriculture sector is the major contributor to the methane emissions and the food sector is becoming the industry that is polluting water the most, our findings suggest interlinkages between air and water pollution. Water pollution will likely be intensified by the increasing demand for biomass-derived fuels for transportation biofuel needs, because large quantities of water are needed to grow the fuel crops, and water pollution is exacerbated by agricultural drainage containing fertilizers, pesticides, and sediment. Potentially, there are major spillovers in environmental degradation across countries, and across air and water pollution levels. As Olmstead (2010) claims, water pollution in transboundary settings is still a challenge since our analysis find an aggregate increase in water pollution even though some countries pollute less over time as relatively lower levels of water pollution in these countries could be due to free-riding. In other words, even though some countries' direct contribution to water pollution is decreased (due to their production levels), their indirect contribution (*i.e.*, due to increased consumption) might have led to an aggregate increase in water pollution levels.

Source: Agliardi E, Pinar M, Stengos T. Air and water pollution over time and industries with stochastic dominance[J]. Stochastic Environmental Research & Risk Assessment, 2016:1–20.

References

[1] Agliardi E (2011) Sustainability in uncertain economies. Environ Resource Econ 48:71–82.

[2] Agliardi E, Pinar M, Stengos T (2015) An environmental degradation index based on stochastic dominance. Empir Econ 48:439–459.

[3] Anderson G (2004) Making inferences about the polarization, welfare and poverty of nations: a study of 101 countries 1970–1995. J Appl Econ 19:537–550.

[4] Arrow K, Dasgupta P, Goulder L, Daily G, Ehrlich P, Heal G, Levin S, Ma¨ler K-G, Schneider S, Starrett D, Walker B (2004) Are we consuming too much? J Econ Perspect 18:147–172.

[5] Barrett GF, Donald SG (2003) Consistent tests for stochastic dominance. Econometrica 71:71–104.

[6] Baumol WJ (1972) On taxation and the control of externalities. Am Econ Rev 62:307–322.

[7] Bernauer T, Kuhn PM (2010) Is there an environmental version of the Kantian peace? Insights from water pollution in Europe. Eur J Int Relat 16:77–102.

[8] Besag J, York J, Mollie A (1991) Bayesian image restoration, with two applications in spatial statistics. Ann Inst Stat Math 43:1–20.

[9] Blanchet D, Fleurbaey M (2013) Beyond GDP: measuring welfare and assessing sustainability. Oxford University Press, New York.

[10] Boden TA, Marland G, Andres RJ, (2013), Global, regional, and national fossil-fuel CO2 emissions, Carbon Dioxide Information Analysis Center, Oak Ridge National Laboratory, U.S. Department of Energy, Oak Ridge, TN, USA, On line at: http://dx.doi.org/10.3334/CDIAC/00001_V2013.

[11] Cocchi D, Greco F, Trivisano C (2007) Hierarchical space-time modelling of PM10 pollution. Atmos Environ 41:532–542.

[12] Dahlman CJ (1979) The problem of externality. J Law Econ 22:141–162.

[13] Dasgupta PS (2000) Human well-being and the natural environment. Oxford University Press, Oxford.

[14] Davidson R, Duclos J-Y (2000) Statistical inference for stochastic dominance and for the measurement of poverty and inequality. Econometrica 68:1435–1464.

[15] Dominguez-Faus R, Powers SE, Burken JG, Alvarez PJ (2009) The water footprint of

biofuels: a drink or drive issue? Environ Sci Technol 43:3005–3010.

[16] Duclos J-Y, Makdissi P, Wodon Q (2005) Poverty-efficient transfer programs: the role of targeting and allocation rules. J Dev Econ 77:53–74.

[17] Duclos J-Y, Makdissi P, Wodon Q (2008) Socially-improving tax reforms. Int Econ Rev 49:1507–1539.

[18] Gans J, King S, Mankiw GN (2012) Principles of microeconomics, 2nd edn. Cengage Learning, Melbourne.

[19] Hettige H, Mani M, Wheeler D (2000) Industrial pollution in economic development: Kuznets revisited. J Dev Econ 62:445–476.

[20] Law J, Quick M, Chan PW (2014a) Bayesian spatio-temporal modeling for analysing local patterns of crime over time at the small area level. J Quant Criminol 30:57–78.

[21] Law J, Quick M, Chan PW (2014b) Analyzing hotspots of crime using a Bayesian spatiotemporal modeling approach: a case study of violent crime in the Greater Toronto Area. Geogr Anal 47:1–19.

[22] Li G, Haining R, Richardson S, Best N (2014a) Space–time variability in burglary risk: a Bayesian spatio-temporal modelling approach. Spat Stat 9:180–191.

[23] Li Q, Song J, Wang E, Hu H, Zhang J, Wang Y (2014b) Economic growth and pollutant emissions in China: a spatial econometric analysis. Stoch Env Res Risk Assess 28:429–442.

[24] Makdissi P, Wodon Q (2004) Robust comparisons of natural resource depletion indices. Econ Bull 9:1–9.

[25] Marks J, Tabereaux A, Pape D, Bakshi V, Dolin EJ (2013) Factors affecting PFC emissions from commercial aluminum reduction cells, In: Bearne G, Dupuis M, Tarcy G (eds) Essential readings in light metals: aluminum reduction technology, Wiley, Hoboken, NJ, On line at: http://dx.doi.org/10.1002/9781118647851.ch151.

[26] Natvig B, Tvete IF (2007) Bayesian hierarchical space-time modelling of earthquake data. Methodol Comput Appl Probab 9:89–114.

[27] Olivier JGJ, Van Aardenne JA, Dentener F, Ganzeveld L, Peters JAHW (2005) Recent trends in global greenhouse gas emissions: regional trends and spatial distribution of key sources. In: Van Amstel A (ed) Non-CO2 Greenhouse Gases (NCGG-4). Mill-press, Rotterdam, pp 325–330.

[28] Olmstead SM (2010) The economics of water quality. Rev Environ Econ Policy 4:44–62.

[29] Orda's Criado C, Valente S, Stengos T (2011) Growth and pollution convergence: theory and evidence. J Environ Econ Manage 62:199–214.

[30] Ors E, Palomino F, Peyrache E (2013) Performance gender-gap: does competition matter? J Labor Econ 31:443–499.

[31] Paruolo P, Murphy B, Janssens-Maenhout G (2015) Do emissions and income have a common trend? A country-specific, time-series, global analysis, 1970–2008. Stoch Env Res Risk Assess 29:93–107.

[32] Persson TA, Azar C, Lindgren K (2006) Allocation of CO2 emission permits–economic incentives for emission reductions in developing countries. Energy Policy 34:1889–1899.

[33] Peters GP, Hertwich EG (2008) CO_2 embodied in international trade with implications for global climate policy. Environ Sci Technol 42:1401–1407.

[34] Pinar M, Stengos T, Topaloglou N (2013) Measuring human development: a stochastic dominance approach. J Econ Growth 18:69–108.

[35] Pinar M, Cruciani C, Giove S, Sostero M (2014) Constructing the FEEM sustainability index: a Choquet integral application. Ecol Ind 39:189–202.

[36] Rezai A, Foley DK, Taylor L (2012) Global warming and economic externalities. Econ Theor 49:329–351.

[37] Sahu SK, Gelfand AE, Holland DM (2007) High resolution spacetime ozone modeling for assessing trends. J Am Stat Assoc 102:1212–1220.

[38] Sang H, Gelfand AE (2009) Hierarchical modeling for extreme values observed over space and time. Environ Ecol Stat 16:407–426.

[39] Scotto M, Guedes Soares C (2007) Bayesian inference for long-term prediction of significant wave height. Coast Eng 54:393–400.

[40] Sigman H (2002) International spillovers and water quality in rivers: do countries free ride? Am Econ Rev 92:1152–1159.

[41] Singh RK, Murty HR, Gupta SK, Dikshit AK (2012) An overview of sustainability assessment methodologies. Ecol Ind 15:281–299.

[42] Sivakumar B, Christakos G (2011) Climate: patterns, changes, and impacts. Stoch Env Res Risk Assess 25:443–444.

[43] Smit B, Wandel J (2006) Adaptation, adaptive capacity and vulnerability. Glob Environ Change 16:282–292.

[44] Sullivan CA (2011) Quantifying water vulnerability: a multi-dimensional approach. Stoch Env Res Risk Assess 25:627–640.

[45] Tamazian A, Chousa JP, Vadlamannati KC (2009) Does higher economic and financial development lead to environmental degradation: evidence from BRIC countries. Energy Policy 37:246–253.

[46] Tol RSJ (2009) The economic effects of climate change. J Econ Perspect 23:29–51.

[47] Vanem E (2011) Long-term time-dependent stochastic modelling of extreme waves. Stoch Env Res Risk Assess 25:185–209.

[48] Velders GJM, Fahey DW, Daniel JS, McFarland M, Andersen SO (2009) The large

contribution of projected HFC emissions to future climate forcing. Proc Natl Acad Sci USA 106:10949–10954.

[49] Waller LA, Carlin BP, Xia H, Gelfand AE (1997) Hierarchical spatio-temporal mapping of disease rates. J Am Stat Assoc 92:607–617.

[50] Wikle CK (2003) Hierarchical models in environmental science. Int Stat Rev 71:181–199.

[51] Wikle CK, Berliner LM, Cressie N (1998) Hierarchical Bayesian space-time models. Environ Ecol Stat 5:117–154.

[52] World Bank (2012) World Development Indicators 2012. World Bank, Washington.

[53] Xepapadeas A (2011) The economics of non-point source pollution. Ann Rev 3:355–373.

[54] Xepapadeas A, Vouvaki D (2008) Changes in social welfare and sustainability: theoretical issues and empirical evidence. Ecol Econ 67:473–484.

[55] Yunfeng Y, Laike Y (2010) China's foreign trade and climate change: A case study of CO2 emissions. Energy Policy 38:350–356.

Appendix

Table A1. List of countries used in each respective analysis.

Country name	Country code	Water pollution	CO_2 emissions	Methane, nitrous oxide and other GHG emissions
Afghanistan	AFG		x	
Albania	ALB	x	x	x
Algeria	DZA		x	x
Andorra	ADO		x	
Angola	AGO		x	x
Antigua and Barbuda	ATG		x	
Argentina	ARG	x	x	x
Armenia	ARM		x	x
Aruba	ABW	x	x	
Australia	AUS		x	x
Austria	AUT	x	x	x
Azerbaijan	AZE	x	x	x
Bahamas, The	BHS	x	x	
Bahrain	BHR		x	x
Bangladesh	BGD	x	x	x
Barbados	BRB		x	
Belarus	BLR		x	x
Belgium	BEL	x	x	x
Belize	BLZ		x	
Benin	BEN		x	x
Bermuda	BMU		x	
Bhutan	BTN		x	
Bolivia	BOL	x	x	x
Bosnia and Herzegovina	BIH		x	x

Continued

Botswana	BWA	x	x	x
Brazil	BRA		x	x
Brunei Darussalam	BRN		x	x
Bulgaria	BGR	x	x	x
Burkina Faso	BFA		x	
Burundi	BDI		x	
Cambodia	KHM	x	x	x
Cameroon	CMR		x	x
Canada	CAN	x	x	x
Cape Verde	CPV		x	
Cayman Islands	CYM		x	
Central African Republic	CAF		x	
Chad	TCD		x	
Chile	CHL	x	x	x
China	CHN	x	x	x
Colombia	COL	x	x	x
Comoros	COM		x	
Congo, Dem. Rep.	ZAR		x	x
Congo, Rep.	COG		x	x
Costa Rica	CRI		x	x
Cote d'Ivoire	CIV		x	x
Croatia	HRV	x	x	x
Cuba	CUB		x	x
Cyprus	CYP	x	x	x
Czech Republic	CZE	x	x	x
Denmark	DNK	x	x	x
Djibouti	DJI		x	

Continued

Dominica	DMA		x	
Dominican Republic	DOM		x	x
Ecuador	ECU	x	x	x
Egypt, Arab Rep.	EGY		x	x
El Salvador	SLV		x	x
Equatorial Guinea	GNQ		x	
Eritrea	ERI	x	x	x
Estonia	EST	x	x	x
Ethiopia	ETH	x	x	x
Faeroe Islands	FRO		x	
Fiji	FJI		x	
Finland	FIN	x	x	x
France	FRA	x	x	x
French Polynesia	PYF		x	
Gabon	GAB		x	x
Gambia, The	GMB	x	x	
Georgia	GEO		x	x
Germany	DEU	x	x	x
Ghana	GHA		x	x
Gibraltar	GIB		x	x
Greece	GRC	x	x	x
Greenland	GRL		x	
Grenada	GRD		x	
Guatemala	GTM		x	x
Guinea	GIN		x	
Guinea-Bissau	GNB		x	
Guyana	GUY		x	

Continued

Haiti	HTI	x	x	x
Honduras	HND		x	x
Hong Kong SAR, China	HKG		x	x
Hungary	HUN	x	x	x
Iceland	ISL		x	x
India	IND		x	x
Indonesia	IDN	x	x	x
Iran, Islamic Rep.	IRN	x	x	x
Iraq	IRQ		x	x
Ireland	IRL	x	x	x
Israel	ISR	x	x	x
Italy	ITA	x	x	x
Jamaica	JAM		x	x
Japan	JPN	x	x	x
Jordan	JOR	x	x	x
Kazakhstan	KAZ	x	x	x
Kenya	KEN		x	x
Kiribati	KIR		x	
Korea, Dem. Rep.	PRK		x	x
Korea, Rep.	KOR	x	x	x
Kuwait	KWT		x	x
Kyrgyz Republic	KGZ	x	x	x
Lao PDR	LAO		x	
Latvia	LVA	x	x	x
Lebanon	LBN		x	x
Lesotho	LSO	x		
Liberia	LBR		x	
Libya	LBY		x	x
Lithuania	LTU	x	x	x

Continued

Luxembourg	LUX	x	x	x
Macao SAR, China	MAC		x	
Macedonia, FYR	MKD	x	x	x
Madagascar	MDG	x	x	
Malawi	MWI	x	x	
Malaysia	MYS	x	x	x
Maldives	MDV		x	
Mali	MLI		x	
Malta	MLT	x	x	x
Marshall Islands	MHL		x	
Mauritania	MRT		x	
Mauritius	MUS	x	x	
Mexico	MEX		x	x
Micronesia, Fed. Sts.	FSM		x	
Moldova	MDA	x	x	x
Mongolia	MNG	x	x	x
Montenegro	MNE		x	
Morocco	MAR	x	x	x
Mozambique	MOZ		x	x
Myanmar	MMR		x	x
Namibia	NAM		x	x
Nepal	NPL		x	x
Netherlands	NLD	x	x	x
New Caledonia	NCL		x	
New Zealand	NZL	x	x	x
Nicaragua	NIC		x	x
Niger	NER		x	
Nigeria	NGA		x	x
Norway	NOR	x	x	x

Continued

Oman	OMN	x	x	x
Pakistan	PAK		x	x
Palau	PLW		x	
Panama	PAN	x	x	x
Papua New Guinea	PNG		x	
Paraguay	PRY	x	x	x
Peru	PER		x	x
Philippines	PHL	x	x	x
Poland	POL	x	x	x
Portugal	PRT	x	x	x
Qatar	QAT	x	x	x
Romania	ROM	x	x	x
Russian Federation	RUS	x	x	x
Rwanda	RWA		x	
Samoa	WSM		x	
Sao Tome and Principe	STP		x	
Saudi Arabia	SAU		x	x
Senegal	SEN	x	x	x
Serbia	SRB		x	x
Seychelles	SYC		x	
Sierra Leone	SLE		x	
Singapore	SGP	x	x	x
Slovak Republic	SVK	x	x	x
Slovenia	SVN	x	x	x
Solomon Islands	SLB		x	
Somalia	SOM		x	
South Africa	ZAF	x	x	x
Spain	ESP	x	x	x
Sri Lanka	LKA		x	x

Continued

St. Kitts and Nevis	KNA		x	
St. Lucia	LCA		x	
St. Vincent and the Grenadines	VCT		x	
Sudan	SDN		x	x
Suriname	SUR		x	
Swaziland	SWZ		x	
Sweden	SWE	x	x	x
Switzerland	CHE		x	x
Syrian Arab Republic	SYR	x	x	x
Tajikistan	TJK	x	x	x
Tanzania	TZA	x	x	x
Thailand	THA	x	x	x
Timor-Leste	TMP		x	
Togo	TGO		x	x
Tonga	TON	x	x	
Trinidad and Tobago	TTO	x	x	x
Tunisia	TUN		x	x
Turkey	TUR	x	x	x
Turkmenistan	TKM		x	x
Turks and Caicos Islands	TCA		x	
Uganda	UGA	x	x	
Ukraine	UKR	x	x	x
United Arab Emirates	ARE		x	x
United Kingdom	GBR	x	x	x
United States	USA	x	x	x
Uruguay	URY		x	x
Uzbekistan	UZB		x	x
Vanuatu	VUT		x	
Venezuela, RB	VEN		x	x
Vietnam	VNM	x	x	x
West Bank and Gaza	WBG		x	
Yemen, Rep.	YEM	x	x	x
Zambia	ZMB		x	x
Zimbabwe	ZWE		x	x

Table A2. Variable definitions and sources.

Variable	Definition and sources
CO2 emissions, emissions from different consumption types and emissions by sectors.	Carbon dioxide emissions are those stemming from the burning of fossil fuels and the manufacture of cement. They include carbon dioxide produced during consumption of solid, liquid, and gas fuels and gas flaring. Detailed data set is obtained from the Carbon Dioxide Information Analysis Center, Environmental Sciences Division, Oak Ridge National Laboratory, Tennessee, United States. The data set can be accessed from: http://cdiac.ornl.gov. All emission estimates are expressed in thousand metric tons of carbon, where total emissions and emissions from different types of consumptions can be accessed: http://cdiac.ornl.gov/ftp/ndp030/nation.1751_2011.ems Data on carbon dioxide emissions by sector are from IEA electronic files: http://www.iea.org/stats/index.asp which are also reported from the World Bank's World Development Indicators (World Bank 2012) can be accessed from http://data.worldbank.org/data-catalog/world-development-indicators/wdi-2012
Methane emissions (thousand metric tons of CO_2 equivalent) and sub-sector contributions	Methane emissions are those stemming from human activities such as agriculture and from industrial methane production. Total methane emissions and sector contributions to methane emission can be accessed from the European Commission, Joint Research Centre (JRC)/Netherlands Environmental Assessment Agency (PBL). Emission Database for Global Atmospheric Research (EDGAR): http://edgar.jrc.ec.europa.eu/ or http://edgar.jrc.ec.europa.eu/overview.php?v=42 which can be also accessed from can be accessed from http://data.worldbank.org/indicator/
Nitrous oxide emissions (thousand metric tons of CO_2 equivalent) and sub-sector contributions	Nitrous oxide emissions are emissions from agricultural biomass burning, industrial activities, and livestock management. Nitrous oxide emissions and sector contributions to nitrous oxide emissions can be accessed from the European Commission, Joint Research Centre (JRC)/Netherlands Environmental Assessment Agency (PBL). Emission Database for Global Atmospheric Research (EDGAR): http://edgar.jrc.ec.europa.eu/ or http://edgar.jrc.ec.europa.eu/overview.php?v=42 can be also accessed from can be accessed from http://data.worldbank.org/indicator/
Other greenhouse gas emissions: perfluorocarbon (PFC), hydrofluorocarbon (HFC), and sulfur hexafluoride (SF6) (thousand metric tons of CO_2 equivalent)	HFC emissions are mostly due to use of refrigeration, air-conditioning, and insulating foam products. PFC emissions are mainly due to aluminum production and SF6 emissions are due to leakage and venting from the electricity sector, magnesium production, and other minor contributions, which can be accessed from the European Commission, Joint Research Centre (JRC)/ Netherlands Environmental Assessment Agency (PBL). Emission Database for Global Atmospheric Research (EDGAR): http://edgar.jrc.ec.europa.eu/ or http://edgar.jrc.ec.europa.eu/overview.php?v=42 can be also accessed from can be accessed from http://data.worldbank.org/indicator/
Water pollution and sector contributions	It is measured by biochemical oxygen demand (BOD) which is the amount of oxygen that bacteria in water will consume in breaking down waste. All the data sets are categorized and taken from the World Bank's World Development Indicators (World Bank 2012). Industry shares of emissions of organic water pollutants are emissions from manufacturing activities as defined by two-digit divisions of the International Standard Industrial Classification revision 3. The detailed data on water pollution could be accessed through http://data.worldbank.org/data-catalog/world-development-indicators/wdi-2012

Chapter 24

Identifying the Impediments and Enablers of Ecohealth for a Case Study on Health and Environmental Sanitation in Hà Nam, Vietnam

Vi Nguyen[1,2,3], Hung Nguyen-Viet[3,4,5], Phuc Pham-Duc[3], Craig Stephen[6,7], Scott A. McEwen[1]

[1]Department of Population Medicine, Ontario Veterinary College, University of Guelph, 2509 Stewart Building (#45), Guelph N1G 2W1, ON, Canada
[2]Public Health Risk Sciences Division, Laboratory for Foodborne Zoonoses, Public Health Agency of Canada, 160 Research Lane, Unit 206, Guelph, ON N1G 5B2, Canada
[3]Center for Public Health and Ecosystem Research, Hanoi School of Public Health (HSPH), 138 Giang Vo Street, Hanoi, Vietnam
[4]Swiss Tropical and Public Health Institute (Swiss TPH), and International Livestock Research Institute (ILRI), Socinstrasse 57, CH-4002 Basel, Switzerland and Hanoi, Vietnam
[5]Swiss Federal Institute of Aquatic Science and Technology (ESWAG), Sandec - Department of Water and Sanitation in Developing Countries, P.O. Box, CH-8600, Dübendorf, Switzerland
[6]Department of Ecosystem and Public Health, Faculty of Veterinary Medicine, University of Calgary, TRW 2D26 3280 Hospital Drive, NW, Calgary, Alberta T2N 4Z6, Canada

[7]Centre for Coastal Health, 900 Fifth Street, Nanaimo, British Columbia V9R 5S5, Canada.

Abstract: Background: To date, research has shown an increasing use of the term "ecohealth" in literature, but few researchers have explicitly described how it has been used. We investigated a project on health and environmental sanitation (the conceptual framework of which included the pillars of ecohealth) to identify the impediments and enablers of ecohealth and investigate how it can move from concept to practice. Methods: A case study approach was used. The interview questions were centred on the nature of interactions and the sharing of information between stakeholders. Results: The analysis identified nine impediments and 15 enablers of ecohealth. Three themes relating to impediments, in particular— integration is not clear, don't understand, and limited participation—related more directly to the challenges in applying the ecohealth pillars of transdisciplinarity and participation. The themes relating to enablers—awareness and understanding, capacity development, and interactions—facilitated usage of the research results. By extracting information on the environmental, social, economic, and health aspects of environmental sanitation, we found that the issue spanned multiple scales and sectors. Conclusion: The challenge of how to integrate these aspects should be considered at the design stage and throughout the research process. We recommend that ecohealth research teams include a self-investigation of their processes in order to facilitate a comparison of moving from concept to practice, which may offer insights into how to evaluate the process.

Keywords: Ecohealth, Evaluation, Health, Sanitation, Case Study, Vietnam

1. Multilingual Abstracts

Please see Additional file 1 for translations of the abstract into the six official working languages of the United Nations.

2. Background

"Ecohealth can be defined as systemic, participatory approaches to understanding and promoting health and wellbeing in the context of social and ecological interactions"[1]. It has been built upon the approach of improving human health through integrated management of ecosystems and the understanding that health is integral to systems at different biological scales, from the individual to the biosphere[2][3]. There is currently no consensus for an overarching paradigm or a particular set of techniques for ecohealth practice[1][4]-[6]. Forget and Lebel's[2] discussion on the history and evolution of the paradigm encompassed and elaborated on the descriptions presented above. Ecohealth is useful to address complex problems that span multiple disciplines and sectors, like many other integrated approaches, such as the Population Health Approach, the Global Health Research Initiative, the Millennium Ecosystem Assessment, and the One Health Initiative[7]-[10]. Recently, there has been an increasing use of the term 'ecohealth' in literature, yet many researchers who have used this approach have not explicitly described how they applied it[5]. A scoping review on ecohealth found that only two primary research papers explained their processes, making it difficult to review the utility of ecohealth in practice from the existing body of literature[11][12].

Monitoring and evaluating the process of ecohealth research and its outcomes are important components of ecohealth[13]. There has been, however, relatively little published research on the evaluation of ecohealth projects, including in-progress evaluation, to determine their consistency with ecohealth concepts[14]. While Boischio and colleagues discussed the challenges and opportunities of ecosystem approaches in the prevention and control of dengue and Chagas disease, the discussion concerned their experience with the Canadian International Development Research Centre's (IDRC) Ecohealth Program Initiative rather than being a project evaluation per se[15]. The IDRC has emphasized outcome mapping for ecohealth evaluation, however, this mapping is difficult to apply to projects in-progress when there is usually inadequate time to achieve project outcomes[13][16]. Thus, a case study involving a midterm examination of the processes used in an integrated approach may provide useful insights for understanding ecohealth's concepts and practices. The case study approach is a well-recognized methodology in qualitative research and is useful for in-depth investigation[17].

The specific challenges and opportunities for implementing ecohealth in practice will be affected by contextual factors such as culture, national policies, infrastructure, and the nature of the problem(s) being examined. However, the implementation issues encountered when working across disciplines, using participatory approaches, ensuring equity in the process, and building capacity for the sustainability of interventions may apply more generally across ecohealth projects. A recent scoping review of the peer-reviewed literature on ecohealth revealed that the practical aspects of applying ecohealth concepts have received relatively little attention[5]. While the present investigation focused specifically on ecohealth, other integrated approaches (that are not limited to "one health") have similar aims and also address health challenges that lie at the human, animal, and environment interface. Thus, these approaches can also benefit from the findings in this paper[18]. Zinsstag *et al.*[18] have discussed these issues through the history of integrative thinking in human and animal health, the evolution of "one medicine" towards "one health", and the emergence of ecohealth over the past few decades in response to broader thinking in global health.

Through the Swiss National Centre for Competence in Research North-South Program (NCCR North-South), a conceptual framework for environmental sanitation assessment to improve human health and environmental sustainability was developed and tested in different settings in Southeast Asia and West Africa[19]. The project in Vietnam aimed to assess the risk of the reuse of human waste and wastewater for agriculture, environmental sanitation, and human health[19]-[24]. The conceptual framework for that project incorporated the following pillars of ecohealth: sustainability, participation, equity, and transdisciplinarity, as defined by the Community of Practice in Ecosystem Approaches to Health - Canada (CoPEH-Can)[25]. We aimed to identify the impediments and enablers of ecohealth in practice for a project on health and environmental sanitation and assess how well the research process fits with the concepts of ecohealth. This was accomplished by examining the nature of the interactions among stakeholders, investigating how knowledge was shared, and identifying which themes were consistent with ecohealth themes in literature and which were unique to this case.

3. Methods

3.1. Study Approach

This research followed a case study structure which included: case and boundary identification, finding and assessing sources of information for data collection, and context description[17]. Our approach examined the nature of interactions among stakeholders and how information was shared through the research process. A stakeholder was defined as a person or a group of people that was affected by the issue of environmental sanitation in the project site and/or involved in the research process. Involvement was defined as participation in problem definition, establishing partnerships/collaborations, research planning, execution, analysis, or results sharing. For our purposes, the researchers were also considered to be stakeholders.

3.2. Study Design

3.2.1. Identification of the System Being Studied

The system being studied was confined to the research project of the NCCR North-South research team in Vietnam and the stakeholders involved. All of our case study data were collected in Vietnam by the first author. Initially, sources of information included some project documents in English and meetings with the NCCR North-South research team.

3.2.2. Selection and Recruitment of Participants

We selected participants by identifying the categories and identities of stakeholders through an interview with the NCCR North-South project lead. All of the four graduate student investigators identified by the project lead described the general roles of the project participants when we interviewed them. We chose the Head of the Health Station and a few health station workers and village health workers from both communes as participants because they provided population health information and have previously conducted interviews with commune residents (see **Table 1**). Project participants were selected from a list of all community members; they were the project's intended beneficiaries. Female participants were purposively selected, as they were primarily responsible for family health, sanitation, and agricultural work in their villages. To capture a diversity of perspectives, they were selected from different villages by convenience sampling, depending on the availability of participants.

Table 1. Case study data collection methods, languages of delivery, and purposes of questions, by stakeholder group.

Category	Stakeholder group			
	Project lead (n = 1)	Graduate student	Head of health station &	Village health workers &
		researchers (n = 4)	health station workers (n = 6)	Community members (n = 22)
Data collection method	Key informant interviews	Key informant interviews	Key informant interviews	Focus groups
Language	English	All Vietnamese except Part 1 with PhD student	Vietnamese	Vietnamese
Purpose of questions	Stakeholder role (1*, 3**)	Stakeholder role (1*, 3**)	Respondent information (2*, 0**) - 3 for health station workers	Involvement in this research (1*, 4 **)
	Understanding the research problem (1*, 5**)	Interaction between the research team (1*, 6**)	Participation in the research (11*, 0**)	Thoughts on the research topic (1*, 14**)
	Establishing collaborations (2*, 8**)	Research objectives (2*, 4**)	Results sharing (4*, 0**)	Researchers' approaches (1*, 8**)
	Research planning (2*, 0**)	Sharing of information (3*, 5**)	Using research results (6*, 0**)	Issues important to the community (1*, 5**)
	Conducting research (2*, 1**)	Understanding the research problem (2*, 4**)		Learning from participation (1*, 4**)
	Analyzing/interpreting results (1*, 0**)	Successes & challenges (2*, 0**)		
	Results sharing (4*, 0**)	Contribution to the community members (1*, 1**)		
	Beneficiaries of the research (3*, 0**)	Beneficiaries of the research (3*, 0**)		
	Research objectives (1*, 0**)	Research approach (9*, 0**)		
	Research approach (15*, 6**)			

*number of questions. **number of probes for each question.

3.2.3. Data Collection

The case study data were collected between January and May 2010. **Table 1** lists the data collection methods, languages, and purposes of questions by stakeholder group. The entire list of interview questions with each stakeholder group was too lengthy to report here, but it is available upon request to the corresponding author. Open-ended questions solicited information on the nature of interactions among project stakeholders and how knowledge was shared. Eight participants were invited to each focus group. Eight and six participants participated in the first and second focus groups in the Nhât Tân Commune, respectively. Five and three participants participated in the first and second focus groups in the Hoang Tay Commune, respectively. All interviews and focus groups were designed to last between 1 to 1.5 hours. We conducted a total of four focus groups.

All questions were drafted in English, and then translated into Vietnamese prior to the interviews. Most interviews were conducted in Vietnamese with the assistance of a translator, while a few were conducted in English with those proficient enough in the language. Interviews were digitally recorded and responses were translated and transcribed directly into English by the translator, then checked by the primary author during analysis. Data collection commenced after approval from the University of Guelph Research Ethics Board (REB# 10JA017) and the Hanoi School of Public Health Ethical Review Board (Decision No. 010-005/DD-YTCC) was obtained.

3.2.4. Translation, Transcription, and Analysis

All responses were analyzed using a modification of the analysis method framework; the first step was adapted to provide guidance on coding themes and writing memos (see **Table 2**)[26]. After each interview, initial themes were identified by listening to the interview recordings directly after rather than waiting for the translation and transcription. The remaining steps of the analysis method framework were implemented for all transcripts following the data collection, and translation and transcription. This was managed using qualitative data analysis software, ATLAS.ti 6.1 (ATLAS.ti GmbH, Berlin, Germany).

Table 2. Steps in the analysis method framework used for the analysis of interview and focus group responses.

Step	Explanation
1	Identifying initial themes by reading the document, writing memos about the data, and creating a coding list with definitions.
2	Labeling or tagging data by theme by applying the coding list to other documents and iteratively making revisions to the coding list for new themes that emerge.
3	Sorting data by theme, each in a separate matrix that allows the reader to clearly see the data and the document from which it came.
4	Summarizing and synthesizing data in another similar matrix that only captures the content and context.
5	Identifying elements and dimensions, refining categories, classifying data in another matrix by reading the matrices from the previous steps and labeling the data to suggest what it represents.
6	Detecting patterns by searching within and then across documents for linkages and repetition.
7	Developing explanations by giving reasons that relate to the patterns found in the previous step.

4. Results

4.1. Description of the Case and Its Context

The NCCR North-South was one of 20 programs initiated in 2001 by the Swiss National Science Foundation for sustainable development research[27]. The purpose of this 12-year program was to build research capacity in partnerships between northern and southern institutions in nine regions of Asia, Africa, Latin America, and Switzerland, while also establishing a formal institutional network in these countries. This case study was limited to Phase 2 of the research, in particular, health and environmental sanitation. The conceptual framework, developed by the NCCR North-South researchers (see **Figure 1**), was tested in Southeast Asia and West Africa[19]. The subject of our case study was the research process in Vietnam (part of NCCR North-South project), which assessed the risk that the reuse of human excreta and wastewater in agriculture and aquaculture poses to the environment, health, and socio-economics (hereafter referred to as "the problem").

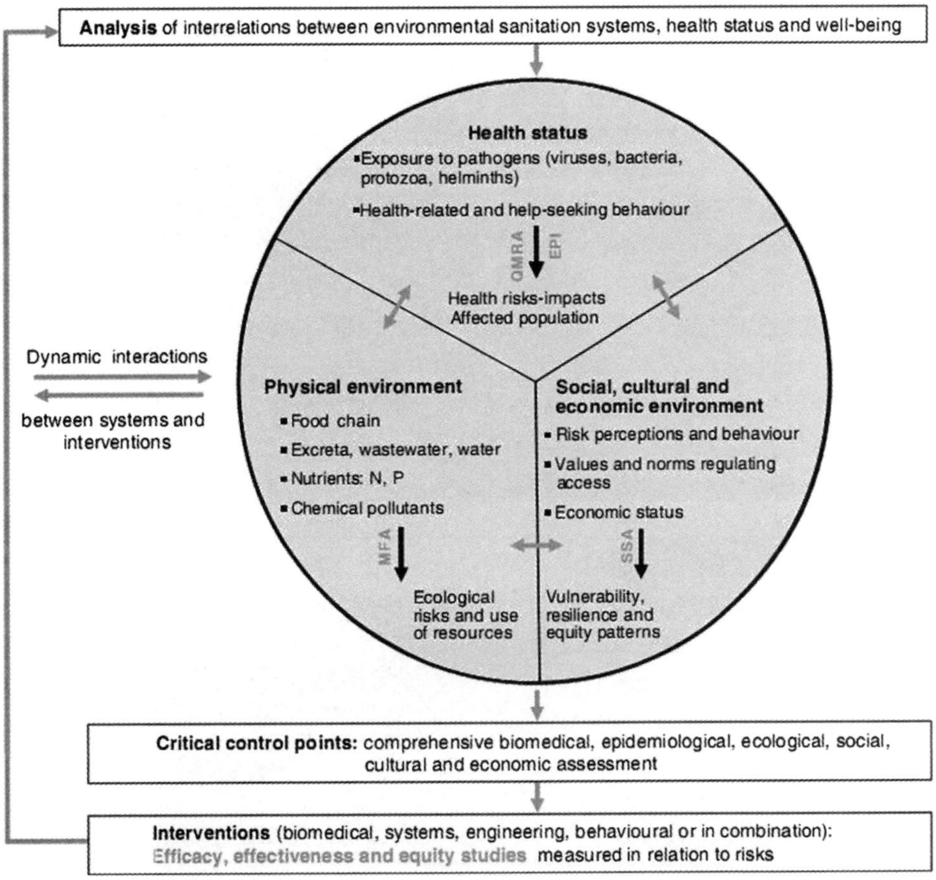

Figure 1. Conceptual framework of the combination of health and an environmental risk assessment for health and environmental sanitation planning. This was the framework of the project that we investigated. EPI: Epidemiology, QMRA: Quantitative Microbial Risk Assessment, MFA: Material Flow Analysis, SSA: Social Science Analysis.

The project was conducted in a peri-urban area, approximately 60 km south of Hanoi, in the Nhât Tân and Hoang Tay Communes, Kim Bang District, Hà Nam Province, Vietnam. Both are typical northern Vietnamese communes, with poor services for sanitation, wastewater drainage, and solid waste management[23] (see **Figure 2**). Household effluent is discharged untreated and flows through dykes that end up in the Nhue River, which flows through the commune. This river, being the only agricultural irrigation source for the communes, also receives untreated effluent from Hanoi[28]. At the time of the study, there was no place for garbage disposal and as a result, rubbish often ended up on the side of the commune roads, where it was often burned. The major land uses are residential, aquaculture,

Figure 2. Open drainage system (top) and Nhue River containing untreated wastewater flowing from Hanoi (bottom) in Hoang Tay Commune, Kim Bang District, Hà Nam Province, North Vietnam. Photo: Vi Nguyen, 2010.

and agriculture (rice cultivation and vegetables); the latter being the main source of livelihood (see **Figure 2**).

Figure 3 shows a broad overview of the environmental, social, economic, and health aspects of the problem, the details of which were extracted from project documents. The project stakeholders included institutions (Hà Nam Centre for Preventive Medicine, National Institute of Hygiene and Epidemiology, and the

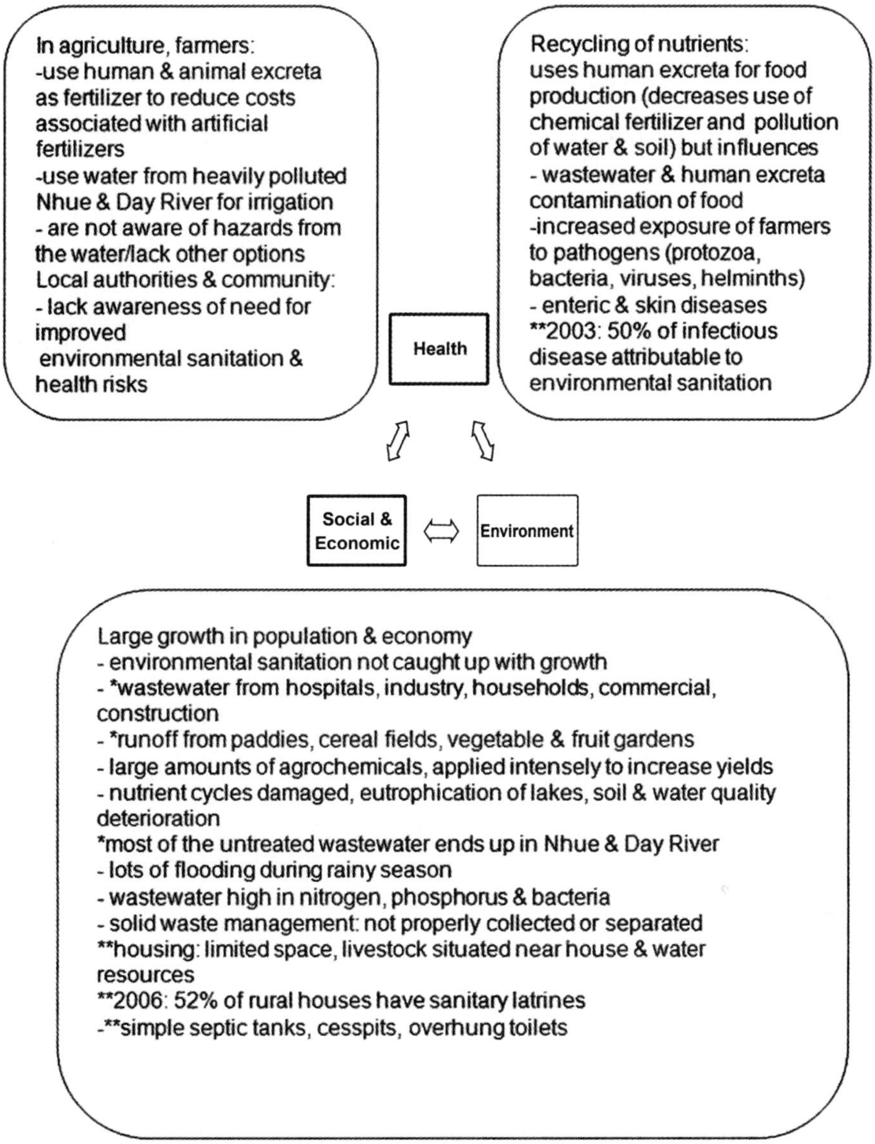

Figure 3. Environmental, social, economic, and health aspects of the problem from a research perspective (*from Hanoi, **in rural areas).

Hanoi School of Public Health), local authorities (Communal Head of the Health Station, health station workers, Communal People's Committee, District Level Health Services, Women's Union, and village health workers), and the NCCR North-South research team and their research participants (community members from both communes who responded to household surveys). The project involved four graduate students working on sub-projects in the same study sites. The general study details of each sub-project are shown in **Table 3**.

Table 3. Description of the major elements of sub-projects within the health, social, and environmental research components.

	NCCR research project component			
Category	Health		Social	Environmental
Degree (number of students), discipline	PhD (1), Epidemiology	MPH (1)	MPH (1)	MSc (1), Environmental Engineering & Management
Title	Health risks of wastewater & excreta reuse in agriculture & aquaculture in northern Vietnam	QMRA1 of exposure to wastewater & excreta in agriculture in Hà Nam, Vietnam	Assessment of human behaviors of reusing wastewater & excreta in agriculture based on PMT2 Framework	Assessing nutrient flows by MFA3 in Hà Nam, Vietnam
Objective(s)	Determine prevalence of infections of helminths, E. histolytica, C. parvum, G. lamblia, & Cyclospora, incidence & risk factors of diarrheal disease	Assess exposure to wastewater & excreta in agriculture & determine the risk of infection by C. parvum, G. lamblia	Examine perception & behavior related to the use of wastewater & excreta (health risk, coping appraisal, intention to act) based on PMT, develop a questionnaire to assess this, validate the questionnaire	Quantify nutrient (N4 & P5) flows in an agricultural & environmental sanitation system, develop scenarios to reduce the N or P discharge into the environment at all critical control points
Data collection dates	June-October 2008, April-June 2009, August-July 2010	October 2008-October 2009	October 2008-October 2009	August 2008-January 2009
Methodologies	Epidemiology Microbiology, Parasitology	QMRA Microbiology, Parasitology	PMT	MFA
Data sources & collection methods	Household surveys, human feces sampling	Wastewater sampling	Qualitative: in-depth interview, focus group discussions with farmers, field observation, quantitative surveys	Annual reports, primary research studies, working group papers, statistical records, maps, field observation, key informant/expert interviews, household surveys

[1]QMRA: Quantitative Microbial Risk Assessment; [2]PMT: Protection Motivation Theory; [3]MFA: Material Flow Analysis; [4]N: nitrogen; 5P: phosphorous.

4.2. Interviews and Focus Groups

All of the interview themes were identified through questions about the nature of interactions and the sharing of information among stakeholders. The analysis identified nine impediments and 15 enablers of eco-health, as shown in **Table 4** and below. The themes presented in-text are not presented in the table to avoid repetition of data.

Table 4. Themes categorized as enablers and impediments of ecohealth for this case study.

Category	Theme	Explanation	Selected quotations
Impediments	Lack of acceptance	People did not want to change their conventional ways of doing research	"For this school, if you look at the topic of Master's thesis, almost all topics were done in a classical way: epidemiological urvey, s cross-sectional study…and what they [students] don't want is to design a study, going to the field, taking samples like [our MSc student] to do analysis. Because [the students] are already staff in different institution so they have a database… to analyse".
	Not comfortable talking to highly educated researchers	Differing education levels and professional backgrounds impeded communication among some stakeholders	"They [the researchers] are nice and enthusiastic but just our ability is limited. When we [Village Health Workers] meet them [we don't feel very comfortable] because we are not highly educated, we can't keep up with them".
	Terminology	Lack terminology in their native language which made it hard to express ecohealth concept for others to understand	"Actually it [the Vietnamese language] doesn't have it [the ecohealth concept] now. I, myself, can't find any Vietnamese word for researchers to understand it clearly. Maybe if someone can combine all the ideas of those people [perspectives of ecohealth], the definition of ecohealth can be clearer".
	Past history of extractive research	Community members expressed frustration with years of research and seeing no changes.	"The people hope that after the research is done, [researchers] will soon have solutions so that they know the situation [in our commune]. If you just come and ask many times without results, they will say 'they come here and ask many times, take the water samples but we haven't seen any results'".

Continued

	Lack of interaction	Difficult to maintain a relationship with stakeholders with whom they didn't have a lot of direct interaction with	"We go regularly to meet them to update about the work… the outputs of the research… I'm talking about the health worker level because in the end you can't have a lot of relationship with the participants from the community".
	Differing priorities	Research that was relevant for what researcher's deemed important did not match the nature of the problem	"For the project objective, we had to make sure it was an environmental health problem. The community's main health problems were skin problems and diarrhoea. Microbiologists are more concerned about the chemicals -heavy metals in wastewater but our background is in the health, about the diarrheal diseases and parasitic infections. Our study objective and the main problem in the study site did not match".
Enablers	Consensus	Agreement among groups	"Need to find compromise between you [researcher], the community, and policy-makers [to plan interventions]. But when you implement, I think we need the strong willingness of the Communal People's Committee, Health Station, other mass organizations, and the community".
	Equity	Accounted for differences among different groups (gender, stakeholder level, social status, etc.)	"It's mainly the Women's Union. If they have their meeting, I would like to have a meeting in this commune about environmental sanitation. Because they [women] are in charge of housework and going to the field. I would like to have a meeting with them because they mainly clean the road. The men don't do it. The custom is like that".
	Evidence	The research provided evidence that the community could use	"The people knew before that there was pollution, but now through the researchers, the main influences have been discovered. Why they are infected with helminths? Or where does the diarrhoea come from? They can be aware of that now. It was vague before".
	Free to express concerns	Health Station Workers and community members were free to ask researchers questions if they didn't understand the survey questions	"When they [the researchers] come, they often ask if we have any concerns [regarding research]. If yes, we will discuss with them so that it's easier to do".

Continued

Funding	Financial contributions from collaborators	"We need financial support to clean and rebuild the facilities so that the environment can be improved. Without funding, the drains would never be clean".
A channel for concerns	Through the Health Station, the community could voice opinions to the Communal People's Committee	"We will give our opinions to the Head of Health Station in a monthly meeting. The Health Station will collect all the opinions and submit them to the upper levels".
Networks	Must be well-known among those working in the area; offers access to other opportunities	"I would go to approach them [policy-makers] once I have more evidence and in particular, a bigger network...people working in the Ministry [of Health]... Environment, in the University, in the nstitute. We can have some kinds of recognition when we can talk with them".
Pluralism	Multiple methods and perspectives, included multiple stakeholders at different levels	"With one person, the problem can't be seen comprehensively but a group of researchers with the same idea about improving environment for health, there will be many researchers joining and thus, many ideas contributed from many sides. About research with community's participation, if we have the participation of the community, the information will be more reliable and timely".
Research in partnership	Decisions on research made together among partners involved in the research	"We discuss together, identify the problem together and we will do research together with the resources we already have. We are also willing to discuss with people to find other funds, other support to support our common interest".
Sharing process	The responsibility for interventions, the data, and results should be shared by stakeholders; each person has a part	"Because when all unions and department co-operate, they can advocate widely to people, the people can follow, and keep good sanitation. It can't work if just one does it. They can't go to each person".
Commitment to ongoing testing and monitoring	The desire for project commitment to addressing sanitation beyond data collection and research outputs	"I also want the people from the environment section to come here and take the [water] sample for testing so that we can know. Or when you do research, you know the information and you will share information with us so that we can learn from experience".
Sharing knowledge gained through research	Village Health Workers shared what they have learned through the research with others in their community	"By talking, for example, with the women here (Village Health Workers) or the neighbours talk with each other or when we have a [Women's Union] meeting".

Three impediment themes in particular—integration is not clear, don't understand, and limited participation—related more directly with the challenges in applying the ecohealth pillars of transdisciplinarity and participation. When asked about how and what was integrated in the research, a project team member explained that "the concepts were developed with the expectation that we would integrate information for the three components... So we did it [the research]. But the integration is not clear...we need to explore further to see the link between the three components".

In a discussion of how the community could directly use the research results, a village health worker said that "if the [community members] didn't participate and just attended to listen to the results, they wouldn't understand them. When the researchers came to present the results, they presented very briefly". Community members and health station workers explained that they wanted to participate in interventions to mitigate the problem as much as they could, but they felt limited by their knowledge, abilities, time, resources, and funding (for example: "The Health Station just advocates. We have to depend on many things. We don't have any funding. We just advocate by using loudspeakers or through the village health workers. We have also launched campaigns to collect garbage and general campaigns, but that's all we can do. It mainly depends on the Communal People's Committee").

On the other hand, the enabler themes—awareness and understanding, capacity development, and interactions—facilitated usage of the research results. Village health workers echoed that "regarding the waste in the Nhue River, we do know about it [its effects on health], but we don't know the percentage of the infection or pollution, whether it is too high, without the [research] results". A project team member said that "NCCR North-South focuses on partnership with Vietnam's institutes...by cooperating with foreign countries, they improve research capacity [of researchers and supporters]...learn new methods and knowledge. NCCR North-South wants them to be active in research so [they] don't need to wait for any external support". Another researcher noted that there has been "more contact with them [health station workers] every time we go [to the study site]... health station workers have much more contact and good relationships with community members. Researchers can't cover everything".

Discussions with community members about solutions, community roles, and signs of improvement in health and environmental sanitation yielded input that spanned not only the health sector, but also the environmental, social, and economic aspects of the issue (see **Table 5**). We felt that this discussion was necessary in order to get community input on what was necessary to enable the next steps since ecohealth is so action oriented[5].

We assessed the project's consistency with ecohealth concepts identified in the scoping review[5] (see **Table 6**). The comparison with project details and interview themes revealed that the main challenges were related to limited participation and how to integrate research components. The strengths of the project were: the timeframe, which showed a long-term commitment (from 2008 and continuing through to 2013 and beyond) to health and environmental sanitation in the community, and that multiple disciplines and research questions examining the different aspects of the issue attempted to address its complexity.

Table 5. Community members' input on the solutions, roles, and signs of improvement for health and environmental sanitation.

Community-identified ideal solutions or community roles in environmental sanitation	Community-identified signs of improvements in health and environmental sanitation
Use a biogas oven (converts waste into fuel)	Cleaner roads (no more garbage thrown randomly)
Burn garbage	Everyone gathers household garbage for a garbage collector; identified the need for regulations
Treat excreta to get rid of smell or compost it properly	Economic status is better
Lead by example by making changes and other people will follow if they see changes working	Improved health means we can do anything
Need funding	Reduction in diseases and conditions they perceived to result from poor sanitation (diarrheal diseases, skin diseases, cancer)
Need awareness & understanding	No smell (from garbage, animal carcasses thrown into the river, and the wastewater itself)
Need a clean water system and wastewater treatment system	No wastewater visible (for human exposure)

Table 6. Assessment of the case study's consistency with ecohealth components identified in the scoping review of ecohealth.

Ecohealth component	Component explanation	Corresponding project elements	Source of information
Participation	- from the beginning, stakeholders (including affected population) collaborate on various research stages using local knowledge and addressing some of their priorities; also refers to participatory action research	- participation from member of local institutions and community members consisted of providing information for the researchers' project and helping them collect data	- interview theme: "limited participation" (**Table 4**)
System	- understanding the whole and its parts (issues, interactions, key actors, components, and interrelationships); includes systems science	- not be evaluated at the time of this study1	N/A
Multidisciplinary	- more than two disciplines working together in their traditional roles	- More than one discipline was involved (epidemiology/public health, environmental engineering) but all were allied health professions	- project documents (**Table 3**)
Action-oriented	- results in something done to solve or mitigate the research problem under study	- no interventions or changes were planned at the time of this study but they intended to address this in the next phase of research	- interview with project lead (interview transcript, not shown here)
Complexity	- made up of many interrelated parts; where ecohealth is best applicable	- the project was designed to address several dimensions of the sanitation problem and made efforts to share results and perspectives across disciplines and stakeholders	- project documents (**Table 3** and **Figure 3**)
Long-term	- ecohealth requires a time-commitment; improvements/ outcomes might only be seen in the future; difficult to contain within a single project	- data collection started in 2008; next phase of research was expected to last until 2013 - project involved multiple components	- project documents (**Table 3**)

Continued

Indicators	- measures used for study outcomes and monitoring should be developed by involved stakeholders and may be different according to each group	-community-identified indicators had not been discussed with the researchers or addressed at the time of this project	- "community identified signs of improvement" **(Table 5)**
Adaptive management	- an iterative learning process with stakeholder participation involving monitoring, evaluating, and adjusting the plan based on the information generated in the process	- could not tell at the time of this study1	N/A
Transdisciplinarity	- collaboration between researchers and practitioners from complimentary disciplines/sectors and/or other stakeholders on a problem; uses multiple methods/tools that facilitate the generation of new frameworks, concepts, methods, institutions, etc. from the knowledge sharing and/or interaction	- integration of research components was not clear; integration of results was anticipated, but how this will happen was not clear	- interview theme: "integration is not clear" **(Table 4)**
Equity	- addresses differences between groups affected by research problem; gender (roles, responsibilities), power (decision making, access to resources), and trade-offs (who benefits)	- statistical analysis of data had been stratified by gender	- interview with PhD student on health research component (interview transcript, not shown here)
Sustainability	- meeting the needs of current generations without compromising the needs of future generations; the outcome or goal of ecohealth, also refers to sustainability of the environment and/or of interventions/projects	- could not tell at the time of this study1	N/A

Continued

Socio-ecological	- understanding the human and environmental components of a problem and their interaction	- health component quantifie human health risks and exposure - social component examines perceptions & behaviours - environmental component quantifies nutrient flows in agricultural & sanitation system - the interaction between components not addressed yet, as integration is not clear	- project document (**Table 3**) - interview theme: "integration is not clear" (**Table 4**)
SOHOs (self-organizing, holarchic open system)	- characterized by holarchy (interactions between nested hierarchies), feedback loops (consequences for another part of the system - positive or negative), self-organization (combination of feedback, boundaries, and openness)	- could not tell at this point in the project1	N/A
Negotiate	- a process in which the decisions on objectives, methods, and indicators are made with stakeholders	- the research was conducted according to researchers' priorities, mainly driven by a conceptual framework developed a priori	- interview theme: "priorities" (**Table 4**) - project document (**Figure 1**)

5. Discussion

Overall, examining the factors that helped or hindered the research team to reach an ecohealth process during the first three years of the project allowed us to identify some enablers and impediments that can help turn the theoretical components of ecohealth into practice. The project we examined was still in-progress during our study period, therefore, our findings do not reflect the entire project. While the case study project faced several challenges in implementing a number of ecohealth concepts, its conceptual framework corresponded quite strongly to ecohealth. This was evident in the design and preliminary documents, where concepts of integration, multi-stakeholder participation, and an understanding of the system were stressed. The main challenges were related to fully realizing a transdisciplinary and participatory approach, and sustaining research efforts. If our assessment was treated like a checklist, then the project could be consistent with most of the pillars of ecohealth. However, when taking in an assessment of 'if ' or 'how' these components were implemented, the project faced challenges in fully realizing these themes in practice.

In terms of enablers of the research approach, an important aspect that we didn't consider initially was the baseline to which we would compare this project. If we consider the pillars of ecohealth as defined by the IDRC as the gold standard but we don't clearly know what that gold standard looks like in practice (in terms of methods and tools), then the best we can do is compare the research approach to a baseline of how research linking environment and health had previously been done in similar contexts, and then document the progress. That being said, the NCCR North-South research project did make efforts to address the sanitation issue from the perspective of other disciplines, to present research results back to the local institutions and community participants, and showed continued commitment to the issue and the particular study sites (see **Table 6**, enabler themes presented in our Results, and the Ecohealth Field Building Initiative discussed below). It is also important to consider this progress in the context of the history of ecohealth in the region. Ecohealth is relatively new in Southeast Asia compared to Latin America, for example, in terms of the development of a community of practice and research capacity[29][30].

The case study showed that the integration aspect of transdisciplinarity was difficult to achieve. The NCCR North-South researchers collected data from different sectors, but they faced challenges integrating these data. This is a common problem for ecohealth research[6]. By extracting information on the environmental, social, economic, and health aspects of environmental sanitation, we found that the issue was not confined to a particular scale or sector, but was interconnected and spanned multiples scales (local, regional, and national) and sectors (health, social, economic, and environment). This complexity is typical of many public health problems when their multi-dimensional natures are adequately taken into account[12]. The need to accommodate multiple scales and sectors is a common feature of complex public health problems. For example, Marko *et al.* developed and applied a framework for analyzing the impacts of urban transportation in Edmonton, Canada and illustrated the economic, socio-cultural, infrastructural, and political factors that affected or were affected by transportation[31]. Murray and Sanchez-Choy conducted research on improving health in rural Amazonian communities, and found that in order to make connections between ecosystem variables, use of resources, and health, it was necessary to analyze the issues at the ecosystem, community, and household levels[32]. While it is acknowledged that complex problems span multiple scales and/or sectors, research should include the collection of data from the scales and sectors influencing the issue being studied. However, as illustrated by this study, there remain significant challenges in developing acceptable and effective means to integrate across disciplines and scales. Recently, Wilcox *et al.*[33] have summarized and described identifiable components of an integrative research project in the context of conservation medicine, which included: making integration part of the project; a clear research question and project goal; inclusion of disciplines; an integrative theory, model, or approach; an operational efficacy; an institutional environment conducive to collective learning; and a project plan (see **Table 2.2** in their paper).

The response "don't understand" reflects that affected stakeholders might have not been equally involved. This lack of understanding could have affected their capacity to learn from and use the research results. This response also highlights that the use of disciplinary methods (e.g. epidemiological surveys) may have limited the participation (another theme) of many stakeholders to help the researchers collect data and provide research inputs. This may have long-term consequences of "research fatigue" if the desired outcomes and expectations are not met. Tools and group processes to facilitate integration, including participatory methods that are not specific to a particular discipline, sector, or education level, may help to overcome this impediment in practice. These may include creating rich picture maps[11], or issue and influence diagrams[12] to develop a shared understanding of the issue being studied. Similar to Mertens *et al.*, eco-health practitioners should strive for collaborative (jointly determining priorities) and collegial participation (knowledge exchange yielding new understandings and locally-controlled action plans) by negotiating research priorities during planning phases and sharing research progress more regularly so that community members can participate in robust results dissemination planning in their own communities[34][35].

The themes "awareness and understanding", "capacity development" at the institutional level, and increased "interactions" among stakeholders highlight some of the challenges of achieving sustainability of the research efforts. These features of research impact are often not captured as research outcomes, as publications generally focus on the technical aspects of the research. Outcome mapping, an evaluation tool promoted and used by the IDRC for programs, projects, and organizations, could be used to capture these other features of ecohealth research[36][37]. At the time of writing this paper, the research team in Vietnam was undertaking the Ecohealth Field Building Leadership Initiative (FBLI) in Southeast Asia, which was focused on research, training, policy, and networking (personal communication with HNV, principal investigator of this initiative). Their research focus was on human health issues associated with agricultural intensification, with research activities in Vietnam focused on the same study site as the NCCR North-South. Their intention was to build on past efforts and lessons learned, which showed a continued commitment to addressing the issues (linking health and the environment) affecting the community. They have implemented a field intervention examining how the combination of human and animal excreta composting influences helminth egg die-off in excreta, while maintaining its nutrient value[38]. The intervention aimed to improve the current storage practices of human excreta and to identify the best option for the safe use of excreta in agriculture. The preliminary results have been reported by Nguyen-Viet *et al.* in[38]. In addition, the NCCR North-South research was the basis from which to launch Vietnam's One Health-Ecohealth Newsletter, as well as Vietnam's One Health University Network (VOHUN) and FBLI.

Negotiation, as a component of ecohealth, included negotiating indicators of the successes of the research[5]. The input from community members on solutions, roles, and signs of improvement, with respect to the problem of sanitation, showed that their participation in interventions required the involvement of multiple sectors and a holistic view of health (see **Table 5**). This broader view of health was evident in the case study as the signs of improvement encompassed many determinants of health that lie outside of the health sector, such as economic status and the physical environment[39]. There were differences in priorities across these various determinants of health. For example, on the one hand, public health professionals have traditionally viewed improvements in health in terms of morbidity or mortality indicators (for example, reduction of diarrheal diseases). On the other hand, communities seemed more interested in cleaner roads and improved economic statuses, as identified in our case study (see **Table 5**). Therefore, indicators of improvements in the problem being studied need to be negotiated in eco-health research, as our scoping review found[5].

Our study was one of few that examined how a research project could implement ecohealth components. Insights from this work could be used to inform other ecohealth projects in their planning and implementation phases. We used our synthesized interpretation of eco-health, which was informed by a scoping review of the literature on ecohealth to assess the case study project's consistency with ecohealth concepts[5]. This was strongly influenced by the IDRC's position on ecohealth, as most of the published research was supported by this funder or they cited use of IDRC's approach to ecohealth[5]. There is currently no consensus on ecohealth concepts among fields that have similar initiatives of working towards more holistic, integrated approaches (e.g. "one health" initiatives, global health research, conservation medicine, and ecosystem management), and application of these concepts is often context-specific[10][40]–[42]. As a result, the understanding of what is meant by ecohealth and its implementation is varied; this particular finding was also cited by the authors of an external review of the IDRC's Ecohealth Program[43]. An explanation of the process as it was implemented is required, as it is not intuitive, to give readers the ability to understand and evaluate a study that is classified as ecohealth. Future research should concentrate on the reporting and evaluation of processes to more rigorously guide ecohealth to develop from concept to practice.

6. Conclusion

Our case study offered insights into the operational challenges that occurred when attempting to implement ecohealth. Three impediment themes in particular— integration is not clear, don't understand, and limited participation—related more directly with the challenges in applying the ecohealth pillars of transdisciplinarity and participation. The enabler themes—awareness and understanding, capacity development, and interactions— facilitated usage of the research results. As there are many integrated approaches with similar aims to ecohealth, these challenges may apply more generally to interventions for health problems that arise at the human, animal, and environment interface. Components of ecohealth should not be treated as a checklist for inclusion. Monitoring processes and progress may also offer insights into how to evaluate ecohealth research, as it would emphasize articulation of the research approach and how implementation corresponds with concepts. Further research stemming from these lessons and insights for research design would contribute to the development of the field of ecohealth.

Additional File

Additional file 1: Multilingual abstracts in the six official working languages of the United Nations. http://www.biomedcentral.com/content/supplementary/2049-9957-3-36-S1.pdf.

Abbreviations

IDRC: International Development Research Centre; NCCR: North-South National Centre for Competence in Research North-South Program; CoPEH-Can: Community of Practice in Ecosystem Approaches to Health-Canada.

Competing Interests

The authors declare that they have no competing interests.

Authors' Contributions

VN conceived the study and its design, collected and analyzed the data, and drafted the manuscript. VN was an external reviewer in the NCCR North-South project. HNV was involved in the study design, revised the manuscript, and provided intellectual input to the interpretation of the findings. PDP was involved in the study design. CS and SAM conceived the study and its design, revised the manuscript, and provided intellectual input to the interpretation of the findings. All authors read and approved the final manuscript.

Acknowledgements

The authors would like to thank the NCCR North-South research team for their support, which ranged from supervision, mentoring, facilitating the research partnerships, and obtaining research approval, to helping with fieldwork activities, translation, and transcription: Jakob Zinsstag, Tu Van Vu, Khuong Nguyen Cong, Nga Thu Do, Nhung Hong Nguyen, and Nhi Truong Thi Phuong; the Canadian Community of Practice in Ecohealth and the International Development Research Centre for funding this research; the Hanoi School of Public Health (Department of Environmental Health), the National Institute for Hygiene and Epidemiology, the Swiss Tropical and Public Health Institute, and the International Livestock Research Institute's Ecosystem Approaches to Better Management of Zoonotic Infectious Diseases project for their in-kind contributions to this research; the research participants from the community and local institutions in the study sites for their insights into the research process; and the Public Health Agency of Canada for providing stipend support to VN.

Source: Nguyen V, Nguyen-Viet H, Pham-Duc P, *et al.* Identifying the impediments and enablers of ecohealth for a case study on health and environmental sanitation in Ha Nam, Vietnam[J]. Infectious Diseases of Poverty, 2014, 3(1):36-36.

References

[1] Waltner-Toews D: Eco-health: a primer for veterinarians. Can Vet J 2009, 50:519–521.

[2] Forget G, Lebel J: An ecosystem approach to health. Int J Occup Env Heal 2001, 7:S3–36.

[3] Lebel J: Health: An Ecosystem Approach. International Development Research Centre: Ottawa; 2003.

[4] Webb J, Mergler D, Parkes MW, Saint-Charles J, Spiegel J, Waltner-Toews D, Yassi A, Woollard RF: Tools for thoughtful action: the role of ecosystem approaches to health in enhancing public health. Can J Public Health 2010, 101:439–441.

[5] Nguyen V: Understanding the Concept and Practice of Ecosystem Approaches to Health Within the Context of Public Health. In MSc Thesis. University of Guelph; 2011.

[6] Charron DF: Ecohealth Research in Practice: Innovative Applications of an Ecosystem Approach to Health, Insight and Innovation in Development. Springer and International Development Research Centre: New York and Ottawa; 2012.

[7] CIHR: Global health-healthy Canadians in a healthy world. Canadian institutes of health research (CIHR). http://www.cihr-irsc.gc.ca/e/35878.html.

[8] Corvalan C, Hales S, McMichael A: Ecosystems and human wellbeing: Health synthesis. Geneva: World Health Organization; 2005.

[9] PHAC: What is a population health approach? http://www.phac-aspc.gc.ca/ph-sp/approach-approche/index-eng.php.

[10] PHAC: One World One Health™: From ideas to action. Report of the expert consultation. March 16-19, 2009 - Winnipeg, Manitoba, Canada. Public Health Agency of Canada (PHAC); 2009. [http://www.aitoolkit.org/site/DefaultSite/filesystem/documents/OWOH% 20Winnipeg%20July%202009% 20version.pdf].

[11] Bunch M: Soft systems methodology and the ecosystem approach: a system study of the cooum river and evirons in Chennai, India. Environ Manage 2003, 31:182–197.

[12] Neudoerffer RC, Waltner-Toews D, Kay JJ, Joshi DD, Tamang MS: A diagrammatic approach to understanding complex eco-social interactions in Kathmandu. Nepal Ecol Soc 2005, 10:12.

[13] IDRC: Evaluation strategy (2005-2010). [https://idl-bnc.idrc.ca/dspace/bitstream/10625/26668/1/122276.pdf].

[14] Sherwood S, Cole D, Crissman C: Cultural encounters: Learning from cross-disciplinary science and development practice in ecosystem health. Dev Pract 2007, 17:179–195.

[15] Boischio A, Sánchez A, Orosz Z, Charron D: Health and sustainable development: challenges and opportunities of ecosystem approaches in prevention and control of dengue and Chagas disease. Cad de Saúde Pública 2009, 25:S149–S154.

[16] IDRC: Findings Brief - External Review of the Ecosystem Approaches to Health Program International Development Research Centre (IDRC). [http://www.idrc.ca/EN/Documents/External-Review-of-the-Ecosystem-Approaches.pdf].

[17] Creswell J: Qualitative Inquiry & Research Design: Choosing Among Five Approaches. 2nd edition. Thousand Oaks: Sage Publications Ltd.; 2007.

[18] Zinsstag J, E S, Waltner-Toews D, M T: From "one medicine" to "one health" and systemic approaches to health and well-being. Prev Vet Med 2011, 101:148–156.

[19] Nguyen-Viet H, Zinsstag J, Schertenleib R, Zurbrügg C, Obrist B, Montangero A, Surkinkul N, Koné D, Morel A, Cissé G, Koottatep T, Bonfoh B, Tanner M: Improving environmental sanitation, health, and well-being: a conceptual framework for integral interventions. Ecohealth 2009, 6:180–191.

[20] Minh HV, Nguyen-Viet H, Thanh NH, Jui-Chen Y: Assessing willingness to pay for

improved sanitation in rural Vietnam. Prev Med: Environ Health; 2013.

[21] Pham-Duc P, Nguyen-Viet H, Hattendorf J, Zinsstag J, Cam PD, Odermatt P: Ascaris lumbricoides and Trichuris trichiura infections associated with wastewater and human excreta use in agriculture in Vietnam. Parasitol 2013, 62(2):172–180.

[22] Pham-Duc P, Nguyen-Viet H, Hattendorf J, Zinsstag J, Cam PD, Odermatt P: Risk factors for Entamoeba histolytica infection in an agricultural community in Hanam province. Vietnam Parasit Vectors 2011, 4:102.

[23] Do-Thu N, Morel A, Nguyen-Viet H, Pham-Duc P, Nishida K, Kootattep T: Assessing nutrient fluxes in a Vietnamese rural area despite limited and high uncertainty data. Resour Conserv Recy 2011, 55:849–856.

[24] Vu-Van T, Pham-Duc P, Nguyen NH, Tamas A, Zurbrügg C: Improving farmers' wastewater handling practice in Vietnam. Sandec News 2010, 11:24.

[25] CoPEH-Can: Canadian Community of Practice in Ecosystem Approaches to Health (CoPEH-Can).[http://www.copeh-canada.org/index_en.php].

[26] Spencer L, Ritchie J, O'Connor W: Analysis: practices, principles, and processes. In Qualitative research practice: A guide for social science researchers. Edited by Ritchie J, Lewis J. Thousand Oaks: Sage Publications Ltd; 2003.

[27] NCCR North-South: Research partnerships for sustainable development in Southeast Asia: Highlights of the National Centre for Competence in Research North-South (NCCR North-South) Program in Southeast Asia, 2005-2009. NCCR North-South, Joint Areas of Case Studies Southeast Asia: Pathumthani; 2009.

[28] MONRE, ICEM: Improving Water Quality in the Day/Nhue River Basin: Capacity Building and Pollution Sources Inventory. Report No. ADB/MARD/MONRE/ Project 3892-VIE. Hanoi: Vietnamese Ministry of Natural Resources and the Environment (MONRE) and International Centre for Environmental Management (ICEM); 2007.

[29] CoPEH-LAC: What is CoPEHs-LAC? (Community of Practice in Ecosystem Approaches to Health - Latin America and the Caribbean). [http://www.copehlac.una.ac.cr/index.php?option=com_content&view=article&id=48& Itemid=98].

[30] VSF-VWB: Community of Practice in Ecohealth - South and Southeast Asia CoPEH-SSEA. Veterinarians Without Borders - Vétérinaires Sans Frontières (VSF-VWB). [https://sites.google.com/site/veterinairessansfrontie res/about-copeh].

[31] Marko J, Soskolne CL, Church J, Francescutti LH, Anielski M: Development and application of a framework for analysing the impacts of urban transportation. Ecohealth 2004, 1:374–386.

[32] Murray TP, Sanchez-Choy J: Health, biodiversity, and natural resource use on the amazon frontier: an ecosystem approach. Cad de Saúde Pública 2001, Supplement 17:181–191.

[33] Wilcox BA, Aguirre AA, Horwitz P: Ecohealth: Connecting Ecology, Health and

Sustainability. In New Directions in Conservation Medicine. Edited by Aguirre AA, Ostfeld RS, Daszak P. Oxford, New York: Oxford University Press; 2012:17–32.

[34] Biggs S: Resource-Poor Farmer Participation in Research: A Synthesis of Experiences from Nine National Agricultural Research Systems. The Hague: International Service for National Agricultural Research; 1989.

[35] Mertens F, Saint-Charles J, Mergler D, Passos CJ, Lucotte M: Network approach for analysing and promoting equity in participatory ecohealth research. Ecohealth 2005, 2:113–126.

[36] IDRC: Outcome mapping. [http://www.idrc.ca/en/ev-26586-201-1-DO_TOPIC.html].

[37] Anonymous: Outcome mapping learning community. [http://www.outcomemapping.ca/].

[38] Nguyen-Viet H, Pham-Duc P, Nguyen V, Tanner M, Vu-Van T, Van-Minh H, Zurbrüg C, Schelling E, Zinsstag J: Chapter B5: A One Health Perspective for Integrated Human and Animal Sanitation and Nutrient Recycling. In One Health: The Theory and Practice of Integrated Health Approaches. Edited by Zinsstag J, Schelling E, Whittaker M, Tanner M, Waltner-Toews D. London: CABI; in press.

[39] WHO: The determinants of health. World Health Organization (WHO). [http://www.who.int/hia/evidence/doh/en/index.html].

[40] Brown K, Mackensen J, Rosendo S: Chapter 15: Integrated Responses. In Ecosystems and human well-being: Policy Responses, Volume 3. Washington: Island Press; 2005:425.

[41] Stephen C, Daibes I: Defining features of the practice of global health research: an examination of 14 global health research teams. Glob Health Action 2010, 3:5188.

[42] Tabor G: Defining Conservation Medicine. In Conservation medicine: Ecological health in practice. Edited by Aguirre AA, Ostfeld RS, Tabor CM, House C, Pearl MC. Oxford and New York: Oxford University Press; 2002.

[43] Finkelman J, MacPherson N, Silbergeld E, Zinstaag J: External review of the IDRC Ecohealth Program Initiative: Final report; 2008.

Chapter 25

Improved Stove Interventions to Reduce Household Air Pollution in Low and Middle Income Countries: A Descriptive Systematic Review

Emma Thomas[1], Kremlin Wickramasinghe[1], Shanthi Mendis[2], Nia Roberts[3], Charlie Foster[1]

[1]British Heart Foundation Centre on Population Approaches for Non-Communicable Disease Prevention, Nuffield Department of Population Health, University of Oxford, Oxford, UK

[2]Chronic Disease Prevention and Management, World Health Organization, Geneva, Switzerland

[3]Bodleian Health Care Libraries, University of Oxford, Oxford, UK

Abstract: Background: Household air pollution (HAP) resulting from the use of solid fuels presents a major public health hazard. Improved stoves have been offered as a potential tool to reduce exposure to HAP and improve health outcomes. Systematic information on stove interventions is limited. Methods: We conducted a systematic review of the current evidence of improved stove interventions aimed at reducing HAP in real life settings. An extensive search of ten databases commenced in April 2014. In addition, we searched clinical trial registers and websites for unpublished studies and grey literature. Studies were included if they reported on an improved stove intervention aimed at reducing HAP resulting from solid fuel use in a low or middle-income country. Results: The review identified 5,243 records. Of these, 258 abstracts and 57 full texts were reviewed and 36 studies identified which met the inclusion criteria. When well-designed, implemented and monitored, stove interventions can have positive effects. However, the impacts are unlikely to reduce pollutant levels to World Health Organization recommended levels. Additionally, many participants in the included studies continued to use traditional stoves either instead of, or in additional to, new improved options. Conclusions: Current evidence suggests improved stove interventions can reduce exposure to HAP resulting from solid fuel smoke. Studies with longer follow-up periods are required to assess if pollutant reductions reported in the current literature are sustained over time. Adoption of new technologies is challenging and interventions must be tailored to the needs and preferences of the households of interest. Future studies require greater process evaluation to improve knowledge of implementation barriers and facilitators. Review registration: The review was registered on Prospero (registration number CRD42014009796).

Keywords: Improved Stoves, Systematic Review, Indoor Air Pollution, Solid Fuel Smoke, TIDieR Checklist

1. Background

The health impact of indoor air pollution among low-and middle-income countries (LMIC) is considerable and primarily results from solid fuel smoke[1]. Solid fuels such as dung, coal, wood and agricultural residues are used by approximately half of the world's population for cooking and heating[2][3]. The burning of these fuels in open fires or inadequate stoves results in harmful pollutants being emitted into the household atmosphere[4].

The 2010 Global Burden of Disease Study ranked household air pollution (HAP) as the third highest global risk factor[5]. In South Asia and sub-Saharan Africa, HAP accounted for the first and second highest risk factor for burden of disease respectively. The study attributed 3.5 million deaths and 4.3 per cent of global disability-adjusted life years (DALYs) in 2010 to HAP from solid fuels[5]. The two major health outcomes associated with HAP are acute lower respiratory infections (ALRI) in children under five years of age and chronic obstructive pulmonary disease (COPD) in adults over 20 years[6]. Women and young children are frequently at greater risk due to longer hours spent indoors[7].

Improved (*i.e.* high-efficiency and low emission) stoves have been offered as a potential tool to reduce exposure to indoor air pollution, improve health outcomes and decrease greenhouse gas emissions and deforestation[8]. During the 1970s higher oil prices, increasing deforestation and concerns of a "fuelwood crisis" created additional pressure on governments and non-government organisations (NGOs) to act[9]. Many NGOs and governments then facilitated the wide-scale distribution of stoves. Initial enthusiasm about stoves was often supported by laboratory-based experiments performed in highly controlled contexts[9][10]. The intervention impact in real world contexts was unrealized. Many organisations believed the improved efficiency of the stoves would be enough to facilitate their widespread adoption. However, traditional, "three-stone" biomass stoves have additional benefits such as heating, protection from insects, and wide variety of fuel flexibility[9]. Additionally, improved stoves must be adopted and maintained by households in order to achieve intended benefits. Despite NGO-led practices of stove distribution and improved epidemiological surveillance, what remains unknown are the best ways of implementing improved stoves.

We aimed to conduct a systematic review of stove interventions that aim to reduce household air pollution in LMIC. In-depth understanding of these interventions is required in order to facilitate policy and funding decisions. Additionally, information on the type, quality and distribution of stove interventions is required to facilitate successful replication and scale-up.

2. Methods

2.1. Search Strategy

Our review was registered on Prospero (registration number CRD 42014009796) and followed the Preferred Reporting Items for Systematic Reviews and Meta-Analyses (PRISMA) review process[11] (Additional file 1). In April 2014, we searched the following ten databases: CINAHL (EBSCOHost) [1982-present], Cochrane Central Register of Controlled Trials (Cochrane Library, Wiley) [Issue 4, 2014], Embase (OvidSP) [1974-present], Global Health (OvidSP) [1973-present],

Ovid MEDLINE (R) In-Process & Other Non-Indexed Citations and Ovid MEDLINE(R) (OvidSP) [1946-present], PsycINFO (OvidSP) [1967-present], Science Citation Index (Web of Science, Thomson Reuters) [1945-present], Global Health Library-Regional Indexes & WHOLIS http://www.globalhealthlibrary.net/php/index.php and Pubmed http://www.ncbi.nlm.nih.gov/pubmed/. The search imposed no limit on study design or date of publication. The Embase search strategy is provided in Additional file 2 as an example of the search terms used. In addition, we searched clinical trial registers and websites for unpublished studies and grey literature.

2.2. Eligibility Criteria

Many studies measure pollutant outcomes over a series of times and as such non-randomised studies (e.g. interrupted time series and before and after studies) are common. Therefore, in order to gain an in-depth understanding of the scope of stove interventions, we included primary intervention studies regardless of study design. Such designs included: individually randomised trials, cluster-randomised trials, controlled before-and-after studies, interrupted time series and project evaluations. Eligible study participants were exposed to HAP from solid fuels such as dung, wood, agricultural residues and coal for cooking and heating; laboratory-based studies were excluded. Interventions were required to take place in LMIC where HAP has the greatest health consequences and systematic information is limited. As such, interventions in high income economies (as per the World Bank[12]) were excluded. Included studies aimed to reduce pollutant emission/exposure through the use of improved stoves. No limit was imposed on the reported outcome as a reduction in any pollutant or health outcome would be important to capture for future studies and scale-up options.

2.3. Study Selection

One reviewer removed obviously irrelevant studies and assessed all remaining titles and abstracts for inclusion. A 10% sample of abstracts were independently assessed by a second reviewer and crosschecked with 90% agreement reached. The same two reviewers also discussed any 'unsure' abstracts. Articles obtained in full text were then reassessed for inclusion.

2.4. Quality Assessment

The quality of the included randomised controlled trails (RCTs) was assessed using the Cochrane Collaboration Risk of Bias Tool. The tool is not appropriate for non-RCT designs and as such was limited to RCTs only. The tool covers six domains of bias: selection bias, performance bias, attrition bias, reporting bias and other bias[13]. For each domain a set of criterion determines if the study is at high risk, low risk or unknown risk.

2.5. Synthesis of the Literature

The Intervention Description and Replication (TIDieR) Checklist[14] was used as a foundation for the synthesis of the literature. This checklist is an extension of the CONSORT 2010 statement (item 5) and the SPIRIT 2013 statement (item 1) and provides key areas of the intervention that should be reported to enhance replication and implementation of interventions[14]. The TIDieR checklist items include the following elements of the intervention: brief name, why, what, who provided, how, where, when and how much, tailoring, modifications and how well. The TIDieR guide provides an explanation and elaboration of each item[14]. The guidance is intended to apply across all evaluation study designs and reviews.

3. Results

Our systematic review identified 5243 potential articles. After duplicates were removed, title screening occurred on 3772 studies of which 258 were further screened on abstract and 57 full texts retrieved. A total of 36 studies were found to meet the full inclusion criteria (see **Figure 1** for a flow chart of study selection). Studies were excluded on the grounds of: study type (only stove intervention or evaluation of stove intervention studies were included); source of air pollution (populations exposed to non-solid fuels only such as tobacco, radon or outdoor sources of air pollution were excluded), the study setting (intervention occurring in non-natural settings such as laboratory-based or non-residential settings such as occupational settings were excluded); the study country (only studies from LMIC were included).

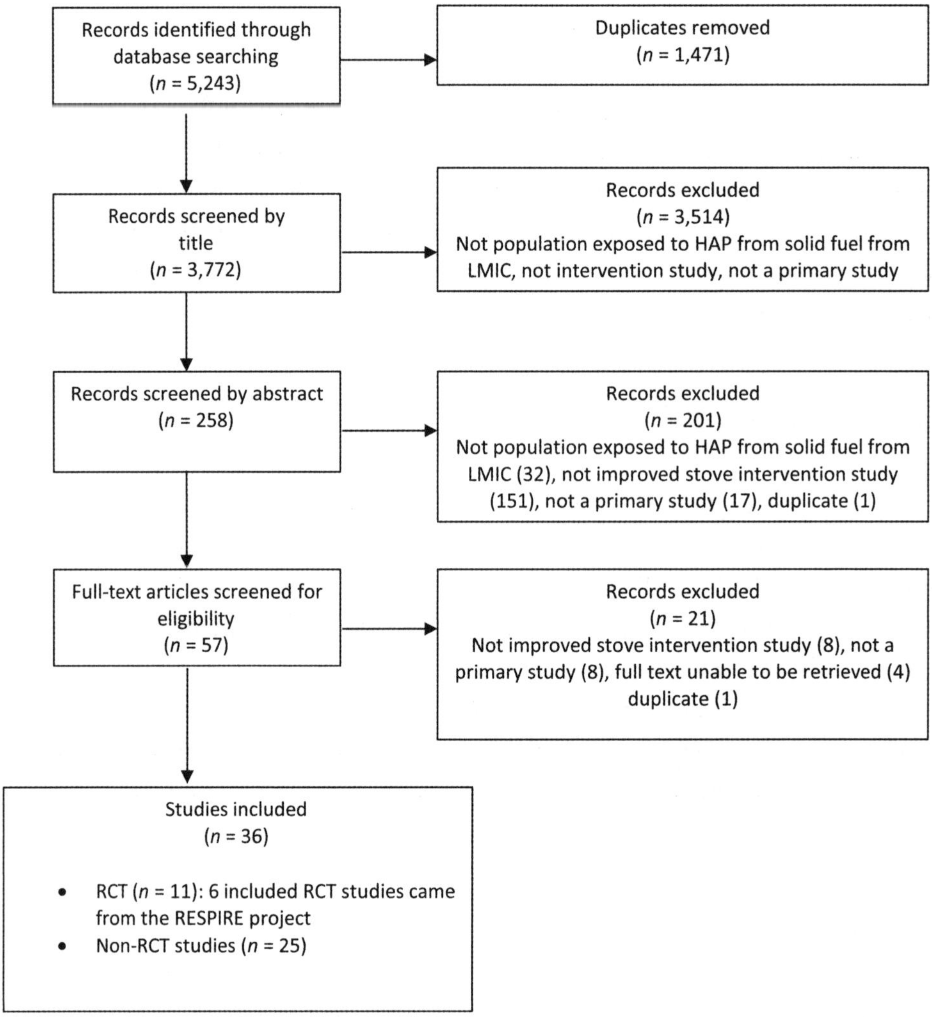

Figure 1. Flow diagram of study selection.

3.1. Reported Effect of Stove Intervention

We examined the effect of stove interventions as reported by study authors. The majority of authors reported a positive reduction of HAP after the installation of an improved stove. Primarily, the main pollutants measured were carbon monoxide (CO) and particular matter (PM). The time frames of measuring pollutant concentration differed greatly across each study from hourly to seven-day measurements. A meta-analysis of the results was not possible due the disparity between pollutant types, methods, and timing of measurement. While pollutant reductions were reported, frequently, these reductions were not enough to meet WHO air quality recommendations. The study by Hanna *et. al.* (2012) which has the longest follow-up (4 years) showed no improvement post one year. Beltramo *et al*'s.[15] solar oven intervention group was higher than the control group (8.09 ppm/h compared to 6.50ppm/h respectively). The authors reported this unexpected increase in CO exposure was largely due to smaller house-hold size in the intervention group.

A wide range of health outcomes were reported across the studies. Self-report measures within studies reported a reduction of respiratory (e.g. cough, phlegm, wheeze, chest tightness)[16][17], non-respiratory (e.g. eye discomfort, headache, backache)[18][19] and sleep symptoms (e.g. snoring, nasal congestion)[17][20] in intervention groups. However, objective measures of pulmonary function were less conclusive. The RESPIRE study did not significantly improve women's lung function or reduce physician-diagnosed pneumonia for children younger than 18 months after 12–18 months of improved stove use[16][21]. Authors of the RESPIRE study suggest that stove or fuel interventions with lower average emissions than the plancha chimney stove may be required for communities with such high exposure to air pollution[21]. Similarly, Clark *et. al.*[22] found no evidence of association between stove type and lung function. The RESPIRE study did report a non-statistically significant reduction in low birth weight in the intervention groups[23], evidence of a reduction in blood pressure[24] and reduced occurrence of non-specific ST-segment depression[25] suggesting improved stove interventions may potentially affect cardiovascular health.

3.2. Description of Study Characteristics Using the TIDieR Checklist

The TIDieR checklist was completed for each included study. Extracted information included: brief name of the intervention (Item 1); the rationale and goal (Item 2), the stove type and educational material provided (Item 3); who provided the intervention (Item 4), the mode of delivery (Item 6); where the intervention occurred (Item 7); the intervention schedule (Item 8); whether the intervention was tailored (Item 9) or modified during the course of the study (Item 10); and whether intervention adherence and fidelity was assessed (Item 11 and 12).

3.3. Item 1 & 2: Brief Name & Why

A brief description of each study can be seen in **Tables 1-3**. Nine studies (6 RCTs, 3 non-RCTs) were affiliated with the Randomized Exposure Study of Pollution Indoors and Respiratory Effects (RESPIRE) study[16][18][19][21][23]–[27]. This was the first RCT to investigate the health effects from solid fuel use[19]. The study occurred in Guatemala and aimed to assess the impact of improved stoves (planchas) on exposure and health outcomes in a rural population reliant on wood fuel. An additional five RCTs were identified, all of which investigated the impact of improved stoves on either exposure outcomes (n = 2) or both exposure and health outcomes (n = 3). The majority of identified non-RCTs were prepost studies investigating the impact on exposure-related health outcomes or exposure reduction of various stove types. Some studies also assessed traditional cooking practices and the acceptability of stoves to the local community members.

3.4. Item 3 & 4: What (Materials & Procedures)

Across the studies more than 15 different stove types were used. The most commonly reported stove was the plancha (largely due to the RESPIRE study), which is an improved chimney woodstove typically built into the home. Other stoves were portable and delivered to the home (e.g. the Eco-Stove[28]) or provided in multiple pieces and built by the household with provided instructions (e.g. the Juntos National Program[29]). In Hanna *et al.*[8], households were responsible for providing mud for the stove base, labour and payment of about US$0.75 to pay the mason who assisted in building and maintaining the stove. One study investigated the use of a solar oven stove (the HotPot)[15].

3.5. Item 5: Who Provided

Studies were largely led by University and NGO collaborations. The most wide-spread dissemination of stoves (n = 40,000) was led by the Joint UNDP/ World Bank Energy Sector Management Assistance Program[30]. This program

Table 1. Randomised control trials from the Randomized Exposure Study of Pollution Indoors and Respiratory Effects (RESPIRE) study.

First author of study, year	Brief name	Study design	Study country	N	Age of partici-pants (years)	Sex	Control group (Y or N)	Pollutant outcome	Health Outcome	Follow-upperiod (post stove installation)	Reported effect of stove use(positive effect (+); negative effect (−); no effect(/))
Diaz, 2008 [18]	RESPIRE: self-rated health among women in the RESPIRE trial	RCT (subsample)	Guatemala	169 (80 Ix; 89 control)	Adult	Female	Y	NA	Self-report of health	Approx. 18 months	+
Diaz, 2007 [19]	RESPIRE: eye discomfort, headache and back pain	RCT (subsample)	Guatemala	504 (259 Ix; 245 control)	Adult	Female	Y	e-CO	NA	12 - 18 months	+
Smith, 2010 [20]	RESPIRE: trial of woodfire chimney cook stoves	RCT	Guatemala	515 infants; 532 mothers	Infants (0 - 18 months); mothers (15 - 55 years)	Female & children	Y	CO	In separate papers	Every 3 months >until the children Reached 18 months	+
Smith, 2011 [21]	RESPIRE: effect on childhood pneumonia	RCT	Guatemala	534 households (269 intervention; 265 control)	Infants (0 - 18 months); mothers (15 - 55 years)	Female & children	Y	CO	Childhood pneumonia	Every 3 months until the children reached 18 months	+
Smith - Sivertsen, 2009 [22]	RESPIRE: Effect on women's respiratory symptoms and lung function	RCT	Guatemala	504 women	15 - 55 years	Female	Y	CO	Chronic respiratory symptoms and lung function	Every 3 months until the children Reached 18 months	+
Thompson, 2011 [23]	RESPIRE: impact of reduced maternal exposure on new born birth weigh	RCT (Subgroup of RESPIRE)	Guatemala	174 infants (69 from Ix; 105 from control)	Infant s	Both	Y	CO	Birth weight	Until birth	+

CO Carbon Monoxide, e-CO exhaled CO, Ix intervention.

Table 2. Additional randomised control trials (non-RESPIRE studies).

First author of study, year	Brief name	Study design	Study country	N	Age of partici-pants (years)	Sex	Control group (Y or N)	Pollutant outcome	Health Outcome	Follow-up period (post stove installation	Reported effect of stove use (positive effect (+); negative effect (−); no effect(/))
Beltramo 2012 [15]	Provision of solar oven + training + education	RCT	Senegal	790 participants (465 Ix; 325 control)	Mean 23 years	Female	Y	CO	NA	6 months	/
Hanna, 2012 [8]	Household behaviour on the impact of improved cook stoves	RCT (stepped wedge)	India	2651 house-holds	Unknown	Female	Y	e-CO, proxy PM	Exposure-related health complaints and health checks	4 years	/after first year
Jary, 2014 [31]	Feasibility of RCT of cook stove interventions	Pilot parallel RCT	Malawi	50	Adults	Female	Y	e-CO	Symptom burden, oxygen saturation	7 days	Feasible
Romieu, 2009 [36]	Improved biomass stove intervention in rural Mexico	RCT	Mexico	552 women	Adult	Women	Y	CO, PAH	Respiratory & lung function measurements, blood samples & health questionnaire	10 months	+
Rosa, 2014 [35]	Impact of water filters and improved cook stoves on drinking water and HAP	RCT (parallel household - randomised RCT)	Rwanda	566 households (HAP sampling in 121 households)	All	Both	Y	PM2.5	NA	5 months	+

CO Carbon Monoxide, e-CO Exhaled CO, PM particular matter, PAH polycyclic aromatic hydrocarbons, Ix intervention group.

Table 3. Non-randomised controlled trials included in review.

First author of study, year	Brief name	Study design	Study country	N	Age of participants (years)	Sex	Control group (Y or N)	Pollutant outcome	Health outcome	Follow-up period (post stove installation)	Reported effect of stove usea
					Before and after studies						
Accinelli, 2014 [17]	Impact of biomass fuel stoves on respiratory and sleep symptoms in children	Before-and-after study	Peru	82	<15	Both	N	NA	Respiratory & sleep symptoms	2 years	+when exclusive use of stove
Castaneda, 2013 [20]	Effect of improved stoves on sleep apnoea in childre	Before-and-after study	Peru	59	<15	Both	N	NA	Sleep symptoms	1 year	+
Clark, 2013 [28]	Impact of cleaner stoves on blood pressure	Before-and-after study	Nicaragua	74	Adults	Female	N	CO, PM2.5	Blood pressure	9 months - 1 year	+
Cynthia, 2008 [40]	Reduction of PM and CO as a result of the Patsari cookstove	Before-and-after study	Mexico	60 households	Adult	Female	N	CO, PM2.5	NA	1 month	+
Fitzgerald, 2012 [41]	Cookstove interventions in Peru	Before-and-after comparative study	Peru	57 house-holds :30 (stove1); 27 (stove 2)	18 - 45	Female	N	CO, PM2.5	NA	3 weeks	+
Li, 2011 [29]	Exposure reduction of stove intervention	Before-and-after comparative study	Peru	Program A) 30; Program B) 27 house-holds	18 - 45	Female	N	CO, PM2.5	Urinary OH-PAH levels	3 weeks	+

Continued

Mukhopadhyay 2012 [34]	Exploratory study of cookstoves to inform large-scale interventions	Before-and-after feasibility study	India	32 house-holds	All	Both: focus on primary cooks	N	CO, PM2.5	NA	12 weeks	NA
Oluwole, 2013 [42]	Effect of stoves on HAP and respiratory health in Nigeria	Before-and-after pilot study	Nigeria	59 mother-child pairs	Mother (20 - 60); child (6 - 17)	Female & children	N	CO, PM2.5	Exposure-related health complaints	1year	+
Pennise, 2009 [43]	Air quality of improved stoves in Ghana and ethanol stove in Ethiopia	Before-and-after comparative study	Ghana and Ethiopia	Ghana: 36 households; Ethiopia 33 households	All	Both	N	CO, PM2.5	NA	Unclear	+
Riojas-Rodriguez [44]	Impact of Patsari improved stoves on PAHs and CO (subproject of Romieu *et al*'s RCT)	Before-and-after study	Mexico	63 women	Adult	Women	Y	CO, PAH	Measured in Romieu's 2010 study	10 months	+
Singh, 2012 [33]	Mud improved stove in Nepal	Before-and-after study	Nepal	47 households	All	Primary cooks (mainly female)	N	CO, PM	Exposure-related health questionnaire	3 & 12 months	+
Torres-Dorsal, 2008 [32]	Evaluation of risk reduction program using biomarkers of exposure and effect	Before-and-after study	Mexico	20 participants	Children (5 - 17); adult (20 - 35)	Both	N	COHb	Urinary1-OHP levels And DNA damage	Unknown	+
Zuk, 2007 [45]	Impact of improved wood stoves in rural Mexico	Before-and-after study	Mexico	53 households	All	Both	N	PM2.5		2 - 3 months	+
				Cross-sectional study							

Continued

Bruce, 2004 [46]	Impact of improved stoves, house construction & child location on IAP levels	Cross-sectional	Guatemala	204 house-hold	<1.5	Both	Y	CO, PM3.5	NA	2 - 3 years	+
Clark, 2009 [22]	Impact of improved stoves on IAP and health	Cross-sectional	Honduras	79	Adult	Female	Y	CO, PM2.5	Pulmonary function, respiratory symptoms, CRP concentrations	NA	+
Guarnieri, 2014 [27]	RESPIRE: airway inflammation	Cross-sectional (within RCT)	Guatemala	45 (19 Ix; 26 control)	Adult	Female	Y	CO, e-CO	Spirometry & induced sputum for cell counts, gene expressions & protein concentrations	18 - 24 months	+
Hartinger, 2013 [47]	Chimney stoves compared to traditional open stoves	Cross-sectional (within RCT)	Peru	93 house-Holds (43 Ix; 48	All	Both	Y	CO, PM2.5	NA	7 months	/unless restricted to full
Henkle, 2010 [48]	Honduras stove project	Cross-sectional	Honduras	34 homes	2 - 84	Both: female (56.4%)	N	CO, TSP	Respiratory surveys, PEFR	NA	Feasible
Cohort study											
Chapman, 2005 [49]	Improved stoves impact on COPD	Retrospective cohort study	China	20,453	Born 1917 - 51	Focus on farmers	Y	NA	COPD diagnosis	Average 12.8 years	+
					Marketing andcampaign						
Joint UNDP [30]	Energy Sector Management Assistance Program (ESMAP): Niger improved stoves project	Marketing and campaign	Niger	40,000 stoves sold	All	Focus on women	N	NA	NA	NA	Successful marketing and sale of stoves

involved training of metalsmith workers, establishment of commercial networks and sensitisation campaigns. Multiple studies recruited bilingual community members or community health workers as field staff. Local brick masons or metalsmiths were frequently used to assist in the building of stoves. Some studies identified community members who used and maintained stoves correctly and employed them as stove inspectors or promoters within communities[8].

3.6. Item 6: How

Table 4 provides a summary of key components of interventions. The majority of studies focused on stove provision only (including education and training on stove use). Very few studies have combined stove provision with additional interventions to reduce HAP such as improving the living environment (e.g., improved kitchen design and ventilation) or modifying user behaviour (e.g., using pot lids, removing children from cooking area).

3.7. Item 7: Where

Figure 2 shows the distribution of study countries identified in the review. In-keeping with the inclusion criteria, only studies from LMIC were included. Fifteen different study countries were identified across Central and South America, Africa and Asia. **Figure 2** groups study countries according to the Global Burden of Disease Regions. The countries are colour-coded to highlight areas of high burden

Table 4. Key study components of included studies.

Intervention	Articles[a]
Stove provision	29
Comparison of 2 or more stove types	5
Stove + behavioural intervention	3
Stove + changes to home environment	1
Marketing campaign	1
Combined stove +other environmental intervention	1

[a]Some studies fall into 2 categories.

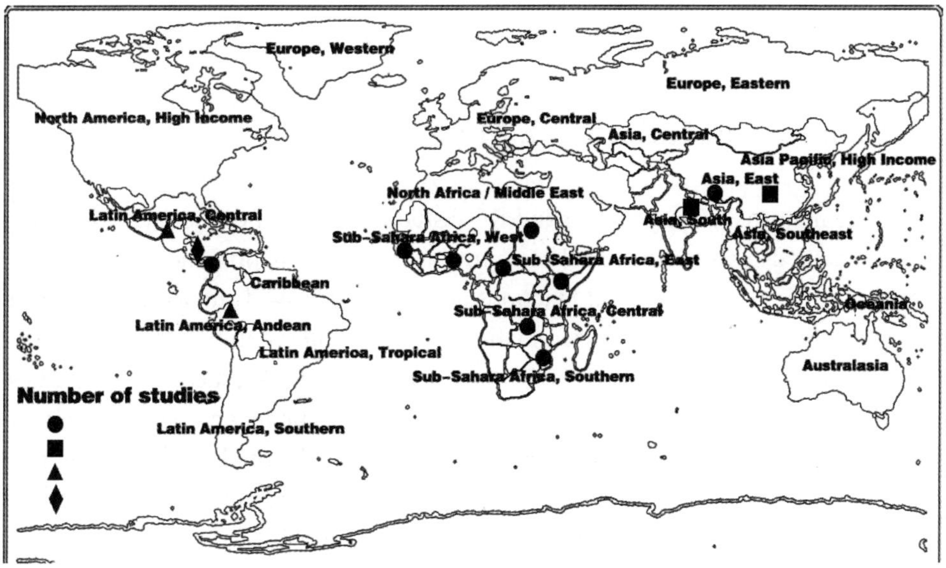

Figure 2. The location of included studies. Countries are grouped as per the Global Burden of Disease regions and colour coordinated in terms of burden of disease attributable to HAP from solid fuels. The numbers of studies in each region are illustrated by the size of the circular marker. High to low burden of disease attributable to HAP from solid fuels as per Lim *et. al.*[5] is represented by signifying highest levels of disease burden to signifying lowest levels of disease burden. This figure was created by the authors using ArcInfo 10.2.1.

of disease attributable to HAP from solid fuels as per Lim *et. al.*[5]. Studies have occurred across a range of settings and locations. Importantly, a range of studies have occurred in South Asia and Sub-Saharan Africa where HAP from solid fuels has the highest disease burden.

3.8. Item 8: When and How Much

Most commonly, studies performed baseline assessments and then provided (or installed) the stove with instructions on stove use. Limited interaction between stove installation and post-intervention follow-up was reported. If stoves were damaged or malfunctioning, participant were often responsible for contacting an appropriate person. Uniquely, the RESPIRE study provided weekly maintenance checks.

Post-intervention follow-up ranged from seven days[31] to four years[8]. In Hanna *et. al.*'s[8] study no meaningful reduction of HAP was seen beyond the first year. No other studies provided follow-up periods beyond two years.

3.9. Item 9 & 10: Tailoring & Modifications

The majority of large-scale and RCT studies included preliminary questionnaires and needs assessments to determine baseline information and cooking practices. Impressions and observed reactions about the improved stoves were reported to impact upon design and dissemination of some interventions[26][30]. However, once the intervention was designed, limited tailoring took place across studies. In-built stoves were designed around the house requirements and occasionally adjustments were individualised to each study house[32]. Mostly, however, stove provision was standardised across households and limited adaption took place. In one study, abnormally sized or shaped kitchens were excluded[33].

3.10. Item 11 & 12: How Well (Adherence & Fidelity)

Many studies reported difficulties with adherence and adoption of stoves. 'Stacking', the use of traditional stoves in conjunction with improved stoves, was a frequently reported issue. However, while commonly reported as a potential issue, few studies (6/36) actively measured and reported adherence or fidelity data. One study objectively measured stove use with the Stove Use Monitoring System (SUMS)[34] which is fixed to the stove and records the temperature profile over time. Random spot checks by Rosa *et. al.*[35] and Romieu *et. al.*[36] reported 64.1% and 50% of checked households to be exclusively using the improved stoves. Romieu *et. al.*[36] conducted a predictive model with longitudinal data to assess factors influencing adoption. The authors reported no clear effect of socio-economic status or education level on stove adoption. Importantly, previous use of a similar stove type was a predictor of improved stove adoption and as such, the authors concluded that greater reinforcement and training of stove use is required in future studies[36]. Much higher adoption rates were reported by the Joint UNDP/ World Bank Energy Sector Management Assistance Program in Niger[30]. This program reported that 96% of consumers in inspected homes were using the stoves properly, 76% maintaining it correctly and 93% had decided to replace it when necessary[30]. This large scale program utilised a sensitisation campaign and publicity to inform both women and men about the existence and advantages of the stove as well as create a market demand. Training on stove use was also provided. Additionally, the design of both the campaign and stove were piloted and modified based on contextual needs.

3.11. Study Quality

To date, the RESPIRE studies (**Figure 3**) have the highest study quality as per the Cochrane Collaboration Risk of Bias Tool. Great variation exists between the additional RCTs (**Figure 4**) with no study achieving all of the study criteria.

4. Discussion

4.1. Findings and Comparison to the Literature

We found evidence of well designed and implemented interventions. We were not able to make firm conclusions on the relative impact of the intervention due to the differences in their outcome measures. Potentially, the impact of the interventions may decrease over time. Findings from one study with a 4-year follow-up period reported a significant drop in HAP reduction beyond one year. However, the majority of studies have follow-up periods less than 18 months. Success of stove interventions are heavily dependent on how well households adopt the intervention and exclusively use the improved stove above traditional options. These difficulties are echoed in Barnes *et al.*'s 1994 comparative review of international stove programs which concluded that "no matter how efficient or cheap the stove, individual households have proved reluctant to adopt it if it is difficult to install and maintain or less convenient and less adaptable to local preferences than its traditional counterpart"[9]. Sensitisation campaigns and publicity such as that used by the Joint UNDP/World Bank in Niger may enhance adoption[30]. Without an in-depth understanding of contextual drivers for implementation success, interventions aimed at reducing HAP cannot be expected to succeed.

4.2. Comparison to Who Air Quality Targets

We found a disconnection between the relative impact of studies and the

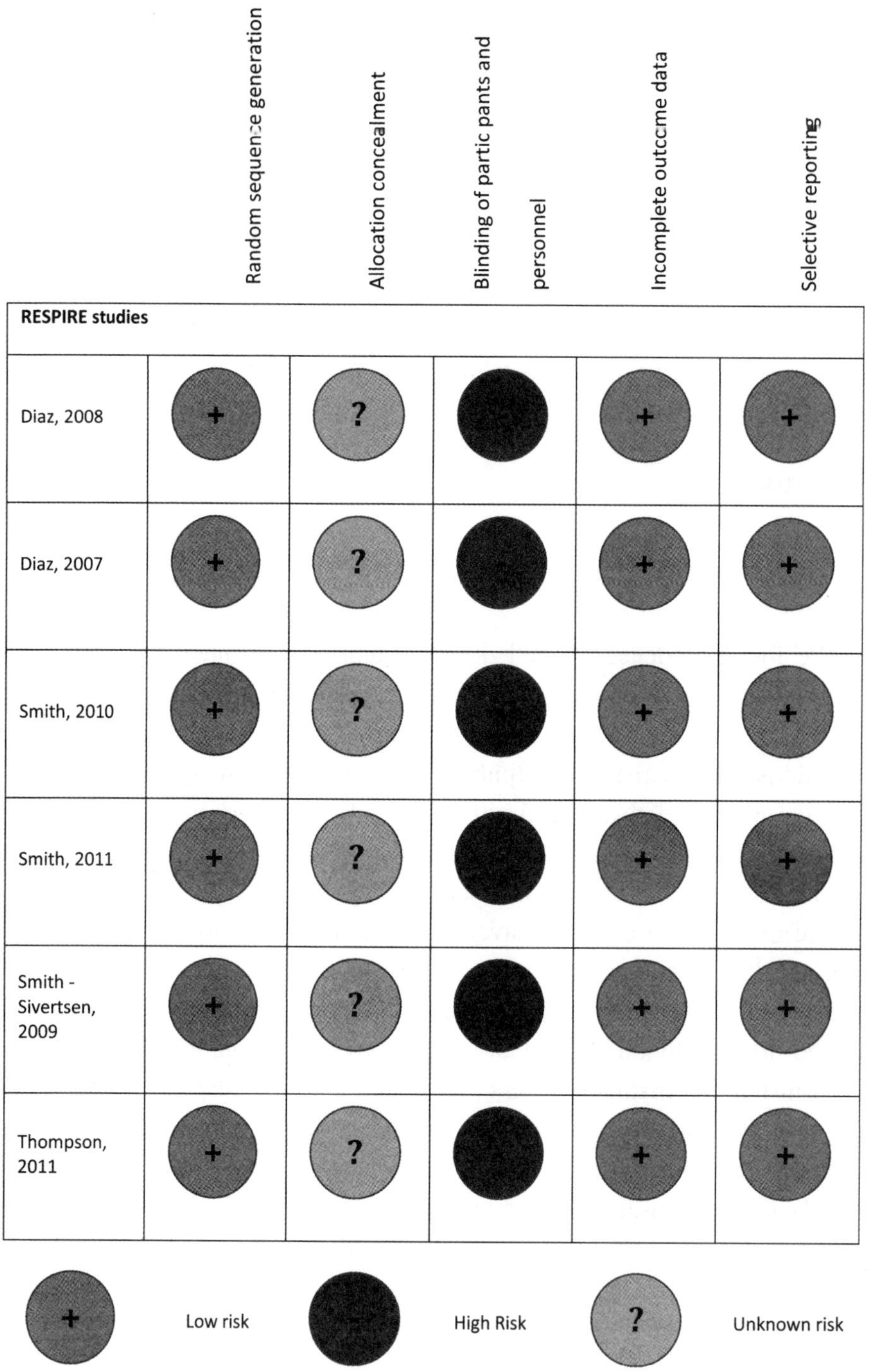

Figure 3. Quality Assessment using the Cochrane Collaboration Risk of Bias Tool of the RESPIRE studies.

Non-RESPIRE studies	Random sequence generation	Allocation concealment	Blinding of participants and personnel	Incomplete outcome data	Selective reporting
Beltramo, 2012	-	?	-	-	?
Hanna, 2012	+	-	-	+	+
Jary, 2014	+	+	-	-	+
Romieu, 2009	+	-	-	+	+
Rosa, 2014	+	-	-	+	+

+ Low risk   - High Risk   ? Unknown risk

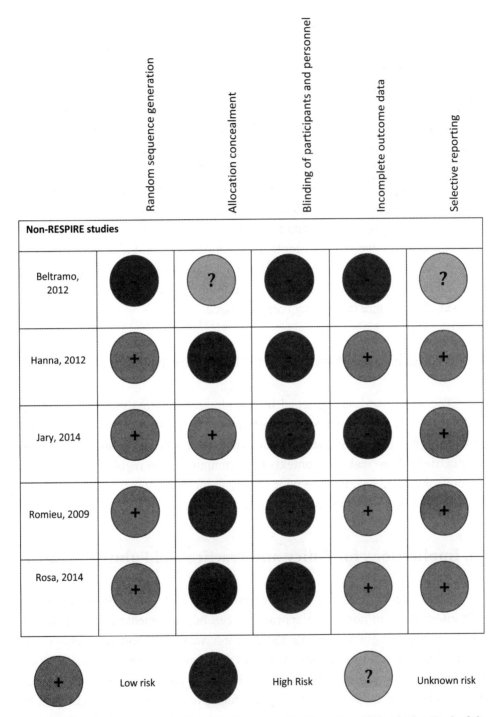

Figure 4. Quality Assessment using the Cochrane Collaboration Risk of Bias Tool of the non-RESPIRE studies.

targets in the WHO Air Quality Guideline[37]. Pollutant concentrations were measured over different time frames for each included study varying from hourly to 7-day measurements. In general, these did not align to timeframes used in WHO guidelines making comparisons problematic. The WHO targets also appear unachievable for some contexts given the exceptionally high baseline pollutant concentrations. For example, the study by Rosa *et. al.*[35] report the mean $PM_{2.5}$ concentrations as 48% lower in the intervention group than the control. However, the levels remained more than six times higher than the WHO interim target for $PM_{2.5}$ (485μg/m^3 compared to 75μg/m^3 respectively). The authors also noted that even the outdoor cooking areas had concentration levels well above recommended targets (243μg/m^3 compared to 75μg/m^3 respectively). Given the extreme levels of HAP reported in some studies, single interventions (e.g. provision of a stove) are unlikely to independently reduce pollution to meet recommended standards. As such, complex interventions involving multiple components may be required, however, the literature in this area remains scarce.

4.3. Strengths and Weaknesses of the Review

The review was strengthened by a wide search across multiple databases. Additionally, the inclusion of a wide variety of study designs enabled the vast contribution of non-randomised study in this field to be included. Further, all studies were carried out in 'real life' contexts in communities who regularly use solid fuels for cooking and heating. The wide variety in included studies interventions and outcome measures made comparison between studies difficult.

4.4. Implications for Policy and Practice and Future Research

Stove interventions can reduce exposure to HAP. However, we have very little information on how these approaches can be implemented in a sustainable way to enhance long-term use. We also lack information on how interventions can be scaled up and what supporting structures are required to assist in their success. As repeatedly advocated throughout the literature, there is a need for practice-based evidence of adoption[38]. We suggest future research evaluate and report the process of implementation along with outcome evaluation to enhance the knowledge of how interventions are carried out. Additionally, use of alternative study designs such as natural experimental studies may assist in understanding the population-level impact of large scale stove interventions[39]. In the interim, implementers should ensure an in-depth understanding of the needs and preferences of consumers and the social, financial and environmental context in which they live. Only through active engagement and involvement of the targeted communities can interventions be adequately tailored to meet their needs and be expected to succeed.

5. Conclusion

When well designed, implemented and monitored, current evidence suggests stove interventions can reduce HAP. However, the intervention impacts are unlikely to reduce pollutant levels to WHO recommended levels. Studies have reported a significant reduction in exposure-related health complaints. However, objective measures of lung function have not shown statistically significant improvements in stove interventions groups. Studies with longer follow-up periods are required to assess if pollutant reductions reported in the current literature are sustained over time. Adoption of new technologies is challenging and interventions must be tailored to the needs and preferences of the households of interest and must repeatedly reinforce stove use and benefits. We suggest that future studies give greater emphasis to process evaluation and consider natural experimental designs to increase understanding of population-level impact of stove interventions in real world contexts.

Additional Files

Additional file 1: PRISMA 2009 Checklist. http://www.biomedcentral.com/content/supplementary/2049-9957-3-36-S1.pdf.

Additional file 2: Embase search terms. http://www.biomedcentral.com/content/supplementary/s12889-015-2024-7-s2.pdf.

Abbreviations

ALRI: Acute lower respiratory infections; CO: Carbon monoxide; COPD:

Chronic obstructive pulmonary disease; DALYs: Disability-adjusted life years; HAP: Household air pollution; LMIC: Low and middle-income countries; NGOs: Non-government organisations; PM: Particulate matter; RCTs: Randomised controlled trails; TIDieR Checklist: The Intervention Description and Replication Checklist; WHO: World Health Organization.

Competing Interests

The authors declare that they have no competing interests.

Authors' Contributions

ET designed the review, screened titles, abstracts and full texts and drafted the manuscript. KW helped conceive the study and provided input into the review design. SM provided input in the conception and design of the review. NR assisted with the development of key terms and ran the database search. CF assisted with conception and design of the review, development of inclusion criteria, screening of abstracts and full texts and contributed to the manuscript. All authors read and approved the final manuscript.

Acknowledgements

This review was commissioned by the World Health Organization. CF and KW are funded by the British Heart Foundation.

Source: Thomas E, Wickramasinghe K, Mendis S, *et al.* Improved stove interventions to reduce household air pollution in low and middle income countries: a descriptive systematic review[J]. Bmc Public Health, 2015, 15(1):1–15.

References

[1] Fullerton DG, Bruce N, Gordon SB. Indoor air pollution from biomass fuel smoke is a major health concern in the developing world. Trans R Soc Trop Med Hyg.

2008;102(9):843–51.

[2] Kim KH, Jahan SA, Kabir E. A review of diseases associated with household air pollution due to the use of biomass fuels. J Hazard Mater. 2011;192(2):425–31.

[3] Smith K, Mehta S, Maeusezahl Feuz M. Indoor air pollution from household use of solid fuels, in Comparative quantification of health risks: global and regional burden of disease due to selected major risk factors. Trans R Soc Trop Med Hyg. 2004;102(9):843–51.

[4] Eisner MD, Balmes JR. Indoor and Outdoor Air Pollution. In: Mason RJ, Broaddus VC, Martin T, King Jr T, Schraufnagel D, Murray JF, Nadel JA, editors. Murray and Nadel's Textbook of Respiratory Medicine: 2-Volume Set. 5th ed. Philadelphia: Elsevier Health Sciences; 2010.

[5] Lim SS, Vos T, Flaxman AD, Danaei G, Shibuya K, Adair-Rohani H, et al. A comparative risk assessment of burden of disese and injury attributable to 67 risk factors and risk factor clusters in 21 regions, 1990–2010 : a systematic analysis for the Global Burden of Disease Study 2010. Lancet. 2012;380:2224–60.

[6] Mehta S, Shahpar C. The health benefits of interventions to reduce indoor air pollution from solid fuel use: a cost-effectiveness analysis. Energy Sustain Dev. 2004;8(3):53–9.

[7] Bruce N, Rehfuess E, Mehta S, Hutton G, Smith K. Indoor Air Pollution. In: Jamison DT, Breman JG, Measham AR, Alleyne G, Claeson M, Evans DB, Jha P, Mills A, Musgrove P, editors. Disease Control Priorities in Developing Countries. 2nd ed. Washington (DC): World Bank; 2006.

[8] Hanna R, Greenstone M. Up in smoke: the influence of household behavior on the long-run impact of improved cooking stoves. Natl Bur Econ Res. 2012. http://papers.ssrn.com/sol3/papers.cfm?abstract_id=2039004.

[9] Barnes DF. What makes people cook with improved biomass stoves?: a comparative international review of stove programs. Washington, D.C.: In. Edited by Bank W; 1994.

[10] Smith KR, Khalil MAK, Rasmussen RA, Thorneloe SA, Manegdeg F, Apte M. Greenhouse gases from biomass and fossil fuel stoves in developing countries: a Manila pilot study. Chemosphere. 1993;26(1–4):479–505.

[11] Moher D, Liberati A, Tetzlaff J, Altman D, The PRISMA Group. Preferred reporting items for systematic reviews and meta-analyses: the PRISMA statement. PloS Med. 2009;6:7.

[12] Country and lending groups. The World Bank.(2015). Country and Lending Groups. Accessed online. http://data.worldbank.org/about/country-and-lending-groups.

[13] Higgins JP, Altman DG, Gotzsche PC, Juni P, Moher D, Oxman AD, et al. The Cochrane Collaboration's tool for assessing risk of bias in randomised trials. BMJ. 2011;343:d5928.

[14] Hoffmann TC, Glasziou PP, Boutron I, Milne R, Perera R, Moher D, et al. Better re-

porting of interventions: template for intervention description and replication (TIDieR) checklist and guide. BMJ. 2014;348:g1687.

[15] Beltramo T, Levine DI. The effect of solar ovens on fuel use, emissions and health: results from a randomised controlled trial. J Dev Effectiv. 2013;5(2):178–207.

[16] Smith-Sivertsen T, Diaz E, Pope D, Lie RT, Diaz A, McCracken J, et al. Effect of reducing indoor air pollution on women's respiratory symptoms and lung function: the RESPIRE Randomized Trial, Guatemala. Am J Epidemiol. 2009;170(2):211–20.

[17] Accinelli RA, Llanos O, Lopez LM, Pino MI, Bravo YA, Salinas V, et al. Adherence to reduced-polluting biomass fuel stoves improves respiratory and sleep symptoms in children. BMC Pediatrics. 2014;14:12.

[18] Diaz E, Bruce N, Pope D, Diaz A, Smith KR, Smith-Sivertsen T. Self-rated health among Mayan women participating in a randomised intervention trial reducing indoor air pollution in Guatemala. BMC Int Health Hum Rights. 2008;8:7.

[19] Diaz E, Smith-Sivertsen T, Pope D, Lie RT, Diaz A, McCracken J, et al. Eye discomfort, headache and back pain among Mayan Guatemalan women taking part in a randomised stove intervention trial. J Epidemiol Commun Health. 2007;61(1):74–9.

[20] Castaneda JL, Kheirandish-Gozal L, Gozal D, Accinelli RA, Pampa Cangallo Instituto de Investigaciones de la Altura Research G. Effect of reductions in biomass fuel exposure on symptoms of sleep apnea in children living in the peruvian andes: a preliminary field study. Pediatr Pulmonol. 2013;48(10):996–9.

[21] Smith KR, McCracken JP, Weber MW, Hubbard A, Jenny A, Thompson LM, et al. Effect of reduction in household air pollution on childhood pneumonia in Guatemala (RESPIRE): a randomised controlled trial. Lancet. 2011;378(9804):1717–26.

[22] Clark ML, Peel JL, Burch JB, Nelson TL, Robinson MM, Conway S, et al. Impact of improved cookstoves on indoor air pollution and adverse health effects among Honduran women. Int J Environ Health Res. 2009;19(5):357–68.

[23] Thompson LM, Bruce N, Eskenazi B, Diaz A, Pope D, Smith KR. Impact of reduced maternal exposures to wood smoke from an introduced chimney stove on newborn birth weight in rural Guatemala. Environ Health Perspect. 2011;119(10):1489–94.

[24] McCracken JP, Smith KR, Diaz A, Mittleman MA, Schwartz J. Chimney stove intervention to reduce long-term wood smoke exposure lowers blood pressure among Guatemalan women. Environ Health Perspect. 2007;115(7):996–1001.

[25] McCracken J, Smith KR, Stone P, Diaz A, Arana B, Schwartz J. Intervention to lower household wood smoke exposure in Guatemala reduces ST-segment depression on electrocardiograms. Environ Health Perspect. 2011;119(11):1562–8.

[26] Smith KR, McCracken JP, Thompson L, Edwards R, Shields KN, Canuz E, et al. Personal child and mother carbon monoxide exposures and kitchen levels: methods and results from a randomized trial of woodfired chimney cookstoves in Guatemala (RESPIRE). J Expo Sci Environ Epidemiol. 2010;20(5):406–16.

[27] Guarnieri MJ, Diaz JV, Basu C, Diaz A, Pope D, Smith KR, et al. Effects of woodsmoke exposure on airway inflammation in rural guatemalan women. PLoS ONE [Electronic Resource]. 2014;9(3):e88455.

[28] Clark ML, Bachand AM, Heiderscheidt JM, Yoder SA, Luna B, Volckens J, et al. Impact of a cleaner-burning cookstove intervention on blood pressure in Nicaraguan women. Indoor Air. 2013;23(2):105–14.

[29] Li Z, Sjodin A, Romanoff LC, Horton K, Fitzgerald CL, Eppler A, et al. Evaluation of exposure reduction to indoor air pollution in stove intervention projects in Peru by urinary biomonitoring of polycyclic aromatic hydrocarbon metabolites. Environ Int. 2011;37(7):1157–63.

[30] Program JUWBESMA. Niger: Improved stoves project. 1987.

[31] Jary HR, Kachidiku J, Banda H, Kapanga M, Doyle JV, Banda E, et al. Feasibility of conducting a randomised controlled trial of a cookstove intervention in rural Malawi. Int J Tuberc Lung Dis. 2014;18(2):240–7.

[32] Torres-Dosal A, Perez-Maldonado IN, Jasso-Pineda Y, Martinez Salinas RI, Alegria-Torres JA, Diaz-Barriga F. Indoor air pollution in a Mexican indigenous community: evaluation of risk reduction program using biomarkers of exposure and effect. Sci Total Environ. 2008;390(2–3):362–8.

[33] Singh A, Tuladhar B, Bajracharya K, Pillarisetti A. Assessment of effectiveness of improved cook stoves in reducing indoor air pollution and improving health in Nepal. Energy Sustain Dev. 2012;16(4):406–14.

[34] Mukhopadhyay R, Sambandam S, Pillarisetti A, Jack D, Mukhopadhyay K, Balakrishnan K, et al. Cooking practices, air quality, and the acceptability of advanced cookstoves in Haryana, India: an exploratory study to inform large-scale interventions. Glob Health Action. 2012;5:1–13.

[35] Rosa G, Majorin F, Boisson S, Barstow C, Johnson M, Kirby M, et al. Assessing the impact of water filters and improved cook stoves on drinking water quality and household air pollution: a randomised controlled trial in rwanda. PLoS ONE[Electronic Resource]. 2014;9(3):e91011.

[36] Romieu I, Riojas-Rodriguez H, Marron-Mares AT, Schilmann A, Perez-Padilla R, Masera O. Improved biomass stove intervention in rural Mexico: impact on the respiratory health of women. Am J Respir Crit Care Med. 2009;180(7):649–56.

[37] World Health Organization. WHO Air quality guidelines for partinculate matter, ozone, nitrogen dioxide and sulfur dioxide. Global update 2005. Geneva: Switzerland WHO Press; 2006.

[38] Lewis JJ, Pattanayak SK. Who adopts improved fuels and cookstoves? A systematic review. Environ Health Perspect. 2012;120(5):637–45.

[39] Craig P, Cooper C, Gunnell D, Haw S, Lawson K, Macintyre S, et al. Using natural experiments to evaluate population health interventions: new Medical Research

Council guidance. J Epidemiol Commun Health. 2012, jech-2011.

[40] Cynthia AA, Edwards RD, Johnson M, Zuk M, Rojas L, Jimenez RD, et al. Reduction in personal exposures to particulate matter and carbon monoxide as a result of the installation of a Patsari improved cook stove in Michoacan Mexico. Indoor Air. 2008;18(2):93–105.

[41] Fitzgerald C, Aguilar-Villalobos M, Eppler AR, Dorner SC, Rathbun SL, Naeher LP. Testing the effectiveness of two improved cookstove interventions in the Santiago de Chuco Province of Peru. Sci Total Environ. 2012;420:54–64.

[42] Oluwole O, Ana GR, Arinola GO, Wiskel T, Falusi AG, Huo DZ, et al. Effect of stove intervention on household air pollution and the respiratory health of women and children in rural Nigeria. Air Qual Atmosphere Health. 2013;6(3):553–61.

[43] Pennise D, Brant S, Agbeve SM, Quaye W, Mengesha F, Tadele W, et al. Indoor air quality impacts of an improved wood stove in Ghana and an ethanol stove in Ethiopia. Energy Sustain Devel. 2009;13(2):71–6.

[44] Riojas-Rodriguez H, Schilmann A, Marron-Mares AT, Masera O, Li Z, Romanoff L, et al. Impact of the improved patsari biomass stove on urinary polycyclic aromatic hydrocarbon biomarkers and carbon monoxide exposures in rural Mexican women. Environ Health Perspect. 2011;119(9):1301–7.

[45] Zuk M, Rojas L, Blanco S, Serrano P, Cruz J, Angeles F, et al. The impact of improved wood-burning stoves on fine particulate matter concentrations in rural Mexican homes. J Expo Sci Environ Epidemiol. 2007;17(3):224–32.

[46] Bruce N, McCracken J, Albalak R, Schei MA, Smith KR, Lopez V, et al. Impact of improved stoves, house construction and child location on levels of indoor air pollution exposure in young Guatemalan children. J Expo Anal Environ Epidemiol. 2004;14 Suppl 1:S26–33.

[47] Hartinger SM, Commodore AA, Hattendorf J, Lanata CF, Gil AI, Verastegui H, et al. Chimney stoves modestly improved Indoor Air Quality measurements compared with traditional open fire stoves: results from a small-scale intervention study in rural Peru. Indoor Air. 2013;23(4):342–52.

[48] Henkle J, Mandzuk C, Emergy E, Schrowe L, Sevilla-Martir J. Global health and international medicine: Honduras Stove Project. Hispanic Health Care Int. 2010;8(1):36–46.

[49] Chapman RS, He X, Blair AE, Lan Q. Improvement in household stoves and risk of chronic obstructive pulmonary disease in Xuanwei, China: retrospective cohort study. BMJ. 2005;331(7524):1050.

[50] Albalak R, Bruce N, McCracken JP, Smith KR, De Gallardo T. Indoor respirable particulate matter concentrations from an open fire, improved cookstove, and LPG/open fire combination in a rural Guatemalan community. Environ Sci Technol. 2001;35(13): 2650–5.

[51] Baris E, Ezzati M. Household energy, indoor air pollution and health: a multisectoral intervention program in rural China. Washington, D.C: Assistance ESM; 2007.

[52] Zhou Z, Jin Y, Liu F, Cheng Y, Liu J, Kang J, et al. Community effectiveness of stove and health education interventions for reducing exposure to indoor air pollution from solid fuels in four Chinese provinces. Environ Res Lett. 2006;1(1):014010.

Chapter 26

Legal Protection Assessment of Different Inland Wetlands in Chile

Patricia Möller[1,2], Andrés Muñoz-Pedreros[3]

[1]Facultad de Ciencias, Universidad Austral de Chile, Casilla 567, Valdivia, Chile
[2]Programa de Humedales, Centro de Estudios Agrarios y Ambientales, Casilla 164, Valdivia, Chile
[3]Núcleo de Estudios Ambientales NEA, Escuela de Ciencias Ambientales, Facultad de Recursos Naturales, Universidad Católica de Temuco, Casilla 15-D, Temuco, Chile

Abstract: Background: Inland wetlands are well represented ecosystems in Chile that are subjected to various pressures affecting conservation. Protection means legal and administrative initiatives which promote the protection and/or preservation of a wetland, either in its entirety, considering their areas of influence or its components. Results: The aim of this work is to develop a methodology for estimating the value of protection of different types of inland wetlands in Chile. For this purpose: (a) the Chilean regulations in relation to the issues of biodiversity, wetlands and water resources were compiled; (b) such legislation and its application were analyzed, (c) protection and restriction values of each legal standard was estimated, and then total protection value of standards applied to six types of wetlands in study. 47 legal rules related to protection of inland wetlands and eight directly or indirectly affecting conservation and wise use of wetlands were identified. Conclusions: In Chile there is no specific statutory rules or regulations on wetlands. Current legal standards do not protect equally the different types of inland wetlands, being swamp forests, peatlands and brackish Andean lakes less protected.

To improve wetlands conservation, incentives promoting wetlands destruction must be eliminated and promulgate specific regulations for proper management and conservation.

Keywords: Inland Wetlands, Legal Standards, Real Protection, Wetland Protection

1. Background

Chile is highly diverse in limnic systems mostly recognized as wetlands and defined as 'areas of marsh, fen, peatland or water, whether natural or artificial, permanent or temporary, static or flowing, fresh, brackish or salt waters, including areas of marine water where tide depth does not exceed the six meters' (Ramsar 2000). This long definition includes very heterogeneous areas, such as swamps, rivers, lakes, seashores, and others (Mitsch and Gosselink 2000). Wetlands are highly productive ecosystems (Novitzki *et al.* 1996) characterized by multiple roles, e.g. hydrological, biogeochemical, habitat conservation, and food webs (Woodward and Wui 2001), as well as providing goods and services relevant to human society (Barbier *et al.* 1997). In Chile, wetlands are estimated to cover 4,498,060, 7ha equivalent to 5.9% of national territory (CONAF/CONAMA 1997).

Wetlands support high biodiversity and are presently recognized as the most threatened systems by human activities (Marín *et al.* 2006). It affects various types of wetlands in Chile (Muñoz-Pedreros 2004, Peña-Cortés *et al.* 2006, Zegers *et al.* 2006, Figueroa *et al.* 2007). Particularly, freshwater aquatic fauna conservation is mostly affected, such as fishes (Vila *et al.* 2006), amphibians (Díaz-Paéz and Ortiz 2003, Veloso 2006), molluscs (Valdovinos *et al.* 2005), and decapod crustaceans (Bahamonde *et al.* 1998, Pérez-Losada *et al.* 2002). Inland wetland conservation is a global priority (Abell 2002, Dudley 2008). However, their particular territorial location and the difficulty to apply current protection categories make a difficult management as protected areas. As traditional ecosystem conservation methods are not well implemented in inland aquatic environments, different conceptual approaches are suggested (e.g. freshwater focal area, critical management zone, and catchment management zones (Abell *et al.* 2007)).

Governance concept applied to natural resources, mainly in waters, has become important (Iza and Rovere 2006). Governance is understood as the economic, political, and administrative practice to manage every country affair. This includes mechanisms, processes, and institutions by which citizens express their interests, exercise their rights, meet their requirements, and mediate their differences. In this context, conservation of wetlands should be addressed under various environmental management tools as a permanent process, where various stakeholders and public, private, and civil society develop specific efforts to preserve, maintain, restore, and make a sustainable use of environment. Environmental management uses different and diverse origin instruments, which can be classified in four main categories (sensu Rodríguez-Becerra and Espinoza 2002): direct regulation, administration, economics, and education (including research, technical assistance, and environmental information). Direct regulatory tools which are also known as command and control regulations prevail in the environmental management and consist of mandatory regulations and standards, which establish environmental quality targets as well as management and preservation of renewable natural resources of environment. Legislation creates legal tools and standards to comply with principles and reach the aims (Asenjo 2006).

Brañes (2000) defines the environmental regulation as a set of standards dealing with legal protection of those conditions making every life to be possible, considering for this purpose relationships among many biotic and abiotic elements in the environment, such as system or ecosystem. To Fernández (2004), the environmental law gathers standards, regulations, and principles recognized as legally protected, safeguarding environmental systems, in a global and inclusive perspective that differ from merely legislation of environmental incidence.

It may be understood here that legal protection of wetlands is provided by all legal and administrative initiatives aimed to protect and/or preserve them. There are many legal standards that apply to inland wetlands in Chile, many of them are sectoral standards related to their components, ecological functions, and biodiversity, which would favor or regulate their status and permanence. Selection and analysis of these regulations will allow to provide foundations, which make possible assessment of legal degree protection that they provide to these systems and to integrity as ecological systems. In Chile, regulations addressed to the conservation of water resources as a whole have been mainly aimed to develop economic activities related to the exploitation of a natural resource. Until 1994, prior to the Environmental Basis Act (Act No. 19,300) enactment, there were only sector legal standards, without global environmental protection objective, where legal rules were mainly aimed to protect health of human life and only incidental protection of nature (Olivares 2010). Act No. 19,300 started regulation process of environmental standards, as well as the creation of institutions providing the state of management tools in this area (Rojas 2011).

However, Hermosilla (2004) estimates that while keeping current protection system and guarantees of property rights in Chile over the common well, effective protection to natural systems including wetland ecosystems will not be possible.

Prior this research, there was no compilation and analysis of rules applied to wetland systems which allows to determine real legal protection in taking care of conservation, but there are legal standards which should be assessed under this context, like those regulations enacted which damage preservation and rational usc of inland wetlands. The rational use concept is defined by the Ramsar Convention as the maintenance of ecological features, achieved through the implementation of ecosystem-based approach, within sustainable development context.

Under the Chilean legal system, environmental heritage conservation is the rational use and exploitation or repair if any, of environment components, to assure sustainability and regeneration capacity (Act No. 19,300; Article 2b); however, it does not point out which are those components.

Some environmental components, those considered to be relevant issues for this study among them, are noted by Fernández (2004): a) land or sea waters, surface water or groundwater, streams or standing waters; b) land, soil and subsoil, including beds, bottom and subsoil of ground waters; c) flora and wildlife, land or water; d) microflora and fauna of land, soil and subsoil, streams or water bodies and beds, bottoms and subsoil of these streams or water bodies; e) genetic diversity and patterns and factors regulating flow; f) natural scenic beauty and rural or urban landscape; g) essential ecological processes. These basic components of the environment can be damaged when misused which results in extinction or serious damage, prevents regeneration, and causes environmental damage determined in law as 'any loss, decrease, impairment, or significant impairment associated to environment or to one or more of its components' (Act No. 19,300: Art. 2e).

Some of the elements or factors which could damage or degrade the environment are (to Fernández 2004): a) any kind of pollution; b) erosion, salinization, alkalinization, infestation, flooding, sedimentation, and desertification of soil and land; c) logging or unreasonable and uncontrolled destruction of trees and shrubs as well as extractive forest use and other vegetation destruction; d) monocultures, overgrazing, and, as a whole, any cultural practice with harmful effects on the environment; e) sedimentation of water streams and lakes; f) harmful alterations of natural water flows; g) adverse changes and misuse of water beds; h) wild flora and fauna over-exploitation; i) elimination, destruction, or degradation of endangered species from flora and fauna habitat; j) wetland eutrophication origins; k) introduction or distribution of exotic plants or animals coming from a national biogeographical province different to natural; l) introduction or spread of animal diseases or plant pests; m) use of non-biodegradable products or substances; n) the accumulation of waste or inappropriate waste disposal; o) visual landscape destruction or alteration, and, as a whole, any act or omission affecting negatively the basic composition, behavior, and natural potential of environment; threatening land genetic viability or affecting life, health, integrity, or development of man, plants, or animals. These environmental components and environmental degradation factors were considered in the analysis of rules to determine protection concept granted by the Chilean law to the rational use of wetlands. Thus, the aim of this work is to develop a multicriteria methodology to quantify the level of legal protection that different types of inland wetlands have in Chile.

2. Methods

2.1. Compilation and Analysis of Current Standards

Compilation and analysis of current legislation related to wetlands and water resources, including legal standards with description and characterization, was carried out, consulting different sources of reference (e.g., CONAMA 1996a, 1996b, 1997a, 1997b, 2009, Castillo 1994, Gallardo 1985 Hermosilla 2004, Ortiz 1986, 1990, Valenzuela 1994, Fernández 2004, Bravo 2010). Also, the website of the Library of Congress of Chile was reviewed (www.leychile.cl), which has a compendium of environmental legal standards and acts classified by topics. The review covered Chilean regulations until April 2012.

The identification of the corresponding legal standards was based on the interest of this study (e.g., fresh water fauna, water resources, wetlands, watersheds), selecting standards aimed at protecting and preserving aspects and components involved in direct or indirect conservation of inland wetlands. Legal standards in force which restrict conservation and protection of inland wetlands, as well as issues related to functionality and biodiversity, were also identified. Selected rules were hierarchically ordered according to Gallardo (1985).

2.2. Multicriteria Method for Legal Protection Assessment

Wetland legal protection will tend to protect and/or conserve, either entirely, in components or considering the influence areas. This protection was estimated with a multicriteria methodology developed according to the following procedure (**Figure 1**): (a) compilation and analysis of existing legislation related to inland wetland issues (aspects related to biodiversity conservation and rational use included); also, current regulations which restrict conservation and rational use of inland wetlands were identified; (b) estimation of protection value (PV) provided by each legal regulation that considers applicability and standard hierarchy; (c) estimation of restriction value (RV) provided by each legal regulation instrument that also considers applicability and legal standard hierarchy; and (d) finally, estimation of real protection value (RPV) by summing protection values (PVs) of each legal standard applied to wetland of interest and restriction values (RVs) applied to the same type of wetland.

2.3. Protection Value (PV) Calculation

Protection value considers applicability and legal standard hierarchy (**Figure 1**). Applicability (A) is given as expressly enacted for wetland conservation purposes (value 3); or if secondarily, it could be considered as it refers to conservation of some environmental components emphasizing the aquatic system as a natural system (value 2), or if it is only tangential or generically involved with conservation of any component without reference to the natural system (value 1). Legal standards applicability was assessed, considering the reference which protects or preserves. It was classified in three levels: a) wetland system, including reference to 'wetlands' 'aquatic environment', 'aquatic ecosystems', or 'biodiversity' concepts

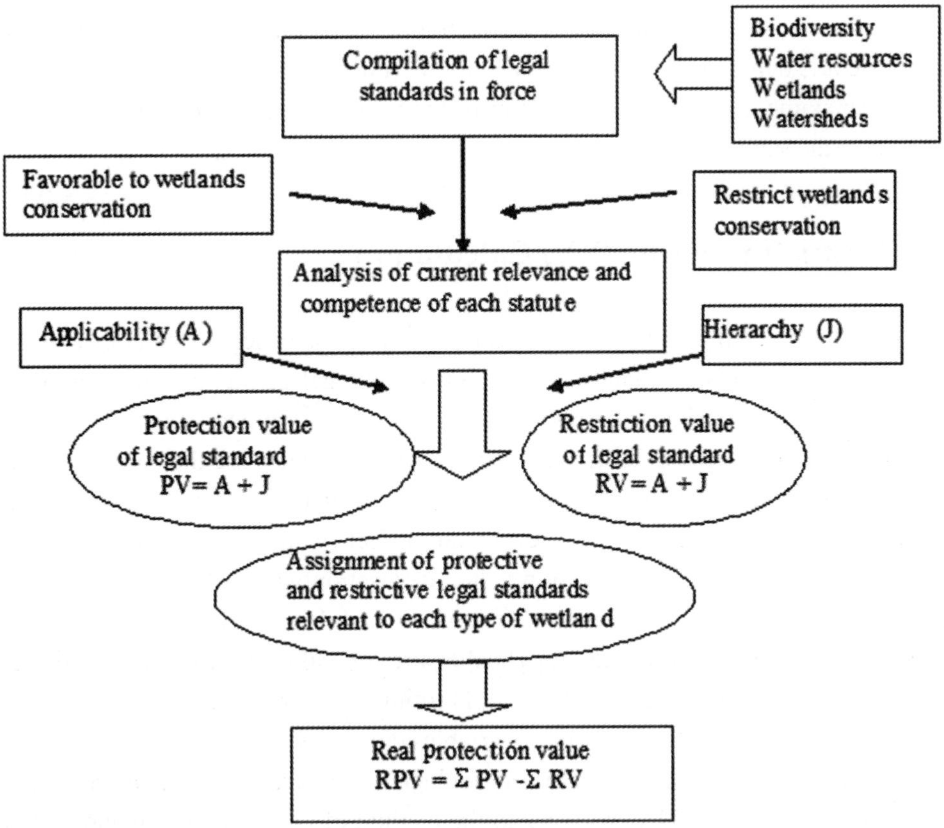

Figure 1. Methodological flow diagram to estimate value of legal protection for wetlands in Chile.

(in ecosystem level); b) components, e.g., water, flora, and plants, with reference to aquatic system (e.g., river or lake); and c) component, e.g., water, wildlife, flora, or vegetation, without reference to aquatic system. These three levels were assigned with values 3, 2, and 1 respectively, getting applicability of each of the legal regulations. Hierarchy (H) was assessed by assigning the maximum value (3) to legal regulations, an average hierarchy (value 2) to decrees with force of law and law decrees, and lower hierarchy (value 1) to standard rules (e.g., regulations, simple decrees, instructions, supreme decrees, resolutions). For details of this hierarchy, see Additional file 1.

Legal protection estimated value (PV) was calculated using the following formula: PV = A + J, where PV = legal protection value of each legal rule, A = applicability of standard to each wetland type, and J = hierarchy that the standard has in the law of Chilean legal system. Thus, the protection value (PV) of each legal instrument is deployed between 2 (PV = 1 + 1) (minimum protection) and 6 (PV = 3 + 3) (maximum protection). The sum of the variables considers that the weight of each is the same.

2.4. Restriction Value (RV) Calculation

It is estimated by analyzing protection restrictions of legal standards and the rational use of diverse wetlands. Applicability and legal hierarchy standards (**Figure 1**) were also considered. Applicability (A) was determined, considering what and how does it affect wetlands. The implementation effects are the complete wetland system destruction or any structural component (value 3); collateral effect (value 2) and whether it may only eventually affect (value 1). Hierarchy (H) was valued in samc way as thc protcction value. Restriction value (RV) was calculated as: RV = A + J, where RV = restriction value of each legal rule, A = statute applicability, and J = standard hierarchy. Thus, restriction value (RV) to each legal regulation is deployed between 2 (VR = 1 + 1) (minimum constraint) and 6 (RV = 3 + 3) (maximum constraint). The sum of variables considers the weight of each one to be the same. Integration of parameters included and their subsequent validation was done by a team of 11 experts using the method of Delphi (Linston and Turoff 1975). The group of experts included professionals from different disciplines, with over five years of professional experience, who in spite of not having legal formation, they were linked to these subjects. Criteria considered specialists in aquatic resources (4), relation with public institutions (3), and environmental management experience (5) (see shaping panel at Additional file 1). Work methodologies were workshops, documents, and questionnaire previously associated, which were duly submitted to experts to review and compile observations by using templates. Observations from experts were systematized and incorporated to a document which was forwarded for assent repeating procedure described to obtain consensus. The algorithm was applied to different types of inland wetlands of Chile.

2.5. Wetlands Studied

Six types of wetlands were selected based on representativeness and wetland they represented. Representation considered frequency, defined as abundance, in a survey of 1,215 inland wetlands (CEA/FIP 2010), with Scott and Carbonell records (1986); Schlatter *et al.* (2001); and Lopez-Lanus and White (2005), which was refined and grouped according to the type by Dugan (1992) and considerations by Ramsar (2000) and Ramirez *et al.* (2002). Based on the above, selected wetlands correspond to fluvial systems, lakes, and palustrine areas, which are: (a) rivers, (b) lagoons, (c) brackish Andean lakes, (d) highaltitude Andean peatlands called 'vegas' and 'bofedales', (e) freshwater swamp forests, and (f) permanent freshwater marshes and swamps also called locally as 'bañados' and 'ciénagas'.

3. Results

3.1. Legal Standards for Wetland Protection

Standards were selected either since they address wetland as entirety natural system or they refer to some constituting component, such as water (in quality), aquatic life, wildlife, and others, as well as the environmental and aesthetic roles. Same way, those referred to any basin component with impact on aspects above mentioned (conservation of watershed vegetation, woodland springs, and riparian zones) were also included. Those standards which currently or potentially damage wetlands or any of their components were also included. Rules were considered from current or potential application point of view. Forty-seven legal rules directly or indirectly related to the protection of wetlands were identified (for each rule purpose, see Additional file 1).

3.2. Protection Value (PV) of Legal Standards

Each legal standard protection value is shown in **Table 1**, which presents 47 legal rules with each potential protection value, according to applicability and hierarchy. It can be observed that they mostly (30 standards) have protection values 2 or 3, it means low; and 11 standards have intermediate protection value (value 4) while only six standards have higher protection values (values 5 and 6). The total protection value given by adding all protection standards is 144.

Table 2 presents legal protection value provided by legal standards to different types of inland wetlands in Chile. Zero values indicate that standard does

Table 1. Applicability (A) of legal standards affecting Chile's inland wetlands depending on what it protects or preserves, hierarchy (J) and protection value (PV) of each legal rule, as well as it total value protection.

Legal standard	What protects/preserves?	A	J	PV
Act N°19,300	Environment	3	3	6
DS N° 531 Washington Convention	Flora and fauna in protected area systems	1	1	2
DS N°141 CITES	Components (flora and fauna)	1	1	2
DL N° 3,485 Ramsar Convention	Systems (wetland)	3	2	5
DS N° 868 Migration of species	Component (fauna)	1	1	2
Decree N° 1,963 Convention in biodiversity	Biological diversity	3	1	4
Act N° 11,402 Barrier works and standardization of bank and channel	Reforestation river basin (headwaters)	2	3	5
Act N° 17,288 National Monuments	Sites (Research interests)	1	3	4
Act N° 18,892 Fishery and Agricultural Standards	Components (fauna, such as hydro-biological resources (fisheries and aquaculture))	2	3	5
Act N° 18,902 Sanitary health superintendency creation; liquid industrial residues control	Component (water quality for human purposes)	1	3	4
Act N° 19,473 Hunting Law	Components (fauna)	1	3	4
DFL N° 208 Provisions for development of fishing activities	Components (water quality, fish, and seafood)	1	2	3
DFL N° 701 About forest development	Components (soil for quality purposes/water quantity for irrigation and drinking)	1	2	3
Act N° 20.283 Native Forest Recovery and Forestry Development rule	Components (native forest associated to springs, natural bodies, and water courses)	3	3	6
Act N° 20,256 Recreational fishing standards	Ecosystem	3	3	6
Act N° 20,411 Prevents constitution of water rights	Components (aquifer)	1	3	4

Continued

DFL N° 725 Sanitary Health Legal Standards Art. 73	Components (quality/water quantity and human health)	1	2	3
DFL N° 1,122 Water Legal Standards	Components (water supply quantity for human use)	2	2	4
DS N° 1 Regulations for water pollution control	Components (water supply quantity for human use)	1	1	2
DS N° 5 Hunting law approval	Components (fauna)	1	1	2
DS N° 29 Approval of classification standards according to wild species conservation conditions	Components (wild species)	1	1	2
DS N° 30 Law 19,300	Systems (wetland)	3	1	4
DS N° 46 Groundwater emission standards	Components (water supply quantity for human use)	1	1	2
DS N° 33 Classification process of species according to conservation condition	Components (wild species)	1	1	2
DS N° 41 Classification process of species according to conservation condition	Components (wild species)	1	1	2
DS N° 42 Classification process of species according to conservation condition	Components (wild species)	1	1	2
DS N° 51 Classification process of species according to conservation condition	Components (ichthyofauna species)	1	1	2
DS N° 82 Soil, water, and wetlands legal standards Act N° 20,283	Systems (bodies and natural waterflows)	3	1	4
DS N° 90 Standards for liquid residues discharge into surface freshwater	Components (water quality in four types of use)	1	1	2
DS N° 95 Environmental Quality Standards Law	Components (water quality for human and environmental health)	2	1	3
DS N° 93 (2009) Native Forest Recovery Law	Components (native forest with soil, water, and wetland protection areas)	3	1	4
DS N° 193 Law Decree 701 General rules	Components (soil, courses and water bodies, flora and fauna)	1	1	2

Continued

DS N° 210 Reforestation and recreational fishing standards	Components (ichthyofauna)	1	1	2
DS N° 236 Standards for disposals not discharged to sewage network	Components (water quality for human consumption)	1	1	2
DS N° 238 Park and marine reserves standards (including inland waters)	Systems (wetland)	3	1	4
DS N° 320 Aquaculture environmental standards	Systems (river and lake)	2	1	3
DS N° 430 Reviewed document for fisheries and aquaculture Law	Systems (river and lake)	2	1	3
DS N° 594 Health and work place environmental condition standards	Components (groundwaters and surface waters)	1	1	2
DS N° 351 Industrial liquid residues discharge depuration and neutralization standards	Components (groundwaters and surface waters)	1	1	2
DS N° 609 Standards for industrial liquid residues to sewerage	Components (surface water quality)	1	1	2
DS N° 4,363 Forest law approval	Components (native plants associated to wellsprings)	2	1	3
DS N° 2,374 Standards for forest exploitation in forest watersheds and dams	Components (forests in watersheds)	1	1	2
Decree N° 878 Prohibition to extractive native fish in inland waters	Components (ichthyofauna)	1	1	2
NCh N° 1,333/87 Chilean standard for water quality	Components (water quality for aquatic life)	2	1	3
Secondary standards for environmental quality	Components (water quality for aquatic life)	3	1	4
Resolution N° 425 Groundwater exploration and exploitation standards	Components (water quality for aquatic life)	2	1	3
Resolution N° 197 Loa River reduction to new exploitation feeding	Components (water)	2	1	3
Total protection value				144

DS Supreme Decree, DL Law Decree, DFL Decree with Force of Law, NCh Chilean Standard, CITES Convention on International Trade in Endangered Species of Wild Fauna and Flora.

Table 2. Protection value and number of legal standards to different types of inland wetlands in Chile.

Legal standards/ protection value	VP potencial wetlands	Rivers	Brackish Andean lakes	Lagoons	High-altitude Andean peatlands	Freshwater swamp forests	Permanent freshwater marshes and swamps
Act N° 19,300	6	6	6	6	6	6	6
DS N° 531 Washington Convention	2	2	2	2	2	2	2
DS N° 141 CITES	2	2	2	2	2	2	2
DL N° 3,485 Ramsar Convention	5	5	5	5	5	5	5
DS N° 868 Migratory species (Bonn Convention)	2	2	2	2	2	0	2
Decree N° 1,963 Convention on Biodiversity	4	4	4	4	4	4	4
Act N° 11,402 Barriers and river banks and streams	5	5	0	0	0	0	0
Act N° 17,288 National Monuments	4	4	4	4	4	4	4
Act N° 18,892 Aquaculture and Fishing Law	5	5	0	5	0	0	0
Act N° 18,902 Sanitary health superintendency creation; industrial liquid residues control	4	4	4	4	4	4	4
Act N° 19,473 Hunting Law	4	4	4	4	4	4	4
DFL N° 208 Provisions to fishing development	3	3	0	3	0	0	0

Continued

DFL N° 701 About forest development	3	3	0	3	0	3	3
Act N° 20,283 Native Forest and Forest Development Recovering Law	6	6	0	6	0	6	6
Act N° 20,256 Recreative fishing standards	6	6	0	6	0	0	0
Act N° 20,411 Prevent constitution water rights (Arica/Parinacota a O'Higgins)	4	4	4	4	4	4	4
DFL N° 725 Sanitary CodeArt. 73	3	3	0	3	0	0	0
DFL N° 1,122 Water Code	4	4	4	4	4	4	4
DS N° 1 Aquatic pollution control rules	2	2	2	2	0	0	0
DS N° 5 Hunting law standard approval	2	2	2	2	2	2	2
DS N° 29 Approval of classification of species standards based on conservation conditions	2	2	2	2	2	2	2
DS N° 30 Act 19,300 Standards	4	4	4	4	4	4	4
DS N° 46 Groundwater discharge standards	2	2	2	2	2	2	2
DS N° 33 Classification process of species according to conservation condition	2	2	0	0	0	0	0
DS N° 41 Classification process of species according to conservation condition	2	2	0	2	2	2	2

Continued

DS N° 42 Classification process of species according to conservation condition	2	2	0	2	0	2	2
DS N° 51 Classification process of species according to conservation condition	2	2	2	2	2	0	2
DS N° 82 Act No. 20,283 Soil, water, and wetland regulations	4	4	4	4	4	4	4
DS N° 90 Standards to liquid residue discharge into surface fresh water	2	2	2	2	0	0	0
DS N° 93 (1995) Primary and secondary environmental quality standards	3	3	3	3	3	3	3
DS N° 93 (2009) Native Forest Recovery and Forestry Development rules	4	4	0	4	0	4	4
DS N° 193 DL N° 701 General rules	2	2	0	2	0	2	2
DS N° 210 Regulation of reforestation and planting for recreational fishing	2	2	0	2	0	0	0
DS N° 236 Regulation not discharged sewage disposal to sewer networks	2	2	2	2	2	2	2
DS N° 238 Regulation for parks and marine reserves (including inland waters)	2	2	2	2	2	2	2
DS N° 320 Environmental aquaculture rules	2	2	0	2	0	0	0

Continued

DS N° 430 Consolidated, coordinated, and systematized document on fishery law	3	3	0	3	0	0	0
DS N° 594 Rules for health and environmental conditions in work places	2	2	2	2	2	2	2
DS N° 351 Rules for neutralization and depuration of liquid industrial residues	2	2	2	2	2	2	2
DS N° 609 Standard for sewerage industrial liquid residues discharge	2	2	0	2	2	2	2
DS N° 4,363 Approves final text forestry law	3	3	0	3	0	3	3
DS N° 2,374 Regulation to forest exploitation in forest watersheds and dams	2	2	0	2	0	2	2
Decree N° 878 Native fish extraction prohibition in inland waters	2	2	2	2	2	2	2
NCh N° 1,333/87 Chilean water quality standards	3	3	0	3	3	3	3
Secondary environmental quality standards	4	4	0	4	0	0	0
Resolution N° 425 Establish groundwater exploration and exploitation standards	3	3	3	3	3	3	3
Resolution N° 197 Declares Loa River reduction and feeders for new farms	3	0	3	0	3	0	0
Total standards	47	46	27	44	28	32	34
Total legal protection value	144	141	80	134	83	98	102

DS Supreme Decree, DL Law Decree, DFL Decree with Force of Law, NCh Chilean Standard, CITES Convention on International Trade in Endangered Species of Wild Fauna and Flora.

not apply. From 47 standards identified, 46 of them apply to rivers, providing total protection value of 141; 27 apply to brackish Andean lakes, providing total protection value of 80; 44 apply to lagoons, providing total protection of 134; 28 apply to peatlands ('vegas' and 'bofedales') with total protection of 83; 32 apply to swamp forests, providing total protection of 98; and 34 apply to marshes ('bañados' and 'ciénagas'), giving full protection of 102. As shown in **Table 2**, greater legal protection is provided in descending order: to rivers, lagoons, marshes, swamp forests, and finally, peatlands and brackish Andean lakes have the lowest legal protection values.

3.3. Restriction Value (RV) of Legal Standards

Eight legal restriction standards considered affecting conservation, and rational use of wetlands (**Table 3**) was identified. Some of these standards are also included in the protection regulations since their articles incorporate both protection and restriction issues for wetland conservation. Legal restriction applicability was assessed considering what and how it affects the wetland system. In this case, unlike the rules of protection, three applicability levels were not identified, since every effect caused by standards are supposed to affect the entire system or biological diversity, which was also considered at ecosystem level, so all were assigned with value 3. **Table 3** also shows restriction value to conservation or sustainable use that provides laws to wetlands, and **Table 4** shows their application to different wetland types considered in this study. Zero values indicate that standard does not apply.

Restriction to sustainable use of wetlands established by legal standards to different types of inland wetlands reveals that the eight identified standards provide a total restriction value of 42. Wetlands mostly affected are rivers and swamp forests, both with five standards, and restriction values of 28 and 25, respectively, followed by lagoons with three rules and restriction value of 18 and marshes with two rules and a restriction value of 12. Least affected are brackish Andean lakes and peatlands to which a single standard applies, giving restriction value 6.

3.4. Real Protection Value (RPV)

Real protection value (RPV) was estimated once the number of legal rules

Table 3. Legal restriction standards which affects the rational use of inland wetlands in Chile, applicability (A) in terms of how and what affect to wetlands, hierarchy (J) and restriction value (RV) of each legal standard and total restriction value.

Legal standard	Standard objective	How affect	What affect	A	J	RV
Act N° 18,248 Mining Law	Property acquisition of a mining concession enables to request the mining right's constitution and water right ownership. Written permission by governor is required for mining operations at lower distance than 50 yards in flood barriers, waterways, and lakes for public use.	Intervention to wetland systems, which may completely affect structure	Every type of wetlands (rivers) and lakes	3	3	6
Act N° 20,256 Standards for recreational fishing	Development of recreational fishing activity, and the economic and associated tourist activities.	Introduction of alien species affects biodiversity and system as a whole	Rivers and lakes	3	3	6
Act N° 11,402 Barriers and standardization of banks and river flows	Barrier works as well as standardization of banks and beds of rivers, lagoons and tidelands carried out with legal involvement can only be executed and planned by sanitary authorities of Public Works Ministry.	Regularization of channels and gravel mining extraction affecting the aquatic system structure	Rivers, lakes, and streams	3	3	6
	Extraction of gravel and sand from beds of rivers and streams should be carried out with municipality permits, which may charge duties or subsidies.					
Act N° 18,450 Rules for private irrigation and drainage development	Cost analysis benefits, construction, and rehabilitation of irrigation as well as drainage works to enable agricultural soils with poor drainage.	Drainage of wetlands results in complete destruction system	Swamp forests, marshes, and swamps	3	3	6
DFL N° 235	Incentive system to degraded soil recovering, proposing regulation of channels as management practices.	Regulation of channels affects the aquatic system structure	Water flows (rivers)	3	2	5

Continued

DFL N° 701	Afforestation development to soil recovery, making economic exploitation available.	Forest development results in swamp forest destruction (replacement by exotic species)	Swamp forests	3	2	5
Decree N° 193 General standards	Assess lands suitable to forestry for quality forest development benefits, those corresponding to Ñadis soils.	Development of forest Ñadis soils results in destruction, affecting swamp forest	Swamp forests	3	1	4
DFL N° 701		(replacement by exotic species)				
Decree N° 98 Act N° 18,450	Defined as drainage of structures, elements, and tasks aimed to evacuate soil waters excess on surface or subsurface with restriction factor to culture development.	Wetland draining results in complete destruction of system	Swamp forests	3	1	4
TOTAL RESTRICTION VALUE						42

DFL Decree with Force of Law.

applied to each type of wetland is established (**Figure 1**). This value is obtained by summing protection values (PVs), given by all legal standards applied to each type of wetland; and subtracting restriction values (RV), given by all legal standards that apply to the same wetland, as follows: RPV = ΣPV − ΣRV. Distribution of values is grouped into three ranges, giving following nominal values: value 1 to less protected wetlands, value 2 to intermediate protected ones, and value 3 to the best legally protected wetlands. The real protection value (RPV) given by the set of laws applied to different types of wetlands in the study is presented in **Table 5**. This (RPV) is deployed between 73 and 116. These values were grouped into ranges to where numeric values between 1 and 3 (**Table 6**) were assigned, being 3 the highest protection value and 1 the lowest protection value.

Real protection provided by laws is higher (value 3) in rivers and lagoons, intermediate (value 2) in marshes, and low (value 1) in brackish Andean lakes, peatlands, and swamp forest. So, the best protected wetlands, according to the

Table 4. Restriction value and number of legal standards to different inland wetlands in Chile.

Legal standards	Wetland potential	Rivers	Brackish Andean lakes	Lagoons	High-altitude Andean peatlands	Freshwater swamp forests	Permanent freshwater marshes and swamps
Act N° 18,248 Mining Code	6	6	6	6	6	6	6
Act N° 20,256 Recreational fishing standards	6	6	0	6	0	0	0
Act N°11,402 Barriers and works for bank and channel standardization	6	6	0	6	0	0	0
Act N° 18,450 Standards for private investment in watering and drainage systems	6	0	0	0	0	6	6
DFL N° 235	5	5	0	0	0	0	0
DFL N° 701 Forest development	5	0	0	0	0	5	0
Decree N° 193 General rules DL N° 701 About forest development	4	0	0	0	0	4	0
Decree N° 98 Act N° 18,450 Development of private investment in watering and drainage systems	4	0	0	0	0	4	0
Total number of standards	8	5	1	3	1	5	2
Total value of legal restrictions	42	28	6	18	6	25	12

DL Decree Law, DFL Decree with Force of Law.

Table 5. Nominal and numerical real protection granted by legal rules to different types of inland wetlands in Chile.

Type of wetland	Tota protection value	Protection value	Total restriction standards	Restriction value	Real protection value	Numeric protection value
Wetlands	47	144	8	42	102	3
Rivers	46	141	5	28	113	3
Brackish Andean lakes	27	80	1	6	74	1
Lagoons	44	134	3	18	116	3
High-altitude Andean peatlands	28	83	1	6	77	1
Swamp forests	32	98	5	25	73	1
Permanent freshwater marshes and swamps	34	102	2	12	90	2

Table 6. Nominal and numerical protection value granted by law to different types of wetlands.

Nominal value	Rate	Real protection number value	Type of wetland
Low protection	≤87	1	Brackish Andean lakes
			High-altitude Andean peatlands
			Swamp forests
Intermediate protection	88–102	2	Permanent freshwater marshes and swamps
High protection	≥103	3	Rivers and lagoons

number and legal hierarchy affecting direct or indirect protection, are rivers and lagoons. Wetlands with intermediate protection category are marshes and peatlands. Finally, the least protected are brackish Andean lakes, peatlands, and swamp forests.

4. Discussion

General review of legal standards affecting conservation of wetlands in

Chile reveals that before year 2012, there were no laws addressed to wetlands as ecosystem, since they only involved some components (water, flora, wildlife). The first reference to wetland concept in the Chilean law, except acts related to Ramsar Convention adhesion enacted in 1981, appears in 1994 with Act No. 19,300, one time mentioned.

Same way, wetland definition arises for the first time in 2011 by act of Supreme Decree N°82, Soil, water, and Wetlands Regulations Act No. 20,283. This is the evidence of the lack of consideration and the late inclusion of these ecosystems in the country regulations.

There is a different legal protection to different types of inland wetlands since current legislation in Chile does not provide equal protection to the different types of wetlands. Lagoons and rivers have the highest legal protection values due to the large number of regulations applied favorably to these systems and fewer standards that affect them adversely. For marshes, the intermediate value of legal protection comes from that; in spite, there are a number of regulations favoring conservation, and there are a few restrictions affecting them.

Andean systems (brackish lakes and peatlands) present different condition; since in spite a few restriction standards affecting them, they are poorly protected by current legislation. Swamp forest is the most dramatic case; which in spite of important existing regulatory standards that should protect them, they are the most affected ones by the restriction rules. The greatest impacts are Act No. 18,450 which promotes irrigation and drainage and Decree Law No. 701 about forest development. These regulations subsidize up to 75% of activities which enable the poorly drainaged lands to become agricultural territory, which means the main threat to swamp forests of coastal watersheds in La Araucanía region (Urrutia 2005, García 2005) and its evolution towards greater fragmentation in recent years (Peña-Cortés *et al.* 2011).

Legal standards above mentioned are clear perverse incentives and disincentives to the rational use of wetlands in Chile.

One aspect considered relevant in the Ramsar Convention was to assess effectiveness of legislative and institutional measures related to promote the conservation and the rational use of wetlands, are sectoral legal and institutional measures affecting wetlands such as financial and tax incentives to convert them. Nelson (1986) noted that wetland policies in the US and UK had administrative divisions with antagonistic competences. So, while the US Department of Agriculture provided incentives to drainage, the Interior Department was promoting conservation of wetlands, same way that the Ministry of Agriculture, Forestry, and Fisheries and the Nature Preservation Council faced each other in the UK. This problem was addressed in the US by removing some incentives for wetland drainage, which also included other conservation and education measures (Dahl 2000), reducing by 80% the wetland loss tendency between 1986 and 1997 compared to previous decade. Apparently, some Latin American countries have also advanced on this subject. For example in Ecuador, the legislation review by Echeverría (2008) about wetland management does not identify measures that indirectly support the loss or degradation of wetlands through negative incentives.

Review of some national reports presented by countries as contracting party in the last conference of the Ramsar Convention, held in Bucharest, Romania in 2012[a] (COP11), allowed to know measures that countries reported as implemented regarding incentives which promote the conservation and the rational use of wetlands, as well as those intended to remove perverse incentives which discourage them.

USA is the American country with the highest amount of initiatives on this subject. Report to COP11[b] indicates several provisions that discourage conversion of wetlands to cropland. The federal Swampbuster policy, provision of Food Security Law dated 1985, has eliminated the policy incentives and other mechanisms which have technically and economically allowed the destruction of wetlands, acting and separating those who make it from other agricultural policy benefits. Federal efforts to restore wetlands have increased from 1987, with legal standards of critical preservation and restoration with two programs which stopped and even reversed loss as Wetland Conservation (WC) provisions of Agricultural Law in 1985 and the Wetlands Reserve Program (WRP) of Agricultural Law in 1990 (Coperland 2010).

[a]http://www.ramsar.org/library/field_documents%253Afield_language/english-1/field_tag_body_event/cop11-bucharest-2012-415.
[b]http://www.ramsar.org/sites/default/files/documents/pdf/cop11/nr/cop11-nr-usa.pdf.

In national report, Canada[c] indicates that measures to eliminate perverse incentives in some provinces consist of wetlands filling prevention rules which enable marginal agricultural soils and balance benefits provided by wetland drain restoration projects with cleaning, storage, and discharge of water functions fulfilled by wetlands. The UK[d] works reforms to agricultural policy to have a positive impact on marsh wetlands used for grazing by agri-environment patterns, management agreements, and other keeping administration systems.

The China[e] report indicates that diverse level governments and their departments have a strict control over every kind of activity that damages wetlands by enacting 11 provincial (autonomous regions) wetland conservation regulations, which prohibit several activities and perverse incentives. People convocation has taken the surveillance and control in the implementation of those laws at various levels. Australia[f] has already indicated in the COP10 national report 2008 about the National Water Initiative, Water for Future, and Water Law 2007 developed to stop perverse incentives in the water management. Same way, New Zealand[g] declared in national report that they have eliminated subsidies to land development.

In South American countries[h], Bolivia, Colombia, Peru, and Chile state, they did not take measures to promote incentives to encourage conservation and rational use of wetlands. Countries showing incentives are Costa Rica by promoting private refuges, territory tax payment exemption, and payment for environmental services; Ecuador by applying tax exemption to rural land containing wetlands and monetary incentives to private owners and community which preserve Andean wetlands; and Argentina and Venezuela by providing financial incentives for conservation and planning in land management.

Regarding measures to remove perverse incentives, Bolivia, Brazil, Costa Rica, and Peru declare that they have not taken them. Ecuador and Venezuela note that they count with these measures. Uruguay, Colombia, and Chile indicate that these measures are being planned and involve payments for environmental services and penalties for damage to the ecosystem. They consider compensatory wet-

[c]http://www.ramsar.org/sites/default/files/documents/pdf/cop11/nr/cop11-nr-canada-e.pdf.
[d]http://www.ramsar.org/sites/default/files/documents/pdf/cop11/nr/cop11-nr-uk.pdf.
[e]http://www.ramsar.org/sites/default/files/documents/pdf/cop11/nr/cop11-nr-china.pdf.
[f]http://www.ramsar.org/sites/default/files/documents/pdf/cop11/nr/cop11-nr-australia.pdf.
[g]http://www.ramsar.org/sites/default/files/documents/pdf/cop11/nr/cop11-nr-newzealand.pdf.
[h]http://www.ramsar.org/library/field_documents%253Afield_language/espa%C3%B1ol-4/field_tag_body_event/cop11-bucharest-2012-415/field_tag_countries/neotropics-15?search_api_views_fulltext=&items_per_page=20.

land restoration. Analysis of measures reveals that these have to do with the creation of new incentives rather than eliminating those perverse incentives which favor, fill, and convert wetlands to agricultural soils or urban with bonus or by other means (see national reports of countries mentionedi).

Based on the formerly mentioned, three actions must be implemented in Chile to improve conservation and rational use of wetlands: 1) Repeal rules which promote destruction of wetlands through incentives. 2) Discuss a special law for proper management and conservation of wetlands, including tax incentives and the corresponding regulation to operate. 3) Prioritize conservation actions for wetland types with low protection value such as swamp forests, peatlands, and brackish Andean lakes.

5. Conclusions

In Chile practically, there are no specific legal regulations on wetlands; there are only some decrees which applied indirectly to these ecosystems, addressed only to some components (water quality and quantity and wild species), without ecosystem approach. Current legal regulations in Chile do not allow a proper protection of wetlands and rational use as mandated by the committed country as a member of the Ramsar Convention. In Chile, to improve wetland conservation conditions, every incentive promoting wetland destruction must be eliminated; promulgate specific regulations for proper management and conservation of wetlands including tax incentives and corresponding operating regulations; and favor conservation actions to less protected wetland, such as swamp forests, peatlands, and brackish Andean lakes.

Additional File

Additional file 1: Complementary background. http://revchilhistnat.biomedcentral.com/articles/10.1186/s40693-014-0023-1.

Competing Interests

The authors declare that they have no competing interests.

Authors' Contributions

PM developed the methodology, conducted legal review and drafted the manuscript. AM participated in the design of the methodology, in writing and revision of the manuscript. Both authors read and approved the final manuscript.

Acknowledgements

We thank to projects: management and restoration measures of wetland ecosystems in southern Chile, Ministry of Environment (2011-2012); identification of potential areas to establish reserves aimed to the protection of native fauna of freshwater aquatic species, FIP No. 2008-58, Fisheries Research Fund (2009-2010); analysis of biodiversity in Antofagasta Region (2007-2008) CONAMA, FNDR, all of them executed by Center for Agricultural and Environmental Studies. Also, we thank Charif Tala (Ministry of Environment) for his comments and suggestions to methodology. AMP appreciates the contribution of the General Direction for Research and Graduate Studies at the Catholic University of Temuco, DGIPUCT project No. CD2010-01 and Project 0804 MECESUP UCT.

Source: Möller P, Muñoz-Pedreros A. Legal protection assessment of different inland wetlands in Chile[J]. Revista Chilena De Historia Natural, 2014, 87(87):1–13.

References

[1] Abell R (2002) Conservation biology for the biodiversity crisis: a freshwater follow-up. Conserv Biol 16:1435–1437.

[2] Abell R, Allan JD, Lehner B (2007) Unlocking the potential of protected areas for freshwaters. Biol Conserv 134:48–63.

[3] Asenjo R (2006) Institucionalidad pública y gestión ambiental en Chile. Expansiva, serie. In: Foco., p 91.

[4] Bahamonde N, Carvacho A, Jara C, López M, Ponce F, Ma Retamal E, Rudolph E (1998) Categorías de conservación de decápodos nativos de aguas continentales de Chile. Bol Mus Nac Hist Nat 47:91–100.

[5] Barbier E, Acreman M, Knowler D (1997) Valoración económica de los hume-

dales. Guía para decidores y planificadores. Oficina de la Convención de Ramsar, Gland, Suiza.

[6] Brañes R (2000) Manual de Derecho Ambiental. Fundación Mexicana para la educación Ambiental, Editado por Fondo de Cultura Económica, 2° edición. Ciudad de México.

[7] Bravo D (2010) Conservación y preservación de los humedales en Chile. Justicia ambiental: Revista de Derecho Ambiental de Fiscalía del Medio Ambiente/FIMA 2:91–158.

[8] Castillo M (1994) Régimen jurídico de protección del medio ambiente. Aspectos generales y penales. Segunda edición. Ediciones Bloc, Santiago de Chile.

[9] CEA/FIP (2010) Estudio identificación de áreas potenciales para establecer reservas destinadas a proteger la fauna nativa de especies hidrobiológicas de agua dulce Centro de Estudios Agrarios y Ambientales/ Fondo de Investigación Pesquera, Informe Final.

[10] CONAF-CONAMA (1997) Catastro y evaluación de los recursos vegetacionales nativos de Chile. Editado por CONAF/CONAMA, Chile.

[11] CONAMA (1996a) Aire, flora, fauna, áreas silvestres protegidas, procedimientos administrativos ambientales, Documento 13, Serie Jurídica.

[12] CONAMA (1996b) Diversidad biológica, identificación y diagnóstico preliminar del ordenamiento jurídico aplicable a la protección de la diversidad biológica, Documento 16, Serie Jurídica.

[13] CONAMA (1997a) Áreas silvestres protegidas, legislación punitiva y sancionatoria ambiental, Documento 27, Serie Jurídica.

[14] CONAMA (1997b) Áreas silvestres protegidas, legislación sancionatoria ambiental, Documento 28, Serie Jurídica.

[15] CONAMA (2009) Ley sobre bases generales del medio ambiente y sus reglamentos. Editado por Comisión Nacional del Medio Ambiente, Santiago, Chile Coperland C (2010) Wetlands: an overview of issues. Congressional Research Service Reports. Paper 37.

[16] Dahl Te (2000) Status and trends of wetlands in the conterminous United States 1986 to 1997. U.S. Department of the Interior, Fish and Wildlife Service, Washington, D.C.

[17] Díaz-Paéz E, Ortiz J (2003) Evaluación del Estado de conservación de los anfibios en Chile. Rev Chil Hist Nat 76:509–525.

[18] Dudley N (ed) (2008) Directrices para la aplicación de las categorías de gestión de áreas protegidas. UICN Gland, Suiza.

[19] Dugan P (1992) Conservación de Humedales. Un análisis de temas de actualidad y acciones necesarias. UICN, Suiza.

[20] Echeverría H (2008) Proyecto Capacitación en base de una Revisión de la Legislación e Institucionalidad relacionada con la Gestión de Humedales en el Ecuador (WFF/06/EC/2), Centro Ecuatoriano de Derecho Ambiental/ Ministerio del Ambiente del Ecuador.

[21] Fernández P (2004) Manual de Derecho Ambiental Chileno. Editorial Jurídica de Chile.

[22] Figueroa R, Palma A, Ruiz V, Niell X (2007) Análisis comparativo de índices bióticos utilizados en la evaluación de la calidad de las aguas en un río mediterráneo de Chile: río Chillán, VIII Región. Rev Chil Hist Nat 80(2):225–242.

[23] Gallardo E (1985) Legislación sobre protección de la flora arbórea y arbustiva nativa chilena. In: Benoit IL (ed) Libro rojo de la flora terrestre de Chile (Primera Parte). Corporación Nacional Forestal, Santiago de Chile.

[24] García A (2005) Análisis estructural y estado de conservación de los rodales costeros de temu y pitra, entre Imperial y Queule, propuestas para su conservacion. Tesis, Universidad Católica de Temuco, Chile.

[25] Hermosilla J (2004) Tratados Internacionales vigentes en Chile en materia de protección a la Biodiversidad y su relación con la legislación interna. Facultad de Ciencias Jurídicas y Sociales. Escuela de Derecho Universidad Austral de Chile, Valdivia, Chile.

[26] Iza AO, Rovere MB (2006) Gobernanza del agua en América del Sur: dimensión ambiental. UICN, Gland, Suiza y Cambridge, Reino Unido.

[27] Linstone H, Turoff M (1975) The Delphi Method. Techniques and Applications. Addison-Wesley, Boston, United States of America.

[28] López-Lanús B, Blanco DE (2005) El Censo Neotropical de Aves Acuáticas 2004. Global Series No. 17. Wetlands International, Buenos Aires, Argentina.

[29] Marín V, Delgado L, Vila I (2006) Sistemas Acuáticos, ecosistemas y cuencas hidrográficas. In: Vila I, Veloso A, Schlatter R, Ramírez C (eds) Macrófitas y vertebrados de los sistemas límnicos de Chile. Editorial Universitária, Santiago de Chile.

[30] Mitsch WJ, Gosselink JG (2000) Wetlands third edition. John Wiley, New York.

[31] Muñoz-Pedreros A (2004) Los humedales del río Cruces y la Convención de Ramsar. Un intento de protección fallido. Gest Amb 10:11–26.

[32] Nelson RW (1986) Wetlands policy crisis: United States and United Kingdom. Agr Ecosyst Environ 18:95–121.

[33] Novitzki R, Smith RD, Fretwell JD (1996) Wetland Functions, Values and Assessment. In: Fretwell JD, Williams JS, Redman PJ (comp). National water summary on wetland resources. United States Geological Survey. Water Supply Paper 2425. Washington, D.C.

[34] Olivares A (2010) El nuevo marco institucional ambiental en Chile. Revi Cat Dret Amb 1(1):1–23.

[35] Ortiz S (1986) Legislación vigente en el ámbito del manejo de cuencas hidrográficas. Corporación Nacional Forestal Octava Región, Concepción, Chile.

[36] Ortiz S (1990) La actividad forestal y su regulación ambiental. Seminario CORMA sobre Legislación forestal y su dimensión en el derecho ambiental, Concepción, Chile.

[37] Peña-Cortés FA, Gutierrez PL, Rebolledo GD, Escalona MA, Bertran CE, Schlatter R, Tapia J (2006) Determinación del nivel de antropización de humedales como criterio para la planificación ecológica de la cuenca del lago Budi, Chile. Rev Geog Nor Gr 36:75–91.

[38] Peña-Cortés FA, Pincheira-Ulbrich J, Escalona-Ulloa M, Rebolledo G (2011) Cambio de uso del suelo en los geosistemas de la cuenca costera del río Boroa (Chile) entre 1994 y 2004. Rev Fac Cienc Agr Univ Nac Cuyo 43(2):1–20.

[39] Pérez-Losada M, Jara CG, Bond-Buckup G, Crandall KA (2002) Conservation phylogenetics of Chilean freshwater crabs Aegla (Anomura, Aeglidae): assigning priorities for aquatic habitat protection. Biol Conserv 105:345–353.

[40] Ramírez C, San Martín C, Rubilar H (2002) Una propuesta para la clasificación de los humedales chilenos. Rev Geog Valp 33:265–273.

[41] Ramsar (2000) Marcos para manejar humedales de importancia internacional y otros humedales. Con comprendidos de los Lineamientos adoptados por la Conferencia de las Partes Contratantes en sus reuniones 4a.,5a.,6a., y 7a. Oficina de la Convención de Ramsar, Gland, Suiza.

[42] Rodríguez-Becerra M, Espinoza G (2002) Gestión ambiental en América Latina y el Caribe: evolución, tendencias y principales prácticas. In: David W (ed) Inter Division de Medio Ambiente, Banco Interamericano de Desarrollo.

[43] Rojas M (ed) (2011) Cuadernos de Análisis Jurídico, Colección Derecho Ambiental Programa de Derecho y Política Ambiental. Facultad de Derecho. Universidad Diego Portales, Santiago de Chile.

[44] Schlatter RP, Espinosa LA, Vilina Y (2001) Coasts of central and southern Chile (Region 15).Chapter 3.3.3. Los Humedales de América del Sur, una agenda para la conservación de su Biodiversidad y las Políticas de Desarrollo. In: Canevari P, Davidson I, Blanco DE, Castro G, Bucher EH (eds) Resumen Ejecutivo. Wetlands International: 14 y CD. Book.

[45] Scott D, Carbonell M (eds) (1986) Inventario de Humedales de la Región Neotropical. IWRB y UICN, Reino Unido.

[46] Urrutia O (2005) Estado de conservación de los bosques pantanosos y su relación con los ecosistemas asociados en el borde costero, entre Imperial y Queule, IX Región. Tesis, Universidad Católica de Temuco, Chile.

[47] Valdovinos C (2005) Moluscos terrestres y dulceacuícolas de la cordillera de La Costa chilena. In: Smith-Ramírez C, Armesto JJ, Valdovinos C (eds) Historia, biodiversidad y ecología de los bosques costeros de Chile. Editorial Universitaria,

Santiago de Chile.

[48] Valenzuela R (1994) Diagnóstico preliminar del ordenamiento jurídico vigente, en lo que se refiere a la protección de la diversidad biológica, Contrato N° 01-0007-001 preparado para la Comisión Nacional del Medio Ambiente, Santiago de Chile.

[49] Veloso A (2006) Batracios de las cuencas hidrográficas de Chile: origen, diversidad y estado de conservación. In: Vila I, Veloso A, Schlatter R, Ramírez C (eds) Macrófitas y vertebrados de los sistemas límnicos de Chile. Editorial Universitaria, Santiago de Chile.

[50] Vila I, Pardo R, Dyer B, Habit E (2006) Peces límnicos, diversidad, origen y estado de conservación. In: Vila I, Veloso A, Schlatter R, Ramírez C (eds) Macrófitas y vertebrados de los sistemas límnicos de Chile. Editorial Universitaria, Santiago de Chile.

[51] Woodward RT, Wui YS (2001) The economic value of wetland services: a meta-analysis. Ecol Econ 37:257–270.

[52] Zegers G, Larraín J, Díaz F, Armesto J (2006) Impacto ecológico y social, de la explotación de pomponales y turberas de Sphagnum en la isla Grande de Chiloé. Rev Amb Des 22:28–34.

Chapter 27

Industrial Air Pollution in Rural Kenya: Community Awareness, Risk Perception and Associations between Risk Variables

Eunice Omanga[1], Lisa Ulmer[2], Zekarias Berhane[2], Michael Gatari[3]

[1]Impact Research and Development Organization, Kisumu, Kenya
[2]School of Public Health, Drexel University, Philadelphia, USA
[3]Institute of Nuclear Sciences and Technology, University of Nairobi, Nairobi, Kenya

Abstract: Background: Developing countries have limited air quality management systems due to inadequate legislation and lack of political will, among other challenges. Maintaining a balance between economic development and sustainable environment is a challenge, hence investments in pollution prevention technologies get sidelined in favor of short term benefits from increased production and job creation. This lack of air quality management capability translates into lack of air pollution data, hence the false belief that there is no problem. The objectives of the study were to: assess the population's environmental awareness, explore their perception of pollution threat to their health; examine the association between specific health hazards. Methods: A cross-sectional study was implemented by gathering quantitative information on demographic, health status, environmental perception and environmental knowledge of residents to understand their view of pollution in their neighborhood. Focus group discussions (FGDs) allowed for corroboration of the quantitative data. Results: Over 80% of respondents perceived industrial pollution as posing a considerable risk to them despite the fact that the economy of the area largely depended on the factory. Respondents also argued that they had not been actively involved in identifying solutions to the environmental challenges. The study revealed a significant association between industrial pollution as a risk and, perception of risk from other familiar health hazards. The most important factors influencing the respondents' pollution risk perception were environmental awareness and family health status. Conclusion: This study avails information to policy makers and researchers concerning public awareness and attitudes towards environmental pollution pertinent to development and implementation of environmental policies for public health.

Keywords: Environmental, Perception, Industrial, Air, Pollution, Risk, Rural

1. Background

Most developing countries, especially in sub-Saharan Africa, do not have air quality management systems because of inadequate legislation, budgetary constraints and lack of political will, among other things[1]. In these regions, there is a major challenge in maintaining a balance between economic development and a sustainable environment; hence investments in pollution prevention technologies like emission controls are commonly outweighed by the short-term benefits that accrue from increased production and job creation. The lack of air quality management capabilities in these regions translate into lack of air pollution data, which in most cases gives the false belief that it is not a problem. This however, is not the case and only further conceals a major public health crisis in the developing world.

More and more people are concerned about environmental hazards and the resultant adverse health effects on humans and the environment at local, regional and global levels[2]-[5]. For example, since the publication of Rachel Carson's[6] book Silent Spring, public fear and concern over cancer from chemicals such as pesticide residues in food are on the increase. Pollution (air and water) adversely impact the environment and the effects frequently spread well beyond geographical borders. Since the establishment of the United Nations Environment program (UNEP) in 1972 with its headquarters in Nairobi, Kenya, national and international focus has been on environmental health effects of pollution, especially water pollution. While progress has been made in reducing sanitation related diseases like diarrhoea in the developing world, little has been done to combat the negative health effects of increased industrialization and resultant pollution.

A 2007 World Health Organization (WHO) report revealed huge inequalities on environmental impact on health in addition to demonstrating how public health could be improved by reducing environmental threats such as pollution, occupational hazards as well as climate and ecosystem changes[7]. This report revealed that globally, up to 13 million deaths could be averted annually through better environmental management and in some countries over 30% of the disease burden could be prevented. The WHO estimate of the burden of disease due to air pollution puts premature deaths attributed to air pollution at over 2 million, with residents of developing countries bearing half of this burden[8].

Pollution from industries negatively impacts the health of employees and neighbouring communities and the potential for adverse health outcomes is heightened when the industries are located in rural areas where the bulk of the population is vulnerable because of limited information about their rights and limited capacity to defend themselves or influence policy decisions. Rural communities are often overlooked by businesses and sometimes even the government. The deficient environmental health awareness coupled with lack of sustainable environmental health programs is a major challenge in most developing countries[8]. While majority of studies have been on urban populations, air pollution in rural communities is on the increase[9].

To date, estimates of air pollution health effects in the developing world rely mainly on data from research conducted in the developed world, specifically North America & Europe which is then extrapolated, yet the characteristics of the ambient pollutants and the environmental conditions, in most cases differ a great deal[10]. Besides, the prevailing health conditions of the population in the developing world differ from those in the developed world and influence therefore health outcomes are influenced differently. In Kenya, a few studies have shown how air pollution has, and continues to adversely affect human health and the built environment (**Figure 1**) and the ecosystem in general[1][11]-[14]. This is because for a long time there was no specific administrative or legislative framework within which to articulate and execute air quality management in Kenya. With the passage of the

Figure 1. Corrosion on corrugated iron roof sheets close to factory due to air pollutions.

Environmental Conservation and Management Act of 1999[15], and the creation of the National Environmental Management Authority (NEMA), Kenya's potential to manage air quality has improved to a limited extent. However, the NEMA administrative framework and professional capacity continues to be expanded hence the urgent need for locally generated scientific data.

This study examined the perception of residents neighbouring a factory situated in a rural township in Kenya. It contributes to the literature on environmental risk perception by addressing environmental risks in a local cultural and social context. Respondents reflected on their perceived health threats and relations to the manufacturing industry in their neighbourhood. This research avails useful information to all the stake holders (community members, industry and government regulators) who may be involved in discussions over existing (or non-existing) environmental threats. It presents the situation as seen by the communities and therefore provides a framework for further research with regard to protection of the public health from industrial pollution in Kenya. Perception studies reveal unique elements such as cultural or local contexts of certain environmental issue, thereby expanding the range of sources for decision making in assessing risk factors and possible mitigation. The use of participatory approach combined with respondents' interest in the study topic, helped to increase trust in this research work and confidence in the data collection process.

2. Methods

2.1. Study Location and Motivation

Four factors motivated the selection of the town for this study: (i) of the extensive media publicity on both air and water pollution including damage to the built environment (**Figure 1**); (ii) the presence of the factory and its being the largest employer in the area and therefore a backbone of the region's economy; (iii) it represented a typical example of a rural based large manufacturing industry and therefore presented an opportunity to survey the neighboring residents' perception of pollution to guide the development of applicable and appropriate policy options; (iv) no community level studies on pollution have been carried out before in the area. It is prudent to point out that even though the study was designed to gather residents' views while the factory was operational, the factory closed down while ethics approval was being awaited (four months before the study commenced) due to liquidity problems.

2.2. Characteristics of the Industry

About 90% of toxic emissions from paper mills go into the atmosphere while 10% ends up in the water[16]. Sources of air pollutants from the industry include: power generation plant, boilers, bleaching plants and caustic soda/chlorine plant. Pollutants include particulate matter, chlorine, sulfur dioxides, hydrogen sulfides, carbon dioxide, carbon monoxides and nitrous oxides. Hydrogen sulfide (HS) in addition to being toxic has a very foul odor. Air pollutants of concern are particulate matter, hydrogen sulfide as well as sulfur and nitrate oxides. The vulnerability and health risks for communities neighboring industrial facilities are compounded by the possibility of an industrial accident, as was experienced in Bhopal, India in 1984 and Chernobyl, USSR in 1986 among others.

2.3. Study Design

This was a descriptive cross-sectional study in one of the seven Provinces of Kenya where residents living within a five-kilometer radius of the paper mill were interviewed on their pollution and health perceptions. The sample was stratified and proportionally allocated to include key informants of the stake holders in the community as well as the general population. Community gatherings (barazas) were used as the main forum for informing the community about the study. Such a system was successfully utilized by Mwanthi and Kimani[17] in their study of agrochemical handling and response in a rural community in Kenya. The purpose of the study and the community's role in this research was first explained to the key leaders and then to the community at large. Prospective respondents were informed of the study and assured of confidentiality.

2.4. Sample Size Determination and Sampling Procedure

The sampling frame was a random sampling of households within a 5-km radius of the industry and the formula below was used to determine the minimum sample size:

$$n = \frac{t^2 \times pq}{d^2}$$

where: n = the desired sample size (if the target population is greater than 10,000),

t = critical value for the desired confidence level (alpha),

p = proportion in the target population estimated to have characteristics being measured,

$q = 1 - p$, and d = desired precision.

Since no estimate was available for the expected proportion of the target population which had the characteristics of interest then, 50% (p = 0.5) was used and confidence level was set at with 95% for which t = 1.96 resulting in sample size of 384.

An additional 10% was added to take care of nonrespondents that resulted with final sample size of 423.

The study area comprised of five sub-locations (clusters) which formed the strata. Proportionate stratified sampling was used to ensure the five clusters were adequately represented in the sample.

Rural Kenyans have structures through which official and community information is disseminated. The smallest unit is the village headed by a village elder. A number of villages form a clan and in some cases a group of ten villages form an administrative group known as mji kumi (ten homes). Village or clan leaders are usually elected by the clan members and are expected to play the intermediary role between the community and the administration. Respondents were selected using cluster sampling. Each mji kumi or a village elder's area of jurisdiction formed a cluster for this study. A skip pattern was used to select households for the 423 quantitative questionnaires while participants for the focus group discussions (FGDs) comprised of all stakeholders (community leaders, political and religious leaders).

2.5. Methodology

The study employed both qualitative and quantitative methods. Qualitative data facilitated interpretation of quantitative results in the context of the respondents' daily lives and the role their knowledge, attitude and practices play in shaping their responses to industrial pollution. All respondents provided written informed consent before the questionnaires were administered. For the FGDs, group oral informed consent was obtained from participants; an oral statement was read to them and their agreement was indicated by a check mark on the consent form before commencing the discussion. We opted for oral consenting because we believed the topic of discussion involved no more than minimal risk to the FGD participants.

The qualitative interviews conducted after completion of the survey provided data that permitted corroboration and evidence on the impact of the factory closure. We took comprehensive notes of each FGD and they served as the textual basis for qualitative analysis.

2.6. Data Collection

The quantitative survey instrument was a questionnaire developed primarily by reviewing and adapting questionnaires from literature. For environmental perception, literature reviewed included previous risk perception, environmental health and environmental psychometric studies/surveys within and outside Kenya which were then modified for relevance[3][17]–[30]. The survey questionnaire consisted of a list of items reported in literature as known or potential environmental health hazards from industrial facilities and environmental health perception indicators. Questions were developed based on variables representing aspects of perceived health risks identified from the above literature pertinent to the study area and population.

2.7. Quantitative Data Management and Analysis

The survey asked respondents to identify and describe the main risks from the paper mill in their neighborhood. Additionally respondents were asked to list and characterize the specific health and environmental risks they believed were associated with the paper mill.

Perception predictor variables were based on Vlek and Stallen's personal decision-making model of risk acceptability and Riechard & Peterson perception questionnaire[26][31]. Found within that model are 32 "aspects of risk" categorized and believed to influence risk perception. Some elements like global warming, radon gas and asbestos were omitted based on the fact that majority of the respondents were likely to be unfamiliar with them. Also, Riechard & Peterson's original questionnaire contained single 6-point Likert-type scale of environmental perception where as the scale here has been reduced to four or three-point (latter collapsed to two for logistic regression) and modified to scales of perceptions of exposure, benefits and control. The final list contained 20 common environmental variables.

The perceptions of control questions were adapted from the works of Riechard & Peterson, Schmidt & Gifford, Slovic and Westmoreland[26][27][32][33]. The same hazards were used for perception of exposure, severity and control. Benefit items were arrived at after reviewing the works of Gregory & Mendelsohn, Hallman and Hallman & Wandersman[23][24][34]. Thus literature on factors influencing the perception of (health) risks associated with exposure to environmental contaminants informed the selection of these variables. Respondents' general risk perception was measured by asking them to rank their perception of risk for common environmental and health hazards in their neighborhood and other known environmental pollutants and health threats.

To "asses the residents' awareness and beliefs concerning environmental risk especially their perception of air pollution health risk relative to other public health and environmental issues in their neighborhoods", respondents' responses to the 20 common environmental variables were compared across sub-locations. The 20 public health and environmental perception and control items were adopted from other environmental risk perception studies but modified for relevance and application to the study population[26][31]–[33]. Respondents were asked to rank their perception of risk (and control) of the 20 common risks on Likert-type scales; 1 (low risk), 2 (some risk) and 3 (high risk) that represented the amount of risk of harm/control they perceived for themselves, family and/or community from each hazard.

Raw data was cleaned by sorting out, grouping and coding of the completed questionnaires. The new data was cleaned by sorting out, grouping and coding of the completed questionnaires. Questionnaires were scrutinized for dishonesty and disregarded as necessary and answers to some interview questions helped in cross checking the consistency of responses. For example, in the early part of the interview respondents were asked the compositions of their household members (number of children and adults) and then in subsequent sections, they responded to another question on whether there were children with chronic illnesses in the household. If a respondent who had declared that there were no children in the household latter indicated that there were children with chronic illnesses in the household (as it happened in a few cases), then such data was deemed inconsistent and as a result the questionnaire was excluded from the analysis.

The quality of data was ensured through cross checking and computation before analysis. Respondent's rating on each of the environmental hazard list was recorded.

All quantitative analyses were done using SPSS (Statistical Package for the Social Sciences) version 18. Bivariate analysis was performed to compare variables across sub-locations and to identify characteristics related to perceptions. Chi-square ($\chi 2$) test was used to test for the association, as all variables were categorical. Statistical levels of significance were set at 0.05. Multivariate analysis revealed factors associated with risk perception of the study population. Variables that were significantly associated with perceptions of risk and control were included in the final logistic regression analysis to confirm the most important factors influencing risk perception and environmental awareness in the communities while accounting for confounding.

Logistic regression was used to study the association of the original categorical independent variables with each of the five primary perception outcome (binary variables): if respondent believes PPM has exposed him/her to hazardous chemicals; if the respondent felt he/she was coughing, breathless or wheezing due to something in the air; if the respondent believes the industry will affect children born in the future in the community; if the respondent believes the industry will expose him/her to hazards if it opens in the future; if the respondent worries about getting health problems in the future because of a polluted environment.

Whilst dichotomization is valid, it involves loss of information and there are fears that it may alter the findings. However, in an analysis of self-rated health and lifetime social class data, Manor *et al.*,[35] compared dichotomized variable using logistic regression with alternative methods for ordered categorical variables and similar results and conclusions emerged from the five statistical approaches.

The covariates entered in all models included the following: the seven demographic factors, four household health indicators, six respiratory symptoms, two health risks characteristics, three benefit variables, four environmental awareness/knowledge variables and six information source variables (total 32 items) as predictors for air pollution as a health risk. All the above independent variables assumed to influence perception of health effects of air pollution were tested, following the stepwise procedure and the predictors that significantly improved the model were the only ones kept in the final step.

A total of 423 questionnaires were given out to respondents from the six sub-locations within the two districts (District A and District B) which fall within the five kilometer radius of the factory, the focus of this study. Three hundred and eighty two questionnaires were returned filled out and after thorough scrutiny 15 were discarded and the remaining 367 were entered in SPSS.

2.8. Qualitative Data Management

Using Atlas.ti (version 5.2) software, the transcripts were coded, categorized and then grouped to generate larger environmental risk themes and subsequently correlated with the quantitative data analysis.

2.9. Ethics Statement

This study was conducted according to the principles expressed in the Declaration of Helsinki. The study was approved by the Institutional Review Board of Kenyatta National Hospital, Nairobi, Kenya, Ref: KNH/UON-ERC/A/249.

3. Results and Discussion

3.1. Results

3.1.1. Descriptive Statistics

Table 1 and **Figure 2** show demographic variables across the sub-locations, including the chi-square tests. They reveal differences in terms of household income, the presence of a child in the household, distance from the study factory and a family member having worked for the factory. This is expected as the sub-locations vary in distance from the town (where the factory is located) and population characteristics. The closure of the factory changed the population especially in the immediate vicinity of the town.

No statistically significant difference was observed for gender, education and duration of stay of less than 20 years. Only 28% of respondents were able to estimate their household income.

3.1.2. Gender

There were slightly more males than females in all the sub-locations except in Sub-location 1 where it was nearly balanced (51% males and 49% females).

Table 1. Selected demographic characteristics of the study population.

District/ Sub-location		Overall (n %)	District A			District B		p value
			S-location 1	S-location 2	Slocation 3	S location 4	S-location 5	
Questionnaires	Qs given	423	143	50	50	110	70	
	Qs Compltd	367 (87%)	111 (78%)	50 (100%)	40 (80%)	108 (98%)	58 (83%)	
Gender	M	205 (56%)	54 (49%)	33 (66%)	26 (65%)	57 (53%)	35 (60.5%)	0.073
	F	154 (42%)	57 (51%)	17 (34%)	14 (35%)	49 (45%)	17 (29%)	
	Missing	8 (2%)				2 (2%)	6 (10.5%)	
Education (% Completed high school)		162 (44%)	56 (51%)	24 (49%)	18 (45%)	37 (34%)	27 (47%)	0.154
Employment (% Employed)		81 (22%)	28 (25%)	13 (26%)	9 (23%)	20 (19%)	11 (19%)	0.694
Household income	Up to 5,000	44 (12%)	5 (5%)	6 (12%)	1 (3%)	22 (20%)	10 (17%)	0.004*
	5,001–10,000	17 (5%)	4 (4%)	4 (8%)	0 (0%)	7 (6%)	2 (3%)	
	>10,000	21 (6%)	8 (7%)	2 (4%)	1 (3%)	4 (4%)	6 (10%)	
	Missing	266 (72%)	92 (83%)	38 (76%)	38 (95%)	58 (54%)	40 (69%)	
	NA	19 (5%)	2 (2%)	0 (0%)	0 (0%)	17 (16%)	0 (0%)	
Family member worked for the factory		70 (19%)	36 (32%)	10 (20%)	2 (5%)	18 (18%)	4 (7%)	<0.001*
Child in household (%)		312 (85%)	80 (72%)	44 (88%)	34 (85%)	100 (93%)	54 (93%)	<0.001*
Distance from the factory	0–2km	147 (40%)	93 (84%)	16 (32%)	10 (25%)	17 (16%)	11 (16%)	<0.001*
	2–5km	220 (60%)	18 (16%)	34 (68%)	30 (75%)	91 (84%)	47 (84%)	
Lived there for >20 years		213 (59%)	57 (52%)	28 (56%)	34 (87%)	62 (57%)	32 (59%)	0.004*
Lived >10 years		286 (83%)	75 (79%)	39 (78%)	36 (92%)	92 (85%)	44 (82%)	0.315

Legend: Qs given: No of Questionnaires given out. Qs Compltd: No. of Questionnaires completed. *p-value ≤ 0.05.

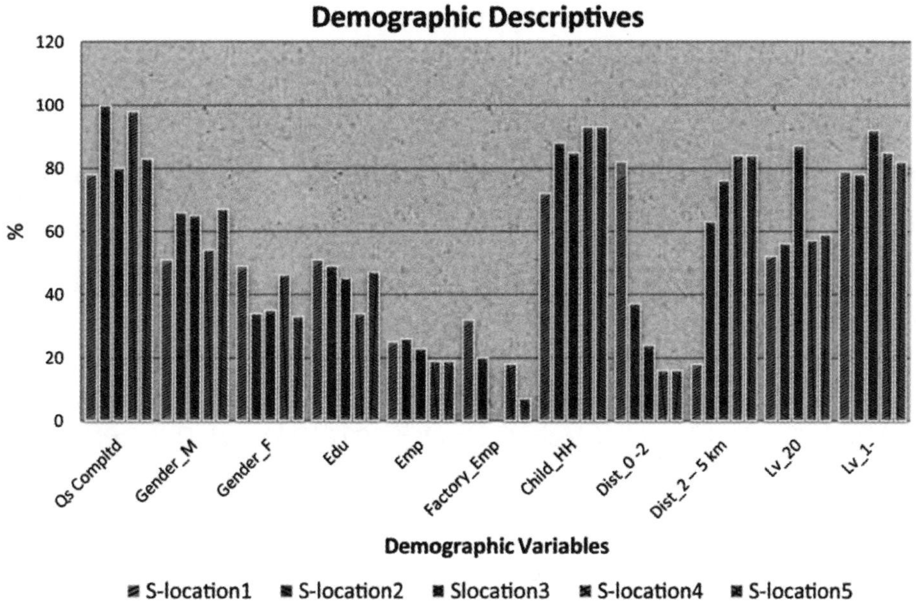

Figure 2. Demographic variables by sub-locations.

3.1.3. Employment Status

Overall only 81 (22%) respondents reported being in employment at the time of the survey with the highest employment rate reported in Sub-location 3 (26%) and Sub-location 1 (25%). Sub-location 4 and Sub-location 5 had the lowest at 19%.

3.1.4. Occupation

Occupations included small scale/informal businesses, teaching, clerical jobs and health professionals (doctors and nurses) with majority of respondents (21%; n = 75) being small scale farmers (peasant).

3.1.5. Employment with the Factory

Only 70 (19.1%) respondents had worked for the factory or had a member of their families who did. This was confirmed by the focus group discussions (FGDs). Surprisingly a very small proportion of the respondents had worked for the factory and even then those who had worked did so through subcontractors. Apparently very few employees were employed directly by the factory; instead the practice was to use subcontractors (out sourcing) as service providers to the factory through which a small number of residents were employed. The factory was situated in sub-location 1 and therefore this location had the highest proportion of respondents (and relatives) who were former factory employees (**Figure 3**).

3.1.6. Education

44% of the study respondents had completed high school and above compared to 22% of all Kenyans according to the 2008/09 Kenya Demographic and Health Survey KDHS[36].

3.2. Environmental Awareness and Risk Perception

The summary of proportions of all respondents (combined) scores for medium to high levels of risk perceptions and control for the items are presented in **Table 2**. More than half of the respondents perceived all the elements as medium to high risk (with the exception of flooding).

3.2.1. Risk Perception

Air pollution, industrial dust/smoke, cigarette smoking, HIV/AIDS and water

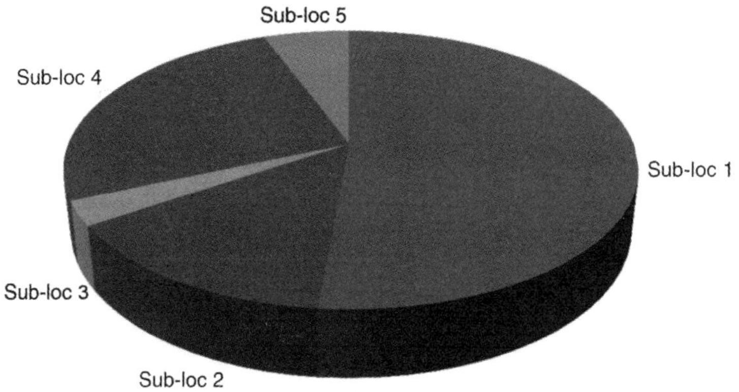

Figure 3. Employment with the study factory by sub-location.

Table 2. Perception and control of 20 environmental health risks and association (perception and control) of variables with air pollution perception.

Risk factors	Perception of risk and control		Association with air pollution perception (p-value)	
	Reporting medium or high risk	Reporting medium or high control	Perception of risk	Perception of control
Overpopulation	218 (63%)	212 (62%)	<0.001*	0.185
Destruction of Forests	254 (75%)	211 64%)	0.028*	0.451
Water Pollution	235 (68%)	214 (64%)	<0.001*	0.118
Air Pollution	275 (82%)	156 (48%)	-	<0.001*
Industrial Noise	246 (73%)	144 (44%)	<0.001*	0.526
Industrial Odor	271 (81%)	157 (48%)	<0.001*	0.453
Industrial Dust/Smoke	271 (82%)	168 (53%)	<0.001*	0.288
Motor Vehicle Accidents	233 (70%)	180 (57%)	0.040*	0.288
Motor Vehicle Pollution	197 (61%)	176 (55%)	<0.001*	0.561
Poor Housing	209 (65%)	200 (63%)	<0.001*	0.036*
Cooking with Firewood	191 (57%)	178 (55%)	0.113	0.515
Agricultural Chemicals	187 (56%)	188 (58%)	<0.001*	0.454
Waste/Plastic Bags	215 (66%)	200 (63%)	0.003*	0.230
Cigarette smoking	270 (80%)	235 (72%)	0.019*	0.560
Working Conditions	194 (62%)	167 (55%)	<0.001*	0.284
HIV/AIDS	269 (80%)	217 (67%)	0.002*	0.320
Drought	189 (58%)	169 (55%)	0.341	0.297
Famine	205 (63%)	183 (59%)	0.010*	0.057
Flooding	104 (33%)	171 (57%)	0.106	0.507
Soil Erosion	166 (51%)	214 (68%)	0.055	0.513

*p-value ≤ 0.05.

pollution were perceived as the greatest risks. Flooding was perceived as the least risk and this is because the study area hardly experiences flooding.

3.2.2. Perception of Control

In all the 20 environmental and public health variables, the majority of the respondents reported low levels of control. Between 44% and 72% respondents felt they had medium/high level of control over the range of listed environmental factors with majority being in the 50s/60s. Respondents had the highest perception of control over cigarette smoking followed by perception of control over soil erosion (68%). On the other hand respondents reported the least level of control over industrial noise (44%), odor and air pollution (48%).

3.2.3. Association between Risk Variables

The second set of analysis involved bivaraite analysis to test the associations between air pollution perception (dependent) with the other environmental and public health variables. **Table 3** shows these analysis results.

1). Association of Air Pollution Perception with Perceptions of other 19 Environmental Risk variables Significant associations were found between perception of air pollution as a risk and all the listed environmental threats except for perceptions of drought, flooding and soil erosion as risks. Thus fifteen of the 19 perception variables were found to be significantly associated with air pollution with nine of them being highly significant; $p < 0.001$.

2). Air Pollution Perception versus Perception of Control of other variables.

Only the association between perception of air pollution and perception of control over air pollution was found to be highly significant ($p < 0.001$). The remaining perception variables were insignificant except for perception of control of famine with borderline significance, $p = 0.0507$. Overall respondents perceived air pollution as a risk. For example out of the respondents who thought overpopulation was a high risk, the majority (74%, $n = 78$) perceived air pollution as a risk and even those who perceived overpopulation as a low risk, majority still perceived

Table 3. Final logistic regression models.

		Believes PPM has exposed him/her to hazardous chemicals		Felt he/she was coughing, breathless or wheezing due to something in the air		Believes the industry will affect children born in the future in the community		Believes the industry will expose him/her to hazards if it opens in the future		Worries about getting health problems in the future because of a polluted environment	
		OR	95% CI	OR	95% CI	OR 9	5% CI	OR	95% CI	OR	95% CI
Demographic	Distance from PPM	2.747*	1.199 - 6.295	1.039‡	0.560 - 1927	1.132‡	0.502 - 2.554				
	Presence of a child in the HH	34.769*	1.390 - 869.96								
	Respondent completed HS	3.105*	1.394 - 6.918							2.553*	1.086 - 6.004
	Employment Status					0.434‡	0.188 - 1.004	0.383*	0.167 - 0.879	0.365*	0.150 - 0.886
	Perception of health	0.527‡	0.239 - 1.162	0.479*	0.261 - 0.878	0.299*	0.135 - 0.660	0.237**	0.108 - 0.517	0.332*	0.149 - 0.738
HH health Indicators	Somebody in the HH has one of the conditions (heart, respiratory or skin disease)	1.788‡	0.755 - 4.237	1.980*	1.031 - 3.802	2.192‡	0.963 - 4.990			2.594*	1.096 - 6.143
Respiratory symptoms	Coughed phlegm daily for ≥2 months, 2 yrs in a row			2.907*	1.127 - 7.498						
	Had an attack of whistling or noisy sound in the chest when breathing			2.881*	1.402 - 5.920						

Continued

Role of Industrial plant	Family depended on industry					2.287*	1.001 - 5.228	2.792*	1.234 - 6.319		
	Community benefited from industry									2.289*	0.997 - 5.257
Environmental awareness/ knowledge	Ever actively looked for environmental health information			1.883*	1.002 - 3.537						
	Willing to participate in environmental campaigns			2.297‡	0.713 - 7.396	2.985*	1.031 - 8.645			3.127*	1.100 - 8.888
Main Information source	Friends/relatives	2.553*	1.198 - 5.438			0.411*	0.180 - 0.937	0.451*	0.207 - 0.981		
	Church/Community leaders					3.148*	1.303 - 7.602	2.368*	1.090 - 5.147	2.447*	1.089 - 5.499
Omnibus Tests of Model Coefficients	Model (Sig.)		<0.001		<0.001		<0.001		<0.001		<0.001
Correctly predicted observations on average			85.6%		79.7%		84.6%		86%		84.4%

‡p > 0.05; *p < 0.05; **p < 0.01; HH - Household: HS - High school.

air pollution as a high risk (48%, n = 60). The same general trend (*i.e.* majority perceiving air pollution as a high risk) was observed across all the twenty perception variables with a couple of exceptions drought and flooding—both of which are typically not a problem in the study area.

3.3. Multivariate Analysis

A summary of the logistic regression models predicting respondents' perceptions of the health effects of PPM pollution and the main predictor factors of three of these perceptions are summarized in **Table 3**. The table shows the confidence intervals, the odds ratios and p values of each remaining predictor in the respective models as well as the percentages of correctly predicted observations for each model.

Based on the variables that emerged significant in this model, factors that influence the residents' belief that PPM had exposed them to hazardous chemicals are; distance from PPM, presence of a child in the HH, perception of health (though not significant) education level, presence of a smoker in the HH, perceived benefits from PPM and friends/relatives being the main source of information.

The odds ratio showed that compared to those who live far, those who live near the factory were almost 3 times more likely to believe that PPM had exposed them to hazardous chemicals. Similarly, families with children are 35 times were more likely to believe they had been exposed; those who had completed high school 3 times more likely to believe they had been exposed; those who rely on friends/family as the information source are 2.5 times more likely to believe PPM had exposed them to hazardous chemicals.

The 'Respondent believes PPM has exposed him/her to hazardous chemicals' model was able to correctly predict 85.6% of observations, from the Omnibus Tests of Model Coefficients table the model significance level was <0.001 showing that hat the final model predicted the dependent variable. The results are consistent with the descriptive and bivariate results showing that respondents generally perceive the paper mill as a major source of pollution.

For the perception: Respondent felt he/she was coughing, breathless or wheezing due to something in the air, the overall model was able to correctly predicted 79.7% of observations. The model significance level was <0.001, showing that the independent variables are associated with the dependent variable. Respondents had somebody in the household with one of the conditions (respiratory, heart or skin conditions), those who coughed phlegm for at least two months for two years, those who had experienced an attack of whistling or noisy breathing and those who had looked for environmental health information were more likely to have sensed respiratory irritants in the air (*i.e.* participant felt he/she was coughing, breathless or wheezing due to something in the air). On the other hand, respondents who view their health as good/excellent (compared to those who perceive their health as poor/fair) are less likely to say they had sensed respiratory irritants in the air.

Respondents who perceive their health is excellent/ good were less likely to say the industry will affect children born in the future in the community. On the other hand, respondents who viewed their health as poor/fair; were willing to participate in environmental campaigns, those whose families benefited on PPM, and those who rely on the church as the main source of information were more likely to believe the industry will affect children born in the future. From the Omnibus Tests of Model Coefficients table the final model for 'Respondent believes the industry will affect children born in the future in the community' had a p-value < 0.001 and was able to correctly predict 84.6% of observations.

Participants who were employed and those who viewed their health as good/excellent were less likely to believe the industry will expose them to hazards if it opens in the future. The model was able to correctly predict 86% of observations.

Likewise, employed participants and participants who view their health as good/excellent were less likely to worry about getting health problems in the future due to a polluted environment. On the other hand participants who had completed high school, knew somebody who is affected, were willing to participate in environmental campaigns and those who believed the community benefited from PPM and relied on community leaders for their information were more likely to worry about getting health problems in the future due to a polluted environment.

The model is highly significant, showing that the independent variables predict the dependent variable well and was able to correctly predict 84.4% of observations.

3.4. Discussion

The qualitative interviews conducted after completion of the quantitative survey provided data that permitted corroboration of quantitative data and evidence about the impact of the factory closure.

A large proportion of respondents were unemployed hence the closure of the factory may have negatively affected the economy of the study area. It is worth noting that majority of respondents (81%, n = 297) had not worked for the factory nor their family members and only 19% (n = 70) of respondents had a family member whom at one time worked for the factory. This sentiment was confirmed during the FGDs where respondents claimed the factory had mostly benefited foreigners; that the locals were rarely employed under the pretext that the locals were uneducated. While most respondents said the factory had not employed locals, the presence of the factory supported other economic activities in the region as confirmed by the FGDs, e.g. ready markets for their agricultural products and income from rental houses among others.

There were mixed reactions to the environmental effects of the factory closure. While some participants said there was no difference, most of them felt the environment (and agricultural production) had improved;

"… since the factory closed, I planted cassava and am seeing prospects of having a good cassava harvest, compared to when the factory was operating… even vegetables" a village elder from Sub-location 5 said.

3.4.1. Awareness and Perception

In this study, efforts were made to examine perceptions of risks around actual hazards that respondents had experienced or at least could relate to. Other researchers have also examined perceptions of risk around specific hazardous activities/technologies in affected communities[24][37]. Respondents in this study rated the risk of harm from a list of common environmental and public health hazards by indicating the extent to which they believed these hazards posed serious health threats to themselves, their family members' and the community. The results revealed high risk perception for the hazards that the respondents are familiar with and commonly exposed to. With the exception of flooding, overall majority (over 50%) off the respondents perceived all the listed environmental/health factors as medium or high risk. Air pollution, industrial dust/smoke (82%), industrial odor (81%) and HIV/AIDS & cigarette smoking (80%) were perceived to have the highest risk factors in the list (**Table 2**).

Risks considered to be involuntary like industrial hazards are generally dreaded because they are viewed as being more dangerous than those that are more familiar or voluntary[38]–[40]. In many previous environmental studies, respondents were asked to rank hazards, some of which had no relevance to them and therefore posed no threat as far as they were concerned. This study only included threats applicable and known to the study community.

The hazards like HIV/AIDS, air and water pollution and motor vehicle accidents that were ranked highly in this study, have already been identified in other studies to be perceived as high risk by Kenyans and are often mentioned as major challenges facing Kenya as a country[41][42].

Generally the risk perception of all respondents reflected a universal concern for pollution. This finding contradicts the widely predicted view that low income individuals/communities have more pressing problems and do not have the 'luxury' or the time and knowledge to worry about pollution of their environment[43][44]. As revealed by the FGDs, residents of the study area did not just accept the status quo. On the contrary, there had been many attempts to address the pollution problem but unfortunately lack of proper representation by local leaders, personal interests of some leaders and political interference were identified as some of the barriers to solving the pollution problem. As a result of the perennial non-response to their many complaints, the residents have developed some apathy and belief that nobody cares about their problems (pragmatic acceptance). This came out repeatedly in the FGDs. Some environmental perceptions studies have shown that low income populations are well aware of their increased vulnerability to environmental hazards but feel disempowered to act appropriately and with time, they embrace their disadvantaged circumstances[45][46].

Other perception studies have revealed different views on environmental issues by rural and urban residents with rural populations having a tendency to focus on community economic benefits of activities/technology over environmental management and protection of natural resources[47]. This study revealed a scenario where the economic benefit of the factory to the residents was mainly derived from business prospects and not from direct employment opportunities as in the studies reviewed. This may explain why they were not shy to express their fears about the negative effects of the factory on probing.

Generally cigarette smoking is rare in rural areas in Kenya hence its high perception of as a health risk in this study. According to WHO (2002) annual cigarette consumption per person in Kenya is less than 500[48]. Accordingly, respondents reported the highest level of control (72%) over this environmental risk. Similarly the high perception of HIV/AIDS as a health risk is in line with the disease epidemic in Kenya; currently 1.4−1.6 million Kenyans are estimated to be living with HIV/ AIDS[35][49].

The perception of flooding as a low risk can be explained by the fact that flooding is a very rare phenomenon in the study area. Consequently respondents reported a relatively high level of control over this risk (67%) but reported least levels of control over air pollution; industrial noise and industrial odor, and perceived them as high risks.

The findings above demonstrate relatively high levels of awareness and concern for environmental pollution. Compared to other environmental risks, air pollution is definitely a major concern for the study town residents where over 80% of the residents perceived it as a high risk.

Only 34% residents perceived flooding as a risk and the perception association between flooding and of air pollution was insignificant. This can be explained by the fact that flooding is not a common occurrence in this region and therefore not one of the problems they worry about. About half of the respondents (51%) perceived soil erosion as a risk but 68% felt they had control over the hazard. Perception association of this hazard with air pollution was insignificant.

3.4.2. Association between Perception of Air Pollution and Perception of Control

Literature on risk perception reveals that the ability of individuals to influence circumstances that affect them is closely tied to perceptions of risk and if they feel they have control over a risk then they perceive it as a low risk and vice versa[50][51]. This phenomenon is contradicted by the bivariate analysis of perception of air pollution and perception of control over the same. Of the respondents who perceived air pollution as a high risk, majority (about 55%) reported having some control over it and 51% of respondents who reported low risk felt they had low level of control.

It is also worth noting that out of those who felt they had a high level of control over air pollution (48%, n = 156), majority (70.5%, n = 109), perceived air pollution as a high risk. So, generally air pollution was perceived as a high risk irrespective of the respondents' perception of control. The association between these two measures of perception was highly significant. The association of air pollution with perceived ability to deal with the problem has been demonstrated in other studies[52][53]. The public has a tendency to feel their actions have little or no impact as far as reducing air pollution is concerned and this perceived low level of control over environmental problems may discourage affected populations from becoming part of the solution.

From the FGDs there was a general belief that other stakeholders (political leaders, management of industries, government officials) were unable to act responsibly to alleviate the pollution problems and some respondents further indicated that it was not their responsibility because they did not have the capacity to effectively act to protect their environment. Most of the respondents had lost faith in the government regulatory agencies, and were of the opinion that the government officers receive bribes to defend the interests of the factory management.

Health status and distance from a polluting facility have been shown to be a strong predictor of risk perception. A couple of studies done in India a developing country revealed that persons neighboring industrial settings are twice more likely to experience respiratory symptoms/illness compared to those living far away[6][31]. Obviously people who live close by bear the burden of the pollution (odor, smoke, dust, etc.) and naturally believe it is affecting their health. Participants were able to clearly narrate the history of the paper mill during the FGDs. Those close to the factory reported that the arrival of the factory marked the beginning of diminishing agricultural outputs. However in the logistic regression models, it was a modest a predictor of environmental risk perception as one of the models predicted that people who live closer to the factory are over twice more likely to believe PPM had exposed them to hazardous chemicals. Also from the logistic regression models, presence of a child in the household was a predictor for one perception variable; belief of PPM was a source of exposure to hazardous chemicals. Existing literature has confirmed that the connection between environmental exposure and possible health effects is mostly attributed to concerns over children's health[20][54][55].

The logistic regression models further confirmed the MLR finding that family health status was the best predictor of environmental risk perception. Family health status variables were predictors in four out of the six models-belief that PPM exposed residents to hazardous chemicals, experience of coughing, breathless or wheezing due to something in the air, belief that the industry will affect children born in the future in the community, the belief that the industry will expose him/her to hazards if it opens in the future and worrying about getting health problems in the future because of a polluted environment.

Out of the health status variable, only the perception variable *'respondents' perceptions of his/her health'* turned out to be a strong predictor for environmental risk perception in this study. The other family health status variables like presence of an unwell child in the household did not come out clearly as predictors for environmental risk perception. The respondent's view of health was a predictor for four out of the seven dependent risk perception variables (the respondent believing the industry will affect children born in the future in the community; the respondent sensing respiratory irritants in the air; the respondent believing the industry will expose them to hazards if it opens in the future and respondent worrying about getting health problems in the future because of a polluted environment). People who perceived their health were found to be less likely to have a high perception of environmental hazards. This can only be attributed to the fact that because they feel healthy, they do not seem to see the danger around them. And have a tendency to be complacent.

4. Conclusion

Study respondents clearly demonstrated their awareness and concern over negative effects of air pollution as well as other environmental and agricultural activities on their health. Although public perceptions are influenced by many factors, the concerns are consistent and call for involvement of the affected individuals—their social status notwithstanding—in environmental management and policy formulation. This study presented a quantitative approach to environmental risk perception. It examined pertinent environmental hazards to generate fundamental information on environmental beliefs within the study community through both closed and open ended questions. With increased awareness, individual involvement and support, participation by rural communities could be central to achieving environmentally friendly and sustainable industrial development in emerging economies.

The findings will enable the stakeholders, (the researcher, the public, the industry and the policy makers) to focus on what is important in mediating the pollution risks in the community as environmental policies are developed.

4.1. Limitations of the Study

This was one of the very few studies in Kenya that attempted to assess community perception of pollution; however, it had following limitations:

- The study area was limited to 5km radius of the paper mill.
- The factory closed down due to financial problems implying views of the long-term residents and regular workers of the factory were not captured.
- Lack of health facility data to confirm or contradict respondents' claims.
- The fact that most respondents felt the factory did not benefit them as much as 'foreigners' may have biased the study.

4.2. Abbreviations

AIDS: Acquired Immunodeficiency Syndrome; FGD: Focus group discussions; HIV: Human immunodeficiency virus; KDHS: Kenya Demographic and Health Survey; MoH: Ministry of health; NASCOP: National AIDS & STI Control Program; NEMA: National Environment Management Authority; UNEP: United Nations Environment program; WHO: World Health Organization.

Competing Interests

The authors declare that they have no competing interests.

Authors' Contributions

EO: Conceived the study and designed; carried out the field data collection and data entry; performed statistical analysis; drafted the manuscript. LU: Guided the conception and design, supervision of the research, helped with manuscript draft and revision. ZB: Guided statistical analysis, helped with manuscript draft. MG: Guided field data collection, helped with manuscript draft. All authors have given final approval of this version to be published.

Authors' Information

EO (DrPH): Head of Research, Impact Research and Development Organization.

LU (PhD): Chair, Community Health and Development Department, Drexel School of Public Health.

ZB (PhD): Prof. Epidemiology and Biostatistics department, Drexel School of Public Health.

MG (PhD): Senior Lecturer, Institute of Nuclear Science and Technology, University of Nairobi.

Acknowledgements

Dr. Omanga specifically wishes to acknowledge funding received from the Margaret McNamara Memorial Fund (MMMF) which enabled her to complete the field data collection exercise.

The authors wish to acknowledge the intellectual contributions towards the research from both Dr. Arthur Frank and Dr. Rene Turchi who were research committee members.

Source: Omanga E, Ulmer L, Berhane Z, *et al.* Industrial air pollution in rural Kenya: community awareness, risk perception and associations between risk variables[J]. Bmc Public Health, 2014, 14(14):1−14.

References

[1] Mulaku GC, Kariuki LW: Mapping and Analysis of air Pollution in Nairobi, Kenya. In International Conference on Spatial Information for Sustainable Development. Nairobi, Kenya: Institution of Surveyors of Kenya; 2001.

[2] Jones RE, Dunlap RE: The social bases of environmental concern: have they changed over time? Rural Sociol 1992, 57:28–47.

[3] Adeola FO: Endangered community, enduring people: toxic contamination, health, and adaptive responses in a local context. Environ & Behav 2000, 32(2):207–247.

[4] Kunzli N, Kaiser R, Medina S, Studnicka M, Chanel O, Filiger P, Herry M, Horak F Jr, Puybonnieux-Texier V, Quénel P, Schneider J, Seethaler R, Vergnaud JC, Sommer H: Public-health impact of outdoor and traffic-related air pollution: a European assessment. Lancet (North American Edition) 2000, 356(9232):795–801.

[5] Kempton W, Boster JS, Hartley JA: Environmental Values in American Culture. Cambridge, MA: MIT Press; 1995.

[6] Carson R: Silent Spring. Boston, MA: Houghton Mifflin; 1980.

[7] WHO: Quantificating Environmental Health Impacts. Geneva, Switzerland: WHO; 2007.

[8] WHO: New Country-by-Country Data Show in Detail the Impact of Environmental Factors on Health. Geneva, Switzerland: World health Organization; 2007.

[9] Byrd T, VanDerlslice J, Peterson S: Attitudes and beliefs about environmental ha-

zards in three diverse communities in Texas on the border with Mexico. Panam Salud Publica 2001, 9(3):154–160.

[10] Cohen AJ, Anderson HR, Ostro B, Pandey KD, Krzyzanowski M, Künzli N, Gutschmidt K, Pope CA III, Romieu I, Samet JM, Smith KR: Urban air Pollution. In Comparative Quantification of Health Risks: Global and Regional Burden of Disease Attributable to Selected Major Risk Factors. Edited by Ezzati M. Geneva, Switzerland: World Health Organization; 2004:1153–1433.

[11] Boleij JSM, Ruigewaard P, Hoek F: Domestic air pollution from biomass burning in Kenya. Atmos Environ 1998, 23:1677–1681.

[12] Gatebe CK, Kinyua AM, Mangala MJ, Kwach R, Njau LN, Mukolwe EA, Maina DM: Determination of suspended particulates matter of major significance to human health using nuclear techniques in Kenya. J Radioanalytical and Nuclear Chemistry 1996, 203(1):125–134.

[13] Gatari MJ, Boman J, Maina DM: Inorganic elemental concentrations in near surface aerosols sampled on the northwest slopes of Mount Kenya. Atmos Environ 2001, 35: 6015–6019.

[14] Gatari M, Wagner A, Boman J: Elemental composition of tropospheric aerosols in Hanoi, Vietnam and Nairobi, Kenya. Sci Total Environ 2004, 341:241–249.

[15] Kenya Government: The environmental Management and Coordination Act (No. 8. of 1999), Judiciary. Nairobi: NEMA; 1999.

[16] UNEP/UNDP/DUTCH: Report on the development and harmonization of environmental standards in East Africa. Nairobi, Kenya: UNEP; 1999.

[17] Mwanthi MA, Kimani VN: Patterns of agrochemical handling and community response in central Kenya. J Environ Health 1993, 55(7):11–16.

[18] Dunlap R, Van Liere K: The new environmental paradigm: a proposed measuring instrument and preliminary results. J Environ Educ 1978, 9(4):10–19.

[19] Dunlap R, Van Liere K, Merig A, Jones R: Measuring endorsement of the new ecological paradigm: a revised NEP scale. J Social Issues 2000, 56(3):425–442.

[20] Elliott SJ, Taylor SM, Walter S, Stieb D, Frank J, Eyles J: Modelling psychosocial effects of exposure to solid waste facilities. Soc Sci Med 1993, 37(6):791–804.

[21] Freudenberg N: Action for environmental health: report of a survey of community organizations. Am J Public Health 1984, 74:444–448.

[22] Greenberg MR: Concern about environmental pollution: How much difference Do race and ethnicity make? A New Jersey case study. Environ Health Perspect 2005, 113(4):369–373.

[23] Gregory R, Mendelsohn R: Percieved risk, dread and benefits. Risk Anal 1993, 13:258–264.

[24] Hallman W: Coping with an environmental stresor: perception of risk, attribution of responsibility and psychological distress in a community living near a harzadous waste facility. University of South Carolina: Columbia; 1989.

[25] Lichtenstein S, Slovic P, Fischhoff B, Layman M, Combs B: Judged frequency of lethal events. J Exp Psychol Hum Learn 1978, 4:551–578.

[26] Riechard DE, Peterson SJ: Perception of environmental risk related to gender, community socioeconomic setting, age, and locus of control. J Environ Educ 1998, 30(1):11–19.

[27] Schmidt FN, Gifford R: A dispositional approach to hazard perception: preliminary development of the environmental appraisal inventory. J Environ Psychol 1989, 9:57–67.

[28] Slovic P, Fischhoff B, Lichtenstein S: Facts and Fears: Understanding Perceived Risk. In Societal Risk Assessment: How Safe is Safe Enough? Edited by Schwing RC, Albers WA. New York: Plenum Press; 1980:p. 181–p. 216.

[29] Taylor-Clark K, Koh H, Viswanath K: Perceptions of environmental health risks and communication barriers among Low-SEP and racial/ethnic minority communities. J Health Care Poor Underserved 2007, 18(4):165–183.

[30] Torres EB, Subida RD, Gapas JL, Sarol JN, Villarin JT, Vinluan RJ, Ramos BM, Quirit LL: Public Health Monitoring of the Metro Manila Air Quality Improvement Sector Development Program. Manila, Philippines: World Health Organization (WHO), Asian Development Bank (ADB), Philippines Department of Health (DoH); 2004:156.

[31] Vlek CJH, Stallen PJM: Rational and personal aspects of risk. Acta Psychologica 1980, 45:273–300.

[32] Slovic P: Perception of Risk: Reflections on Psychometric Paradigm. In Social Theories of Risk. Edited by Krimsky S, Golding D. Westport, CT: Praeger Publishers; 1992:117–152.

[33] Westmoreland G: Perception of Risk from Environmental Hazards: A Comparative Study of Undergraduates in the United States and Costa Rica. Atlanta, GA: Emory University; 1994.

[34] Hallman WK, Wandersman AH: Perception of Risk and Toxic Hazards. In Psychosocial Effects of Hazardous Waste Disposal on Communities. Edited by Peck D. Springfield, IL: Charles C Thomas; 1989.

[35] Manor O, Matthews S, Power C: Dichotomous or categorical response? Analysing self-rated health and lifetime social class. Int J Epidemiol 2000,29:149–157.

[36] KNBS: Kenya National Bureau of Statistics 1999 Kenya population census report. Nairobi: Government Printer; 2000.

[37] MacGregor DG, Slovic P, Morgan MG: Perception of risks from electromagnetic fields: a psychometric evaluation of a risk-communication approach. Risk Anal 1994,

14(5):815–828.

[38] Sandman PM: Hazard versus Outrage in the Public Perception of Risk. In Effective Risk Communication: The role and responsibility of government and non-government organizations. Edited by Covello VT, McCallum DB, Pavlova MT. New York: Plenum Press; 1989:45–49.

[39] Flynn J, Slovic P, Mertz CK: Gender, race, and perception of environmental health risks. Risk Anal 1994, 14:1101–1108.

[40] Miller M, Solomon G: Environmental risk communication for the clinician. Pediatrics 2003, 112(1):211–217.

[41] Khayesi M: Liveable Streets for Pedestrians in Nairobi: The Challenge of Road Traffic Accidents. In The Earthscan Reader on World Transport Policy and Practice. Edited by Whitelegg J, Haq G. London: Earthscan; 2003:p. 35–p. 41.

[42] Lamba D: The forgotten half; environmental health in Nairobi's poverty areas. Environ Urban 1994, 6(1):164–173.

[43] Dunlap RE, Mertig AG: Global environmental concer: an anomaly for pstmaterialism. Soc Sci Q 1997, 78(1):23–29.

[44] Inglehart R: Public support for environmental protection: objective problems and subjective values in 43 societies. Pol Soc Sci 1995, 28(1):57–72.

[45] Adeola FO: Environmentalism and risk perception: empirical analysis of black and white differentials and convergence. Soc Nat Resour 2002, 17(10):911–939.

[46] Satterfield TA, Mertz CK, Slovic P: Discrimination, vulnerability, and justice in the face of risk. Risk Anal 2004, 24(1):115–129.

[47] Freudenberg WR: Rural–urban differences in environmental concern: a closer look. Sociol Inq 1991, 61:167–198.

[48] WHO: World Health Report: Reducing Risks, Promoting Healthy Life. Geneva, Switzerland: World Health Organization; 2002.

[49] NASCOP: Ministry of Health, Kenya AIDS Indicator Survey (KAIS) 2007: Final Report. Nairobi, Kenya: NASCOP; 2008.

[50] Bickerstaff K, Walker G: Public understandings of air pollution: the 'localisation' of environmental risk. Glob Environ Chang 2001, 11:133–145.

[51] Walker G, Simmons P, Wynne B, Irwin A: Public Perception of Risks Associated with Major Accident Hazards. Associated with Major Accident Hazards. In Research Report Series 194/98. Sudbury: HSE Books; 1998.

[52] Geller ES: Actively caring for the environment: an intergration of behaviorism and humanism. Environ & Behav 1995, 27(4):184–195.

[53] Zeidner M, Schechter M: Psychological responses towards air pollutiion: some personality and demographic correlates. J Environ Psychol 1988, 8:191–208.

[54] Cutter S: Living with Risk. New York: Routledge; 1993.

[55] Elliott SJ, Cole DC, Krueger P, Voorberg N, Wakefield S: The power of perception: health risk attributed to Air pollution in an urban industrial neighborhood. Risk Anal 1999, 19(4):621–634.

Chapter 28

Spatial Heterogeneity of the Relationships between Environmental Characteristics and Active Commuting: Towards a Locally Varying Social Ecological Model

Thierry Feuillet[1,2], Hélène Charreire[1,2], Mehdi Menai[1], Paul Salze[3], Chantal Simon[4], Julien Dugas[4], Serge Hercberg[1], Valentina A Andreeva[1], Christophe Enaux[3], Christiane Weber[3], Jean-Michel Oppert[1,5]

[1]University of Paris 13, Equipe de Recherche en Epidémiologie Nutritionnelle (EREN), UMR U1153 Inserm/U1125, Centre de Recherche en Epidémiologie et Biostatistiques Sorbonne, Paris Cité, Bobigny, France
[2]University of Paris Est, Lab'Urba, Urban Institute of Paris, UPEC, Créteil, France
[3]University of Strasbourg, Laboratoire "Image Ville Environnement" UMR 7362 CNRS, Strasbourg, France
[4]CARMEN, Institut National de la Santé et de la Recherche Médicale U1060, University of Lyon 1, Institut National de la Recherche Agronomique U1235, CRNH Rhône-Alpes, Lyon, France
[5]Service de Nutrition GH Pitié-Salpêtrière (AP-HP), Pierre and Marie Curie University, Institut Cardiométabolisme et Nutrition (ICAN), Paris, France

Abstract: Background: According to the social ecological model of health-related behaviors, it is now well accepted that environmental factors influence habitual physical activity. Most previous studies on physical activity determinants have assumed spatial homogeneity across the study area, *i.e.* that the association between the environment and physical activity is the same whatever the location. The main novelty of our study was to explore geographical variation in the relationships between active commuting (walking and cycling to/from work) and residential environmental characteristics. Methods: 4,164 adults from the ongoing Nutrinet-Santé web-cohort, residing in and around Paris, France, were studied using a geographically weighted Poisson regression (GWPR) model. Objective environmental variables, including both the built and the socio-economic characteristics around the place of residence of individuals, were assessed by GIS-based measures. Perceived environmental factors (index including safety, aesthetics, and pollution) were reported by questionnaires. Results: Our results show that the influence of the overall neighborhood environment appeared to be more pronounced in the suburban southern part of the study area (Val-de-Marne) compared to Paris inner city, whereas more complex patterns were found elsewhere. Active commuting was positively associated with the built environment only in the southern and northeastern parts of the study area, whereas positive associations with the socio-economic environment were found only in some specific locations in the southern and northern parts of the study area. Similar local variations were observed for the perceived environmental variables. Conclusions: These results suggest that: (i) when applied to active commuting, the social ecological conceptual framework should be locally nuanced, and (ii) local rather than global targeting of public health policies might be more efficient in promoting active commuting.

Keywords: Walking, Cycling, Active Transportation, Geographically Weighted Regression, Spatial Non-Stationarity

1. Background

Promoting active transportation (walking and cycling) has become a priority in public health policies. The practice of daily active transportation has been shown to provide significant economic and health benefits. It contributes substantially to improving household budgets (reducing car-related expenditure[1]), limiting gas emissions, and decreasing other negative externalities, e.g. congestion, noise and pollution[2]. In addition, active transportation contributes to overall physical activity, which has been shown to have a protective effect against major chronic diseases (cardiovascular disease, type 2 diabetes, and certain cancers, according to the Physical Activity Guidelines for Americans[3]). However, active transportation represents only a small proportion of daily commuting. In France in 2008, 23% of commuting was by walking and 2.7% by cycling[4]. In the US, the corresponding figures were 10.5% and 1% in 2009, according to the National Household Travel Survey[5].

One major concern in developing relevant policies is the need for a better understanding of walking and cycling, which are complex behaviors, and their multiple determinants. Based on the socio-ecological conceptual framework, interacting factors include those related to the individual-level sphere and those associated with the social and physical environment[6][7]. In turn, social and physical environmental factors comprise both perceived and objective dimensions[8]. In this paper, a geographical approach was used to identify spatial variations in the associations between environmental characteristics and active transportation. Thus, the study could aid the development of more focused or locally adapted public policies and planning choices. For the purpose of this research, data on active commuting and explanatory factors were collected from a large sample of French adults and an innovative local-targeted regression technique was applied.

Associations between active commuting behaviors and social support (e.g.[9][10]), different physical environment dimensions, e.g. walkability and bikeability, land use, public transportation availability, safety, aesthetics, etc., in residential and/or work neighborhoods are documented in the literature (e.g.[11]–[24]). However, previous studies have largely been based on the implied and strong assumption that the relationship between individual/environmental factors and active commuting is spatially homogeneous, *i.e.* that pertinent factors operate in a similar manner everywhere. This is a necessary condition for the use of global regression models. Yet, non-stationarity, referring to the variation in relationships across space[25][26], is a very common phenomenon in any geographical dataset like place-level factors. For instance, a global positive association can be found between walking to work and overall walkability, but is this really the case everywhere throughout a city or a neighborhood?

In recent years, an innovative local-based regression technique has been gaining popularity for exploring spatial non-stationarity among data: the Geographically Weighted Regression model (GWR,[25]). GWR allows the parameter estimates to vary locally, unlike in global models where they remain constant. GWR fits local regression at each location by applying a weighting scheme (based on a kernel function) which gives more weight to neighboring locations[27][28]. The results emphasize the spatial patterning of relationships. The GWR technique has been successfully applied in a few health-related studies[29]-[37]. For instance, Chen and Truong[34] used GWR to highlight that township disadvantages increased obesity prevalence only in certain areas in Taiwan. Similarly, Chalkias *et al.*[37] showed that only certain zones in Athens (Greece) constituted an obesogenic environment.

To our knowledge, GWR has not yet been applied to explore the potential non-stationarity of environmental correlates of active commuting. The main objective of our study was thus to investigate whether several objective and perceived environmental determinants of active commuting behaviors varied across space, ceteris paribus (*i.e.* after adjusting for individual characteristics), in Paris and its immediate suburbs. The underlying idea was to question the relevance of using a general conceptual framework, such as the socio-ecological model, when analyzing spatial data including potential non-stationary processes. The study was thus positioned to raise questions about the potential gain of a more geographically nuanced theoretical model, taking into account area-specific attitudes and behaviors. To explore this issue, a multivariate geographically weighted Poisson regression (GWPR) model was used, based on data from French adults.

2. Methods

2.1. Study Area and Population

Information about active commuting behaviors and some of the explanatory factors was derived from the NutrinetSanté study, an ongoing web-based cohort launched in France in May 2009, which focuses on relationships between nutrition and health[38]. Briefly, participants aged 18y. or older completed a set of questionnaires assessing demographic and socio-economic characteristics, as well as physical activity and perceived residential environment (response rate of 48.5%). Residential addresses were obtained from all participants, geocoded to the parcel or street levels and implemented as a shapefile in a geographical information system (GIS). This study was conducted according to the guidelines laid down in the Declaration of Helsinki, and all procedures were approved by the Institutional Review Board of the French Institute for Health and Medical Research (IRB Inserm n° 0000388FWA00005831) and the Commission Nationale Informatique et Libertés (CNIL n° 908450 and n° 909216). All participants gave their written electronic informed consent to take part in the study.

The area covering Paris and its three surrounding départements called the "Petite Couronne" (**Figure 1**), was targeted because it is the most populated region

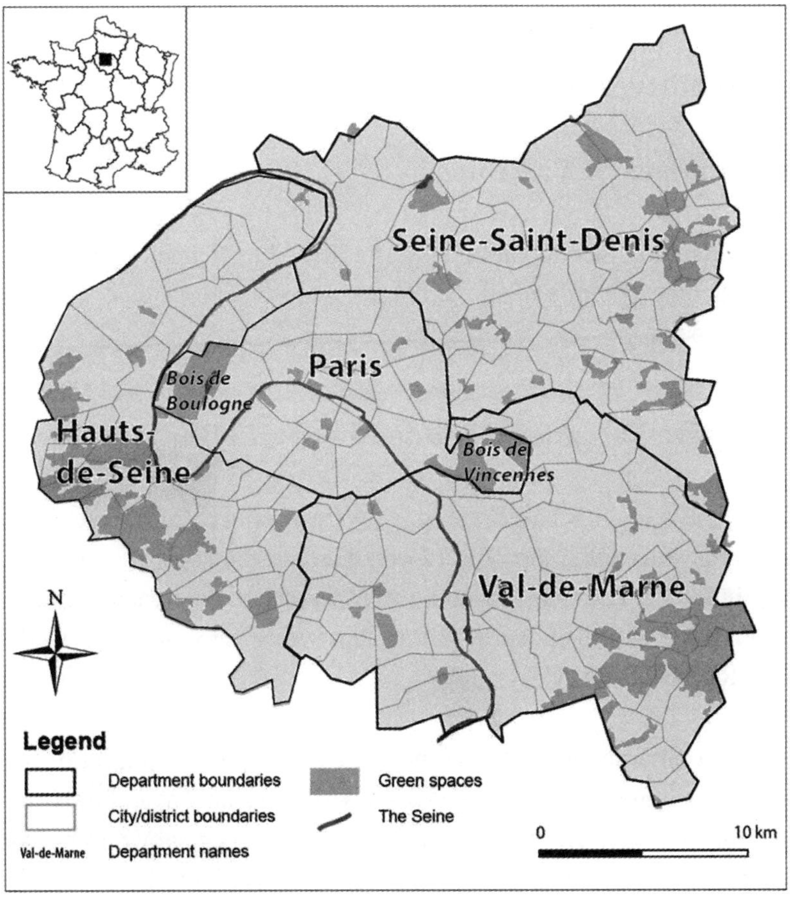

Figure 1. Location map of the study area showing Paris and its three immediate suburbs.

in France, giving access to a large sample of participants. Indeed, this urban area sprawls over 762km^2 and has more than 6.6 million inhabitants (2010 French Census) with a population density of approximately 8,700hab./km^2. More than 2.2 million individuals live in the city of Paris alone, leading to a population density of 21,347hab./km^2.

2.2. Outcome Variable: Active Commuting (Walking and Cycling to and from Work)

Participants reported their time spent on active commuting. For each subject, the variable used was the mean of the hours spent walking and biking to/from work per week during the past 4 weeks.

2.3. Explanatory Variables

2.3.1. Environmental Variables

A set of environmental variables was assessed to characterize the residential neighborhood of each individual. Variables were categorized as objective (*i.e.* GIS-based) or perceived (*i.e.* reported via a questionnaire).

2.3.2. Objective Variables Obtained from GIS (Built and Social)

Overall, fifteen GIS-based variables were obtained from different sources. They were related to either the built (7 variables) or the social (8 variables) environment (**Figure 2**). The GIS procedure used for calculating each variable is illustrated in **Figure 2**. All the geoprocessing steps were performed with ArcGIS 10.1 (ESRI Inc., Redlands, CA, USA).

The built environment set encompassed 7 variables, representing 3 distinct groups (land use and facilities, level of bikeability, and availability of public transportation). These were chosen because they have been shown to be associated with transportation-related physical activity[39]-[41]. According to this literature, expected relationships between walking for transportation and proximity facilities

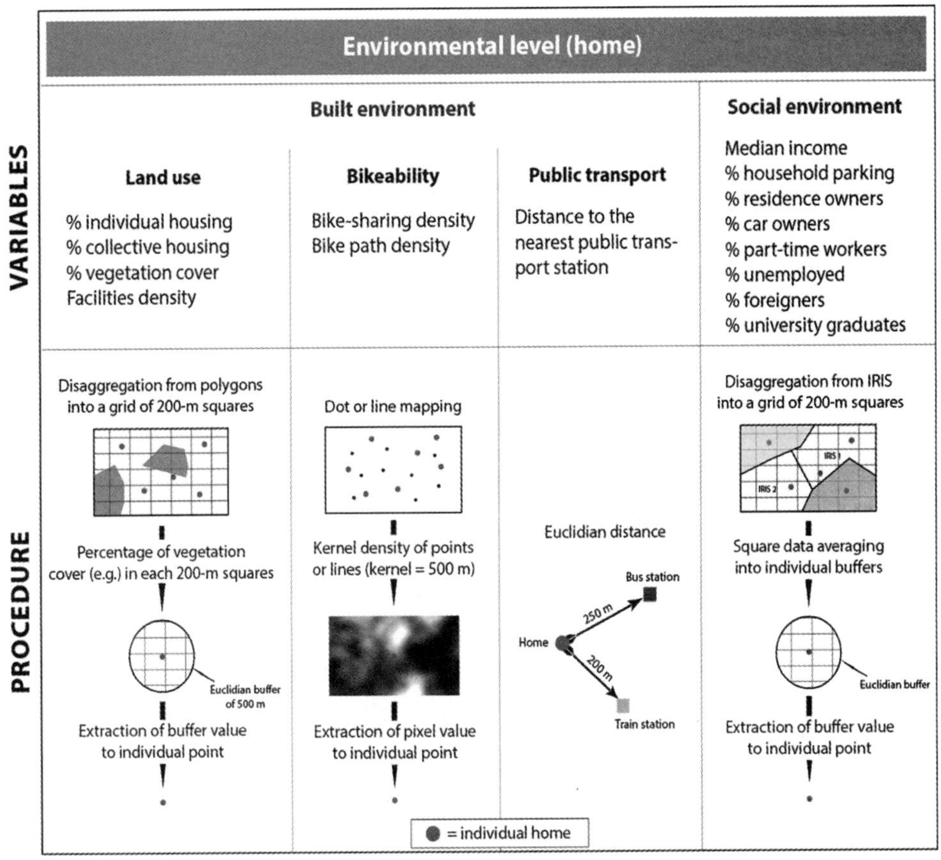

Figure 2. GIS-based schematic procedure for the calculation of objective (both built and social) environmental variables.

and land use mix are positive, even if the specifics of these associations are less clear.

(i) Land use and facilities

Land use included the percentage of area covered by individual housing, collective housing and vegetation, respectively, as well as proximity facilities. Data on housing and vegetation were provided as shapefiles (polygons) by the Paris Region Urban Planning & Development Agency (IAU Île-de-France, 2008, scale: 1:5000). Data on proximity facilities were given as the number of 26 different types of facilities, e.g. banks, bakeries, drugstores, restaurants, etc., by the IRIS Census unit. IRIS areas (acronym for "Aggregated Units for Statistical Information"), provided by the French National Institute of Statistics and Economic Studies (INSEE, www.insee.fr), represent the smallest unit for dissemination of French infra-municipal data. They include an average of 2000 residents per unit and are homogeneous in terms of housing and socio-economic conditions. Data on land use and facilities were then disaggregated into a grid of 200 × 200m cells, in order to obtain a spatially homogeneous data net. Finally, these data were linked to each individual within a Euclidian buffer of 500m around the residential address (mean values of the cells included in the buffer), as this distance is commonly used in accessibility studies[42][43].

(ii) Level of bikeability

Bikeability (positive expected association with active commuting), including bike-sharing facilities and bike path densities, as assessed by fixed kernel density estimation (KDE) with a bandwidth of 500 m, which is equivalent to a Euclidian buffer for polygon-shaped variables. KDE is a smoothing geostatistical technique to transform a point or a line pattern into a continuous surface map of density (raster), with an estimated value for each cell. This method has been extensively used in other accessibility studies (e.g.[31][44]–[46]). Bike-sharing accessibility was assessed by calculating kernel density estimation on the 1,230 bicycle stations of the Parisian bicycle-sharing system, called "Vélib". Next, the value of the overlapping density raster cell (for bike-sharing and bike path densities) was assigned to each individual residential address.

(iii) Public transportation availability

Public transportation availability (positive expected association with active commuting) was assessed by calculating the distance to the nearest subway, bus or train station (provided by IAU) from each individual home.

(iv) Socioeconomic environment

The socioeconomic level of the residential environment has been considered since it is expected to be positively associated with overall physical activity, including active commuting[47]. Eight neighborhood-level socioeconomic variables, from the Census database (www.insee.fr), were included and disaggregated into the 200-m square grid (following the same procedure as used for land use and facilities) (see **Figure 2**): percentage of foreign residents, unemployed, part-time workers, university graduates, homes occupied by their owners, car owners, households with a parking space, and median income.

2.3.3. Statistical Analyses of Objective Environmental Variables

The 7 physical and 8 socio-environmental variables were substantially multicol-linear. As a valid regression analysis requires the explanatory variables to be independent, a suitable solution to this statistical issue is to transform the set of correlated variables into synthetic uncorrelated variables, named principal components (PC). In PC analysis (PCA), the principal components retained are those that explain the maximum amount of variance of the original data. This method was used in our study to avoid dropping any explanatory variable and to keep as much information as possible. PC were retained as new explanatory variables when eigenvalues were greater than one. PC coordinates were assigned to individuals and then mapped to facilitate the interpretation. The PCA was carried out with varimax rotation to represent the linear proximity among variables. It was conducted only with objective variables since we were interested in investigating separately the spatial variation in the influence of objective and perceived variables on active commuting.

The PCA results revealed that the first five principal components accounted for 84% of the total inertia of the original variables. The first two PC explained 64% of the variance, whereas the following three PC had an eigenvalue below 1. For this reason, only the two first PC were kept as independent variables in the main analysis. The highest eigenvectors for the first component (PC1) were associated with low rates of car ownership, individual housing, and households with a parking space, but with high rates of home ownership, collective housing, and high facility and bike-sharing densities (**Table 1**). This first synthetic variable was thus related to the built environment density. The highest eigenvectors for PC2 were associated with high median income, a high proportion of university graduates, and low proportions of foreign residents and unemployed. This second synthetic variable was related to the socio-economic environment (high values indicate well-to-do areas).

Table 1. Results of the principal component analysis conducted with the fifteen objective environmental variables.

Variable	Principal component 1 Densely built-up areas & facility availability	Principal component 2 Socio-economics: well-off neighborhoods
Median income	−0.01	0.51
% households with a parking spot	−0.36	0.04
% home owners	0.32	0.08
% car owners	−0.38	0.04
% individual housing	−0.33	−0.06
% collective housing	0.34	0.13
% part-time workers	0.17	−0.21
% unemployed	0.11	−0.49
% foreign residents	0.17	−0.42
% university graduates	0.20	0.44
% vegetation cover	−0.20	0.17
Facility density	0.30	0.05
Distance to public transportation	−0.15	−0.03
Bike-sharing density	0.34	0.10
Bike path density	0.10	0.05
Overall KMO score = 0.78; Bartlett's test $p < 0.001$		

Values in bold font are greater than |0.3|.

2.3.4. Perceived Environment Variables

Three perceived residential neighborhood variables were defined by specific questions in a self-administered questionnaire. These three questions were derived from the ALPHA questionnaire designed to measure the relationship between physical activity and the environment in a European context[48]. Briefly, the three questions were related to (i) bike safety in road traffic ('cycling is unsafe because of the traffic'), (ii) pollution ('there is too much pollution in my neighborhood'), and (iii) aesthetics ('my neighborhood is not clean and not well-maintained').

Responses included five modalities based on a Likert-type scale (strongly agree, somewhat agree, neither agree nor disagree, somewhat disagree, strongly disagree). To account for substantial multicollinearity among these three variables, an index was built using principal component analysis. This index accounted for 48% of the total variance. The factor loadings were 0.72 for bike safety, 0.69 for pollution and −0.68 for aesthetics. Therefore, a high value in this synthetic index indicates individuals who reported a low level of pollution, a feeling of safety for cycling and a low level of aesthetics. A positive relationship between this synthetic variable and active commuting is expected[49].

2.3.5. Individual Variables

Individual data included age, gender and education (divided into two categories, <high school, ≥high school), number of motor vehicles per household, number of bikes per household, and an indicator related to the possession of a transit pass (yes or no). Finally, two work-related variables were added: self-reported commuting time (divided into tertiles) and availability of a parking space at the workplace (yes or no).

2.4. Statistical Modeling

Given the non-Gaussian, zero-inflated distribution of the outcome variables (walking and cycling to and from work), each value was rounded to the nearest half-unit (0.5). This procedure of discretization enabled the variable to be modeled with a Poisson regression. First, a global Poisson model (GPR) was carried out, *i.e.* parameter estimates were kept constant to explore global relationships between environmental variables and active commuting, while adjusting for individual variables. Secondly, a geographically weighted Poisson regression[28][30] was conducted to account for the possible spatial non-stationarity of these relationships. The statistical specifications of GPR are described in Appendix A.

2.5. Local Model (GWPR)

Geographically weighted regression aims to capture spatial non-stationarity, *i.e.* spatially varying relationships, in a regression model by allowing regression parameters $\beta_0, \cdots, \beta_k$ to vary with location. To do this, GWPR incorporates spatial coordinates in the model. Since this study assumes that only environmental factors have spatial effects and not individual ones, we performed a semiparametric GWPR, *i.e.* we kept fixed the coefficients of the individual factors[30]. The resulting equation of the semiparametric GWPR we used is expressed as follows:

$$\log \lambda_i = \sum_k \beta_k \left(u_i, v_i\right) x_{ik} + \sum_m r_m x_{im} \tag{1}$$

where $\beta_k(u_i, v_i)$ are local model parameters associated with environmental factors and specific to residential location of subject i, (u_i, v_i) denoting the coordinates of residential location of subject i, x_{ik} is the value of the kth environmental variable at residential location of subject i, and γ_m are model parameters associated with the individual variables x_{im}, not assumed to depend on geographical location.

A key step in the development of GWPR consists of calibrating the model by a kernel regression method in order to estimate smoothed geographical variations in the parameters with a distance-based weighting scheme[30]. GWPR uses a spatial kernel since it is assumed that observations near point i have more influence on the estimation of parameter $\beta_k(u_i, v_i)$ than do observations located farther from i. In other words, GWPR integrates multiple local regressions within an overall frame-work, as illustrated in **Figure 3**. The estimation of the parameters is described in Appendix B. Next, parameters at location i are estimated by maximizing the geographically weighted log-likelihood[30]. Thereby, the geographical weight structure can be based on one of two types of kernel function, Gaussian or bi-square[25][28]. The kernel's bandwidth can be set as fixed (based on metric distance) or adaptive (based on a constant number of neighbors considered in each regression calculation). Adaptive kernels are suitable when the units of analysis are irregularly distributed across space, which is the case here: respondents are sparse near the study area boundaries and absent from the two large green spaces of Paris (Bois de Boulogne and Bois de Vincennes, see **Figure 1**). For this reason, the adaptive kernel method was used, making sure that each local regression encompassed enough regression points irrespective of the location, coupled with a bisquare weighting scheme, which gave better results than the Gaussian one alone (see details in Appendix C).

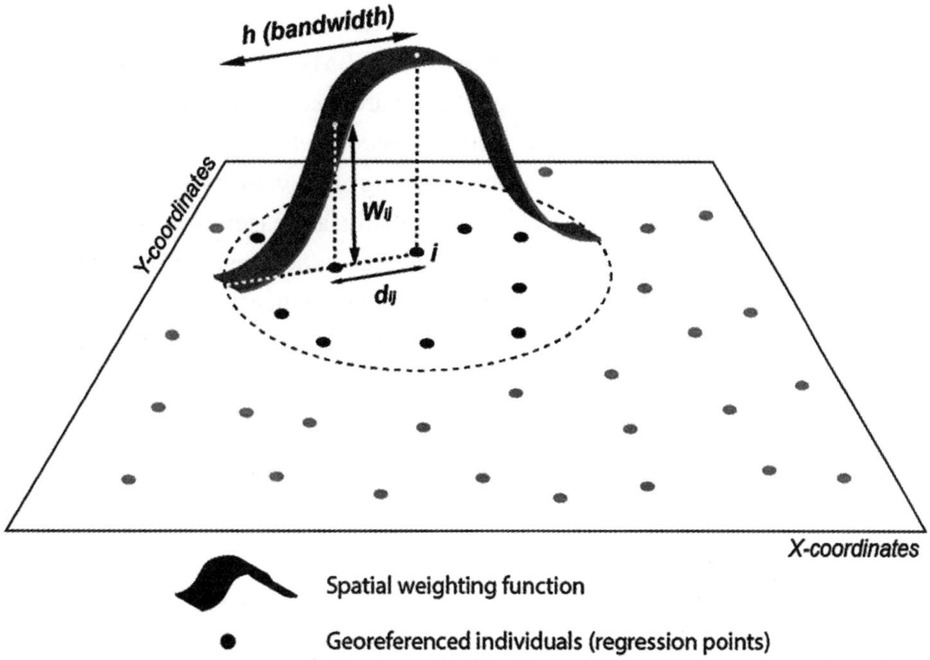

Figure 3. Schematic representation of the geographically weighted regression and its spatial parameters.

The maximum number of neighbors in each regression model was determined by minimizing, in an iterative way, the corrected Akaike's Information Criterion (AICc), which is a statistic based on the log-likelihood of the model, weighted by the actual number of parameters[50]. In this study, the number of neighbors minimizing the AICc was 800 (after testing down from 4,000 to 50, every 50). This amounted to an adaptive bandwidth size varying from 3,150m to 19,250m and a mean distance of 5,650m, knowing that the study area spreads over approximately 30,000m.

AICc was also used to compare the performance of both global and local models. The model with the smallest AICc should be selected as an optimal model called MAICE (minimum AIC estimator, see[30]. A difference in AICc values of more than 2 is considered substantial.

Global models were performed with SAS 9.3 software (SAS Institute Inc., Cary, NC, USA) and the academic piece of software GWR 4.0 was used to calibrate and run geographically weighted models. This software is a tool for modelling varying relationships among variables by calibrating GWR and Geographically Weighted Generalized Linear Models (GWGLM) with their semi-parametric variants (https://geodacenter.asu.edu/gwr_software, see Nakaya *et al.*[51] and Nakaya[52] for additional details). Maps with continuous values were based on an interpolation procedure (inverse distance weighting) and were processed with ArcGIS 10.1 (ESRI Inc., Redlands, CA, USA). Finally, a measure of statistical significance (pseudo t-value) of GWPR estimates was added visually as isolines for each local term map. A value greater than |1.96| indicates a p-value < 0.05. Mapped GWPR estimates are log-odds, with negative and positive values meaning negative and positive relationships, respectively.

3. Results

3.1. Descriptive Statistics

3.1.1. Characteristics of the Study Population

The participation rate for the questionnaire was 48.5%. 61.7% of the valid questionnaires were kept, resulting in a sample of 4,164 respondents. **Table 2** shows that they are 43.6 years-old on average, 78% of them are women, 83% have an educational level above high school and approximately 46% have a transit pass. 41% of respondents do not practice active commuting, while approximately 95% walk or cycle less than 5 h/week. The mean time for active commuting among participants reporting any such activity (59%) is 2.3 h/week. Regarding the perception variables, around 47% of the respondents report too much pollution, approximately 40% report that biking is unsafe in their neighborhood, and 16% find that their neighborhood is not clean (**Table 2**).

3.1.2. Characteristics of the Environment

For the first principal component (PC1) related to the built environment density, the highest values are located within Paris and its immediate suburbs, as expected, whereas values tend to decrease further outwards [**Figure 4(a)**]. For the second principal component (PC2) related to the socio-economic environment, the highest values are found in neighborhoods mostly encompassing the southern and

Table 2. Descriptive statistics of the variables used in the analysis.

Variable (N = 4164)	Mean (or %)	Min	Max	STD
Active commuting (outcome)				
% No	41	/	/	/
% Yes	59	/	/	/
If yes (h/week)	2.34	0.1	10.8	1.8
Individual variables				
Age	43.6	19	87	13.3
Gender				
% Men	22.0	/	/	/
% Women	78.0	/	/	/
Education				
% < high school	16.8	/	/	/
% ≥ high school	83.1	/	/	/
Parking at work				
% Yes	36.7	/	/	/
% No	63.3	/	/	/
Transit pass				
% Yes	45.8	/	/	/
% No	54.3	/	/	/
Commuting time				
% 1st tertile	36.2	/	/	/
% 2nd tertile	30.6	/	/	/
% 3rd tertile	33.1	/	/	/
Number of motor vehicles owned	1.03	0	8	0.95
Number of bikes owned	1.30	0	4	1.38
GIS-based environmental variables				
PC1 (densely built-up areas)	0	−8.6	5.3	2.5
PC2 (well-to-do areas)	0	−8.5	4.0	1.8
Perceived environmental variables		**% agree**	**% neither agree nor disagree**	**% disagree**
Too much pollution		47.3	24.6	28.1
Neighborhood is not clean		15.8	14.3	69.9
Biking is unsafe		34.4	25.7	39.9

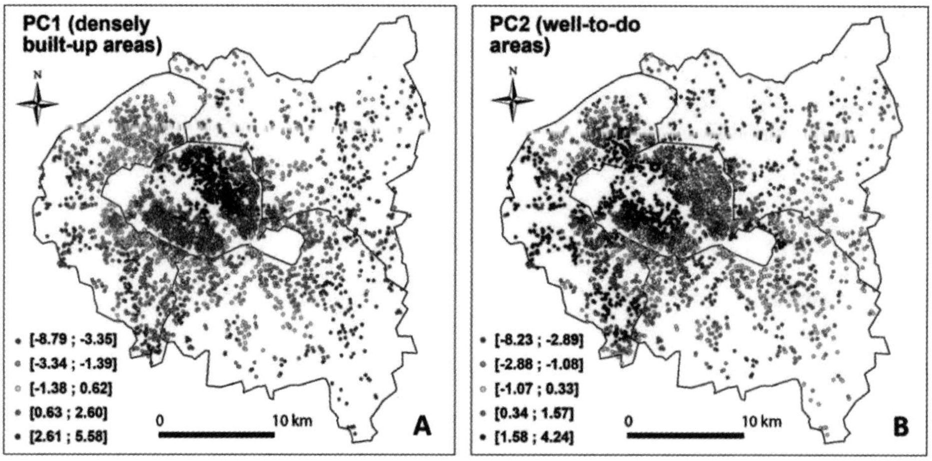

Figure 4. Map view of the first two components derived from the principal component analysis and kept as explanatory variables. (a) The first component (PC1) refers to the built environment (densely built-up areas and facility density) and is characterized by high values in Paris. (B) The second component (PC2) is related to the socio-economic environment (high values indicate well-to-do areas).

western parts of Paris and the surrounding suburbs, as well as areas surrounding the Bois de Vincennes in the eastern part of the city [**Figure 4(b)**]. The lowest values of PC2 are located in the northern part of the area, especially the western part of the Seine-Saint-Denis département.

3.1.3. Global Relationships between Active Commuting and Environmental Variables

Total deviance explained by the global regression model is 24% with an AICc of 6,733 (**Table 3**). Results of the parameter estimations via the global Poisson regression are summarized in **Table 4**. Nagelkerke's R2 for the full model is 0.317 (meaning that approximately 32% of the total variance is explained by the model), with a value of 0.297 for the model including only individual-level variables, *i.e.* the individual level accounts for around 94% of the full model pseudo R2.

Regarding the environmental variables, after controlling for individual ones, the results show that active commuting is positively associated with the first principal component (OR = 1.05, 95% CI 1.03−1.07, implying that an increase of one unit of the densely-built level of the neighborhood is associated with an increase in

Table 3. Overall performances of both global and local (bi-square kernel) regressions used for modeling associations between active commuting behaviors and individual and environmental factors.

Model (N = 4164)	Deviance (D)	Effective number of parameters (k)	AICc (D + 2*k)	Deviance explained
GPR	6733	12	6757	0.24
GWPR (800 neighbors)	6207	167	6556	0.30

Table 4. Parameter estimations from the global Poisson regression model.

	Variable	Log-odds	OR	Wald 95% CI		p-value
Individual level	Intercept	−0.10	0.91	0.68	1.21	0.498
	Age	0.00	1.00	1.00	1.01	0.085
	Gender (ref = male)	0.04	1.04	0.96	1.12	0.393
	Education (ref = < high school)	−0.12***	0.89	0.82	0.96	<.0001
	Parking at work (ref = yes)	−0.29***	0.75	0.69	0.80	<.0001
	Transit pass (ref = yes)	−0.05	0.95	0.88	1.03	0.148
	Commuting time (ref = 1st tertile)	0.67***	1.96	1.87	2.05	<.0001
	Number of vehicles owned	−0.19***	0.83	0.79	0.87	<.0001
	Number of bikes owned	0.09***	1.10	1.07	1.13	<.0001
Environmental level	PC1 (densely built-up areas)	0.05***	1.05	1.03	1.07	<.0001
	PC2 (well-to-do areas)	0.01	1.01	0.99	1.03	0.496
	Neighborhood perception (too much pollution and not clean) (0 = strongly agree; 5 = strongly disagree)	0.04**	1.04	1.02	1.06	0.001

***p < 0.001, **p < 0.01.

active commuting), but no global relationship was detected with the second principal component (socio-economic level of the neighborhood). Finally, the perception of the neighborhood environment is significantly associated with active commuting (OR = 1.04, 95% CI 1.02–1.06).

3.1.4. Spatial Variations in the Relationships

Total deviance explained by the GWPR model is 30% with an AICc of 6,556 (**Table 3**), which is a better goodness-of-fit diagnostic than the global regression model. Although the mean ORs are relatively close to those of the global regression (**Table 5**), the range of GWPR OR estimations shows the non-stationarity of the relationships between the environmental variables and active commuting in the study area.

For the three environmental variables, and while controlling for individual-level covariates, ORs are spread on both sides of 1, meaning that the relationships are sometimes negative, sometimes positive, and sometimes non-significant according to the location in the study area (**Figure 5**). Relationships between active commuting and the built environment vary substantially, with ORs ranging from 0.84 to 1.25 (**Table 5**). The relationships are significant and positive in the southern part (département of Val-de-Marne) and the northeastern part of the study area [**Figure 5(a)**]. Elsewhere, the relationships are mostly non-significant, except at specific locations, such as in a small area in Paris where the relationships are negative. While the global model shows no significant associations between active commuting and the socio-economic environment, GWPR indicates some local nuances, as small parts of the area (extreme north and south) exhibit significantly positive Ors [>1.10, **Table 5** and **Figure 5(b)**].

Perceived environmental variables also show some non-stationarity [**Figure 5(c)**]. In northern and southern central parts of the area, associations are significantly

Table 5. Parameter estimations from the semiparametric geographically weighted Poisson regression model (after adjusting for individual variables).

Variable	Mean log-odds	Mean ORs	STD log-odds	Min ORs	Max ORs	Range ORs
Intercept	0.01	1.01	0.12	0.75	1.43	0.68
PC1 (densely built-up areas)	0.03	1.03	0.06	0.84	1.25	0.41
PC2 (well-to-do areas)	−0.00	0.99	0.03	0.89	1.13	0.24
Neighborhood perception (too much pollution and not clean) (0 = strongly agree; 5 = strongly disagree)	0.03	1.03	0.06	0.85	1.27	0.42

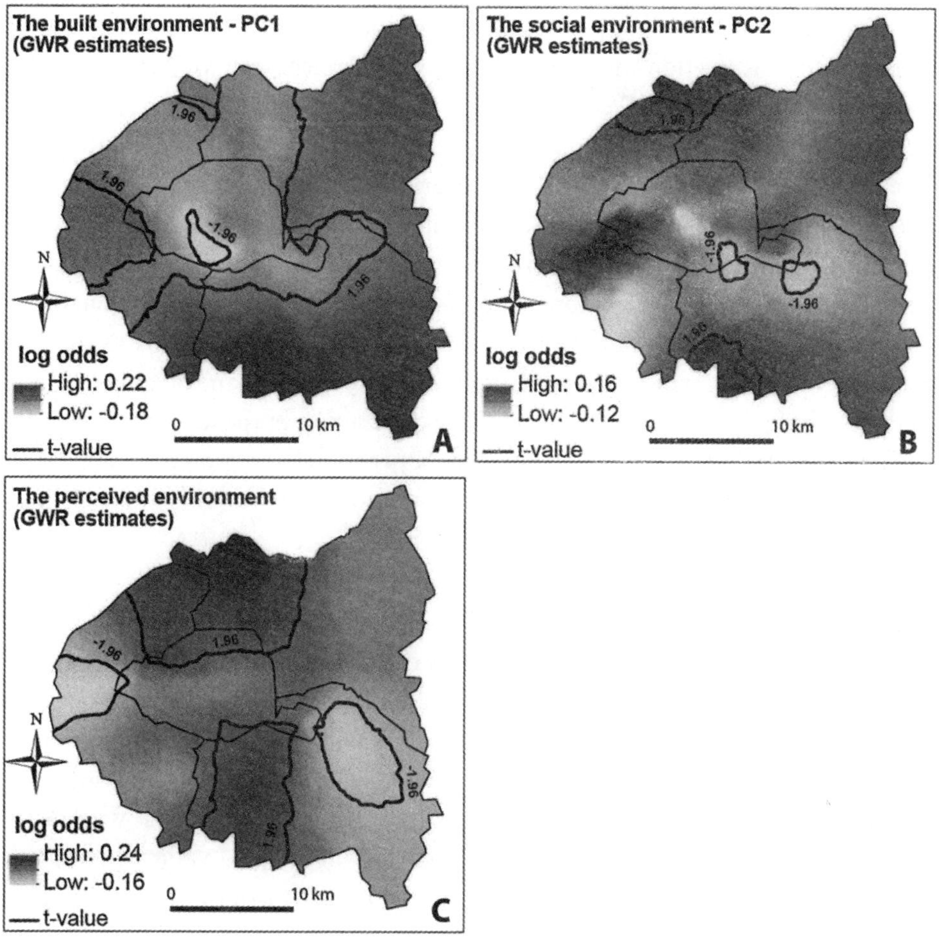

Figure 5. Map results of the geographically weighted Poisson regression parameters (log odds) for the built (A), the social (B) and the perceived (C) environment. Positive values of the log-odds (in red) indicate positive relationships between the respective explanatory variable and active commuting, and negative values of the log-odds (in yellow) indicate negative relationships. A pseudo t-value > |1.96| shows significant associations ($p < 0.05$).

positive, whereas they are non-significant elsewhere, or even negative in some parts of the Hauts-de-Seine and Val-de-Marne départements.

4. Discussion

Using a GWPR, we have clearly demonstrated some spatial heterogeneity within the relationships across our study area. This represents the main novel result of our study. To the best of our knowledge, there is currently no other study on spatially varying relationships between the environment and walking or cycling for commuting purposes. We have shown that the associations of environmental factors with active commuting are area-specific. In other words, the relative influence of the specific environmental characteristics of the neighborhood (built, social and perceived) differs by location. For instance, the southern part of the study area (département of Val-de-Marne) is generally characterized by significant positive associations between environmental characteristics and active commuting. In contrast, in the major part of Paris or in the northern part of the Hauts-de-Seine département, environmental characteristics and perception of the neighborhood are not associated with walking and cycling for commuting purposes. Our results also show the complexity of the relationship between the environment and active commuting. At the same location, certain environmental variables may be associated with the outcome, while others may not.

Regarding the global relationships, we have highlighted that both perceived and objective environmental factors are significantly associated with active commuting, as expected. The importance of both the individual and the environmental level for active transportation is in line with existing evidence (e.g.[14][17][24][49][53]–[57]) based on the social ecological framework[8][58]. Although some environmental characteristics have significant relationships with active commuting, they only weakly contribute to explaining its total variance (as shown by the respective Nagelkerke's pseudo R^2 of the nested models[R^2 = 0.30 without the environment level, and 0.32 with it)]. This corroborates findings by Giles-Corti and Donovan[59] in Australia, and Ogilvie *et al.* in the UK[55]. For instance, Ogilvie *et al.*[55] highlighted that in urban neighborhoods of Glasgow, 18.7% of the total variance in active travel was explained by personal correlates, and 20.1% when environmental-level variables were added. These authors concluded that including environmental characteristics did not substantially modify the influence of the personal characteristics on the associations studied.

4.1. Interpretation of the Observed Spatial Non-Stationarity

The interpretation of the spatial patterning of the relationships needs further and deeper investigation. Three potential causes of parametric instability of the regression parameters have been identified by Fotheringham *et al.*[28]:

(1) The non-stationarity could be due to random sampling variations and hence not related to any underlying spatial process.

(2) The relationships might be intrinsically different across space, in other words "there are spatial variations in people's attitudes or preferences or there are different administrative, political or other contextual issues that produce different responses to the same stimuli over space".

(3) The non-stationarity could also indicate that the model suffers from major misspecification or omission of key variables (or representation by an incorrect functional form).

The demarcation between the second and third potential causes of parametric instability may seem blurred in some cases. Can relationships really be intrinsically different across space? As discussed by Blainey[60], space may represent "merely a proxy for societal factors which are not captured by the model". The crucial issue here is that such space-related societal factors are sometimes very difficult to define, and hence hard to quantify in a model.

However, in some instances, varying relationships seem to be driven both by the level of and the variance in the explanatory variable involved. For example, regarding the variable related to densely-built areas, the absence of a relationship in the major part of Paris could be explained by the fact that facility and building density is already extremely high in the city, so that one additional unit (of density or facility availability) would have a limited marginal effect on active commuting behavior. In contrast, in the less densely-built areas surrounding Paris (south and east), an increase of one unit regarding this variable would have a more noticeable impact. In that case, the instability of the parameters associated with the facility availability would only be due to a threshold effect of the variable itself and not directly linked to another contextual effect. The same rationale was used by Lu *et al.*[61] to explain the heterogeneity of relationships between the number of buildings and non-motorized traffic in Burlington, Vermont, USA. The number of buildings was associated with non-motorized traffic in the suburbs (with low building density), but not in the city center (with high building density).

Local associations with the perceived environment also show interesting spatial patterns. Unlike the objective environmental variables, the perceived ones do not follow any obvious spatial distribution (meaning that they are not correlated with objective measures). This means that spatial non-stationarity associated with these variables could be due to the omission of variables in each local context drawn through the GWPR estimates [**Figure 5(c)**]. For instance, a local confounding factor may have been omitted from the model, leading to a spurious association. In some locations, reporting a low level of pollution and a strong feeling of safety for cycling is associated with a decrease in active commuting, which is counter-intuitive. We can hypothesize that in such places, these perceived neighborhood characteristics may be correlated with exposure to a limited traffic volume, while such little traffic may indicate a less walkable/bikeable environment in terms of infrastructure density[62].

Small single patterns of non-stationarity often remain difficult to interpret in detail, partly because the possibly omitted local variables are often not easily quantifiable, being the results of complex local interactions. In particular, the strength of social interactions within a place may lead to a homogenization of the relationships[63]. However, exploring these interactions, opening "the black boxes of places"[64], needs further and deeper quantitative and qualitative investigations. It is still advisable to keep an overview of the overall spatial patterning of non-stationarity, since this enables some boundaries between different local contexts to be drawn, wherein, for complex reasons, relationships converge.

4.2. Implications of the Findings

The non-stationarity of spatial datasets has been demonstrated in other studies dealing with health-related outcomes, such as obesity[37], cardiovascular mortality[34], and health care system organization[35]. Such observations lead to questions about the relevance of using global models, which tend to smooth the effects of one or another variable across the entire area, whereas these effects are in fact area-specific. Considering local models over global ones has potential implications for public health policies. As emphasized by Yang and Matthews[35], GWR-based analyses in health-related research could be used as a tool for place-specific targeting and/ or tailoring of public health interventions. For example, potential interventions by local authorities regarding bike safety in traffic, e.g. installing separate bike paths away from roads, might have a stronger impact on cycling behaviors in the southern and northern parts of our study area than in the eastern part

[**Figure 5(c)**]. In addition, increasing bike-sharing stations and facility density (*i.e.* contributing to increasing PC1) would be more useful in the Val-de-Marne děpartement than in the Hauts-de-Seine děpartement, or in Paris [**Figure 5(a)**], in terms of active commuting. Such place-level prioritizing could be not only more efficient than whole-area interventions for promoting active transportation, but could also ensure substantial cost-efficiency in planning policies.

4.3. Theoretical Considerations: Toward a Locally Varying Social Ecological Model

According to Sallis *et al.*[58], the first principle of ecological models is that multiple levels of factors, including individual and environmental ones, influence health behaviors. Our results on active commuting behaviors, namely the spatial variability of the relative influence of individual and environmental factors, suggest the importance of considering the local context. In other words, the social ecological framework needs to be locally adapted, according to the spatial patterning of the relationships. In certain areas, policy-makers might achieve better results by acting on one particular level rather than on multiple ones. This also suggests expanding the fourth principle proposed by Sallis *et al.*[58]: ecological models are most efficient when they are not only behavior-specific, but also area-specific.

Beyond the physical activity domain, the idea of spatially varying relationships also fits the theoretical framework developed by Lytle[65] for eating behaviors. This author hypothesized that the relative influence of individual, environmental and social factors on the proportion of variance explained in eating behaviors varied as a function of the level of restriction of the environment, specifically: "the more restricted an environment is with regard to availability and accessibility of healthy, inexpensive options, the more influence the physical environment may have with regard to food choices that are made"[65]. Our GWPR analyses provide empirical evidence for the application of Lytle's assumption to active commuting behaviors.

4.4. Strengths and Limitations

One major advantage of GWR modeling, which is based on individual locations, is its ability to reveal spatial variations beyond the actual administrative units, and therefore to highlight spatial patterning. Regression parameters can then be seen as new continuous variables, which can lead to a data-based spatial clustering of the relationships for each explanatory variable. As previously shown, GWR also identifies significant associations at the local level that do not appear when fitting the usual global models. Finally, the mapped results are easily readable and can be considered turnkey products for policy-makers. Despite these advantages, GWR has some limitations. First, there are border effects inherent in the concept of spatial kernels. Local regressions for individuals located near the boundary of the study area do not follow exactly the same weighting scheme as those for individuals located in the center, since the former do not have neighbors all around them. This leads to a larger adaptive spatial kernel for these individuals. Second, GWR can lead to local multicollinearity among the explanatory variables, even if the variables are not collinear at a global scale. These facts can affect the validity of the model and the results and therefore require careful consideration.

From a statistical point of view, we performed GWPR models by rounding the outcome variable values to the nearest half-unit (*i.e.* 30 minutes) in order to get integer values. We also run models using other discretization procedures (rounding to the nearest unit and double unit) to check for statistical stability and findings were essentially unaltered, with unchanged spatial patterning of relationships (data not shown). We also run a geographically weighted logistic regression model by separating active commuters from non-active ones, followed by a second Gaussian GWR model only applied to active commuters, and results again were very similar in terms of direction, intensity and spatial structure of the relationships (data not shown).

In addition, a limitation of our study is the fact that the outcome variable, walking and cycling to/from work, was self-reported. This may be a source of potential misclassification, knowing that physical activity usually tends to be over-reported[18][66][67]. Second, we only focused on active transportation for commuting, but some studies have shown that relationships can be inversed when dealing with walking for leisure or errands. For instance, in four Japanese cities, Inoue *et al.*[68] showed the expected associations between neighborhood aesthetics and walking in the neighborhood, walking for leisure and walking for daily errands, while no relationship was found with walking for commuting purposes. Finally, this study did not take into account the environmental characteristics of the workplace or the commuting routes, which may also be associated with active transportation behaviors[10][69].

5. Conclusion

After showing global, significant associations between individual/environmental factors and active commuting (walking and cycling to/from work) in a French web-cohort, this study implemented a geographically weighted Poisson regression to investigate possible non-stationarity among these associations. At a local scale, GWPR-based analyses enable nuances to be understood by clearly highlighting the spatial heterogeneity of the relationships. For instance, the influence of the overall neighborhood environment appears to be more pronounced in the southern part of the study area (département of Val-de-Marne) than in Paris, whereas more complex patterns were revealed elsewhere. We also showed that socio-economic level is significantly and positively associated with the outcome in the extreme northern and southern parts of the area. On the contrary, in some locations, the built environment appears to be non-significantly, or even correlated, in an unexpected way with active commuting (for example in Paris). Perception-based variables are also subject to non-stationarity. Specifically, a better perception of bike safety in traffic is mainly associated with an increase in walking and cycling, except in the northwestern part of the area (Seine-Saint-Denis) where the relationships were inversed. This non-stationarity in the relationships has two main implications. First, from a practical point of view, our results suggest that public policies should follow the spatial patterning of the relationships in order to strengthen their efficiency. GWPR modeling, and its easily readable associated maps, can be a useful tool to guide the design of tailored and area-targeted public policies promoting physical activity for health. Second, from a theoretical point of view, our data suggest that ecological models of health behavior should be not only population- and behavior-specific but also location-specific.

Competing Interests

The authors declare that they have no competing interests.

Authors' Contributions

TF, HC, PS and JMO conceptualized the research idea, TF and MM performed the statistical analysis, and the manuscript was written by TF, under the supervision of JMO, and reviewed by VA, SH, CW, CS, JD and CE. All authors read and approved the final manuscript.

Acknowledgements

This work is part of the ACTI-Cités project (coordinator: J.M. Oppert) carried out with financial support from the French National Cancer Institute (Institut National du Cancer, INCa) through the Social sciences and humanities and public health programme (2011-1-PL-SHS-10).

The NutriNet-Santé cohort study is funded by the following public institutions: Ministère de la Santé, Institut de Veille Sanitaire (InVS), Institut National de la Prévention et de l'Education pour la Santé (INPES), Fondation pour la Recherche Médicale (FRM), Institut National de la Santé et de la Recherche Médicale (INSERM), Institut National de la Recherche Agronomique (INRA), Conservatoire National des Arts et Métiers (CNAM) and Paris 13 University.

The authors also wish to thank the two anonymous referees for their constructive comments.

Source: Feuillet T, Charreire H, Menai M, *et al.* Spatial heterogeneity of the relationships between environmental characteristics and active commuting: towards a locally varying social ecological model[J]. International Journal of Health Geographics, 2015, 14(1):1−14.

References

[1] Garrard J. Active Transport: Adults. An Overview of Recent Evidence. Melbourne: Victorian Helth Promotion Foundation; 2009.

[2] Commission of the European Communities. Green Paper. From the Commission to

the Council, the European Parliament, the European Economic and Social Committee and the Committee of the Regions. Adapting to Climate Change in Europe – Options for EU Action. Brussels Commission of the European Communities; 2007.

[3] USDHHS. Physical Activity Guidelines for Americans. Washington DC: USDHHS; 2008.

[4] Department for observation and statistics. National Survey on Transport and Mobility (in French). Paris: French ministry of sustainable development; 2010.

[5] Pucher J, Buehler R, Merom D, Bauman A. Walking and cycling in the United States, 2001–2009: evidence from the National Household Travel Surveys. Am J Public Health. 2011;101 Suppl 1:S310–7.

[6] Sallis JF, Owen N, Fisher EB. Ecological models of health behavior. In: Glanz K, Rimer BK, Viswanath K, editors. Health Behavior and Health Education: Theory, Research, and Practice. 4th ed. San Francisco: Wiley: Jossey-Bass; 2008. p. 465–82.

[7] Richard L, Gauvin L, Raine K. Ecological models revisited: their uses and evolution in health promotion over two decades. Annu Rev Public Health 2011;32:307–26.

[8] Saelens BE, Sallis JF, Frank LD. Environmental correlates of walking and cycling: findings from the transportation, urban design, and planning literatures. Ann Behav Med Publ Soc Behav Med. 2003;25:80–91.

[9] De Bourdeaudhuij I, Teixeira PJ, Cardon G, Deforche B. Environmental and psychosocial correlates of physical activity in Portuguese and Belgian adults. Public Health Nutr. 2005;8:886–95.

[10] Titze S, Stronegger WJ, Janschitz S, Oja P. Environmental, social, and personal correlates of cycling for transportation in a student population. J Phys Act Health. 2007;4:66–79.

[11] Nelson A, Allen D. If you build them, commuters will use them: association between bicycle facilities and bicycle commuting. Transp Res Rec J Transp Res Board. 1997;1578:79–83.

[12] Trost SG, Owen N, Bauman AE, Sallis JF, Brown W. Correlates of adults' participation in physical activity: review and update. Med Sci Sports Exerc. 2002;34:1996–2001.

[13] Dill J, Carr T. Bicycle commuting and facilities in major U.S. cities: if you build them, commuters will use them. Transp Res Rec. 2003;1828:116–23.

[14] Troped PJ, Saunders RP, Pate RR, Reininger B, Addy CL. Correlates of recreational and transportation physical activity among adults in a New England community. Prev Med. 2003;37:304–10.

[15] Humpel N, Owen N, Iverson D, Leslie E, Bauman A. Perceived environment attributes, residential location, and walking for particular purposes. Am J Prev Med. 2004;26:119–25.

[16] De Geus B, De Bourdeaudhuij I, Jannes C, Meeusen R. Psychosocial and environmental factors associated with cycling for transport among a working population. Health Educ Res. 2008;23:697–708.

[17] Titze S, Stronegger WJ, Janschitz S, Oja P. Association of built environment, social-environment and personal factors with bicycling as a mode of transportation among Austrian city dwellers. Prev Med. 2008;47:252–9.

[18] Charreire H, Weber C, Chaix B, Salze P, Casey R, Banos A, *et al.* Identifying built environmental patterns using cluster analysis and GIS: relationships with walking, cycling and body mass index in French adults. Int J Behav Nutr Phys Act. 2012;9:59.

[19] Dalton AM, Jones AP, Panter JR, Ogilvie D. Neighbourhood, route and workplace-related environmental characteristics predict adults' mode of travel to work. PLoS One. 2013;8:e67575.

[20] Hino AAF, Reis RS, Sarmiento OL, Parra DC, Brownson RC. Built environment and physical activity for transportation in adults from Curitiba, Brazil. J Urban Health Bull N Y Acad Med. 2014;91:446–62.

[21] Panter J, Desousa C, Ogilvie D. Incorporating walking or cycling into car journeys to and from work: the role of individual, workplace and environmental characteristics. Prev Med. 2013;56:211–7.

[22] Panter J, Griffin S, Dalton AM, Ogilvie D. Patterns and predictors of changes in active commuting over 12 months. Prev Med. 2013;57:776–84.

[23] Adams MA, Frank LD, Schipperijn J, Smith G, Chapman J, Christiansen LB, *et al.* International variation in neighborhood walkability, transit, and recreation environments using geographic information systems: the IPEN adult study. Int J Health Geogr. 2014;13:43.

[24] Bopp M, Child S, Campbell M. Factors associated with active commuting to work among women. Women Health. 2014;54:212–31.

[25] Brunsdon C, Fotheringham AS, Charlton ME. Geographically weighted regression: a method for exploring spatial nonstationarity. Geogr Anal. 1996;28:281–98.

[26] Brunsdon C, Fotheringham S, Charlton M. Geographically weighted regression. J R Stat Soc Ser Stat. 1998;47:431–43.

[27] Fotheringham AS, Brunsdon C, Charlton M. Quantitative Geography: Perspectives on Spatial Data Analysis. London: SAGE; 2000.

[28] Fotheringham AS, Brunsdon C, Charlton M. Geographically weighted regression: the analysis of spatially varying relationships. Chichester, England, Hoboken, NJ, USA: Wiley-Blackwell; 2002.

[29] Congdon P. Modelling spatially varying impacts of socioeconomic predictors on mortality outcomes. J Geogr Syst. 2003;5:161–84.

[30] Nakaya T, Fotheringham AS, Brunsdon C, Charlton M. Geographically weighted

Poisson regression for disease association mapping. Stat Med. 2005;24:2695–717.

[31] Maroko AR, Maantay JA, Sohler NL, Grady KL, Arno PS. The complexities of measuring access to parks and physical activity sites in New York City: a quantitative and qualitative approach. Int J Health Geogr. 2009;8:34.

[32] Chen VY-J, Wu P-C, Yang T-C, Su H-J. Examining non-stationary effects of social determinants on cardiovascular mortality after cold surges in Taiwan. Sci Total Environ. 2010;408:2042–9.

[33] Comber AJ, Brunsdon C, Radburn R. A spatial analysis of variations in health access: linking geography, socio-economic status and access perceptions. Int J Health Geogr. 2011;10:44.

[34] Chen D-R, Truong K. Using multilevel modeling and geographically weighted regression to identify spatial variations in the relationship between place-level disadvantages and obesity in Taiwan. Appl Geogr. 2012;32:737–45.

[35] Yang T-C, Matthews SA. Understanding the non-stationary associations between distrust of the health care system, health conditions, and self-rated health in the elderly: a geographically weighted regression approach. Health Place. 2012;18: 576–85.

[36] Broberg A, Salminen S, Kyttä M. Physical environmental characteristics promoting independent and active transport to children's meaningful places. Appl Geogr. 2013;38:43–52.

[37] Chalkias C, Papadopoulos AG, Kalogeropoulos K, Tambalis K, Psarra G, Sidossis L. Geographical heterogeneity of the relationship between childhood obesity and socio-environmental status: empirical evidence from Athens, Greece. Appl Geogr. 2013;37:34–43.

[38] Hercberg S, Castetbon K, Czernichow S, Malon A, Mejean C, Kesse E, *et al.* The Nutrinet-Santé study: a web-based prospective study on the relationship between nutrition and health and determinants of dietary patterns and nutritional status. BMC Public Health. 2010;10:242.

[39] Saelens BE, Handy SL. Built environment correlates of walking: a review. Med Sci Sports Exerc. 2008;40(7 Suppl):S550–66.

[40] Bergeron P, Reyburn S. L'impact de L'environnement Bâti Sur L'activité Physique, L'alimentation et Le Poids. Québec: Direction du développement des individus et des communautés. Institut national de Santé Publique du Québec; 2010.

[41] Knuiman MW, Christian HE, Divitini ML, Foster SA, Bull FC, Badland HM, *et al.* A longitudinal analysis of the influence of the neighborhood built environment on walking for transportation. The RESIDE study. Am J Epidemiol. 2014;180:453–61.

[42] Larsen K, Gilliland J. Mapping the evolution of "food deserts" in a Canadian city: supermarket accessibility in London, Ontario, 1961–2005. Int J Health Geogr. 2008;7:16.

[43] Larsen K, Gilliland J, Hess P, Tucker P, Irwin J, He M. The influence of the physical environment and sociodemographic characteristics on children's mode of travel to and from school. Am J Public Health. 2009;99:520–6.

[44] Gatrell AC, Bailey TC, Diggle PJ, Rowlingson BS. Spatial point pattern analysis and its application in geographical epidemiology. Trans Inst Br Geogr. 1996;21:256.

[45] Fan Y, Khattak AJ. Urban form, individual spatial footprints, and travel: examination of space-use behavior. Transp Res Rec J Transp Res Board. 2008;2082:98–106.

[46] Buck C, Pohlabeln H, Huybrechts I, De Bourdeaudhuij I, Pitsiladis Y, Reisch L, *et al.* Development and application of a moveability index to quantify possibilities for physical activity in the built environment of children. Health Place. 2011;17: 1191–201.

[47] McNeill LH, Kreuter MW, Subramanian SV. Social environment and physical activity: a review of concepts and evidence. Soc Sci Med. 2006;63:1011–22.

[48] Spittaels H, Verloigne M, Gidlow C, Gloanec J, Titze S, Foster C, *et al.* Measuring physical activity-related environmental factors: reliability and predictive validity of the European environmental questionnaire ALPHA. Int J Behav Nutr Phys Act. 2010;7:48.

[49] Panter JR, Jones A. Attitudes and the environment as determinants of active travel in adults: what do and don't we know? J Phys Act Health. 2010;7:551–61.

[50] Johnson JB, Omland KS. Model selection in ecology and evolution. Trends Ecol Evol. 2004;19:101–8.

[51] Nakaya T, Fotheringham AS, Charlton M, Brunsdon C. Semiparametric Geographically Weighted Generalised Linear Modelling in GWR4.0, Proceedings of geocomputation. 2009.

[52] Nakaya T. Geographically weighted generalised linear modelling. In: Brunsdon C, Singleton A, editors. Geocomputation: A Practical Primer. London: Sage Publication; 2015. p. 217–20.

[53] Wendel-Vos GCW, Schuit AJ, de Niet R, Boshuizen HC, Saris WHM, Kromhout D. Factors of the physical environment associated with walking and bicycling. Med Sci Sports Exerc. 2004;36:725–30.

[54] Wendel-Vos W, Droomers M, Kremers S, Brug J, van Lenthe F. Potential environmental determinants of physical activity in adults: a systematic review. Obes Rev Off J Int Assoc Study Obes. 2007;8:425–40.

[55] Ogilvie D, Mitchell R, Mutrie N, Petticrew M, Platt S. Personal and environmental correlates of active travel and physical activity in a deprived urban population. Int J Behav Nutr Phys Act. 2008;5:43.

[56] Bopp M, Kaczynski AT, Besenyi G. Active commuting influences among adults. Prev Med. 2012;54:237–41.

[57] Bopp M, Kaczynski AT, Campbell ME. Social ecological influences on work-related active commuting among adults. Am J Health Behav. 2013;37:543–54.

[58] Sallis JF, Owen N, Fisher E. Ecological Models of Health Behavior. San Francisco: Jossey-Bass; 2008.

[59] Giles-Corti B, Donovan RJ. The relative influence of individual, social and physical environment determinants of physical activity. Soc Sci Med. 2002;54:1793–812.

[60] Blainey SP. Forecasting the use of new local railway stations and services using GIS. phd. PhD thesis, University of Southampton; 2009.

[61] Lu GX, Sullivan J, Troy A. Impact of ambient built-environment attributes on sustainable travel modes: a spatial analysis in Chittenden county, Vermont. 91st Annual Meeting of the Transportation Research Board. Washington, DC. 2012;21.

[62] Reis RS, Hino AAF, Parra DC, Hallal PC, Brownson RC. Bicycling and walking for transportation in three Brazilian cities. Am J Prev Med. 2013;44:e9–17.

[63] Parkes A, Kearns A. The multi-dimensional neighbourhood and health: a cross-sectional analysis of the Scottish Household Survey, 2001. Health Place. 2006;12:1–18.

[64] Macintyre S, Ellaway A, Cummins S. Place effects on health: how can we conceptualise, operationalise and measure them? Soc Sci Med 1982. 2002;55:125–39.

[65] Lytle LA. Measuring the food environment: state of the science. Am J Prev Med. 2009;36(4 Suppl):S134–44.

[66] Shephard R, Vuillemin A. Limits to the measurement of habitual physical activity by questionnaires. Br J Sports Med. 2003;37:197–206.

[67] Andrews GJ, Hall E, Evans B, Colls R. Moving beyond walkability: on the potential of health geography. Soc Sci Med 1982. 2012;75:1925–32.

[68] Inoue S, Ohya Y, Odagiri Y, Takamiya T, Ishii K, Kitabayashi M, *et al.* Association between perceived neighborhood environment and walking among adults in 4 cities in Japan. J Epidemiol. 2010;20:277–86.

[69] Karusisi N, Thomas F, Méline J, Brondeel R, Chaix B. Environmental conditions around itineraries to destinations as correlates of walking for transportation among adults: the RECORD cohort study. PLoS One. 2014;9: e88929.

Appendix A. Global Poisson Regression Model (GPR)

GPR is a kind of generalized linear model and is typically used for the modeling of count data. The Poisson probability distribution of the number h of occurrences of an event is expressed as follows:

$$p(h \setminus \lambda) = \frac{e^{-\lambda} \lambda^{h}}{h!} \quad \text{for } h = 0, 1, 2, \cdots \text{ and } \lambda > 0 \tag{A1}$$

where λ is the only Poisson parameter, as the distribution is equidispersed (*i.e.* the mean and variance of Poisson distribution are both equal to λ_i). In the Poisson regression model, the expected value λ is the result of the exponential function of the linear combination of the explanatory variables. Indeed, the log-linear model does not contain negative values of λ. Hence, GPR takes the following form:

$$\text{Log}\lambda_i = \beta_0 + \beta_1 X_{i1} + \beta_2 X_{i2} + \cdots + \beta_k X_{ik} \tag{A2}$$

where β_0, $\cdots$, β_k are the parameters (or coefficients) of the model and x_1, $\cdots$, k_x are the predictors (individual and environmental variables). Regression parameters are then estimated by maximizing the log-likelihood in an iterative manner. Because the dependent variable is log-transformed, parameters are interpreted as odds ratios (ORs), just as in logistic regression: each parameter β_i is the estimated increase in the log-odds of the outcome per unit increase in the value of the predictor x_i.

Appendix B. GWPR Parameter Estimation

According to Nakaya *et al.*[30], the local parameters can be estimated with a modified local Fisher scoring procedure, a form of iteratively reweighted least squares:

$$\hat{\beta}(u_i, v_i) = \left(X'W(u_i, v_i)A(u_i, v_i)X\right)^{-1} X'W(u_i, v_i)A(u_i, v_i)y \tag{B1}$$

where X is the design matrix of explanatory variables, $W(u_i, v_i)$ is the diagonal spatial weights matrix calculated for each calibration residential location of subject i,

$A(u_i, v_i)$ denotes the variance weights matrix associated with the Fisher scoring for each residential location of subject i and y is the $n \times 1$ vector of adjusted dependent variables.

Appendix C. GWPR Bi-Square Weighting Scheme

The bi-square function is expressed as follows[28]:

$$w_{ij} = \begin{cases} \left(1-\left(d_{ij}/G_{i(k)}\right)^2\right)^2 & d_{ij} < G_{i(k)} \\ 0 & d_{ij} < G_{i(k)} \end{cases} \tag{C1}$$

where w_{ij} is the geographical weight of the jth observation at the ith regression point, d_{ij} is the Euclidean distance between i and j and $G_{i(k)}$ is the adaptive bandwidth size defined as the kth nearest neighbor distance.

Chapter 29

Spatiotemporal Patterns of Particulate Matter (PM) and Associations between PM and Mortality in Shenzhen, China

Fengying Zhang[1,2,3*†], Xiaojian Liu[4*†], Lei Zhou[1], Yong Yu[1], Li Wang[2], Jinmei Lu[5], Wuyi Wang[3], Thomas Krafft[2,6]

[1]China National Environmental Monitoring Centre, Beijing 100012, China
[2]CAPHRI School of Public Health and Primary Care, Maastricht University, Maastricht, The Netherlands
[3]Institute of Geographic Sciences and Natural Resources Research, Chinese Academy of Sciences, Beijing 100101, P. R. China
[4]Shenzhen Center for Disease Control and Prevention, Shenzhen 518055, China
[5]Department of Engineering and Safety, University of Tromsø, N-9037 Tromsø, Norway
[6]Institute of Environment Education and Research, Bharati Vidyapeeth University, Pune, India

Abstract: Background: Most studies on air pollution exposure and its associations with human health in China have focused on the heavily polluted industrial areas and/or mega-cities, and studies on cities with comparatively low air pollutant concentrations are still rare. Only a few studies have attempted to analyse particulate

[*]Fengying Zhang and Xiaojian Liu are co-first authors.
[†]Equal contributors

matter (PM) for the vibrant economic centre Shenzhen in the Pearl River Delta. So far no systematic investigation of PM spatiotemporal patterns in Shenzhen has been undertaken and the understanding of pollution exposure in urban agglomerations with comparatively low pollution is still limited. Methods: We analyze daily and hourly particulate matter concentrations and all-cause mortality during 2013 in Shenzhen, China. Temporal patterns of PM ($PM_{2.5}$ and PM_{10}) with aerodynamic diameters of 2.5 (10) μm or less [or less (including particles with a diameter that equals to 2.5 (10) μm] are studied, along with the ratio of $PM_{2.5}$ to PM_{10}. Spatial distributions of PM_{10} and $PM_{2.5}$ are addressed and associations of PM_{10} or $PM_{2.5}$ and all-cause mortality are analyzed. Results: Annual average PM_{10} and $PM_{2.5}$ concentrations were 61.3 and 39.6μg/m^3 in 2013. $PM_{2.5}$ failed to meet the Class 2 annual limit of the National Ambient Air Quality Standard. $PM_{2.5}$ was the primary air pollutant, with 8.8% of days having heavy $PM_{2.5}$ pollution. The daily $PM_{2.5}/PM_{10}$ ratios were high. Hourly $PM_{2.5}$ concentrations in the tourist area were lower than downtown throughout the day. PM_{10} and $PM_{2.5}$ concentrations were higher in western parts of Shenzhen than in eastern parts. Excess risks in the number of all-cause mortality with a 10μg/m^3 increase of PM were 0.61% (95% confidence interval [CI]: 0.50–0.72) for PM10, and 0.69% (95% CI: 0.55–0.83) for $PM_{2.5}$, respectively. The greatest ERs of PM_{10} and $PM_{2.5}$ were in 2-day cumulative measures for the all-cause mortality, 2-day lag for females and the young (0–65 years), and L02 for males and the elder (>65 years). $PM_{2.5}$ had higher risks on all-cause mortality than PM_{10}. Effects of high PM pollution on mortality were stronger in the elder and male. Conclusions: Our findings provide additional relevant information on air quality monitoring and associations of PM and human health, valuable data for further scientific research in Shenzhen and for the on-going discourse on improving environmental policies.

Keywords: Temporal-Spatial Patterns, Particulate Matter, Mortality, Shenzhen

1. Background

Airborne particulate matter (PM) consistently associated with adverse health effects at current levels of exposure in urban populations[1]–[4]. Air pollution has serious direct and indirect effects on public health in China[2][5]–[8]. PM with aerodynamic diameters less than 2.5μm ($PM_{2.5}$) has become the fourth prominent threat to the health of Chinese people[9].

The range of adverse health effects of air pollution is broad[2][10][11]. Susceptibility to pollution may vary depending on overall health condition and age[5][6][12]-[14]. Risk of various effects has been shown to increase with exposure, but there is little evidence to suggest a threshold below which no adverse health effects can be anticipated[15][16]. The lowest concentration at which such effects begin to manifest is not much greater than the background concentration, which has been estimated at 3–5μg/m^3 for $PM_{2.5}$ in the United States and western Europe[15]. Most studies on air pollution exposure and its effects on human health in China have focused on heavily polluted cities or mega-cities[8][17]-[19], whereas studies on cities with relatively low air pollutant concentrations are rare.

Shenzhen is a major coastal city with a population of some 15 million. It is situated within the Pearl River Delta (PRD) and Guangdong Province, immediately north of Hong Kong. Shenzhen has become China's most crowded city and is the fifth most densely populated city in the world, with a population density of 17,150 per square kilometre. Shenzhen is listed as the fourth most important economic centre among Chinese cities. As China's first and still one of the most successful special economic zones the city has an important position in the PRD region and the country. Compared with other cities in China, air quality in Shenzhen is high. Nevertheless, the city has been experiencing elevated levels of PM pollution in recent years because of rapid economic development[20]. As one of the first-stage cities implementing the National Ambient Air Quality Standard (GB3095-2012) in 2013, Shenzhen provided real-time hourly monitoring concentrations of air pollutants to the general public since January 1 2013. According to air quality monitoring data from the China National Environmental Monitoring Center (CNEMC), respective annual average concentrations of PM_{10} and $PM_{2.5}$ were 61.3 and 39.6μg/m^3 in Shenzhen in 2013. The annual average PM_{10} concentration was higher than in 2012 (52μg/m^3). However, comprehensive studies on PM in Shenzhen have been rare and there have been no systematic investigations of PM spatiotemporal patterns in Shenzhen.

We carried out a time-series analysis on daily and hourly PM concentrations and daily number of all-cause mortality (excluding accidental deaths) during the first year (2013) of National Ambient Air Quality Standard implementation in

Shenzhen. Daily and hourly patterns of $PM_{2.5}$ and PM_{10} were summarized and the daily $PM_{2.5}/PM_{10}$ ratio was calculated. Spatial distributions of PM_{10} and $PM_{2.5}$ were investigated. Associations of PM and all-cause mortality were analysed and the susceptibility differentiated according to gender and age were addressed. The objectives were to provide daily/hourly PM_{10} and $PM_{2.5}$ monitoring information for Shenzhen during 2013 to the general public and scientific researchers, investigate spatiotemporal characteristics of PM_{10} and $PM_{2.5}$, evaluate changes of $PM_{2.5}/PM_{10}$ ratio, and to discover potential relationships between daily exposure to $PM_{2.5}$ or PM_{10} and all-cause mortality.

Rang of health effects of air pollution was broad[2][10][11]. Susceptibility to pollution may vary with health or age[5][6][12]–[14]. Risk of various effects has been shown to increase with exposure, but there is little evidence to suggest a threshold below which no adverse health effects can be anticipated[15][16]. The lowest concentration at which such effects begin to manifest is not much greater than the background concentration, which has been estimated at 3–5μg/m^3 for $PM_{2.5}$ in the United States and western Europe[15]. Most studies on air pollution exposure and its effects on human health in China have focused on heavily polluted cities or megacities[8][17]–[19], whereas studies on cities with relatively low air pollutant concentrations are rare.

2. Methods

2.1. Study Area

Shenzhen is in southern China, 113°46–114°37E and 22°27–22°52N, with an area of 1991.64km^2. There is a subtropical oceanic climate, with warm temperatures and abundant rainfall. Annual average temperature is 22.4°C. The monthly average temperature in January is 15.4°C, and 28.9°C in July.

2.2. Data Sources

2.2.1. Mortality Data

All non-accidental mortality data for calendar year 2013 were obtained from death certificates recorded at the Shenzhen Center for Disease Control and Prevention. In the death registry, causes are coded by the International Classification of Disease revision 10 (ICD10).

2.2.2. Air Pollutant Monitoring Data

Daily air quality monitoring data were provided by the Shenzhen Environmental Monitoring Center and CNEMC. Daily PM_{10} and $PM_{2.5}$ concentrations were derived from the average of available hourly data measured at 11 state-controlled monitoring stations across Shenzhen, the locations of which are presented in **Figure 1**. According to Technical regulation for ambient air quality assessment (on trial) (HJ 663–2013), when calculating daily means of a city, at least 75% hourly concentrations from the monitoring stations of the city had to be available in a single day. If more than 25% of the data in a monitoring station was missing in the whole study period, the entire station would be excluded. According to technical guidelines of the Chinese government, these locations must not be in the immediate vicinity of traffic intersections or major industrial polluters, and should be sufficiently distant from any other emission sources. Thus, the monitoring data reflect the general background urban air pollution level in our study area.

To discern spatiotemporal changes of hourly $PM_{2.5}$ and PM_{10} concentrations,

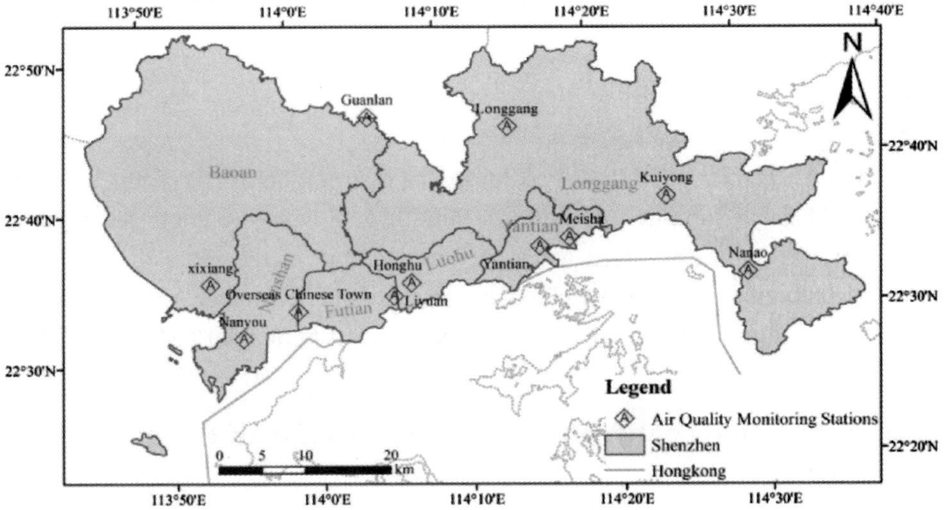

Figure 1. Distribution of 11 state-controlled air quality monitoring sites in Shenzhen.

we acquired hourly monitoring data from 1 January through 30 November 2013 at two state-controlled monitoring stations. These stations were downtown (Huaqiao-cheng, HQC) and in a tourist area (Nan'ao, NA). Hourly $PM_{2.5}$ and PM_{10} monitoring data were from the National Real-Time Air Quality Monitoring Data Publishing Platform developed by CNEMC, which is publicly accessible via the website http://113.108.142.147:20035/emcpublish/.

2.2.3. Meteorological Data

To control for effects of weather on mortality, meteorological data (temperature, relative humidity, barometric pressure and wind speed) were obtained from the Meteorological Bureau of Shenzhen Municipality. The weather data was monitored at a weather station belonging to that bureau. The monitoring standard is consistent with the international WMO (World Meteorological Organization) standard. There were no missing meteorological data.

2.3. Data Analysis

2.3.1. Statistical Analysis

Spearman correlation coefficients were used to reflect the relationship between PM and meteorological factors during the study period.

2.3.2. Spatial Analysis

In the Macroscopic regional scale, spatial distribution of $PM_{2.5}$ concentration follow the basic assumption of 'the first law of geography', namely the regional concentrations in nearby areas are more similar than in the more distant areas. Therefore, inverse distance weighted model (Inverse Distance Weighted, IDW) interpolation analysis was used to analyze spatial distributions of $PM_{2.5}$ and PM_{10}.

2.3.3. Associations between Daily Concentration of PM and Mortality

Consistent with other time-series studies[21][22], we used a generalized additive model (GAM) with penalized splines to analyze mortality, PM, and confounding factors (calendar time, day of week, temperature, barometric pressure, wind speed and humidity). Because the daily mortality number was small and typically followed a Poisson distribution[23][24], the core analysis was via a GAM with log link and Poisson error that accounted for smooth fluctuations of that number.

In preparation for conducting the model analyses, we conducted two steps in the procedure of the model building and model fit: development of the best base model (without a pollutant) and development of the main model (with a pollutant). The latter is achieved by adding the PM to the final cause-specific best base model, assuming a linear relationship between the logarithmic mortality number and PM concentration.

First, we constructed the basic pattern of mortality number excluding PM. We incorporated smoothed spline functions of time and weather conditions, which can include non-linear and non-monotonic links between mortality and time/weather conditions, offering a flexible modelling tool. Day of the week was also included in the basic models.

After we established the basic models, we introduced the PM and analyzed their associations with mortality. To compare the relative quality of the mortality predictions across these non-nested models, Akaike's Information Criterion (AIC) was used as a measure of how well the model fitted the data. Smaller AIC values indicate the preferred model. Briefly, we fitted the following log-linear generalized additive models to obtain the estimated pollution log-relative rate β in the study district:

$$\log\left[E(Yt)\right] = \alpha + \sum_{i=1}^{q} \beta(Xi) + \sum_{j=1}^{p} fj(Zj, df) + Wt(week)$$

Here $E(Yt)$ represents the expected number of mortality at day t; β represents the log-relative rate of mortality associated with an unit increase of PM; Xi indicates the concentrations of pollutants at day t; $Wt(week)$ is the dummy variable for day of the week $\sum_{j=1}^{p} fj(Zj, df)$ is the non-parametric spline function of calendar time, temperature, barometric pressure, wind speed and humidity. A detailed introduction to GAM is given in Wood[24]. We initialized the df as 7 df/year for time,

3 df for temperature, barometric pressure, wind speed and humidity[25].

Results were expressed as excess risk (ER) in mortality number per 10μg/m^3 increases in PM concentrations (ER = ($e^{\beta x \Delta C}$ − 1) × 100, where ΔC is the incremental PM amount, which was 10μg/m^3 here for comparison with similar studies in other locations of China).

Values of $p < 0.05$ were considered statistically significant.

We also examined PM effects with different lag (L) structures of single-day (distributed lag; L0–L3) and multi-day (moving average lag; L01–L03) lags. Here, a lag of 0 day (L0) corresponds to current-day pollution and a lag of 1 day to the previous-day concentration. In multi-day lag models, L03 corresponds to a 4-day moving average pollutant concentration of the current and previous 3 days[26][27]. Meteorological factors used in the lag models (distributed and moving average) were from current-day data. While running the models we also considered lags of more than three days for each of the pollutants, but very few associations were identified and these results have been excluded from further analyses.

2.3.4. Software Used

Temporal changes of $PM_{2.5}$ and PM_{10} were summarized by Origin 9.0 software, and their spatial differences were presented by ArcGIS 10.2 using Inverse distance weighted (IDW). Other statistical analyses were conducted in R3.1.0, and MGCV package in R3.1.0 was used for the GAM analysis.

3. Results

3.1. Descriptive Results

Table 1 summarizes annual means and percentages of daily mortality number, PM_{10} and $PM_{2.5}$ concentrations, and meteorological factors for Shenzhen in 2013.

During the study period, mean daily temperature and humidity were 23.1˚C and 74.8%, respectively. Mean daily temperature was 9.8˚C–31.2˚C and mean

Table 1. Statistical characteristics of air pollutants, meteorological factors, and daily mortality number.

Items	Average	SD	Min	25%	Mid	75%	Max
All-cause mortality[a]	32.7	6.5	9	28	32	36	51
Male[a]	20.5	4.9	5	17	20	24	36
Female[a]	12.1	3.7	1	10	12	14	25
Young (0–65 years)[a]	17.6	4.5	3	15	17	20	32
Elder (>65 years)[a]	15.1	4.4	4	12	15	18	29
Temperature (°C)	23.1	5.2	9.8	19.4	24.2	27.7	31.2
Humidity (%)	74.8	15.6	24	67	78	87	100
Pressure (hPa)	1005.2	6.2	986.8	1000.5	1005.1	1010.8	1019.2
Wind speed (m/s)	2.1	0.8	0.3	1.6	2	2.5	5.5
PM_{10} ($\mu g/m^3$)	61.3	33.5	10	34	53	81	179
$PM_{2.5}$ ($\mu g/m^3$)	39.6	24.8	9	20	35	52	135
$PM_{2.5}/PM_{10}$ (%)	62.4	10.7	37	54.4	63.2	69.6	100

[a]Daily number. Pressure is barometric.

daily humidity ranged from 24%–100%, reflecting the subtropical oceanic climate of Shenzhen.

Annual average concentrations were 61.3$\mu g/m^3$ for PM_{10} and 39.6$\mu g/m^3$ for $PM_{2.5}$. Averages were higher than median values for the two air pollutants. $PM_{2.5}/PM_{10}$ ratios ranged from 37.0% to 88.3%, with a mean of 62.4%.

A total mortality of 11,919 people from all causes was observed in 2013. Among these, 7494 were male and 4425 female. There were 6421 people in the 0–65 year age group and 5498 in the 65+ group. The daily number of all-cause mortality was between 9 and 51.

Spearman correlation coefficients for PM and meteorological factors are presented in **Table 2**. Significant positive correlations were found between PM and barometric pressure. Temperature was negatively and significantly correlated with

Table 2. Spearman correlation coefficients between air pollutants and meteorological factors.

	Temperature	Humidity	pressure	Wind speed	PM10	PM2.5
Temperature	1	0.250[a]				
Humidity	0.250[a]	1				
Pressure	−0.831[a]	−0.516[a]	1			
Wind speed	0.006	−0.048	−0.054	1		
PM10	−0.475[a]	−0.624[a]	0.578[a]	−0.188[a]	1	
PM2.5	−0.559[a]	−0.556[a]	0.624[a]	−0.170[a]	0.973[a]	1

[a]Correlation significant at 0.01 level (2-tailed test). Pressure is barometric.

PM. Similar patterns of correlations were also found for humidity and wind speed.

3.2. Temporal Changes

3.2.1. Daily Concentrations of PM_{10} and $PM_{2.5}$

Figure 2 shows temporal characteristics of PM_{10} and $PM_{2.5}$. Daily PM_{10} and $PM_{2.5}$ concentrations showed significant similar temporal trends, with relatively high levels during October-December and low levels for May-September.

Daily PM_{10} concentrations were 10−179μg/m^3 with an average of 61.3μg/m^3, and $PM_{2.5}$ concentrations were 9−135μg/m^3 with an average of 39.6μg/m^3. PM_{10} and $PM_{2.5}$ concentration ranges were wide, and their maxima were twice the Class 2 limits of National Ambient Air Quality Standard.

3.2.2. Ratios of $PM_{2.5}$ to PM_{10}

Temporal characteristics of daily $PM_{2.5}/PM_{10}$ ratios are shown in **Figure 2.** The ratios peaked during December-February and April-May, with low values from June to August, which mean high fine particulate ratio in December-February and April-May. The ratios ranged from 37.0% to 88.3%, with an average of 62.4%.

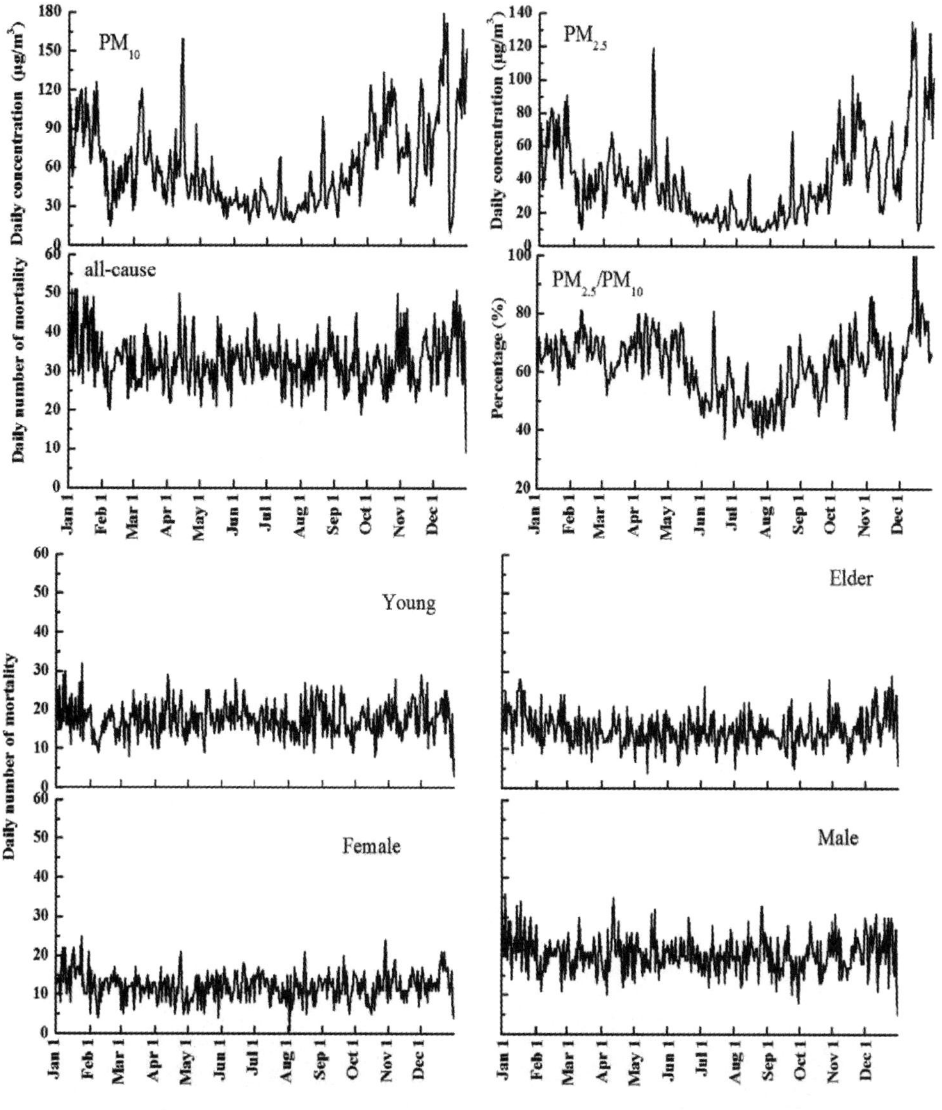

Figure 2. Daily concentrations of PM_{10} and $PM_{2.5}$ and daily number of mortality.

3.2.3. Temporal Trends of Daily Mortality Number

Daily trends on number of mortality for all-cause, male, female, young and elder are also summarized in **Figure 2**. Daily mortality number for all-cause was 9 −51 with an average of 33, 5−36 for male with an average of 21, 1−25 for female with an average of 12, 3−32 for young with an average of 18, 4−29 for elder with an average of 15.

3.3. Spatial Differences

Figure 3 presents spatial distributions of PM_{10} and $PM_{2.5}$ in Shenzhen during 2013. PM_{10} and $PM_{2.5}$ concentrations were higher in western parts of Shenzhen

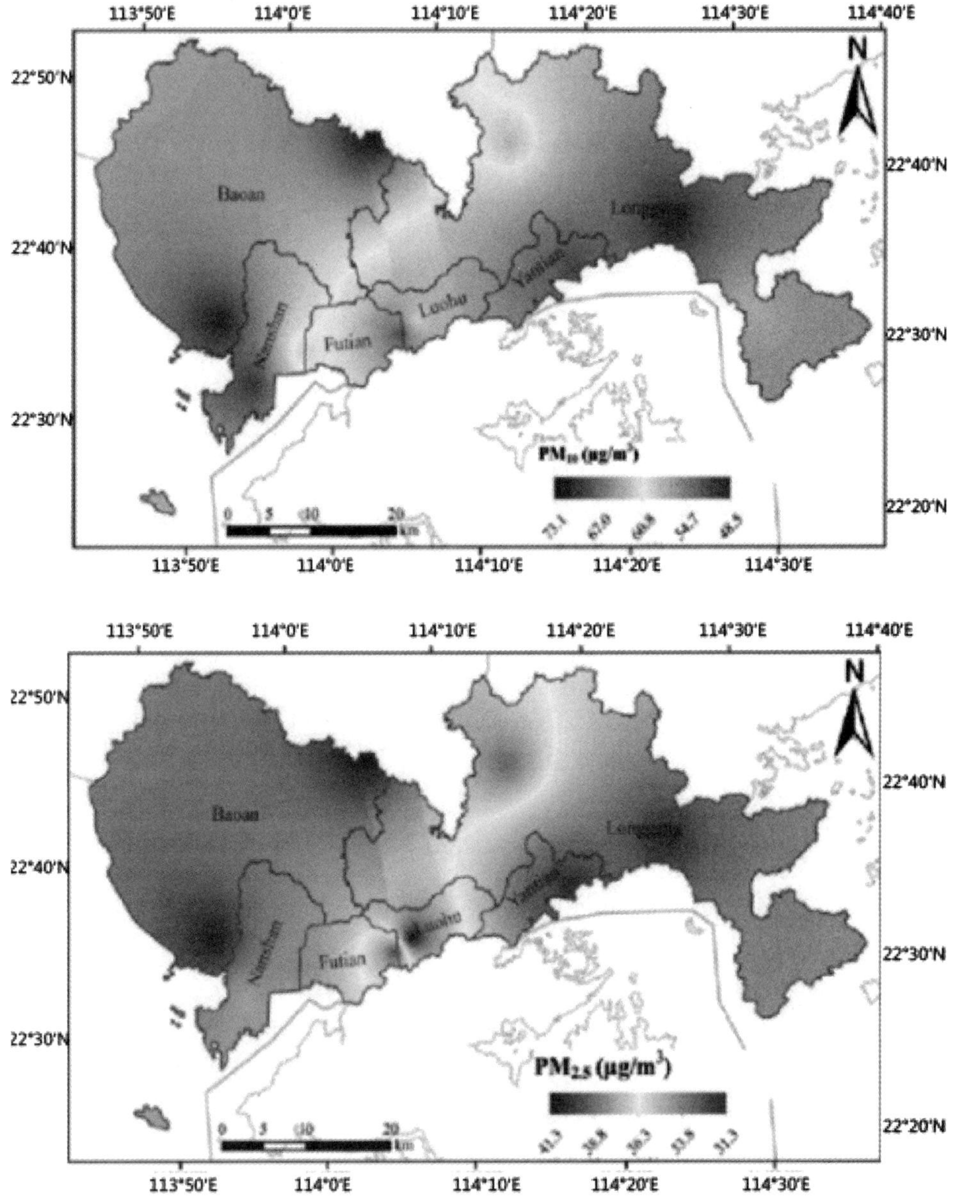

Figure 3. Spatial distributions of PM_{10} and $PM_{2.5}$ in Shenzhen.

than in eastern parts. According to National Ambient Air Quality Standard, annual average $PM_{2.5}$ concentrations at five monitoring stations named Nan'ao, Liyuan, Kuiyong, Meisha and Yantian were within Class 2 limits (annual average $PM_{2.5}$ concentration < 35μg/m^3), but exceeded these limits at the other six stations.

To represent spatial differences of PM in Shenzhen more directly, we analysed monthly and hourly concentrations of PM_{10} and $PM_{2.5}$ at HQC and NA sites, in the downtown and tourist areas of Shenzhen, respectively. These two parts of the city serve distinct and quite different urban functions, and differences in air quality might indicate that they are affected by different pollutant emission sources.

3.3.1. Monthly Differences

Figure 4 shows average monthly PM_{10} and $PM_{2.5}$ concentrations at HQC and NA from January through November. The results show that PM_{10} and $PM_{2.5}$ had similar hourly trends at HQC, which had higher concentrations in January and October and lower concentrations in July. At the NA monitoring station, both PM_{10} and $PM_{2.5}$ had higher concentrations in January and October, and lower concentrations in May for PM_{10} and June for $PM_{2.5}$. Concentrations of PM_{10} at HQC were higher than at NA during January-May and September-November, and lower than at NA during May-September. Concentrations of $PM_{2.5}$ at HQC were higher than at NA during January-June and September-November, and lower than at NA during June-September.

3.3.2. Hourly Differences

Hourly average PM_{10} and $PM_{2.5}$ concentrations at HQC and NA in 2013 are also presented in **Figure 4**. Changes in hourly concentrations of PM_{10} and $PM_{2.5}$ had similar patterns at NA, which had maxima at 20:00 and secondary maxima at 9:00. Minima were from 3:00 to 4:00 Changes in hourly concentration of PM_{10} and $PM_{2.5}$ did not show patterns common with HQC, which may be related to changes of pollution source over a day in the downtown area. Hourly concentrations of $PM_{2.5}$ at NA were lower than at HQC throughout the day, but the hourly PM_{10} concentration did not show this pattern.

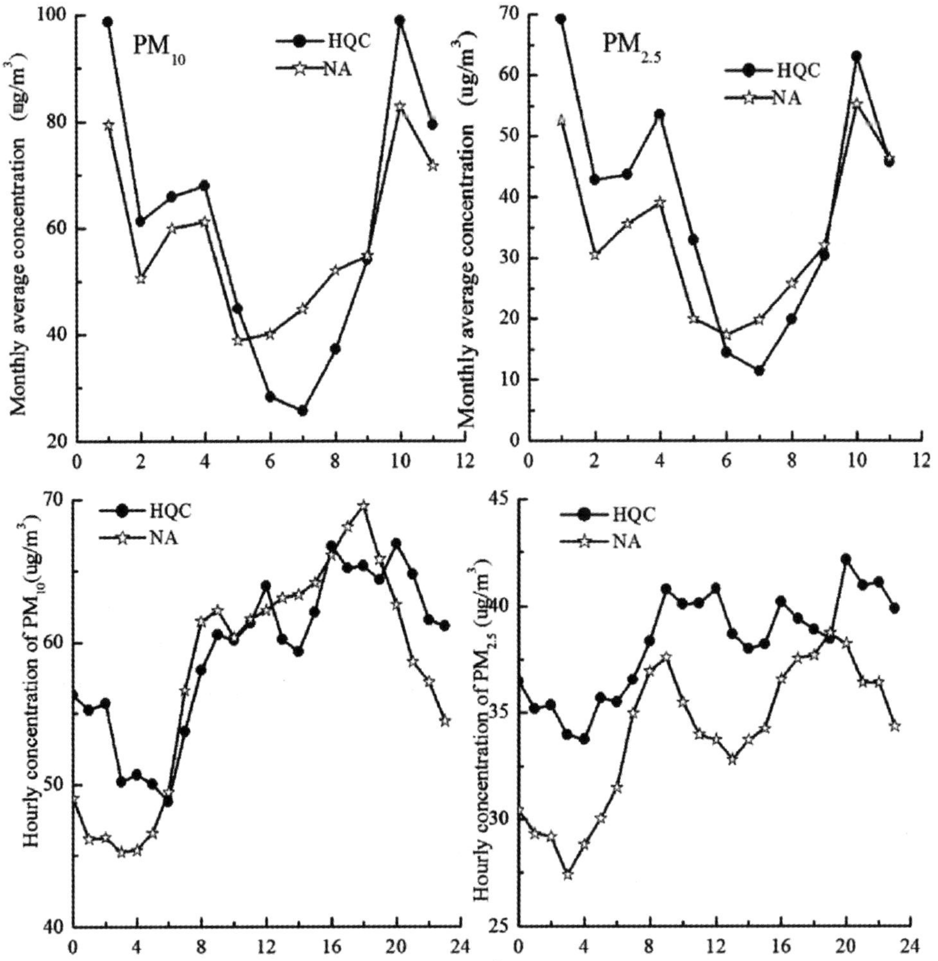

Figure 4. Monthly/hourly PM_{10} and $PM_{2.5}$ concentrations at HQC and NA.

3.4. Associations with All-Cause Mortality

Table 3 presents ER percentages(ERs) (95% confidence interval [CI]) of daily all-cause mortality number with every 10μg/m^3 increase in PM_{10} or $PM_{2.5}$ concentration.

To identify possible time delay of PM_{10} or $PM_{2.5}$ pollution exposure and daily mortality number, we analyzed lag effects of air pollutants. ER in the all-cause mortality number with a 10μg/m^3 increase of pollutants for single-day measures, 1–3 days prior to mortality (L0–L3), and moving averages from day 0 and day 1 to

Table 3. Excess risk (ER) percentage for daily all-cause mortality number with every 10-μg/m^3 increase in PM concentration.

Items		All-cause ER (95 % CI)	Female ER (95 % CI)	Male ER (95 % CI)	Elder ER (95 % CI)	Young ER (95 % CI)
PM_{10}	L0	0.22 (0.12 ~ 0.32)	−0.32 (−0.48 ~ 0.17)	0.54 (0.42 ~ 0.67)	1.35 (1.21 ~ 1.49)	−0.79 (−0.92 ~ 0.65)
	L1	0.37 (0.28 ~ 0.46)	0.33 (0.19 ~ 0.47)	0.40 (0.29 ~ 0.51)	1.05 (0.92 ~ 1.17)	−0.24 (−0.36 ~ 0.12)
	L2	0.51 (0.43 ~ 0.60)	0.33 (0.21 ~ 0.46)	0.62 (0.52 ~ 0.73)	0.85 (0.74 ~ 0.97)	0.21 (0.10 ~ 0.33)
	L3	−0.02 (-0.10 ~ 0.06)	−0.47 (−0.59 ~ 0.34)	0.25 (0.15 ~ 0.35)	0.41 (0.29 ~ 0.53)	−0.41 (−0.52 ~ 0.29)
	L01	0.42 (0.31 ~ 0.52)	0.05 (−0.12 ~ 0.22)	0.64 (0.51 ~ 0.77)	1.54 (1.39 ~ 1.69)	−0.57 (−0.71 ~ 0.42)
	L02	0.61 (0.50 ~ 0.72)	0.23 (0.06 ~ 0.40)	0.85 (0.71 ~ 0.98)	1.55 (1.4 ~ 1.71)	−0.22 (−0.37 ~ 0.07)
	L03	0.45 (0.34 ~ 0.57)	−0.07 (−0.25 ~ 0.10)	0.77 (0.64 ~ 0.91)	1.39 (1.24 ~ 1.55)	−0.38 (−0.54 ~ 0.23)
$PM_{2.5}$	L0	0.12 (−0.01 ~ 0.24)	−0.35 (−0.54 ~ 0.15)	0.40 (0.24 ~ 0.56)	1.67 (1.48 ~ 1.85)	−1.26 (−1.44 ~ 1.09)
	L1	0.46 (0.35 ~ 0.58)	0.76 (0.58 ~ 0.94)	0.29 (0.15 ~ 0.44)	1.39 (1.22 ~ 1.55)	−0.37 (−0.53 ~ 0.21)
	L2	0.68 (0.57 ~ 0.79)	0.75 (0.57 ~ 0.92)	0.65 (0.51 ~ 0.78)	1.13 (0.97 ~ 1.28)	0.28 (0.12 ~ 0.43)
	L3	−0.23 (−0.34 ~ 0.12)	−0.63 (−0.80 ~ 0.45)	0.01 (−0.13 ~ 0.15)	0.55 (0.39 ~ 0.71)	−0.94 (−1.1 ~ 0.79)
	L01	0.41 (0.28 ~ 0.55)	0.31 (0.10 ~ 0.52)	0.48 (0.31 ~ 0.65)	1.92 (1.73 ~ 2.11)	−0.92 (−1.11 ~ 0.73)
	L02	0.69 (0.55 ~ 0.83)	0.62 (0.40 ~ 0.84)	0.74 (0.56 ~ 0.91)	1.97 (1.77 ~ 2.16)	−0.45 (−0.64 ~ 0.25)
	L03	0.44 (0.29 ~ 0.58)	0.19 (−0.04 ~ 0.42)	0.59 (0.41 ~ 0.76)	1.82 (1.62 ~ 2.03)	−0.80 (−1.00 ~ 0.60)

day 3 prior to the mortality are also listed in **Table 3**. When running the models, lag effects of more than 3 days for PM_{10} and $PM_{2.5}$ were also considered. However, as little to no relationship was found, the results of that analysis were not included. Gender and age differences were also considered. Unlike cities in northern China, temperature differences in Shenzhen were not significant. Therefore, we did not run seasonal models.

The results showed that the greatest ERs of PM_{10} and $PM_{2.5}$ were in 2-day cumulative measures (L02) for the all-cause mortality group, 2-day lag (L2) for females and the young (0–65 years), and L02 for males and the elder (>65 years). The greatest ERs in the mortality number with a 10-μg/m^3 increase of PM_{10} were 0.61%, 0.33%, 0.85%, 1.55% and 0.21% for the all-cause mortality group, females, males, elder and young, respectively. The greatest ERs in the mortality number with a 10-μg/m^3 increase of $PM_{2.5}$ were 0.69%, 0.76%, 0.74%, 1.97% and 0.28% for the same respective groups. ERs of males with increases in PM_{10} or $PM_{2.5}$ concentration were greater than those of females, and ERs of the elder were greater than the young with concentration increases of PM_{10} or $PM_{2.5}$.

4. Discussion

This study focused on spatiotemporal patterns and possible associations of PM_{10} and $PM_{2.5}$ with all-cause mortality during the first year (2013) of National Ambient Air Quality Standard implementation in Shenzhen, a relatively clean city compared to other cities in China. The objectives were to provide 2013 PM monitoring information of Shenzhen to the general public, to discover possible associations between PM and mortality in a comparatively clean city, and to provide scientific results to researchers in other areas. We also intend to encourage health services and public health policymakers in Shenzhen to consider ideas for real-time public health alerts for air quality, so that vulnerable groups and others affected by air pollution can be appropriately advised. The present study was unique in the following aspects: 1) we analysed daily patterns of PM and air quality during the first year (2013) of National Ambient Air Quality Standard implementation in Shenzhen; 2) based on reliable data sources, hourly/monthly patterns of PM in two functional areas of Shenzhen were addressed; 3) spatial patterns of PM were determined; and 4) to our knowledge, the study is the first to investigate associations between PM and all-cause mortality in Shenzhen.

During 2013, annual average PM_{10} and $PM_{2.5}$ concentrations were 61.3 and 39.6μg/m^3, respectively; averages were higher than the median values for the two air pollutants. According to National Ambient Air Quality Standard and the Technical Regulation on Ambient Air Quality Index (on trial) (HJ633-2012), $PM_{2.5}$ was the major air pollutant in Shenzhen, with 104 days as the primary pollutant and 32 days as a "non-attainment" pollutant. The latter indicates 32 days with heavy $PM_{2.5}$

pollution. Annual average PM_{10} and $PM_{2.5}$ concentrations were 108 and 89μg/m^3 in Beijing, and 72 and 53μg/m^3 in Guangzhou[28]. Compared with heavy PM- polluted cities in China (Beijing, Guangzhou, and others), Shenzhen has good air quality[9][19][29]. Shenzhen was ranked 7th among 74 first-stage cities, but its annual average $PM_{2.5}$ concentration exceeded the Class 2 limit of National Ambient Air Quality Standard[28][30].

The annual average ratio of $PM_{2.5}$ to PM_{10} was 62.4%, which indicates a high percentage of $PM_{2.5}$ in ambient air pollution of Shenzhen. $PM_{2.5}/PM_{10}$ maximized in December-February and April-May, with lower values in June-August. Compared with Beijing the $PM_{2.5}/PM_{10}$ ratio in Shenzhen was higher than Beijing autumn normal days and lower than haze days and winter normal days in Beijing; the average $PM_{2.5}/PM_{10}$ ratios in Beijing were correspondingly 0.63, 0.32, 0.70, and 0.66 in autumn haze, autumn normal, winter haze and winter normal days, respectively[31]. The $PM_{2.5}/PM_{10}$ ratio is 0.575 in Taiwan[32]. These findings may be related to meteorological conditions and pollution sources in the city. $\mathbf{PM_{2.5}}$ concentrations can be affected by both local emissions and contributions of meso-scale origin[33]. Further studies on concentrations and ratios of PM at intercity level should be conducted.

Hourly concentrations of PM_{10} and $PM_{2.5}$ had similar patterns in the tourist area (NA monitoring station) but did not have any patterns in common in the downtown area. This may be related to changes of pollution source like traffic emission in downtown over a single day. Hourly concentrations of $PM_{2.5}$ in the tourist area were lower than downtown throughout the day, which may be attributed to more intensive human activities downtown. PM_{10} and $PM_{2.5}$ concentrations were higher in western parts of Shenzhen than eastern parts, which may be related to land use, pollution sources, industrial structure, traffic conditions, and other factors[34]. There should be further study of relationships between driving factors (e.g., spatial distribution of pollutant emissions, pollutant emission intensity, and regional industrial structure) and pollutant concentrations.

Time-series studies estimate that a 10μg/m^3 increase in mean 24-hour $PM_{2.5}$ concentration increases the ERs of daily cardiovascular mortality by ~0.4 to 1.0%[35]. Consistent with other studies[2][11][12][36][37], we found a statistically significant association between PM_{10} or $PM_{2.5}$ and daily mortality number. There were lag effects in all the study groups, and ERs of $PM_{2.5}$ were greater than PM10 in all study groups with concentration increases. ERs in the all-cause mortality number with a 10-μg/m^3 increase of PM_{10} and $PM_{2.5}$ were 0.61% (95% CI: 0.50%–0.72%) and 0.69% (95% CI: 0.55%–0.83%), respectively. A study in Beijing for 2005–2009 showed that a 10μg/m^3 increase in $PM_{2.5}$ was associated with a 0.65% rise in all-cause mortality, whereas the same increase in PM_{10} was associated with an increase of 0.15%[38]. During 2006–2009 in Guangzhou, increments of 10μg/m^3 in PM_{10} were associated with a ER of 1.26% for total non-accidental deaths, and 1.79% for cardiovascular deaths[39]. During 2007 to 2009 in Tianjin, the effect estimates per 10μg/m^3 increase in PM_{10} concentrations at the moving average of lags 0 and 1 day in high temperature level were 0.62% for non-accidental mortality[40]. ERs in our study were greater than those reported for Beijing and smaller than those for Guangzhou. Such inter-city variability in ER estimates may have been influenced by a number of factors, such as demographic and socioeconomic variables, culture, air pollution sources, and geographical and weather conditions[39]. Both temperature and particulate air pollution are associated with increased death risk; and extreme high temperature increased the associations of PM_{10} with daily mortality[41]. ERs in the all-cause mortality number with 10-μg/m^3 increase of PM_{10} and $PM_{2.5}$ were 0.33 and 0.76% for females, respectively, 0.85 and 0.77% for males, 1.55 and 1.97% for the elder, and 0.21 and 0.28% for the young. Males were more sensitive to PM_{10} or $PM_{2.5}$ concentration changes than females. The elder appeared to be more affected than the young by PM_{10} or $PM_{2.5}$ concentration increase. These findings indicate that PM effects on mortality were stronger among the elderly and on male. Because the seasonal difference was not significant in Shenzhen, we did not consider seasonal associations among PM and mortality in this study. The present study has certain limitations. We considered the target population to be relatively homogeneous and did not consider residence or work location of deaths, owing to a lack of data. Pollutant exposure levels were derived from 11 fixed-site monitoring stations. However, because air pollution varies spatially within a city, averages drawn from these stations may not reflect actual exposure levels. Accurate exposure assessment and a homogeneous target population are important factors to consider in future studies estimating mortality risk from air pollution[39]. Further in-depth studies should require air pollutant composition, pollution emission sources, pollutant emission patterns, time-series of human activity, individual exposure to pollutants, social economy, and human health at the city level.

5. Conclusion

During 2013, annual average PM_{10} and $PM_{2.5}$ concentrations were 61.3 and 39.6μg/m^3 in Shenzhen. $PM_{2.5}$ failed to meet the Class 2 annual limit of National Ambient Air Quality Standard and was the major air pollutant, with 8.8% of days having heavy $PM_{2.5}$ pollution. The annual average $PM_{2.5}/PM_{10}$ ratio was 62.4%. Hourly $PM_{2.5}$ concentrations in the tourist area were lower than downtown throughout the day. PM_{10} and $PM_{2.5}$ concentrations were higher in western parts than eastern parts. ERs in the all cause mortality number increased with PM_{10} and $PM_{2.5}$. $PM_{2.5}$ had higher risks than PM_{10}. PM effects on mortality were stronger among male and the elderly. Our findings provide additional information on air quality monitoring and associations between PM and all-cause mortality, and valuable data for scientific research in Shenzhen. It also contributes to the discussion on further developing environmental health policies in urban China.

Competing Interests

The authors declare that they have no competing interest.

Authors' Contribution

FYZ led the study, carried out the time-series studies, analyzed the data and wrote the first draft of the manuscript. XJL participated in all-cause mortality date in Shenzhen. LZ and YY conducted PM monitoring/ meteorological data collecting and spatial analysis. LW and JML assisted with statistical analysis and language editing. WYW and TK helped to conceptualize the study, provided intellectual advice, contributed to data interpretation and helped to revise various drafts of the manuscript. All authors read and approved the final manuscript.

Acknowledgements

The authors wish to thank all the staff members at the Shenzhen Centre for Disease Control and Prevention for their strong support of this study. We thank the

Shenzhen Environmental Monitoring Center and Meteorological Bureau of Shenzhen Municipality for providing data. The present study was supported by the National Natural Science Foundation of China (NO. 41401101 & NO.41371118).

Source: Zhang F, Liu X, Zhou L, *et al.* Spatiotemporal patterns of particulate matter (PM) and associations between PM and mortality in Shenzhen, China[J]. Bmc Public Health, 2016, 16(1):1−11.

References

[1] Dominici F, Greenstone M, Sunstein CR. Particulate Matter Matters. Science. 2014;344(6181):257−9.

[2] Shang Y, Sun Z, Cao J, Wang X, Zhong L, Bi X, et al. Systematic review of Chinese studies of short-term exposure to air pollution and daily mortality. Environ Int. 2013;54:100−11.

[3] Bell ML, Zanobetti A, Dominici F. Evidence on vulnerability and susceptibility to health risks associated with short-term exposure to particulate matter: a systematic review and meta-analysis. Am J Epidemiol. 2013;178:kwt090.

[4] Kim K-H, Kabir E, Kabir S. A review on the human health impact of airborne particulate matter. Environ Int. 2015;74:136−43.

[5] Tang D, Li TY, Chow JC, Kulkarni SU, Watson JG, Ho SSH, et al. Air pollution effects on fetal and child development: A cohort comparison in China. Environ Pollut. 2014;185:90−6.

[6] Zhou M, Liu Y, Wang L, Kuang X, Xu X, Kan H. Particulate air pollution and mortality in a cohort of Chinese men. Environ Pollut. 2014;186:1−6.

[7] Dong G-H, Qian ZM, Xaverius PK, Trevathan E, Maalouf S, Parker J, et al. Association between long-term air pollution and increased blood pressure and hypertension in China. Hypertension. 2013;61(3):578−84.

[8] L-w Z, Chen X, Xue X-d, Sun M, Han B, Li C-p, et al. Long-term exposure to high particulate matter pollution and cardiovascular mortality: A 12-year cohort study in four cities in northern China. Environ Int. 2014;62:41−7.

[9] Chen Z, Wang J-N, Ma G-X, Zhang Y-S. China tackles the health effects of air pollution. Lancet. 2013;382(9909):1959−60.

[10] Pope CA, Dockery DW. Health effects of fine particulate air pollution: Lines that connect. J Air Waste Manage Assoc. 2006;56(6):709−42.

[11] Wagner JG, Allen K, Yang H-y, Nan B, Morishita M, Mukherjee B, et al. Cardiovas-

cular Depression in Rats Exposed to Inhaled Particulate Matter and Ozone: Effects of Diet-Induced Metabolic Syndrome. Environ Health Perspect. 2014;122(1):27–33.

[12] Goldberg MS, Burnett RT, Stieb DM, Brophy JM, Daskalopoulou SS, Valois M-F, et al. Associations between ambient air pollution and daily mortality among elderly persons in Montreal, Quebec. Sci Total Environ. 2013;463:931–42.

[13] Mahiyuddin WRW, Sahani M, Aripin R, Latif MT, Thuan-Quoc T, Wong C-M. Short-term effects of daily air pollution on mortality. Atmos Environ. 2013;65: 69–79.

[14] Burnett RT, Pope III CA, Ezzati M, Olives C, Lim SS, Mehta S, et al. An Integrated Risk Function for Estimating the Global Burden of Disease Attributable to Ambient Fine Particulate Matter Exposure. Environ Health Perspect. 2014;122(4):397–403.

[15] WHO Regional Office for Europe. Air quality guidelines, global update 2005: particulate matter, ozone, nitrogen dioxide, and sulfur dioxide. World Health Organization; 2006. http://www.euro.who.int/en/health-topics/environment-and-health/air-quality/publications/pre2009/airquality-guidelines.-global-update-2005.-particulate-matter,-ozone,-nitrogen-dioxide-and-sulfur-dioxide.

[16] Brunekreef B, Holgate ST. Air pollution and health. Lancet. 2002;360(9341): 1233–42.

[17] Huang W, Cao J, Tao Y, Dai L, Lu S-E, Hou B, et al. Seasonal Variation of Chemical Species Associated With Short-Term Mortality Effects of PM2.5 in Xi'an, a Central City in China. Am J Epidemiol. 2012;175(6):556–66.

[18] Cao J, Xu H, Xu Q, Chen B, Kan H. Fine Particulate Matter Constituents and Cardiopulmonary Mortality in a Heavily Polluted Chinese City. Environ Health Perspect. 2012;120(3):373–8.

[19] Ma Y, Chen R, Pan G, Xu X, Song W, Chen B, et al. Fine particulate air pollution and daily mortality in Shenyang, China. Sci Total Environ. 2011;409(13):2473–7.

[20] Dai W, Gao J, Cao G, Ouyang F. Chemical composition and source identification of PM2.5 in the suburb of Shenzhen, China. Atmos Res. 2013;122:391–400.

[21] Zhang FY, Wang WY, Lv JM, Krafft T, Xu J. Time-series studies on air pollution and daily outpatient visits for allergic rhinitis in Beijing, China. Sci Total Environ. 2011;409(13):2486–92.

[22] Bhaskaran K, Gasparrini A, Hajat S, Smeeth L, Armstrong B. Time series regression studies in environmental epidemiology. Int J Epidemiol. 2013;42(4):1187–95.

[23] Box GE, Jenkins GM, Reinsel GC. Time series analysis: forecasting and control. 4th ed. New York: John Wiley & Sons; 2008.

[24] Wood S. Generalized additive models: an introduction with R. CRC Press; 2006. https://www.crcpress.com/Generalized-Additive-Models-An-Introduction-with-R/Wood/9781584884743.

[25] Yang Y, Cao Y, Li W, Li R, Wang M, Wu Z, et al. Multi-site time series analysis of acute effects of multiple air pollutants on respiratory mortality: A population-based study in Beijing, China. Sci Total Environ. 2015;508:178–87.

[26] Gasparrini A, Armstrong B, Kenward M. Distributed lag non - linear models. Stat Med. 2010;29(21):2224–34.

[27] Zhang F, Krafft T, Ye B, Zhang F, Zhang J, Luo H, et al. The lag effects and seasonal differences of air pollutants on allergic rhinitis in Beijing. Sci Total Environ. 2013;442:172–6.

[28] China National Environmental Monitoring Center. Environmental Quality Report of China in 2013. 2014.

[29] Huang D, Xu J, Zhang S. Valuing the health risks of particulate air pollution in the Pearl River Delta, China. Environ Sci Pol. 2012;15(1):38–47.

[30] Ministry of Environmental Protection of the People's Republic of China M. Report on the State of the Environment in China 2013. 2014.

[31] Gao J, Tian H, Cheng K, Lu L, Zheng M, Wang S, et al. The variation of chemical characteristics of PM2.5 and PM10 and formation causes during two haze pollution events in urban Beijing, China. Atmos Environ. 2015;107:1–8.

[32] Chu H-J, Huang B, Lin C-Y. Modeling the spatio-temporal heterogeneity in the p M10-PM2.5 relationship. Atmos Environ. 2015;102:176–82.

[33] Marcazzan GM, Vaccaro S, Valli G, Vecchi R. Characterisation of PM10 and PM2.5 particulate matter in the ambient air of Milan (Italy). Atmos Environ. 2001;35(27):4639–50.

[34] Zhang J, Ouyang Z, Miao H, Wang X. Ambient air quality trends and driving factor analysis in Beijing, 1983-2007. J Environ Sci (China). 2011;23(12):2019–28.

[35] Brook RD, Rajagopalan S, Pope III CA, Brook JR, Bhatnagar A, Diez-Roux AV, et al. Particulate Matter Air Pollution and Cardiovascular Disease An Update to the Scientific Statement From the American Heart Association. Circulation. 2010;121(21):2331–78.

[36] Atkinson RW, Kang S, Anderson HR, Mills IC, Walton HA. Epidemiological time series studies of PM2.5 and daily mortality and hospital admissions: a systematic review and meta-analysis. Thorax. 2014;69(7):660–5.

[37] Krall JR, Anderson GB, Dominici F, Bell ML, Peng RD. Short-term exposure to particulate matter constituents and mortality in a national study of US urban communities. Environ Health Perspect. 2013;121(10):11–48.

[38] Li P, Xin J, Wang Y, Wang S, Shang K, Liu Z, et al. Time-series analysis of mortality effects from airborne particulate matter size fractions in Beijing. Atmos Environ. 2013;81:253–62.

[39] Yu ITS, Zhang Y, San Tam WW, Yan QH, Xu Y, Xun X, et al. Effect of ambient air

pollution on daily mortality rates in Guangzhou, China. Atmos Environ. 2012;46: 528–35.

[40] Li G, Zhou M, Cai Y, Zhang Y, Pan X. Does temperature enhance acute mortality effects of ambient particle pollution in Tianjin City, China. Sci Total Environ. 2011; 409(10):1811–7.

[41] Meng X, Zhang Y, Zhao Z, Duan X, Xu X, Kan H. Temperature modifies the acute effect of particulate air pollution on mortality in eight Chinese cities. Sci Total Environ. 2012;435–436:215–21.

Chapter 30

Study on Wastewater Toxicity Using ToxTrak™ Method

Ewa Liwarska-Bizukojc[1], Radoslaw Ślęzak[2], Małgorzata Klink[3,4]

[1]Institute of Fermentation Technology and Microbiology, Lodz University of Technology, Wolczanska 171/173, 90-924 Lodz, Poland
[2]Department of Bioprocess Engineering, Lodz University of Technology, Wolczanska 213, 90-924 Lodz, Poland
[3]Institute of Environmental Engineering and Building Installations, Lodz University of Technology, Al. Politechniki 6, 90-924 Lodz, Poland
[4]Water Supply System and Sewer-Zgierz Ltd., ul. A. Struga 45, 95-100 Zgierz, Poland

Abstract: ToxTrak™ method is an analytical tool for the measurement of toxicity of drinking water, wastewater and natural water. It is based upon the estimation of the inhibitive effect on bacterial respiration processes. The main aim of this work was to test the applicability of ToxTrak™ method in the assessment of wastewater toxicity in a full-scale WWTP in Poland. In order to achieve it, the study was divided into two parts. First, the validation of ToxTrak™ method was performed. Second, wastewater toxicity was monitored in the long- and short-term campaigns. Validation of ToxTrak™ method revealed that the indigenous biomass (mixed cultures of activated sludge microorganisms) was more sensitive than *Escherichia coli* for both materials (wastewater and phenol) tested. The values of degree of inhibition determined for phenol towards indigenous biomass and *E. coli* were close to each other, and no statistically significant difference between them was found. It confirmed the reliability of the results obtained with the help of ToxTrak™ test. The toxicity of the effluent was always lower than that of the influent and the linear correlation between them was found. Despite, the decrease of wastewater toxicity in the WWTP, the effluents were ranked as toxic or highly toxic according to the classification of wastewater based upon the acute toxicity.

Keywords: Activated Sludge, Monitoring, Toxicity, ToxTrak™ Method, Validation, Wastewater

1. Introduction

In the last decade, not only the removal of macropollutants but also that of micropollutants from wastewater was intensively investigated. Wastewater is a complicated matrix of compounds, some of them are present in very small amounts (below 1mg l) and, what is more important, their presence does not influence on the values of such combined indicators of contamination as chemical oxygen demand (COD). It suggests that the standard physicochemical determinations used for the characterisation of wastewater composition are not sufficient nowadays. Modern chromatographic methods coupled to mass spectrometry allow for the determination of the individual components of wastewater (*i.e.* selected pharmaceuticals, pesticides or detergents); however, they do not quantify the effect of wastewater on living organisms and do not assess the environmental risk. For this purpose, the application of toxicity tests is desirable. It is regarded that the evaluation of toxicity is necessary to complement the physicochemical measures of wastewater quality (Hernando *et al.* 2005; Libralato *et al.* 2010a). Such holistic attempt is particularly important in the countries facing water scarcity, in which water reuse is widely implemented.

A variety of tests were used over the last years in order to estimate the toxicity of wastewater. The most common were bioluminescence tests with Vibrio fischeri and growth inhibition tests with Pseudomonas putida, the acute immobilization tests with Daphnia magna and algae growth inhibition tests (Hernando *et al.* 2005; Ra *et al.* 2007; Libralato *et al.* 2010b; Vasquez and Fatta-Kassinos 2013; Tobajas *et al.* 2015). Also, other organisms like diatom Phaeodactylum tricornutum Bohlin (Libralato *et al.* 2016) and bivalve molluscs (Libralato *et al.* 2010a) were applied in order to evaluate wastewater toxicity and/or efficiency of wastewater treatment processes. It shows that wastewater toxicity was measured towards pure cultures of organisms representing different trophic levels, *i.e.* producers, consumers and decomposers. These tests were usually made in agreement with the OECD standard procedures or the ISO norms.

In parallel, several classifications of wastewater toxicity were developed. They were described in detail elsewhere (Libralato *et al.* 2010b). One of the most commonly used is the classification proposed by Persoone *et al.* (2003), which distinguished five classes of wastewater toxicity dependent on the value of toxicity units (TU).

Literature review revealed that there was a lack of data concerning wastewater toxicity towards mixed cultures of organisms, while the mixed cultures are exposed to wastewater in biological treatment processes. Also the number of scientific papers presenting the variations of wastewater toxicity in the long-term period and/or seeking for the correlation between the physicochemical measures and toxicity is very limited (Ra *et al.* 2007; Yi *et al.* 2009; Vasquez and FattaKassinos 2013; Xiao *et al.* 2015).

In this work, ToxTrak™ toxicity test using the indigenous biomass (mixed cultures) was a tool for the measurement of wastewater toxicity. The main aim of the study was to test the applicability of ToxTrak™ method in the assessment of wastewater toxicity in a full-scale WWTP.

In order to achieve this aim, the work was divided into two parts. First, the validation study for ToxTrak™ method was performed. Second, for better assessment of wastewater toxicity, its variability was monitored in the long-term and short-term campaigns. The second part of this work was also aimed at the evaluation of toxicity reduction and seeking for the correlations between physicochemical measures of wastewater composition and its toxicity.

2. Materials and Methods

2.1. Description of the WWTP

The object of this study was the WWTP Zgierz (Poland) that generally treats municipal wastewater. The contribution of industrial wastewater usually varies from 7% to 15%. The average pollutant load to the plant corresponds to approximately 94,000 PE. The biological stage consists of one five-zone bioreactor and secondary clarifier run in the Phoredox process configuration. The total volume of bioreactor is 24,000m^3. The scheme of the biological system working in the WWTP Zgierz including the sampling points is depicted in **Figure 1**. Hydraulic retention time of wastewater in the WWTP studied was about 48 h; thus, the effluent (sampling point no. 2) was each time sampled with 2-day delay in order to relate it to the influent properly. The sampling dates presented in the figures and tables correspond with the date of the sampling of influent.

In the period of study, *i.e.* from August 2013 to August 2014, the average inflow of wastewater was 8725m^3·day^{-1}, reaching the maximum value of 17,300m^3·day^{-1} in October 2013. BOD5 of the influent varied from 425 to 6520mg O_2 l^{-1}, while COD was in the range from 632 to 8020mg O_2 l^{-1}. BOD5/COD ratio varied from 0.202 to 0.696. The concentration of total nitrogen in the influent was in the range from 67.8 to 98.3mg·l^{-1}, while the total phosphorus varied from 8.9 to 22.4mg·l^{-1}. More detailed characteristics of the influent is presented in **Table 1**. In the period of study, the levels of carbon, nitrogen and phosphorus removal from wastewater required by Polish legislation were achieved in this plant. There were no significant disturbances in the operation of the WWTP under study.

2.2. ToxTrak™ Test

ToxTrak™ methodorated to be an analytical tool for the measurement of

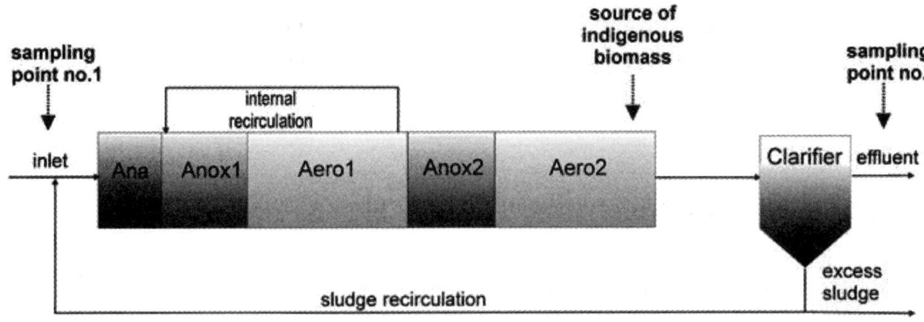

Figure 1. Scheme of the activated sludge system of the WWTP Zgierz including the sampling points.

Table 1. Characteristics of the influent and effluent in the WWTP studied.

Characteristic feature	Short-term campaign January 2014		Short-term campaign June-July 2014		Long-term campaign	
	Influent (sampling point no. 1)	Effluent (sampling point no. 2)	Influent (sampling point no. 1)	Effluent (sampling point no. 2)	Influent (sampling point no. 1)	Effluent (sampling point no. 2)
pH (−)	7.0−7.7	7.0−7.3	7.3−7.6	7.1−7.2	7.0−7.8	7.0−7.3
COD (mg O_2 l^{-1})	632−3932	34−59	1240−3760	24−84	632−8020	19−84
BOD5 (mg O_2 l^{-1})	440−1200	2−10	425−980	2−11	425−6520	2−12
Conductivity (μS cm^{-1})	1587−2740	1502−2220	1748−2420	1694−1946	1505−2740	1461−2220
Ammonium (mg N-NH_4^+)	57.63−79.59	0.42−0.65	66.92−70.21	0.29−0.79	57.53−79.59	0.28−0.79
Ntot (mg N l^{-1})	67.81−98.26	5.06−5.71	75.19−84.59	3.94−6.14	65.38−98.26	5.06−7.42
Ptot (mg P l^{-1})	8.9−22.4	0.21−0.40	9.5−14.4	0.30−0.69	8.9−23.3	0.21−0.69

toxicity of drinking water, wastewater and natural water. The method is based on the reduction of resazurin, a redox-active dye, by bacterial respiration. When it is reduced, resazurin changes its colour from blue to pink. The presence of toxic substances in the sample decreases the rate of resazurin reduction, which can be measured colorimetrically. The endpoint of ToxTrak™ method is the inhibition of bacterial respirometric activity. The results of this test (toxicity scores) were expressed as the degree of inhibition (DI) in a percentage (%). In order to make them more universal and easier comparable to other toxicity data, the values of inhibition concentration (IC50) were determined by the linear regression between wastewater concentration (diluted samples of wastewater) and the degree of inhibition in a logarithmic coordinates system. Then, the toxicity units (TU) were calculated (Swedish EPA 1997; Libralato *et al.* 2010a; Vasquez and Fatta-Kassinos 2013).

In this work, ToxTrak™ test was used to measure the toxicity of municipal wastewater from the WWTP Zgierz (Poland). Inoculum was basically the indigenous biomass taken from the second aeration chamber of the Zgierz WWTP

(**Figure 1**) and prepared in agreement with the guidelines delivered by HACH Company. Absorbance was measured with the use of spectrophotometer DR 6000 at λ = 603nm. The test was made in accordance with the guidelines for ToxTrak™ (Toxicity ToxTrak™ Method 10017, HACH LANGE Manual. http://www hach.com/toxtrak-toxicity-reagent-set-25-49-tests/product-downloads?id=7640273 469). Each sample was made in five replications. If necessary, additional replications were made in order to obtain reliable results.

2.3. Physicochemical Analyses

Apart from toxicity, the physicochemical indicators of wastewater (influent and effluent) were also determined in agreement with the standard methods. These were pH, COD, BOD5, conductivity, ammonium and total nitrogen and total phosphorus (APHA-AWWA-WEF 2012). Their values for the influent are included in **Table 1**.

2.4. Validation Study

The application of any toxicity test in new conditions or for a new object requires the validation of the experimental procedure to evaluate its sensitivity and precision. Here, the validation of ToxTrak™ method comprised two stages (**Figure 2**). First, ToxTrak™ test was conducted with two different inocula, *i.e.* the indigenous biomass (mixed culture) from the WWTP Zgierz (**Figure 1**) and pure culture of *Escherichia coli* from the collection DSM 30083 (**Figure 2**). Activated sludge used in the tests was characterised by high biodiversity and its biotic index varied from 6 to 8 (Madoni 1994). Second, two types of tested materials were used. These were raw wastewater (influent) from the WWTP Zgierz and the solution of phenol. Phenol is recommended as the organic reference substance for testing of toxicity towards bacteria (NaleczJawecki *et al.* 2010; Microtox® Acute Toxicity Test 1998).

2.5. Monitoring of Wastewater Toxicity

The monitoring study was performed from August 2013 to August 2014 in

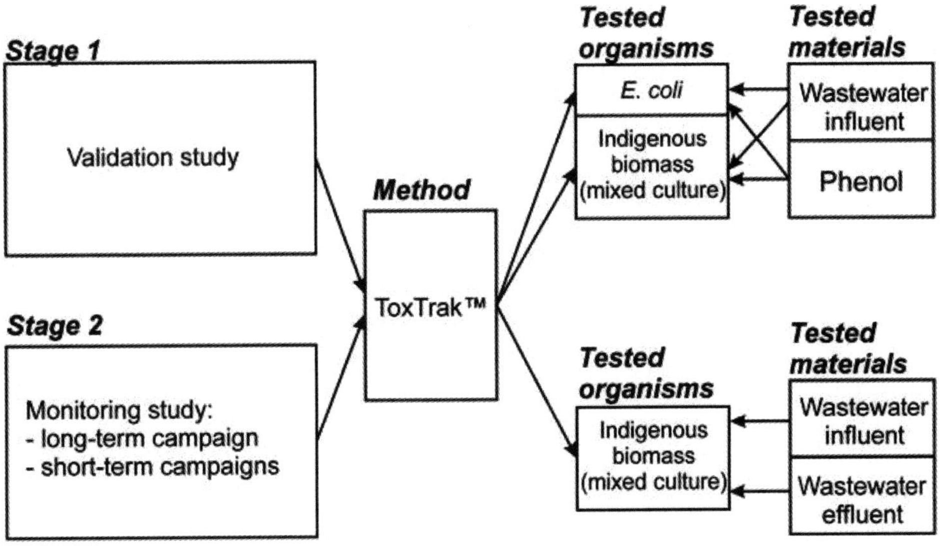

Figure 2. Experimental scheme for the wastewater toxicity study.

the WWTP Zgierz (**Figure 2**). Toxicity of raw wastewater (influent) and treated wastewater (effluent) was determined (**Figure 1**). The averaged diurnal samples of the influent and effluent were taken from the WWTP Zgierz under dry weather conditions and tested within 2 h.

Two types of the measurement campaigns, *i.e.* long and short term, were conducted. In the long-term campaign, toxicity was measured once a month over 13 months. During the short-term campaigns, the samples were taken three times a week. Each short-term campaign lasted 15 days and comprised seven samples. One short-term campaign was conducted in winter (January 2014) and the other one in summer (June and July 2014).

2.6. Data Analysis

Toxicity score in ToxTrak™ method was expressed as the percent inhibition degree. In accordance with the guidelines of the test (Method 10017, HACH, Loveland, CO, USA), each sample was tested in five replications. The results of tests were subjected to the basic statistical analysis that comprised the calculation of mean values, standard deviations and relative standard deviations (RSD) of the measured degrees of inhibition. Linear regression (R2) and Pearson's coefficients were used in order to find the correlation between toxicity and physicochemical indicators of wastewater or operation parameters of the WWTP studied. Moreover, linear regression was applied to seek for the correlation between influent and effluent toxicity. Both basic statistical analysis and correlation analysis were made with the use of MS Excel. The confidence level of 95% was each time assumed. Results (from **Figures 3–6**) were presented as mean values with the standard deviations.

Additionally, one-way analysis of variance (ANOVA) was applied to evaluate whether the degrees of inhibition in the validation study were equal. The null hypothesis was that they were equal. The confidence level of 95% was assumed, too. ANOVA implemented in MS Excel (Analysis ToolPak) software was used.

3. Results and Discussion

3.1. Validation Study

In the validation study, the toxicity of raw wastewater from the WWTP

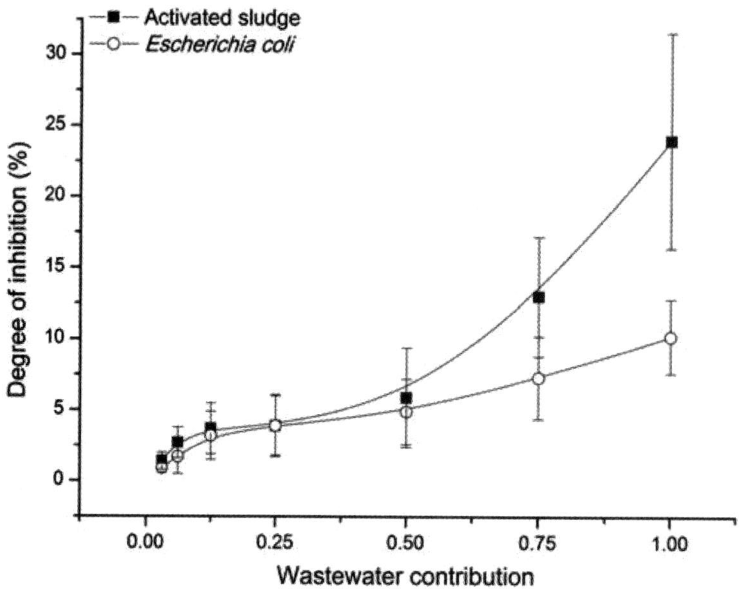

Figure 3. Wastewater toxicity of towards activated sludge microorganisms and *Escherichia coli*.

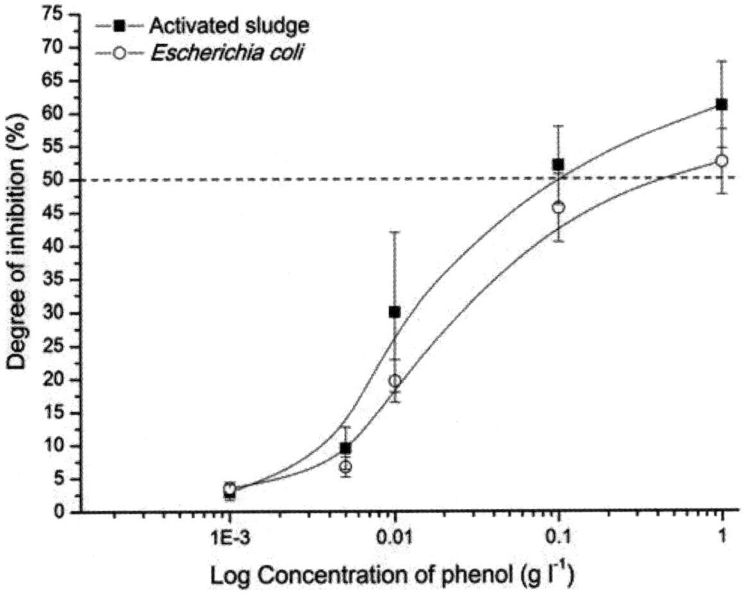

Figure 4. Toxicity of phenol towards activated sludge and *Escherichia coli*.

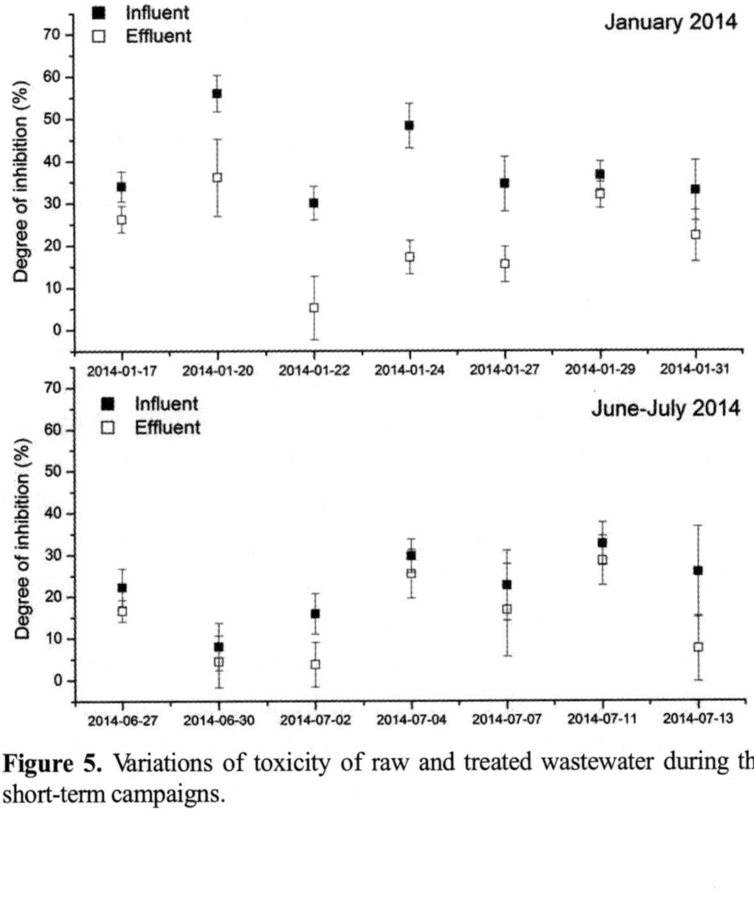

Figure 5. Variations of toxicity of raw and treated wastewater during the short-term campaigns.

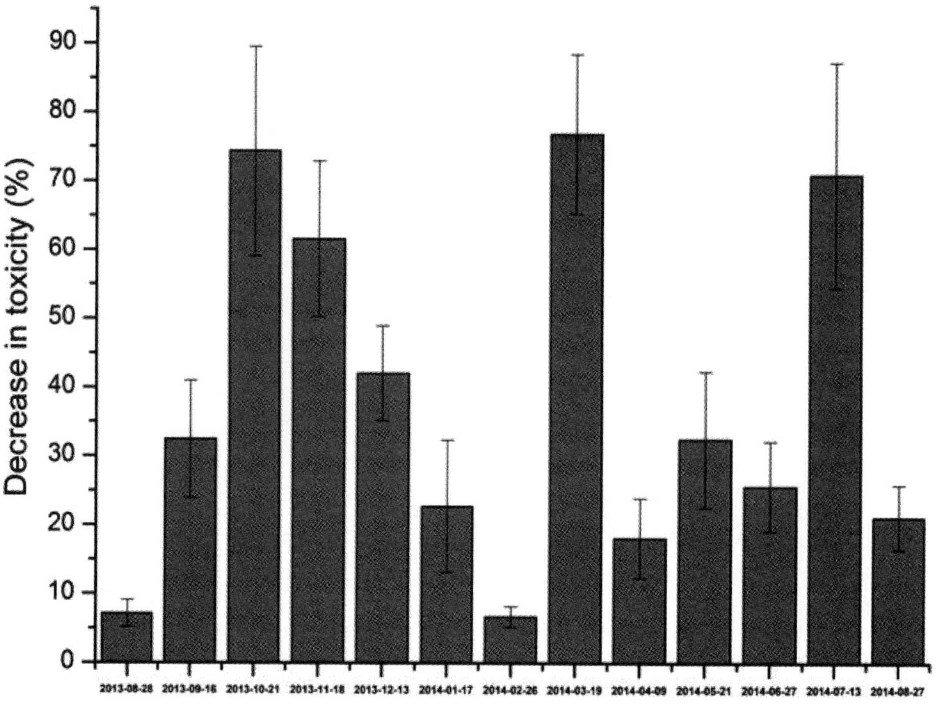

Figure 6. Variations of the decrease of wastewater toxicity during the long-term campaign.

Zgierz towards mixed culture (activated sludge microorganisms) and pure culture (*E. coli*) was tested. It was also made for phenol solutions (**Figure 2**). The results of these tests are presented in **Figure 3** and **Figure 4**, respectively. The curves illustrating the changes of the degree of inhibition dependent on wastewater dilution or phenol concentration had the same shape for both types of biomass (mixed and pure culture) used. The values of the degree of inhibition determined for activated sludge and *E. coli* were often close to each other. Taking the fact that the ranges of standard deviations calculated for the degree of inhibition towards mixed and pure culture covered themselves into account, it could be claimed there was no statistically significant difference between them. It concerned the values of the degree of inhibition estimated for phenol and for more diluted wastewater (wastewater contribution up to 0.5v/v). The exceptions were undiluted or slightly diluted (wastewater contribution 0.75v/v) raw wastewater (**Figure 3**), when the discrepancy between the degree of inhibition for activated sludge and the one for *E. coli* reached even 200%. The most probable reason was the material tested, *i.e.* wastewater, which was the complicated matrix of compounds. In the tests with phenol, such discrepancies between the degrees of inhibition calculated in the tests towards activated sludge and *E. coli* were not observed (**Figure 4**). The results of one-way ANOVA test confirmed that there was no sufficient evidence to reject the null hypothesis that the group means were equal (P value $>\alpha$, where $\alpha = 0.05$) with regard to both experiments, *i.e.* the tests with phenol (P value = 0.334) and the tests with wastewater (P value = 0.673).

The reproducibility of ToxTrak™ tests, calculated as the relative standard deviation (RSD), varied from 9.3% to 52.6%. The lowest values up to 24.2% were found in the tests, in which toxicity of phenol towards *E. coli* was measured. They were in agreement with the results obtained by Hernando *et al.* (2005), where RSD was in the range from 5% to 22.3% in the toxicity tests with the use of Vibrio fisheri, Selenastrum capricornotum and D. magna. The application of activated sludge biomass in place of *E. coli* in the experiments with phenol caused to the decrease of reproducibility of the ToxTrak™ test (higher RSD up to 39%). It was the most probable associated with the biodiversity of activated sludge biomass used in the tests. Nevertheless, the application of ToxTrak™ test using indigenous biomass in a WWTP can help to recognise the impact of influent on the biomass, which is responsible for the microbiological removal of pollutants from wastewater. As a result, the operators can avoid disturbances in functioning of the biological part of a WWTP.

In ToxTrak™ tests, the indigenous biomass as well as pure culture can be used as the inoculum. The comparison of these two types of inocula made here showed that activated sludge microorganisms were more sensitive than *E. coli*, irrespective of the tested material (wastewater or phenol solution) (**Figure 3** and **Figure 4**). The agreement between these results confirmed the correctness of the laboratory work performed. The inhibition concentration IC50-45 min estimated for phenol was $49.3 mg \cdot l^{-1}$ towards activated sludge biomass and $373 mg \cdot l^{-1}$ in the tests with *E. coli*. The value of IC50-5 min for phenol towards V. fischeri in Microtox® Acute Toxicity Test should be included in the range from 13 to $26 mg \cdot l^{-1}$ (Modern Water Microtox Acute Toxicity and Modern Water 1998).

Lower value of IC50-45 min determined in the tests with the use of indigenous biomass compared to these with *E. coli* indicated on the higher resistance of *E. coli* towards tested materials (wastewater or phenol solution). Vasquez and Fatta-Kassinos (2013) found that wastewater samples were less toxic to V. fischeri than Pseudokirchneriella subcapitata or D. magna. But it must be remembered that activated sludge is a mixture of organisms consisting mainly of bacteria and protozoa and its sensitivity may vary to a higher extent than it may in the case of pure cultures.

Basically, the organisms used in the toxicity tests should be sensitive to a variety of chemicals (U.S. EPA 2002). Thus, activated sludge organisms were selected as the inoculum for the further tests made within the monitoring study. The second reason of this choice was the fact that these organisms were exposed to raw wastewater in the biological step of the WWTP and the results of toxicity tests allowed for the prediction, whether the influent could exert any effect on biological treatment processes. As a result, the relationships between toxicity of raw wastewater and the efficiency in removal of C, N and P from wastewater could be found.

3.2. Monitoring Study

The toxicity of raw wastewater expressed as the degree of inhibition varied significantly from 10.2% to 59.8%, whereas for the treated ones from 3.3% to 35.6%. Such variability of results concerned also other measures as COD or BOD5 and is typical for wastewater being the mixture of various compounds. What is more, the composition of this mixture varied in time. The variations in wastewater toxicity were observed in the long as well as in the short-term campaigns. Yi *et al.* (2009) and Vasquez and Fatta-Kassinos (2013) found the great changes in the toxicity of wastewater in the monitoring study of the WWTPs, too. Furthermore, Vasquez and Fatta-Kassinos (2013) observed significant variations of toxicity dependent on the season and the species tested.

In this work, variations of wastewater toxicity in respect to the season were also apparent. The results of the short-term campaigns revealed that raw wastewater were more toxic in winter than in summer (**Figure 5**). The mean value of DI in the whole winter campaign was equal to 38.9%, while in the summer one, it was 22.4%. The reason was the most probably connected with the activity of small businesses including factories, from which wastewater was delivered by sewer system to the studied WWTP. Their activity was higher in January in comparison to the end of June and July, when many of them reduced or even stopped their manufacturing due to the holidays. It was confirmed by the contribution of industrial wastewater in the influent, which was at the level of 12% in the winter campaign and 7% in the summer one. Ra *et al.* (2007) observed higher toxicity of municipal wastewater in winter than in summer, too. At the same time, Vasquez and Fatta-Kassinos (2013) found the opposite relation. According to their work, municipal wastewater toxicity was higher in summer than in winter due to lower dilution of wastewater in the summer period. However, it should be added that in the study performed by Vasquez and Fatta-Kassinos (2013), four samples per year were taken, whereas in this work, each shortterm campaign comprised seven samples. It makes the results presented here more reliable with regard to the seasonal variations of toxicity in the WWTP Zgierz.

Irrespective of the sampling date, the toxicity of the influent was higher than the toxicity of the effluent (**Figure 5**). It indicated that the biological treatment system reduced wastewater toxicity. However, in some cases, the difference between the toxicity of raw and treated wastewater was relatively small. The decrease of toxicity in the WWTP studied varied widely from 5.8% to 76.6% (**Figure 6**). There were several possible reasons of the variability of these results. The most probable was the composition of raw wastewater and operational parameters of the activated sludge system. Yi *et al.* (2009) also observed the decrease of wastewater toxicity after biological step and secondary clarifier; however, after Fenton process, the toxicity increased again. There was no correlation between the decrease of toxicity and the biodegradability of the influent expressed as BOD5/COD (R2 below 0.200). Also, such operational parameters as sludge loading rate, sludge age or temperature in the bioreactor did not have any unequivocal effect on the decrease of wastewater toxicity (R2 below 0.200). Thus, it is difficult to say, which operational conditions favour the decrease of wastewater toxicity.

At the same time, the correlation between the toxicity of the influent and effluent was found. The coefficient of determination (R2) was equal to 0.772 (**Figure 7**). Generally, the higher inlet toxicity is, the higher toxicity of the effluent can be expected.

Additionally, it was checked whether the toxicity of influent was correlated with physicochemical indicators of raw wastewater or degree of removal of organic compounds (COD), nitrogen or phosphorus. The values of Pearson's coefficient

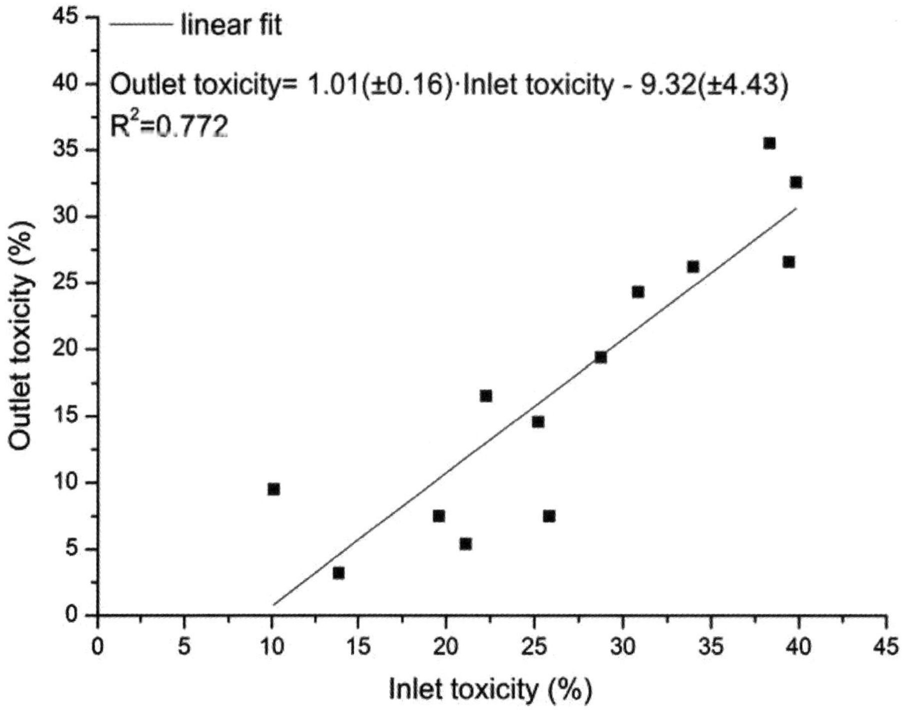

Figure 7. Correlation between toxicity of influent and effluent.

and R-squared are presented in **Table 2**. Unfortunately, the obtained results did not allow for the statement that the toxicity of raw wastewater was correlated with any of physicochemical indicator or biodegradability. The values of R-squared were usually below 0.200 (Pearson's coefficients below 0.400). According to Postma *et al.* (2002) ammonium and conductivity were considered as "confounding factors" that may interfere with the biological effects of micropollutants removal. It was found elsewhere that the conductivity of raw wastewater was correlated with the toxicity to D. magna (Vasquez and Fatta-Kassinos 2013). Therefore, the correlations between toxicity and these two parameters were checked carefully. Only a weak correlation was found between the conductivity and toxicity of raw wastewater in the summer campaign (R^2 = 0.521) (**Table 2**). At the same time in winter and long-term campaigns, such correlation was not observed (R2 below 0.100) (**Table 2**). Toxicity of raw wastewater, if it was in the measured range (DI from 10.2% to 59.8%), did not interfere with the efficiency of organic carbon (COD), nitrogen and phosphorus removal. As written earlier, the levels of carbon, nitrogen and phosphorus removal from wastewater required by Polish legislation were achieved in the WWTP studied in each measurement campaign.

Table 2. Pearson's and R-squared coefficients for the correlations between toxicity and physicochemical indicators of raw wastewater (influent) and between toxicity of raw wastewater and efficiency of removal of organic compounds and nutrients.

Correlation	Pearson's coefficient/R^2		
	Short-term campaign January 2014	Short-term campaign June-July 2014	Long-term campaign
Toxicity vs. COD	0.231/0.0534	−0.388/0.151	$0.0701/4.92 \times 10^{-3}$
Toxicity vs. BOD5/COD	0.171/0.0292	0.371/0.138	0.138/0.0191
Toxicity vs. conductivity	$0.0537/2.88 \times 10^{-3}$	0.721/0.521	0.231/0.0534
Toxicity vs. ammonium concentration	0.189/0.0357	−0.177/0.0313	0.301/0.0906
Toxicity vs. removal of COD	0.191/0.0365	0.358/0.128	$-0.0592/3.50 \times 10^{-3}$
Toxicity vs. removal of nitrogen	−0.403/0.162	$-0.0359/1.29 \times 10^{-3}$	−0.238/0.0566
Toxicity vs. removal of phosphorus	0.405/0.164	−0.389/0.151	0.454/0.206

What is more, no strong linear correlation between toxicity of raw wastewater and degree of removal of COD, Ntot and Ptot was found. The values of R-squared and Pearson's coefficients were from 1.29×10^{-3} to 0.206 and from −0.403 to 0.454, respectively (**Table 2**). A weak or moderate positive correlation was found between toxicity of wastewater and phosphorus removal (**Table 2**). A weak negative correlation was observed between toxicity of wastewater and nitrogen removal too but it concerned the winter campaign only (**Table 2**).

Although the toxicity of effluent was always lower than that of influent, it did not mean that the treated wastewater was environment friendly. According to the classification of wastewater toxicity based upon the acute toxicity proposed by Persoone *et al.* (2003), the effluent was toxic (class III) or highly toxic (class IV). Out of 25 samples of the treated wastewater, two belonged to class III and 23 samples were ranked as class IV. The results of toxicity tests performed to V. fischeri indicated that the effluent of the WWTP tested by Vasquez and Fatta-Kassinos (2013) was usually classified as class III or IV, *i.e.* similarly as in this work. Toxic properties of the treated wastewater meant that its introduction to water bodies could adversely affect the organisms living there as well as functioning of the whole ecosystem. Furthermore, it may have the influence on public health.

In order to decrease the risk of water toxicity and to protect water environment, the additional treatment after biological step should be implemented in wastewater treatment plants, particularly in the countries facing water scarcity problems.

4. Conclusions

The validation study revealed the usefulness of Toxtrak™ test in the measurement of wastewater toxicity. Both indigenous biomass from wastewater treatment plants and *E. coli* can be successfully applied in the measurement of wastewater toxicity. Indigenous biomass was more sensitive than *E. coli* for both materials (wastewater and phenol solution) tested. The values of the degree of inhibition determined for phenol solutions towards activated sludge microorganisms and *E. coli* were close to each other and no statistically significant difference between them was found. It confirmed the reliability of the results obtained with the help of Toxtrak™ test.

The toxicity of wastewater in the short- as well as longterm campaigns varied widely. In respect to the season, lower toxicity of raw wastewater was observed in summer than in winter. The toxicity of the effluent was always lower in comparison to the influent. The linear correlation between the toxicity of the influent and effluent was found. At the same time, any strong linear correlation between the toxicity of raw wastewater and physicochemical parameters (pH, COD, ammonium nitrogen, total nitrogen, total phosphorus) or biodegradability (BOD5/COD) was not observed. Despite the fact that the WWTP effluents obeyed all regulations in Poland with respect to the physicochemical properties, they were not biologically safe. They were ranked as toxic or highly toxic according to one of the classifications of wastewater based upon the acute toxicity.

This work confirms that apart from traditional, physicochemical measures of wastewater composition, also, biological toxicity-based monitoring is necessary in order to make the effluent safer to the aquatic environment.

Acknowledgements

This work was made within the project NR14-0004-10 financed by the National Centre for Research and Development, Republic of Poland.

Source: Ewa L B, Ślęzak Radoslaw, Małgorzata K. Study on wastewater toxicity using ToxTrak^{TM} method:[J]. Environmental Science & Pollution Research, 2016, 23:1−9.

References

[1] APHA-AWWA-WEF (2012) American Public Health Association/American Water Works Association/Water Environment Federation. Standard methods for the examination of water and wastewater 22nd edn. APHA/AWWA/Water Environment Federation, Washington DC.

[2] EPA US (2002) Methods for measuring the acute toxicity of effluents and receiving waters to freshwater and marine organisms, 5th edn. U.S. EPA, Washington DC.

[3] Hernando MD, Fernandez-Alba AR, Tauler R, Barcelo D (2005) Toxicity assays applied to wastewater treatment. Talanta 65:358−366. doi:10.1016/j.talanta.2004.07.012.

[4] Libralato G, Ghirardini AV, Avezzù F (2010a) Toxicity removal efficiency of decentralised sequencing batch reactor and ultra-filtration membrane bioreactors. Wat Res 44:4437− 4450. doi:10.1016/j.watres.2010.06.006.

[5] Libralato G, Ghirardini AV, Francesco A (2010b) How toxic is toxic? A proposal for wastewater toxicity hazard assessment. Ecotoxicol Environ Saf 73:1602−1611. doi:10.1016/j.ecoenv.2010.03.007.

[6] Libralato G, Gentile E, Ghirardini AV (2016) Wastewater effects on Phaedodactylum tricornutum (Bohlin): setting up a clasification system. Ecol Indic 60:31−37. doi:10.1016/j.ecolind.2015.06.014.

[7] Madoni P (1994) A sludge biotic index (SBI) for the evaluation of the biological performance of activated sludge plants based on the microfauna analysis. Wat Res 28(1):67-75. doi:10.1016/0043-1354(94)90120-1.

[8] Modern Water Microtox Acute Toxicity Overview, Modern Water N.D. (1998.) http://www.coastalbio.com/images/Acute_Overview.pdf. Accessed 29 January 2015.

[9] Nalecz-Jawecki G, Baran S, Mankiewicz-Boczek J, Niemirycz E, Wolska L, Knapik J, Piekarska K, Bartosiewicz M, Pietowski G (2010) The first Polish interlaboratory comparison of the luminescent bacteria bioassay with three standard toxicants. Environ Prot Eng 36(3):95−102.

[10] Persoone G, Marsalek B, Blinova I, Törökne A, Zarina D, Manusadzianas L, Nalecz-Jawecki G, Tofan L, Stepanova N, Tothova L, Kolar B (2003) A practical and us-

er-friendly toxicity classification system with microbiotests for natural waters and wastewaters. Environ Toxicol 18(6):395–402. doi:10.1002/tox. 10141.

[11] Postma JF, De Valk S, Dubbeldam M, Maas JL, Tonkes M, Schipper CA, Kater BJ (2002) Confounding factors in bioassays with freshwater and marine organisms. Ecotoxicol Environ Saf 53(2):226–237. doi: 10.1006/eesa.2002.2195.

[12] Ra JS, Kim HK, Chang NI, Kim SD (2007) Whole effluent toxicity (WET) tests on wastewater treatment plants with Daphnia magna and Selenastrum capricornutum. Environ Monit Assess 129:107–113. doi:10.1007/s10661-006-9431-2.

[13] Swedish EPA (1997) Characterisation of discharges from chemical industry—The stork project. Swedish Environmental Protection Agency. Report no. 4766, Stockholm.

[14] Tobajas M, Verdugo V, Polo AM, Rodriguez JJ, Mohedano AF (2015) Assessment of toxicity and biodegradability on activated sludge of priority and emerging pollutants. Environ Technol 11:1–9. doi:10. 1080/09593330.2015.1079264.

[15] Vasquez MI, Fatta-Kassinos D (2013) Is the evaluation of "traditional" physicochemical parameters sufficient to explain the potential toxicity of the treated wastewater at sewage treatment plants? Environ Sci Pollut Res 20:3516–3528. doi:10.1007/s11356-013-1637-6.

[16] Xiao Y, De Araujo C, Sze CC, Stuckey DC (2015) Toxicity measurement in biological wastewater treatment processes: a review. J Hazard Mater 286:15–19. doi:10.1016/j.jhazmat.2014.12.033.

[17] Yi X, Kim E, Jo H-J, Schlenk D, Jung J (2009) A toxicity monitoring study on identification and reduction of toxicants from a wastewater treatment plant. Ecotoxicol Environ Saf 72:1919–1924. doi:10. 1016/j.ecoenv.2009.04.012.